digitalSTS

digitalSTS

A Field Guide for Science & Technology Studies

EDITED BY

Janet Vertesi & David Ribes

CO-EDITED BY

Carl DiSalvo
Yanni Loukissas
Laura Forlano
Daniela K. Rosner
Steven J. Jackson
Hanna Rose Shell

PRINCETON UNIVERSITY PRESS / PRINCETON & OXFORD

Published by Princeton University Press
41 William Street, Princeton, New Jersey 08540
6 Oxford Street, Woodstock, Oxfordshire OX20 1TR

press.princeton.edu

LCCN 2018955221
ISBN 978-0-691-18707-5
ISBN (pbk.) 978-0-691-18708-2

British Library Cataloging-in-Publication Data is available

Editorial: Eric Crahan, Pamela Weidman, Kristin Zodrow
Production Editorial: Terri O'Prey
Production: Jacquie Poirier
Publicity: Alyssa Sanford, Julia Hall
Copyeditor: Joseph Dahm

This book has been composed in IBM Plex Serif

Printed on acid-free paper. ∞

Printed in the United States of America

10 9 8 7 6 5 4 3 2 1

Contents

Preface
The digitalSTS Community

The contributions to this volume are the outcome of a five-year community-building endeavor under the title *digitalSTS*: an effort to theorize the *next* generation of STS encounters with digital artifacts, environments, and interactions. These published contributors are only a small selection of all those participants who—across multiple conference panels, workshops, and online submissions and reviews—helped to produce an emerging understanding of digital studies rooted in the fields and commitments that constitute science and technology studies (STS). This edited collection is a product of that emergent, experimental, and participant-driven process. These events deserve an introduction as they shaped the voices in this collection and beyond.

digitalSTS got its start at the Society for Social Studies of Science (4S) meeting in Cleveland in the fall of 2011 when over 80 people gathered for a participant-driven conversation about the challenges that digital tools, practices, and platforms posed for theory and practice, for the career trajectories of scholars identifying with digital scholarship, and for communicating our findings to STS and beyond. Eight panelists, ranging from junior to senior scholars and drawn from anthropology, history, communication, sociology, STS, and other interdisciplinary fields, issued brief provocations to a packed room; two hours of lively discussion followed. Among the many inspiring comments from panelists were those of anthropologist Gabriella Coleman, who exhorted the assembled STS scholars in the audience not to cede public discourse on digital topics to media pundits, where revolutionary rhetoric would dominate the subject. This prompt and the remarkable energy and vivacity in the room inspired us to turn a one-time event into the platform for a broader discussion, to open avenues for addressing "the digital" within STS, and to serve as a springboard to push the conversation forward.

Over the subsequent five years we convened four workshops, deployed an online platform, hosted events at 4S meetings and elsewhere, and set about building a community for scholarship on digital topics in STS. At 4S in Copenhagen in 2012, 40 workshop participants convened to specify the challenges that digital studies and scholarship posed to STS. This discussion was the source of the idea for a published "fieldguide" of cases, tools, and born-digital projects. Realizing that "making and doing" needed further elaboration within STS, co-editors Laura Forlano and Yanni Loukissas took the lead organizing a three-day follow-up workshop at the Arnold Arboretum of Harvard University in 2013. This hands-on meeting gathered STS scholars and makers (broadly construed) to explore the field's specific intersections with design and to more strongly theorize what might set STS making apart from other forms of DIY or digital intellectual endeavors. Later, as projects began to crystallize and develop, we established an online open platform for peer-to-peer discussion of abstracts and papers, and workshopped precirculated drafts in person. Many of these meetings were sponsored by the Community for Sociotechnical Systems Researchers (CSST—www.sociotech.net), by the National Science Foundation (NSF) Office of Cyberinfrastructure, and by the NSF Science, Technology and Society program, with additional event sponsorship from the Sloan Foundation and Microsoft Research.

We embraced the concept of open scholarship and digital publishing from the outset, for instance by deploying practices from collaborative peer production and online communities. The first round of review, editing, and feedback for this volume was conducted on a platform (adapted from the alt.CHI community of human-computer interaction researchers) that allowed anyone to register, submit a paper, and review a colleague's work. Those whose interest was piqued by an article posted reviews, cross-talk among reviewers was encouraged, and neither authors nor reviewers were blinded. This fostered a collaborative reviewing environment as authors sought to help each other, point to key references or theories, and offer constructive critiques in preparation for the traditional blind review process conducted by Princeton University Press.

As the result of considerable bottom-up work, this volume stands as an achievement of many complementary community-building aims. We sought to bring together those scholars with one foot in STS and the other in information technology fields to encourage more cross-talk and to incorporate lessons learned from programming, design, and development into STS scholarship and practice. We also hoped to open a space for STS scholars studying digital topics to develop their contributions, and for those in relevant neighboring fields to identify with and contribute to STS scholarship through the events and engagements of *digitalSTS*. We purposefully integrated opportunities for junior academics to work together and alongside mid- and senior-career scholars in an effort to elaborate and develop relevant conceptual vocabulary and themes in the STS register.

Community building can easily produce exclusions instead of inclusions, or homogeneity instead of heterogeneity. Our continued interest in developing the diversity of participation sets this volume apart. Our goal was not to definitively theorize "the digital" or to identify an overarching perspective on digital work that all must share. Instead, the shape and scope of the topic were defined by the self-selected community, with the editorial board serving as organizers and shepherds along the way. Hence, it was conversations among community members that actively shaped the volume and its approach from the outset, with topics emerging from among the many active and interactive discussions. It is our hope that this will ensure a volume with broad scope and reach across STS and its sister disciplines.

Fortunately, digital scholarship no longer seems such a marginal topic in STS as it did in 2011. Perhaps this reflects a changing public consciousness, as data, algorithms, infrastructure, and a host of other digital objects have taken on a heightened relevance in the press and in politics. It may also reflect a changing membership in STS, as those who have grown up with devices always in hand or in pocket seek to reflect that ubiquity in their scholarly work. The question of whether or not "digitalSTS" marked something truly unique haunted our discussions from the very first. However, at our final workshop, held in advance of the 2015 4S meeting in Denver, Matt Ratto noted that perhaps the term "digital" "marked time" for this project, capturing a sensibility toward this emerging scholarship between 2011 and 2016. The contributions to this volume are simply "doing STS," albeit with digital subject matter, tools, and products. If not carving out a new intellectual domain, then, the moniker of digitalSTS serves instead as a rallying call to push our field forward, to seize the public conversation surrounding "the digital revolution," to engage with policy and design practice, and to knit together a robust, thoughtful community of scholars dedicated to both the core principles and the continued future of STS in the world.

digitalSTS

Introduction

The field of science and technology studies (STS) was born well into the era of computation. One would be hard-pressed to find an investigation from the 1970s or 1980s of a contemporary scientific or technological site that did not include a recognizable computational artifact: a machine in the corner of a lab, an instrument dependent on algorithms, an expert system at center stage. Studies of infrastructure, such as the work of Thomas Hughes or Susan Leigh Star, were inflected by systems theory, cybernetics, and even the budding milieu of artificial intelligence. Some of STS's early central concepts originate from intellectual movements entangled with computation: Thomas Hughes and Bruno Latour's black box (a systems theory concept), Leigh Star's boundary object that she described as influenced by "both its computer science and pragmatist senses" (Leigh Star 2010, 603), or Donna Haraway's Cyborg Manifesto (1991).

If STS has long been (in some way) digital, then why should this volume call for a digitalSTS? Since another long-standing intellectual commitment of STS is "situatedness" (Suchman 1987; Haraway 1988), we must admit that our situations have changed. To paraphrase Latour and Woolgar (1979), rats are no longer being turned (only) into paper; now they are transformed into PowerPoint files and archival document placeholders. Laboratory work today involves distant collaborations enacted through communication tools, incompatible file formats, and a dizzying array of software analytic tools. Lay-expert groups find solidarity on online fora or social network sites, coordinate via Twitter feeds, and deploy bots for community management and mediation. Microchips are no longer confined to hefty machines in the corner of the laboratory or even the workplace desktop: they are in our homes, our pockets, our clothing, sometimes under our skin. In short, the textures of scientific and daily life at the beginning of the 21st century are suffused with online platforms and heterogenous informational environments.

Scholars drawing from STS are well placed to analyze this contemporary turn of events and to inspect their long-arc historical trajectories. The field maintains a commitment to unpacking the layered, social, and gradual aspects of scientific and technical change, undermining common accounts of revolution, disruption, or inevitable progress. STS scholarship provides tools for locating the politics in technical and scientific decision making, for examining the global yet unevenly distributed tools of computing, and for unearthing the power-laden absences and silences in small and large-scale systems alike. Examinations of science and technology written from the STS perspective have long focused on topics like quantification, standardization, classification, and representation—all themes that sustain an importance to the digital—and have done so with close attention to the practical, local, situated elements of knowledge construction and technological development. STS scholarship ultimately addresses how what we consider to be universal, ubiquitous, or inevitable—such as a contemporary digital lifestyle—is accomplished in a certain time and place, by specific actors and with particular materials at hand.

The chapters in this volume sustain these commitments and bring them to the encounter with the digital anew. They do so through exploring once-familiar sites reconfigured by digital technologies, by investigations that grapple with novel tools, and through efforts that tackle design and making rather than treating technology as (only) an object of social and humanistic investigation. In doing so, these essays pose new questions for scholarship in STS. What is revealed about laboratory life when the "shop floor" goes digital? What can we learn about expertise when new constituencies of laypeople come together online at great distances? How do we incorporate computational artifacts like bots, algorithms, and hidden taxonomies into our extant concepts of distributed or hybrid agency? And how do our theories of materiality map onto digital objects and practices, and vice versa?

Responding to these questions, the chapters in this volume examine digital field sites and mediated interactions as opportunities for theory building, speaking to the field's core assumptions about the construction, shaping, or hybridity of knowledge, objects, and expertise. The authors deploy the emerging tools of digital scholarship, such as critical making and large-scale data analytics, to enhance the analysis of core concepts like representation, quantification, and materiality. The volume as a whole aims to populate our scholarly toolkit by adding contemporary examples to our field's repertoire of cases, theories, and methods. These examples aim to sit alongside, rather than replace, the classic "evocative objects" (Turkle 2007) of our field in our course syllabi and beyond: for instance, the bicycle (Pinch and Bijker 1987), the speed bump (Latour 1992), or the reactor beam (Traweek 1988). They also complement several already-influential studies of digital systems in STS such as the DVD player (Gillespie 2007), the trader's screen (Knorr Cetina and Bruegger 2002; MacKenzie 2006), the configured user (Woolgar 1990), or the "always on" worker (Wajcman 2015).

digitalSTS as Departures

Calling something digitalSTS does not denote a standalone branch of STS, or a bracing theoretical departure, but it does mark a series of disruptions for STS scholarship, even while sustaining intellectual continuities. Let us examine the differences first.

The study (and sometimes making) of digital systems is alive and thriving in sister fields: certainly from the engineering and information sciences, but also always already in sociology and anthropology, communication, and the digital humanities. Classic STS theories and cases have played a role formulating many of these approaches as our concepts and exemplary objects have traveled. For instance, information scientists use "boundary objects" (Star and Griesemer 1989) or "infrastructural inversion" (Bowker 1994) to describe the social work of data management, while studies of algorithmic inequality among Internet researchers have frequently been inspired by Langdon Winner's (1986) account of how the Long Island Expressway overpasses prohibited bus-riding access by poorer citizens. This speaks to an existing exchange of ideas across these disciplines.

But an imbalance persists with STS ideas influencing "outward" rather than welcoming concepts, topics, and framings "inward." This, despite the fact that there is much to learn from how digital scholarship, artifacts, and systems are evolving, changing, and challenging assumptions in sister fields. If there is novelty in the concept of a digitalSTS, it is in **bilateral bridge building between STS and fields that have embraced digital studies and making**. Some of these fields are already known to STS: prior studies of public understanding of science established ties with communications and media studies (Lewenstein 1995; Kirby 2011), while classic ethnographies of machine work engaged with computer-supported cooperative work and other design-oriented enterprises (Orr 1996; Suchman 1987). Other fields are new configurations, such as the digital humanities (Gold 2012), digital sociology (Daniels et al. 2017; Marres 2017), and scholarship in the emerging information schools.

This volume therefore extends this conversation and broadens our field's scope to include voices from scholars engaged in digital studies across many fields. The chapters combine approaches, topics, artifacts, and literatures from multiple disciplines to demonstrate their relevance for STS scholarship, embracing them as subjects for STS analysis. At the same time, however, the volume is grounded in the core concepts and literature of STS. The authors herein not only address digital topics as core matters of concern but also speak back to classic concepts and cases as they do so, developing new theoretical tools for further analysis. The essays therefore demonstrate a way forward for digital studies writ large that take the primary tenets of STS seriously. We hope that this perspective will prove valuable to scholars engaging with digital topics within, across, and beyond STS.

Second, **digital methods are animating scholarship across the academy**. STS has been slow to take up these techniques largely due to our resistance to the unreflexive use of systematizing or formal tools and methods (e.g., Law 2004; Lury and Wakeford 2012). Another goal for this volume is to provide examples of a kind of reflexive, digital methodological inquiry. Inspecting how such systems are constitutive and destructive, revealing and blinding, or powerful here but weak there is an essential feature of STS work. In sum, the sentiment is that STS cannot incorporate the configurations of epistemic tools and methods offered by digital media—from network analysis to data mining and topic modeling—without some (even if inevitably incomplete) archaeology of these tools' assumptions, methods, and roles in situated knowledge production. The contributions herein deploy such tools while inspecting with care the emerging technique, technology, internal logic, rhetoric, or broader milieu of digital methods. They also describe connections and challenges to classic STS theories, to inspire the next generation of critique.

Another thread of contributions has taken on **the challenge of "making"** by combining the design and use of digital tools, artifacts, or methods in tandem with explicit reflection and inquiry on the approach itself. When we began our explorations of the relationship between STS, making, and design in 2012, the field had only the most tenuous of spaces dedicated to these modes of inquiry. We assert that design techniques not only have a place in STS: they are an important locus for the exploration of concepts and analysis, and a site for scholarly intervention with technologists and publics. An online curated collection of notable STS maker projects accompanies this volume to stimulate the scholarly imagination with examples of objects, experiences, or software tools designed with the intention to "make durable" STS concepts through design, participation, and critical making. These contributions demonstrate what is possible when we embrace digital methods of knowledge production—not as revolutionary or straightforwardly objective, but rather as a matter of inquiry involving varying doses of agnosticism, reflexivity, symmetry, and critical perspectives on knowledge construction itself.

Finally, the reader may notice that we do not offer any singular definition or criteria for "the digital" in the volume. Throughout this project, we purposefully suspended any propositions for a universal definition or methodology for the digital. Participants in the workshops that led to this volume were especially concerned not to draw rigid boundaries or produce residual categories by enforcing strict definitions. The volume therefore embraces a dynamic and grounded approach to the study of digital systems, treating the category of the digital as an emergent feature among communities of users, designers, and maintainers, and inspecting sociotechnical architectures in long-arc development trajectories. In essence, the category "digital" as it appears here was also itself emergent through the process that led to the volume (see the preface). Although this may appear disruptive for a collection of papers about digitality, the collected papers demonstrate **the methodological value of digital agnosticism, pluralism, or symmetry** as a point of departure for STS studies of digital phenomena.

digitalSTS as Continuities

In addition to new ground, the volume sustains important continuities. Arguably, what sets this volume apart is its continued commitment to core STS principles as deployed in the analysis of digital systems and interactions, broadly construed. Such commitments will be familiar to STS scholars; their novelty lies in their considered, thorough application to the digital spheres of action, and vice versa. However, these principles may be less familiar to the wider group of scholars who explore digital topics, especially those who may be newcomers to STS. We therefore review these animating commitments here for both sets of readers, outlining their contours for those who may be unfamiliar with STS themes and intellectual traditions, and demonstrating their connection to digital topics for those grounded in these traditions. This list is not exhaustive, but taken together, these renewed commitments demonstrate the unique voice that STS writ large brings to digital scholarship.

First, the contributions in this volume examine **digital objects and practices in sociohistorical context**, locating them in time and place and demonstrating their grounded, emergent contingencies. In line with foundational scholarship by Thomas Kuhn, Ludwig Fleck, and others, STS scholars do not embrace any notion of drastic social change effected by a "digital revolution" (Kuhn 1962; Fleck 1979; see also Shapin 1996). Instead, scholars look for historical continuities and attend to everyday cultural practices, revealing the "normal science" side of digital and technical work. This also means turning away from lofty "great man" stories, teleological narratives of discovery (aka Whig histories), and discussions of the exceptional nature of scientific or technical work. STS scholars of the digital continue to advance the field's claim that knowledge and technology alike are responsive to, in dialog with, and reflective of social and political context—but that they do not, on their own, drive social or political change.

Further, digitalSTS contributors remain committed to the principle that there is no equal contest of ideas in which the best ideas or inventions win out because they are true. Truth is a consequence rather than an antecedent. STS studies scholars in the 1980s adopted **the principle of "symmetry"** to underline that "truth" and "falsity" are not determining inputs or preconditions, but are rather outcomes of the work called science, research, scholarship, and so on (Bloor 1991; Collins 1985;

David 1985; Cowan 1984). This same sense of symmetry and historicism must illuminate any STS analysis of digital systems—today the site of considerable hyperbolic rhetoric about their transformational, disruptive, and revolutionary potential—if we are to uncover the social and historical mechanisms that give rise to ubiquitous systems, devices, and infrastructures.

In tandem with this commitment to contextual and historically grounded factors is our continued investigation of the **practical, situated, local, and grounded nature of digital work**. This commitment animated the accounts of sociologists and anthropologists who first ventured into laboratories to observe the work of scientists and engineers for themselves. STS scholars do not hold scientific and technical work in isolation from the pressures of society or cultural norms, but instead show how fields "advance" stepwise due to their messy work, culturally laden communities, practical activity, and everyday achievements. Whether in the preparation of protein gels or papers for publication (Lynch 1985; Latour and Woolgar 1979), or the local temporalities that govern scientific activity (Traweek 1988), contingent social practices are the order of the day (Collins 1985; Pickering 1995), with scientists responding to pressures of funding, publication, and reputation instead of lofty ideals such as replicability or skepticism (Merton 1942; Mitroff 1974). This sensibility also suffuses technology studies, where approaches such as the Social Construction of Technology (SCOT) program, social shaping (MacKenzie and Wajcman 1999), and theories of technological politics (Winner 1986; Jasanoff and Kim 2015) investigate the culturally situated and power-laden battles over knowledge and artifacts alike. While STS theories have certainly been elaborated since these early days of the field, the work herein reflects commitments that tie knowledge production and technical achievement to observable action and interaction. Contributions to this volume, then, frequently rely on ethnographic experience and attenuation to lived, enacted, embodied work with digital systems on the ground, whether the mundane software tools of coordination work (Vertesi), the transformation of reefs into bits (Parmiggiani and Monteiro) or the simple act of navigating via GPS (Singh et al.).

An increasing number of studies of digital technologies and online life deploy the tools of the social construction of technology or social shaping to show the grounded nature of technical artifacts. But perhaps more unique to STS is the sensibility toward **networked agency** that we bring to our digital objects (and subjects!) of study. Not to be confused with theories of digital networks or networked publics, this line of thinking instead stems from several related strands of STS theory related to materiality and its intersection with the social world. A formative instance is actor-network theory (ANT; see Latour 2005), which posits that agency does not arise from singular objects, devices, or individuals but rather that people, technologies, and scientific objects act by virtue of being embedded in a network composed of humans and nonhumans (Callon 1986). This makes digital devices or software tools inherently unstable and unable to act or circulate freely on their own; indeed, a small shift in the network can affect an object's—or an individual's!—ability to act. In a classic case, a light bulb kit made in France for deployment in West Africa cannot follow its "script" and reliably provide light when placed in the local contexts of, say, village generator ownership or taxation through electric bills (Akrich 1992). This lesson continues to hold in the context of digital device ecosystems, whether in the favelas of Brazil or the One Laptop Per Child project in Peru (Nemer and Chirumamilla; Chan). As networked thinking takes hold in a variety of academic fields (e.g., social network analysis and networked publics among them), this also raises methodological questions about the overlaps with actor-networks (Venturini et al.).

Objects are also fluid and hybrid when considered through the lenses of feminist theory and new materialism. These prominent theoretical approaches reject any clear line between "the social" and "the technical" (or between "science" and "technology"), and instead look to where and how such objects and categories are constructed in action (Haraway 1997; Barad 2007; Suchman 2011). Scholars who work in this vein resist dichotomous vocabulary, preferring portmanteau words such as "sociotechnical" or "technoscience" or hybrid figures such as the cyborg (Haraway 1991) to denote their analytical inseparability. They also describe how practices "enact" objects into being and "entangle" both matter and meaning (Moll 2002; Barad 2007). This strand of thinking is influential in this volume. It animates our own editorial agnosticism about the nature of digital materials and bounded nature of digitality as a concept. It is also present in essays that examine where actors themselves draw the boundaries around "the material" and "the digital," for instance in the case of digitizing musical recordings (Camus and Vinck), data or specimens (Ribes), or locally encoded software ontologies as representational taxonomies (Allhutter). And it is present in provocative "maker" products in this volume that inspire contemplation on our place in the environment (Calvillo, Winthereik et al.) or our interaction with archival data (Loukissas).

Boundaries themselves are long-standing themes for STS scholarship that have developed tools for **investigating the production and maintenance of closed spaces** like laboratories or expert communities, and the circulation between and across these sealed spaces. Boundaries can be a source of strength, though they always exact costs. Classic STS studies of laboratories and experiments, for instance, noted that it was only within the effortfully ordered spaces of a laboratory that anthrax could be isolated from cows (Latour 1988), genomes from organisms (Lynch 1985), plans from actions (Suchman 1987). To this end, much of the work herein is not concerned with public spheres—e.g., use of social media or search engines writ large—but examines particular groups and their use of digital tools to effect or efface boundaries. Not only the objects but also the subjects of such spaces are inspected: scientists and technologists may cultivate particular and changing forms of objectivity (Daston and Galison 2007), cast divides between basic science and applied engineering (Gieryn 1999), or distinguish findings from policy concerns (Jasanoff 1987). This boundary work may generate useful products such as the vaccine and the treatment, but the isolation involved can also produce myopic overgeneralization such as the standard human (Epstein 2007), or facilitate distancing the design of technologies from their consequences in use (Abbate 2012). Similarly, pieces herein examine the boundary work of drawing women in and out of computing (Kerasidou, Dunbar-Hester; see also Light 1999; Ensmenger 2010; Abbate 2012), circumscribing migrant use of digital systems (Hawthorne), and identifying system "misuse" in opposition to innovation (Latzko-Toth et al).

One result of boundary drawing is the **inclusion and exclusion** of different groups in the production of objects or knowledge, a process that also accompanies socially produced categorization and standardization regimes. For instance, Steve Epstein (2007) has tracked the coalitions that formed in the 1980s and 1990s to overturn the white male as the standard human for biomedical investigations in favor of studies inclusive of women, racial minorities, the elderly, and the young. The specter of exclusion animates branches of STS, including both infrastructural categories (Bowker and Star 1999) and the systematic production of ignorance (Proctor and Schiebinger 2008; Oreskes and Conway 2010), as well as how such systematic un-incorporations often take place along existing lines of power. The initiative to surface the excluded reverberates across the volume, but especially animates studies of digital systems in transnational, racialized, or gendered contexts: for instance, the distributed invisible laborers in the Global South for whom voice technologies that standardize accents become part of their embodied practice (Poster) or the software workers who craft boundaries around source code, thereby excluding and occluding female participation (Couture).

Inclusions, exclusions, boundary making, and boundary crossings also actively draw the line around **sanctioned forms of expertise** as credible knowledge. This long-standing thread of work in STS has inspired scholars to take stock of lay knowledge as well as its productive and antagonistic intersections with professionals and other experts (Wynne 1992; Collins and Evans 2009). For instance, Ruha Benjamin (2013) and Alondra Nelson (2016) describe complex intersections between race and medicine in the context of sickle cell anemia or DNA ancestry, describing the fraught participation of racialized medical subjects in research and practice. Benjamin argues for a radical, participatory approach to scientific and medical work: an approach resonant with work in critical design that invites lay publics to participate in knowledge making and analysis (see Balsamo 2011; Vertesi et al. 2016). Such concerns are visible in this volume too, such as with the making of digital tools for visualizing controversies and publics (Munk et al.) or coordinating activists via digital tools (Ilten and McInerney).

Finally, inclusion, exclusion, and expertise recall the importance of **visibility and invisibility to technopolitics.** Infrastructure scholars like Paul Edwards (2013), Christine Borgman (Borgman et al. 2014), and Susan Leigh Star (1999) have been keen to retrieve the otherwise invisible architectures of the digital—software, servers, cables, technical laborers, crowdworkers, data points—all of which are the products of human labor in their creation and maintenance but otherwise backgrounded, rendered as infrastructure ready to hand. Recent work has also surfaced the role of algorithms in the production of social control (Schüll 2017; Noble 2017; Eubanks 2018). Several chapters in the volume therefore follow Geoffrey C. Bowker's call for infrastructural inversion (1994), seeking to re-reveal the work, debates, and decisions that subtend digital action, as well as the accompanying role of visibility in the workplace. Scholarship in this volume also plays with the relationship between the visible and the invisible, surfacing the work of keeping systems up and running effectively (Cohn, Sawyer et al.), deploying digital systems to visualize and trace interactions or experience energy lines (Cardoso Llach, Salamanca, Winthereik et al.), or revealing the work of scholarly tools such as affect (Stark) or algorithms (Seaver). This theme, especially relevant to the study of infrastructures, continues to play a key role in digital knowledge and object production today.

How to Read This Volume

This volume is organized into six sections, each of which analyzes a different topic that speaks to long-standing issues in STS: infrastructure, gender, global inequalities, materiality, visualizing the social, and software. An editorial essay at the start of each section lays out the continuities and departures in that section, situating the contributions among relevant literatures. As an effort to mitigate the dangers of residual categories, the essays themselves are also tagged in the online volume according to different cross-categorizations, some related to theories in use and others topics of interest: the mobility of objects, standardization, expertise, hybridity, or the role of instrumentation, to name a few. The print volume offers one pathway through the book, the online version others; we welcome readers to forge their own.

In addition to these thematic sections, the chapters align with a typology that emerged from the workshops that led to this volume. These different types of essays were each curated by a different group of editors and speak to a different set of concerns. This includes **case studies** that develop robust theoretical insights (edited by Janet Vertesi and Steve Jackson), research that brings reflexive perspectives to **digital methods** (edited by David Ribes and Daniela Rosner), and examples of critical **making** (curated by Laura Forlano, Yanni Loukissas, Carl DiSalvo, and Hanna Rose Shell). Among the last category, an online gallery complements the essays in the volume. Annotated in the text, each of these types of contributions stakes out a different approach to the question of how digital studies meets STS and collectively offer a "fieldguide" to the diversity of approaches that enliven the field today.

Far from abandoning our commitments to STS or calling for a reenvisioning of the field, then, the present volume argues that paying explicit attention to digital sites, environments, and methods requires returning to STS's classic orienting theories and scholarship, while developing new articulations. Bringing information technologies into view as they are embedded in multivariate contexts of use, of practice, of development, and of knowledge making presents an exciting opportunity to bring the agenda of the field forward, to continue our productive conversation with emerging disciplines, and to develop new pedagogical tools. Ultimately, the moniker of digitalSTS reminds us that there are many, many opportunities in the analysis of digital systems to return to our core commitments, while at the same time pushing the boundaries of our field.

Works Cited

Abbate, Janet. 2012. *Recoding Gender: Women's Changing Participation in Computing*. History of Computing. Cambridge, MA: MIT Press.

Akrich, Madeline. 1992. "The De-scription of Technological Objects." In *Shaping Technology/Building Society*, edited by Wiebe E Bijker and John Law, 205–24. Cambridge, MA: MIT Press.

Balsamo, Anne. 2011. *Designing Culture: The Technological Imagination at Work*. Durham, NC: Duke University Press.

Barad, Karen Michelle. 2007. *Meeting the Universe Halfway: Quantum Physics and the Entanglement of Matter and Meaning*. Durham, NC: Duke University Press.

Benjamin, Ruha. 2013. *People's Science: Bodies and Rights on the Stem Cell Frontier*. Palo Alto, CA: Stanford University Press.

Bloor, David. 1991. *Knowledge and Social Imagery*. 2nd ed. Chicago: University of Chicago Press.

Borgman, Christine L., Peter T. Darch, Ashley E. Sands, Jillian C. Wallis, and Sharon Traweek. 2014. "The Ups and Downs of Knowledge Infrastructures in Science: Implications for Data Management." In *Proceedings of the 14th ACM/IEEE-CS Joint Conference on Digital Libraries*, 257–66. Piscataway, NJ: IEEE Press. http://dl.acm.org/citation.cfm?id=2740769.2740814.

Bowker, Geoffrey C. 1994. *Science on the Run: Information Management and Industrial Geophysics at Schlumberger, 1920–1940*. Inside Technology. Cambridge, MA: MIT Press.

Bowker, Geoffrey C., and Susan Leigh Star. 1999. *Sorting Things Out: Classification and Its Consequences*. Cambridge, MA: MIT Press.

Callon, Michel. 1986. "Some Elements of a Sociology of Translation: Domestication of the Scallops and the Fishermen of St. Brieuc Bay." In *Power, Action and Belief: A New Sociology of Knowledge*, edited by John Law, 196–233. London: Routledge & Kegan Paul.

Collins, Harry. 1985. *Changing Order: Replication and Induction in Scientific Practice*. London: Sage.

Collins, Harry, and Robert Evans. 2009. *Rethinking Expertise*. Chicago: University of Chicago Press.

Cowan, Ruth Schwartz. 1984. *More Work for Mother: The Ironies of Household Technology from the Open Hearth to the Microwave*. New York: Basic Books.

Daniels, Jessie, Karen Gregory, and Tressie McMillan Cottom. 2017. *Digital Sociologies*. Chicago: Policy Press.

Daston, Lorraine, and Peter Galison. 2007. *Objectivity*. New York: Zone Books.

David, Paul A. 1985. "Clio and the Economics of QWERTY." *American Economic Review* 75 (2): 332–37.

Edwards, Paul N. 2013. *A Vast Machine: Computer Models, Climate Data, and the Politics of Global Warming*.

Ensmenger, Nathan. 2010. *The Computer Boys Take Over: Computers, Programmers, and the Politics of Technical Expertise*. History of Computing. Cambridge, MA: MIT Press.

Epstein, Steven. 2007. *Inclusion: The Politics of Difference in Medical Research*. Chicago Studies in Practices of Meaning. Chicago: University of Chicago Press.

Eubanks, Virginia. 2018. *Automating Inequality: How High-Tech Tools Profile, Police and Punish the Poor*. New York: St. Martin's.

Fleck, Ludwig. 1979. *Genesis and Development of a Scientific Fact*. Edited by T. J. Trenn and Robert K. Merton. Chicago: University of Chicago Press.

Gieryn, Thomas F. 1999. *Cultural Boundaries of Science: Credibility on the Line*. Chicago: University of Chicago Press.

Gillespie, Tarleton. 2007. *Wired Shut: Copyright and the Shape of Digital Culture*. Cambridge, MA: MIT Press.

Gold, Matthew K., ed. 2012. *Debates in the Digital Humanities*. Minneapolis: University of Minnesota Press.

Haraway, Donna J. 1988. "Situated Knowledges: The Science Question in Feminism and the Privilege of Partial Perspective." *Feminist Studies* 14 (3): 575–99. https://doi.org/10.2307/3178066.

———. 1991. *Simians, Cyborgs, and Women*. New York: Routledge.

———. 1997. *Modest_Witness@Second Millenium. FemaleMan_Meets_OncoMouse: Feminism and Technoscience*. New York: Routledge.

Jasanoff, Sheila. 1987. "Contested Boundaries in Policy-Relevant Science." *Social Studies of Science* 17 (2): 195–230.

Jasanoff, Sheila, and Sang-Hyun Kim, eds. 2015. *Dreamscapes of Modernity: Sociotechnical Imaginaries and the Fabrication of Power*. Chicago: University of Chicago Press.

Kirby, David A. 2011. *Lab Coats in Hollywood: Science, Scientists, and Cinema*. Cambridge, MA: MIT Press.

Knorr Cetina, Karin, and Urs Bruegger. 2002. "Global Microstructures: The Virtual Societies of Financial Markets." *American Journal of Sociology* 107 (4): 905–50. https://doi.org/10.1086/341045.

Kuhn, Thomas S. 1962. *The Structure of Scientific Revolutions*. Vol. 2, 2nd ed. Foundations of the Unity of Science. Chicago: University of Chicago Press.

Latour, Bruno. 1988. *The Pasteurization of France*. Cambridge, MA: Harvard University Press.

———. 1992. "Where Are the Missing Masses? The Sociology of a Few Mundane Artifacts." In *Shaping Technology/Building Society*, edited by Wiebe E. Bijker and John Law, 225–58. Cambridge, MA: MIT Press.

———. 2005. *Reassembling the Social: An Introduction to Actor-Network-Theory*. Oxford: Oxford University Press.

Latour, Bruno, and Steve Woolgar. 1979. *Laboratory Life: The Construction of Scientific Facts*. Princeton, NJ: Princeton University Press.

Law, John. 2004. *After Method: Mess in Social Science Research*. New York: Routledge.

Leigh Star, Susan. 2010. "This Is Not a Boundary Object: Reflections on the Origin of a Concept." *Science, Technology, & Human Values* 35 (5): 601–17. https://doi.org/10.1177/0162243910377624.

Lewenstein, Bruce. 1995. "Science and the Media." In *Handbook of Science and Technology Studies*, edited by Sheila Jasanoff, Gerald Markle, James Peterson, and Trevor Pinch, 343–60. Thousand Oaks, CA: Sage.

Light, Jennifer. 1999. "When Computers Were Women." *Technology and Culture* 40 (3): 455–83.

Lury, Cecelia, and Nina Wakeford, eds. 2012. *Inventive Methods: The Happening of the Social*. Culture, Economy and the Social. New York: Routledge.

Lynch, Michael, ed. 1985. *Art and Artifact in Laboratory Science: A Study of Shop Work and Shop Talk in a Research Laboratory*. Studies in Ethnomethodology. London: Routledge & Kegan Paul.

MacKenzie, Donald A. 2006. *An Engine, Not a Camera: How Financial Models Shape Markets*. Inside Technology. Cambridge, MA: MIT Press.

MacKenzie, Donald A., and Judy Wajcman, eds. 1999. *The Social Shaping of Technology*. 2nd ed. Buckingham: Open University Press.

Marres, Noortje. 2017. *Digital Sociology: The Reinvention of Social Research*. Malden, MA: Polity.

Merton, Robert. 1942. "The Normative Structure of Science." In *The Sociology of Science: Theoretical and Empirical Investigations*, edited by Norman W. Storer, 267–78. Chicago: University of Chicago Press.

Mitroff, Ian. 1974. "Norms and Counter-norms in a Select Group of the Apollo Moon Scientists." *American Sociological Review* 39:579–95.

Moll, Annemarie. 2002. *The Body Multiple: Ontology in Medical Practice*. Durham, NC: Duke University Press.

Nelson, Alondra. 2016. *The Social Life of DNA: Race, Reparations, and Reconciliation after the Genome*. Boston: Beacon.

Noble, Safiya Umoja. 2017. *Algorithms of Oppression: How Search Engines Reinforce Racism*. New York: New York University Press.

Oreskes, Naomi, and Erik M. Conway. 2010. *Merchants of Doubt: How a Handful of Scientists Obscured the Truth on Issues from Tobacco Smoke to Global Warming*. London: Bloomsbury.

Orr, Julian E. 1996. *Talking about Machines: An Ethnography of a Modern Job*. Ithaca, NY: Cornell University Press.

Pickering, Andrew. 1995. *The Mangle of Practice: Time, Agency, and Science*. Chicago: University of Chicago Press.

Pinch, Trevor, and Wiebe E. Bijker. 1987. "The Social Construction of Facts and Artifacts: Or How the Sociology of Science and the Sociology of Technology Might Benefit Each Other." In *The Social Construction of Technological Systems: New Directions in the Sociology and History of Technology*, edited by Trevor Pinch, Wiebe E. Bijker, and Thomas P. Hughes, 17–50. Cambridge, MA: MIT Press.

Proctor, Robert, and Londa Schiebinger, eds. 2008. *Agnotology: The Making and Unmaking of Ignorance*. Palo Alto, CA: Stanford University Press.

Schüll, Natasha. 2017. *Keeping Track*. New York: Farrar, Straus and Giroux.

Shapin, S. 1996. *The Scientific Revolution*. Chicago: University of Chicago Press.

Star, Susan Leigh. 1999. "The Ethnography of Infrastructure." *American Behavioral Scientist* 43 (3): 377–91. https://doi.org/10.1177/00027649921955326.

Star, Susan Leigh, and James R. Griesemer. 1989. "Institutional Ecology, 'Translations,' and Boundary Objects: Amateurs and Professionals in Berkeley's Museum of Vertebrate Zoology, 1907–39." *Social Studies of Science* 19 (3): 387–420.

Suchman, Lucy. 1987. *Plans and Situated Actions: The Problem of Human-Machine Communication*. Cambridge: Cambridge University Press.

———. 2011. "Subject Objects." *Feminist Theory* 12 (2): 119–45. https://doi.org/10.1177/1464700111404205.

Traweek, Sharon. 1988. *Beamtimes and Lifetimes: The World of High Energy Physicists*. Cambridge, MA: Harvard University Press.

Turkle, Sherry, ed. 2007. *Evocative Objects*. Cambridge, MA: MIT Press.

Vertesi, Janet, David Ribes, Laura Forlano, Yanni A. Loukissas, and Marisa Cohn. 2016. "Engaging, Making, and Designing Digital Systems." In *Handbook of Science and Technology Studies*, 4th ed., edited by Ulrike Felt, Rayvon Fouché, Clark A. Miller, and Laurel Smith-Doerr, 169–94. Cambridge, MA: MIT Press.

Wajcman, Judy. 2015. *Pressed for Time: The Acceleration of Life in Digital Capitalism*. Chicago: University of Chicago Press.

Winner, Langdon. 1986. "Do Artifacts Have Politics?" In *The Whale and the Reactor: A Search for Limits in an Age of High Technology*, edited by Langdon Winner, 19–39. Chicago: University of Chicago Press.

Woolgar, Steve. 1990. "Configuring the User: The Case of Usability Trials." *Sociological Review* 38 (S1): 58–99. https://doi.org/10.1111/j.1467-954X.1990.tb03349.x.

Wynne, Brian. 1992. "Misunderstanding Misunderstanding: Social Identities and Public Uptake of Science." *Public Understanding of Science* 1 (3): 281–304.

Introduction
Materiality

Laura Forlano

STS is well known for its unique approaches to the study of materiality, a central topic for inquiry since the founding of the field. The late 1980s saw the development of actor-network theory (ANT): an approach that provocatively assumed the analytical equivalence ("symmetry") of human and nonhuman actors (Callon 1986, 1987; Latour 1996, 2005; Law and Hassard 1999). Microbes, electrons, plants, animals, test tubes, people, and laboratory equipment therefore all played a role in scientific discoveries and technological developments. For ANT scholars, agency is not embedded in a single device, object, or person, but emerges from its distribution among this network of human and nonhuman actors. By the early 21st century, STS scholars regularly took such objects, devices, and tools as "matters of concern" (Latour 2004) by examining how they incorporate mixed agencies and politics: in other words, examining hybrid ontologies (Woolgar and Lezaun 2013).

A similar focus on materiality appeared in the work of feminist technoscience around the same time. This strand of research focuses on the ways in which gender and identity are constructed in and through science and technology, while emphasizing how knowledge is situated, embodied, and localized in such a way as to exclude minority voices (Haraway 1988). The recent development of new materialism synergizes this approach with ontologies research by paying attention to the hybrid agencies and ethics of living and nonliving things (Barad 2007; Bennett 2009; Dolphijn and van der Tuin 2012). We are asked to "meet the universe halfway" by viewing objects like quarks as imbued with both social understandings and material agencies. New materialism often takes up questions related to human relations with other beings in order to contribute to questions around climate change and the environment and has been particularly influential in multispecies anthropology (Kirksey 2014; Haraway 2008). These core approaches in STS define the field's attention to materiality not as a static, obdurate, or objective constraint upon social life, but as hybrid matter constituted through the arrangements of people and things, talk, and practice.

Along with text-based scholarship about materiality in STS, scholars have explored the topic through research methodologies that offer new ways of thinking about and engaging publics around complex sociotechnical issues (DiSalvo 2009; Michael 2012). These projects make representations and visualizations; design things, prototypes, and experiments; create opportunities for intervention and participation; and explore the topic through art and performance (Felt et al. 2016; Latour and Weibel 2005). Examples include Natalie Jeremijenko's Environmental

Health Clinic and Feral Robotic Dog project (Bratton and Jeremijenko 2008; Lane et al. 2006), Trevor Paglen's critical geography (2009a, 2009b; Paglen and Thompson 2007), Carl DiSalvo's adversarial design and speculative civics (2012, 2016; DiSalvo et al. 2016), Natasha Myers and Joseph Dumit's gaming and visualizations (Burri and Dumit 2008; Dumit 2014; Myers and Dumit 2011), Matt Ratto and Garnet Hertz's critical making (Hertz and Parikka 2012; Ratto 2011), and Hanna Rose Shell's films (2012a, 2012b). Within this tradition, the making of digital technologies and systems is of particular interest to this volume (Vertesi et al. 2016). As evidence for the growing interest in these more inventive and engaged forms of scholarship within STS, we have also explored design and making at workshops (Forlano et al. 2012; Loukissas et al. 2013) and in a "Making and Doing" exhibition at the Society for the Social Studies of Science conference since 2015.

Despite this focus on materiality in STS over the past several decades, emerging scholarship on the digital and the social in the 1990s initially emphasized the dematerialized, virtual nature of online human relations, rejecting earlier materialist theory. Drawing upon media studies or communication theory, the digital and material were essentialized and separated into discrete units. Digital essentialism still haunts many studies of emerging technology today, in part due to the linguistic difficulties of articulating the mutual shaping and interdependence of the material and the digital. Still, more recent work on digital systems in STS and related fields has gravitated toward more complex, even hybrid understandings of digital materiality (Blanchette 2011; Dourish and Mazmanian 2011; Pink et al. 2016). Such scholars explore the ways in which the digital can be understood to be material (Dourish 2017) or explore digital work as practical action. They also reclaim the material, social, and environmental conditions of digital production, use, and discard through investigations into maintenance (Graham and Thrift 2007), repair (Jackson 2014), failure and breakdown (Rosner and Ames 2014; Rosner and Fox 2016), and care (Mol 2008).

The chapters in this section serve to advance and deepen our understanding of digital materiality. Rather than offering generalizations about the properties of materiality or digitality, the essays explore how digital materialities emerge in their sites of inquiry. Alexandre Camus and Dominique Vinck, for example, offer an ethnographic account of the digitization of the extensive concert archives of the Montreux Jazz Festival over the past 50 years. For the archives, "becoming digital" is a dynamic and interactive process of digital craftsmanship that requires the embodied, material labor (physical, cognitive, visual, and aural), time, and effort of engineers. In this case, the digital materiality of music takes on specific qualities such as tangibility and fluidity as well as textures such as softness, thickness, weight, boundaries, spatiality, relations, and networks. As the concerts are digitized—rather than becoming immaterial—they are rematerialized into new forms such as individual songs, playlists, and setlists. These new material forms are clickable, taggable, searchable, and indexable; as such, they have new associations with one another as well as with networks. The authors thus illustrate the ways in which the digital is distributed into networks that "do and undo the concerts," thereby saving the archives and allowing them to circulate. They also demonstrate how the digital and the material are not discrete categories or properties, but emerge locally in dynamic relation to each other.

Yanni Loukissas's interactive visualization and essay, which is presented as an online "data documentary," complements this volume. The piece investigates the "life and death of data" through engagement with the plant collection of Harvard

University's Arnold Arboretum, which includes 71,250 accessions between 1872 and 2012. This project draws on social and cultural research as well as the making of a longitudinal digital visualization in order to inquire into the social, material, and institutional histories of data. It also questions how to study the institutions that create, maintain, and share large digital collections. The visual and literal dimensions of this piece inspire us to examine data differently, developing notions of "hybrid materialities." Specifically, Loukissas highlights the tension between the virtual, ancillary, freely accessible, open, and transparent qualities that are often attributed to digital data on the one hand, and their materiality, centrality, locality, and situated significance on the other.

David Ribes considers the methodological implications of studying digital materiality by drawing on four intellectual traditions: ethnomethodology, actor-network theory, the anthropology of classification, and historical ontology and epistemology. Specifically, he asks, "How do we approach studies of things, objects, stuff, and materials, their agencies and interrelations, in action and across time?" Drawing on empirical cases from two large ethnographic studies, the Multicenter AIDS Cohort Study (MACS) and Long-Term Ecological Research (LTER), Ribes argues that researchers must "discover" the digital and the material through fieldwork. In these cases, blood and water samples as well as related datasets exhibit tremendous flexibility, allowing them to be understood as either digital or material in nature depending on the specific context. As a result, Ribes argues that interpretations about the nature of materiality must be read through multiple, sometimes competing, theoretical traditions.

Nerea Calvillo deploys digital visualizations as an inventive research method (Lury and Wakeford 2012) for "thinking with the environment." Across two projects—*In the Air* and *Pollen In the Air*—Calvillo describes the process of hands-on collaborative making of visualizations as a means of investigating the materiality of invisible gases and the politics around public air quality datasets. The visualizations are speculative in that they bring to life new imaginaries and worlds that engage with environmental issues. The resulting airscapes—aerial maps that present "air as a landscape that can be inhabited"—are a form of ethnographic engagement with the air. Calvillo's visualizations do not merely represent scientific data about air, but also reimagine the relations between humans and gaseous nonhumans through embodied, affective experiences. Most significantly, these projects reconfigure the politics around "air as a harm" to humans in favor of a feminist, multispecies encounter predicated on collective values and environmental justice.

As a group, then, these essays offer a perspective upon digital encounters that embraces the material without essentializing its properties. This requires engaging with digital materiality as hybrid, shifting, and situated; an emic category to be analyzed in context; and a property to be played with and ultimately troubled.

Works Cited

Barad, Karen. 2007. *Meeting the Universe Halfway: Quantum Physics and the Entanglement of Matter and Meaning*. Durham, NC: Duke University Press.

Bennett, Jane. 2009. *Vibrant Matter: A Political Ecology of Things*. Durham, NC: Duke University Press.

Blanchette, Jean-François. 2011. "A Material History of Bits." *Journal of the American Society for Information Science and Technology* 62 (6): 1042–57.

Bratton, Benjamin H., and Natalie Jeremijenko. 2008. *Suspicious Images, Latent Interfaces*. New York: Architectural League of New York.

Burri, Regula Valérie, and Joseph Dumit. 2008. "Social Studies of Scientific Imaging and Visualization." In *The Handbook of Science and Technology Studies*, edited by Edward J. Hackett, Olga Amsterdamska, Michael Lynch, and Judy Wajcman, 297–317. Cambridge, MA: MIT Press.

Callon, Michel. 1986. "The Sociology of an Actor-Network: The Case of the Electric Vehicle." In *Mapping the Dynamics of Science and Technology*, edited by Michel Callon, John Law, and Arie Rip, 19–34. London: Macmillan.

——. 1987. "Society in the Making: The Study of Technology as a Tool for Sociological Analysis." In *The Social Construction of Technological Systems: New Directions in the Sociology and History of Technology*, edited by Wiebe E. Bijker, Thomas P. Hughes, and Trevor Pinch, 83–103. Cambridge, MA: MIT Press.

DiSalvo, Carl. 2009. "Design and the Construction of Publics." *Design Issues* 25 (1): 48–63.

——. 2012. *Adversarial Design*. Cambridge, MA: MIT Press.

——. 2016. "The Irony of Drones for Foraging: Exploring the Work of Speculative Interventions." In *Design Anthropological Futures*, edited by J. H. Rachel Clark and Kapser Tang Vangkilde, 139–54. New York: Bloomsbury.

DiSalvo, Carl, Tom Jenkins, and Thomas Lodato. 2016. "Designing Speculative Civics." Paper presented at the CHI Conference on Human Factors in Computing Systems, San Jose, CA.

Dolphijn, Rick, and Iris van der Tuin, eds. 2012. *New Materialism: Interviews and Cartographies*. Ann Arbor, MI: Open Humanities Press.

Dourish, Paul. 2017. *The Stuff of Bits: An Essay on the Materialities of Information*. Cambridge, MA: MIT Press.

Dourish, Paul, and Melissa Mazmanian. 2011. "Media as Material: Information Representations as Material Foundations for Organizational Practice." Paper presented at the Third International Symposium on Process Organizational Studies, Corfu, Greece.

Dumit, Joseph. 2014. "Writing the Implosion: Teaching the World One Thing at a Time." *Cultural Anthropology* 29 (2): 344–62.

Felt, Ulrike, Rayvon Fouché, Clark A. Miller, and Laurel Smith-Doerr, eds. 2016. *The Handbook of Science and Technology Studies*. Cambridge, MA: MIT Press

Forlano, Laura, Dehlia Hannah, Kat Jungnickel, Julian McHardy, and Hannah Star Rogers. 2012. "Experiments In (and Out of) the Studio: Art and Design Methods for Science and Technology Studies." https://www.hastac.org/opportunities/4seasst-workshop-experiments-and-out-studio-art-and-design-methods-science-and.

Graham, S., and N. Thrift. 2007. "Out of Order: Understanding Repair and Maintenance." *Theory, Culture & Society* 24 (3): 1–25.

Haraway, Donna. 1988. "Situated Knowledges: The Science Question in Feminism and the Privilege of Partial Perspective." *Feminist Studies* 14 (3): 575–99.

——. 2008. *When Species Meet*. Minneapolis: University of Minnesota Press.

Hertz, Garnet, and Jussi Parikka. 2012. "Zombie Media: Circuit Bending Media Archaeology into an Art Method." *Leonardo* 45 (5): 424–30.

Jackson, Steven J. 2014. "Rethinking Repair." In *Media Technologies: Essays on Communication, Materiality, and Society*, edited by Tarleton Gillespie, Pablo Boczkowski, and Kirsten Foot, 221–39. Cambridge, MA: MIT Press.

Kirksey, Eben. 2014. *The Multispecies Salon*. Durham, NC: Duke University Press.

Lane, Giles, Camilla Brueton, George Roussos, Natalie Jeremijenko, George Papamarkos, Dima Diall, . . . Karen Martin. 2006. "Public Authoring & Feral Robotics." *Proboscis*, no. 11:1–12.

Latour, Bruno. 1996. "On Actor-Network Theory: A Few Clarifications." *Soziale Welt* 47:369–81.

——. 2004. "Why Has Critique Run Out of Steam? From Matters of Fact to Matters of Concern." *Critical Inquiry* 30 (2): 225–48.

——. 2005. *Reassembling the Social: An Introduction to Actor-Network-Theory*. Oxford: Oxford University Press.

Latour, Bruno, and Peter Weibel, eds. 2005. *Making Things Public: Atmospheres of Democracy*. Cambridge, MA: MIT Press.

Law, John, and John Hassard. 1999. *Actor Network Theory and After*. Oxford: Blackwell.

Loukissas, Yanni, Laura Forlano, David Ribes, and Janet Vertesi. 2013. "digitalSTS and Design." http://stsdesignworkshop.tumblr.com.

Lury, Celia, and Nina Wakeford. 2012. *Inventive Methods: The Happening of the Social*. New York: Routledge.

Michael, Mike. 2012. "'What Are We Busy Doing?' Engaging the Idiot." *Science, Technology, & Human Values* 37 (5): 528–54.

Mol, Annemarie. 2008. *The Logic of Care: Health and the Problem of Patient Choice*. London: Routledge.

Myers, Natasha, and Joseph Dumit. 2011. "Haptic Creativity and the Mid-embodiments of Experimental Life." In *A Companion to the Anthropology of the Body and Embodiment*, edited by Frances E. Mascia-Lees, 239–61. New York: Wiley-Blackwell.

Paglen, Trevor. 2009a. *Blank Spots on the Map: The Dark Geography of the Pentagon's Secret World*. New York: Penguin.

——. 2009b. "Experimental Geography: From Cultural Production to the Production of Space." *Brooklyn Rail*, March 6.

Paglen, Trevor, and Adam Clay Thompson. 2007. *Torture Taxi: On the Trail of the CIA's Rendition Flights*. London: Icon Books.

Pink, Sarah, Elisenda Ardevol, and Debora Lanzeni, eds. 2016. *Digital Materialities: Design and Anthropology*. New York: Bloomsbury.

Ratto, Matt. 2011. "Critical Making: Conceptual and Material Studies in Technology and Social Life." *Information Society* 27 (4): 252–60.

Rosner, Daniela K., and Morgan Ames. 2014. "Designing for Repair? Infrastructures and Materialities of Breakdown." Paper presented at the 17th ACM Conference on Computer Supported Cooperative Work & Social Computing, Baltimore.

Rosner, Daniela K., and Sarah E. Fox. 2016. "Legacies of Craft and the Centrality of Failure in a Mother-Operated Hackerspace." *New Media & Society* 18:558–80. doi:10.1177/1461444816629468.

Shell, Hanna Rose. 2012a. "Blind: The Phenomenology of Camouflage." *Sensate*, December. http://sensatejournal.com/hanna-rose-shell-blind-on-the-phenemology-of-camouflage/.

——. 2012b. "Locomotion in Water." *Journal of Short Film*, Spring.

Vertesi, Janet, David Ribes, Laura Forlano, Yanni Loukissas, and Marisa Cohn. 2016. "Engaging, Designing and Making Digital Technologies." In *The Handbook of Science and Technology Studies*, edited by Ulrike Felt, Rayvon Fouché, Clark A. Miller, and Laurel Smith-Doerr, 169–94. Cambridge, MA: MIT Press.

Winner, Langdon. 1986. "Do Artifacts Have Politics?" In *The Whale and the Reactor: A Search for Limits in an Age of High Technology*, edited by Langdon Winner, 19–39. Chicago: University of Chicago Press.

Woolgar, Steve, and Javier Lezaun. 2013. "The Wrong Bin Bag: A Turn to Ontology in Science and Technology Studies?" *Social Studies of Science* 43 (3): 321–40.

Unfolding Digital Materiality
How Engineers Struggle to Shape Tangible and Fluid Objects

Alexandre Camus and Dominique Vinck

Grasping a Matter-Network

In this chapter we address two issues related to the notion of digital materiality. The first one concerns the fluidity: On what rests the fluidity of the digital? And how can we qualify the process leading to the production of objects that can circulate? The second one concerns the *mode of existence* (Souriau [1954] 2009; Latour 2013) of digital artifacts: What does becoming digital imply for an artifact? How to qualify the dynamic of becoming digital? We deal with these questions, keeping distance from an understanding of digital as a matter, which can be defined according to an opposition container/content or matter/form. We do not propose a new minimal unit that is supposed to reduce the composition of a digital matter. On the contrary, we chose to tackle the digital materiality from the more dynamic angle of the material, which we define as the matter mobilized in activity and which is transforming as the interactions between the craftsman and the materials unfold. First, a constantly growing number of people are involved in the production of information and digital craftsmanship: these interactions remain to be studied. Second, following the moves of the material will enable us to overtake the vain quest to establish the minimal unit of the digital.

STS scholars have given renewed importance to the materiality of information, revealing the machines and invisible workers (Star 1991; Denis and Pontille 2012) that allow them to function—as well as the protocols and organizations that support them (Bowker et al. 2010). The attempt to conceive information materiality in terms of material infrastructure and extended embedded artifacts led to the birth of a subfield that Bowker et al. (2010) call "Information Infrastructure Studies" (IIS). As a major result IIS have gone *behind the screen*, revealing the *invisible infrastructures* (Star and Ruhleder 1996; Star and Strauss 1999) that produce information.[1] In a nutshell, these works have led us to think about information and the digital as a complex and heterogeneous *assemblage*. These works also question the understanding of information and the digital as an intrinsically fluid content that is ontologically distinct from and superior to materiality. In the wake of laboratory ethnographies (Latour and Woolgar 1979; Lynch 1985; Pinch 1986), IIS conceptualize information as the result of concrete operations from which it cannot be dissociated. The question *"what is digital information made of?"* can be reformulated into another question: *"how is digital information produced?"* The answer rests on the study of the sociotechnical networks that participate in shaping information and constitute its eminently material being (Denis and Pontille 2012).

Digital information is indeed difficult to grasp. Understanding it often means understanding the pipes, machines, invisible work, protocols, classifications, and standards through which it is produced and flows, with particular attention to collective data shaping. In IIS, the emphasis was put on aggregated entities such as collaborative research systems (Star and Ruhleder 1996), transnational databases (Bowker 2000; Millerand 2012; Heaton and Proulx 2012; Heaton and Millerand 2013; Van Horn et al. 2001), health information systems (Hanseth and Monteiro 1997), and digital library systems (Gal et al. 2004). These very interesting works can lead to the strange impression that everything depends on the infrastructure. ISS already suggested that interactions between operators and the material are important to study, but we still lack a close look at the practical way the shaping is engaged.

In order to extend ISS results with a closer look at how information is produced, this chapter draws on the analysis of a process of digitization, that of the concert recordings of the Montreux Jazz Festival. Within this ambitious project of digitization, we focus on a specific activity, that of indexing events (speeches, musical performances, applause, etc.) in the digitized files. To complete this work, operators must "tag" the events that occurred onstage at the time of recording. In order to do so, the lab team is often confronted with the need to remove a number of uncertainties that "primary digitization" (conversion of the analog signal into a digital signal) has not dispelled: What is recorded on the digitalized tapes? In what state are they? What does their video and video signal say? What songs were played that evening onstage?

By working on the case of soft and very light artifacts—the shaping of digital pieces of music from recorded live concerts—we challenge the "hardness/softness" of things and look at specific textures of the digital such as digital spatiality, relations and embedded networks, enabling and constraining. The challenge here is that of working on a seemingly double immateriality: music and digital. Following Hennion (2015b), we propose to consider digital (music) as something that has no proper matter of its own while only manifesting its presence to those who interact concretely with it. It is made of matter in the sense of Latour (2013), that is to say, "the set of all the other beings upon which any given entity depends" (Mode of Existence n.d.).[2] This conception of matter as a network of relations allows to free the question "what is digital made of?" from the quest for an essence by providing a relational understanding of materiality. The *matter-network* proposed by Latour relates to a sociotechnical network, that is to say a set of associations between heterogeneous entities, human or not (Callon 1986), an *assemblage* (Latour 2005) that makes beings exist. Since a thing is made through movements and the process of associating all the elements that constitute it, there is no reason to seek which of these elements is the most important or ontologically superior, nor even of what "ingredients" it is composed. Understanding what a thing is made of thus equates to unfolding the operations that shape and maintain the sociotechnical network that underlies the existence of this thing.

The aim of the analysis is then to describe this network of associations, paying attention to all the actions and actors that contribute to the *assemblage* that is the thing. Since the network is not defined nor delimited a priori, the concept of matter-network does not provide any limit to the associations that are to be taken into account; this means that we have to *follow the associations* (Latour 1987). This points to the relational dimension of the notion of materiality and to the interactions that are needed for the matter-network to exist. Such an analysis is empirically grounded in the description of observable interactions between the elements of this matter-network.

Through this study, we aim to contribute to theoretical and methodological debates about the ways in which STS can help to think about and engage with digitality. Considering the digital shaping of music recordings helps to theorize the question of materiality and the ways in which it can be taken into account through a variety of practices studied by STS scholars, such as producing knowledge or shaping a theory. Our main question is, what does "becoming digital" imply for the digitized artifacts? In fact, following ISS we consider that *doing* digitality takes material work, time, and effort by human beings, but we also propose to add an analysis of what *making* digital things implies. We consider digital things as a contingent result of a process of association, which is, at the same time, precisely aiming for reducing the contingence of the existence of digital productions. Then, the key point is to account for this crucial issue by proposing an analysis of the dynamic of reducing the contingence of digital things.[3]

An Ethnography Inside the Shaping of Cultural Digital Beings

In this chapter we describe how digital information is produced through a case study in the field of digitization of cultural heritage. Carrying out ethnographic fieldwork in the lab of Ecole Polytechnique Fédérale de Lausanne (EPFL) in charge of this archive, we follow the process of digitizing the archives of the Montreux Jazz Festival. In this lab, a few dozen researchers, engineers, and technicians work daily on the digital files of 10,000 hours of audiovisual recordings. Ethnographic investigation allows us to grasp the process of shaping digital entities and, as participating observers, to take part in it and to understand its phenomenological implications. Through this inquiry, we show how the methodology and results of IIS can be completed and deepened not only by going *behind the screen* (Jaton 2017) but also by taking a closer look at, and accounting for, the experiences made *inside and with the screen*. In order to do so, we have personally experienced the set of tasks of shaping digital beings. Alongside the lab team for six months, we took part in the digitization process. This experience, together with informal interviews and observations of work situations over one year, allows us to provide a precise account of the constitution of *digital beings*.[4] Then the work of information is grasped in its materiality: stages, operators, and tools, and some tangible (computers, keyboards, and mouse) but also more evanescent ones, such as a signal processing software. This work remains difficult to describe, as undemonstrative as a mouse click that positions a cursor on the screen. However, the click's almost silent action has irreversible consequences on the inherently material destiny of information.

In the first part of the chapter, we briefly present the history of the recordings of the festival concerts. We describe the evolution of the network that supports the conservation of the tapes, that is, their forced rescue through digitization. We set the context, which explains why an engineering university took into its hands the destiny of a segment of cultural heritage. It then becomes possible to understand the switch from conservation to promotion in the process of transforming heritage into raw material for innovation. In the second part, we enter the lab where digital files are monitored, digitized, and processed, starting with the investigation initiated by the lab, to determine what there is on the digital files corresponding to the tapes. We will see how the team builds objects that can be manipulated and qualified. We describe how they interact with physical media, tools, and interfaces, producing *holds* (Bessy and Chateauraynaud 2003; Hennion 2015a) on the digital files.[5] We show that the existence of the informational material becomes more and more certain as they build its *equipment* (Vinck 2011). We account for the negotiations between these operators and the "pieces of concerts" they manipulate, leading to the transformation of the archive into an original artwork. We focus on the process of establishing the boundaries for the musical pieces identified within the concerts. These operations of building *pieces of digital music* often face objects that resist the definition of their boundaries. In the last section, we discuss some of the results of this study in the light of works dealing explicitly with the issue of materiality and of a nonessentialist understanding of materiality.

Part 1: Digitization to Rescue Concerts

Tapes Threatened by an Eroding Network

Since its foundation, the Montreux Jazz Festival's organizers have recorded concerts. The archive now represents about 5,000 hours of audio and video recordings. Endorsed as Memory of the World by UNESCO in 2013, this unique collection covers 50 years of jazz, blues, and funk music. As famous US artists and politicians (e.g., Stevie Wonder, Herbie Hancock, Hillary Clinton) have participated, the Montreux archive is one of the most valuable testimonies to American contemporary music. The records were made in the most innovative formats of the time: stereo multitracks since 1975, HDTV since 1991, 3-D experiments in the 2010s. According to the leaders of the digitization project, the decision to put together the archive dates back to 1988, when a manager of the festival made a request to the national television network, which was supposed to preserve the recordings. He asked them to provide a given recording and discovered that the tape had been reused to record a football match between two small towns. There was no other copy of the recording; the memory of that concert was therefore lost. Worried, the festival's founder decided to retrieve the records and construct a building on his property to house safely the tapes whose number increased year by year.

However, the "bunker," as aficionados call it, was quick to show its limits. Seemingly safe in Compactus storage that housed them for 20 years, the tapes were again threatened, this time by oxidation. In order to be read, some formats required technologies that were becoming scarce (two-inch audio and U-matic for example). Some machines could no longer be found; others were very bulky. In addition, the knowledge required to run them was not easily available; it was sometimes concentrated in a few specialized companies, sometimes possessed by only a few people no longer working actively in the field. The 18 different formats used during 50 years for recordings and the rarity of some of them give value to the archive and, at the same time, make the perennial conservation of the archive difficult. Technology and the knowledge on which the archive depended were endangered.

Other elements that support the existence of the tapes and their contents are stable and constitute good allies in the preservation of the recordings: the bunker of the archive (regularly maintained and built according to the best standards of the time); up-to-date inventories; contracts with artists who rely on the wide network of the copyright; amateurs who continue to listen to edited or pirated copies; collectors who maintain their collection and sometimes enrich it with tapes acquired when organizations that held copies let them go. All those elements were thereby compensating somewhat for the erosion of the network of organizations and actors supposed to ensure the preservation. However, these mediators associated with the archive who converge on "concerts that have taken place" were not enough to ensure the sustainability of their recordings. Gradually, their mediation weakened and the network supporting the recordings crumbled in places.

This threat highlights the network-constitutive mediations of memory and illustrates the relational dimension of maintaining in existence the concerts. If the recorded concerts (their materiality) depend on such a network to exist and not to disappear, the question may arise, *what is their content?* Thus formulated, the question assumes, however, a dichotomy between material form (media) and content that we believe should be abandoned in order fully to grasp the relational dimension of materiality (Law and Mol 1995; Hennion 2003; Ingold 2011; Latour 2013). Examining the network of associations that constitutes the existence of the "concerts that have taken place" also allows us to understand how this existence is threatened and the dynamics of the interaction between the elements of the network that do and undo the concerts.

From Conservation to Promotion and Exploitation: Rescuing the Concerts by Making Them Fluid and Able to Circulate

The destiny of the concert recordings of Montreux brings together actors whose collaboration may be surprising. People originating from the worlds of culture and technology agree on an operation to rescue the recordings. Because of its concern for preserving this heritage and awareness of the importance of the technical and financial investment required for the sustainability of the archive, the responsible Heritage Foundation has signed a partnership with EPFL, which seeks to develop its practice in the field of the digitization of cultural heritage. Access to large digitized corpora and to an already shaped *data territory* (Vinck and Camus, forthcoming) will provide resources for the development of its laboratories.

In 2011, on the basis of these contacts and agreements, an ambitious and large-scale project of digitization of audiovisual archives was born. Over five years, more than 15 million Swiss francs were allocated to digitize 10,000 tapes covering the 50 years of the festival in order to preserve this heritage. But the aim was also to promote the festival and its archives and to generate added value (academic publications, design of products and services, creation of start-up). At the time of the survey, although 75% of the tapes were digitized and available on the five-petabyte storage system hosted in a secure room beside the computing machinery and other sensitive data from the engineering university, the destiny of the archive remains uncertain. Conservation does not happen by itself; our interlocutors state that there is no permanent solution other than the systematic and regular transfer of the archive from one format to another following their evolution, coupled with a copy in at least three different media and locations. To cope with the uncertainty of conservation, the hypothesis formulated by those responsible for the lab—which was, in fact, created at the occasion of this digitization project—is that the sustainability of the archive will be achieved through its promotion and exploitation, including by offering it as *raw material for innovation* to different laboratories of the university. Thus, with the material attached to new sociotechnical networks, the risk of its disappearance would be reduced, including the risk of missing out on new computer formats, through the monitoring of technological developments by researchers from the university for their own needs. The lab thus achieves a translation (Callon 1986) between the preservation of a cultural archive and the scientific life of a state-of-the-art lab network. The digitization of the festival archive is therefore driven by a goal of promotion of the labs involved and of the university that hosts these labs. The association of the archive with this network of labs should also produce a dynamic of innovation from which archive promotion should benefit.

Between 2011 and 2014, five labs and over 30 researchers worked on the digitized archive, resulting, for example, in the development of a musical suggestion application and automatic playlist generation (*Genezik*) and in an application allowing one to browse the festival's archives on an iPad (*Archive Discovery Application*). These devices form new sociotechnical extensions of the archive and thus extend its existence. They feed on the digitization of the archive and, at the same time, guide the treatment of the digital archive according to their specific needs. They depend, in particular, on a music content classification system used by contemporary media players: the playlist. To serve these applications, recorded and digitized concerts must be formatted so as to generate a playlist. Through this way of processing digital files, the suggestion algorithm is able to offer a "musical journey" based upon the analysis of the first piece selected by the user. From this first selection, the application developed by researchers in signal processing analyzes the musical content (they call it the "musical DNA") and calculates the supposed tastes of the user so as to offer song transitions as "soft" as possible according to "audio only" parameters (rhythmic structure, timbre, harmonic progression).[6] The application thus distinguishes itself from its competitors whose algorithmic recommendations depend on metadata (for example, qualification of the "style")—an unreliable criterion according to the researchers of the project. The application allows users to discover pieces they had forgotten and should nevertheless enjoy. However, it assumes that the digital files it handles are equipped with tags transforming concerts into a world of independent musical pieces, that is, clearly marked with a beginning and an end. For this application, an entire concert is not suitable material for deployment; it needs pieces. The lab accordingly shapes digital files, allowing the archived concerts to be deployed in new sociotechnical networks, at the price of a *musical-material translation* of concerts into playlists. The *Archive Discovery Application* offers users the possibility to navigate the festival concerts on an iPad by selecting a year, a concert, or a song. The concert as a digital entity is not a problem as long as it is equipped with tags that allow it to be read and translated as if it were a playlist of the concert. This playlist allows the user to switch from one song to another.

Both applications briefly presented here highlight the fact that the rescue of endangered tapes involves the creation of new networks of relations between entities that are themselves also new. The association of the archive with these technological artifacts offers it new conditions for digital existence; however, this association involves a series of transformations of the archive and concerts. The temporality of recordings and concerts does not remain intact; it does not become what it would have been in a classical conservation process of archives where the integrity of the document is an end in itself. Embedded in a dynamic of digitization, the archive is put to the test, employed and transformed in order to participate in its promotion, which is posed as a condition of its preservation. The renewed vitality of the concerts, through the sociodigital treatment they undergo, suggests that they are on track to be saved. Digitization is the opportunity to decouple threatened concert tapes and to embed them into a new network under construction. The survival of the festival concerts depends on the success of this construction and association with new entities that are, from now on, protective of the henceforth digital existence of the recordings.

Part 2: Multiplying Mediators in Order to Obtain Tangible Material

In this part of the chapter, we continue the investigation of the digital deployment of the music archive by entering the lab in which this digital music material, which is intended to circulate in other labs, is prepared. We begin by operating an "infrastructural inversion" (Bowker 1994) to highlight the operations that make and unmake the digital material. We situate our analysis upstream of the circulation outside the lab, at the time when a collection of digitized tapes acquires circulation potential. This preparation for circulation involves manipulations of digitized records and qualification operations of what there is "in" the files.

Getting Acquainted with the Digital Material

In order to create mobile units and tag concert sequences (*songs* of *playlists*), researchers transform digital files into a material with which interactions become possible. They begin by looking for a way to manipulate the material and "catch the tapes." Their investigation consists in qualifying what these tapes may *be* and what they might be *made of.*

At its arrival at the lab, the tape is digitized and copied onto two durable storage media.[7] A unique ID and a digitizing report available in the project database are already associated and linked to the magnetic tape data storage. These are the basic handles available for researchers to locate files in the collection under construction. They also provide them with a set of clues concerning their probable state: if the digitization has gone well, the signal is probably in good condition. This is precisely what they will begin to verify. The investigation they initiate shows that digitization means not dematerialization but rather another materialization The primary digitization transforms a single magnetic tape into six files stored on two physical supports (LTO cartridges). To each file corresponds a unique ID that takes the formalism of the analog archive inventories, to which is added the file name extension (e.g., WAV, AVI). These indications enable them to manage the unloading of the files from one of the LTO cartridges into the storage system. The primary, "predigital" archive is the copy of the original magnetic tape on a new physical support and magnetic tape. This archive is kept as such on its new support, duplicated and stored in a safe, waiting to migrate to the next generation of the support. The secondary (II) and tertiary (III) versions of the digital archive are uploaded into the storage system. The secondary archive will provide the basis for continuing the process due to its good value in relation to weight (about 50 GB per hour of recording) according to the researchers.

On the basis of this version, they transform the files to prepare pieces on which they will run tests. They then undertake a series of tests to learn more about the tapes. The first test concerns the signal. To qualify the audio signal, they use specialized software (*Adobe Audition*), which is their main tool for audio signal processing. For the video signal, they use other specialized software (*Adobe Premiere*) and tools present in the "reference monitors" with advanced signal processing functions.

The tests and inspections of signals constitute an introduction. The signal is then taken as an intermediary through which researchers catch "what there is on the tape." This contact is limited to the most experienced members of the lab. In concrete terms, the researcher opens the file corresponding to the digital copy of the tape in its secondary archive version with the specialized software. Her workstation is equipped with two screens. On the top of the right screen appears the moving image of the video recording, under which are displayed the different signal channels: video is represented by a sequence of still *frames*, audio by the graphical representation of the *waveform* corresponding to each stereo channel. Audio can be listened to through headphones, but most of the time researchers only see it on the screen.

The concert—both its sound and image—is thus visually deployed on multiple screens. Two screens are used for this operation described as *quality assessment*. A third screen, known as the "reference screen," is put to use from time to time for some operations. On the right-hand screen, the video is displayed in play mode via the software interface; on the second, left-hand screen, the researcher varies visualizations of the video signal corresponding to different approaches of the signal (see figure 1). The lab has previously referred to local experts, who have long been related to the recording of the festival concerts, to help establish acceptance criteria, such as chrominance (part of the signal concerning the color) and luminance (level of light). They have agreed to set the luminance of the video between 0.3 and 0.9–1.1. The researcher explores the digitized tapes to value indicators that allow her to estimate the quality of the digitization performed by the subcontractor.

The researcher involved in this process of testing the signal is constantly navigating between the two main screens that combine different approaches of the recordings through specific visualizations, each corresponding to different types of tests. To chrominance and luminance is added the detection of errors and defects, such as *dropouts* (small losses of analog signal due to digitization or originating in the recording itself; see figure 2).

During this first stage, browsing in the video is done randomly, through trial and error. The mouse cursor moves and clicks on the timeline, which is located on the right-hand screen, allowing circulation in the concert file. The representations of the signal coexist with the video. Moving forward on the *timeline*, the researcher goes ahead in the recording and in the discovery of the digitized tape, seeking to verify that there is still something that looks like a concert, assessing the state of the signal and detecting possible defects.

If indicators of the video signal delivered on the screen do not show values outside the standards, the read-discovery-analysis continues linearly, moving

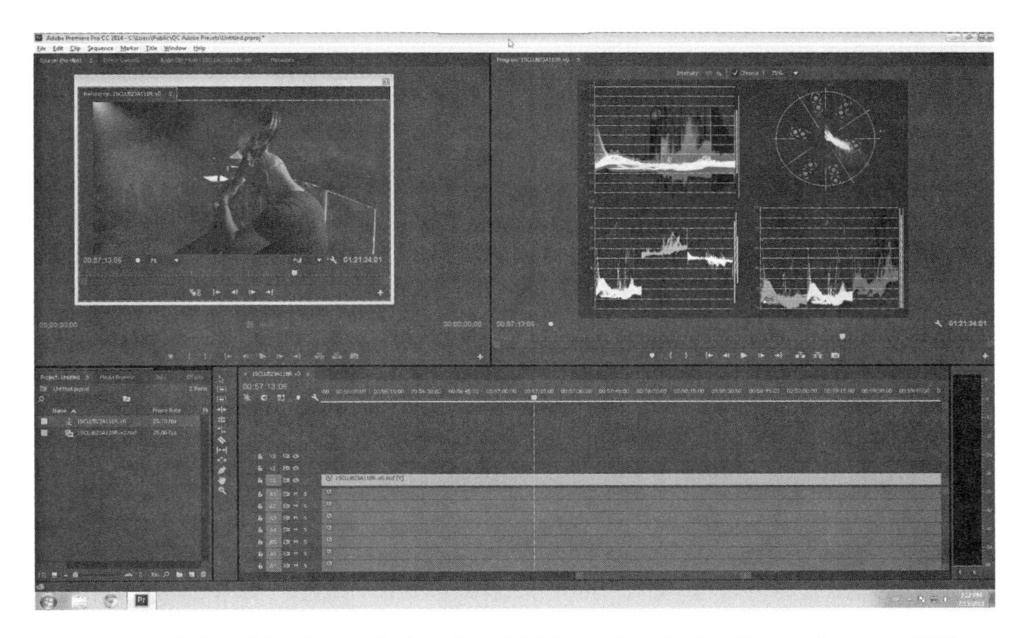

FIGURE 1: Variation of the signal in the interface of *Adobe Premiere*. Credits: Metamedia Center, EPFL.

FIGURE 2: Dropouts (see on the hear). Credits: Metamedia Center, EPFL.

forward in the recording. However, when an anomaly is detected, scrolling is immediately stopped and the researcher goes back. Too high luminance or chrominance leads her to a more detailed examination: Is it certain? By how much? Under what conditions? Which color exactly? The dropouts also interrupt the reading because their perception involves finding the incriminated frames. As a first step, she looks back on her perception and aligns it on a part of the video signal. She identifies the sequence (zone) where the dropout appeared. Then she zooms inside the frame sequence represented by a series of rods. Finally, she changes tools. Leaving the mouse, she moves to the keyboard and uses the arrows, scrolling through the frames one by one until she finds and notes the timecode (to the hundredth of a second) that matches the defect.[8] This operation can be tricky because the matching of perception and the sequence that contains the defective images is not easy. Sometimes the perceived dropout is not found and the reading-analysis resumes.

Giving Birth to Musical-Digital Intermediary Objects

This first meeting between researchers and digitized tapes is an important step in understanding the recordings because *holds* are built in order to qualify them. Knowledge is channeled through the signal that translates the digital file into an object that can be handled with clues, indicators, and concepts that are known and tested by the specialists of signal processing. The signal appears on the screen, and it is "here" or "there," on its graphical-digital representation, that the researchers identify relevant information, such as a remarkable value of the signal or images containing *dropouts*. The tapes are then recognized as containing content that the researchers can qualify, at least the small portions that they can qualify and with which they build an object of knowledge. They thus begin to know "what the recordings are made of."

Approaching the tape through its audio or video signal also allows them to connect the contents of the tape with instruments mastered by the researchers. With their conceptual and instrumental equipment, they capture and translate a portion of what is on the tape and spread it over two or even three 22-inch screens. Even deployed in this way, the qualification of the signal remains difficult; the material is not fully captive. Some reactions to the signal testing are furtive and sometimes difficult to catch even with a trained eye.

The result of this first encounter is materialized by a report (*quality assessment report*), another computer file that reports on the meeting with and the behavior of the tape grasped through its video signal. Researchers formalize *the competence deduced from observed performances* (Latour 2004). This process leads to descriptions that increase knowledge while multiplying the elements supposed to be constitutive of digitized records (signal, color, light, frame, timeline, timecode, etc.). These elements appear as *intermediary objects* (Vinck 2011) at the interface level (screens, headphones) and are sometimes *inscriptions* (Latour and Woolgar 1979), due to operations carried out by researchers, their computers, and algorithms. This set of successive operations of transforming the recording make it an object of knowledge and a component of the information infrastructure constituted by the digital archive.

Preparing the Recorded Concert for New Encounters: Networking the Content with Other Files

We have just seen a foundational stage of the process that helps dispel initial uncertainty about the state of digitized tapes and their capacity to be grabbed. The trial also involves many researchers, their techniques, and their skills. This first encounter verifies whether it is possible to go further in relation to the recording. The sociotechnical network constitutive of the material has significantly increased by connecting with instruments and scientific skills of the lab that multiply the material properties. However, we still have to consider many operations before the digitized tapes become equipped concerts that will drive musical playlists. Now, we briefly account for the deployment of the material before stopping at an important moment of interaction between the operators and digital-musical material.

The reconstitution of a concert involves a detailed examination of the recordings on which their existence depends. During this review, the researchers seek to answer questions such as these: What happened that night on the stage? What was played? In doing so, they "reveal" the concert that is supposed to be "contained" in the tapes. By putting to work these recordings as well as inventories and other traces that have been kept since the days of video recording (labels on the case of the tape and ID), researchers link tapes to concerts and identify objects (such and such piece of music) that are supposed to exist in the file of the digitized concert. Up until this point, the researchers have revealed a signal proving the existence of a concert. The next part of the process is about bringing into existence its content and what this content is made of. To do so, they use archives from the legal files of the festival, which, in the negotiation of contracts with the artists, declare the song list supposed to have been played that night. So these songs have a legal existence. This *setlist* provided by the festival organizers (*Festival-setlist*) introduces in the lab a potential *hold* on the content of the concerts. Considering together the tape IDs attached to the corresponding digital files, the presence of a signal attested by first tests and the *Festival-setlist* leads to the idea that there indeed are pieces of music in the digital file corresponding to the concert, pieces whose list is given a priori. This new step then consists in testing this hypothesis, verifying whether what is described on the *setlist* is really to be found in the digital files and avatars of the tapes received from the subcontractor. The *Festival-setlist* looks like a relatively cursory Excel document; it is, for now, the most accurate description available to researchers and temporary employees (who generally are equally male and female engineering students) about the events that may have been preserved in the digitized tapes. This description of the concert does not necessarily correspond to a tape on which these pieces may be because in the world of copyright, the reference unit is not the tape but the concert. The next task is to establish a matching of concerts setlists and tapes (often two tapes for a concert).

The descriptive file of the concert is included in a file directory that gradually gathers a set of files to be investigated in order to determine what concerts are made of. The directory is named after a concert and collects lightened copies of audio files (WAV versions slightly compressed after extracting the audio stream of the archive II), video files (highly compressed MP4 versions of the archive II) of each tape related to the concert, the setlist provided by the legal service of the festival (*Festival-setlist*), and finally a list of songs to be completed (*Indexing-setlist*) as and when the *Festival-setlist* is validated by an exploration of audio and video files.

In this exploration of the digitized concert, researchers and temporary employees favor the audio file over the video file. The audio signal is said to be lighter; especially the *waveform* visualization and the possibility to zoom into the signal make the audio signal an object considered to be manageable and easy to handle. Comparatively, the video signal representation is said to be hard to grasp. So it is on the audio file that temporary employees place markers that distinguish the events constituting the concert. The video file, considered to be peripheral, is there "just in case"; the user guide explains to temporary employees that in general it will not be used. However, this MP4 version of the archive is that which is included on the iPad and its Archive Discovery App.

Changing the Properties of the Material: Obtaining a Reasonably "Clickable" Material

The network of entities gathered around the tapes is already dense. New people and machines intervene in the handling of the files that researchers now call "concert files." As the network grows, the archive corpus becomes a more and more tangible material. The deployment of the network gives thickness to the digitized records by associating inscriptions and accumulating clues used to qualify the material.

At this stage, temporary employees have access to material that is just a click away. The sum of the files gathered in their working directory becomes the ground on which they prepare to take action and draw the outlines of what could be the concert they are in charge of. The recorded concert is from now on a set of *concert files* constituted of various computer files. The copresence of these files and clues reinforces the thickness of what is now the concert. By their participation in the evidential base (Ginzburg 1984) that makes the object, their role as evidence provides them also with a role as mediator. These files are the mediators of an indexed concert in the making, with which temporary employees should be able to interact.

This *proliferation of mediators* (Hennion 2003) ensures a possible and tangible interaction with the corpus of digitized recordings: the concert is distributed and expands into a series of new objects. Its *matter* (Latour 2013) extends and is populated with elements that can be put to work in future interactions. The mediators built by engineers enable them to multiply *holds* for future interactions with the concert that can now be manipulated and modified.

Part 3: The Struggle for Fluidity: Building Mobile Pieces

Transforming Concerts into Collections of Small Pieces: Songs and Playlists

The digitized recordings transformed into *concert files* pass through the hands of the temporary employees who ask, What happened that night onstage? What was actually played? The lab team thus progresses in its knowledge of digitized tapes and becomes prepared to index recording files by identification of events in order to be able to find them easily or even to extract and build them as *songs* that "may be played individually in a music software playlist" (as written into the *Indexing user guide*). This indexing operation adds tags to digital files, allowing, for example, the *Archive Discovery Application* to explore videos of concerts in the media player *playlist* of an iPad, but also allowing *Genezik* to analyze musical similarity of pieces.

On their arrival at the lab, the dozen new temporary employees are trained for a week and receive 13 pages of guidelines that remind them of the goals and steps to follow:

- "Locate transitions between songs and events, placing markers in [the audio edition window] and naming them from the provided concert setlist;
- Build the setlist associated to the .WAV file in progress, copying information from the concert setlist. If mismatch is observed, compared with the audio, apply the corrections; and,
- Generally speaking, consider that each song will be isolated and played back alone, or as a part of a playlist, making use of fade out and fade in transitions." (*Indexing user guide*, lab internal document, p. 9)

They must scrutinize the whole tape in order to describe the content and mark events. To do this, they have a list of typical events:

- Intro: the whole recorded area before the first song, including silence, applause, speeches
- Song
- Interlude: short piece of music, instrument tests
- Applause: includes "thank you" and song introduction speeches by the artists
- Speech: something more than just simple song introduction, including applause
- Come back: long applause time before come back of an artist, may include speeches
- Silence: rarely used
- To check: in case something does not fit the previous types (*Indexing user guide*, lab internal document, p. 9)

The guidelines for indexing, setlists, and the graphical representation of the waveform on the screen are used as support for the temporary employees to guide their work of qualifying file content but always need to be interpreted. This work leads to the emergence of *quasi-objects*, including songs that are not separate entities and detached from the flow of the concert but that anticipate their potential extraction and their establishment as new objects. At this stage, they are temporary objects made for the construction of new objects, which include songs and playlists.

Feeling and Touching the Sound Material before Modifying It: The Songs as Quasi-Objects

The first of these supports is the setlist, listing the songs as so many music-digital objects supposed to have been recorded and stored on tape. It orients the work of the temporary employees who, without this support, cannot a priori know what to look for in the recording and might put off their investigation by just leaving a note saying "no setlist." Without the *Festival-setlist*, the investigation stops. If the *Festival-setlist* and the *Indexing-setlist* to be completed appear in the workspace, temporary employees open the file from the audio version (WAV) of the recording. It is with this file that they will interact. The markers they place on it will testify to the fact that an event is recorded at this precise location.

At this point, the tape is no longer simply an audio signal; it begins to exist as a concert. With the setlist, it even starts to become a potential playlist. All the temporary employees need, in principle, is to recognize the pieces assumed to be in the recording; then, with a click, they set new tags to delimit the songs. This is a first decisive but sometimes difficult step toward fluidization. Sometimes, they find the songs easily, especially when the song's title words are regularly quoted in the lyrics; they may have more difficulty tracking and identifying others, including instrumentals, which jazz is particularly fond of.

Having opened the audio file, mouse in hand, they move the cursor from a "moment" of the concert to another on the graphical representation of the sound wave displayed on the screen. First, they scan visually the shape of the waveform recording to identify its structure. With their headphones on, they spot graphical wave packets mainly from visual cues on the screen, clicking the mouse for a brief stop at the beginning of each graphical wave packet to check whether it is music or not. With this visual scan of sound, in some way a warm-up for the real work of identification that will follow, they seek what they expect to find according to the setlist. They get an idea of what looks like the concert and the way in which the pieces are connected and the transitions done. They count as music tracks the graphical wave packets a priori identified, hoping that the setlist displays the same number. This identification of the relevant graphical wave packets does not happen by itself; it takes the engineering students a while to learn how to determine the difference, at a glance, between a graphical wave packet that shows a "song" and another packet that represents a "speech." By doing this work for a few weeks, they become apprentices or even experts, able to distinguish a piece of music from a speech at a single glance.

Interacting with digital recordings, armed with a mouse, a headset, and especially a screen, with their eyes and fingertips, they explore the structure of the concert. Doing so, they are already transforming the continuous flow of the concert into a playlist whose labels are those of the *Festival-setlist*. The concert is seen on the screen, mainly on the audio file. The amount of clues and evidence gathered around it grants *thickness* to the audio file (Ginzburg 1984). Through the visualization of the audio file, the recorded concert becomes a space in which it is possible to move freely, in small steps or giant leaps, thus shaping quasi-objects (Lécaille 2003) preparing the installation of a markup. Once these (reversible) bollards are installed, the pieces take on more consistency. They are now defined by boundaries under construction.

Sketching of the Piece

Indexing consists of adding tags to digitized files in order to mark the beginning and the end of each identified event on the audio avatar of the tape recording (WAV file). The time code of these tags can then be used to guide the extraction

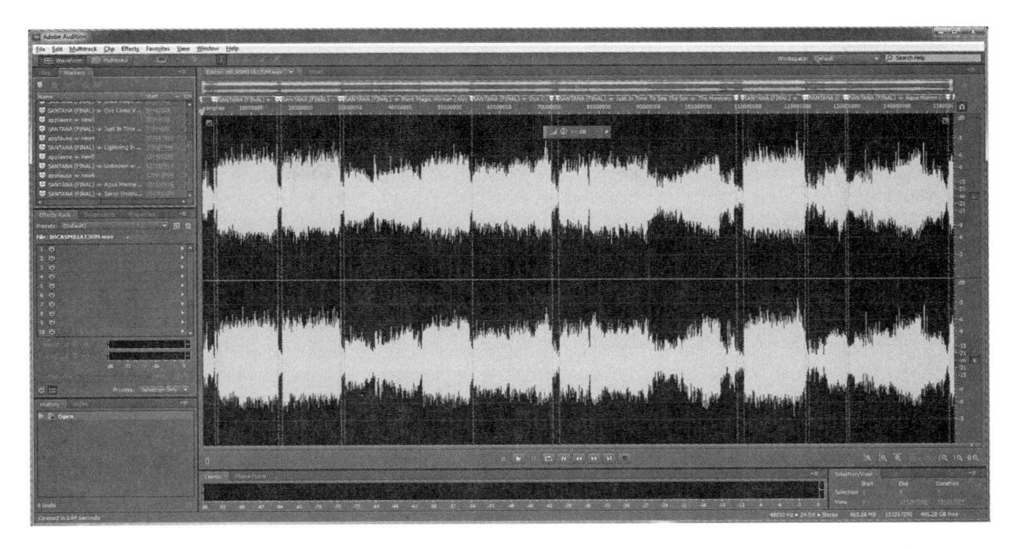

FIGURE 3: The graphical wave packets in the interface of *Adobe Audition*. Credits: Metamedia Center, EPFL.

of tracks and to facilitate searching in and playing concerts that have become playlists. To reach this goal, the engineering students have to build songs. They do this by starting from the *Festival-setlist* and a first visual exploration of the sound file. Then comes the moment to determine the physical limits of these songs.

Having spotted on the audio file the first song corresponding to the setlist, they position the cursor and set a tag at the end of the graphical wave packet. This end point is the exact beginning of the next transition, that is, an "applause" sequence, whose end is at first fixed temporarily. It becomes final once the start of the second piece has been clearly identified; for the moment this is somewhere between the possible end of the "applause" and the following graphical wave packet (see figure 3). In this still unclear area, the temporary employee walks the cursor with small touches on the graphical representation of the audio signal; each stop generates the corresponding sound in the headphones. She thus tests some possible beginnings until she marks one provisionally; automatically opening a new time range, she extends to the end of the graphical wave packet that is about to become the second piece. The end marker is positioned roughly toward the end of the graphical wave packet. She then moves the cursor within the package that could be the second piece, with some hops in order to proceed through a quick listening. This ensures that the piece does not contain anything unexpected that she could not otherwise "see."

The wave packet is now becoming a song. The temporary employee then returns to its borders to define them more accurately. She retrieves the start marker and zooms in on this provisional start. The software then produces a detailed representation of the waves in order to sort out the sounds. She moves around and tests two or three plausible places, working on a different scale from the one used at the time of the provisional setting. She now targets a small portion, that is, a sample, defined at a 48,000th of a second (sampling rate of 48 kHz) on the audio file. Her eye guides the operation, as a slowed listening would distort the sound too much; the temporary employees act on the audio signal through its visualization and the mouse controlling the cursor position.[9] Decisions on sound are thus taken visually.

On the Way to the First Note: Boundary Work Struggle

How do engineering students determine the most appropriate sample to materialize the beginning of the song? The *Indexing user guide*, of which they always have a handy copy, insists on the fact that a good song beginning should be located in the first half second before the first note. This rule refers explicitly to the music industry and particularly to the edition of live CDs. In this interval, they must take the decision to define the small piece of waveform on which to fix the tag. At the scale of the sample 1/48,000th of a second, half a second is actually very large and contains a mixture of various noises: audience applause, whistles, shouts, and so on, with which the words of an artist can overlap (thanks or announcements of the next song). Cases of relative silence are rare; there is no tabula rasa on which they can fix the tag arbitrarily. This is nearly always applause that articulates two pieces. Zooming becomes strategic for indexers, in order to choose the sample that will host the tag of the beginning of the song; a second, stretched by the zoom, then spreads on a quarter of the 22-inch screen.

Described as such because it hinders action, noise is also a resource for indexers. They treat it as a buffer zone, a material martyr in which they can cut without mercy. This buffer zone forces them to zoom in and turn the time lapse of noise into a subsequence with many possibilities. They scrutinize it to know its composition and detail one by one the sounds that could interfere with the beginning of the first note. For example, they wait for a cry or whistle to fade and place the marker then, to avoid giving the impression of starting on something that is ending. It is therefore not a straightforward task to get closer and closer to the first note. Indexers use many tricks, knowledge of editing software, their sharpened ear, and body control that ensures fine coordination between hand, mouse, cursor, graphical-numerical representation, and sight. This work also puts to use their acoustic knowledge learned on the job about the behavior of the signal: the velocity of each sound, the need to wait for a sound to finish in order to be able to isolate it without disturbing listening. This indexing work, in fact, opens a new set of uncertainties. The division into distinct sequences does not impose by themselves as evident. The indexers must negotiate with the sound stream and dynamics of the concert. Nothing is obvious; the sound must be very finely qualified in order for the indexer to be able to decide where to put its click markup.

Once the cursor has almost found its final place after a series of trials and hesitations, the indexer checks once again, listening to what happens by mimicking the start of the song as if it were in a playlist. She then repeats listening once or twice by pressing the buttons on the built-in player of the editing software. Sometimes dissatisfied, she shifts again slightly the cursor to skip a tiny passage finally judged to be bad. She starts listening again in the manner of the player until she is satisfied with the result; the piece has therefore a beginning and an end, at least temporarily. This work will be taken up later in the file biography, during a step called "quality control," consisting of verifying the quality of boundaries, particularly to ensure nobody has missed notes before the start of the song, that they have left enough, but not too much, applause at the end of the song, and have not forgotten real "silences" in the song.

Shaping the Object, Equipping Its Boundaries

Transforming heavy musical events like concerts into more fluid materials like songs goes through boundary work consisting of identifying what a song is and what its boundaries are. But it also goes through an equipping work (Vinck 2011) of these boundaries. In order to give more consistency to these boundaries, they are associated with new elements, which give them more weight and paradoxically contribute to the fluidization of the concerts. So, once the sequence has precise limits, indexers describe it in the *Indexing-setlist*. They assign a name using the generic formalism of *Festival-setlist* (when the song actually exists in the recording). They also include parentage with respect to the concert of songs that are on their way to become autonomous entities:

E.g.: artist -x- title -x- album -x- year -x- track number -x- ID

B.B. King -x- Strung Out -x- The Jazz Festival archive -x- 2015 -x- 1 -x- 173

From the *Indexing-setlist*, which is destined to replace the *Festival-setlist*, the temporary employees copy the corresponding line in the list of events that is built using the editing software (*Audition*); the "piece" that is shaped by the work of boundary markers has no name yet but only timecodes (the two samples hosting the start and end of a sequence). These timecodes are embedded in *Audition* and in the indexed audio file it produces. They are then extracted from the indexed audio file with another software, a freeware that structures a text file (.TXT), a description that reproduces the formalism of the example and adds timecodes. The sound file is not attached to this timecode.txt file, which is designed to serve as a basis for playing audio and video files and to provide time stamps that will be transformed into reading marks in the player of the *Archive Discovery Application*.[10] The indexing operation ends with file saving. The *Indexing-setlist* is saved in two formats including *XML*, which "dialogues" more easily with the general database. This *Indexing-setlist* may now replace the list provided by the festival, which was the reference to this point. However, indexing does not end here. The songs now constituted as tagged entities are still embedded in the recording of the concert. The issue is then to extract them and add another set of equipment to them in order to ensure their new and autonomous life.

With the indexed audio file still open, a new directory named *SONGS* is created and placed in the same directory as the other files being processed. The engineering student selects musical events she has just indexed via the interface of editing software, more precisely in its descriptive list on the left of the waveform representation. With a right-click, she selects *export*, sets the format (WAV 24 bits 48 kHz, the same as the one with which she just tagged pieces of concert) and introduces the destination file: *SONGS*. She repeats the operation by changing the format for copying songs to MP3. She then reduces the editing software window and goes to the *SONGS* directory where she sorts out what she will keep. *Audition* automatically generates a third set of files in *PKF* format. This proprietary format saves an image of the graphical representation of the signal. It is an image of wave images that evolves according to signal processing; it is always up-to-date because it is produced in the background; it thus avoids having to generate a graphical representation each time the file is opened. These files, however, are the first to be deleted, indicating that the lab has no plans to reopen the tracks in the editing software. To do this, the "SONGS" files are sorted by "type"; all *PKF* files are selected and then deleted. Then, WAV and MP3 files are sorted by name. Events that are not songs, such as *speech*, *applause*, and others, are also deleted in order to leave only songs in the SONGS folder.

Then the temporary employees structure song metadata using the *Tag & Rename* program. In concrete terms, they take all the songs in MP3 version, drag and drop them into the program, click on "edit tag," and the metadata that was in the name of the file itself is structured in columns. Continuing the example from BB KING:

artist -x- title -x- Album -x- year -x- track number -x- ID

B.B King -x- Strung Out -x- The Montreux Jazz Festival archive -x- 2011 -x- 1 -x- 173

The term "-x-" separates the variables. Artist, Title, Album, Year, Track-Number now constitute columns of metadata that media players are supposed to recognize and read. Engineering students then add to each piece the poster of the year of the concert. The pieces, whose parentage remains attached in the metadata, are now recognizable by any media player that will "play" them as any other MP3 song. In order to become autonomous, the song files are equipped with data that keep track of their parentage and refer to documents confirming their origin and copyrights associated with the concert.

Discussion: Digitalization as Rematerialization

A Produced Fluidity

During digitization, the material of the recordings has been greatly deployed through multiple mediators and associations established by the lab team. In a first step, the digitized tapes are literally multiplied in different places: a safe for the two conservation cartridges with the uncompressed files, then a storage system containing the uncompressed files and the lower quality version that will be used for the rest of the process. Then their general condition is checked before they are engaged in a space of action in which they are grasped via the mediation of several objects, supposed to help unravel the mystery of their condition and their constitution. In fact, these mediators increase the number of elements on which the existence of the digitized concert rely. They became related as constitutive elements of the matter-network. It is not a unique *essence*, which irrigates a network in construction, but rather a *material*, which is constantly transforming as its existence is deploying.

Researchers and temporary employees interact with digital recordings, sometimes very closely, for example in negotiating the marks these artifacts can accept. These artifacts are worked on in a process through which researchers and temporary employees get to know their materiality and composition (including the sequences that make up the concerts). Conversely, progressively built objects lead researchers and temporary employees to act with caution, taking the materiality of these artifacts into account. While this digital-musical material has become tangible (reasonably clickable), this does not mean the material has become docile. Even the transitions that should help extract the "musical content" are negotiated and inscribed with difficulty. Researchers and temporary employees go through a struggle that results in an acceptable compromise at some point on samples of 1/48,000th of a second. They negotiate those object boundaries piece by piece. The result, written, described, and recopied in several places, unfolds into multiple traces, all of which are certificates and containers of what has just been built and is still attached to the digitized tapes.

The indexing process we described is the most complete attempt to find and reveal what there is "on the tapes." Through indexing, knowledge of the concert is crystallized into a number of entities (inscriptions, equipment, intermediary objects that become material mediators of a matter-network) that are associated with the object (i.e., part of the audio file) that has been built through these operations. This produced object becomes a known object (e.g., a song); the materiality of the produced object (e.g., the piece of file) is considerably extended with respect to that of the object to be built (i.e., songs and playlists). In this sense, our account for indexing process is an attempt to reveal the distribution of being digital. Think of what one should do to move a very small marker materializing the beginning of one song. The number of elements related to the markers—the matter of the marker—could itself be discouraging since it is very difficult to redo the indexing process. The whole chain of mediators must be activated once again and the intermediary objects (timecodes, up-to-date setlists, indexed audio files, independent songs, etc.) have to be rewritten and reproduced. While the process of digitization aims at producing and shaping *digital facts*, questioning them would require a new inquiry. Our original question on the destiny of the matter of recordings when it is confronted with a dynamic of digitization leads us to show that the deployment of matter and the dynamic of multiplication of mediators come with enhanced vitality of the *musical fact*. The new life of festival concerts materializes in two ways: on the one hand, concerts can be yet again read with contemporary devices. New means of navigation are made possible thanks to the reading tags on pieces. On the other hand, bits of concerts, the "songs," are equipped in order to become autonomous and to circulate in a new sociotechnical network, that of *media players*.

The fluidity that has been acquired through the digitization process is grounded in new associations on which, from now on, the recordings of the festival concerts depend. In this sense, this new potential for circulation depends on the capacity of the constitutive elements of the new matter-network to become active in order to allow interactions when necessary. This emphasizes the amount of work accomplished to build a potential of fluidity that can rely on the media readers that are widely available around the world. The fluidity, or potential for circulation, results from the building of a material with which transformation and thus gain of fluidity is possible. The material that is at stake had to become tangible and modifiable in order to become fluid. This is a crucial point that our ethnography can shed light on. On the way to becoming a tangible material one can work with, these very hybrid digital pieces (signal, music, bit, waveform, sample, timecodes, songs, etc.) modify their constitution. These transformations are particularly visible when operators are building for themselves a set of *holds* in order to stabilize a material with which they will be able to interact. These *holds* are forged in interaction with the material and modify its properties by becoming a constitutive part of it. These both practical and conceptual elements show with accuracy that reducing tangibility and contingence are at the heart of the process of becoming digital. In our understanding, this is also a key point for the conceptualization of digital materiality where equipping work (Vinck 2011) is playing a central role.

Following on material properties, the example of copyrights is revealing. While becoming digital, concerts have become richer and have received constitutive elements from the elements used to construct them. While concerts benefit from the emergence of pieces, their existence previous to songs is modified and enhanced by this new presence, which is tested and testified on the indexed recording. The copyright information that has been transformed as one of the prior holds in order to produce "surfable" concerts fosters the emergence of entities that have strong mobility potential. The "same" list of legal objects changes the status of the pieces once they have been created. There is no longer the possibility to pretend these objects don't exist and are not the subject of a contract. Concerts are weighed down by the presence of new pieces and copyright turns back on an absolute and ubiquitous conception of digital fluidity. Law is used as a way to handle the tapes and is inscribed in the objects it refers to. It limits circulation when it has taken part in building the conditions for possible circulation. This composite matter contains elements that may tend to conflicting paths. The pieces that have been built in order to become autonomous may well not be able to exit the servers that host them because of copyright issues. This point encourages avoiding presuming matter is univocal: the apparently "same" constitutive ingredient can shape and reshape the destiny of the material.

Rescuing Materiality: The STS Contribution

Looking briefly at the history of the notion of materiality in the humanities and social sciences, we can trace back to the original temptation of reducing digital to a matter of pure semantics. For example, the fields of anthropology and archaeology, while traditionally sensitive to cultural artifacts and the materiality of the social world, have long contributed to conceptualize a sort of materiality they think of as material culture or material traces of social activity in which, ultimately, the material dimension is secondary. If analyzing materiality means understanding what there is *in* artifacts, this equates to constructing them as social objects that lose their material properties to become objects of meaning and interpretation. Thus, the literature explicitly dealing with the materiality of things in the social world (Godelier 1986; Miller 1998; Toren 1999; Graves-Brown 2000) has given material objects the status of tabulae rasae, unimportant foundations on which the social world and culture are built. Material objects are thus understood to participate passively in social life by providing their material form for the construction of a semantic layer, the higher activity specific to human societies (Godelier 1986). This conception of materiality is based on a generally obvious and implicit hierarchy between humans (who are superior through their use of intentionality and meaning for example) and nonhumans (taken to be inert or transparent). This *Great Divide* (Latour 1993) supports the idea of materiality taken in an essentially semantic sense and leads to a conceptual impasse where materiality and materials are opposed (Ingold 2007).

STS has played an important role in rescuing materiality from the dead end of the absolute exteriority of material things with regard to human beings and what they do. In this respect, we have made extensive use of foundational notions such as inscriptions, instruments, intermediary objects, and mediations in our analysis, which could also have been called an analysis of the *construction of a digital fact*, a distortion of the title of the famous 1979 book by Latour and Woolgar. STS has been particularly productive in describing and building concepts around the materiality of scientific facts and technical artefacts, whose analysis became more relevant once it was freed from the mental sphere (Godelier's *idéel* sphere) where it was locked up by positivist epistemology. Facts are stabilized by a set of operations from which they cannot be fundamentally dissociated.

On a parallel front, the foundational works of what were to become the Infrastructure Information Studies led by Star and Bowker showed that study of information and its production were not contained in its semantic dimension. Information was thus successfully displaced from being reduced to its semantic content to become a collective act, itself constituted of actions, movements, power, organizations, and so on. In a nutshell, thanks to this work, the notion of information has gained thickness and intelligibility. If we want to understand how it is made, we can bring in the notion of infrastructure. The matter of information is made of all the relations that partake in its production, between human beings or with artifacts.

Conclusion: Toward a Nonessentialist Materiality

The attempt to qualify digital materiality in terms of material infrastructure and extended embedded artifacts has given extremely interesting results, such as the deployment of numerous artefacts involved conditioning the digital. However, digital materiality remains hard to grasp; furthermore, emphasizing digital embeddedness might appear to paradoxically posit digital as escaping materiality, as something nonmaterial flowing inside the cables and processed by humans and machines. But, at the same time, catching digital through the computational infrastructure, which is definitely part of its matter-network, helped also to reintroduce a fundamental possibility to understand digital materiality, avoiding any dualism between digital and material. In the fight against digital essentialism inherited from the late 1990s claim for a "digital world,"[11] one of the seminal STS pieces had addressed a now famous question: "If bits are not made of atoms, what could they possibly be made of?" (Blanchette 2011). This statement might give to the atoms a material reality, which reflects only some types of interaction with them and conduces to a materialist understanding of information turned into what we propose to call a *hard materialism* with unexpected consequences as it seems to oppose one essentialism to another. This hard materialism leads to looking for a canonical element to understand digital materiality and still contribute to the old quest of *essence*, even if we conclude to a material one. Another representation of the atoms, depending of other types of interaction, would show them as vacuum, distributed presence, wave, and fuzzy entities with moving boundaries. With this quantic representation of atoms, they are material in the sense of hard and weak interactions, which means a materiality not so different from bits. This relational understanding opens up another materialism coherent with the approach we developed in this chapter.

In recent years, the question of materiality has acquired new relevance as scholars have grappled with the issue of the materiality of the social world, with the result that debates have extended to several subfields of human and social sciences: organizational studies (Orlikowski 2007; Carlile et al. 2013), situated action (Suchman 2007, 2011; Hutchins 1995, 1999), or gender studies (Braidotti 2011a, 20011b; Frost 2011). STS play a special role in this debate. During the 1980s, ANT was central in reviving the debate on materiality and offered new pathways to rescue matter from centuries of philosophical (and theological) essentialism. In this chapter, we have tried to demonstrate the relevance of the notion of matter as recently formulated by Latour as "the set of all the other beings upon which any given entity depends" (2013). This concept seems particularly relevant to account for the unfolding process that is characteristic of digital becoming.

In other fields closely related to STS, exemplified by works on *care* (Mol 2008; Mol et al. 2010; Puig de la Bellacasa 2010) and *repair and maintenance studies* (Henke 2000; Dant 2010; Gregson et al. 2009; Edensor 2011; Denis et al. 2015), in-depth dialogue with the studies gathered under the plural name of *new materialisms* (Barad 2003; Bennett 2004; DeLanda 2006; Pels et al. 2002) has taken place. This ongoing fruitful discussion builds an increasingly relational understanding of matter and materiality, and rejects any kind of exteriority of matter. In this movement toward taking into account *posthumanist performativity* (Barad 2003), matter can be considered as: "an active participant in the iterative process of the world's becoming. Matter is neither a given resource, nor a mere effect of human action. Rather, materials move, transform, damage, mutate, form alliances in a more or less durable way and are constitutive parts of animated things called humans, made of water, bones, blood, hair. . . . As protein source of being and 'energetic forces' (Bennett 2004), they are essential features of human agency" (Denis and Pontille 2015, 351). In a complement to these studies, we have shown that issues relative to materiality and material ecologies are not issues that can be restricted to the field of metaphysics. Our aim here was not so much to speculate on the composition of matter, but to account for the dynamic composition of the world, following the unfolding of the existence of few digital beings. Our ethnography of the work of digital material has allowed us to ground in practice the issue of digital materiality, considering the material in the sense of the matter of interaction, accounting for engaged mediations both material and corporeal.

Following the transformation of the material during the shaping process largely overflows the material infrastructure. For example, the scenes we described stage the bodily commitment of the operators. The crucial digital fluidity is conditioned by the abilities of the digital craftsmen to build a material to work with and all the negotiations between operators and the material to become are mediated through perceptions. When aligning the material to bodily perceptions is at stake, the uncertainty of the result overflows material determinism. These digital artifacts in the making bear the marks of all of those negotiation results that can be considered as *compromises with the material*. Tracking the moves of the material in interaction seems to be an effective means to qualify digital materiality. This probably relies on the close ethnography and the experimenting of the work of this particular material. This text is the account of our encounter with this material for which the fluidity is not given but a hardly satisfied condition of existence, and maybe its only viable mode of existence. This kind of engaged investigation should be reiterated in order to fulfill our knowledge of the digital beings that are constantly being deployed around us.

Acknowledgments

Translated by Dinah Gross.

Notes

1. "The daily work of one person is the infrastructure of another" (Star and Ruhleder 1996).
2. See www.modesofexistence.org/inquiry/?lang=en#a=SEARCH&s=0&q=matter.
3. For a synthetic presentation of different approaches of digital materiality, see Ribes (this volume), especially on the "additive approach."
4. We use the term "digital being" to refer to a variety of entities such as data, files, algorithms, databases, digital pieces of music, visualizations, graphical-numerical objects (Lécaille 2003).
5. A hold or take (in the climbing vocabulary) is a detail into an object, which receives its relevance only in the interaction with this object. The hold allows an actor to grasp the object in order to integrate it into its action. The hold can be created intentionally into the object in prevision of the action, but it could also be an accident, which receives its relevance at the moment of the action and through the interaction (e.g., the hand of the climber looking for an accident into the rock that could became a hold for him).
6. The calculation uses a method of clustering by similar pieces according to the audio variables.
7. The "primary digitization," consisting of the conversion of the analog signal to a digital one, is outsourced to one of the few specialized firms in Europe. The storage medium is Linear Tape Open (LTO) cartridges—a technology that stores a digital signal on a magnetic tape. The life expectancy of these tapes is estimated by experts to range between 15 and 30 years. This has been the most used long-term storage system since the beginning of the 2000s.
8. "A timecode is a sequence of numeric codes generated at regular intervals by a timing synchronization system" (https://en.wikipedia.org/wiki/Timecode).
9. The manipulation of the video signal, which is split into frames that can be scrolled through by using the arrow keys of the keyboard, is very different because the arrows impose a different sequence of cursor movement than the manipulation of the mouse.
10. The videos are not sampled at the same time scale as audio files. A sample in kHz must be translated into the target frame, which means that the marked sample must be converted to correspond to a frame of the sequence of 25–30 frames per second video.
11. For a canonical example of this claim, see Negroponte (1995).

Works Cited

Barad, Karen. 2003. "Posthumanist Performativity: Toward an Understanding of How Matter Comes to Matter." *Signs: Journal of Women in Culture and Society* 28 (3): 801–31.

Bennett, Jane. 2004. "The Force of Things: Steps toward an Ecology of Matter." *Political Theory* 32 (3): 347–72.

Bessy, Christian, and Francis Chateauraynaud. 2003. *Experts et Faussaires: Pour une Sociologie de la Perception*. Paris: Métaillé.

Blanchette, Jean-François. 2011. "A Material History of Bits." *Journal of the American Society for Information Science and Technology* 62 (6): 1042–57.

Bowker, Geoffrey. C. 1994. *Science on the Run: Information Management and Industrial Geophysics at Schlumberger, 1920–1940*. Cambridge, MA: MIT Press.

——. 2000. "Biodiversity Datadiversity." *Social Studies of Science* 30 (5): 643–83.

Bowker, Geoffrey C., Karen Baker, Florence Millerand, and David Ribes. 2010. "Toward Information Infrastructure Studies: Ways of Knowing in a Networked Environment." In *International Handbook of Internet Research*, edited by Jeremy Hunsinger, Lisbeth Klastrup, and Matthew Allen, 97–117. Dordrecht: Springer.

Braidotti, Rosi. 2011a. *Nomadic Subjects. Embodiment and Sexual Difference in Contemporary Feminist Theory*. 2nd ed. New York: Columbia University Press.

——. 2011b. *Nomadic Theory. The Portable Rosi Braidotti*. New York: Columbia University Press.

Callon, Michel. 1986. "Some Elements of a Sociology of Translation: Domestication of the Scallops and the Fishermen of St Brieuc Bay." In *Power, Action and Belief: A New Sociology of Knowledge?*, edited by John Law, 196–223. London: Routledge.

Carlile, Paul R., Davide Nicolini, Ann Langley, and Haridimos Tsoukas, eds. 2013. *How Matter Matters: Objects, Artifacts and Materiality in Organization Studies*. Oxford: Oxford University Press.

Dant, Tim. 2010. "The Work of Repair: Gesture, Emotion and Sensual Knowledge." *Sociological Research Online* 15 (3). doi:10.5153/sro.2158.

DeLanda, Manuel. 2006. *A New Philosophy of Society. Assemblage Theory And Social Complexity.* London: Continuum.

Denis, Jérôme, and David Pontille. 2012. "Workers of Writing, Materials of Information." *Revue d'Anthropologie des Connaissances* 6 (1): 1–20. doi:10.3917/rac.015.0001.

——. 2015. "Material Ordering and the Care of Things." *Science, Technology, & Human Values* 40 (3): 338–67.

Denis, Jérôme, Alessandro Mongili, and David Pontille. 2015. "Maintenance and Repair in Science and Technology Studies." *Tecnoscienza* 6 (2): 5–15.

Edensor, Tim. 2011. "Entangled Agencies, Material Networks and Repair in a Building Assemblage: The Mutable Stone of St Ann's Church." *Transactions of the Institute of British Geographers* 36 (2): 238–52.

Frost, Samantha. 2011. "The Implications of the New Materialisms for Feminist Epistemology." In *Feminist Epistemology and Philosophy of Science: Power in Knowledge*, edited by Heidi E. Grasswick, 69–83. Dordrecht: Springer.

Gal, Uri, Yoo Youngjin, and Richard Boland. 2004. "The Dynamics of Boundary Objects, Social Infrastructures and Social Identities." Working Papers in Information Systems. http://sprouts.aisnet.org/4-11/.

Ginzburg, Carlo. 1984. "Morelli, Freud, and Sherlock Holmes: Clues and Scientific Method." In *The Sign of Three: Dupin, Holmes, Peirce*, edited by Umberto Eco and Thomas Sebeok, 81–118. Bloomington: Indiana University Press.

Godelier, Maurice. 1986. *The Mental and the Material: Thought, Economy and Society.* London: Verso.

Graves-Brown, Paul, ed. 2000. *Matter, Materiality and Modern Culture.* London: Routledge.

Gregson, Nicky, Alan Metcalfe, and Louise Crewe. 2009. "Practices of Object Maintenance and Repair: How Consumers Attend to Consumer Objects within the Home." *Journal of Consumer Culture* 9 (2): 248–72.

Hanseth, Ole, and Eric Monteiro. 1997. "Inscribing Behaviour in Information Infrastructure Standards." *Accounting, Management and Information Technologies* 7 (4): 183–211.

Heaton, Lorna, and Florence Millerand. 2013. "La Mise en Base de Données de Matériaux de Recherche en Botanique et en Écologie." *Revue d'Anthropologie des Conaissances* 7 (4): 885–913. doi:10.3917/rac.021.0885.

Heaton, Lorna, and Serge Proulx. 2012. "La Construction Locale D'Une Base Transnationale de Données en Botanique." *Revue D'Anthropologie des Connaissances* 6 (1): 141–62. doi:10.3917/rac.015.0179.

Henke, Christopher R. 2000. "The Mechanics of Workplace Order: Toward a Sociology of Repair." *Berkeley Journal of Sociology* 44:55–81.

Hennion, Antoine. 2003. "Music and Mediation: Towards a New Sociology of Music." In *The Cultural Study of Music: A Critical Introduction*, edited by Martin Clayton, Trevor Herbert, and Richard Middleton, 80–91. London: Routledge.

——. 2015a. "Paying Attention: What Is Tasting Wine About?" In *Moment of Valuation: Exploring Sites of Dissonance*, edited by Ariane Berthoin Antal, Michael Hutter, and David Stark, 37–56. Oxford: Oxford University Press.

——. 2015b. *The Passion of Music: The Sociology of Mediation.* Aldershot: Ashgate.

Hutchins, Edwin. 1995. *Cognition in the Wild.* Cambridge, MA: MIT Press.

——. 1999. "Cognitive Artifacts." In *The MIT Encyclopedia of the Cognitive Sciences*, edited by Robert Wilson and Frank Keil, 126–28. Cambridge, MA: MIT Press

Ingold, Tim. 2007. "Materials against Materiality." *Archeological Dialogues* 14 (1): 1–16.

——. 2011. *Redrawing Anthropology: Materials, Movements, Lines.* Aldershot: Ashgate.

Jaton, Florian. 2017. "We get the algorithms of our ground truths: Designing referential databases in digital image processing." *Social Studies of Science* 46(7): 811–40.

Latour, Bruno. 1987. *Science in Action: How to Follow Scientists and Engineers through Society.* Cambridge, MA: Harvard University Press.

——. 1993. *We Have Never Been Modern.* Cambridge, MA: Harvard University Press.

——. 2004. *Politics of Nature: How to Bring the Sciences into Democracy.* Cambridge, MA: Harvard University Press.

——. 2005. *Reassembling the Social: An Introduction to Actor-Network Theory.* Oxford: Oxford University Press.

——. 2013. *An Inquiry Into Modes of Existence.* Cambridge, MA: Harvard University Press.

Latour, Bruno, and Steve Woolgar. 1979. *Laboratory Life: The Social Construction of Scientific Facts*. Prince ton, NJ: Princeton University Press.

Law, John, and Annemarie Mol. 1995. "Notes on Materiality and Sociality." *Sociological Review* 43 (2): 274–94.

Lécaille, Pascal. 2003. "La Trace Habilitée: Une Ethnographie des Espaces de Conception Dans un bureau D'Etudes de Mécanique: L'Echange et L'Equipement des Objets Grapho-Numériques entre Outils et Acteurs de la Conception." Doctoral dissertation, Université de Grenoble. www.theses.fr/2003GRE21021.

Lynch, Michael. 1985. *Art and Artifact in Laboratory Science: A Study of Shop Work and Shop Talk in a Research Laboratory*. London: Routledge & Kegan Paul.

Miller, Daniel, ed. 1998. *Material Cultures. Why Some Things Matter*. Chicago: University of Chicago Press.

Millerand, Florence. 2012. "La Science en Réseau: Les Gestionnaires D'Information 'Invisibles' Dans la Production D'Une Base de Donnnées Scientifiques." *Revue d'Anthropologie des Connaissances* 6 (1): 163–90. doi:10.3917/rac.015.0201.

Mode of Existence. N.d. "Matter." http://modesofexistence.org/inquiry/#a=SET+VOC+LEADER&c[leading]=VOC&c[slave]=TEXT&i[id]=#vocab-289&i[column]=VOC&s=0.

Mol, Annemarie. 2008. *The Logic of Care: Health and the Problem of Patient Choice*. London: Routledge.

Mol, Annemarie, Ingunn Moser, and Jeannettte Pols, eds. 2010. *Care in Practice: On Tinkering in Clinics, Homes and Farms*. London: Transcript.

Negroponte, Nicholas. 1995. *Being Digital*. New York: Knopf.

Orlikowski, Wanda J. 2007. "Sociomaterial Practices: Exploring Technology at Work." *Organization Studies* 28 (9): 1435–48.

Pels, Dick, Kevin Hetherington, and Frederic Vandenberghe. 2002. "The Status of the Object: Performances, Mediations, and Techniques." *Theory, Culture & Society* 19 (5/6): 1–21.

Pinch, Trevor. 1986. *Confronting Nature: The Sociology of Solar-Neutrino Detection*. Dordrecht: Springer.

Puig de la Bellacasa, Maria. 2010. "Matters of Care in Technoscience: Assembling Neglected Things." *Social Studies of Science* 41 (1): 85–106.

Souriau, Etienne. [1954] 2009. *Les Différents Modes D'Existence*. Paris: Presses Universitaires de France, Coll. MétaphysiqueS.

Star, Susan L. 1991. "The Sociology of the Invisible: The Primacy of Work in the Writings of Anselm Strauss." In *Social Organization and Social Process: Essays in Honor of Anselm Strauss*, edited by David Maines, 265–83. Hawthorne, NY: Aldine de Gruyter.

Star, Susan L., and Karen Ruhleder. 1996. "Steps toward an Ecology of Infrastructure: Design and Access for Large Information Spaces." *Information Systems Research* 7 (1): 111–34.

Star, Susan L., and Anselm Strauss. 1999. "Layers of Silence, Arenas of Voice: The Ecology of Visible and Invisible Work." *Computer Supported Cooperative Work* 8:9–30.

Suchman, Lucy. 2007. *Human-Machine Reconfigurations. Plans and Situated Actions*. 2nd ed. Cambridge: Cambridge University Press.

——. 2011. "Lecture Practice and Its Overflows: Reflections on Order and Mess." *Tecnoscienza* 2 (1): 21–30.

Toren, Christina. 1999. *Mind, Materiality and History*. London: Routledge.

Van Horn, John Darrell, Jeffrey S. Grethe, Peter Kostelec, Jeffrey B. Woodward, Javed A. Aslam, Daniela Rus, Daniel Rockmore, and Michael S. Gazzaniga. 2001. "The Functional Magnetic Resonance Imaging Data Center (fMRIDC): The Challenges and Rewards of Large-Scale Databasing of Neuroimaging Studies." *Philosophical Transactions of the Royal Society of London Series B* 356 (1412): 1323–39.

Vinck, Dominique. 2011. "Taking Intermediary Objects and Equipping Work into Account in the Study of Engineering Practices." *Engineering Studies* 3 (1): 25–44.

Vinck, Dominique, and Alexandre Camus. Forthcoming. "The Role of Corpora in the Innovation Dynamics in the Field of Digital Cultures and Humanities." In *Public Policy for Innovation and Development*, edited by Igor Rivera. London: Routledge.

The Life and Death of Data

Yanni Loukissas

This interactive essay is available at the companion website: https://digitalsts.net.

Arnold Arboretum Plant Accessions (Living and Dead) from 1872-2012

1872 1877 1882 1887 1892 1897 1902 1907 1912 1917 1922 1927 1932 1937 1942 1947 1952 1957 1962 1967 1972 1977 1982 1987 1992 1997 2002 2007 2012

Jan Feb Mar Apr May Jun Jul Aug Sep Oct Nov Dec

The Life and Death of Data ?

0 5 10 15 20 25 30 35 40 45 50 55 60 65 70

The Life and Death of Data

by Yanni Alexander Loukissas
Digital Media / LMC, Georgia Tech

Credits:
The interactive features of this project were designed and implemented in collaboration with Krystelle Denis. Our work was supported by The Lasky-Barajas Dean's Innovation Fund for Digital Arts and Humanities at Harvard University and made possible by members of metaLAB and the Arnold Arboretum.

00. About

In collections of scientific and cultural history that are too big to see, data act as virtual handles for rare and delicate artifacts from the past. At the Arnold Arboretum, a long-lived collection of trees, vines and shrubs managed by Harvard University, landscapes from around the world and across time are stitched together by data. However, data are worthy of study themselves. Created in varied social and technological eras, they register the organizational structures and values of their time. Through a combination of data visualization and interviews with Arboretum staff, this data documentary illuminates what data can teach us about their own social and material histories, as well as how to study institutions of collecting digitally.

A screenshot from the interactive essay, *The Life and Death of Data,* about the accession records of Harvard University's Arnold Arboretum, accessible online. Image by Yanni Loukissas and Krystelle Denis.

Materiality Methodology, and Some Tricks of the Trade in the Study of Data and Specimens

David Ribes

In recent years we have witnessed a revitalized interest in the study of materiality within social, organizational, and humanistic scholarship. Programmatic calls to attend to materials, things, and objects, and their entanglement with practice have been issued in a range of disciplines. These reexaminations of materiality often position themselves as a corrective to approaches that overemphasized rhetoric and discourse, or that treated "the social" as an abstraction or idealism. For STS, however, materiality is not altogether new, but rather a long-standing theoretical and methodological topic that investigators have wrestled with since the founding of the field. In the past three decades, STS has examined concepts such as *objects with agency*, has followed the *turn to practice*, and has considered *artifacts that have politics*. Such propositions have always been accompanied by lively critique from, for example, social constructivists or adherents of the strong program who, in various ways, have accused materialist analyses of falling into the traps of naïve realism or technological determinism. Throughout these discussions, and on both sides of the fence, the best STS work has retained a staunchly empirical commitment even as it has engaged these thorny conceptual questions.

STS has come to some useful conclusions, contentious though they may still be, about materiality along the way, particularly in the realm of methodology; that is, how do we approach studies of things, objects, stuff, and materials, their agencies and interrelations, in action and across time? In this chapter, I will exposit four methodological threads that have practically influenced my approach to studies of materiality, and from these I distill four "tricks of the trade" that have helped me investigate materiality as encountered in the field. Recounting tricks of the trade is never a comprehensive endeavor, and there are always more tricks to be exchanged. In this essay there is room for only four, one for each intellectual thread. Thus, the goal of this essay is neither exhaustive nor theoretically synthetic, but rather the beginning of a conversation for how to facilitate empirical investigations.

The overall argument will be that materials must be encountered specifically as one goes about research. I use the word "materiality" because it is the current term of art, but in actuality it makes me somewhat uncomfortable, particularly as the banner for a movement or common field of inquiry. This is because I feel there is very little that can be said about the topic of "matter," as such. There is just so much of it, it comes in so many forms, and it plays so many changing roles over time. And yet, a great deal of current social and philosophical scholarship about materiality is pitched in a high and abstract register, often foreclosing further study with sweeping declarations about the nature of materiality. Rather, materiality is an ongoing interest within and among various topics of study; its investigation should be specific, situated and tied to research questions at hand rather than cast as an investigation of materiality in general; and our methodologies should allow for the surprises of fieldwork to emerge rather than foreclose the discussion with received categories for the role of the material.

This is why this chapter revisits the craft and method for how to study materiality: How to make it something available for inspection in the social and humanistic disciplines? How to recount findings meaningfully? And, when should materiality be foregrounded in an analysis rather other topics? This chapter will outline methodological approaches from four intellectual threads: ethnomethodology (Garfinkel et al. 1981); the anthropology of classification, similarity, and difference (Douglas 1986); actor-network theory (Latour 1988); and historical ontology and epistemology (Hacking 2004; Daston 2000). It is not possible to recap these traditions or debates in this chapter, and so instead I will draw out threads from classic works that have informed my approach to materiality—works that have also served as key points of dialogue between these traditions over the years. At key junctures these traditions have been in dialogue with each other, sometimes in a relationship of critique, and at other times of theoretical elaboration. Each has offered a useful orientation to investigation (methodology), and from each I will pluck a maxim, or what Howard Becker called a "trick of the trade" (Becker 1998). Posed as method and methodology, these "tricks" will not cohere as a theory—that is, I am not proposing an analytic or theoretical synthesis of ethnomethodology, ANT, and the other traditions. Rather, I emphasize the utility of a syncretic toolset, and the value of having multiple handles to investigate the issues of materiality. Theoretical coherence (or not) is a matter for the downstream arranging and recounting of findings.

Some of these tricks were formulated by their authors as principles, or even metaphysical postulates, but here I present them as methodological aids. Many of the tricks specifically "reset" the researcher, enabling an escape from received understandings of the material, or from a predefined role for their agencies; thus, some of the tricks have the role of encouraging an *ethnomethodological indifference* or an agnosticism to what materials "really are," and to how they must be approached through investigation. The goal of starting from an agnostic position (itself a trick) is to enable the investigator to find unique configurations of the material, the natural, the technological, the social, and their hybrid or entangled combinations in the research object at hand.

At issue in this digitalSTS volume, and more broadly in studies of materiality, is the question of how to investigate the digital. Is it something new, requiring inventive theoretical formulations, methods, and a reorientation to our objects of study? Or is it a continuation of long-standing themes, such as agency, standardization, and immutable mobility? For STS, the recent interest in materiality, particularly materiality of information or the digital, is a contemporary inflection to long-standing theoretical and methodological topics. This chapter won't begin from the premise that "the digital" and/or "the material" are particular kinds of concerns for STS. Instead, I argue that the investigator must "discover" the digital and/or material through fieldwork.

Throughout the chapter I will draw from examples of the production, storage, and use of blood and river water samples and their entangled data as they circulate among scientific practitioners. I have found that sometimes data are "quite digital," stored in hard drives and transferred across networks, but at other times data never become computational, preserved only as chicken-scratched notes on paper forms archived in file folders. Mostly, specimens display properties we associate with physical objects, such as archived blood aliquots that are treated as finite resources to be used sparingly in research; but sometimes blood is enacted in ways that display qualities we associate with digital materials, such as DNA amplification or creating immortalized cell lines, that is, a potentially indefinite "copying" of materials. Rather than making general claims about the digital or material (theory), instead I will outline the results of treating the materiality of data and specimens as something to be discovered relative to empirical investigations (methodology).

A final point before diving in: One of the leading recent invigorations of materialist thinking (but not, I think, research) has emerged from a variety of brands of *new materialist* philosophy. While occasionally I have gained some philosophical insight from reading these new materialist texts, I have never gained a methodological insight, and so I say nothing more about these lines of thought herein. Here, I side with Latour's response to Graham Harman's demand for metaphysical generalizations beyond specific research: "the empirical is not disposable" (Latour et al. 2011). This is an assertion that *if* there is a metaphysics to be uncovered it is immanent rather than transcendent. Materialist thinking and research, such as within organizational theory (Orlikowski 2007), has been more germane to my interests with its commitment to concrete research. But it has also been plagued by hard and fast assertions about materiality and its agencies, often limiting investigations to "technology," for instance: "As people approach technological artifacts they form particular goals (human agency) and they use certain of the artifact's materiality to accomplish them (material agency)" (Leonardi 2012). The tricks I outline here are intended to forestall such starting declarations; they instead open avenues for situated and specific research rather than closing down investigations with predefined categories, agencies, or roles for materials.

Data, Blood, and Water

I approach these questions through my recent investigations of data and sample archives within scientific research investigations. Should we make a distinction between these two fundamental materials of science: specimens and data? I draw from two case studies that I have written about extensively: the Multicenter AIDS Cohort Study (MACS) and Long-Term Ecological Research (LTER). For simplicity these are characterized here as two longitudinal data and sample collection ventures, but in other texts I have written of the MACS and LTER as "research infrastructures" (Ribes and Polk 2015; Ribes 2014): these organizations do a great deal more than build long-term archives of data and specimens, but I will only focus on those aspects here.

Since 1984, the MACS has been generating data and specimens from cohorts of gay and bisexual men; every six months the men return to a clinic to undergo a medical examination, fill in behavioral and demographic questionnaires, and donate multiple specimens. I will mostly focus on blood, but many other specimens are also collected. The data are collated in archives, and the specimens are kept in cold storage—for both, MACS members seek to render these resources available for reuse in ongoing or new scientific investigations. Similarly, since 1980, LTER has been collecting data and samples from a distributed network of ecological sites across the globe. I will focus on the collection of stream water data, along with stream water samples from their Baltimore County site in Maryland. These data and samples are also placed in archives that are made available for scientists to repurpose in their investigations.

I "cut my research teeth" on ventures that collected no specimens, but rather were exclusively focused on data, their interoperability, representation, or manipulation. This is a central feature of the "cyberinfrastructure" projects that were my empirical cases during the 2000s: all of which defined themselves as primarily concerned with the circulation and reuse of data (Ribes and Lee 2010). Consequently, when I began my investigations of the MACS and LTER, I initially treated those organizations as data-centric enterprises too. It took many months of fieldwork for me to attune myself to the sample archive that they had been building alongside data repositories for many decades. As an STS researcher, I had certainly been aware of the material practices of sample collection in each venture, but my tendency was to treat blood and water as steps in a long translation chain that would ultimately end with their representation as the (never quite) immutable mobiles called data (Latour 1999), that is, I followed my actors to the sites where river water was collected in bottles, then trucked to a laboratory where silt was filtered, and trucked to another laboratory where they were assayed for the presence and concentration of various chemicals, ultimately recorded as data. And certainly this is the case for a significant portion of the blood and water that is collected: within hours or days these materials are subjected to one assay or another, and recorded and preserved as data—for example, salinity, calcium, HIV serostatus, white blood cell counts, and so on. But the samples are *not only* a step in a translation chain to data. For these ecologists and biomedical researchers, samples are an invaluable resource unto themselves. Ecologists travel to the vast cold rooms of the Cary Institute in Millbrook, New York, to visit these repositories in search of decades-old water samples: a vial of water collected from Gwynns Falls in 1995, or perhaps a longitudinal cross-section of samples from 2001 to 2009 from Herring Run. They may seek out these samples for many reasons, such as to search for a chemical or organism not yet recorded in their databases, or to reexamine the samples using a more sensitive or reliable instrument. In 1985, following the discovery of HIV and availability of the antibody test, MACS scientists returned to their archive of blood to search the stored serum for signs of the virus. No scouring of the databases could have revealed this new entity in the data archive; only by returning to the sample repositories with new instruments could these new ontological entities be enacted across the past.

The rest of this chapter will exposit four methodological threads in STS, and for each thread return to these two scientific materials—data and specimens—to illustrate an empirical examination of materiality. Are digital data and material specimens fundamentally different things? Or, perhaps, are they fundamentally the same if considered differently? Is one more material than the other, presumably specimens, or should we approach them as equally material? Or, should we consider each to have its own distinct materiality? Though these are good starting points, ultimately I reject the premises of all these questions. I argue instead for a situated, speculative, and historical approach to materiality. I have found that some actors approach data and specimens quite similarly, and that others treat them completely differently; that the treatment of their material nature has changed drastically for both over the years; and that both data and specimens reveal emergent, open-ended trajectories for their use, limits, and replenishment as resources for scientific investigations.

Ethnomethodology: Interactionism at the Emergence of an Object

Our starting point will be Garfinkel, Lynch, and Livingston's ethnomethodological study of the "discovery" of a pulsar (Garfinkel et al. 1981). An arbitrary starting point for certain, but one that has been particularly impactful within STS for investigating the emergence of new phenomena and their enactment as objects. Here, scientific objects are the hard-won outcome of actors' practical work and negotiation, with materials as the pliable but constraining resources that enable them to do so. The authors do not insist on any particular status or role for materials, those are topics for the actors to practically work over.

Using a tape recording that had been running during the 1969 process of "first time through" discovery Pulsar NP 0532 in the Crab Nebula, the authors outlined how scientists inspected data, evaluated the positioning and calibration of instruments or modeling tools, and over time came to shape an object that could be accounted for as distinct from the circumstances of its discovery. Garfinkel, Lynch, and Livingston likened the activity to a figure-ground gestalt shift in which, from an image filled with foliage, an animal was extracted; or perhaps in a more apt metaphor, they recount the discovery of the pulsar as a "potter's object" which is slowly, methodically, but not deterministically crafted from a simultaneously constraining and enabling material.

A more nuanced term they use for "discovery" of the pulsar is "first time through" (Garfinkel et al. 1981, 134), referring to the challenge of trying to find when one does not quite know what is sought, or how it will manifest. While a pulsar had been theorized in astronomy, and research had already indicated where to look for some of its features using investigative techniques and instruments, no one had yet actually found a pulsar by working their way through scientific materials. How will such an object manifest itself *as* data? Their answer is that the pulsar is not an object at all at the beginning of the night but rather that "somehow it was 'evolved' from an evidently-vague 'it' which was an object-of-sorts with neither demonstrable sense nor reference, to a 'relatively finished object' over the period of the night" (135). The "somehow" is what the analysts explore as a matter of the practical, local, and negotiated inspection that the astronomers engaged in using the materials at hand, including the data, the positioning of the instrument (i.e., the telescope), various visualization tools such as the oscilloscope, modeling tools and their settings.

Only near the end of the audio recording are Garfinkel et al. willing to concede that the pulsar is "in hand" as a distinct object. Early in the evening the object is "witnessably vague"—that is, it is discussed and posited but remains an "it" rather than "the pulsar"—whereas by the end of the night it is a "relatively finished object" (157). By the time of publication of the scientists' findings, "the work of the optically discovered pulsar's local historicized production is rendered as the properties of an independent Galilean pulsar" (134), that is, an autonomous thing existing prior to any method or activity of its detection and that is causing its inscription as data through the astronomical instruments. Rather than "first time through," it was now possible to render an account of the pulsar as "the-exhibitable-astronomical-analizability-of-the-pulsar-again" (135).

The term "material" appears several times in this study of the pulsar's discovery, but never in the sense of *stuff and things out there*; rather, materials refer to the practical at-hand resources that are, in time, enacted as visible, workable, and accountable features of the pulsar in the lab. That is, printouts of data, diagrams, the various instruments, and their calibrated arrangement—these are some of the local materials upon which the pulsar's enactment in situ, and eventually in formal print, depend. This is a definition of materials later adopted by the actor-network theorists (Latour 1986; see esp. Camus and Vinck in this volume). The pulsar of Garfinkel et al.'s study was achieved through practical manipulation of many local resources: data and visualizations, instruments, and the coordinated interpretive work of multiple scientists. In the final scientific publication, these remain only in the traces scientists provide in their formal methods write-up, while the pulsar is presented as independent object in the world. But in the lab that Garfinkel et al. inspected, the pulsar could be "found" only by placing the various materials in a just-so relationship to the other, and at the beginning of the night, in its "first time through" it was unclear exactly how this should be done. Both data and object were emergent, their relationship was interactionally built-up and built-together in the process of discovery.

Example: The (Seemingly) Endless Discovery of the Value of Data and Specimens

I turn now to the data of the ecological research infrastructure LTER. Rather than being the materials for crafting a single object, some of these data have played that role multiple times for distinct scientific objects. These data have in various ways been enacted to stand in, or mean, different things for different kinds of studies. Hidden in the past, these activities remain only in traces such as publications, but at some past point data, instruments, and vials of water were enacted as the accountable and observable objects of ecology. It is by placing materials in different relations to each other that LTER has supported the investigation of thousands of distinct objects over its thirty years, often using "the same materials" for highly heterogeneous purposes (this is a key quality of what it means to approach LTER *as a research infrastructure*).

No datum or specimen is meaningful on its own: they gain and maintain their value by the enacted and sustained links to other materials. For instance, by combining temperature data from the Gwynns Falls watershed and assays of river samples, ecologists have examined "how stream temperature affect phosphorus concentrations" (Kaushal et al. 2010). By combining *that same river temperature data* with those from other catchments they have studied "riparian ecologies" (Groffman et al. 2003). And by using stream flow data along with the *same temperature dataset* they have modeled "impervious surface drainage" (Kim 2007). What objects can be crafted from the data and specimens, and how those objects will be revealed from those materials are ongoing questions for these ecological researchers. Over time new uses are developed for old data and specimens, an unfolding "purpose" for LTER's data and specimen archives.

Some of these data and materials come packaged together, or entangled: it is worthless that a data point for a stream reads as "27 degrees Celsius" without the additional data that tell us what from what catchment (Gwynns Falls) and when (April 7, 2011) it was collected. Much more is collected at the same time—observations about the smells, the turbidity, and the height of the water—and all of these must be kept together in the paper tables that ecologists use to generate their data. As these on-paper data points wind their way to digital databases, these relations must be preserved. There are several check points to ensure this is the case, including, beginning in the late 1980s, the automated alerts that appear if values are entered outside specified ranges, or if a database field is left blank. Similarly, a sample of water is always entangled with the data (or "metadata") that tell of when it was collected and from where. Through those key points, a sample is linked to the vast array of data that were collected with it, or that come from "the same" source over time. No matter how well physically preserved a sample of water may be, its value also depends on sustaining those links of reference: a desirable and necessary entanglement that information managers and specimen archivists seek to preserve.

Approached in the abstract, there is little to say about the materiality of LTER's data or samples, in part because that materiality is at stake in their use. For instance, salinity is not a stand-alone material property. That is, whether a particular data point can be treated as a direct stand-in for salinity, or whether that datum must be modeled relative to broader temporal patterns of salinity, or questions about whether the instrument used from that reading was properly calibrated (and so on) are matters for the actors to debate, negotiate, and come to some form of agreement (or not). Across the decades of data collected on salinity in Gwynns Falls, and the scores of uses for those data, we will find many enactments of the materiality they call "salinity."

Trick of the Trade I: Place Yourself to Observe People Interact While New Materials, Objects, and Things Are Emerging, Negotiated, and Agreed upon (or Not)

For research technologists and scientists, new things, materials, and objects are "hard won" (Daston 2000) and those wins are very often not once-and-for-all; they may recur in later uses of those materials, or through reapproaches to those objects. That said, even the most hard-won objects or materials are often later treated as a given, or blackboxed. Scholars of objects, materials, or things should place themselves "where the action" is of discovering, negotiating, and renegotiating materiality. This can be done through participant observation, or through clever archival reconstructions, but they are rarely visible in the tidy texts of scientific publications or interview-based reconstructions.

Scientists, engineers, and others kinds of actors put a great deal of effort into discovering and understanding materials and objects. Following from the general ethnographic maxim to "respect your actors," the first trick of the trade is to approach the work of those you are investigating as interactional, and the materials as emergent and relational. The cases of the discovery of the pulsar and the reuse of data and specimens in LTER are two examples where actors are involved in an ongoing process of investigating the properties of materials and crafting objects. The investigator need not decide "what are materials," and is usually not in a good position to do so; the actors themselves are working away on the topic, using the tools of their own trades. They do so in time, in practice, and in changing relations with each other and the materials at hand. As we have seen in the case of the pulsar, materials are often left behind in favor of sought-after "facts or findings" of investigations, but as we have seen with LTER, actors may return to those data and specimens as they "recur" in new investigations (Rheinberger 2000). Approached in this manner, no material can be taken as having a forever closed meaning, property, or affordance; in principle they can be, and often are, revisited, repurposed, or rediscovered (leading to the fourth trick of the trade: historicism; see below). For the investigator, the key is to be at the right place and the right time, whether that be in person or establishing copresence through documentation.

Anthropology of Classifications: Similarity and Difference Are Institutions

As Mary Douglas wrote, "Nothing else but institutions can define sameness. Similarity is an institution" (Douglas 1986, 55). She was writing specifically of categories, but the point can be extended for the "sameness" of materials, practices, or the fittings of technologies, whether that of plugs or ports, or code and categories (Bowker and Star 1999). Difference too is an institution: things are kept apart and distinct as a matter of routinized activity, or through the sustained activity of classificatory devices. More vivaciously, Henri Bergson puts to us "it's not enough to shout 'Vive the multiple!'; the multiple has to be done" (quoted in de la Bellacasa 2012). Finally, similarity and difference are often a matter of nuanced and situated commensuration (Espeland 1998), as things may be the same for some purposes and different for another.

Example: Can Robots Step in the Same River Twice?

In "Data Bite Man" (2013), Steve Jackson and I recounted how LTER ecologists attempted to transition the by-hand collection of water specimens to an automated robotic system. Their hope was that by installing devices that regularly "sipped" river water, ecologists could cut out the weekly trek to collect these specimens, thus reducing labor, cost, and time, and perhaps adding a layer of objectivity by removing human variation (Daston and Galison 1992). However, they found that the automated system collected water samples in a very different way than people, that is, by pooling weekly samples of water rather than keeping each sample separate as ecologists had done since the inception of their project. This key difference in the practice of a human and a robot rendered the comparison of a decades old legacy archive with the new samples challenging, if not impossible. A deliberative process among these scientists deemed the two methods to be too different, and the loss of historical commensurability with their sample and data archive unacceptable. Today these ecologists collect water both by hand *and* via robot, now two longitudinal sample and data repositories that serve distinct scientific purposes.

The institutions of difference and similarity may be sunk deep into the architectures of data and sample production. For the analyst this may mean attending to the seemingly trivial work of maintenance that sustains similarity, such as metrological practices that make sure temperature data are the same via regularized instrument calibration (O'Connel 1993). When I accompanied and participated in water temperature measurement, even more mundane than calibration, I was instructed to stand in the middle of river and to position my body downstream of the thermometer to ensure my own body heat would not warm the water and distort the reading. This is a colloquial, embodied standardization that any graduate student who collects river samples and data in LTER learns so as to ensure that all materials are collected in the same way, over time (Goodwin 1994). Such a production of sameness is not recorded in any ecological paper I have read, but when I have recounted this story to ecologists, they sagely nod in recognition.

The generation of water samples can be compared to those of blood specimens, but at a granular scale of analysis the practices are unique. The MACS is geographically distributed across four US sites: Baltimore, Chicago, Los Angeles, and Pittsburgh. At each site they have a lab for the analysis of their locally collected biological specimens. However, the practices and technologies at their labs are forever threatening to diverge; for example, samples are left out of the fridge at differing times or technicians may handle them differently. To ensure that the results of an assay are the same at all sites, they often purchase the same brand, make, or model of reagents and instruments (a further proliferation of materials, see irreducibility below), and share protocols for their use across the distributed geographic locations. On occasion, the same specimen of blood is circulated across these geographically distributed sites to be tested by each instrument assembly there—results are compared at the level of practice, instrument calibration, and assay outputs. The "same" blood specimen—sustained as such by packing it on ice in its travels from Baltimore to Chicago—serves to evaluate practices and instruments that ongoingly diverge, but by calibrating them in this manner assays can thereafter be taken to produce comparable results of different blood samples. Here, the blood specimen that is circulated across sites ceases to be an object of inquiry (our usual understanding of the role for specimens), and instead becomes part of the instrument assembly, ensuring calibration, or "sameness."

Each of these practical and technical regimes seeks to ensure particular "samenesses" or mark meaningful differences. Tied as they are to distinct interests and concerns of scientists, one scientist's routine for ensuring sameness is often not sufficient for another. In a common *infrastructural inversion* (Bowker 1994) scientists and technicians may return to inspect the protocols of data and specimen collection and preservation to evaluate these according to their needs, and on occasion these protocols are changed to meet the criteria of emergent instruments and objects.

Trick of the Trade II: When There Are Claims That Things Are the Same or Different, Seek Out the Work and Technologies That Make Them So

When things fit together—whether interoperable data or physical modules—seek out the practices and protocols that hold together these fittings. Often this requires digging into the histories of standardization or interoperability. Similarity is always sustained in the present, but the complications of its routinization may have been worked out in the past (Ribes 2017). This general methodological point is valuable for the inspection of data, widgets, or classifications, though how similarity and difference are established and sustained is always by specific means that demand a close practical and sociotechnical investigation.

If similarly and difference are sufficiently institutionalized, one may also find that the actors have developed ways of testing or revisiting them: for example, as we saw above, the MACS has developed an extensive calibration regime that, every few years, seeks to confirm the similarity of practices and technologies that lead to the generation of data and specimens. All institutionalized regimes of standardization display such occasional infrastructural inversions, and, as with any such inversion, tends to produce a significant paper trail. Whether as an ethnographer following these inversions live or historically reconstructing these practices (ideally, both), inversions offer opportune moments (i.e., observable practice, and paper trails) for the investigation of the production of sameness and difference.

The Irreducibility Principle

As Latour states (rather than finds), "Nothing is, by itself, either reducible or irreducible to anything else" (Latour 1988, 153). No entity on its own can substitute for another, such as electromagnetic waves for the experience of a sunset. In some circumstances, the electromagnetic explanation *stands in* for the world, but when that is the case one will also find a great deal of (often backgrounded) work, and many artifacts to make that the case and to sustain the link of reference. Latour quotes Whitehead in support: "For natural philosophy everything perceived is in nature. We may not pick and choose. For us the red glow of the sunset should be as much part of nature as are the molecules and electric waves by which men of science would explain the phenomenon. It is for natural philosophy to analyze how these various elements of nature are connected" (Whitehead 2013, 29). The electromagnetic explanation is certainly not "wrong," per se, but it also does not substitute for an experience of the sunset—together, the scientific explanation and the experience of the sunset add to the richness of the world. For the philosopher, like Whitehead, exploring how the elements of nature are connected becomes his professional task; for the actor-network theorist, the empirical task is investigating how links between the scientific and experiential spheres are created (or not), and sustained.

The maxim is known as the "irreducibility principle" and is akin more to a metaphysical assertion than a methodological approach. But I have found irreducibility a useful starting point for investigations, serving to sensitize me to the production of additional materials and the work of generating and sustaining more links of reference. As with sameness and difference of materials, the reduction of one material to another requires work and technique, and that work is inspectable as practice and action.

Example: A Generative Approach to Materials

Irreducibility leads to a distinct exploration of data and specimens. Data are not (only) "representations" of something else: a reduced or more essential capturing of the world that is thereafter able to substitute for it (as with recordings of electromagnetic waves from a sunset). Instead, each data point and each specimen can also be considered as something new in the world: *a generative model of data and specimens*. Rather than (only) reducing materials, they proliferate them. In and across vials, refrigerators, files, or disks, these entities thread through their own

lives. Data and specimens have properties unto themselves *qua* data and specimens, and they must also be sustained as such along with their ties of reference.

In any longitudinal collection endeavor, as with LTER, each data point or specimen is not a replication of the previous data or specimen, it is a new collection, capturing a river that is warmer or colder, with a changing salinity, and potentially a new chemical or biological composition. Each is a novel temporal slice. This is one reason that these scientists continuously return to the river to collect new materials: if data could fully capture a river's properties they would not keep samples; if samples could stand in for the river they would not return weekly to the river for more.

Similarly with the MACS: blood, on the one hand, is certainly not the whole of the person, but through blood—its analysis as a specimen or its preservation as an aliquot—one can do innumerable new things that cannot be done with the person. Scientists can store it for later retrieval to search for things today that could not be found then, they can return to reinspect it for errors in analysis, or manipulate it to produce altogether new materials such as serum, plasma, or a concentrated "pellet" of *peripheral blood mononuclear cells* (PBMCs). Such is also the case with subject data, which are clearly not the totality of the person or any aspect of them, but which can allow new procedures and analysis, such as longitudinal aggregation to generate a view on an individual's life course, and combined with other people's data can produce "a population."

As Latour tells us, in demography the statistician loses the crying baby but gains a nativity rate (Latour 1987, 234). Data, unlike people, afford these aggregations to produce longitudinal, comparative views. Contrary to any naïve reductionist formulation, in which less and less will explain more and more of the world, what I have found in any ongoing collection endeavor is a proliferation of materials and data in all fields, each entangled with their own specializations, methods, apparatuses, and instruments.

While data or specimens are not reductions of the world on their own, in combinations of instruments, arguments, procedures, and visualizations, they are used to stand in for the world on many occasions. A sample of river water from May 5, 1995, "is from" the river at that time, the accompanying data "are" its salinity, temperature, and nitrogen levels on that day (with a reminder of the caveat above, that such properties are negotiated, rather than essential). And so, in addition to the data and specimens, we will also find a battery of work and materials dedicated to sustaining the relation of reference between data, specimens, and the river they came from. In inspecting the activities that seek to sustain a relation between data, specimens, and where they came from, we will find a further explosion of materials and artifacts: the sterilized vials for storing water and blood, the labels that mark their accession, the forms that are used to record data, the reference thermometer that is used to calibrate the one used in the field or the clinic. All of these are needed to ensure that scientists can continue to say that these data can stand in for this river at a particular point in time.

Trick of the Trade III: Seek Out New and More Materials, and Track the Work of Sustaining Ties of Reference (and Reduction) across Them

Overall, in the study of materials, an approach informed by irreducibility leads one to attend to the generation of new and more materials rather than the elimination of the world by the substitution of complex entities by simpler or fundamental ones.

A dangerous thread of recent materialist thinking has taken as its task to promote particular reductionisms, often casting all activity, things, and objects as fundamentally material. For instance, in an otherwise well-researched and argued investigation of bits, Jean-François Blanchette seems resolutely set on the task of displaying forever more material foundations to computation, asserting "if bits are not made from atoms, then what?" (Blanchette 2011). This approach, more theoretical assertion than methodological approach, takes the materialist argument one step too far, shifting from a practical investigation to an ontological assertion, inevitably leading us back to many old and polarizing problems that may culminate in discussions about whether there is such a thing as the experience of being a bat (Nagel 1974).

I find this tendency troubling for many reasons, but I will focus on the methodological difficulties that emerge: By espousing a particular material reductionism (data are fundamentally ordered atoms), the investigator may be blocked from an empirical investigation of the actors' epistemologies and ontologies, or more specifically, their relevant form of reduction. In calling data ordered atoms, what has occurred is an adoption of a particular reductionism, in the case, the reductionism of electrical engineering. That reductionism is real, instituted in innumerable texts and technologies; however, it is not the reductionism of ecologists and biomedical researchers. In studying their data practices, I have never seen them enact data as reducible to arranged atoms. They may very well think this is the case, but it is not enacted in their practice, nor particularly important for it. Instead, the important reductions are those of, in this case, ecological and biomedical science: in that field a thermometer reading from a river or a body "is" (and then immediately "was") its temperature, thereafter recorded *as data*. As a cool river or warm body is reduced to its temperature, we can observe the generation of something new: data. After this process we have a (presumably) differently cool river or warm body, along with a new thing, data, that these scientists seek to keep "the same" across all its future trajectories. Atoms and molecules may still play a role, such as for stream chemists interested in phosphorous content. Here the river is reduced to its chemical contents as new data are generated, but certainly no ecologist is thereafter reducing that data to the ordered atoms of electrical engineering.

The important practices that sustain these reductions for ecologists include calibration of their instruments, or careful body positionings relative to the river (see above). Sustaining ties of reference (e.g., this sample is from *this* river) and crafting compelling reductions (e.g., these data *are* the river's temperature) can be inspected as the practical work and technical armature of domains, disciplines, or actors. In doing so, the social analyst need neither commit nor oppose the links of reference or reductions.

Instead of reinterpreting the world as one set of fundamental materials, a material methodology gives the tools to recognize the situated and specifically textured nature of reductions and generations, as well as the importance of material agencies when they are encountered. In this sense, rather than casting materiality as an ontological assertion to be enacted across the board, materiality is an additional sensitizing concept along with those that draw our attention to the processuality of, say, practice, documents and archives, collaboration, power, and so on.

Historicist Materiality

The final trick of the trade is drawn from the nuanced empiricism of historical epistemology and ontology. Michel Foucault, who innovated the term "historical ontology" (more commonly known as the genealogical method), sought to examine the historical conditions of possibility for researchers to take interest in a "thing." In an example quite relevant to the case of the MACS, Foucault unearthed the conditions of possibility that paved way for doctors, psychologists, sociologists, and others in the 18th and 19th centuries to become interested in, and, more importantly, concerned with, the health of the population. Sex, or rather the regulation of sexual acts, came to be a central concern in the 19th century when the reproduction of the population became synonymous with a healthy economic and political environment; "There emerged the analysis of modes of sexual conduct, their determinations and their effects, at the boundary line of the biological and economic domains" (Foucault 1990, 26). Over time, such new objects of investigation were materialized by instrumentation that, for example, tracked the number of sexual partners or the tumescence of the penis (Waidzunas and Epstein 2015).

The work of historical epistemologists such as Lorraine Daston has additionally drawn our attention to the ways in which objects exit the repertoire of reality (2000). Things do not simply become objects of scientific concern and are thereafter explored "linearly," forever becoming better understood. Rather, the interest of scientists in certain objects can repeatedly wax and wane, and they can also cease to be of interest to science altogether. Thus, for Daston and others exploring historical epistemology, the focus is not only the things in themselves but also their social life within the sciences, and then without. A favored example in STS has been "phlogiston" (White 1932), a substance that for a time organized the attention of scientists interested in combustion before eventually falling out altogether from the pantheon of reality.

Rheinberger's work has emphasized the emergent or unfolding role of materials as sources of surprise, recalcitrance, and recurrence. He tracks a circuitous role for materials in his investigation of "cytoplasmic particles" as they were, first, epistemic objects of scientific investigation, but later those same entities became the tools or instruments for further investigations (Rheinberger 2000). It is common for scientific materials to play both roles in parallel, as with the case below where the Epstein Bar virus is used as the tool for immortalizing the cell lines even as that virus also remains the object of research for other scientists who continue to investigate its genetics, prevalence, transmission, and so on.

Example: Immobilized Data and Replicating Materials

A received view of data, as with any form of information, is that it can be copied indefinitely, while the correlate view of specimens is that they are finite. This is largely how they are treated in the MACS, resulting in different *regimes of valuation* for both (Dussauge et al. 2015). In the MACS, the demographic, behavioral, or medical data that have been generated about the participants have been reused hundreds, perhaps thousands of times in studies of many kinds. No matter how many times the data are used, they can still can be repurposed once again (as we saw above with the case of stream temperature data in LTER). In contrast, the specimens of blood are each unique and finite, with only a certain quantity collected and preserved from each man at each time point. To maximize their usage, the six vials that are collected at each visit are pipetted (or "alliquotted") into dozens of smaller vials before being placed in cold storage. Each aliquot can be defrosted, used in part, and the remaining materials frozen again—detailed instructions accompany each specimen to ensure investigators use only what they need and then properly return whatever is left over to the archive. Despite such care, ultimately, each use depletes the vial, and eventually for a given sliver of time there will be none left. And so, differently than data, the use of blood specimens is carefully deliberated, that is, weighing the prospective value of a particular study against the quantity of stored blood that will be used. Proposals for a study are occasionally rejected on this basis.

But there are circumstances where these received truths do not hold: data are not all organized to encourage sharing and replication, and materials need not always be finite. Unlike the demographic, behavioral, or biomedical data, personally identifying data that are held about the MACS participants, whether HIV positive or negative, are never shared. These data are kept separate from the rest (under "lock and key" for the first decades of the project, and now "behind password and encryption") and only a select group of staff and investigators have access to the men's real names, addresses, and contact information. In principle these data can be copied indefinitely, never depleting their archives, but in practice the MACS has evolved a complex privacy regime to ensure this is never the case. They do so at some expense, developing systems that keep personally identifying information in secured sites, and at a cost to their scientific enterprise (i.e., imagine the wealth of research that could be conducted if their vast troves of behavioral data could be linked to social media traces). Treated as data, identifying information can be copied and shared. But, inspected in relation to the sociotechnical system that sustains them, these data are best understood as part of an operation dedicated to ensuring they do not travel. However, a change in that privacy regime, or the more general subjects' protection guidelines and laws established in the United States, could change how those data circulate. Approached as a matter of the sociotechnical system that sustains privacy (or not), at this time these data have legal, technical, and practical protections that prevent the kind of indefinite copying we usually associate with data. The replicability and mobility of data should be approached not as inherent technical properties but as sociotechnical ones that shift at the intersection of technical capacities, guidelines, and laws.

Similarly, while in general we can say the blood archive is depleted by every use, there are circumstances when this is not the case. For instance, in 1997 MACS scientists isolated white blood cells from the specimens of over 1,900 participants and "immortalized" them through a technique that involves infecting them with Epstein-Barr virus. Thereafter these lymphoblastoid cell lines would reproduce indefinitely, providing a potentially infinite source of "the same" genetic material, a permanent resource for the studies that rely on these materials: "As these samples are used for a wide range of studies and will become limited as more studies related to human disease are performed, the establishment of cell lines as permanent resources of genomic DNA is considered a potential solution" (Herbeck et al. 2009). Again, the properties of these materials—blood, DNA, PBMCs—are not fixed, they operate in their own shifting regimes.

Trick of the Trade IV: Approach Materials Historically: They May Come, Go, or Change

A received understanding of data is that they can be copied indefinitely, whereas material specimens are finite, but inspected more specifically, and placed in relation to their use and stewarding, certain data inhabit complex sociotechnical architectures that are intended to prevent their copying, reuse, or circulation, and certain materials can be copied, or made to copy themselves, indefinitely. In the abstract, data are an infinite resource while specimens are scarce, but inspected historically, and in relation to the techniques and systems they inhabit, at times they display completely different properties.

We must "discover" the agencies and roles of materials in situ, and track their evolution empirically. The procedures for immortalizing cell lines have evolved in biology over the past century. The first immortal cell lines, such as the infamous HeLA (Landecker 2007), were developed almost serendipitously, but today novel techniques allow for the systematic creation of specific cell lines. Thus, across time, the technical capacities of biomedical research have shifted, and so too have the agencies of certain key biological materials—now "immortal."

The specimens of the MACS archive were once depleted by each use—and this is still the case with most materials. But for some very specific materials, it is now possible to "clone," "replicate," "amplify," or "immortalize" them through various techniques. Coupled with those innovations, we find a complex *regime of similarity and difference* that has emerged for lymphoblastoid cell lines that both enables and constrains their use as resources that treated them "as if" they were the same as the original specimens. Today, immortalized lymphoblastoid cell lines operate in a regime of care to avoid significant genetic drift relative to the originals, and even with the best methods, the process of their generation may lead to genotypic discrepancies. Thus, by treating materials historically, we must also return to the additional tricks of the trade discussed above: the regimes of similarity and difference that constrain and limit the use of these materials for generalizable research, or the local and relational enactment of materials. This form of analysis demands an astute and unrelenting historicism from the scholar of materiality: a material can never be treated as investigated, understood, and thereafter blackboxed; rather, it must be resituated historically and practically at each turn.

Overall, placing materials in changing historical and practical circumstances affords an understanding of their plastic and situated roles. Materials may come and go: sometimes in the sense of break-down (as with a gear), wearing away (like a pebble on a beach), or wearing down (like cartilage); but sometimes in the sense that they are no longer considered materials at all, as with phlogiston, a kind of dematerialization that retrospectively rewrites history. Even the most mundane materials can be an emergent source of deep and challenging complexity: so while "mud" has not been an exemplary object for the scientists most revered in history, it is an ongoing research challenge for the civil or automotive engineer (Hacking 1995).

Materiality as One Sensitizing Concept

I have presented four tricks of the trade out of a vast and diverse repertoire that is available to us for the inspection of materials, objects, and things in the STS tradition and beyond. I have not attempted a theoretical synthesis nor recounted a methodological progression. While the tricks appear in this chapter in a roughly chronological order to their association with methodological schools, it is not that the "newer tricks" have superseded the older ones—I continue to draw on all. So too is the case for tricks of the trade that preceded the materialist turn; for here I am advocating for adding tools to our repertoire rather than a revocation of what came before.

A story to exemplify this: In my recent move from Washington, DC to Seattle, I packed up my belongings and my dog Beemo for the final flight. Airlines are particular about how pets travel in their cargo holds and their well-documented guidelines are available online: for example, the transport crate should be made of hard plastic and held together with four metal screws, vaccinations should be current, and the pet should be checked in at least one hour before the flight. I fastidiously followed the extensive rules. But when I arrived at the airport, well early of the flight and with certificates in hand, one single missing screw almost derailed the entire move to Seattle. Somehow, one of the four required bolts that secured the crate had gone missing. The airline attendant told us in no uncertain terms that Beemo could not fly without. My meticulous planning for this move, otherwise smooth as butter, was derailed by a single absent bolt.

Is this case best analyzed in terms of its materiality—that is, of a missing bolt? In this case I tend to think not. Understanding this circumstance would be better served by approaching it in a more (sociologically) conventional matter of organizations and institutions. Beemo could not travel because she was not in line with the rules the airline had set forth for pet travel. Of course, in some extreme cases, that missing bolt could save her life. But that was not what was occurring: what was interrupting Beemo's flight was the enactment of an accountable checklist procedurally tied to pet travel.

What followed corroborates an organizational explanation: the missing bolt sent me, and some airline staff, into a frantic search for a screw that would fit the crate. After failing to find such a fastener, and after much pleading, the original airline attendant called her supervisor, who made a one-time exception to the rule. As with most institutional rule following, the encounter was morally charged: the supervisor did her due diligence by chiding me for my negligence. But Beemo made it on to the flight and off to Seattle we went. This work-around is also best understood institutionally—as a well-executed appeal to authority and situated decision making within an enacted organizational hierarchy. The absence of the fourth bolt played a role in this entire event, not in the material sense conveyed by Latour's speed bump that physically slows a car (Latour 1995), but rather in its more symbolic sense imparted within an institutional ecology.

Making sense of this circumstance as I have done here could be done with the tools supplied by Max Weber over a century ago. Despite the novelty of planes, rules posted online, computer-based checklists, plastic crates, and the luxury of flying dogs, this little vignette is best understood as a matter of organizational hierarchies, accountability, rule following, work-arounds, and other tricks of the trade that follow from the Weberian tradition of organizational analysis.

A materialist approach should not be dogma—a drive for a materialist purity—rather it is a sensitizing tool of the analyst, allowing us to hone in and make sense of the central aspects of the study at hand.

We cannot say anything about materiality in the abstract. Or more precisely, we can't say anything *interesting* about materiality in the abstract. Instead, it is best to say something about how to study materiality, how to make it something available for the inspection of its social, sociotechnical, or phenomenotechnical investigation. A common starting point for analysis of materiality is that past social theory and methodology has elided the material, and consequently has provided few analytic resources to make sense of it. In this respect, the turn to material analyses is a potentially revolutionary addition to our repertoire of study methods. But foreclosing the discussion with declarations about the nature of that materiality is foolhardy. Rather, materiality is our ongoing object of analysis within the manifold phenomena that are of interest to us.

My core point to the scholar interested in studying materiality is that if you really think materiality matters, then stay away from analyzing it in the abstract. If you believe "materiality" is a foundational reorientation of social research—and not simply an intellectual fad—then what we must do is develop the program by extending our ability to investigate materiality adeptly and to recount our findings meaningfully as we broaden our studies to more and more subjects.

Works Cited

Becker, Howard S. 1998. *Tricks of the Trade: How to Think about Your Research While You're Doing It*. Chicago: University of Chicago Press.

Blanchette, Jean-François. 2011. "A Material History of Bits." *Journal of the American Society for Information Science and Technology* 62 (6): 1042–57.

Bowker, Geoffrey C. 1994. *Science on the Run: Information Management and Industrial Geophysics at Schlumberger, 1920–1940*. Cambridge, MA: MIT Press.

Bowker, Geoffrey C., and Susan Leigh Star. 1999. *Sorting Things Out: Classification and Its Consequences*. Inside Technology. Cambridge, MA: MIT Press.

Daston, Lorraine, ed. 2000. *Biographies of Scientific Objects*. Chicago: University of Chicago Press.

Daston, Lorraine, and Peter Louis Galison. 1992. "The Image of Objectivity." *Representations* 40:81–128.

de la Bellacasa, María Puig. 2012. "'Nothing Comes without Its World': Thinking with Care." *Sociological Review* 60 (2): 197–216.

Douglas, Mary. 1986. *How Institutions Think*. Syracuse, NY: Syracuse University Press.

Dussauge, Isabelle, Claes-Fredrik Helgesson, and Francis Lee. 2015. *Value Practices in the Life Sciences and Medicine*. New York: Oxford University Press.

Espeland, Wendy Nelson. 1998. *The Struggle for Water: Politics, Rationality, and Identity in the American Southwest*. Chicago: University of Chicago Press.

Foucault, Michel. 1990. *The History of Sexuality: An Introduction*. Vol. 1. Translated by Robert Hurley. New York: Vintage.

Garfinkel, Harold, Michael Lynch, and Eric Livingston. 1981. "The Work of Discovering Science Constructed with Materials from the Optically Discovered Pulsar." *Philosophy of the Social Sciences* 11 (2): 131–58.

Goodwin, Charles. 1994. "Professional Vision." *American Anthropologist* 96 (3): 606–33.

Groffman, Peter M., Daniel J. Bain, Lawrence E. Band, Kenneth T. Belt, Grace S. Brush, J. Morgan Grove, Richard V. Pouyat, Ian C. Yesilonis, and Wayne C. Zipperer. 2003. "Down by the Riverside: Urban Riparian Ecology." *Frontiers in Ecology and the Environment* 1 (6): 315–21.

Hacking, Ian. 1995. "The Looping Effects of Human Kinds." In *Causal Cognition: An Interdisciplinary Approach*, edited by D. Sperber, D. Premack, and A. Premack, 351–83. Oxford: Oxford University Press.

——. 2004. *Historical Ontology*. Cambridge, MA: Harvard University Press.

Herbeck, Joshua T., Geoffrey S. Gottlieb, Kim Wong, Roger Detels, John P. Phair, Charles R. Rinaldo, Lisa P. Jacobson, Joseph B. Margolick, and James I. Mullins. 2009. "Fidelity of SNP Array Genotyping Using Epstein Barr Virus-Transformed B-Lymphocyte Cell Lines: Implications for Genome-Wide Association Studies." *PLOS ONE* 4 (9): e6915. doi:10.1371/journal.pone.0006915.

Kim, Hyun Jin. 2007. "Temperatures of Urban Streams: Impervious Surface Cover, Runoff, and the Importance of Spatial and Temporal Variations." Master's thesis, University of Maryland, Baltimore County.

Kaushal, Sujay S., Gene E. Likens, Norbert A. Jaworski, Michael L. Pace, Ashley M. Sides, David Seekell, Kenneth T. Belt, David H. Secor, and Rebecca L. Wingate. 2010. "Rising Stream and River Temperatures in the United States." *Frontiers in Ecology and the Environment* 8 (9): 461–66.

Landecker, Hannah. 2007. *Culturing Life: How Cells Became Technologies*. Cambridge, MA: Harvard University Press.

Latour, Bruno. 1986. "Visualization and Cognition: Thinking with Eyes and Hands." *Knowledge and Society* 6:1–40.

——. 1987. *Science in Action: How to Follow Scientists and Engineers through Society*. Cambridge, MA: Harvard University Press.

——. 1988. *The Pasteurization of France*. Cambridge, MA: Harvard University Press.

——. 1995. "The Sociology of the Door-Closer." In *Ecologies of Knowledge*, edited by Susan Leigh Star, 257–80. New York: State University of New York Press.

——. 1999. "Circulating Reference: Sampling the Soil in the Amazon Forest."

Latour, Bruno, Graham Harman, and Peter Erdélyi. 2011. *The Prince and the Wolf: Latour and Harman at the LSE: The Latour and Harman at the LSE*. Alresford: John Hunt.

Leonardi, Paul M. 2012. "Materiality, Sociomateriality, and Socio-technical Systems: What Do These Terms Mean? How Are They Different? Do We Need Them?" In *Materiality and Organizing: Social Interaction in a Technological World*, edited by P. M. Leonardi, B. A. Nardi, and J. Kallinikos, 25–48. Oxford: Oxford University Press

Nagel, Thomas. 1974. "What Is It Like to Be a Bat?" *Philosophical Review* 83 (4): 435–50.

O'Connel, Joseph. 1993. "Metrology: 'The Creation of Universality by the Circulation of Particulars.'" *Social Studies of Science* 23:129–73.

Orlikowski, Wanda J. 2007. "Sociomaterial Practices: Exploring Technology at Work." *Organization Studies* 28 (9): 1435.

Rheinberger, Hans-Jorg. 2000. "Cytoplasmic Particles: The Trajectory of a Scientific Object." In *Biographies of Scientific Objects*, edited by Lorraine Daston, 270–94. Chicago: University of Chicago Press.

Ribes, David. 2014. "The Kernel of a Research Infrastructure." In *Proceedings of the 17th ACM Conference on Computer Supported Cooperative Work and Social Computing*, 574–87. New York: ACM.

——. 2017. "Notes on the Concept of Data Interoperability: Cases from an Ecology of AIDS Research Infrastructures." In *Proceedings of the 2017 ACM Conference on Computer Supported Cooperative Work and Social Computing*, 1514–26. New York: ACM.

Ribes, David, and S. J. Jackson. 2013. "Data Bite Man: The Work of Sustaining a Long-Term Study." In *"Raw Data" Is an Oxymoron*, edited by Lisa Gitelman, 147–66. Cambridge, MA: MIT Press.

Ribes, David, and Charlotte P. Lee. 2010. "Sociotechnical Studies of Cyberinfrastructure and e-Research: Current Themes and Future Trajectories." *Journal of Computer Supported Cooperative Work* 19 (3–4): 231–44.

Ribes, David, and Jessica Beth Polk. 2015. "Organizing for Ontological Change: The Kernel of an AIDS Research Infrastructure." *Social Studies of Science* 45 (2): 214–41. doi:10.1177/0306312714558136.

Waidzunas, Tom, and Steven Epstein. 2015. "'For Men Arousal Is Orientation': Bodily Truthing, Technosexual Scripts, and the Materialization of Sexualities through the Phallometric Test." *Social Studies of Science* 45 (2): 187–213.

White, John Henry. 1932. *The History of the Phlogiston Theory*. London: E. Arnold.

Whitehead, Alfred North. 2013. *The Concept of Nature*. Mineola, NY: Dover.

Digital Visualizations for Thinking with the Environment

Nerea Calvillo

Visualizations have been an ongoing object of research within science and technology studies (STS) since Bruno Latour and Steve Woolgar identified their role in the production of scientific knowledge and followed them in the laboratory ([1979] 1986; see also Coopmans et al. 2014; Burri and Dumit 2008). This chapter draws on this literature but shifts the focus of attention from visualizations as an object of study to digital visualizations as a research method in themselves. It also shifts the setting, moving out of the laboratory and the scientific realm into messier and more distributed environments including domestic spaces, art institutions, or hostel lobbies. This interest in digital visualizations as research tools has a tradition in STS, ranging from controversy analysis, where visualizations are used to make visible the networks involved in a certain controversy (Yaneva 2012; Venturini 2010; Latour et al. 1992); or "issue mapping" (Marres 2007), where digitally produced maps, charts, and graphs are used for sociological research within the wider frame of digital sociology; to visual artifacts produced by researchers with digital images or video to inquire, sustain, and even present practice-based research outputs in visual sociology (Wakeford 2006). In this context, this chapter proposes the production of visualizations not as tools to represent or organize previous knowledge, as in the networks of controversy mapping, nor to identify emerging patterns or research questions by activating datasets. It proposes the design and production of digital visualizations as research starting points to detect and engage with the environment's sociotechnical assemblages and create alternative representations of it. It does this drawing from feminist technoscience and nonrepresentational methods literature, to account for the openness of the method and the capacity of research to interfere in the world (Mol 2002; Haraway 1992; McCormack 2015; Lury and Wakeford 2012; Manning 2015; among many others).

This proposal emerged out of the design and development of the visualization project *In the Air*, used here to illustrate and support the main argument. I initiated *In the Air* in 2008 at a workshop in Medialab Prado in Madrid,[1] and it has been developed collaboratively throughout these years in workshops, exhibitions, and independent research.[2] The overall investigation emerged from a simple question: how can we, as researchers, deal with and get involved with the invisible gases or particles suspended in the air that we live in? The project engaged with Madrid as a case study, and through the visualization process we acknowledged that the challenge of investigating the air is not only about the design or analysis of visualizations, but about the process of visualizing itself. So, by making visualizations in different formats, we started unfolding questions that the mapping process was opening up, such as the political use of air quality information (recognizing the double-sided nature of transparency of public data, for example) or the technical and normative implications of what it means to sense the air (the difficulty in calibrating the sensors, in deciding what to measure and where, etc.). Since then, *In the Air* has enabled us to test, challenge, and advance both in the design of visualizations and in empirical research on air pollution.

Doing visualizations as a research method also aims to contribute to the long tradition in STS engaged in environmental knowledge production. It moves away from the objectivity expected from experimental practices (Shapin and Schaffer 2011), and from mathematical approaches in climate predictive models (Edwards 2001). Closer to ethnographies of toxicity to acknowledge the inapprehensibility of air pollution (Choy 2011; Shapiro 2015), it shows how making a visualization is a form of doing an ethnography of Madrid's air data, as well as creating a new image of it that can circulate outside academic contexts.

So this is a chapter on digital visualizations *in the making*, structured as follows: first it justifies, retrospectively, the ways in which the production of digital applications can be an inventive method of inquiry for STS, and how those applications can help to deal with material environmental conditions, working through the notion of what Latour calls "inscriptions" (1990). Then it focuses on two main contributions of the method: (1) it shows how the object of inquiry's sociotechnical assemblage emerges in the process and (2) it reveals how visualizations are not just descriptions or representations of reality, but can make or enact worlds (Latour 1986) through design strategies, creating affective airscapes.

In the Air

On a warm October morning of 2008, the Visualizar '08: Database City workshop began at Medialab Prado in Madrid. Selected projects from an international open call were presented, and collaborators chose the ones they wanted to be involved in. Nine international participants from different backgrounds (an interaction designer, architects, a teacher, an anthropologist, and a new media researcher) joined *In the Air* and dedicated two intense weeks to work together within the institution's facilities developing a digital application and a "diffuse façade" prototype. The aim of the project was, as stated on the web page, to serve as a "platform for individual and collective awareness and decision making, where the interpretation of results can be used for real time navigation through the city, opportunistic selection of locations according to their air conditions and as a base for political action" (www.intheair.es).

The digital application, still running, scrapes the data measured by the Madrid City Council's 24 monitoring stations, made available on their website (although not in a usable format, so the programmer had to scrape the data from the web page's API), aiming to propose an alternative to the forms in which air quality was displayed by Madrid City Council and in most institutions at that time: providing only the numbers obtained from the stations at their location.

In the Air is a relational tool to compare components in space and time, as well as with the European Union (EU) regulation (figure 1). The five gases and particles

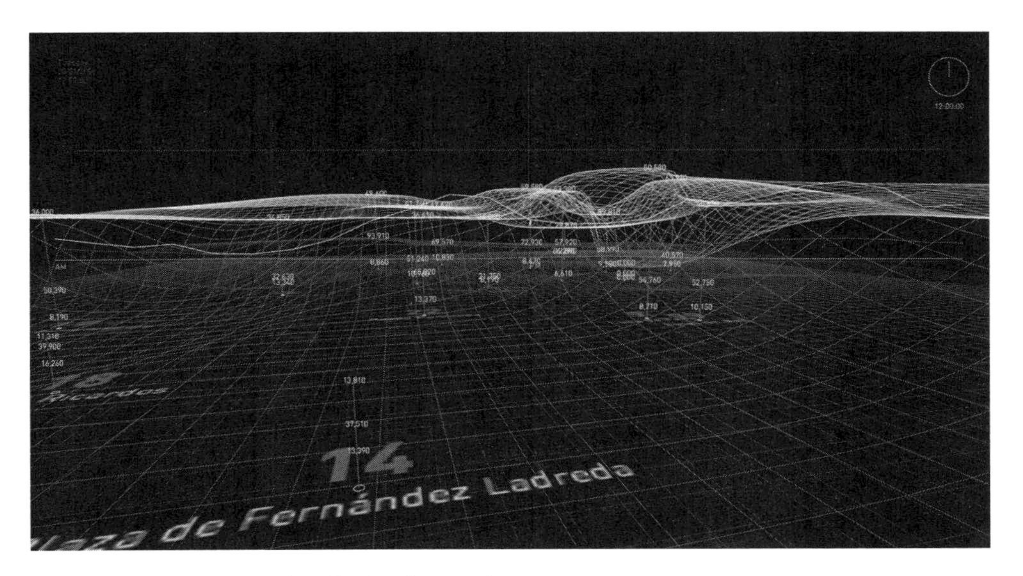

FIGURE 1: *In the Air*. Digital application (2008–12).

that cities have to monitor by EU law are displayed independently but in relation to one other. Instead of air quality averages or different maps for different components, data are interpolated to better understand the differences of pollution across the city. The goal of this interpolation is not to estimate air quality between observing stations (although the formula takes into consideration the distribution of gases), but to visually compare the main differences of pollution across the city. Therefore, it does not take into consideration meteorological data, the built structure of the city, or the fluidity of the air, because the objective is not to understand the performance of the air in its complexity, nor to predict it. The objective is to compare the relationships between some components and different areas of the city and, as data are stored and can be accessed through a timeline, to compare the levels in time. In addition, the interpolation aims to create an image of the air as a landscape that can be inhabited. So it is not a model to represent Madrid's air, but a visualization to present it differently.

In the Air was conceived of as a visualization project to explore forms of making air quality visible. The digital application became the core tool that collected the data and proposed an alternative way of thinking and experiencing air pollution, and was adapted to other cities, like Santiago de Chile and Budapest (figure 2). But other visualizations have been proposed along the years to test what, how, and where different aspects of air pollution are worth making visible.[3] The "digital façade" prototype (figure 3), for instance, used water vapor clouds to display the concentrations of gases and particles in the nearest monitoring station. Through the clouds' different colors, densities, and rhythms, citizens were able to experience air pollution in the public square. The air represented itself (air pollution was seen *in* the air), and at the same time it transformed certain physical conditions of the public space, such as humidity or temperature (Calvillo 2017). Another set of iterations have been developed for urban screens, to help guide passers-by through the city's pollution, collectively and in real time (figure 4).

FIGURE 2: Adaptation of *In the Air* to Budapest.

FIGURE 3: *In the Air*. Digital façade prototype (2008).

Digital Platforms as Inscription Devices

In the Air is not an inscription as an "immutable-mobile" (Latour 1987), as it is not a fixed image that aims to prove a scientific fact. Neither does it intend to tell the truth about a dataset (Tufte 2001). It acts as an epistemic device, whose aim, instead of "stabilizing and representing patterns so that they can be interpreted" (Ruppert et al. 2013), is to expand modes of knowing, perceiving, or engaging with the air. Also, it inscribes not only the object of research (the air), but also its regulations, norms, or politics.

And here is where the digital plays a relevant role. In order to understand the relational features of the permanent fluctuations of the air, a dynamic visualization is required, as well as an automatic processing of the data. Nevertheless,

FIGURE 4: *In the Air* application for Medialab Prado digital façade (2009).

the automatic processing is not addressed to deal with big data and complexity, as the case of Loukissas's Boston arboretum visualization (this volume). Digital software allows the creation of the three-dimensional spaces of the maps and their navigation, creating a more immersive experience with the data, where the user becomes a performer in a new space (Lammes and Verhoeff 2009; Calvillo 2012). The digital also permits to customize the parameters of the visualization: which components are displayed or the color range of the interpolation. So the digital facilitates a responsiveness to the environment and the experience of the data.

Making Visualizations as a Method to Think with the Environment

Thinking *through* things has a long tradition, where objects found throughout research are used as heuristics to develop theories (see Henare et al. 2006 or McCormack 2015). There are also researchers who think by *making* things (DiSalvo 2012; Ratto and Boler 2014; Wilkie and Mattozzi 2014; Dunne and Raby 2013; Jeremijenko 2017; Public Lab 2017; among many others). Their practices are heterogeneous and have different aims and claims, ranging from critical design to critical making or citizen science, yet they share with this chapter an underlying challenge: how to relate research and making. Here, the focus will be not on what the object can do for research, but on how the making of an object (a digital visualization) can be a research method for STS, to explore the materiality of environmental issues, engaging with their sociotechnical assemblages and creating new imaginaries.

Inventive Method

The chapter engages with nonrepresentational approaches to methods, "less (as) a way of articulating a set of practices that are forced to stand up in a particular epistemological theatre of proof, and more a way of going on in the world that allows its different modes of making difference potentially sensed" (McCormack 2015, 93). This method cannot be defined once and for all because the visualizations that I am proposing are all different, "customized" if you will, to the specific object of study. So the method has to take into account the differences and requirements of the object of inquiry, and therefore its protocols cannot be set in advance. Hence, this method also has to be *inventive*. Inventive not because of its newness, but because, as Lury and Wakeford have noted, it "seeks to realize the potential of this engagement, whether this is an intervention, interference or refraction" (2012, 6). Moreover, inventive because the research is organized not only as a description of what is there, but more as a matter of configuring what comes next. Through a hands-on and collaborative process, questions emerge, expanding the realms of inquiry in a nonlinear and unpredictable way, in our case from the air itself to open data, institutional politics, sensors in the public space, or public awareness, highlighting the relevance of unexpected guests such as bees or worms, rain and the wind.

For all these reasons it has to be an open method, more an attitude than a set of protocols, attuned to the messy process of testing and bringing together, taking into account the human and more-than-human support and care involved in these practices. As Erin Manning notes, "Each step will be a renewal of how this event, this time, this problem, proposes this mode of inquiry, in this voice, in these materials, this way. At times, in retrospect, the processes developed might seem like a method. But repeating it will never bring it back, for techniques must be reinvented at every turn and thought must always leap" (Manning 2015, 69).

Even though a method in the natural and social sciences implies a systematic approach, method is here used as "an orientation to investigation" (Ribes, this volume) to highlight the capacity of this process to drive research on an environmental object. As an open and inventive method it has only one clear step: start designing and making the visualization, as a point of departure. However, as an orientation, it involves certain attitudes, like an attunement to the emergences and a critical response to them. It also involves a feminist approach to visualizations (D'Ignazio and Klein 2016), to deal with uncertainty, enable dissent, avoid binaries, and question the power relations embedded in the data.

Material Method

One of the main opportunities for making visualizations is to explore invisible materialities. As a material method, it enables one to "discover" the agencies and roles of materials in situ, and track their evolution empirically" (Ribes, this volume). Because part of the challenge of thinking with the environment is to make visible an environmental condition without trying to impose what we think we know about it, and learning instead about its ontology through the process. In this context, *In the Air* is an exercise in how to represent gases and particles in order to generate a more complex relationship with the air, where instead of knowing about the environment, we can think with it by means of *cosmopolitical experiments* (Hinchliffe et al. 2003). Following Hinchliffe et al.'s case study, this device, like their field books, is a way of "talking around" (and "walking around") rather than "talking about," as a way of knowing and making visible. Not by showing the object of analysis in itself (the air here, the water vole for them), but by producing multiple inscriptions that may allow us to get close, learning to be affected on the way, to produce a more symmetrical relationship between the object of analysis, the nonhuman air, and the humans who research it (producing a different political ontology).

So, instead of representing gases or particles, *In the Air* sets up the conditions to connect to the air. For this purpose, rather than producing an inscription, we had to build an inscription device, a digital application to enable some aspects of the air to reveal themselves, mediated by technology, in real time. As Hinchliffe et al. demonstrate, this inscription device functions as a Deleuzian diagram making the "water vole [or the air] more rather than less real" (2003, 648).

Although this method can be used by many disciplines or contexts, I am after the possibilities it offers to STS. As I have argued elsewhere (Calvillo 2019), it is a collaborative research practice where visualizations work as boundary objects (Henderson 1991) that enable STS practitioners to work with others to act in the world, as many scholars are already doing in experiments in participation or activist research, for instance (Ellis and Waterton 2005). It also helps to understand the networks of an object of inquiry (including its power relations and politics), as well as to create new frameworks or imaginaries, as a form of material ecocriticism (Phillips and Sullivan 2012), which is what the next two sections will focus on.

I am not arguing that all of these objectives cannot be achieved through text. Making visualizations is another way of doing, which adds an additional dimension. As Erin Manning suggests (2015), as a method that might be framed as "research-creation," it produces new forms of experience, where the visualization can function, like the balloon for McCormack for "doing atmospheric things," as "a device for a form of practical aesthetics, in which different qualities of atmospheric spacetimes can be experimented with" (2015, 106).

Emerging Sociotechnical Assemblages

While hurriedly putting together *In the Air*, within the tension and excitement of the workshop, slowly and without even noticing, the sociotechnical assemblages of gases and particles emerged through the making. By paying attention to the difficulties and challenges faced in the process, different agents popped up, revealing their relations.

To illustrate this point, let us take *Pollen In the Air*, the last development of the project, as an example. *Pollen In the Air* makes visible another component of the air proposed in the first iteration: pollen (figure 5). It was produced collaboratively as a commission for the Laboral Centro de Arte y Creación Industrial for the exhibition Datascape, with a smaller group and some financial support. The new visualization shows pollen distribution in Gijón (Spain) as well as the different plant species that produce it.

When constructing the digital application, we had to identify who published the data and how frequently. As we were unable to find the data in the City Council's website, we had to contact the measuring institution, the regional health institute AsturSalud. This institute publishes graphs in PDF format, under the assumption that pollen concentrations are not needed in real time. Obtaining the dataset in a programmable format required institutional collaboration and an email exchange, revealing who manages pollen monitoring and how it is managed. In principle, we were not interested in this, but we had to find out who could provide the data in a usable format. And through our conversations with them, we not only obtained five months of data, but the Pharmacy Medical School and the Open Data Department of the City Council realized their lack of synchronization and decided to change their protocols. We also discovered which institution makes the location of urban trees available online to promote open data. While trying to match these trees with pollen counts we discovered that the trees we had were only the ones taken care of by the City Council. They are "singular" species whose pollen is not counted because it is quantitatively irrelevant, and the plants producing the pollen recognized in the datasets were wild shrubs located on the outskirts of the city, areas that were not mapped. So what we thought would be "just" a technical and design task of translating a dataset into visual form became something where the object of study forced us to connect with the agents involved (many of the institutions in charge of the management of pollen in the area) and activated connections between them (contributing to the rearticulation of their protocols), and we learned not only about pollen, but also about tree species, Gijon's green areas management, pollinators, glass plates, the wind, and so on.

The agents involved did not emerge alone, but also their power relations and what was at stake, or to put it bluntly, their politics. When trying to understand why the data feed from the City Council was not active a week before the opening, following urgent correspondence with different institutions, we learned that there were no actual data because the meter was broken, and due to the economic crisis there was no intention of repairing it. The City Council argued that having one meter in Oviedo (60 km away) was "enough" to comply with regulations (and satisfy citizens). So the publication of the data was politically intended to perform as a "transparency device" (Harvey et al. 2012), where what matters is to say that data are being published rather than actually thinking how they can be mobilized, or how they may condition humans' health and everyday lives.

FIGURE 5: *Pollen In the Air*. Digital application (2014).

The criticality of this approach lies in looking at the process of making with care, in identifying and acknowledging the agents that emerge and taking them into consideration in the design of the visualization (not necessarily represented). The criticality also lies in taking the obstacles found in the process as knowledge production instances and feeding back into the visualization.

Designing Affective Airscapes

What is the aim of making the environment visible: to represent a truth, as most scientific and policy-making maps do (Kitchin et al. 2009)? Or to interfere with the world, as feminists scholars of technoscience claim and Annemarie Mol demands of STS (2002)? In Latour and Woolgar's account of the laboratory, inscriptions have to circulate in order to gain relevance (and to interfere). But the conditions of this project are far from being scientific, first, because it did not emerge from a "center of power," such as a research institute or policy-making center, but on the contrary emerged from a medialab and was designed by nonexperts on air pollution; and second, because the visualization does not intend to succeed by demonstrating a fact, but intends to provide an alternative vision, the construction of new imaginaries of what the air or the environment is or could be. It is, as Haraway suggests, a form of "speculative fabulation, speculative feminism, science fiction, and scientific fact" (2015, 160). Speculation is the only way of engaging with the environment, at the intersection of science, art, and activist practices.

But the process of making visible can also interfere in its reality, as Donna Haraway demonstrated (1988) by looking specifically at the *agency of making visible* itself, and *how* things are made visible: how to decide what is worth seeing, with whom, for whom, under which conditions, how the invisible asks or allows to be made visible, and what our role is, as visualizers, in engaging with the object matter. Yet that is not all; the main point is how different ways of designing an aerial map construct what I have called "airscapes," with completely different implications. Through design decisions a different story is told, because every line or color requires making a decision that forces us to think how to engage and respond to the object of inquiry (Gray et al. 2016), and challenges not only what it is but, most interestingly, what things can become. Therefore, design decisions are propositional world-making techniques, positioning against current critiques to STS and actor-network theory as being an uncritical and nonpropositional approach (for a debate on these critiques, see Farías 2011), and from a feminist perspective, they can be practices of care that hold together the object of inquiry (Bellacasa 2011).

When I speak about design decisions I am not referring to traditions of STS where art has been an instrument for scientists to engage with different publics and allow their work to circulate, due to their aesthetic qualities or their multiplicity of meanings (Galison and Jones 1998; Burri and Dumit 2008). Design decisions are considered as instruments to think with the environment, to explore the ways in which we consider what is relevant of the air and to compare its materiality in space and time. Collaboration is not relegated to the deployment of the visualization (Burri and Dumit 2008), but takes place in the making of it.

In the Air aims to reframe the institutional way of displaying air quality through numbers, and what it cares for are not (only) the concentrations in themselves (the measurements), but the relationships of the aerial components in time, space, with one another, and the city. For this purpose, we used a topographical mesh as a relational entity (figures 6, 7), and the different components were interpolated to see not absolute concentrations, but how their levels are related to the limits admitted by the EU. So it is not about the air as an independent measurement, but the air (and its measurements) as a situated entity. Because location and exposures matter in air pollution. So by representing the data we constructed the space that they generate, aiming to see the space that the air inhabits (Loukissas 2016), but also its own landscape. So what emerged is the *airscape of the toxic*, where one can see the pharmakon capacity of the air, of being good and bad for our bodies according to the concentrations, the weather or the location. This airscape of the toxic enables us to decide how and where we want to see the air, if it is below us, or on top, as a landscape or a depiction, forcing a partial perspective (Haraway 1988). It also enhances a sensorial experience with the air and its visualization by navigating its landscape. If some scholars have been searching for the ways in which the air is felt (Choy 2011; Shapiro 2015), could this be an instrument to expand these modes of feeling the air through digital sensation? Could they be called affective airscapes?

Pollen In the Air brings us to a totally different realm. It shifts the focus of pollen as harmful to humans, as most pollen apps do, toward a more-than-human and ecosystem-based perspective. To do so I draw on Hustak and Myers's work on "affective ecologies" (2012) where they describe how plants and insects have interactions beyond productivity, where they communicate, have bodily (and eventually pleasant) interactions, and so on. So, what is the landscape of a pollen that is in itself a reproductive unit, that serves as food for bees, that pollinates other plants, flies with the wind, and makes some of us sneeze? *Pollen*'s airscape is upside down, reversing the human-centered aerial perspective. Humans are floating with bees or pollen particles. Instead of connecting it to human health, we connected the pollen data to data on the trees in the city, so when pollen activates we can see which trees and plants are blossoming at a precise time and location. So, it is not the landscape of risk or health prediction, but the landscape of plant, wind, and animal intercourse. With this in mind we tested another visualization device, focused not on numbers, but on intensities of things (figure 8). Because demarcations are

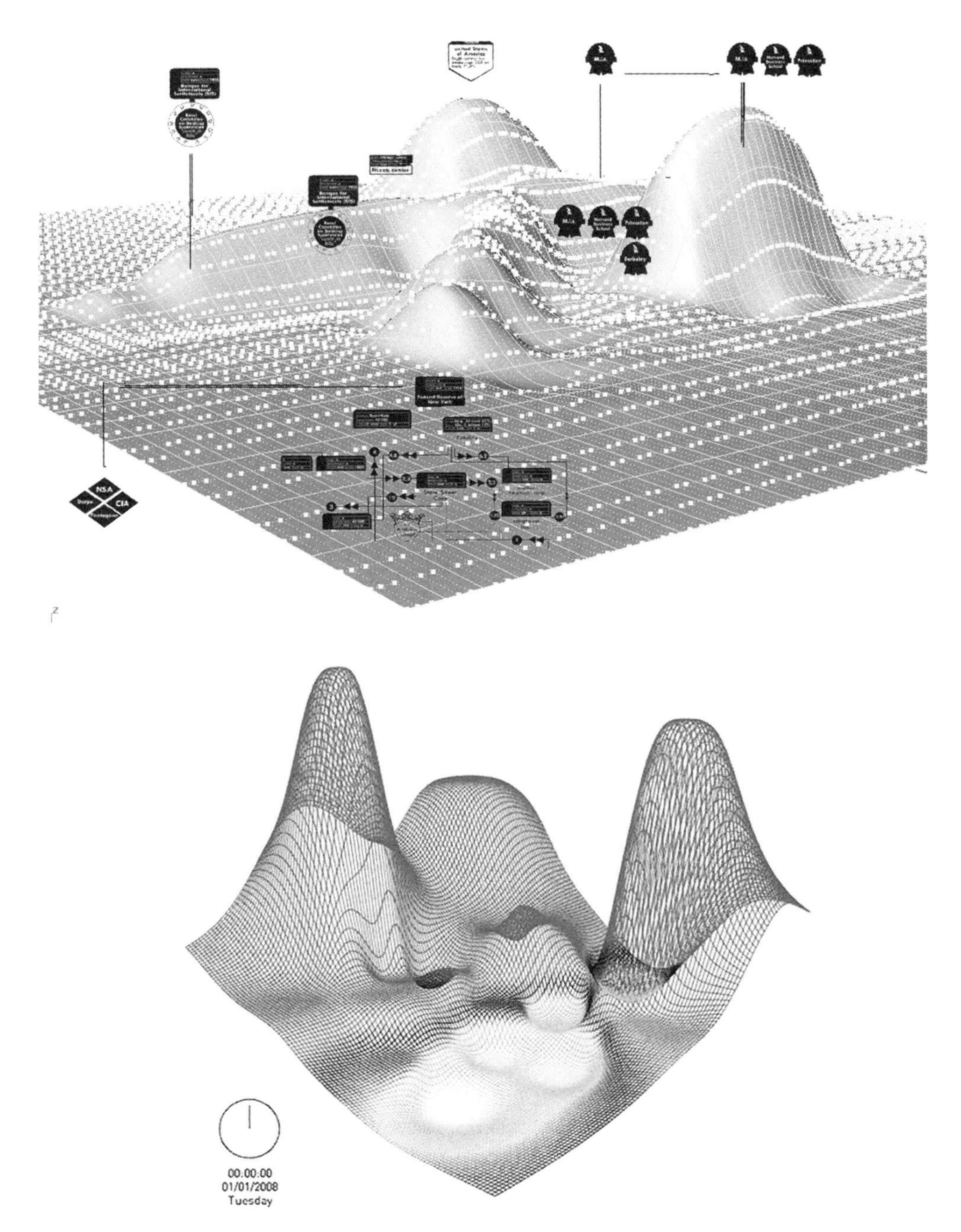

FIGURES 6, 7: *In the Air*. Topographical mesh. Initial concept and first digital prototype, testing the interpolation resolution (2008).

impossible to define in the air, we used particle clouds as the visualization strategy, which from the beginning takes into account the material properties of the air, enacting the air of queer ecologies (Gandy 2012) where multiple species have physical connections that construct various types of affects beyond production and reproduction.

FIGURE 8: *Pollen In the Air*. Particle's clouds. First digital prototype, testing the densities of the pollen clouds (2014).

The criticality of this aspect of the method relies precisely in the design decisions, that permit us to acknowledge the specificity of the materiality of the air in relation to broader issues, such as the implications of its lack of limits in geopolitical terms, as well as being able to frame the air from different perspectives. So instead of re-presenting the air as a harm for individual human health, both visualizations, *In the Air* and *Pollen In the Air*, frame the air of the collective, of environmental justice and multispecies urban relations. There is also a criticality in making the data not only visible, but also experienced (Corby 2008), as it creates awareness and affective engagements with the air.

Conclusions

The exploration of *In the Air* suggests that the production of visualizations can be a research method for STS, particularly suited to examine the invisible materiality of environmental agents and to think with the environment. As an inventive and material method, it makes emerge its sociotechnical assemblages, enabling to interfere in them. Reflecting on the making process from a feminist perspective has also demonstrated how design matters in world making, where speculation in the configuration of the visualization facilitates the construction of new imaginaries in regard to air pollution. This inquiry has also raised epistemic questions about the environment, such as the challenge of representing more-than-humans, proposing that the production of cosmopolitical experiments may be one way of redistributing hierarchies and thinking with the environment.

The critical approach of the method is twofold. It is sensitive to the making process, where the findings and critical positions toward them are built into the visualization. Second, design decisions are considered political decisions that define what is made visible, by whom, how, and many other questions, aligned with feminist visualizations.

The objective of the project is not to create the most efficient, accurate, or realistic representations of the air. Instead, building on Latour and Hermant's proposal in Paris Ville Invisible (1998), it engages with the proliferation of inscriptions, from different perspectives, to get closer to the air. The making of visualizations such as *In the Air* expands the ways, formats, and means by which to think with the air, and test the capacity of design strategies for constructing alternative affective airscapes that connect air concentrations to environmental justice or multispecies urban ecologies.

Acknowledgments

This research would not have been possible without the contributions of all the generous collaborators of *In the Air*. I would especially like to thank Martin Nadal and Marina Fernandez for their support throughout these years, and the curators and institutions who made this collaboration possible. I would also like to thank Ben Dalton, Karen Martin, Jonathan Lukens, and Brit Winthereik for their helpful comments to a first draft of this chapter. Other drafts have been presented at seminars at the STS department of the University of California, Davis, the Royal Anthropology Institute of Great Britain and Ireland Conference on Weather, Anthropology and Climate Change 2016, and at the Monsoons Air Conference 2017 (Westminster University, London). I thank Tim Choy and Lindsay Bremner for their invitation and all the attendants for their feedback.

Notes

1. *In the Air* (www.intheair.es) was initiated at the workshop Visualizar 08, curated by Jose Luis de Vicente. See "Visualizar: Database City," Medialab Prado, http://medialabprado.es/article/visualizar08_database_city_-_lista_de_proyectos_seleccionados_.
2. The collaborators at Visualizar were Sandra Fernández, Carlota Pascual, Greg J. Smith, Guillermo Ramírez, Miguel Vidal, Paco González, Raphaël de Staël, Susanna Tesconi, Victor Viña. For a full list of collaborators, see the project's different development stages on its website.
3. Thanks to curators Jose Luis de Vicente, Valentina Montero, Marília Pasculli, and Benjamin Weil for their support in different phases of the project.

Works Cited

Bellacasa, Maria Puig de la. 2011. "Matters of Care in Technoscience: Assembling Neglected Things." *Social Studies of Science* 41 (1): 85–106.

Burri, Regula Valérie, and Joseph Dumit. 2008. "Social Studies of Scientific Imaging and Visualization." In *The Handbook of Science and Technology Studies*, edited by Edward J. Hackett, Olga Amsterdamska, Michael Lynch, and Judy Wajcman, 297–317. Cambridge, MA: MIT Press.

Calvillo, Nerea. 2012. "The Affective Mesh: Air Components 3D Visualizations as a Research and Communication Tool." *Parsons Journal for Information Mapping* 4 (2): 1–8.

——. 2017. "Air Infrastructures for the Common." In *Imminent Commons: Urban Questions for the Near Future*, 54–59. Barcelona, NY: Actar.

——. 2019. "Slowing Data." In *Transmissions and Entanglements*, edited by Katrina Jungnickel. Cambridge, MA: MIT Press.

Choy, Timothy K. 2011. *Ecologies of Comparison: An Ethnography of Endangerment in Hong Kong*. Durham, NC: Duke University Press.

Coopmans, Catelijne, Janet Vertesi, Michael Lynch, and Steve Woolgar. 2014. *Representation in Scientific Practice Revisited*. Cambridge, MA: MIT Press.

Corby, Tom. 2008. "Landscapes of Feeling, Arenas of Action: Information Visualization as Art Practice." *Leonardo* 41 (5): 460–67.

D'Ignazio, Catherine, and Lauren F. Klein. 2016. "Feminist Data Visualization." Paper presented at the Workshop on Visualization for the Digital Humanities (VIS4DH), Baltimore.

DiSalvo, Carl. 2012. *Adversarial Design*. Cambridge, MA: MIT Press.

Dunne, Anthony, and Fiona Raby. 2013. *Speculative Everything: Design, Fiction and Social Dreaming*. Cambridge, MA: MIT Press.

Edwards, Paul N. 2001. "Representing the Global Atmosphere: Computer Models, Data and Knowledge about Climate Change." In *Changing the Atmosphere*, edited by Clark A. Miller and Paul N. Edwards, 31–65. Cambridge, MA: MIT Press.

Ellis, Rebecca, and Claire Waterton. 2005. "Caught between the Cartographic and the Ethnographic Imagination: The Whereabouts of Amateurs, Professionals, and Nature in Knowing Biodiversity." *Environment and Planning D: Society and Space* 23 (5): 673–93.

Farías, Ignacio. 2011. "The Politics of Urban Assemblages." *City* 15 (3–4): 365–74.

Galison, Peter, and Caroline A. Jones, eds. 1998. *Picturing Science, Producing Art*. New York: Routledge.

Gandy, Matthew. 2012. "Queer Ecologies." *Environment and Planning D: Society and Space* 30:727–47.

Gray, Jonathan, Liliana Bounegru, Stefania Milan, and Paolo Ciuccarelli. 2016. "Ways of Seeing Data: Towards a Critical Literacy for Data Visualizations as Research Objects and Research Devices." In *Innovative Methods in Media and Communication Research*, edited by Sebastian Kubitschko and Anne Kaun, 290–325. London: Palgrave Macmillan.

Haraway, Donna. 1988. "Situated Knowledges: The Science Question in Feminism and the Privilege of Partial Perspective." *Feminist Studies* 14 (3): 575–99.

——. 1992. *Simians, Cyborgs and Women: The Reinvention of Nature*. New York: Routledge.

——. 2015. "Anthropocene, Capitalocene, Plantationocene, Chthulucene: Making Kin." *Environmental Humanities* 6:159–65.

Harvey, Penny, Madeleine Reeves, and Evelyn Ruppert. 2012. "Anticipating Failure: Transparency Devices and Their Effects." *Journal of Cultural Economy* 6 (3): 294–312.

Henare, Amiria, Martin Holbraad, and Sari Wastell, eds. 2006. *Thinking through Things: Theorising Artefacts Ethnographically*. London: Routledge.

Henderson, Kathryn. 1991. "Flexible Sketches and Inflexible Data Bases: Visual Communication, Conscription Devices, and Boundary Objects in Design Engineering." *Science Technology Human Values* 16 (4): 448–73.

Hinchliffe, Steve, Matthew Kearnes, Monica Degen, and Sarah Whatmore. 2003. "Urban Wild Things: A Cosmopolitical Experiment." *Environment and Planning D: Society and Space* 23:643–58.

Hustak, Carla, and Natasha Myers. 2012. "Involutionary Momentum: Affective Ecologies and the Sciences of Plant/Insect Encounters." *Differences* 23 (3): 74–118.

Jeremijenko, Nathalie. 2017. "The Environmental Health Clinic + Lab." www.environmentalhealth clinic.net.

Kitchin, Rob, Chris Perkins, and Martin Dodge. 2009. "Thinking about Maps." In *Rethinking Maps*, edited by Martin Dodge, Rob Kitchin, and Chris Perkins, 1–25. New York: Routledge.

Lammes, Sybille, and Nanna Verhoeff. 2009. "Landmarks: Navigating Spacetime and Digital Mobility." In *Proceedings of ISSEI Language and the Scientific Imagination*, 1–21. Helsinki: University of Helsinki.

Latour, Bruno. 1986. "Visualisation and Cognition: Drawing Things Together." In *Knowledge and Society: Studies in the Sociology of Culture Past and Present*, edited by H. Kuklick, 1–40. New York: Elsevier.

——. 1987. *Science in Action*. Cambridge, MA: Harvard University Press.

——. 1990. "Drawing Things Together." In *Representation in Scientific Practice*, edited by Michael Lynch and Steve Woolgar, 19–68. Cambridge, MA: MIT Press.

Latour, Bruno, and Emilie Hermant. 1998. *Paris Ville Invisible*. Paris: Institut Sythélabo pour le progrés de la connaissance.

Latour, Bruno, Philippe Mauguin, and Geneviève Teil. 1992. "A Note on Socio-technical Graphs." *Social Studies of Science* 22 (1): 33–57.

Latour, Bruno, and Steve Woolgar. [1979] 1986. *Laboratory Life. The Construction of Scientific Facts*. Princeton, NJ: Princeton University Press.

Loukissas, Yanni. 2016. "A Place for Big Data: Reading the Arnold Arboretum through Its Accession Records." *Big Data & Society*, 1–20. http://journals.sagepub.com/doi/pdf/10.1177/2053951716661365.

Lury, Celia, and Nina Wakeford, eds. 2012. *Inventive Methods*. London: Routledge.

Manning, Erin. 2015. "Against Method." In *Non-representational Methodologies*, edited by Phillip Vanni, 52–71. London: Routledge.

Marres, Noortje. 2007. "The Issues Deserve More Credit Pragmatist Contributions to the Study of Public Involvement in Controversy." *Social Studies of Science* 37 (5): 759–80.

McCormack, Derek P. 2015. "Devices for Doing Atmospheric Things." In *Non-representational Methodologies*, edited by Phillip Vanni, 89–111. London: Routledge.

Mol, Annemarie. 2002. *The Body Multiple: Ontology in Medical Practice*. Durham, NC: Duke University Press.

Phillips, Dana, and Heather I. Sullivan. 2012. "Material Ecocriticism: Dirt, Waste, Bodies, Food, and Other Matter." *Interdisciplinary Studies in Literature and Environment* 19 (3): 445–47.

Public Lab. 2017. "Public Lab." https://publiclab.org.

Ratto, Matt, and Megan Boler, eds. 2014. *DIY Citizenship: Critical Making and Social Media*. Cambridge, MA: MIT Press.
Ruppert, Evelyn, John Law, and Mike Savage. 2013. "Reassembling Social Science Methods: The Challenge of Digital Devices." *Theory, Culture & Society* 30 (4): 22–46.
Shapin, Steven, and Simon Schaffer. 2011. *Leviathan and the Air-Pump: Hobbes, Boyle, and the Experimental Life*. Princeton, NJ: Princeton University Press.
Shapiro, Nick. 2015. "Attuning to the Chemosphere: Domestic Formaldehyde, Bodily Reasoning, and the Chemical Sublime." *Cultural Anthropology* 30 (3): 368–93.
Tufte, Edward R. 2001. *The Visual Display of Quantitative Information*. 2nd ed. Cheshire, CT: Graphics Press.
Venturini, Tommaso. 2010. "Diving in Magma: How to Explore Controversies with Actor-Network Theory." *Public Understanding of Science* 19 (3): 258–73.
Wakeford, Nina. 2006. "Power Point and the Crafting of Social Data." *Ethnographic Praxis in Industry Conference Proceedings* 2006 (1): 94–108.
Wilkie, Alex, and Alvise Mattozzi. 2014. "Learning Design through Social Science." In *About Learning and Design*, 196–203. Bozen-Bolzano: Bozen-Bolzano University Press.
Yaneva, Albena. 2012. *Mapping Controversies in Architecture*. Surrey: Ashgate.

Introduction
Gender

Daniela K. Rosner

One of the great debates of our time concerns how gender enfolds with everyday life: how it is enacted through practice, for instance, or how the category prefigures our social lives. STS scholars have embraced both arguments as they have developed tools for analyzing gender in the field. On the one hand, scholars examine the ways in which people produce, sustain, and challenge gender identity within the fields of—or using the tools of—science and technology. Studying science in practice, they trace how gendered norms shape and circumscribe scientific insights (Haraway 1984; Schiebinger 1991; Fausto-Sterling 2000; Milam and Nye 2015). Or, examining technology in context, they reveal how we co-construct gendered identities using technologies or technical orientations (Bardzell 2018; Perez-Bustos 2018; Wajcman 1991; van Oost 2003; Dunbar-Hester 2008), often in connection with alignments of race and class (Amrute 2016; Brock 2011; Gray 2012; Noble 2018; Nyugen 2018; Pham 2015). Influential feminist theories in STS have developed alongside these studies, including the concept of situated knowledge that produces insights into—and alternatives to—the "god's eye view" of scientific "objectivity" (Haraway 1988; Longino 1990; Suchman 2011).

On the other hand, STS scholars also explore how gender categories generate variations in experience that produce—or further entrench—structural inequalities in the STEM fields. Scholars in this vein may chart women's exclusion from developments in science and technology, or aim to recover women's work and contributions to the field (Rossiter 1993; Kline and Pinch 1996; Cowan 2011). Gendered narratives of the history of computer engineering especially haunt and extend STS's ideas of technological belonging: from establishing the Jacquard loom as a precursor to the Babbage Analytical Engine to recalling that the first "computers" were young women (Light 1999; Ensmenger 2010; Abbate 2012). STS scholars have therefore attended to the history of gender in computing in particular, and to undermining the oft-repeated "gendered and normative oppositions between the active and passive audience, from the male wireless amateur versus the distracted housewife in the 1920s, to the degraded 'couch potato' versus the heroic internet surfer of the 1990s" (Boddy 2004, 43, quoted in Hu 2015, 120). Such examples demonstrate how certain gendered norms govern technical participation and how other gendered identities are left out, producing continuing absences in the field.

Across these perspectives, gender is both a powerful analytical tool and a site for analytical work as scholars describe the situated articulation of difference and inequalities, often in the shadow of acclaimed scientific and technical developments.

The chapters in this section deftly apply these multivariant perspectives on gender and technology to the realm of digital culture. In doing so, they bring forward new ways of thinking about digital tools, their circumscriptions, narratives, cultures, and silences. Dunbar-Hester vividly cuts to the heart of gender issues embedded within computing cultures by exploring questions of "diversity" tied to FLOSS and hackerspace projects. Drawing from ethnographic fieldwork within these sites, she reveals a wide array of motivations behind diversity programs. In doing so, she assesses the potential of such projects for political intervention, foregrounding the placement of technology at the center of social empowerment initiatives. The result of this work is a profound recognition of the continued gulf between rhetorics of plurality and claims to social power.

Kerasidou asks, "Whose visions are the field of ubiquitous computing?" Weaving together the futurist discourse of Mark Weiser with a speculative feminist rewriting, Kerasidou examines the theoretical lines drawn (and redrawn) between human, machine, and nature. Her analysis illustrates different possibilities for material and semiotic intervention: retelling stories that, after Donna Haraway, may produce new, potent modes of world making. She ultimately uses her standpoint to recover legacies of ubiquitous computing that dominant historical narratives typically suppress.

Stark offers an extensive review of the literature on emotion in the world of silicon and bits, arguing that human affect is a powerful driver of agency and action. Tracing developments across fields of human-computer interaction, informatics, psychology, and science and technology studies, he highlights how contemporary approaches to the study of emotion offer different analytic tools for examining the interaction between computing and emotion. Introducing the notion of an *emotive actant*, he charts the ways digital actors may in fact intensify the experience and expression of human feelings within the theater of digital culture.

Finally, Couture addresses questions of gender and computing by examining the definitions people give to source code while developing software. He explores how people stabilize and contest those definitions while writing PHP, a language not requiring a compilation phase. Women's work, he argues, becomes less visible and valued in part due to the form it takes within a given project (text, image, or particular programming language, for example). This leads to local definitions of source code that exclude female contributions. He points to this legitimation process as deserving central attention as STS approaches the study of software in development.

Works Cited

Abbate, Janet. 2012. *Recoding Gender: Women's Changing Participation in Computing*. History of Computing. Cambridge, MA: MIT Press.

Amrute, Sareeta. *Encoding Race, Encoding Class: Indian IT Workers in Berlin*. Durham, NC: Duke University Press, 2016.

Bardzell, Shaowen. 2018. "Utopias of participation: Feminism, design, and the futures." *ACM Transactions on Computer-Human Interaction (TOCHI)* 25 (1): 6.

Brock, André. 2011. "Beyond the pale: The Blackbird web browser's critical reception." *New Media & Society*. 13, no. 7: 1085–1103.

Chun, Wendy Hui Kyong. 2011. *Programmed Visions: Software and Memory*. Cambridge, MA: MIT Press.

Cowan, Ruth Schwartz. 2011. *More Work for Mother: The Ironies of Household Technology from the Open Hearth to the Microwave*. New York: Basic Books.

Dunbar-Hester, Christina. 2008. “Geeks, Meta-geeks, and Gender Trouble: Activism, Identity, and Low-Power FM Radio.” *Social Studies of Science* 38 (2): 201–32.
Ensmenger, Nathan. 2010. *The Computer Boys Take Over: Computers, Programmers, and the Politics of Technical Expertise*. History of Computing. Cambridge, MA: MIT Press.
Fausto-Sterling, Anne. 2000. *Sexing the Body: Gender Politics and the Construction of Sexuality*. New York: Basic Books.
Gray, Kishonna L. 2012. “Intersecting oppressions and online communities: Examining the experiences of women of color in Xbox Live.” *Information, Communication & Society* 15 (3): 411–28.
Haraway, Donna. 1984. “Teddy Bear Patriarchy: Taxidermy in the Garden of Eden, New York City, 1908–1936.” *Social Text*, no. 11: 20–64. https://doi.org/10.2307/466593.
——. 1988. “Situated Knowledges: The Science Question in Feminism and the Privilege of Partial Perspective.” *Feminist Studies* 14 (3): 575–99. https://doi.org/10.2307/3178066.
Hu, Tung-Hui. 2015. *A Prehistory of the Cloud*. Cambridge, MA: MIT Press.
Kline, Ronald, and Trevor Pinch. 1996. “Users as Agents of Technological Change: The Social Construction of the Automobile in the Rural United States.” *Technology and Culture* 37:763–95.
Light, Jennifer. 1999. “When Computers Were Women.” *Technology and Culture* 40 (3): 455–83.
Longino, Helen E. 1990. “Values and Objectivity.” In *Science as Social Knowledge: Values and Objectivity in Scientific Inquiry*, 62–82. Princeton, NJ: Princeton University Press.
Milam, Erika, and Robert Nye, eds. 2015. *Scientific Masculinities*. Chicago: University of Chicago Press.
Nakamura, Lisa. 2014. “Indigenous circuits: Navajo women and the racialization of early electronic manufacture.” *American Quarterly*. 66 (4): 919–41.
Noble, Safiya Umoja. 2018. *Algorithms of Oppression: How Search Engines Reinforce Racism*. New York, NY: NYU Press.
Oost, Ellen van. 2003. “Materialized Gender: How Shavers Configure the Users’ Femininity and Masculinity.” In *How Users Matter: The Co-construction of Users and Technology*, edited by Trevor Pinch and Nelly Oudshoorn, 193–208. Cambridge, MA: MIT Press.
Pérez-Bustos, Tania. 2018. “Let Me Show You”: A Caring Ethnography of Embodied Knowledge in Weaving and Engineering.” In *A Feminist Companion to the Posthumanities*: 175–87. Berlin: Springer.
Pham, Minh-Ha T. 2015. *Asians Wear Clothes on the Internet: Race, Gender, and the Work of Personal Style Blogging*. Durham, NC: Duke University Press.
Rosner, Daniela K. 2018. *Critical Fabulations: Reworking the Methods and Margins of Design*. Cambridge, MA: MIT Press.
Rossiter, M. W. 1993. “The Matthew Matilda Effect in Science.” *Social Studies of Science* 23 (2): 325–41. https://doi.org/10.1177/030631293023002004.
Schiebinger, Londa. 1991. *The Mind Has No Sex? Women in the Origins of Modern Science*. Cambridge, MA: Harvard University Press.
Suchman, Lucy. 2011. “Subject Objects.” *Feminist Theory* 12 (2): 119–45. https://doi.org/10.1177/1464700111404205.
Wajcman, Judy. 1991. *Feminism Confronts Technology*. University Park: Pennsylvania State University Press.

If "Diversity" Is the Answer, What Is the Question?

Understanding Diversity Advocacy in Voluntaristic Technology Projects

Christina Dunbar-Hester

In July 2006, a few attendees of the Hackers on Planet Earth (HOPE) Conference in New York City offered for sale homemade silk-screened T-shirts that riffed on a recent gaffe by Senator Ted Stevens (R-AK). Stevens, a senator charged with regulating the Internet, had recently remarked, "The Internet is not something you just dump something on. It's not a truck. It's a series of tubes" (Wired Blogs 2006). This comment was widely circulated online and roundly mocked by many who insisted that Stevens understood neither the technical aspects nor the principles of the regulation he oversaw, which concerned "net neutrality" and whether Internet service providers should be barred from giving delivery priority to favored content.[1]

The activists hawking T-shirts at the HOPE Conference not only swiped at Stevens's lack of support for net neutrality. They also lobbed a separate critique at their own community: their T-shirt featured Stevens's quote "The Internet is a series of tubes" as the caption for an anatomy-book representation of the female reproductive system (figure 1). One woman claimed that she would not sell shirts to men unless they first said out loud to her "uterus" or "fallopian tubes" (Dunbar-Hester 2008a, 119). In other words, these activists creatively and humorously challenged the notion that a technical domain such as the Internet is a masculine one.[2] Significantly, they did so at a conference for computer hackers, an event dominated by male speakers and audience members (participants estimated the ratio of men to women as 40:1, though there is no way of verifying this officially). In this context, asking men to say out loud words related to women's reproductive organs before letting them buy T-shirts was a kind of flag-planting gesture. It also provided fodder for storytelling after the fact, emphasizing both the scarcity of women in spaces like hacker conferences and the fact that this scarcity was not proceeding unchallenged.

And this episode is only one minor, fleeting example: contestations surround who participates in technology production abound in our contemporary historical moment. This chapter uses the empirical site of advocacy around "diversity" in

FIGURE 1: Activist T-shirt, ca. 2006. Courtesy Steph Alarcón.

software and hackerspace communities to ethnographically assess technology and technical practice as a site of purposive political action.[3] This chapter explores multiple framings surrounding the overlapping issues of who participates in amateur technology cultures, to what ends, and with what consequences. It argues that activist engagement with media technologies may challenge elite cultures of expertise that often accompany technology, including "universalist" notions that efface social difference and position in order to present technical practice as universally appealing and attainable. At the same time, presenting technical practice as a main plank in attaining social equality carries risks, including mistaking "technological inclusion" for social power.

Gender advocacy within amateur technology projects like free/libre and open source software (FLOSS) and hackerspaces illustrates how technologies acquire political meanings within technical communities.[4] In this site, we can observe how activists who are concerned with expressing political beliefs do so through engagement with technologies. Geek communities are important because they are situated between "downstream" end users of technology and "upstream" social groups like policy makers and designers. "Geek" as a social identity is constructed around the formation of strong affective relationships with highly specialized pursuits (including fan cultures, though in recent decades "geek" has acquired a dominant meaning related to technology, especially electronics and computers) (Dunbar-Hester 2016a). While geek pursuits may sometimes appear idiosyncratic to those outside their communities, technologically oriented geeks are significant because of the interpretive work they conduct. They mediate between those who build and regulate technology and everyday users of technology. Geeks' interventions into the politics of artifacts have a profound impact on how technology may be built, enabled, or constrained by policy, or taken up by those of us who are not geeks.

Contemporary social studies of technology treat technology neither as wholly socially determined nor as conforming to or flowing from an internal rational logic. Technologies and technical practices are understood as durable (but not immutable) assemblages of social relations and technical artifacts. In keeping with this tradition, but specifically in relation to gender, feminist social studies of technology "conceives of technology as both a source and a consequence of gender relations" (Wajcman 2007). Gender structure and identity are materialized in technological artifacts and practices,[5] and technology is implicated in the production and maintenance of a relational system of gender.[6] Technical domains such as electronics tinkering, including computers and ham radio, have historically served as sites of masculine identity construction (Douglas 1987; Dunbar-Hester 2014; Haring 2006; Edwards 1990; Light 1999; Wajcman 1991). Looking at these issues in their present-day context, scholars have noted that "in spite of the possibility of emancipation from corporeal realities imagined by early theorists and boosters of new media and cyberspace, bodies and social positions are anything but left behind in relationships with computers. . . . It is still the case within the so-called high tech and new media industries that 'what kind of work you perform depends on how you are configured biologically and positioned socially'" (Sara Diamond, quoted in Suchman 2008, 149). In other words, social context and position, including gender, matter greatly as we consider who participates in technical practices and who possesses agency with regard to technology, both historically and in the present.

Having established that geeks act as mediators of technology, and that technology is a site of gender production and maintenance, I discuss methods, and then turn to the empirical case.

Research Methods and Position

Diversity advocacy is multisited and multivocal.[7] My research methods here are informed by an ethnographic sensibility, but lack the "deep hanging out" component that is a hallmark of traditional single-site ethnographic studies (Geertz 1998). Instead, I have sought to mirror the distributed nature of this advocacy, conducting participation observation at a number of sites (North American hackerspaces, fablabs, software conferences, "un-conferences" for women in open technology, corporate events, and software training events/meetups). An alternative approach would be to embed myself and closely attend to a single FLOSS project or hackerspace, but the networked nature of this phenomenon requires that I traverse multiple sites. What this approach loses in granularity and depth at a single site is offset by the benefits of a comparative approach, as discussed by Karin Knorr Cetina, who writes that "using a comparative optics as a framework for seeing, one may look at one [site] through the lens of another. This 'visibilizes' the invisible; each pattern detailed in one [site] serves as a sensor for identifying and mapping (equivalent, analog, conflicting) patterns in the other. A comparative optics brings out not the essential features of each field but differences between the fields" (1999, 4). Since multiple emphases and orientations within diversity advocacy are occurring, comparison is a valuable enterprise, and allows more meaningful analytical points to be made.

One thing to note is the relevance of my own subject position and social identity to this research. As a white, middle-class, highly educated and literate person in North America, these communities and their conversations are relatively accessible to me and hospitable to my presence; my presence requires little justification in most cases. That said, my training, expertise, and commitments are those of the academy, specifically interpretive social science, not computer coding, geeking or hacking, navigating NGOs or start-ups, or feminist activism. Of special importance is my position as a person with a feminine gender identity. Many of these sites are literally closed to people who do not identify as women (though most are explicitly genderqueer and trans* inclusive, some require that people identify as women "in ways that are significant to them"). This means that my gender is implicated in my ability to conduct this research; such strictures draw out quite plainly the fact that the knowledge I make here is situated (as all knowledge is).

Fieldwork and data gathering spanned 2011 to 2016, with continuous attention to listservs and online traffic, and punctuated conference attendance and interviewing. This period is meaningful because it saw several feminist hackerspaces appear as well as growing attention to gender in mainstream open source; at the same time, it presents a snapshot of an unfolding story with both a prehistory and a future that are outside the scope of the present research. It is significant that several initiatives that became research sites were born during this period; while this indicates that I "have my finger on the pulse" of a meaningful social phenomenon, it also means that the objects of study were a moving target and hard to identify before the fact, which creates a methodological challenge.

I have interviewed participants in these activities as well as founders of hackerspaces, open source software projects, and initiatives to promote women's participation in technology (20 formal and informal semistructured interviews to date), mainly in North America but including a few Europeans. And I follow much online activity, lurking on project lists and following social media, which again mirrors the fact that much of these efforts are coordinated and distributed across space, even as they also include local, static components "in real life" such as hacker- and maker-spaces, or project- or programming-language-based meetups. Conferences, of course, are important for participants (and researchers) for the ritual elements that occur when a community comes together for a short time, not only for the information that is transmitted within them (Coleman 2010). (Software and hacker conferences can also be occasions for scandal, including controversy and behavior and boundary policing within a community, which are of anthropological interest.) In weaving together these threads of activity, I gain the ability to map the meaningful (and contested) discourses that surround diversity advocacy, situating them within varying social contexts. It is not an exhaustive or "god's-eye" (Haraway 1991) perspective on these initiatives, but it is not wholly idiosyncratic either; I trace multiple skeins of distinct and interwoven activity in order to draw out meaningful contrasts, and interpret the implications of these varying positions within the space of this advocacy.

Diversity Activism in Open Technology Cultures

In "diversity" advocacy in FLOSS and hackerspaces, self-consciously feminist activists and allies have identified low rates of participation by women in particular in these spaces. Here they confront technical cultures around the issue of "diversity" itself. These initiatives begin with a critique of the liberal Habermasian citizen in how the activists frame and address the problem: they openly admit that there is inequality in their communities, and acknowledge the effects of positionality in producing different rates of participation between men and women. (Not everyone in these technical communities agrees with this assessment, but among the advocates addressing "diversity," it is not controversial.)[8] This is consonant with the acknowledgment by Wendy Faulkner and others that context matters, and "one size does not fit all": "the same measures [to improve gender inclusion in work with communication technologies] may not be effective with different groups or in different settings" (Faulkner 2004, 14; see also Sørensen et al. 2011). Such a framing stands in tension with forms of technologically engaged activism that present technical engagement in universalizing ways (see Suchman 2003; Haraway 1991; Dunbar-Hester 2014; Kerasidou, this volume).

Our contemporary moment is saturated with exhortations for women (and members of other underrepresented groups, but particularly women) to take up participation in science and technology (the common abbreviation is STEM, for science, technology, engineering, and math). Rationales for this push vary, but common ones are national competitiveness and women's empowerment. Both could be found on the Obama White House's website in 2015: (1) "Supporting women STEM students and researchers is . . . an essential part of America's strategy to out-innovate, out-educate, and out-build the rest of the world"; and (2) "Women in STEM jobs earn 33 percent more than those in non-STEM occupations and experience a smaller wage gap relative to men" (Office of Science & Technology Policy n.d.).[9]

Industry, too, often regards increasing women's participation in technical fields as desirable. Google neatly summarizes the corporate agenda surrounding "women in technology" on a web page: "Technology is changing the world. Women and girls are changing technology. . . . We always believed that hiring women better served our users" (Google n.d.-a).[10] In other words, the corporation's full market potential is not being realized without a developer base that can cater to diverse users. On another page, titled "Empowering Entrepreneurs," Google explicates the global reach of its vision and reiterates that "technology" is a route to empowerment: "Archana, an entrepreneur from Bangalore, shows how women are using technology to better their businesses, improve their lives and make their voices heard around the world" (Google n.d.-c). (Note that while my research sites are predominantly North American, Archana is in India; technical work is used to bring people in to globalized capitalism, literally and figuratively [Freeman 2000].)

These agendas reflect the complex social reality within computing and technical fields, in which "what kind of work you perform depends on how you are configured biologically and positioned socially," as noted above. They also provide a backdrop for the object of focus in this project, "diversity" initiatives emanating from FLOSS and hacking communities. Consciousness about diversity (including but not limited to gender) is evident across a wide swath of groups and sectors, including FLOSS development projects, informal hacker groups, and technology-based political collectives (loosely lumped together as free culture or open technology projects). Activists, advocates, and developers are increasingly addressing disparities including gendered divisions of technical labor and the gendered "baggage" of some media and information technologies, including computers and electronics hardware more generally. Indeed, there has been a veritable explosion of interest in holding conversations about the gender implications of work with communication technology.

Reasons for this are complex and varied. As historians of computing have shown, women were programmers of electronic computers in their earliest days, assisting the Allied wartime efforts in Great Britain and the United States (Light 1999; Abbate 2012; see also Misa 2010). Nonetheless, programming was predominantly associated with masculinity within a decade after the war; women's work in computing was effaced (Abbate 2012) and men flooded the growing computer-related workforce and established the academic field of computer science (Ensmenger 2010). In 1991, MIT researcher Ellen Spertus famously asked, "Why are there so few women computer scientists?" By the first decade of the 21st century, women's rate of participation in academic computer science had declined even further in the United States. US Department of Education statistics indicate that in 1985, a few years before Spertus's essay, 37% of computer science majors were women; in 2009 this number had dropped to 18%, and steadily hovered around that percentage during the 2010s.[11]

Beginning in the mid-2000s, the FLOSS community reacted not only to this longer trajectory of men's dominance in computing but to a policy report released by the European Union in 2006. This report showed that while women's presence in proprietary software was around 28%, in FLOSS it was an astonishing 1.5% (Nafus et al. 2006; see also Ghosh 2005). The reasons for this disparity were wide-ranging, probably including such factors as domestic divisions of labor that set up men in heterosexual partnerships to have more leisure time to pursue affective technical passions, wider historical and cultural factors that gendered computing masculine, and the persistent notion that FLOSS projects were liberal, egalitarian spaces where social identity was irrelevant, among others (see, e.g., Lin 2006; Nafus 2012; Reagle 2013; Karanovic 2009).

The numerical breakdown provided by this report served as a rallying cry: this statistic was mobilized to justify increased attention to women's participation in FLOSS. As one person stated in 2009 on a newly launched listserv for women in FLOSS, "There is nothing particularly male about either computers or freedom—and yet women account for fewer than 2% of our [FLOSS] community."[12] (Notably, the agendas of FLOSS and amateur technical projects that seek to promote diversity may exhibit contiguity with, but are not necessarily identical to corporate and policy diversity initiatives.[13] But similarities are rampant: in an online post, one advocate for diversity in open source writes, "Our [diversity imbalance] is reducing our ability to bring the talent we need into our profession, and reducing the influence our profession needs to have on society" [Fowler 2012].)

My project here is distinctly *not* to ask (or answer) questions pertaining to the issue of "why aren't there more women in STEM?" or "how can we bring more women into STEM?," for example. Rather, I uncover a range of motivations behind amateur interventions into diversity questions, in order to evaluate the political potentials and limitations of such projects, including the placement of technology at the center of a project of social empowerment. In other words, the multiple framings of who participates in technology development, and to what end, are taken to be objects of inquiry in their own right. (Note that I do not attempt to define "diversity" myself; I am interested in the work it does as actors identify it as a concern and mount interventions based on this concern [Ahmed 2012].)

In some ways, the diversity advocacy that I examine in this chapter bears similarities to the government and corporate agendas mentioned above. At the same time, unlike White House policy or Google programs, the initiatives I examine are driven by the voluntaristic ethos that surrounds FLOSS. We have to account for why fairly grassroots civil society groups are also pouring their energies into this diversity advocacy, usually as volunteers. Diversity advocacy here is not necessarily identical to corporate or government agendas, though there is certainly overlap. What these sites have in common is that they are not especially institutionalized and are suffused by a voluntaristic ethos.

Many scholars of hacking and tinkering have focused on the fact that these activities often take on meaning as communal and shared actions.[14] Anthropologist Gabriella Coleman has demonstrated that hackers deploy a range of stances including agnosticism and denial of formal politics (exceeding software freedom),[15] though implications for intellectual property in particular are at least implicit and often explicit in the technical and social practices of hacking (Coleman 2012).[16] Scholars have noted that the denial of formal politics makes FLOSS an unlikely site for gender activism, at least historically (Nafus 2012; Reagle 2013). But FLOSS projects are not monolithic, and have matured over time.[17] They are also in dialogue with the wider culture, which is, as noted above, currently awash in "women in tech" discourses (including the publication of and reaction to Facebook COO Sheryl Sandberg's 2013 *Lean In*). The raft of initiatives around "diversity" must be placed within this context, while keeping in mind that geek politics exist along a continuum.

A salient reason that FLOSS participants emphasize diversity is because they believe that free software is emancipatory, and they seek to build a broad commitment to its use, development, and principles (see, e.g., Söderberg 2008, 30). The following quote is a neat summation of this sentiment: "The free software movement needs diverse participation to achieve its goals. If we want to make proprietary software extinct, we need everyone on the planet to engage with free software. To get there, we need people of all genders, races, sexual orientations, and abilities leading the way. That gives the free software movement a mandate to identify under-represented groups and remove their barriers to access" (Free Software Foundation 2012). Here the aspirational goal is nothing less than to have "everyone on the planet" engaged with free software, as the underpinning of an inchoate political agenda tying user empowerment and "freedom" to the ethics and practices of free software. Proponents of FLOSS also express this desire to open up free software user communities as a commitment to furthering affective pleasure, the jouissance that will bind empowered users and user-developers to free software and thus build its reach. One person wrote in a 2009 blog post, "I have strong feelings about Free Software. . . . [One reason to] to improve diversity in FLOSS is to increase contributor retention by increasing joy. . . . [And] the most obvious reason to reach out to groups of people who do not typically contribute is that we can *increase our numbers*" (Laroia 2009, emphasis original). In a similar vein, another advocate writes, "We need more and better software developers to produce valuable software that improves our lives" (Fowler 2012). In general, even when the "why" of FLOSS was underspecified, the reflexive self-importance of participation in this pursuit was unquestioned; in the words of anthropologist Jelena Karanovic, "Many . . . internet professionals ravenously read books by communication theorists on the ways in which the internet [is] transforming sociability and [are] very interested in how their own practices might contribute to realizing the revolutionary potential of the internet" (Karanovic 2009).

For our purposes, it is important to note that like the government and industry agendas discussed above, free software proponents also believe that computing technology is an engine driving society (Smith and Marx 1994) and its use is empowering. Note also that the first quote occurred on the occasion of Ada Lovelace Day (Lovelace was a 19th-century mathematician famed for working on Charles Babbage's difference engine; she is, along with Grace Hopper and Anita Borg, commonly referenced as a figurehead representing women in computing). The second quote, meanwhile, was written by an avid proponent of women's participation in free software who has made an effort to couch his arguments for diversity in broader terms than gender (here, South Asians in free software; elsewhere, "shy people"; etc.).

And of course, within "diversity," gender diversity is commonly understood to be a primary goal, most often expressed as the inclusion of women. Groups with titles like LinuxChix (founded ca. 1998), Debian Women (the Debian operating system project, ca. 2004), Ubuntu Women (2006), the Geek Feminism project, and, more recently, PyLadies (from the Python computer language community, 2011) proliferate, and the list goes on and on.[18] One person on a listserv for women in FLOSS, with a masculine username, addressed the list to recruit women to FLOSS projects in which he was involved:

> I had a look at the projects I'm directly professionally involved in—[Project A] and [Project B]. And, well, they're pretty much your typical F/OSS sausage fests [normative, masculine-dominant spaces], I'm afraid. We do actually have a few women involved, but they're all Red Hat [company] employees; on the volunteer side, it's all men so far.
>
> So I'm hoping to encourage people—women in particular—reading this list to come and get involved with [Project A] and [Project B].[19]

This email represents a banal example of list traffic, and did not generate controversy. (Another list subscriber replied, "Thanks, [Name], for taking the time to make that bid for participants in your project. It was exactly what the world actually needs[,] much more so than almost any other single action.")[20] I include these quotes to illustrate a typical, mundane framing of the issue of "gender diversity" as inclusion of women in free software projects (which, as noted above, should be read in part as a direct reaction to the FLOSSPOLS report).

A more controversial topic on this list, however, did surface: picking a logo for the list. One list subscriber proposed, "If we took the picture of a GNU used by FSF [Free Software Foundation] [and] added lipstick, eye shadow, and mascara, replace the beard by a string of pearls, and replaced the horns by a feminine hat, with a flower sticking up from the hat, I think that would convey the idea."[21] (The GNU symbol she references is the logo of a Unix/Linux-related operating system, a line drawing of the gnu antelope, replete with beard and horns as described in the email; see figure 2.) In other words, the subscriber proposed adorning the GNU with normative markers of femininity. Responses to this suggestion indicated

FIGURE 2: GNU logo. Used with permission under the Creative Commons Attribution-ShareAlike 2.0 License.

discomfort with it. One person commented, "I . . . am not a big fan of this idea. Most women in free software do not adhere to traditionally feminine styles of dress/grooming—I have seen very few wearing makeup let alone pearls at free software events—and I think this sort of appearance would be alienating to many of us."[22] The original poster agreed with this ("You're right. . . . Most of us don't dress over-the-top feminine. I certainly don't.")[23] and added that the original suggestion was intended to be a humorous way of depicting women in FLOSS. Posters to the list struggled with how to represent the presence of women without falling back on representations of normative femininity that many of them found "alienating." (They also touched on race, as one commenter wrote, "I think the gnu is more appealing than the WASP-y noses and dainty lips [in other ideas for logos].")[24]

But they also identified another issue. One person commented, "I think the question of gender identity goes deeper than 'do we all wear pearls here at [Womeninfreesoftware]' to, are we really limiting our reach to 'women' or is there also room for gender queer techies who don't identify with the gender binary?"[25] In other words, using normatively feminine images to represent women in FLOSS was problematic for two reasons. First, these images invoked and threatened to reinscribe a version of femininity that many geek women did not relate to. Second, the emphasis on femininity undermined a commitment to gender diversity common in techie circles, where the prevalence of nonbinary-gendered and trans*-identified people seems relatively high (or is, at least, visible and vocal). Gender diversity did not stop with "women." (As noted above, many projects and hackerspaces with a commitment to gender inclusion explicitly address and include people who identify as queer, nonbinary, and so forth. One representative example is from a hackerspace that describes its community as, "We are intersectional feminists, women-centered, and queer and trans-inclusive" (Double Union n.d.).

Scholars of postfeminism have persuasively argued that much of the cultural work to single out and hail women and girls *as women and girls* in the contemporary moment has to do with constructing feminine, consumerist, individual subject positions within capitalism, aligned with and enacted through neoliberalism (Banet-Weiser 2012; McRobbie 2008). Those insights are useful here, especially as many strains of diversity advocacy align with values of bootstrapping, workplace preparedness, and configuring consumers (often, diverse developers/producers are assumed to better serve consumers). As noted above, reasons for diversity advocacy span a spectrum of political possibility; many do plainly configure subjects for the workplace with an ultimate goal of constructing and serving diverse consumers (difference is mobilized in order to be commodified; see also Dunbar-Hester 2016b). And others leave the "why" of "diversity" underspecified, potentially ripe for appropriation by multiple and possibly incommensurate agendas (e.g., Boston PyLadies, whose website states, "Our goal is to get a larger number of women coding and involved in the open source community"; PyLadies Boston n.d.).

At the same time, some strands of diversity activism exhibit collectivity formation that is more politicized, and often more attuned to structural issues of social inequality. One person wrote on a listserv for women in open technology, "The change I want to see, for mothers, non-mothers, women, people who want a stable balance between work and all the rest of life . . . isn't about leaning . . . anywhere . . . it should be about actually changing the system and inequalities around leave and work environments and people's attitudes."[26] Notably, her reference to "leaning . . . anywhere" is a dig at Sandberg's *Lean In*, which was lambasted by many critics for being insufficiently attuned to structural issues in its exhortation that women "lean in" and take responsibility for perseverance and success at work (see, e.g., hooks 2013). (Sarah Fox et al. [2015] found that feminist hackerspace members used Sandberg's book as a litmus test to understand the degree of politicization of people at their events; they write, "Attitudes toward [*Lean In*] became a gauge by which people could assess each other.") Even so, the poster's emphasis on work-life balance illustrates that she imagines herself and her audience primarily as workers, not as people engaging with technology for other purposes; certainly the jouissance or emancipation sometimes imagined in open technology cultures is not given primacy in her discussion.

By contrast, in 2007, to commemorate International Women's Day, feminist techies based in Europe coordinated a virtual march through Internet Relay Chat channels. They adopted handles associated with women in technology, including Ada Lovelace and Grace Hopper, and other varied feminist figures from history, literature, and pop culture, including experimental novelist Kathy Acker, musician-performer Peaches, and Victorian writer George Eliot, and "marched" through IRC spouting feminist slogans. An excerpt:

<peaches> When men are oppressed, it's a tragedy. When women are oppressed, it's a tradition.
<graceHopper> It's better to act on a good idea than to ask for permission to implement one.
<charlottePgilman> When the mother of the race is free, we shall have a better world
* sistero (sister@ [IP address]) has joined #back chat
<tux> hey sister welcome to #backchat!
<charlottePgilman> happy iwd [International Women's Day] 2007!
<simoneDeBeauvoir> Well-behaved women seldom make history
* [M—] (milena@ [IP address]) has left #backchat
<graceHopper> Bread and Roses
(Genderchangers.org 2007a)

According to the organizers of the march, marchers were kicked out of a number of IRC channels because other users thought they were "bots" due to the coordinated nature of their appearance: "Naturally we were deftly kicked and banned from most servers as a result of our actions. One set of tech operators apologised and lifted the ban when they realised we weren't bots: they found us so co-ordinated they couldn't believe it to be otherwise" (Genderchangers.org 2007b). They make this claim with obvious relish because it signifies the marchers' effectiveness at creating a spectacle. It constitutes storytelling about the marchers' disruptive feminist and feminine presence in a space where hegemonic masculinity tended to reign uncommented upon. Moreover, the marchers' claim that they were assumed to be bots before actual live women users does work to establish the ostensible strangeness of feminine presence here.

Likewise, the Geek Feminism site and wiki (founded 2008) is devoted to providing a community for feminist techies to come together. It conjoins the project of feminism with the culture and aesthetics of geekdom: wiki pages address, for example, fan fiction, "recreational medievalism," and cosplay (dressing up as a character from a story, particularly anime) (Geek Feminism Wiki n.d.): "Things that are on-topic . . . : 1. geeky discussion about feminism; 2. feminist discussion about geekdom; 3. geek feminist discussion of other things" (Geek Feminism n.d.). Crucially, the website also offers a wiki on feminist topics in order to "avoid Feminism 101 discussions"; it assumes conversance in topics such as "privilege, sexism, and misogyny" and recommends visitors to the site unfamiliar with such concepts start by reading as opposed to contributing to discussion (Geek Feminism Wiki n.d.; see Reagle 2016).

The space is moderated, and various forms of behavior are not tolerated, contra the more anarchic and libertarian strands of open source culture where "anything goes" and norms of free expression trump other boundaries, at least rhetorically (Reagle 2013).[27] Geek Feminist actors have advanced a series of critiques of tech cultures, among them lobbying for codes of conduct at tech conferences, establishing a series of "unconferences" for women in open technology (which are explicitly separatist, as people who do not identify as women may not attend), and in general placing priority on the creation of "safer spaces."

Such an emphasis is informed by the Geek Feminists' collective understanding that women in the wider culture (and in tech culture in particular) routinely suffer systemic and gender-based harassment and abuse. One person wrote in a post, "When trying to explain how hostile an environment the geek world can be, I'd tell people, 'I've been attending cons [conferences] of various types since I was thirteen, and I have never, not once, been to a con where I wasn't harassed'" (Geek Feminism 2013). Not infrequently, dark, ugly reports come to light,[28] prompting members of this network to offer support and advice to victims of these incidents. Though digitally distributed and at core united in technological affinity or geek identity, these efforts resemble practices of consciousness-raising and crisis counseling established by feminists in the 1970s.[29] These commitments in some ways run counter to norms of openness upon which open source rests (Reagle 2013), but organizers unapologetically come down in favor of keeping spaces gated and participants vetted.

Moreover, the lesser status and routine mistreatment of women in these communities are assumed to stem from structural inequalities, including different levels of privilege and protection for men and women (again, both in the society at large and in the tech community). At a feminist women*-only hackerspace in San Francisco founded in 2013, the Wi-Fi password when I visited in 2014 was "meritocracy is a joke."[30] This is notable as an identity display within FLOSS or open tech culture, as "meritocracy" has been a shibboleth within that culture since the early days (Reagle 2013),[31] and has been used as an explanation for the lack of representation of women in FLOSS; arguments have been made that if women were more interested in programming, or better at it, more of them would be present.[32] (Relatedly, diversity advocates reject arguments that attention to "diversity" will dilute contributor quality: "A common argument against pushing for greater diversity is that it will lower standards, raising the [unfounded] spectre of a diverse but mediocre group" [Fowler 2015].)

The Wi-Fi password "meritocracy is a joke" confronts "meritocratic" framing of FLOSS head-on and invites reflection upon and agreement with this statement, as the user literally has to enter these words into her computer in order to get online. However, it is inviting only in-group engagement; the hackerspace is not for "everyone" in the tech community. (I myself had to make contact with research informants to gain an introduction and invitation, and, again, would have had a much harder time if I did not identify as a woman.)

The event at the San Francisco hackerspace to which I had to be granted an invitation was a 'zine-making night. Around ten participants sat around a communal table clipping pictures from magazines, chatting and passing back and forth glue sticks and magic markers. Each person was to leave her pages with the 'zine maker who had proposed the event, who was planning to assemble them into a 'zine representing the hackerspace. At another evening meetup for a budding feminist hackerspace in Brooklyn, NY, in August 2015, four young women sat around a table in relative silence, each immersed in her laptop screen, breaking for light chitchat when food for dinner was delivered. One was working on coding the gender drop-down menu for a website, adding options beyond the binary choices of "male" or "female." Another was working on code for an e-reader that ran off a Raspberry Pi (single-board computer). Both had rushed from workplaces to the meetup, and both had forgotten crucial pieces for their projects including power cords and connectors, meaning they could not work long on the projects or accomplish much beyond rudimentary next steps. I include these descriptions in part to show that the activities that occur in women-centered hackerspaces do not all center around electronics, and are not necessarily "productive" in the sense of demonstrable progress being made toward tangible products (see Couture, this volume). What was palpable in both spaces, though, was the sense of a separate space where the sense of being a feminine craft or tech enthusiast—or, more accurately, the fusion of feminine craft *and* tech enthusiast—was displayed and reinforced.

Conclusions

Having sketched the differing impulses guiding "diversity in open technology" initiatives, we can step back and assess them. On the positive side, some strands of this advocacy offer an acknowledgment that—the "openness" ideal of open source notwithstanding—some people have historically been more equal than others when it comes to engagement with these technologies. In this, activists have begun to confront the legacy of electronics and computing as white, elite, masculine domains, as discussed by historians of radio and computing, with an eye to change. The geek feminists' emphasis on the formation of feminist collectivity and safer spaces for people in technical fields and hobbies who have experienced isolation and harassment within these communities are also positive developments.

That said, these initiatives seem to come up short in other ways. First and foremost, the emphasis on gender diversity often misses an opportunity to frame "diversity" more broadly, especially attending to issues of class, race, and disability status.[33] Though exceptions exist, and emerging feminist hackerspaces in particular often gesture toward "intersectionality" (Crenshaw 1991), the dominant discourse is around gender, which critics note has the potential to allow white women to stand in for all women, and to give white, educated women the possibility of forming alliances with and moving into greater positions of power vis-à-vis white, educated men, with little change to technical cultures beyond the relative empowerment of educated white women. This serves to perpetuate the marginalized status of poor white women and women of color in technical cultures (hooks 2013). Furthermore, exhortations that various groups underrepresented in technology fields "learn to code" in order to improve their social position shoulder *individuals* with the onus to bootstrap or lean in.[34] This draws attention away from social and economic policies that contribute to their occupying more marginalized social positions in the first place, and places an immense burden on people most afflicted by conditions of precarity and structural inequality.

In addition, as noted above, gender diversity initiatives struggle with how to represent the presence of women without reinscribing normative femininities. Ironically, the struggle to render women's presence visible means coming into contact with gender stereotypes and symbolism that have been critiqued by both feminists and geek women as problematic for them. Programmatically making women visible is hard to do without inadvertently presenting them as a monolithic class of people. And the problems of how to represent "women" without essentializing them is additionally complicated by the salience of queer, nonconforming, trans*, and other identities within these technical cultures. In addition, representation as a goal has limits as a project of empowerment, as noted by scholars of postfeminism and race such as Sarah Banet-Weiser and Herman Gray (Banet-Weiser 2012; Gray 2013).

Finally, whether technical engagement is empowering in domains exceeding workplace preparedness is largely unexamined and underspecified in these diversity initiatives. It is fair to say that at present, geek communities are struggling with how a formerly marginal and derided social identity (Dunbar-Hester 2008b) is colliding with the exaltation of computers and tech work and the celebration of Silicon Valley as the seat of cultural innovation. In other words, while geeks are enjoying a cultural moment where they are at least as revered as reviled, they have not historically been a monolithically politicized constituency (see, e.g., Wisnioski 2012). Having social power conferred upon their class may not lead them to goals greater than building better products, or taking home better pay under more stable working conditions. Job precarity in tech fields is legion (Turner 2009, 77; Neff 2012). Programmers have struggled to retain their autonomy in the face of managerial control for decades (even as the idea of a looming shortage of workers leading to "software crisis" is also a decades-old discourse) (Ensmenger 2010). Women in these communities are entirely right to suspect that they have it harder than male peers, given the statistics on the wage gap and the punishing conditions of start-up culture, and so on. Yet it is unclear that many of the collectivities of "women in tech" are pressing for more than the opportunity to be valued as workers ticking boxes for corporations that are valorizing diversity as a means to capture a "diverse" consumer market. While job security or value as a worker is hardly something we can fault people for pursuing, the wider emancipatory politics imagined by some who pursue and promote technical engagement is not consistently audible here.

What is so appealing about activism around technology for some is perhaps the way that technology as a focus can seem to skirt or avoid some of the problems that attend movements for social change. At first glance, technology seems more neutral; thorny issues of identity and positionality are not visibly at the fore.[35] But of course they are there, baked into the legacy of electronics tinkering, computer programming, and all manner of technical pursuits. As shown in the above examples, technical communities focused on equality and emancipation are quickly faced with the question, *do we change the dominant culture or start our own space?* The above cases represent partial answers to that charge. In taking their measure, we might conclude that activism around technological participation is useful for changing technologically oriented communities, but it is more limited as a strategy to build a more just social world. Social initiatives centered on participation in technology are likely to reinscribe the placement of men, college-educated people, and whites at the center of social power (see Dunbar-Hester 2014; Wolfson 2014). Gender advocacy in technical cultures challenges the primacy of masculinity, but does little to destabilize other ways in which power and privilege have consolidated around communication technologies.

We should also be very careful with what we mean by participation (or access) in the first place. For example, many poor women and women of color do in fact have plenty of experience with ICTs (Eubanks 2012). But "empowerment" is not a defining feature of their encounters with ICTs. Rather, ICTs are implicated in their surveillance and configuration as wards of the state, or as low-wage, low-status workers. Therefore, they have good reason to regard ICTs with what Virginia Eubanks calls "critical ambivalence" (Eubanks 2012). Such experiences also prime them to reject taking up identities like geek, hacker, or maker. This serves as a potent reminder of why we need to locate technical practices in culture, as suggested by feminist STS. It also underscores the problems with "bringing women in," since gender, class, and race are mutually constitutive.

Recognizing that there are profound differences between participation and social power is all the more important in a context in which "participatory culture" and voluntaristic forms of organization—widely assumed to be key features of digital cultures—are expected to level social inequalities. What this chapter has shown is that it is unlikely that voluntaristic diversity advocates will be adequately equipped to solve these problems. This is not necessarily surprising, nor is it evidence of diversity advocates' shortcomings per se. More fundamentally, it points to how essential it is to be conscientious about what "problems" one is trying to "solve" as a precondition for intervention, voluntaristic or otherwise. Present calls for "diversity in tech" are largely muted in terms of their political potential. A fuller appraisal of what is at stake in FLOSS diversity advocacy calls for greater attention to justice and equity, exceeding the domain of "technology."

Acknowledgments

This research was supported by NSF award 1026818 (Science, Technology & Society). Any opinions, findings, and conclusions or recommendations expressed in this material are those of the author and do not necessarily reflect the views of the National Science Foundation. The author wishes to thank C. W. Anderson, Paula Chakravartty, Cynthia Chris, and Jinee Lokaneeta for writing group feedback, Laura Portwood-Stacer for being an early audience for this project, and a raft of digitalSTS reviewers for feedback on this chapter during its development.

Notes

1. The mockery of Stevens appeared as widely as Jon Stewart's *Daily Show* on the Comedy Central cable network (for example, "Net Neutrality Act" segment, July 19, 2006).
2. Cleverly extending the metaphor into reproductive politics and women's right to choice, the back of the T-shirt read, "Senator Stevens, don't tie our tubes!"
3. An earlier and abbreviated version of this argument appears as Dunbar-Hester (2017), and a fuller one appears as Dunbar-Hester (forthcoming).
4. FLOSS is alternately referred to as free software, Free/Libre software, and open source software production (with each label carrying different emphases); for shorthand this chapter lumps all of these forms of practice into the label "FLOSS." I also include related informal hacker groups and technology-based political collectives, with the acknowledgment that this category of practice is certainly not monolithic.
5. It is widely acknowledged that gender occurs not in isolation but within a matrix of factors that affect social identity, which include class, nationality, ethnicity, and race.
6. In spite of the attention given to gender identity, I mean in no way to discount social structure (along with gender symbolism) as an important site of production of gender (Lerman et al. 2003, 4; see also Faulkner 2007). It is tricky business, but both individual agency (individuals "doing") and social structure (which may act on individuals and groups) are tenets of gender identity.
7. George Marcus discusses "multi-sited ethnography" as a way to adapt to more complex objects of study (1995).
8. For more on hostility to issues of gender parity in FLOSS communities, see Nafus (2012) and Reagle (2013).
9. The page also quotes President Barack Obama as having said in February 2013, "One of the things that I really strongly believe in is that we need to have more girls interested in math, science, and engineering. We've got half the population that is way underrepresented in those fields and that means that we've got a whole bunch of talent . . . not being encouraged the way they need to." (As of this writing, efforts to promote women in STEM had vanished as a White House priority under the Trump administration.)
10. The page "Google Women, Our Work" additionally states, "Our goal is to build tools that help people change the world, and we're more likely to succeed if Googlers reflect the diversity of our users" (Google n.d.-b).
11. As reported in Raja (2014); see also Gelvin (2016).
12. [Womeninfreesoftware] listserv, September 24, 2009. It should be noted that within the United States, women's presence in academic and industry computing fields fell in the 1990s and 2000s. National context matters, and there are significant cultural and national variations in whether women do tech work (see, e.g., Lagesen 2008; Mellström 2009).
13. Nafus et al. write, "The goals of rectifying the loss of a talented labour pool and with it the opportunity to build better technologies is something that is already recognised as a problem within F/LOSS communities, and is far more likely to motivate action than social justice concerns" (2006, 6).
14. See, for example, Coleman (2012).
15. The Free Software Foundation explains, "To use free software is to make a political and ethical choice asserting the right to learn, and share what we learn with others. Free software has become the foundation of a learning society where we share our knowledge in a way that others can build upon and enjoy" (n.d.).
16. Christopher Kelty adds that arguments among geeks about "technical" details are not restricted to technical issues, insofar as technical and political-legal structures are inseparable for these actors: "Techniques and design principles that are used to create software or to implement networking protocols cannot be distinguished from ideas or principles of social and moral order" (2005, 186).
17. The current attention to "diversity" represents a turning point within a collectivity focused on FLOSS as a product, though of course this turn is not universal in FLOSS. See Hess (2005).
18. It is beyond the scope of this chapter to comment on the prehistory of gender activism in FLOSS but it would certainly include WELL and Usenet discussion groups; Systers (a play on "sisters" and "sys," as in "sys admin"), a mailing list for technical women in computing founded in 1987; the Anita Borg Institute's Grace Hopper Celebration (begun in 1994); and various cyberfeminist efforts of the 1990s.

19. [A—] to [Womeninfreesoftware], email, September 28, 2009.
20. [K—] to [Womeninfreesoftware], email, September 28, 2009.
21. [M—] to [Womeninfreesoftware], email, September 24, 2009.
22. [K—] to [Womeninfreesoftware], email, September 24, 2009.
23. [M—] to [Womeninfreesoftware], email, September 24, 2009.
24. [A—] to [Womeninfreesoftware], email, September 24, 2009.
25. [Womeninfreesoftware], email, September 24, 2009.
26. [L—] to Adacamp Alumni, email, March 1, 2015.
27. Coleman argues that free software communities frequently form collective rules, but the norm of individual freedom is extremely salient nonetheless (2012).
28. See Lisa Nakamura's discussion of "glitch racism" (2013).
29. This network is also activated to name and shame abusers; names and details are reported not uncommonly, both to support victims and to offer strategic advice. This also illustrates how these virtual and "real life" spaces are not quite public.
30. Fieldnotes, July 2014, San Francisco.
31. Notably, the coiner of the term "meritocracy" intended it as a satirical concept, which was lost on many who advocated for it in subsequent decades. See "Down with Meritocracy," *Guardian*, June 29, 2001, www.theguardian.com/politics/2001/jun/29/comment. Thanks to Peter Sachs Collopy for directing me to this column.
32. Meritocracy can also be mobilized to argue for initiatives supporting diversity. Martin Fowler writes, "I'm a strong meritocrat, who believes that we should strive for a society where everyone has an equal opportunity to fulfill their potential. A diversity imbalance suggest [*sic*] that there are many women, who would have good careers as programmers, who are not getting the opportunity to do so" (Fowler 2012). Part of what this discussion shows is how strongly ingrained the pro-meritocracy arguments are within FLOSS. Meritocracy also rhetorically links participation in FLOSS to career empowerment.
33. Certain FLOSS projects have imagined the (dis)abilities of users for a long time and include attention to accessibility issues in their practice and rhetoric fairly consistently (e.g., GNOME), while others are less attuned to these topics (and, for example, the FLOSS graphics editor project GIMP [an acronym for GNU Image Manipulation Program] has been criticized for its name).
34. For more on bootstrapping and romantic individualism in the context of the Internet, see Streeter (2010).
35. Thanks to Lucas Graves for discussion on this point.

Works Cited

Abbate, Janet. 2012. *Recoding Gender*. Cambridge, MA: MIT Press.

Ahmed, Sara. 2012. *On Being Included*. Durham, NC: Duke University Press.

Banet-Weiser, Sarah. 2012. *Authentic™: The Politics of Ambivalence in a Brand Culture*. New York: New York University Press.

Coleman, Gabriella. 2010. "The Hacker Conference: A Ritual Condensation and Celebration of a Lifeworld." *Anthropological Quarterly* 83:47–72.

——. 2012. *Coding Freedom*. Princeton, NJ: Princeton University Press.

Crenshaw, Kimberlé. 1991. "Mapping the Margins: Intersectionality, Identity Politics, and Violence Against Women of Color." *Stanford Law Review* 43:1241–99.

Double Union. N.d. www.doubleunion.org.

Douglas, Susan. 1987. *Inventing American Broadcasting*. Baltimore: Johns Hopkins University Press.

Dunbar-Hester, Christina. 2008a. "Propagating Technology, Propagating Community? Low-Power Radio Activism and Technological Negotiation in the U.S., 1996–2006." Doctoral dissertation, Cornell University.

——. 2008b. "Geeks, Meta-Geeks, and Gender Trouble: Activism, Identity, and Low-Power FM Radio." *Social Studies of Science* 38:201–32.

——. 2014. *Low Power to the People: Pirates, Protest, and Politics in Low Power FM Radio*. Cambridge, MA: MIT Press.

——. 2016a. "Geek." In *Digital Keywords*, edited by Benjamin J. Peters, 149–55. Princeton, NJ: Princeton University Press.

——. 2016b. "'Freedom from Jobs' or Learning to Love to Labor? Diversity Advocacy and Working Imaginaries in Open Technology Projects." *Revista Teknokultura* 13:541–66.
——. 2017. "Feminists, Geeks, and Geek Feminists: Understanding Gender and Power in Technological Activism." In *Media Activism in the Digital Age*, edited by Victor Pickard and Guobin Yang, 187–204. New York: Routledge.
——. Forthcoming. *Hacking Diversity: The Politics of Inclusion in Open Technology Cultures.* Princeton, NJ: Princeton University Press.
Edwards, Paul. 1990."The Army and the Microworld: Computers and the Politics of Gender Identity." *Signs* 16:102–27.
Ensmenger, Nathan. 2010. *The Computer Boys Take Over.* Cambridge, MA: MIT Press.
Eubanks, Virginia. 2012. *Digital Dead End.* Cambridge, MA: MIT Press.
Faulkner, Wendy. 2004. "Strategies of Inclusion: Gender and the Information Society." Final report (public version), University of Edinburgh.
——. 2007. "'Nuts and Bolts and People': Gender-Troubled Engineering Identities." *Social Studies of Science* 37:331–56.
Fowler, Martin. 2012. "DiversityImbalance." January 11. http://martinfowler.com/bliki/Diversity Imbalance.html.
——. 2015. "DiversityMediocrityIllusion." January 13. http://martinfowler.com/bliki/DiversityMedio crityIllusion.html.
Fox, Sarah, Rachel Rose Delgado, and Daniela Rosner. 2015. "Hacking Culture, Not Devices: Access and Recognition in Feminist Hackerspaces." In *CSCW Proceedings*, 56–68. New York: ACM.
Freeman, Carla. 2000. *High Tech and High Heels in the Global Economy: Women, Work, and Pink-Collar Identities in the Caribbean.* Durham, NC: Duke University Press.
Free Software Foundation. 2012. "Happy Ada Lovelace Day!" October 16. www.fsf.org/blogs/community /happy-ada-lovelace-day.
——. N.d. "What Is Free Software?" www.fsf.org/about/what-is-free-software.
Geek Feminism. 2013. "That Time I Wasn't Harassed at a Conference." http://geekfeminism.org/2013 /08/15/that-time-i-wasnt-harassed-at-a-conference/.
——. N.d. "About." http://geekfeminism.org/about/.
Geek Feminism Wiki. 2015. http://geekfeminism.wikia.com/wiki/Geek_Feminism_Wiki.
——. N.d. "Feminism 101." http://geekfeminism.wikia.com/wiki/Feminism_101.
Geertz, Clifford. 1998. "Deep Hanging Out." *New York Review of Books*, October 22.
Gelvin, Gaby. 2016. "Study: Middle School Is Key to Girls' Coding Interest." *U.S. News & World Report*, October 20. www.usnews.com/news/data-mine/articles/2016-10-20/study-computer-science-gender -gap-widens-despite-increase-in-jobs.
Genderchangers.org. 2007a. "International Women's Day: Feminist Techies, Female Geeks Take to the Streets of the Internet!" http://genderchangers.org/images/irc_march.pdf.
——. 2007b. "International Women's Day 2007." http://genderchangers.org/march.html.
Ghosh, Rishab. 2005. "Free/Libre/Open Source Software: Policy Support." *FLOSSPOLS: An Economic Basis for Open Standards*, December. www.flosspols.org/deliverables/FLOSSPOLS-D04-openstandards-v6.pdf.
Google. n.d.-a. "Google Women." www.google.com/diversity/women/.
——. n.d.-b. "Google Women, Our Work." www.google.com/diversity/women/our-work/index.html.
——. n.d.-c. "Google Women, Our Future." www.google.com/diversity/women/our-future/index.html.
Gray, Herman. 2013. "Subject(ed) to Recognition." *American Quarterly* 65:461–88.
Haraway, Donna. 1991. "Situated Knowledges." In *Simians, Cyborgs, and Women*, 149–81. New York: Routledge.
Haring, Kristen. 2006. *Ham Radio's Technical Culture.* Cambridge, MA: MIT Press.
Hess, David. 2005. "Technology- and Product-Oriented Movements: Approximating Social Movement Studies and Science and Technology Studies." *Science, Technology, & Human Values* 30:515–35.
hooks, bell. 2013. "Dig Deep: Beyond Lean In." *Feminist Wire*, October 28. http://thefeministwire.com /2013/10/17973/.
Karanovic, Jelena. 2009. "Activist Intimacies: Gender and Free Software in France." Lecture at the American Anthropological Association annual meeting, Philadelphia.
Kelty, Christopher. 2005. "Geeks, Social Imaginaries, and Recursive Publics." *Cultural Anthropology* 20:185–214.
Knorr Cetina, Karin. 1999. *Epistemic Cultures: How the Sciences Make Knowledge.* Cambridge, MA: Harvard University Press.

Lagesen, Vivian. 2008. "A Cyberfeminist Utopia? Perceptions of Gender and Computer Science among Malaysian Women Computer Science Students and Faculty." *Science, Technology, & Human Values* 33:5–27.

Laroia, Asheesh. 2009. "Diversity in Free Software: South Asians as an Example." December 18. http://asheesh.org/note/debian/indians.html.

Lerman, Nina, Arwen Mohun, and Ruth Oldenziel. 2003. *Gender & Technology: A Reader*. Baltimore: Johns Hopkins University Press.

Light, Jennifer. 1999. "When Computers Were Women." *Technology & Culture* 40:455–83.

Lin, Yuwei. 2006. "Women in the Free/Libre Open Source Software Development." In *Encyclopedia of Gender and Information Technology*, edited by Eileen Moore Trauth, 1286–91. Hershey, PA: Idea Group.

Marcus, George. 1995. "Ethnography In/Of the World System." *Annual Review of Anthropology* 24:95–117.

McRobbie, Angela. 2008. *The Aftermath of Feminism: Gender, Culture, and Social Change* London: Sage.

Mellström, Ulf. 2009. "The Intersection of Gender, Race and Cultural Boundaries; or Why Is Computer Science in Malaysia Dominated by Women?" *Social Studies of Science* 39:885–907.

Misa, Thomas, ed. 2010. *Gender Codes*. Hoboken, NJ: John Wiley.

Nafus, Dawn. 2012. "'Patches Don't Have Gender': What Is Not Open in Open Source." *New Media & Society* 14:669–83.

Nafus, Dawn, James Leach, and Bernhard Krieger. 2006. "Free/Libre and Open Source Software: Policy Support (FLOSSPOLS), Gender: Integrated Report of Findings." Cambridge: University of Cambridge.

Nakamura, Lisa. 2013. "Glitch Racism." *Culture Digitally*, December 10. http://culturedigitally.org/2013/12/glitch-racism-networks-as-actors-within-vernacular-internet-theory/.

Neff, Gina. 2012. *Venture Labor*. Cambridge, MA: MIT Press.

Office of Science & Technology Policy. N.d. "Women in STEM." www.whitehouse.gov/administration/eop/ostp/women.

PyLadies Boston. N.d. "Meetup." www.meetup.com/PyLadies-Boston.

Raja, Tasneem. 2014. "Is Coding the New Literacy?" *Mother Jones*, July/August. www.motherjones.com/media/2014/06/computer-science-programming-code-diversity-sexism-education.

Reagle, Joseph. 2013. "'Free as in Sexist?' Free Culture and the Gender Gap." *First Monday* 18 (1). http://firstmonday.org/article/view/4291/3381.

——. 2016. "The Obligation to Know: From FAQ to Feminism 101." *New Media & Society* 18:691–707.

Sandberg, Sheryl. 2013. *Lean In: Women, Work, and the Will to Lead*. New York: Knopf.

Smith, Merritt Roe, and Leo Marx. 1994. *Does Technology Drive History?* Cambridge, MA: MIT Press.

Söderberg, Johan. 2008. *Hacking Capitalism*. New York: Routledge.

Sørensen, Knut Holtan, Wendy Faulkner, and Els Rommes, eds. 2011. *Technologies of Inclusion: Gender in the Information Society.* Trondheim: Tapir Akademisk Forlag.

Spertus, Ellen. 1991. "Why Are There So Few Female Computer Scientists?" MIT Artificial Intelligence Laboratory Technical Report 1315, August.

Streeter, Thomas. 2010. *The Net Effect*. New York: New York University Press.

Suchman, Lucy. 2003. "Located Accountabilities in Technology Production." Lancaster: Centre for Science Studies, Lancaster University. www.comp.lancs.ac.uk/sociology/papers/Suchman-Located-Accountabilities.pdf.

——. 2008. "Feminist STS and the Sciences of the Artificial." In *New Handbook of Science & Technology Studies*, edited by Edward Hackett et al., 139–64. Cambridge, MA: MIT Press.

Turner, Fred. 2009. "Burning Man at Google: A Cultural Infrastructure for New Media Production." *New Media & Society* 11:73–94.

Wajcman, Judy. 1991. *Feminism Confronts Technology.* University Park: Pennsylvania State University Press.

——. 2007. "From Women and Technology to Gendered Technoscience." *Information, Communication & Society* 10:287–98.

Wired Blogs. 2006. "Your Own Personal Internet." *Wired*, June 30. www.wired.com/2006/06/your_own_person/.

Wisnioski, Matthew. 2012. *Engineers for Change*. Cambridge, MA: MIT Press.

Wolfson, Todd. 2014. *Digital Rebellion*. Urbana: University of Illinois Press.

Feminist STS and Ubiquitous Computing

Investigating the Nature of the "Nature" of Ubicomp

Xaroula (Charalampia) Kerasidou

> *Machines that fit the human environment instead of forcing humans to enter theirs will make using a computer as refreshing as taking a walk in the woods.*
>
> **—Mark Weiser on ubiquitous computing (1991, 104)**

In the late 1980s, ubiquitous computing made its first appearance in the labs of Xerox's Palo Alto Research Center (PARC) as the "third wave" in computing (Weiser 1996, 2). Mark Weiser along with his collaborators at Xerox PARC envisioned a "new technological paradigm" that would leave behind the traditional one-to-one relationship between human and computer and spread computation "ubiquitously, but invisibly, throughout the environment" (Weiser et al. 1999, 693). Weiser named this new paradigm *ubiquitous computing*, and the term made its first public appearance in 1991 in an article published by the magazine *Scientific American* under the title "The Computer of the 21st Century." The aim of ubiquitous computing was to integrate interconnected computers seamlessly into the world (Weiser 1991, 1993). In Weiser's words, "Specialized elements of hardware and software, connected by wires, radio waves and infrared, will be so ubiquitous that no one will notice their presence" (1991, 94). Since then, the field has grown and now counts several peer-reviewed journals, professional conferences, and a number of both academic and industrial research centers that, with the help of millions of pounds in research funding, have set out to study the new "post-PC computing" under names such as pervasive computing, ambient intelligence, tangible computing, context-aware computing, the Internet of things, and others.

From the outset, ubiquitous computing was presented as an approach that differed from other contemporaneous computational projects in its explicit focus on the human and on social interactions, rather than on the technical aspects of technology design. As Weiser wrote, ubiquitous computing sought to "concentrate on *human-to-human* interfaces and less on *human-to-computer* ones" (Weiser et al. 1999, 694, emphasis original) and aimed to shift the focus from the personal computer per se to the ways in which it can enrich users' everyday experience. At the same time, ubicomp's early vision expressed a deep-seated nostalgia for an almost lost, implicitly better world that it ought to revive by taking the focus away from the machines and "back to us" in its promise to return us to a natural, instinctive human state that we were all assumed to share. So, talk of "ubiquitous computers, [that] reside in the human world and pose no barrier to personal interactions" was accompanied by dreams of "bring[ing] communities closer together" and of "mak[ing] using a computer as refreshing as taking a walk in the woods" (Weiser 1991, 104).

Reacting to such proclamations of human centeredness, I met the optimistic visions of ubiquitous and hassle-free interactions with a deep skepticism. For a computer science graduate/aspiring STS scholar like myself—admittedly, with strong Luddite-like tendencies that I always tried to manage productively (with various results), but at the same time someone who was and still is unwilling to surrender to dreams of technological utopias—I was troubled by these visions.

A lot has changed since Weiser first articulated these early ubicomp visions, yet such proclamations of the human-centeredness of ubiquitous computing, where the human is figured in broad, universal strokes, along with its alleged uniquely social approach to design, have proven powerful tropes that have become central to the ways that subsequent projects have articulated and continue to articulate their own ambitious visions. And while some theorists within the ubicomp community have sought to "move on" from Weiser's early visions (Rogers 2006), or to propose alternative conceptualizations (such as Dourish and Bell's "ubicomp of the present" (2011)), they too appear to accept this *one thing*, this one ubicomp that is tied to one person, one place, and one vision, even if only to push against it. Efforts such as these demonstrate that foundational stories might be messy but they are worked and reworked and reworked again through and into (hi)stories, and they become powerful and productive in shaping realities and futures in particular ways.

I wish to intervene in this process and tell a different story. This is a story that not only seeks to resist the inevitability of ubicomp's technological vision and challenge the determinism on which it is based, but also allows an exploration and a reflection on questions such as these: Despite their alleged universality, whose visions are these? What kind of worlds do they imagine and then seek to build in "our" name? And how can we intervene in these futures?

Conscious that by focusing on and foregrounding these foundational stories and figures I too would play part to their reproduction, I am up for the challenge. Haraway (1997), after all, has already warned me that things can get messy.

Stories and Figures

Along with others in the fields of feminist technoscience and STS, Haraway has long argued for forms of engagement and critique that do not pretend to gaze objectively and, hence, irresponsibly, but seek to participate messily and partially in the making and the unmaking of the worlds that technoscientific projects seek to bring forth (Haraway 1988, 1991; Kember 2003; Suchman 2007; Traweek 1988). My story is kin to this body of work and presents one suggestion on how such a critical engagement and intervention can be achieved by resisting the ways that ubicomp frames its stories and upsetting instead the naturalization of its figures as they are produced and reproduced within these stories. In simpler words, I want to narrate the past differently in order to open up possibilities for different futures.

Specifically, mobilized by ubiquitous computing's proclaimed human-centeredness as over against the alleged machine-centeredness of the personal computer, I focus on the interaction between human and machine as one of the core issues that arise within ubiquitous computing. But instead of asking traditional HCI questions—*how can we build a better, human-centered ubiquitous computing?*—I take one step back. I trace through some of its early stories the ways that ubiquitous computing figures the human and the machine, and, through the retelling of these stories, I bring to light the imaginaries that have inspired and continue to inspire ubiquitous computing, and the worlds that it works in turn to enact and materialize. In that way, the critical focus changes from assuming the futures and the relations that these technologies project and then considering the consequences for the subjects involved, to, as Suchman (2007, 224n22) writes, the prior and more immediate question of what kinds of relations, ontologies, and agencies are assumed to be desirable, or deemed to be expendable, in these technological worlds.

To this end, this chapter adopts a material-semiotic approach, and focusing on the stories of ubiquitous computing, it explores how specific figurations get constructed and performed within its context and how, in turn, they perform and bring forth specific versions of reality. Stories and figures are two key tools of the material-semiotic approach, an approach that is not foundational in its nature but descriptive. That is, it seeks to tell interesting stories about how things come into being, hold together, or not (Law and Singleton 2000; Law 2000; Law 2009, 141), while it urges us to shift from questions of reference—*what is ubiquitous computing?*—to relational configurations—*how is ubiquitous computing figured in particular practices and knowledges?*

Figuration is a methodological, descriptive tool, developed most explicitly within feminist cultural studies of science, which seeks to both unpack the domains of practice and significance that are built into each figure and articulate the semiotic and material practices involved in the making of worlds (Castañeda 2002). In other words, figuration provides the means to attend to the dual process through which the figure is produced and brought into being (the figure as an effect) at the same time as, in its turn, it brings a particular version of the world into being (the figure having effects). Thinking in terms of effects, attending to the specific and laborious configurations of knowledges, practices, and powers that bring these effects into being, as Castañeda writes, entails "generating accounts of necessarily powerful and *yet still contestable worlds*" (2002, 4, emphasis added) and, hence, leaves space for other possibilities where things are, or could be, configured differently (see Suchman 2007).

In the rest of this chapter, I turn my attention to one of these stories and specifically to ubicomp's vision to return us to a past and more natural world that the personal computer has arguably displaced, and one that ubiquitous computing seeks to revive, investigating the ways that ubiquitous computing figures nature through specific discursive and material practices. This exploration directs me to an entangled knot that tightly ties notions of the natural, the machine, and the human. Resonant with the scholarship that rejects essentialism and endorses relationality, I trace some of the relational entanglements of these three figures as articulated and performed in ubiquitous computing discourses, having as a guide the question, what is the nature of the "nature" that ubiquitous computing invokes, imagines, and performs?

The Nature of "Nature"

Invisibility was at the core of Mark Weiser's vision. As he wrote in 1993, "It was the desire to build technology truer to the possibility of invisibility that caused me to initiate the ubiquitous computing work at PARC five years ago."[1] Since then the idea of the invisible or disappearing computer has made numerous appearances. Donald Norman wrote a book in 1999 titled *The Invisible Computer: Why Good Products Can Fail, the Personal Computer Is So Complex, and Information Appliances Are the Solution*, and the same year the director of MIT's lab for Computer Science, Michael Dertouzos, introduced the *Oxygen* project, which aimed to make computing as pervasive as oxygen (Dertouzos 1999). In 2001, Satyanarayanan identified invisibility as one of the four research thrusts incorporated into the agenda of pervasive computing (Satyanarayanan 2001); the EU, within its initiative on the future of ambient intelligence, co-funded the Disappearing Computer project; and in 2005, *Communications of the ACM* dedicated a special issue to the topic "The Disappearing Computer," where the introduction reads, "It seems like a paradox but it will soon become reality: The rate at which computers disappear will be matched by the rate at which information technology will increasingly permeate our environment and our lives" (Streitz and Nixon 2005, 33).

The question of how we can make computers disappear is being addressed in contemporary computer research in various ways. One of the most prominent of these is the approach of tangible computing, which focuses on the physicality and tangibility of the real world, seeking to build physical interfaces through which the physical and virtual worlds can be bridged. The main advocate of this approach is MIT's "Tangible Media Group," led by Professor Hiroshi Ishii. The group has been working on their vision, which they call "Tangible Bits," for almost two decades now, and in 2009 they were awarded the "Lasting Impact Award" at the ACM Symposium on User Interface Software and Technology (UIST 2009) for their metaDesk project (Ullmer and Ishii 1997). Tangible Bits has positioned itself as the "legitimate" ancestor of ubiquitous computing against various "misinterpretations" and "misuses" of Weiser's concept of ubiquitous computing, and Weiser himself in a personal email to Ishii and his colleague Brygg Ullmer recognized the close affinities between the two projects (see the appendix in Ishii 2004, 1310).

Tangible computing is based on the premise that we inhabit two worlds: the physical world and cyberspace, or as Ishii and Ullmer (1997, 234) put it, the world of atoms and the world of bits. Tangible computing asserts that there is gap between these two worlds that leaves us "torn between these parallel but disjoint spaces." This agrees with Weiser's argument that cyberspace, and specifically the computer, has taken center stage, leaving the real world—the real people, the real interactions—in the background and neglected. Tangible computing seeks to address this problem by "bridg[ing] the gaps between both cyberspace and the physical environment," achieving a seamlessness that "will change the world itself into an interface" (234). Specifically, as Ishii and Ullmer write, "The aim of our research is to show concrete ways to move beyond the current dominant model of GUI [Graphic User Interface] bound to computers with a flat rectangular display, windows, a mouse, and a keyboard. To make computing truly ubiquitous and invisible, we seek to establish a new type of HCI that we call 'Tangible User Interfaces' (TUIs). TUIs will augment the real physical world by coupling digital information to everyday physical objects and environments. . . . Our intention is to take advantage of natural physical affordances to achieve a heightened legibility and seamlessness of interaction between people and information" (235).

In one of his earlier works where he explored the foundations of what he termed "embodied interaction" and the relationship that ties ubiquitous computing and tangible interfaces, computer scientist Paul Dourish (2001a, 232) writes that one of the critical features that tangible computing and ubiquitous computing share is that "they both attempt to exploit our natural familiarity with the everyday environment and our highly developed spatial and physical skills to specialize and control how computation can be used in concert with naturalistic activities." Tangible computing then, as Dourish (2001b, 17) writes, seeks to capitalize on these, now naturalized and unquestioned, skills in order to build computational interfaces that fit seamlessly within our everyday, real world.

The above quotes present a number of themes that I would like to explore further such as ideas of everydayness, familiarity, and naturalness that tangible computing appears to invoke, which are also coupled with notions of directness, transparency, and immediacy. Starting my explorations from the latter, I hope to work my way toward the former. So, anticipating an argument, in the following paragraphs I will try to demonstrate that although, at first glance, the idea that "taking advantage of multiple senses and the multimodality of human interactions with the real world, . . . will lead us to a much richer multisensory experience of digital information" (Ishii and Ullmer 1997, 241) seems to make sense, at a closer look it appears to be reduced to a few sets of simplified opposing dualisms, visual versus tactile, symbolic versus physical, technological versus natural, virtual versus real, mediated versus direct/transparent, where the personal computer comes to embody the first components and the "tangible and ubiquitous ones" the second.

Nature

One conventional way of defining the natural, central to Western thinking, is in opposition to the artificial or the cultural, that is, what is "self-occurring" as opposed to the product of skill or artifice (Soper 1995, 37–38). Tangible computing redefines this distinction, shifting the discourse from issues of productive activity (who produces what) to issues of essence (what is made of what) and introducing a distinction between the world of bits and the world of atoms. The former is occupied by entities such as cyberspace, digital information, and computation. The latter is occupied by everyday familiar material objects such as chairs, tables, bottles, and others that, according to tangible computing, invite tangible and direct interactions. As Dourish (2001b, 16) writes, "A . . . topic of investigation in tangible computing is how these sorts of approaches can be harnessed to create environments for computational activity in which we interact directly through physical artefacts rather than traditional graphical interfaces and interface devices such as mice. . . . So tangible computing is exploring how to get the computer 'out of the way' and provide people with a much more direct-tangible-interaction experience."

Interestingly, in contrast to what Weiser and the tangible computing advocates argue, directness and transparency was what the traditional graphic user interface (GUI) was also striving for; hence, it was based on the idea of "direct manipulation" that sought to "replac[e] the complex command language syntax by direct manipulation of the object of interest" (Shneiderman 1983, 57), and it was described in ways that are uncannily similar to the language used by ubiquitous computing and tangible computing.[2] The use of graphics and visuals, which was the main characteristic of GUI in contrast to the text-based earlier interfaces, and the use of mediatory devices (keyboard, mouse) to manipulate these graphics (which represented virtual objects), led the tangible computing advocates to make the distinction that what we see on our screens is symbolic and the result of mediation and technology, while what we touch is direct and therefore real and natural. "More natural that what, though?" Dourish wonders. And he continues, "More natural, presumably, than the abstract, symbolic styles of representation and interaction that characterize conventional interfaces. Symbolic representation is the traditional core of computational media, and it carries over into interface design, which also relies on symbolic representations. . . . With tangible computing, such symbolism can be displaced by more natural, physical interaction" (2001b, 206). But what does direct and natural actually mean in this context? And does the opposition between symbolic and physical hold when one examines some of the tangible technologies that Ishii and his colleagues have developed?

Let's take for example the *bottles*, a system developed in 1999 by the Tangible Media Group and presented again in 2004 in Ishii's paper "*Bottles*: A Transparent Interface as a Tribute to Mark Weiser." According to Ishii, *bottles* "illustrates Mark Weiser's vision of the *transparent* (or *invisible*) interface that weaves itself into the fabric of everyday life" (2004, 1299). The system, which uses glass bottles as an interface in order to "contain" and "control" digital information, is composed of a table 40 inches tall and 25 inches across made of "rich, luxurious materials"—"The legs were solid aluminum with shelves cut from thick mahogany wood"—while its top, or else the "stage," was made of a layer of frosted glass over a layer of Plexiglas. On the stage, one would find three glass bottles each representing a different instrument. By placing and displacing the tops of the bottles, or else by opening or closing the bottles, the user can start or stop the music of its represented instrument. The system also provides a visual stimulation by three different lights illuminating the three corners of the stage from below. The lights correspond to the manipulation of the bottles in order to provide a more "aesthetically pleasing result" (1304). As Ishii (2004) writes, "The metaphor is a perfume bottle: Instead of scent, the bottles have been filled with music—classical, jazz, and techno music. Opening each bottle releases the sound of a specific instrument accompanied by dynamic coloured light. Physical manipulation of the bottles—opening and closing—is the primary mode of interaction for controlling their musical contents" (1299).

From this we see that again the metaphor is at the center of the interface, even if, instead of the desktop metaphor used in the GUI, we have another metaphor, that of the perfume bottle "that evoke[s] the smell of perfume and the taste of exotic beverages" (Ishii 2004, 1299). Besides (or even apart from) the materiality and tangibility of the glass bottle, it is apparent that Ishii draws from an extensive symbolic history of this object. So he demonstrates, regardless of his insistence on separating the symbolic from the physical, that the two are actually intertwined—intertwined along with the human actor, Ishii himself, with his gendered memories and desires of exotic perfumes and beverages—in a performance that results in the object *itself*.

But this is not the story that Ishii wants to tell. In a way, Ishii and his colleagues perform the glass bottle, or to be more precise, a specific glass bottle, both as an object that has traveled through history and as an object that transcends history. The video that used to accompany the project on the group's website (now available on YouTube)[3] starts with an image of dozens of beautiful glass bottles in an unidentifiable space with no labels and no contents, with different shapes and intricate tops that fill the screen. The bottles are arranged and lit in a way that makes the most of their interesting shapes, colors, and reflections. This image is followed by a black screen with the message "glass bottles have been a part of human culture for thousands of years." The beautiful glass bottle (and not the "ugly" plastic bottle with the screwed top that one is more likely to find in the home) is performed as a stand-alone object without a label, without a context that would ground it or situate it. It is enacted as a universal object that transcends cultures, countries, languages, ages, classes, boundaries and becomes a guarantor of what unites us, a guarantor of our humanness, and in one and the same breath it becomes "nature" as we all know it; maybe some of us find it difficult to open a browser, but we all know how to open a bottle, Ishii (2004) tells us in his article. The screen fades out once again and is followed by another image full of beautifully lit glass bottles. Another message fades in that tells us "glass bottles are tangible and visual, and evoke the smell of perfume and the taste of exotic beverages." Ishii's evocations, memories, desires are performed as universal in a way that we are all made to share a history that is now common and familiar.

This familiarity and commonality then becomes the basis for the natural. Unlike opening a browser, we are told, opening a glass bottle is a common, familiar, and therefore natural action. In one move, nature is reduced to the common, the familiar, and in yet another move, nature is reduced to the direct, the uncomplicated, and the unmediated. And, hence, another dualism is put forth. The natural and the mediated are set in opposition where mediation is deemed complicated, unnatural, and therefore undesirable. This is reflected in Ishii's words where he writes that the origin of his idea to design a bottle interface lies in the concept of a "weather forecast bottle," an idea he intended to develop as a present for his mother. "Upon opening the weather bottle, she would be greeted by the sound of singing birds if the next day's weather was forecasted to be clear. On the other hand, hearing the sound of rainfall from the bottle would indicate impending rain." In these two paragraphs of Ishii's article, the readers are introduced to a nice senior lady who has opened thousands of bottles; "she opened and smelled bottles of soy sauce thousands of times" while cooking for her son and family in her familiar physical environment, that is, her kitchen (2004, 1300). This senior lady, who is made to embody the symbolic alignment between woman, the domestic, and nature (Soper 2000; Rose 1993; Plumwood 1993), "has never clicked a mouse, typed a URL, nor booted a computer in her life." Instead, "my mother *simply* wanted to know the following day's weather forecast. *Why should this be so complicated*?" (2004, 1300, emphasis added).

The idea of a primary set of natural tactile skills appears to come hand in hand with a romantic view of an innocent and long-gone natural world that tangible computing seeks to revive, not only a world in which the personal computer did not fit but a world that the personal computer displaced. Thus, Ishii and Ullmer (1997, 234) write about their decision to start their investigations about the "future of HCI" in the museum of the Collection of Historic Scientific Instruments at Harvard University, where they found "beautiful artifacts made of oak and brass", and again artifice is being overshadowed by essence:

> Long before the invention of personal computers, our ancestors developed a variety of specialized physical artefacts to measure the passage of time, to

predict the movement of planets, to draw geometric shapes, and to compute. . . . We were inspired by the aesthetics and rich affordances of these historical scientific instruments, most of which have disappeared from schools, laboratories, and design studios and have been replaced with the most general of appliances: personal computers. Through grasping and manipulating these instruments, users of the past must have developed rich languages and cultures which valued haptic interaction with real physical objects. Alas, much of this richness has been lost to the rapid flood of digital technologies. We began our investigation of "looking to the future of HCI" at this museum by looking for what we have lost with the advent of personal computers. Our intention was to rejoin the richness of the physical world in HCI. (234)

The idea of our direct experience of the world through our bodily senses along with the romantic view of a past, purer, and better world that the computer threatens and that future technological developments promise point toward what Leo Marx has described as America's "pastoral ideal," a force that according to Marx is ingrained in the American view of life (2000). Balancing between primitivism and civilization, nature and culture, Romanticism and Enlightenment, the pastoral ideal "is an embodiment of what Lovejoy calls 'semi-primitivism'; it is located in a middle ground somewhere 'between,' yet in a transcendent relation to, the opposing forces of civilisation and nature" (Marx 2000, 23). So, unlike Heim's "naïve realists" who rejected the computer fearing the loss of their world to the virtual and perverse reality that cyberspace was introducing and who called for a return to "God's pristine world" (1998, 37), the advocates of tangible and ubiquitous computing seek to find the balance, the "middle state," that the American pastoral ideal sought to achieve. This is a precarious position that managed to reconcile the disfavor and fear of Europe's "satanic mills" and their destructive consequences on England's "pleasant pastures" with an admiration for the technological power of the Industrial Revolution. Or, in other words, a position that managed to reconcile the admiration for technological development with the bucolic ideal of an unspoiled and pure nature.

Machine

But how was such a balance to be achieved? How could the ideal middle state be achieved balancing the opposing forces of technological development and the dream of the return to a serene pastoral existence? According to Leo Marx, for the European colonizers the New World was to provide the answer to this exact question (2000, 101). The American landscape was to become the terrain where old and new, nature and technology harmonically meet to form a libertarian utopia. Technology was seen as "naturally arising" from the landscape as another "natural 'means of happiness' decreed by the Creator in his design of the continent. So, far from conceding that there might be anything alien or 'artificial' about mechanization, [technology was seen] as inherent in 'nature,' both geographic and human" (2000, 160).

Since then, according to Marx (2000), the idea of the "return" to a new Golden Age has been engrained in the American culture and it appears that the power of this idea informs ubiquitous computing's own vision. The idea of a "naturally arising" technology that will facilitate our return to the once lost Garden was to become a dominant and repeating theme within ubiquitous computing discourses. Hence, Weiser envisioned that ubiquitous technologies will make "using a computer as refreshing as taking a walk in the woods" (1991, 104), and twelve years later and writing about the vision of ambient intelligence, Marzano promises that "the living space of the future could look more like that of the past than that of today" (2003, 9).

But while the pastoral defined nature in terms of the geographical landscape, ubiquitous computing defines nature in terms of the objects, tools, and technologies that surround us and our interactions with them. So, while pastoral America defined itself in contradistinction to the European industrial sites and the dirty, smoky, and alienating cityscapes, within ubiquitous computing discourses the role of the alienating force is assigned to the personal computer. And whereas the personal computer with its "grey box" is rejected as the modern embodiment of the European satanic mills, computation is welcomed as a natural technological solution that will infuse the objects that "through the ages, . . . are most relevant to human life—chairs, tables and beds, for instance, . . . the objects we can't do without" (Marzano 2003, 9). Or else, it will infuse the—as we saw earlier, newly constructed—natural landscape, fulfilling the promise that when the "world of bits" and the "world of atoms" are finally bridged, the balance will be restored. But how did these two worlds come into existence? How did bits and atoms come to occupy different and separate ontological spheres?

Far from being obvious or commonsensical, the idea of the separation between bits and atoms has a history that grounds it to specific times and places, and consequently makes those early ubiquitous and tangible computing discourses part of a bigger story that, as Hayles (1999) has documented and as Agre (1997) has argued, started some time ago. This view is endorsed and perpetuated by both ubiquitous and tangible computing and is based on the idea of the separation of computation from its material instantiation, presenting the former as a free floating entity able to infuse our world. As we saw earlier, tangible computing takes the idea of the separation of the two worlds of bits and atoms as an unquestioned fact, which then serves as the basis for its visions and research goals.[4] In this way, the idea that digital information does not *have to* have a physical form, but is *given* one in order to achieve a coupling of the two worlds, not only reinforces the view of digital information as an immaterial entity, but also places it in a privileged position against the material world. In this light, ideas of augmentation—"TUIs will *augment* the real physical world by coupling digital information to everyday physical objects and environments" (Ishii and Ullmer 1997, 2, emphasis added)—or of "awakening" the physical world (Ishii and Ullmer 1997, 3) reinforce the idea of a passive material world that can be brought to life and become worthy and meaningful only through computation, and in that way make ubiquitous computing part of an even bigger and more familiar story. Restaging the dominant Cartesian dualism between the "ensouled" subject and the "soulless" material object, the latter is rendered passive, manipulable, and void of agency, and just like Ishii's bottles, it is performed as a mute, docile *empty vessel* ready to carry out any of its creator's wishes; hold perfumes and beverages, play music, or tell the weather.

At the same time, computation is presented as the force that will breathe life into a mundane and passive world. "As technology becomes hidden within these static, unintelligent objects, they will become subjects, active and intelligent actors in our environment" (Marzano 2003, 8–9). Computation becomes a free-floating, somewhat natural, immaterial entity like oxygen, like the air we breathe (hence MIT's project named *Oxygen*),[5] that can travel unobstructed through any medium, our everyday objects and our environment. But how far does computation's power extend? Or in other words, what sort of agency is granted to it?

It is interesting to note that while computation appears to be foregrounded as a powerful, almost magical, entity that is able to give life and soul to our soulless material world, at the same time it is presented as rather controlled and muted: "This model of technology [referring to ubiquitous computing] stands in stark contrast to most interactive computational technologies whose complexity makes them extremely obtrusive elements of our working environments, to the extent that those environments—working practices, organizational processes and physical settings—need to be redesigned to accommodate computation" (Dourish 2001a, 231). The computational power that will fill our lives, according to ubiquitous computing, will not be alienating, complex, obtrusive, or even noticeable for that matter, and again we come full circle to ubiquitous computing's goal of invisibility. It will be invisible, as its advocates envision, it will leave no traces and bring no radical changes. If anything it will enable us to *reestablish* our humanness and return us to our past, natural state. It will not change us or our lives by introducing something new and unfamiliar, but it will enable us to "remain serene and in control" (Weiser and Brown 1996). Benefit us but not change us. Serve us but not get in our way. Stay invisible without "intrud[ing] on our consciousness" (Weiser 1994, 7). Ubiquitous technologies, as this story goes, are supposed to blend into the environment as harmoniously as the smoky train and the industrial buildings blend into the American landscape in Inness's painting *The Lackawanna Valley* (1855), which "seems to say that 'there is nothing inorganic'" (Marx 2000, 221).

Human

At least since Descartes and the mechanical philosophers of the 17th and 18th centuries, the machine has come to challenge man's ontology, blurring the boundaries between humans and the artificial. The technologies of ubiquitous computing carry on this tradition. Marzano, in the book *The New Everyday*, asks, "We live at a time when many of our traditional certainties are being challenged. What does it mean to be human? . . . Where is the borderline between the natural and the artificial?" (2003, 10). Similarly, the scientists and theorists who held a forum in 2007 titled HCI 2020: Human Values in a Digital Age for "anyone interested in the ramifications of our digital future and in ways society must adjust to the technological changes to come"[6] pose these questions: "What will our world be like in 2020? Digital technologies will continue to proliferate, enabling ever more ways of changing how we live. But will such developments improve the quality of life, empower us, and make us feel safer, happier and more connected? Or will living with technology make it more tiresome, frustrating, angst-ridden, and security-driven? What will it mean to be human when everything we do is supported or augmented by technology?" (Harper et al. 2008, 10).

In the following paragraphs, I seek to join these discussions. However, instead of assuming the futures and the relations that these technologies project and then considering the consequences for the subjects involved, I investigate the prior question of what sort of humanness ubiquitous computing imagines, desires, and naturalizes. And at this point my questions join another technological story. After a close reading of the discourses and practices of projects ranging from traditional AI (1987) to the more recent developments of ALife and situated robotics (2007), and while investigating what it means to be human, Suchman (2007) has unearthed a strong sense of sameness that underpins these projects. She argues that despite the abundance of experiments that investigate and invite crossings of the human-machine boundary, the Euro-American figure of autonomous and rational human agency remains central and uncontested, an agency that these projects then seek to extend to other entities through tactics of mimicry resulting in *humanlike* machines—machines that look, act, think *like* humans. This leads Suchman to conclude, "Reading AI discourses would seem to indicate that the project is less to displace an individualist conception of agency with a relational one so much as to displace the biological individual with a computational one. All else in traditional humanist understandings of the nature of agency seems unquestioned" (240).

Interestingly, ubiquitous computing seems to reverse AI's strategy and, instead of seeking narratives of sameness, it is the differences between human and machine that it strives to bring to the foreground. Influenced by the work of Suchman, an anthropologist employed at the time in Xerox PARC who in 1987 published a groundbreaking critique of AI, Weiser (1994, 8) rejected AI's mimetic tendencies to build machines that can think and act like humans, exclaiming, "Why should a computer be anything like a human being?" Moreover, he explicitly rejected AI's mentalist origins and its eagerness to make things intelligent or smart: "It is commonly believed that thinking makes one smart. But it's frequently the opposite: in many situations, the less you have to think about the smarter you are. . . . Previous revolutions in computing were about bigger, better, faster, smarter. In the next revolution, as we learn to make machines that take care of our unconscious details, we might finally have smarter people" (1996, 8).

So, whereas AI is dreaming of worlds where one will not be able to tell a human from a machine, ubiquitous computing calls for a future that "takes into account the human world and allows the computers themselves to vanish into the background" (Weiser 1991, 94). At first sight it appears that ubiquitous computing seeks to act as a corrective to AI's blindness to human-machine differences by highlighting those differences—"Why should a computer be anything like a human being?" (Weiser 1994, 8). But does that mean that by seeking to foreground human-machine differences, ubiquitous computing seeks to challenge the traditional humanist imaginaries of autonomous, individual agents with essential characteristics? My answer will have to be no. As we will see in the following paragraphs, while AI fixates on issues of sameness by sidestepping the differences between humans and machines, ubiquitous computing brings some differences to the fore, but at the same time reaffirms aspects of the liberal humanist discourse, which identifies a human essence and defines it in terms of its possessive qualities.

According to C. B. Macpherson (1962), the possessive individualism that appeared in the 17th century has been one key characteristic of the subsequent liberal tradition (1). As Macpherson writes, one of the defining propositions that composes possessive individualism states that "what makes a man human is freedom from dependence on the will of others" (263). Or, in other words, "The human essence is freedom from dependence on the will of others and freedom is a function of possession" (3). From this proposition I want to pull the following threads that connect me to ubiquitous computing discourses. One is the specific articulation of the concept of freedom as independence from the will of others, an articulation that defines freedom as the opposite of control and, as we will see, is being challenged by autonomous technology. The other is the articulation of the human essence according to a historically specific concept of freedom. In other words, I would like to attend to the ways that human essence is constructed in specific ways, which then come to be read back into the nature of humanity itself.

Being in Control

Autonomous technology came to highlight the tensions evident in the ways that liberal freedom was being articulated. While on the one hand, according to Hayles who cites Otto Mayr (1989), autonomous technology facilitated the transition from "the centralised authoritarian control that characterised European political philosophy during the 16th and 17th centuries to the Enlightenment philosophies of democracy, decentralised control and liberal self-regulation" (Hayles 1999, 86), on the other, it undermined the latter's very existence. According to Winner (1978), "autonomy" is at heart a political or moral conception that brings together the ideas of freedom and control." And he continues, "To be autonomous is to be self-governing, independent, not ruled by an external law or force. In the metaphysics of Immanuel Kant, autonomy refers to the fundamental condition of free will" (16). Free will, a fundamental characteristic of the liberal humanist subject, is defined as the opposite of control.[7] In this light, autonomous, self-regulating technology threatens the liberal humanist subject as the question is posed, "if technology can be shown to be nonheteronomous [not governed by an external law], what does this say about human will? Ellul is explicit on this point: 'There can be no human autonomy in the face of technical autonomy.' In his eyes there is a one-for-one exchange" (16).[8]

Ubiquitous computing restages this one-for-one exchange between "us" and the personal computer, each struggling for control. Evident particularly in the early ubiquitous computing writings, the post-desktop vision makes its appearance accompanied by clear statements of what humans and machines are, what the former want and why the latter should be kept at bay, placing the two in opposite and antagonistic terrains, as illustrated in Norman's words: "The problem comes about in the form of interaction between people and machines. . . . So when the two have to meet, which side should dominate? In the past, it has been the machine that dominates. In the future, it should be the human" (1999, 140). The tone for this opposition was already set in Weiser's first writings. Weiser (1991) not only envisioned "specialized elements of hardware and software, connected by wires, radio waves and infrared, [which] will be so ubiquitous that no one will notice their presence" (94). He promised a different human-machine interaction with "machines that fit the human environment instead of *forcing* humans to enter theirs" (104, emphasis added).

Within ubiquitous computing discourses, the computer comes to embody a technological menace, the machine that threatens the liberal humanist value of being free and *hence* being in control. As Norman (1999) warns in a book that was characterized as "the bible of 'post-PC' thinking" by *Business Week*, "Today's technology imposes itself on us, making demands on our time and diminishing our control over our lives" (6). And in another point he exclaims, "We have let ourselves to be trapped. . . . I don't want to be controlled by a technology. I just want to get on with my life, enjoy my activities and friends. I don't want a computer, certainly not one like today's PC, whether or not is personal. I want the benefits, yes, but without the PC's dominating presence. So down with PC's; down with computers. All they do is complicate our lives" (72).

The computer is found guilty on the grounds that it has surreptitiously taken control over our lives. As the website of MIT's first ubicomp project *Oxygen* writes, "Purporting to serve us, [computers] have actually forced us to serve them. They have been difficult to use. They have required us to interact with them on their terms, speaking their languages and manipulating their keyboards or mice. They have not been aware of our needs or even of whether we were in the room with them. Virtual reality only makes matters worse: with it, we do not simply serve computers, but also live in a reality they create."[9]

To make things worse, not only is the computer purported to have taken control over our, that is, the users', lives, but it appears to have even escaped the technologists' control. Note for example the following quote where Michael Dertouzos, the director of the MIT Laboratory for Computer Science from 1974 to 2001, describes the feelings of frustration and, more importantly, disempowerment the computer evokes to a group of prominent computer experts:

> Last year a few of us from the Laboratory for Computer Science at the Massachusetts Institute of Technology were flying to Taiwan. I had been trying for about three hours to make my new laptop work with one of those cards you plug in to download your calendar. But when the card software was happy, the operating system complained, and vice versa. Frustrated, I turned to Tim Berners-Lee sitting next to me, who graciously offered to assist. After an hour, though, the inventor of the Web admitted that the task was beyond his capabilities. Next I asked Ronald Rivest, the co-inventor of RSA public key cryptography, for his help. Exhibiting his wisdom, he politely declined. At this point, one of our youngest faculty members spoke up: "You guys are too old. Let me do it." But he also gave up after an hour and a half. So I went back to my "expert" approach of typing random entries into the various wizards and lizards that kept popping up on the screen until by sheer accident, I made it work . . . three hours later. (1999, 52)

Almost like a modern Frankenstein's monster, the computer is performed here as a creature that appears uncontained, unruly, and, therefore, dangerous. It escapes the creator's control and, hence, comes to embody the liberal humanist's nightmare.

Reflecting the paradox it is based upon—flooding our lives with computers while they effectively disappear—ubiquitous computing introduces itself as a technological alternative to our apparently technologically oversaturated and alienated lives. Ubiquitous computing then becomes the solution; the human-centered, somewhat natural approach, which will shift the emphasis away from the machine and bring control back to its legitimate owner, the liberal autonomous human subject. Ubiquitous computing comes to reclaim the control we lost over our machines and becomes the facilitator of our humanness. Its ultimate promise? To enable us to "have more time to be more fully human" (Weiser and Brown 1996). Or, as Dertouzos (1999) puts it, to reestablish our superiority, placing us, once again, at the center of everything that matters. "Perhaps the time has come for the world to consider a fourth revolution, aimed no longer at objects but at understanding the most precious resource on earth—ourselves" (55).

Through these visions and promises a human essence is being invoked, an essence that, apparently, got lost during the computer's reign. This is an essence that "we" all share and that connects us with our true nature. Tangible computing bases its projects on a set of natural, tangible, and universal skills, and ubiquitous computing is supposed to unite us under our alleged frustration with the personal computer, while promising to facilitate the return to our shared humanness. Work, play, and home (Weiser 1993, 77) become the defining human arenas, tables, glasses, and chairs the defining objects, while we are all meant to be united in our desires to escape our windowless offices with their glowing computer screens and take refreshing walks in the woods. Humanity and nature are here reciprocally performed, both united in one front against the technological, and hence unnatural, "other," the personal computer. But here I seek to follow Readings's lead and ask, "Who are we to speak?" (2000, 118).

As Readings (following Lyotard) has argued, the claim to universality where human essence is constructed in specific ways only to be read back into the nature of humanity itself as timeless, universal, natural, and "essential" (hence revealing its tautological nature) is a strategy that liberalism has championed. Under the republican "we," liberalism sought to "build a consensus that defines its community as that of humanity in its freedom" (Readings 2000, 118), while freedom itself was defined as a function of possession. As such, the Jeffersonian democracy of the New World promised a society where "everyone" would be economically independent. As C. B. Macpherson (1962) writes, "[Individualism's] possessive quality is found in its conception of the individual as essentially the proprietor of his own person or capacities, owing nothing to society for them. The individual was seen neither as a moral whole, nor as a part of a larger social whole, but as an owner of himself. The relation of ownership, having become for more and more men the critically important relation determining their actual freedom and actual prospect of realising their full potential was read back to the nature of the individual" (3).

The gendered and racial conceptualizations of the individual in this quote are not symptomatic. The figure of the universal individual is indeed male, white, and free; or, in other words, in possession of his own land and destiny. Jefferson called him the "husbandman" and Jackson the "common man," yet both terms were to capture the "mythical cult-figure" (Empson's term, quoted in Marx 2000, 130) who, according to Leo Marx, can claim a somewhat moral superiority solely on the grounds of his connection with the unspoiled American landscape (131). A mixture of simplicity and sophistication, and with a distaste for the abstract, the intellectual, and the artificial, the "common man" claims a modesty and an earthly wisdom "embod[ying] the values of the middle landscape" (2000, 133). As Jefferson wrote, "State a moral case . . . to a ploughman and a professor. The former will decide it as well, and often better than the latter, because he has not been led astray by artificial rules." The "true American," according to Jefferson's views, is the ploughman, "whose values are derived from his relations to the land, not from 'artificial rules'" (Marx 2000, 130).

Yet, just as the American pastoral ideal seeks to strike the "middle state" by balancing between nature and technology, the American Everyman balances between his love for the land and his "decided taste" (Jefferson quoted in Marx 2000, 134) for business enterprise and progress, as embodied by Jefferson himself who, as Marx writes, perplexed the scholars with the seeming inconsistency of his views (2000, 135). The figure of the American Everyman appears to work in such a way as to reconcile, or better, hold together, these very contradictions.

In their writings, Weiser, Ishii, and Dertouzos appear to adopt this figure of the noble Everyman along with its inherent contradictions. The ubiquitous computing advocate is a person no different from the next person, they tell us, who, his achievements, position, and knowledge notwithstanding, is just like you and me. Dertouzos's earlier story of the four MIT experts with a computer on a plane is here to prove it. Ishii along with the American pastoral farmer aspires to build technologies that will *simply* ease their mother or wife's everyday tasks: "Sometimes, [the pastoral farmer says], I delight in inventing and executing machines, which simplify my wife's labour" (in Marx 2000, 115). And Weiser wants to connect once again with nature and envisions being able to see the traces of the creatures that occupy his neighborhood, yet without leaving the comfort and safety of his own home: "Once woodsmen could walk through the forest and see the signs of all the animals that had passed by in the previous few hours. Similarly, my see-through display and picture window will show me the traces of the neighborhood as faintly glowing trails: purple for cats, red for dogs, green for people, other colors as I request" (1996, 6).

Yet these apparent contradictions get folded and usually remain hidden from view through the evocations of a universal figure, just as Jefferson's idea of the Everyman, of the republican "we," actually excluded and silenced huge numbers of individuals. In the Republic's case, it was the possessive nature of freedom that created a community of "human" subjects under the republican "we" excluding other humans, such as women, Africans, and the Native Americans, who were unable to own productive property in their own right.[10] Indeed, liberal inclusion has always been exclusive. Yet who/what gets excluded or silenced, in the case of ubiquitous computing, is a different question that would lead us to another story, which we have to leave for another time.

Epilogue

Ubiquitous computing is multiple and messy, and done differently in different sites and different stories (Kerasidou 2017; see also Dourish and Bell 2011). Indeed even Mark Weiser, before his death in 1999, had identified two homonymous yet different things under the name ubiquitous computing. One was his own vision, and the other was what "they" had turned it into.[11] Yet, this multiplicity and messiness is worked in such ways as to get folded into and hidden away. The stories of the multiple and, sometimes contradictory, ubiquitous computing*s*, in the plural, get sterilized, reduced, and almost solidified around this one thing, the one dominant story and history of the founding father with the ordered past and the visionary future. And indeed even when dissenting voices emerge (see Rogers 2006; Dourish and Bell 2011), they are positioned only as reactions to some*thing* already there; something that needs to be pushed against in order to be able to articulate what a *better, different, alternative* ubicomp might look like. This is *a* ubicomp, as we are all now meant to agree and repeat, that is tied to a specific time, place, and person, and figures strongly, repeatedly, and, as I have demonstrated elsewhere (Kerasidou 2017), reductively, within the technological stories that the emerging field of ubiquitous computing shares across sites and times. This process then results in a configuration sturdy enough so as to be easily and readily reproducible, and one that can become the basis for other foundational stories, such as stories about nature, hence furthering its dominance.

Haraway (1997, 45) warns us that there is no way out of stories, "We exist in a sea of powerful stories" that weave the technical, social, political, mythic, organic, textual together in their world-making patterns. Yet, she asserts that changing the stories, in both material and semiotic senses, is a modest intervention worth making. Ubiquitous computing's proclaimed human-centeredness along with its alleged uniquely social approach to a "simpler" and more "natural" computational design have indeed proven powerful tropes that have informed, and continue to inform, in various ways the visions of ubicomp's offspring projects, hence making their examination and challenge an important goal worth pursuing.

This is then the modest goal of this chapter: to closely attend to some of the stories and figurations that circulate within ubiquitous computing and then to retell them, tell them *differently* as *my* way of intervening. But make no mistake. Telling stories is no simple matter. Technological stories are not innocent (Law and Singleton 2000). Telling stories, writing histories, performing realities are mingled in the same political and ethical turmoil where my own entanglements and interferences "with other performances of technoscience to prop these up, extend them, undermine them, celebrate them, or some combination of these" (769) cannot but be acknowledged.

So, instead of attempting to *merely* reproduce the stories about nature that circulate within ubiquitous computing (as if even reproduction can ever be complete or innocent), this story is the result of my efforts to consciously and cautiously retell ubicomp's stories in my own way as a way of resisting and critiquing the naturalization of its claims, and as a way of intervening in the future that ubicomp imagines and seeks to build in "our" name.

Acknowledgments

Sections of this chapter have previously appeared in Xaroula Charalampia Kerasidou, "Regressive Augmentation: Investigating Ubicomp's Romantic Promises," *M/C Journal* 16, no. 6 (2013). With thanks to the *M/C Journal*.

Notes

1. http://project.cyberpunk.ru/idb/ubicomp_world_is_not_desktop.html.
2. Rutkowski, for example, used the principle of transparency in 1982 to describe a similar concept as Shneiderman's idea of direct manipulation, writing, "The user is able to apply intellect directly to the task: the tool itself seems to disappear" (quoted in Shneiderman 1983, 63).
3. www.youtube.com/watch?v=U4IYyNL4ld8.
4. http://tangible.media.mit.edu/.
5. http://oxygen.csail.mit.edu/.
6. http://research.microsoft.com/en-us/um/Cambridge/projects/hci2020/default.html.
7. See Chun (2006) for an alternative articulation of the relationship between freedom and control.
8. See also Hayles (1999, 86–87) and Wise (1998, 417–20).
9. www.oxygen.lcs.mit.edu/Overview.html.
10. This exclusion is where C. B. Macpherson (1977) bases his argument that the American democracies as envisioned by Rousseau and Jefferson were only precursors to the theory of liberal democracy

and not liberal democracies themselves. According to his reasoning, the defining characteristic of a liberal democracy is its catering to a class-divided society where class is defined on the basis of a wage relation. The Rousseauean and Jeffersonian democracies, according to the author, were not class-divided societies but one-class societies since their promise was that everyone would be able to own or be in a position to own productive land and capital. The fact that "everyone" was a category that paradoxically excluded women (and others) is explained away, according to Macpherson, on the grounds that the women of the seventeenth century could not be regarded as a class since their labor, unpaid and invisible, did not qualify as wage labor and hence was not regulated by the market (17–22).

11. As is recalled in his obituary, Weiser once told Xerox's chief scientist and PARC's director, John Seely Brown, "they've completely missed the non-technical part of what ubiquitous computing is all about" (quoted in Galloway 2004, 386). These two objects were so different that, apparently, the difference led Weiser to discomfort and frustration and even to an effort to change the name of his vision (see the appendix in Ishii 2004, 1310).

Works Cited

Agre, Philip. 1997. *Computation and Human Experience*. New York: Cambridge University Press.

Castañeda, Claudia. 2002. *Figurations: Child, Bodies, Worlds*. Durham, NC: Duke University Press.

Chun, Wendy Hui Kyong. 2006. *Control and Freedom: Power and Paranoia in the Age of Fiber Optics*. Cambridge, MA: MIT Press.

Dertouzos, Michael. 1999. "The Future of Computing." *Scientific American* 281 (2): 52–55.

Dourish, Paul. 2001a. "Seeking a Foundation for Context-Aware Computing." *Human–Computer Interaction* 16 (2–4): 229–41.

——. 2001b. *Where the Action Is: The Foundations of Embodied Interaction*. Cambridge, MA: MIT Press.

Dourish, Paul, and Genevieve Bell. 2011. *Divining a Digital Future: Mess and Mythology in Ubiquitous Computing*. Cambridge, MA: MIT Press.

Galloway, Anne. 2004. "Intimations of Everyday Life: Ubiquitous Computing and the City." *Cultural Studies* http://www.informaworld.com/smpp/title~db=all~content=t713684873~tab=issues list~branches=18 - v1818 (2–3): 384–408.

Grimes, Andrea, and Richard Harper. 2008. "Celebratory Technology: New Directions for Food Research in HCI." In *CHI'08, Proceedings of the SIGCHI Conference on Human Factors in Computing Systems*, 467–76. New York: ACM.

Haraway, Donna. 1988. "Situated Knowledges: The Science Question in Feminism and the Partial Perspective." *Feminist Studies* 14 (3): 575–99.

——. 1991. "A Cyborg Manifesto: Science, Technology, and Socialist-Feminism in the Late Twentieth Century." In *Simians, Cyborgs and Women: The Reinvention of Nature*, 149–82. London: Free Association.

——. 1997. *Modest_Witness@Second Millenium. FemaleMan_Meets_OncoMouse: Feminism and Technoscience*. New York: Routledge.

Harper, Richard, Tom Rodden, Yvonne Rogers, and Abigail Sellen, eds. 2008. *Being Human: Human-Computer Interaction in the Year 2020*. Microsoft Research. http://research.microsoft.com/en- us/um /Cambridge/projects/hci2020/downloads/BeingHuman_A3.pdf.

Hayles, Katherine. 1999. *How We Became Posthuman: Virtual Bodies in Cybernetics, Literature, and Informatics*. Chicago: University of Chicago Press.

Heim, Michael. 1998. *Virtual Realism*. New York: Oxford University Press.

Ishii, Hiroshi. 2004. "*Bottles*: A Transparent Interface as a Tribute to Mark Weiser." *IEICE Transactions on Information and Systems* 87 (6): 1299–1311.

Ishii, Hiroshi, and Brygg Ullmer. 1997. "Tangible Bits: Towards Seamless Interfaces between People, Bits and Atoms." In *CHI '97, Proceedings of the ACM SIGCHI Conference on Human Factors in Computing Systems*, 234–41. New York: ACM.

Kember, Sarah. 2003. *Cyberfeminism and Artificial Life*. London: Routledge.

Kerasidou, Xaroula Charalampia. 2017. "Figuring Ubicomp (Out)." *Personal and Ubiquitous Computing* 21 (3): 593–605.

Law, John. 2000. "On the Subject of the Object: Narrative, Technology, and Interpellation." *Configurations* 8:1–29

——. 2009. "Actor Network Theory and Material Semiotics." In *The New Blackwell Companion to Social Theory*, edited by S. B. Turner, 141–58. Chichester: Wiley-Blackwell.
Law, John, and Vicky Singleton. 2000. "Performing Technology's Stories: On Social Constructivism, Performance, and Performativity." *Technology and Culture* 41 (4): 765–75.
Macpherson, C. B. 1962. *The Political Theory of Possessive Individualism: Hobbes to Locke*. Oxford: Oxford University Press.
——. 1977. *The Life and Times of Liberal Democracy*. Oxford: Oxford University Press.
Marx, Leo. 2000. *The Machine in the Garden: Technology and the Pastoral Ideal in America*. 35th ed. Oxford: Oxford University Press.
Marzano, Stefano. 2003. "Cultural Issues in Ambient Intelligence." In *The New Everyday: Views on Ambient Intelligence*, edited by E. Aarts and S. Marzano, 8–12. Rotterdam: 010 Publishers.
Norman, Don. 1999. *The Invisible Computer: Why Good Products Can Fail, the Personal Computer Is So Complex, and Information Appliances Are the Solution*. Cambridge, MA: MIT Press.
Plumwood, Val. 1993. *Feminism and the Mastery of Nature*. London: Routledge.
Readings, Bill. 2000. "Pagans, Perverts or Primitives? Experimental Justice in the Empire of Capital." In *Posthumanism*, edited by N. Badmington, 112–28. London: Palgrave Macmillan.
Rogers, Yvonne. 2006. "Moving on from Weiser's Vision of Calm Computing: Engaging Ubicomp Experiences." In *UbiComp 2006, LNCS 4206*, edited by Paul Dourish and Adrian Friday, 404–21. Berlin: Springer.
Rose, Gillian. 1993. *Feminism and Geography*. Cambridge: Polity.
Satyanarayanan, Mahadev. 2001. "Pervasive Computing: Vision and Challenges." *IEEE Personal Communications* 8 (4): 10–17.
Shneiderman, Ben. 1983. "Direct Manipulation: A Step beyond Programming Languages." *IEEE Computer* 16 (8): 57–69.
Soper, Kate. 1995. *What Is Nature?* Oxford: Blackwell.
——. 2000. "Naturalised Woman and Feminized Nature." In *The Green Studies Reader: From Romanticism to Ecocriticism*, edited by L. Coupe, 139–43. London: Routledge.
Streitz, Norbert, and Paddy Nixon. 2005. "The Disappearing Computer." *Communications of the ACM* 48 (3): 32–35.
Suchman, Lucy. 1987. *Plans and Situated Actions: The Human-Machine Communication*. Cambridge: Cambridge University Press.
——. 2007. *Human-Machine Reconfigurations: Plans and Situated Actions*. 2nd ed. Cambridge: Cambridge University Press.
——. 2011. "Anthropological Relocations and the Limits of Design." *Annual Review of Anthropology* 40:1–18.
Traweek, Sharon. 1988. *Beamtimes and Lifetimes: The World of High Energy Physicists*. Cambridge: Harvard University Press.
Ullmer, Brygg, and Hiroshi Ishii. 1997. "The metaDESK: Models and Prototypes for Tangible User Interfaces." In *UIST '97, Proceedings of the 10th Annual ACM Symposium on User Interface Software and Technology*, 223–32. New York: ACM.
Weiser, Mark. 1991. "The Computer for the 21st Century." *Scientific American* 265 (3): 94–104.
——. 1993. "Some Computer Science Issues in Ubiquitous Computing." *Communications of the ACM* 36 (7): 75–84.
——. 1994. "The World Is Not a Desktop." *Interactions* 1 (1): 7–8.
——. 1996. "The Open House." *ITP Review 2.0*. http://makingfurnitureinteractive.files.wordpress.com/2007/09/wholehouse.pdf.
Weiser, Mark, and John Brown. 1996. "The Coming Age of Calm Technology." www.ubiq.com/hypertext/weiser/acmfuture2endnote.htm.
Weiser, Mark, Rich Gold, and John Brown. 1999. "The Origins of Ubiquitous Computing at PARC in the Late 80s." *Pervasive Computing* 38 (4): 693–96.
Winner, Langdon. 1978. *Autonomous Technology: Technics-Out-of-Control as a Theme in Political Thought*. Cambridge, MA: MIT Press.
Wise, MacGregor. 1998. "Intelligent Agency." *Cultural Studies* 12 (3): 410–28.

Affect and Emotion in digitalSTS

Luke Stark

Emotions and Actors in Digital Systems

"'You can't mess with my emotions. It's like messing with me. It's mind control.'" This sort of public reaction confronted Cornell University's Jeffrey Hancock and the other authors of the now-infamous Facebook "emotional contagion" study (Kramer et al. 2014). Hancock had partnered with researchers from Facebook to determine if changing the frequency of positive or negative emotional keywords in a user's Facebook newsfeed tilted that same user's posting in a more positive or negative direction (Kramer et al. 2014). To test the hypothesis, Hancock and his coauthors had altered the flow of newsfeed posts for two groups of Facebook users simultaneously. This process, in which two possible interfaces are shown to social media users in real time for experimental purposes, is known in Silicon Valley as an A/B test (Cristian 2012). When the research describing the experiment was released, media coverage was highly critical: Facebook, in the words of one headline, had "manipulated emotions for science" (Hill 2014).

Human affect and emotion are simultaneously integral and unsettling to our contemporary experience of digital technologies. The furor around Facebook's "emotional contagion" study was exemplary of a public controversy at the sociotechnical intersection of computational media, social and cultural practices, and lived emotional experience (Selinger and Hartzog 2015). Such intersections have proliferated: in April 2017, for instance, Facebook's Australian division was exposed as having worked to target advertising to teenage users based on longitudinal emotional profiles developed out of the company's personal data (Levin 2017). As the computational tools on which we rely to mediate our everyday social lives become increasingly interconnected, pervasive, and sensor-rich, the digitally mediated quantification and expression of human emotions is an increasingly central part of the experience of digitally mediated existence.

As a discipline, science and technology studies (STS) is ideally placed to explore and critique the ways the varied landscape of human emotional experience and expression shifts as a "matter of concern" refracted through diverse digital interfaces, systems, and platforms (Latour 2004). There is strong extant scholarship on the history of emotion as a scientific and technical object. Otniel Dror (1999b, 2001, 2009) has articulated how the development of physiological laboratory cultures of the 19th century was "immanent in the very design of . . . laboratory models of feelings" (2009, 851), while Brenton Malin (2014) describes how new technological apparatuses such as film shaped the emotional terrain of 20th-century social science. Elizabeth Wilson (2010) argues for the centrality of affective response to early computer science and cybernetics pioneers like Alan Turing, and how "computational logic, the building of mechanical devices, and fantastic anticipation were always intimately allied" (39). And two recent edited volumes (Tettegah and Noble 2016; Hillis et al. 2015) have done much to put historical accounts of affect and emotion into conversation with digital media studies.

A wider scholarly focus on human affect and emotion in the history and contemporary development of digital and computational media is nonetheless long overdue. Attention to these realms of human experience opens the way for a broader explication of the human sciences (Rose 1988; Foucault 1994) and "sciences of subjectivity" (Shapin 2012) as "technical" mechanisms in digitally mediated worlds. The designers of digital media platforms and applications draw on the sciences of emotional tracking, classification, and management to build systems through which everyday acts of emotional expression can be made comprehensible as data, and put to use as components of sophisticated digital profiles of the individual (Haggerty and Ericson 2000; Cheney-Lippold 2011; Pasquale 2015; Stark 2018a). Computer science has been heavily influenced by psychology and the behavioral sciences more broadly (Stark 2016), and Facebook's controversial actions point to the ways techniques for mediated affective and emotional management are already being put to use in the service of digital advertising, surveillance, and profiling (Cadwalladr 2018; Rosenberg et al. 2018). The application of these technologies and techniques has gone hand in hand with the design and spread of neoliberal economic models of individualization, and asymmetries of knowledge and power implicating core aspects of how we understand the emotional and reflective self (Illouz 2007).

Even—and I argue especially—in the world of silicon and bits, human emotions are powerful drivers of agency and action. Here I lay out the range of phenomena encompassed by terms such as *affect*, *emotion*, *feeling*, and *mood*. After providing this overview of definitions around affect and emotion, I suggest an analytic frame to help make the impact of these phenomena more legible: the *emotive actant*. Bruno Latour's well-known notion of an actant, or anything "modifying other actors through a series of . . . actions," has widespread currency in STS (Latour 2004, 2005), and Latour's actor-network theory (ANT) is a well-known lens through which to understand the impacts of sociotechnical apparatuses. I argue *emotive actants* are *actants intensifying the experience and expression of human feelings*, and have an increasingly palpable influence within the contours of digitally mediated culture, politics, and social experience.

Definitions are a challenge for scholarship across the fields studying affect and emotion (and associated terms such as feelings, moods, sentiments, sensations, and passions). As William Reddy (2001, 3) observes, "Emotions have been compared to colors. . . . [Both] have a strong subjective or experiential character [and] in both cases, there is no way for an independent observer to check these 'self- reports.'" Despite the fact, as Reddy notes, the "reported experiential qualities display great constancy from one person to another," the inherently subjective experience of emotions makes systematic comparison both of particular genres of feeling and of generalized emotional states difficult enough in everyday life.

Such polysemy is potentially crippling for STS scholars particularly attuned to ways in which scientific practice and discourse are constitutive of scientific "facts," with usages shifting depending on the discipline, subfield, or context of conversation. The language used to describe human feelings is therefore as much an object of study for STS as it is a necessary resource. Resisting the temptation to blur definitions together is key, and the taxonomy I present here should be taken as provisional and partial, reliant on similar efforts by sociologists like Arlie Russell Hochschild (2003b) and Deborah Gould (2010), psychologists such as Jerome Kagan (2007) and Rom Harré (2009), and humanities scholars such as Teresa Brennan (2004).

One central definitional distinction cutting across many disciplines separates *emotion* from *affect*. Jerome Kagan summarizes current psychological consensus around *emotion* as fitting within, and simultaneously constituted by, four interrelated human phenomena: *affect* (or in Kagan's terms, "a change in brain activity to select incentives"), *feeling* or *sensation* ("a consciously detected change in feeling that has sensory qualities"), *emotion* proper ("cognitive processes that interpret and/or label the feeling with words"), and *reaction* ("a preparedness for, or display of, a behavioral response") (2007, 23). Deborah Gould defines *affect* as "nonconscious and unnamed, but nonetheless registered, experiences of bodily energy and intensity that arise in response to stimuli" (2010, 26), and *emotion* as "what from the potential of [affective] bodily intensities gets actualized or concretized in the flow of living" (26). For Gould, such actualization might come through linguistic categorization, gestural performance, or both. Teresa Brennan suggests affect is "the physiological shift accompanying a judgment," inasmuch as affects imply a focus of attention and action toward a particular stimulus presumes adjudication between one potential response and another. The bottom line, as Brennan and others reiterate, is "feelings are not the same as affects" (2004, 5).

The links, causal and otherwise, between affect and the wider range of human felt experience are contested within both the human and biological sciences. The *affective turn* resulted in a rise in interest from scholars in the humanities and social sciences over the past two decades in material, nonconscious, bodily, or somatic factors in the experience of human subjectivity (Gregg and Seigworth 2010)—a turn characterized by "an amalgamation, a revisiting, reconsideration and reorientation of different theoretical traditions" (Hillis et al. 2015, 4); with many such scholars self-identifing as "new materialists." Gould's definition of affect is grounded in Brian Massumi's (2002) theorization of affect as a play of intensities and valences, itself based on the work of philosopher Gilles Deleuze (Deleuze and Guattari 1987). Massumi draws on research from neuroscience (Damasio 1994) and cognitive psychology (Ekman and Rosenberg 2005) to assert affect is invariably precognitive, and is thus dispositive in the shaping of emotional experience. This interpretation has been strongly critiqued both on conceptual grounds and for what its critics assert is a misinterpretation of the underlying empirical evidence (Leys 2011). Ruth Leys's (2017) *The Ascent of Affect* strongly contests the science underpinning the affective turn—and even the concept of affect itself. The affective turn has nonetheless influenced a variety of fields including digital media studies, and provides a set of conceptual springboards for work at the intersections of computational media studies and STS (Sengers et al. 2008; Hillis et al. 2015).

Definitions of emotion proper assume a high degree of cultural specificity in how feelings (in Kagan's sense of the term, consciously detected changes in feeling) are perceived and interpreted by the self and others. Sociologists have been interested in human emotion and its cultural specificity for several decades (Lively and Heise 2004). Contemporary sociological scholarship on emotion grounded in both social psychology (Solomon 2003; Kagan 2007) and affect control theory (ACT) in particular (Lively and Heise 2004; Shank 2010), provides further potential definitional and analytic tools to explore feelings in digital contexts.

Based on her sociological work on gender and labor in the early 1980s, Arlie Russell Hochschild identified two distinct models dominating scholarship on emotion at the time: (1) an *organismic* model formulated in physiology, psychology, and evolutionary biology and (2) an *interactional* model stemming from anthropology and sociology (Hochschild 2003b, 215). These models persist: in general, proponents of an organismic model understand emotions as strongly determined by affective biological processes or drives, and assume by extension basic human emotions are universal across different cultural and sociotechnical contexts. In contrast, interactional models of emotion focus less on underlying affects, and more on the social and cultural factors shaping particular instances of emotive and emotional expression. Psychologist Paul Ekman's research suggesting emotional facial expressions are common across cultures is a well-known example of work grounded in the organismic model (Ekman and Friesen 1971); Lila Abu-Lughod's anthropological work on emotion in the context of North African Bedouin cultures represents an interactional approach (Abu-Lughod and Lutz 1990). The debates between the proponents of each of these models over the degree to which emotions are either universal or culturally specific are a further definitional and sociotechnical variable for STS researchers to keep in view. In *The Managed Heart*, Hochschild argued for an analytic model for emotion synthesizing the organismic and interactional models, what she termed "a new social theory" for emotion. Hochschild suggested emotion is "a biologically given sense, and our most important one . . . unique among the senses [in being] related not only to an orientation toward action but also to an orientation towards cognition" (229). In other words, emotions are more than simply a bridge between physiological and psychological responses to stimuli: they also engage thinking, and by extension purposeful action, in tandem with somatic or instinctual responses.

This capsule introduction to terminological distinctions and controversies around affect and emotion is not meant to slight the extensive and complex debates around how these phenomena are defined and understood across many disciplines. Instead, it is meant to help STS scholars orient themselves amid the thickets of terminological and disciplinary difference, and understand these debates around the vocabulary of affect and emotion as themselves resources for sociotechnical inquiry.

Emotive Actants

Among the language used to describe emotional phenomena is one term, *emotive*, which I argue deserves more prominence in our analyses of digital media. The word *emotive* means, "arousing intense feeling"—whereas the word *emotional*, with which *emotive* is sometimes used synonymously, means "characterized by intense feeling." These terms reflect slightly different models of felt subjective human experience—an emotive response is closer to the aforementioned definitions of *affect* than is a reflexive emotional one, and is characterized by the powerful expression of feelings but not necessarily a fully reflective emotional experience.

In their introduction to *Networked Affect*, Ken Hillis, Susanna Paasonen, and Michael Petit observe, "the need to focus on connections and relations in studies of action and agency" (2015, 10), and how this need is especially acute in the context of digital mediation. Yet affect and emotion, while clearly relational and agentic, have not always been well explicated in relation to ANT approaches (Latour 2005). Hillis and their coauthors stop short at identifying a specific mechanism through which to articulate this relationship—but the term *emotive actant* fits the bill.

I define an emotive actant as an agent intensifying affect, feeling, sensation, and even emotion. Latour observes ANT is interested in tracing the interacting effects of agents; by extension, Latour notes, "if you mention an agent, you have to provide the account of its action, and to do so you need to make more or less explicit which trials have produced which observable traces" (Latour 2005, 53). Emotive actants produce a notable change in expression, a trace explicable by their particular presence, configuration, and influence within a chain of interactions. Latour calls this general process "translation"—a "connection that transports, so to speak, transformations" (2005, 108). Identifying an agent as an emotive actant describes a process, not an end state: many human technologies can be mobilized as emotive actants.[1] Yet in cases such as the Facebook emotional contagion study, a set of digital artifacts—social media classification schemes, algorithmic data analysis, and digitally enabled A/B tests—became mobilized as emotive actants in various deliberate and accidental ways, actively translating and mediating human affective and emotive expressions for particular technoscientific, economic, and political ends.

Explicitly identifying emotive actants as having effects on embodied human mental and physical activity is part of the broader STS project of "rethinking both human and nonhuman actors and how affect is generated and circulated" (Hillis et al. 2015, 10). Emotive actants clarify what Latour describes as the problem of "figuration," proceeding from his observation, "what is doing the action is always provided . . . with some flesh and features that make them have some form or shape, no matter how vague." For Latour, description itself can cloud sociological analysis by limiting the categories of what is considered a legitimate social actor (2005, 53). By focusing on the transformative effects of particular emotive actants in describing agency in the context of digital mediation, I am not diminishing the importance of these actants' figuration—far from it. Instead, the concept of the emotive actant allows a wide range of digital artifacts, and discourses about them, to become comparable within the same category of agency (55).

The notion of an emotive actant also resonates beyond ANT by highlighting the ways human affect and emotion shape the normative value systems underpinning the creation, use, and development of digital media technologies. Values, whether understood as agent-centered or outcome-centered systems of reasoning or activity (Nagel 1979), are central to analyses of sociotechnical systems (Friedman and Nissenbaum 1996). As Deborah Johnson (2007) observes, "Values can be seen as interests, the interests of particular groups struggling over the design or meaning of a technology" (27). Changes in a particular technology's materials, discourses, and practices change the conditions under which values, ranging from individual privacy to transphobia to racial bias, are expressed through it (Nissenbaum 2015; Haimson and Hoffmann 2016; Noble 2018). Notwithstanding the complexity of these questions, scholars in philosophy, information studies, STS, and computer science are increasingly engaged in analyzing the politics of technologies (Winner 1988; Feenberg 1992) through methodological traditions like value sensitive design

(Friedman et al. 2006), reflective design (Sengers et al. 2005), and studying values at play (Flanagan and Nissenbaum 2014). Like ANT, these approaches seek to tease out the mechanisms by which human values influence technologies, and in turn dynamically shape user experiences and sociotechnical milieus.

Digital media's increasingly deleterious effects on democratic processes make interrogating digital systems with an emphasis on affective and emotive values all the more urgent. Digital media manipulation and disinformation within the context of the attention economy (Beller 2006a, 2006b) are enabled by the manipulation of emotive actants. In this societal context, STS scholarship must attend to how affects and emotions—in their presumed salience, their lived experience, and their political resonance—infuse the norms and values of particular digital platforms (Konnikova 2013) and/or shape subjective and social reactions to particular technologies in their myriad transnational and global contexts (Powell 2013; Beer 2016; Stark 2018a).

Emotions in Digital Context

One of the most prescient accounts of computation and its contemporary social contexts comes from Continental philosophy, in the writing of Gilles Deleuze (1990). Deleuze argued the 21st century would be typified by a new form of "control society": one premised on computational media and data analytics as means for powerful institutions to modulate and restrict an individual's smooth flow through and access to social systems. Deleuze's notion of the "control society" describes the broader sociotechnical framework in which data about human emotional expression have been incorporated into systems of algorithmic modulation, management, and control (Crawford et al. 2015; Lee et al. 2015; Hoffmann et al. 2017; Stark 2018a).

Digital media technologies working as emotive actants often convert embodied emotional expression into computationally legible numbers, words, and symbols. Such technologies then reproduce these computational logics as mediating models through which individuals reinterpret and re-present their own subjective feelings, both to themselves and to others. Marisa Brandt (2013), drawing on the work of Jay David Bolter and Richard Grusin (2000), describes this process as *therapeutic remediation*, a cybernetic logic through which computational media can have a transformative effects on the human sense of self. Emotion's entanglement with digital media technologies and platforms has become more and more widespread as these phenomena have been quantified under the purview of clinical psychology, psychiatry, and neuroscience (Damasio 1994). Contemporary quantified models of emotion are grounded in foundational "organismic" works from biology and experimental psychology, including from Charles Darwin's *The Expression of the Emotions in Man and Animals* (Darwin [1872] 2009), in which Darwin postulated emotional expression signaled either the direct action of physiological reflexes, or a habituated response to some external stimuli (Winter 2009; Snyder et al. 2010); and from William James's 1885 "What Is an Emotion?" the basis for what is now known as the James-Lange theory of emotion (James and Lange 1922; Wassmann 2010). The psychoanalytic tradition founded by Sigmund Freud has also influenced contemporary scholarship on emotion and digital media, although in sometimes-oblique ways (Turkle 2004; Liu 2011). Sherry Turkle's work has emphasized the ways in which humans ascribe animation and agency to computational tools depending on the contours of interaction, findings supported by the work of the late Clifford Nass (Reeves and Nass 2003; Robles et al. 2009).

What Otniel Dror (2001) identifies historically as "emotion-as-number"—a scientific discourse equating the range of human emotional experience to physiologically quantifiable metrics—has been a key technical mechanism enabling human felt experience to be simplified, translated, and incorporated into the structures of digital computing. This discourse influenced the first cyberneticists and artificial intelligence pioneers in the late 1940s and early 1950s. Researchers were interested both in behavior and in the notion of physiological feedback (Orr 2006; Kline 2009; Wilson 2010; Pickering 2010). While explicitly cybernetic research trajectories fell out of favor by the 1970s, interest in the digital quantification of emotional responses continued to percolate, such as in the work of Manfred Clynes, who coined the term "cyborg" (Clynes and Kline 1960) and later developed a system to measure human emotion via haptic feedback (Clynes 1989).

Clynes and others in turn helped inspire the contemporary work of computer scientists such as Rosalind W. Picard, who in her seminal work *Affective Computing* (2000) revived the study of emotions via computational data in human-computer interaction (HCI). HCI had long been grounded solely in cognitive psychology (Card et al. 1983), but Picard suggested classifying and quantifying human emotive signals was a first step in learning how to simulate the experience of emotion in machines. The recent expansion of interest in interaction design, machine learning, and artificial intelligence (AI) has prompted growth in the study of social and emotional aspects of AI and HCI as a byproduct of affective computing utility's in translating expressions of emotional experience into quantifiable and machine-legible data (Höök et al. 2010; Höök et al. 2015).

Social and emotional HCI has seen significant growth in the last decade, due to advances in the speed of hardware, the capacity of software, engagement by masses of users in social media, and the increasing public salience of HCI problems to the general public (Shank 2014). These developments in HCI have their genesis in several distinct phases, from early work in human factors computing in the 1970s and 1980s, and a focus on usability in the 1990s (Bødker 2015), to the contemporary proliferation of academic and design work in the affective computing centered on the concept of the "user experience" (UX) of digital information technology (Dourish 2004; Grudin 2012). The temptation to equate all aspects of human emotional expression with physiological data describing bodily activities—information such as heart rate or the rate of blood flow—has been strong in HCI. In a 2007 paper, Kirsten Boehner and colleagues argued for an alternate *interactionist* model for exploring emotion in the context of digital design practice.

As Boehner and her coauthors write, emotion is "an intersubjective phenomenon, arising in encounters between individuals or between people and society, an aspect of the socially organized lifeworld we both inhabit and reproduce" (280). In contrast to research in affective computing and "emotional AI" interested primarily in quantifying and translating physiological responses as emotional signals (Picard 2000; Scheirer et al. 2002; Picard and Daily 2008), the interactionist research paradigm incorporates alternative forms of experiential data as elements of digital design alongside quantitative data (Leahu et al. 2008; Boehner et al. 2007; Boehner, Sengers, and Warner 2007; Sengers et al. 2008), such as those elicited by Katherine Isbister and Kia Höök's sensual evaluation instrument (Isbister et al. 2006). From an STS perspective, however, both of these schools of HCI research implicitly understand digital technologies as emotive actants and not neutral agents: not merely passively transmitting emotional signals between individuals, but modifying, intensifying, and prompting new configurations of feelings in users. Just as, in the words of sociologist Arlie Russell Hochschild, "every emotion has a signal function," so too does every digitally mediated human signal have an emotive and by extension social function in the age of computational media.

Emotive Actants at Work: Clicking, Tracking, Expressing, and Parsing

As examples of some of the ways in which emotive actants might shape sociotechnical processes, consider a few broad categories (though needless to say, the artifacts within these categories often interact and overlap as parts of larger networks). The first category consists of artifacts and systems designed to facilitate *clicking*, wherein scholarship from psychology and neuroscience is brought to bear on intensifying the effects of the UX design for social media platforms. The second category is made up of technologies for *tracking*, encompassing a variety of devices deployed by individuals, institutions, or both to detect, collect, and monitor human emotive states. The third category consists of artifacts designed for correlating and *parsing* data produced by user engagement and expression. Finally, a fourth category of technologies have been developed for *expressing* emotion, enabling us to communicate emotional states online via emoji, emoticons, stickers, and animated GIFs—including expressing cultural and political resistance to the status quo. Across this taxonomy, there is broad scope for new digital STS research describing, contextualizing, and troubling the technical definitions, translations, experiences, and expressions of affect and emotion within the broader social contexts of our networked world.

Clicking

Nikolas Rose (1988, 1996, 2013), Anthony Giddens (1991), and Eva Illouz (2008) have all explored the role of the psychological sciences in disciplining and modulating the modern self; new digital technologies both for expressing emotion socially and for experiencing and managing it therapeutically promise to once again shift our understanding of ourselves. Michael Hardt (1999) has suggested pessimistically, "we increasingly think like computers"; Jeanette Wing, coming to a very different conclusion regarding desirability, has extolled the benefits of such "computational thinking" (Wing 2006). David Golumbia describes this discourse as "the cultural logic of computation," a set of social assumptions about the utility of digital media influencing human social and subjective life (Golumbia 2009).

The business models of many social media platforms entail catching and holding human attention (Beller 2006a, 2006b); engaging human affects and emotions, especially high-intensity or negative ones, helps keep and maintain our engagement across space and time. Psychological and behavioral science has shaped technical strategies to keep users engaged from the 1980s onward (Card et al. 1983; Moran and Card 1982; Newell and Card 1985), and affective or emotional blocks to digital media use, such as computer anxiety, were problematized in the same period (Heinssen et al. 1987; Doyle et al. 2005; Powell 2013). Compelling, even "addictive" UX and interaction design are now understood as major components of Silicon

Valley commercial success (Norman 1989; Katz 2015). With concerns on the rise about the amount of time and attention social media users pay to their digital devices, many critics of digital platforms—and those platforms themselves—have also begun to promote hazy notions of "digital wellbeing" (Stark 2018c).

Advances in digital game design are driving new strategies to engage and channel the affective and emotional impulses of users (Juul 2010a). Digital game designers seek to create compelling effects through design decisions regarding game play and narrative. Decisions around viscerally and emotionally compelling game design are examples of what Ian Bogost terms "procedural rhetoric" (Bogost 2006)—the ability to encode particular persuasive elements into the actions required of players, and by extension by digital media users in general. With the more recent rise in popularity of casual and social games, many of which are played primarily on smartphones, designers have sought to incorporate the same kinds of gamic interface strategies (most notably viscerally compelling feedback within interface graphics) into smartphone applications more broadly. Making a digital interface "juicy," in the words of game theorist Jesper Jull, entails bring colors, quasi-organic movements, and digital objects programmed to be mimic the tactility of their physical counterparts (Juul 2010b). UX design guru Donald Norman coined the term "visceral design" in 2005 to describe these design elements (Norman 2005).

Activating and intensifying visceral, emotional, and noncognitive impulses has a long history in advertising, graphic design, and industrial design (McGrath 2008), and a new scholarly vogue in the study and application of behavioral economics and behavioral "nudges (Thaler 1980; Kahneman 2013; Thaler and Sunstein 2008). The standardization and mobilization of human feelings is also a growth industry for many social media companies. Facebook's 2016 move to expand the "Like" button to include a broader range of emotional "Reaction" icons (Goel 2015) is exemplary of the role these platforms have in encouraging their users to tag and organize their own emotional data, and collect them for a variety of purposes including driving increased use of the site and more targeted advertising decisions (Oremus 2013; Boesel 2013).

STS scholarship has the opportunity to expand into analyses of the technologies and practices of these "creative" industries, exploring how they have incorporated concepts, values, and techniques from the psychological and behavioral sciences, engineering, industrial design, and gaming (Holland et al. 2014; Parisi 2018; Stark 2018b). These projects connect with a recent emphasis on making, doing, and critical design in STS work (Pullin 2011; Dunne and Raby 2001; DiSalvo 2012). Other extant projects examining the historical and contextual complexity of emotion to the technical and design community include cases in which design is mobilized to compel and addict (Schüll 2012), and working with designers and technologists themselves to explore novel practices and strategies around emotion and design (Demir et al. 2009; Stark 2014).

Tracking

Technologies for tracking, recording, collecting, and quantifying human emotions are tightly tied to the development of Western technoscience, and to the more recent rise of so-called "surveillance capitalism" (Zuboff 2015). As Otniel Dror argues, physiology and psychology developed discourses and technologies to produce "emotion-as-number," based on physiological quantification of affects and reactions, as a method of containment for phenomena scientists themselves found unsettling (Dror 1999a, 1992b, 2001, 2009). The 20th century saw increased integration of such quantified data into technoscientific epistemologies of reasoning, experimentation, and evidence. Yet while discourses of emotion-as-number are longstanding, digitally mediated systems of pervasive surveillance are a more recent development. Any external material trace or emanation that can be potentially correlated with an interior affective state is technically traceable (Kerr and McGill 2007): technologies for tracking the physiological expression of affects include facial recognition and movement recognition technologies such as the Facial Action Coding System (FACS; Ekman and Rosenberg 2005); systems gauging affect via vocal tone (Mizroch 2014); mobile monitoring applications for collecting physiological data such as heart rate and skin conductivity (Picard and Klein 2002; Picard and Scheirer 2001); and a range of wearable consumer hardware devices to record movement and sleep patterns, gestures, and gait (Schüll 2016). These technologies are becoming increasingly common both in enterprise applications such as hiring, and in enabling new tools for emotional expression in everyday social media (Stark 2018b).

Digital self-tracking has also increased in popularity over the past decade (Richards and King 2013; Wolf 2010; Nafus and Sherman 2014). Mobile applications such as Mood Panda and MoodScope encourage users to actively track and quantify their mood in the service of self-improvement and sociality. Mood and emotion tracking is just one aspect of the self-tracking ecosystem, with many trackers drawing on techniques from social psychology to assess their feelings longitudinally over time (Carmichael and Barooah 2013). Other devices are pushing techniques of mood management into an ostensibly more direct interface with the body itself, while black boxing the scientific assumptions of their component technologies: such tracking is thus a part of the "neuroscientific turn" (Littlefield and Johnson 2012) connecting brain science to traceable everyday activities.

Parsing

What happens to the data we produce when we click and are tracked? These inadvertent digital traces are parsed and analyzed by institutions ranging from social media platforms and advertising agencies to governmental and security services (Kerr and McGill 2007). Sociologist Ulrich Beck (2009) points to postindustrial societies as framing future decision making within a probability calculus and the language of risk; increasingly, data about emotional reactions and behaviors are included in these calculations. Out of these aggregated patterns of data, personalized profiles of an individual's emotional expressivity (or the set of signals performed by the human body deemed to correspond to it) can be interpreted by the authorities—or the individual herself—as symptomatic of more fundamental aspects of the self: disorder, criminality or culpability, productivity, mindfulness, or well-being (Cheney-Lippold 2011).

Data mining and analytic techniques are often proprietary, opaque, and increasingly incorporated into comprehensive profiles and scores of our online "data doubles" (Haggerty and Ericson 2000; Citron and Pasquale 2014). Such traces can also be linguistic, and thus tracked and analyzed via sentiment analysis techniques counting the number of positive and negative words in a corpus of data (Hu et al. 2013; Andrejevic 2013). Affectiva, a company founded at MIT, promises to deploy a wide range of emotional tracking and monitoring techniques, such as sentiment analysis and facial recognition techniques, to user data in order to assist companies in marketing and advertising (Coldewey 2016).

Parsing human feeling for commercial ends is not new, but digitally mediated parsing can easily be enlisted in the management of labor, consumption, and health (Scholz 2013). Arlie Russell Hochschild (2003a, 2012) explores how flight attendants, retail clerks, nurses, and care workers, often but not exclusively women, are taught to understand, manage, and even shift their own feelings in the service of their jobs. Similar studies on taxi drivers (Facey 2010) and service center workers document how widespread and technologically embedded these practices of emotional self-management have become. For Michael Hardt, the radical challenge of technology's ability to shape our emotional lives for good or ill is posed primarily by this "*affective labor* of human conduct and interaction" (Hardt 1999, 94). By affective labor, Hardt means the energies and passions expended in the service of work—like emotional labor, affective labor relies on human energy, but is more concerned with the constantly renewed drives for sociality and connection humans exhibit online. Affect and emotion also play a particularly central role in the web of contemporary online "knowledge work," and are central to the political stakes around the contemporary experience of digital labor in the so-called "on-demand economy," such as those of Mechanical Turk workers (Irani 2015), Uber drivers (Rosenblat and Stark 2016), and creative workers more broadly (Gregg 2015).

Digital technologies further enable new configurations of surveillance and control across spatial borders. In her ethnographic accounts of call center labor in India and elsewhere, Winifred Poster argues emotional labor is inextricably tied not only to novel shifts in communications technologies (for instance, the adoption of text-based chat programs for customer assistance), but also to already-extant asymmetries of economic power, misogyny, racial and linguistic prejudice, and national chauvinism (Poster 2011, 2013). The digital technologies used by Poster's interlocutors serve to paper over and exacerbate these divisions: by compelling workers to conform to a particular set of linguistic, affective, and stylistic "best practices" around the provision of customer service, these systems and their human proponents promote an invariable standard of human emotional conduct that both suppresses the subtleties of rich human interaction and enlists workers around the world in an forced hermeneutics of monitored, regimented interaction.

Expressing

The parameters for emoting via digital systems are set by those systems' affordances and design; in turn, emoting is a key element of both the computational logics of digitally mediated sociality, and of resistance to hegemonic technoscientific and economic logics. The emoji character set, initially developed as a proprietary feature of Japanese telecom company NTT Docomo's cellular phones, is exemplary of how affective labor is captured through digital mediation, and how these forms of emotional expression are not only social, but also entangled in questions of personal identity and political economy (Stark and Crawford 2015). The enormous popularity of the characters as a means of social expression ultimately forced Apple to provide functionality for emoji on its devices worldwide. Further clamor from users led to the character set's eventual incorporation into the global Unicode technical standard for interoperable digital symbols. Emoji, emoticons, "sticker" pixel images, animated GIFs, and proprietary animations point to a digitally mediated future in which proprietary logics and interfaces drawn from animation increasingly structure the means through which we communicate socially and emotionally (Silvio 2010; Gershon 2015).

The mechanisms of traditional clinical interventions and therapeutic techniques have also changed with the spread of digitally media systems (Mishna et al. 2016). While digitally mediated or e-therapy has a long history ranging from the famous Rogerian therapeutic chatbot ELIZA (Wilson 2010), its increasing use calls into question how practitioners and patients are reshaping the experience of therapy in conjunction with mediating technologies and new business models designed, in many cases to bring market logics of efficiency and cost-effectiveness to the therapeutic process (Alexander and Tatum 2014; Atkins et al. 2014; Brandt 2013). Both Eva Illouz (2007) and Martijn Konings (2015) have observed therapeutic cultures have an ambivalent relationship with capitalist logics, both teaching self-efficacy and autonomy and equipping and acclimatizing individuals as better adjusted and more productive capitalist subjects. This ambiguity characterizes services like Talkspace, a service offering connection with a therapist via text (Cook 2015; Essig 2015): it is unclear how the relationship formed via this type of remote mediation, while broadening the accessibility of therapy, shapes the long-term experience of the patient or the therapist. The increasing penetration of digital technologies into health care settings, including the provision of mental health, is a core area for future STS research (Brandt and Stark 2018).

The categories described above evidently overlap and enable one another, serving as emotive actants in complex ways: for instance, a smartphone-based mood tracking application like Mood Panda functions as an emotive actant by engaging multiple sets of social and technical linkages, and shaping them and the emotional subjectivity of users in turn (Stark 2016). Future STS work on affect, emotion, and digital media will necessarily rely on a multidisciplinary set of literatures exploring both emotions and digital technologies writ large. Yet the political salience of emotion and affect in the context of increasingly ubiquitous digital networks is incontrovertible (Papacharissi 2014), particularly in light of recent public furor over the psychographic profiling performed by British firm Cambridge Analytica in the service of the Donald Trump presidential campaign (Cadwalladr 2018; Stark 2018a).

This extant work in STS provides templates for more broadly sociotechnical assessments of feelings, values, and technologies. Understanding contemporary digital devices as emotive actants provides a mechanism for exploring the effects of these technologies on human feeling, and by extension on human social life. Perhaps most importantly, the notion of an emotional actant actively translating and mediating human emotional expressions highlights political questions of human sociality, equality, and solidarity within ANT, and STS scholarship more broadly, in new, urgent, and productive ways.

Note

1. Indeed, human language is perhaps the ultimate emotive actant, providing a technical means to translate the affective impulses of the body into culturally specific and reflective emotional responses.

Works Cited

Abu-Lughod, Lila, and Catherine A. Lutz. 1990. "Introduction: Emotion, Discourse, and the Politics of Everyday Life." In *Language and the Politics of Emotion*, edited by Catherine A. Lutz and Lila Abu-Lughod, 1–23. Cambridge: Cambridge University Press.

Alexander, Valerie L., and B. Charles Tatum. 2014. "Effectiveness of Cognitive Therapy and Mindfulness Tools in Reducing Depression and Anxiety: A Mixed Method Study." *Psychology* 5 (15): 1702–13. doi:10.4236/psych.2014.515178.

Andrejevic, Mark. 2013. *Infoglut: How Too Much Information Is Changing the Way We Think and Know*. New York: Routledge.

Atkins, David C., Mark Steyvers, Zac E. Imel, and Padhraic Smyth. 2014. "Scaling Up the Evaluation of Psychotherapy: Evaluating Motivational Interviewing Fidelity via Statistical Text Classification." *Implementation Science* 9:49. doi:10.1186/1748-5908-9-49.

Beck, Ulrich. 2009. *World at Risk*. New York: Polity.

Beer, David. 2016. *Metric Power*. London: Palgrave Macmillan.

Beller, Jonathan. 2006a. "Paying Attention." *Cabinet*, no. 24.

——. 2006b. *The Cinematic Mode of Production: Attention Economy and the Society of the Spectacle*. Lebanon, NH: Dartmouth College Press.

Bødker, Susanne. 2015. "Third-Wave HCI, 10 Years Later—Participation and Sharing." *Interactions*, September, 24–31.

Boehner, Kirsten, Rogério DePaula, Paul Dourish, and Phoebe Sengers. 2007. "How Emotion Is Made and Measured." *International Journal of Human-Computer Studies* 65:275–91. doi:10.1016/j.ijhcs.2006.11.016.

Boehner, Kirsten, Phoebe Sengers, and Simeon Warner. 2007. "Interfaces with the Ineffable: Meeting Aesthetic Experience on Its Own Terms." *ACM Transactions on Computer-Human Interaction* 15 (3): 12.

Boesel, Whitney Erin. 2013. "Your Feels as Free Labor: Emoticons, Emotional Cultures, and Facebook." *Cyborgology*, April 11. http://thesocietypages.org/cyborgology/2013/04/11/your-feels-as-free-labor-emoticons- emotional-cultures-and-facebook/.

Bogost, Ian. 2006. "Playing Politics: Videogames for Politics, Activism, and Advocacy." *First Monday*, September. http://firstmonday.org/article/view/1617/1532.

Bolter, Jay David, and Richard Grusin. 2000. *Remediation: Understanding New Media*. Cambridge, MA: MIT Press.

Brandt, Marisa. 2013. "From 'the Ultimate Display' to 'the Ultimate Skinner Box': Virtual Reality and the Future of Psychotherapy." In *Media Studies Futures*, edited by Kelly Gates, 1–22. London: Blackwell.

Brandt, Marisa, and Luke Stark. 2018. "Exploring Digital Interventions in Mental Health: A Roadmap." In *Interventions: Communication Research and Practice* (International Communication Association 2017 Theme Book), edited by Adrienne Shaw and D. Travers Scott, 167–82. Bern: Peter Lang.

Brennan, Teresa. 2004. *The Transmission of Affect*. Ithaca, NY: Cornell University Press.

Cadwalladr, Carole. 2018. "Revealed: 50 Million Facebook Profiles Harvested for Cambridge Analytica in Major Data Breach." *Guardian*, March 17. www.theguardian.com/news/2018/mar/17/cambridge-analytica-facebook-influence-us-election.

Card, Stuart K., Thomas P. Moran, and Allen Newell. 1983. *The Psychology of Human-Computer Interaction*. Hillsdale, NJ: Erlbaum.

Carmichael, Alexandra, and Robin Barooah. 2013. *Getting a Hold on Your Mood: A Quantified Self Approach*. Sebastopol, CA: O'Reilly Media.

Cheney-Lippold, John. 2011. "A New Algorithmic Identity." *Theory, Culture & Society* 28 (6): 164–81. doi:10.1177/0263276411424420.

Citron, Danielle Keats, and Frank Pasquale. 2014. "The Scored Society: Due Process for Automated Predictions." *Washington Law Review* 89:1–33.

Clynes, Manfred. 1989. *Sentics: The Touch of the Emotions*. London: Prism.
Clynes, Manfred, and Nathan S. Kline. 1960. "Cyborgs and Space." *Astronautics* 14 (9): 26–27, 74–76.
Coldewey, Devin. 2016. "Affectiva Partners with Giphy and Opens Its Emotion-Sensing API to Small Businesses." *TechCrunch*, September 13. https://techcrunch.com/2016/09/13/affectiva-partners-with-giphy-and-opens-its-emotion-sensing-api-to-small-businesses/.
Cook, Jordan. 2015. "Talkspace Therapy-by-Text Service Launches Asynchronous Audio, Video Messaging." *TechCrunch*, October 9. http://techcrunch.com/2015/10/29/talkspace-therapy-by-text-service-launches-asynchronous-audio-video-messaging/.
Crawford, Kate, Jessa Lingel, and Tero Karppi. 2015. "Our Metrics, Ourselves: A Hundred Years of Self-Tracking from the Weight Scale to the Wrist Wearable Device." *European Journal of Cultural Studies* 18 (4–5): 479–96. doi:10.1177/1367549415584857.
Cristian, Brian. 2012. "Test Everything: Notes on the a/B Revolution." *Wired*, May 9. www.wired.com/2012/05/test-everything/.
Damasio, Antonio. 1994. *Descartes' Error: Emotion, Reason, and the Human Brain*. New York: Putnam.
Darwin, Charles. [1872] 2009. *The Expression of the Emotions in Man and Animals*. 4th ed. New York: Oxford University Press.
Deleuze, Gilles. 1990. "Postscript on Control Societies." In *Negotiations, 1972–1990*, translated by Martin Joughin, 177–82. New York: Columbia University Press.
Deleuze, Gilles, and Felix Guattari. 1987. *A Thousand Plateaus: Capitalism and Schizophrenia*. Translated by Brian Massumi. Minneapolis: University of Minnesota Press.
Demir, Erdem, Pieter M. A. Desmet, and Paul Hekkert. 2009. "Appraisal Patterns of Emotions in Human-Product Interaction." *International Journal of Design* 3 (2): 41–51.
DiSalvo, Carl. 2012. *Adversarial Design*. Cambridge, MA: MIT Press.
Dourish, Paul. 2004. "Social Computing." In *Where the Action Is: The Foundations of Embodied Interaction*, 55–97. Cambridge, MA: MIT Press.
Doyle, E., I. Stamouli, and M. Huggard. 2005. "Computer Anxiety, Self-Efficacy, Computer Experience: An Investigation Throughout a Computer Science Degree." Paper presented at the 35th ASEE/IEEE Frontiers in Education Conference, Indianapolis.
Dror, Otniel E. 1999a. "The Scientific Image of Emotion: Experience and Technologies of Inscription." *Configurations* 7 (3): 355–401.
——. 1999b. "The Affect of Experiment: The Turn to Emotions in Anglo-American Physiology, 1900–1940." *Isis* 90 (2): 205–37.
——. 2001. "Counting the Affects: Discoursing in Numbers." *Social Research* 68 (2): 357–78.
——. 2009. "Afterword: A Reflection on Feelings and the History of Science." *Isis* 100 (4): 848–51.
——. 2011. "Seeing the Blush: Feeling Emotions." In *Histories of Scientific Observation*, edited by Lorraine Daston and Elizabeth Lunbeck, 326–48. Chicago: University of Chicago Press.
Dunne, Anthony, and Fiona Raby. 2001. *Design Noir: The Secret Life of Electronic Objects*. Berlin: August/Birkhäuser.
Ekman, Paul, and Wallace V. Friesen. 1971. "Constants across Cultures in the Face and Emotion." *Journal of Personality and Social Psychology* 17 (2): 124–29.
Ekman, Paul, and Erika L. Rosenberg. 2005. *What the Face Reveals: Basic and Applied Studies of Spontaneous Expression Using the Facial Action Coding System (FACS)*. 2nd ed. New York: Oxford University Press.
Essig, Todd. 2015. "Talkspace Argues with Talkspace: Conflicting Messages and Clinical Risk." *Forbes*, June 29. http://onforb.es/1JjyALz.
Facey, Marcia. 2010. "'Maintaining Talk' among Taxi Drivers: Accomplishing Health-Protective Behaviour in Precarious Workplaces." *Health & Place* 16 (6): 1259–67. doi:10.1016/j.healthplace.2010.08.014.
Feenberg, Andrew. 1992. "Subversive Rationalization: Technology, Power, and Democracy." *Inquiry* 35 (3–4): 301–22. doi:10.1080/00201749208602296.
Flanagan, Mary, and Helen Nissenbaum. 2014. *Values at Play in Digital Games*. Cambridge, MA: MIT Press.
Foucault, Michel. 1994. *The Order of Things: An Archaeology of the Human Sciences*. New York: Vintage.
Friedman, Batya, Peter H. Kahn, and Alan Borning. 2006. "Value Sensitive Design and Information Systems." In *Human-Computer Interaction in Management Information Systems: Foundations*, edited by B. Schneiderman, Ping Zhang, and D. Galletta, 348–72. New York: M.E. Sharpe.
Friedman, Batya, and Helen Nissenbaum. 1996. "Bias in Computer Systems." *ACM Transactions on Information Systems* 14 (3): 330–47.
Gershon, Ilana. 2015. "What Do We Talk about When We Talk about Animation." *Social Media + Society* 1 (1): 1–2. doi:10.1177/2056305115578143.

Giddens, Anthony. 1991. *Modernity and Self-Identity: Self and Society in the Late Modern Age*. Cambridge, MA: Polity.

Goel, Vindu. 2014. "As Data Overflows Online, Researchers Grapple with Ethics." *New York Times*, August 12. www.nytimes.com/2014/08/13/technology/the-boon-of-online-data-puts-social-science-in-a-quandary.html.

——. 2015. "Facebook to Test Emoji as Reaction Icons." *New York Times*, October 8. www.nytimes.com/2015/10/09/technology/facebook-to-test-emoji-as-reaction-icons.html?ref=technology.

Golumbia, David. 2009. *The Cultural Functions of Computation*. Cambridge, MA: Harvard University Press.

Gould, Deborah. 2010. "On Affect and Protest." In *Political Emotions*, edited by Janet Staiger, Ann Cvetkovich, and Ann Reynolds, 18–44. New York: Routledge.

Gregg, Melissa. 2015. "Getting Things Done: Productivity, Self-Management and the Order of Things." In *Networked Affect*, edited by Ken Hillis, Susanna Paasonen, and Michael Petit, 187–202. Cambridge, MA: MIT Press.

Gregg, Melissa, and Gregory J. Seigworth, eds. 2010. *The Affect Theory Reader*. Durham, NC: Duke University Press.

Grudin, Jonathan. 2012. "A Moving Target—The Evolution of Human-Computer Interaction." In *The Human-Computer Interaction Handbook*, edited by Julie A. Jacko, xxvii–lxi. Boca Raton, FL: CRC Press.

Haggerty, Kevin D., and Richard V. Ericson. 2000. "The Surveillant Assemblage." *British Journal of Sociology* 51 (4): 605–22. doi:10.1080/00071310020015280.

Haimson, Oliver L., and Anna Lauren Hoffmann. 2016. "Constructing and Enforcing 'Authentic' Identity Online: Facebook, Real Names, and Non-normative Identities." *First Monday* 21 (6). doi:10.5210/fm.v21i6.6791.

Hardt, Michael. 1999. "Affective Labor." *Boundary 2* 26 (2): 89–100.

Harré, Rom. 2009. "Emotions as Cognitive-Affective-Somatic Hybrids." *Emotion Review* 1 (4): 294–301. doi:10.1177/1754073909338304.

Heinssen, Robert K., Jr., Carol R. Glass, and Luanne A. Knight. 1987. "Assessing Computer Anxiety: Development and Validation of the Computer Anxiety Rating Scale." *Computers in Human Behavior* 3:49–59.

Hill, Kashmir. 2014. "Facebook Manipulated 689,003 Users' Emotions for Science." *Forbes*, June 28. www.forbes.com/sites/kashmirhill/2014/06/28/facebook-manipulated-689003-users-emotions-for-science/.

Hillis, Ken, Susanna Paasonen, and Michael Petit, eds. 2015. *Networked Affect*. Cambridge, MA: MIT Press.

Hochschild, Arlie Russell. 2003a. *The Commercialization of Intimate Life*. Berkeley: University of California Press.

——. 2003b. *The Managed Heart: Commercialization of Human Feeling*. 2nd ed. Berkeley: University of California Press.

——. 2012. *The Outsourced Self: Intimate Life in Market Times*. New York: Metropolitan Books.

Hoffmann, Anna Lauren, Nicholas Proferes, and Michael Zimmer. 2017. " 'Making the World More Open and Connected': Mark Zuckerberg and the Discursive Construction of Facebook and Its Users." *New Media & Society* 20 (1): 199–218.

Holland, S. P., M. Ochoa, and K. W. Tompkins. 2014. "On the Visceral." *GLQ* 20 (4): 391–406. doi:10.1215/10642684-2721339.

Höök, Kristina, Katherine Isbister, Steve Westerman, Peter Gardner, Ed Sutherland, Asimina Vasalou, Petra Sundström, Joseph "Jofish" Kaye, and Jarmo Laaksolahti. 2010. "Evaluation of Affective Interactive Applications." In *Emotion-Oriented Systems—The Humaine Handbook*, edited by Paolo Petta, Catherine Pelachaud, and Roddy Cowie, 687–703. Berlin: Springer. doi:10.1007/978-3-642-15184-2_36.

Höök, Kristina, Anna Ståhl, Martin Jonsson, Johanna Mercurio, Anna Karlsson, and Eva-Carin Banka Johnson. 2015. "Somaesthetic Design." *Interactions* 22 (4): 26–33.

Hu, Xia, Jiliang Tang, Huiji Gao, and Huan Liu. 2013. "Unsupervised Sentiment Analysis with Emotional Signals." In *Proceedings of the 22nd International Conference on World Wide Web*, 607–18. New York: ACM.

Illouz, Eva. 2007. *Cold Intimacies: The Making of Emotional Capitalism*. Cambridge: Polity.

——. 2008. *Saving the Modern Soul*. Berkeley: University of California Press.

Irani, Lilly. 2015. "The Cultural Work of Microwork." *New Media & Society* 17 (5): 720–39. doi:10.1177/1461444813511926.

Isbister, Katherine, Kristina Höök, Michael Sharp, and Jarmo Laaksolahti. 2006. "The Sensual Evaluation Instrument: Developing an Affective Evaluation Tool." In *Proceedings of the SIGCHI Conference on Human Factors in Computing Systems*, 1163–72. New York: ACM.

James, William. 2003. "From *What Is an Emotion?*" In *What Is an Emotion? Classic and Contemporary Readings*, 65–76. New York: Oxford University Press.

James, William, and C. G. Lange. 1922. *The Emotions*. Baltimore: Williams & Wilkins.

Johnson, Deborah G. 2007. "Ethics and Technology 'in the Making': An Essay on the Challenge of Nanoethics." *Nanoethics* 1 (1): 21–30. doi:10.1007/s11569-007-0006-7.

Juul, Jesper. 2010a. "A Casual Revolution." In *A Casual Revolution: Reinventing Video Games and Their Players*, 1–24. Cambridge, MA: MIT Press.

——. 2010b. "Social Meaning and Social Goals." In *A Casual Revolution: Reinventing Video Games and Their Players*, 121–28. Cambridge, MA: MIT Press.

Kagan, Jerome. 2007. *What Is Emotion? History, Measures, and Meanings*. New Haven, CT: Yale University Press.

Kahneman, Daniel. 2013. *Thinking, Fast and Slow*. New York: Farrar, Straus and Giroux.

Katz, Barry M. 2015. *Make It New: The History of Silicon Valley Design*. Cambridge, MA: MIT Press.

Kerr, Ian, and Jena McGill. 2007. "Emanations, Snoop Dogs and Reasonable Expectations of Privacy." *Criminal Law Quarterly* 52 (3): 392–431.

Kline, R. 2009. "Where Are the Cyborgs in Cybernetics?" *Social Studies of Science* 39 (3): 331–62. doi:10.1177/0306312708101046.

Konings, Martijn. 2015. *The Emotional Logic of Capitalism*. Palo Alto, CA: Stanford University Press.

Konnikova, Maria. 2013. "How Facebook Makes Us Unhappy." *New Yorker*, September 10. www.newyorker.com/online/blogs/elements/2013/09/the-real-reason-facebook-makes-us-unhappy.html.

Kramer, A. D. I., J. E. Guillory, and J. T. Hancock. 2014. "Experimental Evidence of Massive-Scale Emotional Contagion through Social Networks." *Proceedings of the National Academy of Sciences* 111 (24): 8788–90. doi:10.1073/pnas.1320040111.

Latour, Bruno. 2004. "Why Has Critique Run Out of Steam? From Matters of Fact to Matters of Concern." *Critical Inquiry* 30:225–48.

——. 2005. *Reassembling the Social: An Introduction to Actor-Network Theory*. New York: Oxford University Press.

Leahu, Lucian, Steve Schwenk, and Phoebe Sengers. 2008. "Subjective Objectivity: Negotiating Emotional Meaning." In *Proceedings of the 7th ACM Conference on Designing Interactive Systems*, 425–34. New York: ACM.

Lee, Min Kyung, Daniel Kusbit, Evan Metsky, and Laura Dabbish. 2015. "Working with Machines." In *Proceedings of the 33rd Annual ACM Conference on Human Factors in Computing Systems*, 1603–12. New York: ACM. doi:10.1145/2702123.2702548.

Levin, Sam. 2017. "Facebook Told Advertisers It Can Identify Teens Feeling 'Insecure' and 'Worthless.'" *Guardian*, May 1. www.theguardian.com/technology/2017/may/01/facebook-advertising-data-insecure-teens.

Leys, Ruth. 2011. "The Turn to Affect: A Critique." *Critical Inquiry* 37 (3): 434–72.

——. 2017. *The Ascent of Affect*. Chicago: University of Chicago Press.

Littlefield, Melissa, and Jenell Johnson. 2012. *The Neuroscientific Turn: Transdisciplinarity in the Age of the Brain*. Ann Arbor: University of Michigan Press.

Liu, Lydia H. 2011. *The Freudian Robot: Digital Media and the Future of the Unconscious*. Chicago: University of Chicago Press.

Lively, Kathryn J., and David R. Heise. 2004. "Sociological Realms of Emotional Experience." *American Journal of Sociology* 109 (5): 1109–36. doi:10.1086/381915.

Malin, Brenton. 2014. *Feeling Mediated: A History of Media Technology and Emotion in America*. New York: New York University Press.

Massumi, Brian. 2002. *Parables for the Virtual: Movement, Affect, Sensation*. Durham, NC: Duke University Press.

McGrath, Charles. 2008. "The King of Visceral Design." *New York Times*, April 27.

Mishna, Faye, Sophia Fantus, and Lauren B. McInroy. 2016. "Informal Use of Information and Communication Technology: Adjunct to Traditional Face-to-Face Social Work Practice." *Clinical Social Work Journal* 45 (1): 1–7. doi:10.1007/s10615-016-0576-3.

Mizroch, Amir. 2014. "App Tells You How You Feel." *Wall Street Journal*, March 10. www.wsj.com/news/articles/SB10001424052702303824204579421242295627138?mod=WSJ_business_whatsNews&mg=reno64-wsj.

Moran, Thomas P., and Stuart K. Card. 1982. "Applying Cognitive Psychology to Computer Systems: A Graduate Seminar in Psychology." *ACM SIGCSE Bulletin* 14 (3): 34–37.

Nafus, Dawn, and Jamie Sherman. 2014. "This One Does Not Go Up to 11: The Quantified Self Movement as an Alternative Big Data Practice." *International Journal of Communication* 8:1784–94.

Nagel, Thomas. 1979. "The Fragmentation of Value." In *Mortal Questions*, 128–41. Cambridge: Cambridge University Press.

Newell, Allen, and Stuart K. Card. 1985. "The Prospects for Psychological Science in Human-Computer Interaction." *Human-Computer Interaction* 1:209–42.

Nissenbaum, Helen. 2015. "Respecting Context to Protect Privacy: Why Meaning Matters." *Science and Engineering Ethics* 24 (3): 851–52. doi:10.1007/s11948-015-9674-9.

Noble, Safiya Umoja. 2018. *Algorithms of Oppression: How Search Engines Reinforce Racism*. New York: New York University Press.

Norman, Donald A. 1989. "The Psychopathology of Everyday Things." In *The Design of Everyday Things*, 1–33. New York: Currency and Doubleday.

——. 2005. *Emotional Design: Why We Love (or Hate) Everyday Things*. New York: Basic Books.

Oremus, Will. 2013. "Facebook's Cute New Emoticons Are a Fiendish Plot. Don't Fall for It." *Slate*, April 10. www.slate.com/blogs/future_tense/2013/04/10/facebook_emoji_status_update_emoticons_are_bad_for_privacy_good_for_advertisers.html.

Orr, Jackie. 2006. *Panic Diaries: A Genealogy of Panic Disorder*. Durham, NC: Duke University Press.

Papacharissi, Zizi. 2014. *Affective Publics: Sentiment, Technology, and Politics*. Oxford: Oxford University Press.

Parisi, David. 2018. *Archaeologies of Touch: Interfacing with Haptics from Electricity to Computing*. Minneapolis: University of Minnesota Press.

Pasquale, Frank. 2015. "Privacy, Autonomy, and Internet Platforms." In *Privacy in the Modern Age the Search for Solutions*, edited by Marc Rotenberg, Julia Horwitz, and Jeramie Scott, 165–73. New York: New Press.

Picard, Rosalind W. 2000. *Affective Computing*. Cambridge, MA: MIT Press.

Picard, Rosalind W., and Shaundra Bryant Daily. 2008. "Evaluating Affective Interactions: Alternatives to Asking What Users Feel." www.media.mit.edu/publications/evaluating-affective-interactions-alternatives-to-asking-what-users-feel-2/.

Picard, Rosalind W., and Jonathan Klein. 2002. "Computers That Recognise and Respond to User Emotion: Theoretical and Practical Implications." *Interacting with Computers* 14:141–69.

Picard, Rosalind W., and Jocelyn Scheirer. 2001. "The Galvactivator: A Glove That Senses and Communicates Skin Conductivity." Paper presented at the 9th International Conference on Human-Computer Interaction, New Orleans.

Pickering, Andrew. 2010. *The Cybernetic Brain*. Chicago: University of Chicago Press.

Poster, Winifred R. 2011. "Emotion Detectors, Answering Machines, and E-unions: Multi-surveillances in the Global Interactive Service Industry." *American Behavioral Scientist* 55 (7): 868–901. doi:10.1177/0002764211407833.

——. 2013. "Hidden Sides of the Credit Economy: Emotions, Outsourcing, and Indian Call Centers." *International Journal of Comparative Sociology* 54 (3): 205–27. doi:10.1177/0020715213501823.

Powell, Anne L. 2013. "Computer Anxiety: Comparison of Research from the 1990s and 2000s." *Computers in Human Behavior* 29 (6): 2337–81. doi:10.1016/j.chb.2013.05.012.

Pullin, Graham. 2011. *Design Meets Disability*. Cambridge, MA: MIT Press.

Reddy, William M. 2001. *The Navigation of Feeling: A Framework for the History of Emotions*. Cambridge: Cambridge University Press.

Reeves, Byron, and Clifford Nass. 2003. *The Media Equation: How People Treat Computers, Television, and New Media Like Real People and Places*. Palo Alto, CA: Center for the Study of Language and Inference.

Richards, Neil M., and Jonathan H. King. 2013. "Three Paradoxes of Big Data." *Stanford Law Review Online* 66:41–46.

Robles, Erica, Clifford Nass, and Adam Kahn. 2009. "The Social Life of Information Displays: How Screens Shape Psychological Responses in Social Contexts." *Human-Computer Interaction* 24 (1): 48–78. doi:10.1080/07370020902739320.

Rose, Nikolas. 1988. "Calculable Minds and Manageable Individuals." *History of the Human Sciences* 1 (2): 179–200.

——. 1996. *Inventing Our Selves: Psychology, Power, and Personhood*. Cambridge: Cambridge University Press.

——. 2013. "The Human Sciences in a Biological Age." *Theory, Culture & Society* 30 (1): 3–34. doi:10.1177/0263276412456569.

Rosenberg, Matthew, Nicholas Confessore, and Carole Cadwalladr. 2018. "How Trump Consultants Exploited the Facebook Data of Millions." *New York Times*, March 17. www.nytimes.com/2018/03/17/us/politics/cambridge-analytica-trump-campaign.html.

Rosenblat, Alex, and Luke Stark. 2016. "Algorithmic Labor and Information Asymmetries: A Case Study of Uber's Drivers." *International Journal of Communication* 10:3758–84.

Scheirer, Jocelyn, Raul Fernandez, Jonathan Klein, and Rosalind W. Picard. 2002. "Frustrating the User on Purpose: A Step toward Building an Affective Computer." *Interacting with Computers* 14:93–118.

Scholz, Trebor, ed. 2013. *Digital Labor: The Internet as Playground and Factory*. New York: Routledge.

Schüll, Natasha Dow. 2012. *Addiction by Design: Machine Gambling in Los Vegas*. Princeton, NJ: Princeton University Press.

——. 2016. "Data for Life: Wearable Technology and the Design of Self-Care." *BioSocieties* 11 (3): 1–17. doi:10.1057/biosoc.2015.47.

Selinger, E., and W. Hartzog. 2015. "Facebook's Emotional Contagion Study and the Ethical Problem of Co-opted Identity in Mediated Environments Where Users Lack Control." *Research Ethics* 12:35–43. doi:10.1177/1747016115579531.

Sengers, Phoebe, Kirsten Boehner, Shay David, and Joseph "Jofish" Kaye. 2005. "Reflective Design." In *Proceedings of the 4th Decennial Conference on Critical Computing: Between Sense and Sensibility*, 49–58. New York: ACM. doi:10.1145/1094562.1094569.

Sengers, Phoebe, Kirsten Boehner, Michael Mateas, and Geri Gay. 2008. "The Disenchantment of Affect." *Personal and Ubiquitous Computing* 12 (5): 347–58. doi:10.1007/s00779-007-0161-4.

Shank, Daniel B. 2010. "An Affect Control Theory of Technology." *Current Research in Social Psychology* 15 (10): 1–13.

——. 2014. "Technology and Emotions." In *Handbook of the Sociology of Emotions*, vol. 2, edited by J. E. Stets and J. H. Turner, 511–28. Dordrecht: Springer. doi:10.1007/978-94-017-9130-4_24.

Shapin, Steven. 2012. "The Sciences of Subjectivity." *Social Studies of Science* 42 (2): 170–84. doi:10.1177/0306312711435375.

Silvio, Teri. 2010. "Animation: The New Performance." *Journal of Linguistic Anthropology* 20 (2): 422–38. doi:10.1111/j.1548-1395.2010.01078.x.

Snyder, Peter J., Rebecca Kaufman, John Harrison, and Paul Maruff. 2010. "Charles Darwin's Emotional Expression 'Experiment' and His Contribution to Modern Neuropharmacology." *Journal of the History of the Neurosciences* 19 (2): 158–70. doi:10.1080/09647040903506679.

Solomon, Robert C. 2003. *What Is an Emotion? Classic and Contemporary Readings*. New York: Oxford University Press.

Stark, Luke. 2014. "Come on Feel the Data (and Smell It)." *Atlantic*, May 19. www.theatlantic.com/technology/archive/2014/05/data-visceralization/370899/.

——. 2016. "That Signal Feeling: Emotion and Interaction Design from Social Media to the 'Anxious Seat.'" Doctoral dissertation, New York University.

——. 2018a. "Algorithmic Psychometrics and the Scalable Subject." *Social Studies of Science* 48:204–31.

——. 2018b. "Facial Recognition, Emotion and Race in Animated Social Media." *First Monday* 23 (9). doi:10.5210/fm.v23i9.9406.

——. 2018c. "Silicon Valley Wants to Improve Your 'Digital Well-Being'—and Collect More of Your Personal Data along the Way." *Boston Globe*, July 24. https://www.bostonglobe.com/magazine/2018/07/24/silicon-valley-wants-improve-your-digital-well-being-and-collect-more-your-personal-data-along-way/cdw24TGja17KqhfAVMKAkN/story.html.

Stark, Luke, and Kate Crawford. 2015. "The Conservatism of Emoji: Work, Affect, and Communication." *Social Media + Society* 1 (2). doi:10.1177/2056305115604853.

Tettegah, Sharon Y., and Safiya Umoja Noble, eds. 2016. *Emotions, Technology, and Design*. London: Elsevier.

Thaler, Richard H. 1980. "Towards a Positive Theory of Consumer Choice." *Journal of Economic Behavior and Organization* 1:39–60.

Thaler, Richard H., and Cass R. Sunstein. 2008. *Nudge*. New Haven, CT: Yale University Press.

Turkle, Sherry. 2004. *The Second Self: Computers and the Human Spirit*. New York: Simon & Schuster.

Wassmann, Claudia. 2010. "Reflections on the 'Body Loop': Carl Georg Lange's Theory of Emotion." *Cognition & Emotion* 24 (6): 974–90. doi:10.1080/02699930903052744.

Wilson, Elizabeth A. 2010. *Affect and Artificial Intelligence*. Seattle: University of Washington Press.

Wing, Jeannette M. 2006. "Computational Thinking." *Communications of the ACM* 49 (3): 33–35.
Winner, Langdon. 1988. "Do Artifacts Have Politics?" In *The Whale and the Reactor*, 19–39. Chicago: University of Chicago Press.
Winter, Sarah. 2009. "Darwin's Saussure: Biosemiotics and Race in Expression." *Representations* 107 (1): 128–61. doi:10.1525/rep.2009.107.1.128.
Wolf, Gary. 2010. "The Data-Driven Life." *New York Times*, April 28. www.nytimes.com/2010/05/02/magazine/02self-measurement-t.html.
Zuboff, Shoshana. 2015. "Big Other: Surveillance Capitalism and the Prospects of an Information Civilization." *Journal of Information Technology* 30 (1): 75–89. doi:10.1057/jit.2015.5.

The Ambiguous Boundaries of Computer Source Code and Some of Its Political Consequences

Stéphane Couture

This chapter analyzes how the notion of source code remains unstable and contested, and how its different definitions and statuses may be associated with certain types of activities that are valued differently. The notions of computer code and source code are at the core of all digital infrastructures and technologies, and are of increasing interest in academic research. In his famous book *Code: And Other Laws of Cyberspace*, Lessig (1999) puts forth the idea that computer code acts as a kind of law, and that those who write the code—the programmers—are in some ways the Internet's legislators. This approach has given rise to the well-known metaphor "code is law," which has allowed fruitful reflection on the governance capacity of technological artifacts and digital infrastructures (DeNardis 2014). The notion of source code is also at the core of the free and open source movement, whose normative and organizational model depends on open access to source code. The concept of "open source" places the emphasis on the openness of source code. The importance of source code has moreover been put forth in several of the declarations of the movement's actors, who insist for instance on considering access to source code as a form of free speech (Byfield 2006; Coleman 2009). Scholarly literature has also addressed code and source code, in particular through the study of hacker cultures and free and open source movements.

In her book *Coding Freedom*, Coleman (2012) analyzes the rise and political significance of free and open source software, by mostly focusing on hackers' ethics and coding aesthetics, while noting that access to source code has been rearticulated as a matter of free speech (Coleman 2009). Works tied to the emerging fields of software studies (Fuller 2008) and critical code studies (Marino 2010) have focused more directly on the question of software and computer code as the material, or the logic, of new media. These works (in particular those in critical code studies) are usually characterized by a rather formal or hermeneutic approach that focuses on interpreting the meaning of specific pieces of code.[1] Closer to an STS perspective, Mackenzie's book *Cutting Code: Software and Sociality* proposes to study software—and code—as an object that is shaped, articulated, and felt in practice and by different situations (Mackenzie 2006, 7).

Despite this proliferation of discourses and scholarly work addressing directly or indirectly software, code, and source code, these concepts are often taken for granted and considered stable and unproblematic. In particular, very little has been done to critically and empirically assess how the definition of source code may change over time, space, and situation. Using an STS perspective, this chapter addresses the following questions: What are some of the diverse definitions given to source code by actors? Can the way source code is defined have repercussions on the type of activity that is or isn't valued? In particular, which computerized activities are considered—or not—as "coding"? How are the definitions of source code connected to dynamics of visibility and invisibility and to the valuing of work?

My argument is twofold. First, I show that "source code" is not a stabilized category but is rather ambiguously engaged with by actors in a given software development project. These ambiguities relate foremost to the *forms* that source code takes (whether it is a text file or not, if it is written in a particular language), and to the *position* of source code in a particular project. Second, I argue that the way in which source code is defined and considered as such may have consequences in terms of valuing or giving more visibility to one activity over others, thus contributing to the distinct status of actors involved. As I will also explain, this has gendered implications, as women's work in software development tends to concentrate on activities less perceived as "coding" and therefore less visible and valued both inside a given project as well as in the broader developer community (Wajcman 2004; Ghosh et al. 2002; Lin 2006; Haralanova 2010). Throughout this chapter, I will explore different definitions given to source code by actors involved in software development, and also analyze the different forms that source code takes in these projects. In each of the analyses, I will show that the definition of source code and coding is correlated with the form source code takes and its position in the project.

Methodology: Unfolding Source Code, Following the Actors

The analysis is based on a study of source code, which took as fieldwork two free/open source software projects I studied during my doctoral thesis research. One of the first surprises during my research was to notice that there was no consensus around the definition of "source code," and that the very relevance of this notion was contested. I had indeed, at the beginning of the study, a rather clear idea of what constituted source code, but it quickly became obvious that this notion was frankly problematic for some of the people whom I met. Right before the first interview, during an informal discussion, a participant in my research stated that the notion of "source code" did not make sense for the software project I was studying, since there was no notion of *compilation* and *object code*. The classic definition of source code indeed involves the notion of compilation, an operation of translation from a programming language that is humanly comprehensible, to the binary language of the machine. Compilation thus involves translating from source code to object code (also called executable code). However, in the case of PHP—the language in which the projects I studied are programmed—the code is instead "directly" executed and does not require a compilation phase. In the case of written code in the PHP language, my interlocutor preferred to use "script code" instead of "source code."

This episode disturbed me a little, and I worried that I had chosen the wrong fieldwork to address the question of source code (which was formally the object of my study). However, a more detailed analysis of the documents and exchanges between actors showed that the notion of source code was effectively used in the context of the projects, being referred to as such on the project website and on mailing lists. Clearly then, the notion of "source code" had an empirical reality in the studied projects. Nonetheless, this first informal discussion brought me to question myself regarding the definition of source code, to set aside my own conceptions of what this artifact was, and to look at diverging ways to conceive source code.

A quick look at some of the available definitions helps us understand the ambiguity of defining source code. Let's take as an example this definition of source code, given by Krysia and Grzesiek, in the book *Software Studies*: "Source code (usually referred to as simply 'source' or 'code') is the uncompiled, non-executable code of a computer program stored in source files. It is a set of human readable computer commands written in higher level programming languages" (Krysia and Grzesiek 2008, 237). Compare it to this other definition of source code, which is given as a prelude to the GNU General Public License (GPL), used in several free and open source software projects: "The source code for a work means the preferred form of the work for making modifications to it."[2] The spectrum of these definitions is evidently very large. On one hand, Krysia and Grzesiek's definition is very specific, and consists in characterizing source code as something *uncompiled, written, using programming languages, and for making* a computer program. On the other hand, the GPL's definition as the "preferred form of the work for making modifications" doesn't assume that source code is actually in a written form or even that it is used to make software. Following the GPL's definition, an image could have a source code, which would be the Photoshop or GIMP file used to make it or edit it. In the same way, the source code of a PDF document would be the Word or LibreOffice document used to write it. This definition of source code thus covers a wide range of artifacts and activities that are not strictly limited to computer programming.

The analysis is based on the study of two free and open source projects mainly written in the PHP web programming language and used to build interactive websites. The first project, *SPIP*, is a web content management system (CMS) that was initiated in 2001 with the aim to defend both the independent web and freedom of expression on the Internet. SPIP—which was also studied by Demazière, Horn, and Zune (2007a)[3]—is mostly used by NGOs and small associations, but also by some government institutions, especially in France. The second project, *Symfony*, is best described as a web framework, that is, as a set of separate software components that can be assembled to build web applications and complex websites. Symfony is used in the functioning of several large websites such as Daily Motion and Delicious. Both projects could be considered as "midsized," reaching a community of a few hundred people interacting on mailing lists. The two projects diverge by their values, and their linguistic and social groundings. Although both projects were created in France, Symfony's working language is English and thus reaches an international community. On the other hand, the development of SPIP (including coding) is mostly done in French, which restricts its public to France. SPIP is also grounded in activists and "non-programmer communities," while Symfony is much more oriented toward commercial uses and professional programmers. As I will show, this has some consequences for the ways people characterize some artifacts as "source code."

The study used an ethnographic approach and was mostly conducted between 2009 and 2010. It involved studying source code, analyzing online discussions, and attending physical meetings of each of the communities. I also conducted in-depth interviews with 20 people chosen to capture the diversity of engagements within the projects. For instance, I interviewed actors in leadership roles, but also—as I will address later—some people who do not identify as "coders" but who nevertheless participate in meet-ups and discussions. Ten people, including one woman, were interviewed for each of the projects.[4] During these interviews, I systematically asked participants about their definition of source code. I present here some of their responses to show the definitional ambiguities related to this notion.

Throughout this analysis, I will take Krysia and Grzesiek's (2008) definition of source code as a starting point to explore more nuanced and problematic definitions of this notion. Rather than providing an a priori definition of source code, I follow the empirical stance within science and technology studies and look at how this definition is socially constructed and/or contested among actors and how, in turn, these definitions can "do politics" (Suchman 1993; Winner 1980) by valuing certain kinds of activities over others.

The goal of this chapter is not to develop an a priori definition of source code that applies at all times. Instead, I go along with Kerasidou's proposal (in this handbook) to shift from questions of reference—what source code is, in my case—to relational configurations—how source code is defined and apprehended in particular settings and practices. Methodologically, my approach could generally be described as "unfolding source code and letting the actors speak." It follows actor-network theory's injunction to focus on the work of actors as they "make the social." In the context of controversy analysis, Latour recommends following the actors and relying on the categories they themselves mobilize: "The task of defining and ordering the social should be left to the actors themselves, not taken up by the analyst" (Latour 2005, 23). Suchman also insists upon the fact that research done about sociomaterial objects must particularly focus on two aspects: work on the demarcation and breakdown of the network through which the entities themselves have their boundaries set, as well as the localization of said entities within the context of temporal and spatial relations (Suchman 2007, 283). My approach finally participates in Chun's project to escape the "sourcery of source code" (Chun 2008). Chun notes that many research projects, especially in the field of software studies, tend to reduce computer code to source code, and obfuscate the "vicissitudes of execution." To break free of this "sourcery," Chun insists that we should "interrogate, rather than venerate" the logic of software. By problematizing the multiple definitions given by actors to source code and by analyzing these definitions as they relate to power and authority, I hope to contribute to this critical analysis of software and code.

Source Code: An Unstable Notion with Fuzzy Boundaries

Source Code as a Text?

An important ambiguity regarding the definition of source code concerns its presumed *written* nature. As cited before, Krysia and Grzesiek (2008) note that source code is the "set of human readable computer commands *written* in higher level programming languages." Most of the people I met for interviews effectively described source code as a *text*, something that is *written*. For instance, one study participant I interviewed told me: "Source code, it's really a listing, a text" (sf06).[5] However, when interviewed more formally on their definition of source code, study participants added some nuances, for example by noting that within source code, there is documentation, and *several things behind it*:

> Source code, it's really a listing, a text.
> Q: It's only text? Source code is text?
> That's very summarized. Source code, of course, isn't only text, but in the context of this question, of course, if I had to explain to a novice what source code is, I would say that yes, source code is a text, that corresponds to a certain number of instructions. Knowing that after source code, there are several things. There is documentation, there are several things behind it. (sf06)

Other people I met questioned the written nature of source code. A research participant recounted his past experience in a mechanical engineering company where the specification of software was always made in a graphic form, using block diagrams: "For a project like this one, we can consider that the block diagrams used to specify, and that were executed in a given environment, are also a form of source code" (spip11). In this case, source code takes the form of block diagrams, that is, a graphic form, rather than a written one, that simulates an information system. Once the simulation is devised, the model in block diagrams is "manually" translated—that is, by humans—in another computer language that will be more efficient for production usage. An example of a block diagram can be found in figure 1. This description of block diagrams places the emphasis on two crucial points. First, source code is not always a text and can take a graphical form. Second, the translation operation is not always automated and may possibly be done by a human instead of a computer.

For this participant, source code is the *reference code* of which other code ensues, the code that "we will knead" in order to make software evolve: "Actually, I think source code is also the one upon which we will work, which is malleable, and which we will make evolve, which we will knead in order to make the final software evolve. . . . It's the one that is used as reference to produce other code that will serve, and the one that is the basis, well, that we make evolve in order to make software evolve. And of which other code ensues actually" (spip11). This way of defining source code as the code "that we will knead" and that is used to "produce other code" is undoubtedly one of the more general and formal definitions of source code that has been given to me. This definition is actually quite similar to the one made explicit by the GNU General Public License as "the preferred form of the work for making modifications to it."

Source Code as Specification

Another SPIP participant also criticized this presumed characteristic of source code as text. He however went further by stating that source code lies in artifacts that *specify* the functioning of an information system or software, whether it is executed or not: "Source code, true source code, is your specification" (spip01). For

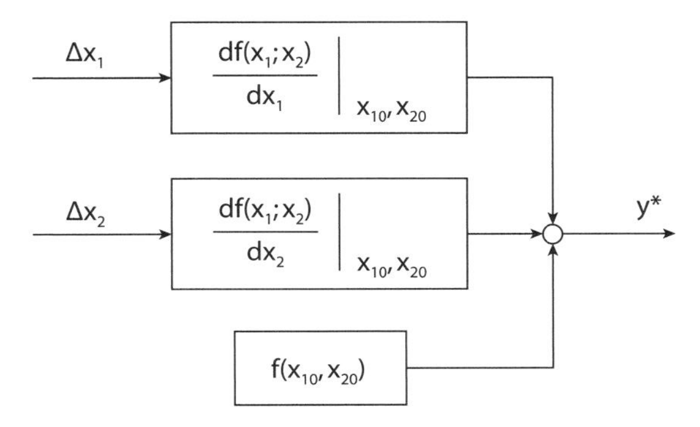

FIGURE 1: Example of a bloc diagram. Source: https://de.wikipedia.org/wiki/Blockdiagramm.

example, in the case of the making of a website or of open source software, source code would be the graphic models or the functional prototypes describing the behavior: "These functional prototypes, once they are conceived, and once she [a participant in SPIP] shows us on the screen, with some kind of model, that if we click here, it does this, if we click there, it does that, etc. . . . The programming source, it's in the functional model. So, this is what has the status of source code" (spip01). Thus, functional prototypes in the form of graphic models constitute a kind of specification for the functioning of software and, as such, act as source code. The same participant went further in this logic, stating that for the case of the Internet as a whole, source code would correspond to *requests for comments* (RFCs):

> Or when it comes to the Internet, it's mostly RFCs.
> Q: RFCs would be source code?
> Well, the Internet source code. Well yeah, that's clear. Because you can purge everything and replace it with something else, as long as they respect the RFCs, that they respect the norms. (spip01)

Let's recall that RFCs, mentioned in the interview extract, are documents that describe some aspects of the Internet. Started by experts and expanded upon jointly with the community, some of those RFCs, but not all, constitute the standards of the Internet. The idea that RFCs are the Internet's source code refers to a much broader definition of source code. Source code is what constitutes the base specification of a software product, of a computer device, even of the Internet. Let's note moreover that this definition does not imply that said specification is directly understood by the machine, that it is executable. Source code is, according to this study participant, "the place where the precise definition of functionalities takes place . . . it's a specific thing that says that I have this button, that form, etc. And when I click on it, it provokes this element, it cues in that algorithm" (spip01). This definition of source code—especially when referred to RFCs—is of course provocative and stretches our common understanding of the notion. It nevertheless shows the extent to which we could approach source code. Indeed, as I will explain in detail later, the goal of this participant was to increase my awareness regarding the social and even gendered consequences of having too narrow of a definition of source code that excludes different forms of participation in the making of software, such as prototyping or graphic design.

This first exploration of the definitions given by some of the participants shows that defining source code is not as simple as it seems to be. Although source code is usually characterized as a text comprising programming instructions—echoing Krysia and Grzesiek (2008) definition—the metaphors given by study participants show a much broader conception of source code that may even include broad specifications of the functioning of a computer system.

The Ambiguities around Forms and Status

Another way of looking at source code is to follow the artifact(s) empirically referred to by this term. For instance, some of the study participants referred to source code as "what is available for downloading." Thus, in the case of Symfony, a participant stated that "when we download Symfony, we download the sources of Symfony" (sf07). As I mentioned earlier, in the case of software or applications written in languages such as PHP, source code is directly executed by the computer, which explains why when one downloads Symfony for instance, one downloads its source code. I did actually download Symfony and SPIP from the projects' web pages, in the form of a .zip archive. The analysis of this archive allows one to grasp the complexity of the studied projects' source code, which comprises hundreds of files of different types and directories. Table 1 shows in numbers the content of these archives—that is, the source code of each of the projects.

These numbers show a definite complexity of the source code artifact, or rather, of all artifacts that make up source code. One may see that source code is composed of thousands of files of different types, and distributed in hundreds of distinct directories. In both cases, the majority of files bear a .php extension, which means that they are effectively *text files*, mainly written in the PHP language, that also contain comments that in theory have no impact on the functioning of the software (I will come back to this). Figure 2 shows an excerpt of one of those files that bears a .php extension in SPIP. This analysis also highlights certain ambiguities regarding source code's definition. Indeed, if text files (of the .php or .html kind) count for the majority of each of the archives' files, they don't make up its entirety. Within these archives are found different images (.gif, .png, .jpg), such as icons for hyperlinks, buttons, or the logo of each of the projects. Furthermore, about a quarter of the lines in PHP files are comments lines. It is worth noting that both of these components (images and comment lines) don't really correspond to the definition given by Krysia and Grzesiek (2008, 237) of source code as "a set of human readable computer commands written in higher level programming languages."

This analysis echoes Robles and Gonzalez-Barahona's (2004) remark that that source code, understood as the components used to produce an executable software version, encompasses more than the "classical" (their own word) conception of source code as text written in a programming language. These components include, for instance, technical documentation, interface specifications, translation modules, and multimedia files, namely images. Even more importantly, in relation to my analysis, Robles and Gonzalez-Barahona note that those files, which are *not* texts written in a programming language, take up more and more space in the case of end-user applications:

The concept of source code, understood as the source components used to obtain a binary, ready to execute version of a program, comprises currently more than source code written in a programming language. Specially when we move apart from systems-programming and enter the realm of end-user applications, we find source files with documentation, interface specifications, internationalization and localization modules, multimedia files, etc. All of them are source code in the sense that the developer works directly with them, and the application is built automatically using them as input. (Robles and Gonzalez-Barahona 2004, 149)

```
171 // Convertir dates a calendrier correct
172 // (exemple: 31 fevrier devient debut mars, 24h12 devient 00h12 du lendemain)
173
174 // http://doc.spip.org/@change_date_message
175 function change_date_message($id_message, $heures,$minutes,$mois, $jour, $annee,
    $heures_fin,$minutes_fin,$mois_fin, $jour_fin, $annee_fin)
176 {
177         $date = date("Y-m-d H:i:s", mktime($heures,$minutes,0,$mois, $jour, $annee));
178
179         $jour = journum($date);
180         $mois = mois($date);
181         $annee = annee($date);
182         $heures = heures($date);
183         $minutes = minutes($date);
184
185         // Verifier que la date de fin est bien posterieure au debut
186         $unix_debut = date("U", mktime($heures,$minutes,0,$mois, $jour, $annee));
187         $unix_fin = date("U", mktime($heures_fin,$minutes_fin,0,$mois_fin, $jour_fin, $annee_fin));
188         if ($unix_fin <= $unix_debut) {
189                 $jour_fin = $jour;
190                 $mois_fin = $mois;
```

FIGURE 2: PHP file excerpt of SPIP's source code. Source: SPIP project.

TABLE 1: Some Numbers That Show Source Code's Complexity*

Quantitative Characteristics	Symfony		SPIP	
Size	14,334 Mb		14,015 Mb	
Number of directories	788		99	
Number of files	2,924		1,305	
Number of files by type (extensions)	.php	2,104	.php	714
	.dat	331	.gif	187
	.yml	157	.png	178
	.xml	102	.html	146
	.png	68	.js	30
	None	30	.txt	17
	.mod	23	.css	15
	.sf	19	.jpg	7
	.txt	17	.xml	6
	.ini	8	.ttf	2
	Other	65	Other	3
Number of lines of computer commands, .php files	302,803		223,161	
Number of comments lines	110,773		23,179	

*The analysis was performed on the source code of Symfony version 1.4.10 and SPIP version 2.1.7 using *Simple Directory Analyser*, *LocMetrics*, and the Windows browser software.

The Status of Configuration Files and the SPIP Language

One important problematic aspect regarding the status of source code is the "configuration files" that in my two cases usually serve to specify the interface of a website. In the Symfony project, these configuration files bear the .yml extension and are written in YAML, a language mostly used to describe data structure. Table 1 shows that, in the case of Symfony, the download archive accounts for 157 files bearing the .yml extension, meaning that they are in YAML format. Is YAML considered source code by Symfony actors? The response of a study participant interviewed on this question is ambivalent. For some, YAML configuration files are not part of the source code, but are rather its *inputs*:

> Q: YAMLs. . . . That, would you consider to be part of source code? Because it does not include loops. . . .
>
> Ha! That's a good question. . . . For me, it configures the source code, but it is not a part of the code itself. Because in fact, those files . . . are read, and then, are transferred in a PHP file, in order to be re-used. . . . They're the *inputs*, ultimately, of the source code. (sf05)

In SPIP, files that are used to configure the appearance of a website are written in a homemade language invented by members of the community and called the *SPIP language* or *skeleton language* (because it is used to build the "skeleton" of a website). This language is a sort of extension of HTML, which includes some additional instructions for interacting with the database. Interestingly, this homemade language has been designed in French, contrary to most programming languages, which are written in English.[6] Figure 3 is an example of a file written in this language.

The SPIP language is seen by actors as an important feature of their project. This is because it is an intermediary between PHP and HTML, which makes it "more accessible than PHP" according to one participant (spip01), and thus allows for broader participation from people who do not have a strong background in computer programming (or are not fluent in English). However, the specificity of the SPIP language creates ambiguity concerning its status as source code. It is seen as more accessible to people, but at the same time—and maybe because of this—it is not always considered as "real code," which is seen as the domain of technologist elites. A participant described the SPIP language as a feature that allows people *who do not code to act as if they were coding*, a statement that emphasizes the ambiguous status of these files as code:

> In SPIP, what is extraordinary is, when you don't code, you will succeed in doing things, *as if you were coding*. . . . Everything was done so that even if you do not understand, you'll be able to get what you want. Because you have a simplified language of loops that can do what you do with PHP, but in simpler. There is a layer of abstraction, in fact, which comes to simplify everything. So, suddenly it becomes accessible, and it is no longer an elitist domain where we must have three years of hard code to understand how to do something. (spip08)

The previous quote demonstrates well the ambivalence toward what is or not considered as (source) code. Files written in the SPIP language seem to fit nicely with Krysia and Grzesiek's definition of source code as "human readable computer

```
1  [(#HTTP_HEADER{Content-type: text/xml[; charset=(#CHARSET)]})]<?xml
2  version="1.0"[ encoding="(#CHARSET)"]?>
3  <rss version="2.0" [(#REM) rss 2.0.9)]
4          xmlns:dc="http://purl.org/dc/elements/1.1/"
5          xmlns:content="http://purl.org/rss/1.0/modules/content/"
6  >
7
8  <channel[ xml:lang="(#LANG)"]>
9          <title>[(#NOM_SITE_SPIP|texte_backend)]</title>
10         <link>#URL_SITE_SPIP/</link>
11         [<description>(#DESCRIPTIF_SITE_SPIP|supprimer_tags|texte_backend)</description>]
12         <language>#LANG</language>
13         <generator>SPIP - www.spip.net</generator>
14
15 [       <image>
16                 <title>[(#NOM_SITE_SPIP|texte_backend)]</title>
17                 <url>(#LOGO_SITE_SPIP||image_reduire{150,150}|extraire_attribut{src}|url_absolue|texte_backend)</url>
18                 <link>#URL_SITE_SPIP/</link>
```

FIGURE 3: File written with the SPIP language. Source: SPIP project.

commands written in higher level programming languages." However, the activity of writing these files is characterized as something "like coding," meaning it is not really coding but only a semblance of coding. This ambivalence is also expressed by diverging responses from participants. For instance, one participant is reluctant to consider the SPIP language as "programming" or "computer code" as it is used for structuring information, rather than manipulating it: "It's not computer code in the sense that. . . . [*thinks*] The SPIP language is a language for structuring information. A computer language more generally is done for full manipulation" (spip09). However, other participants are more assertive about the status of "skeletons":

> Q: The source code of SPIP, what does that mean for you?
> For me it means, uh. . . . [*thinking*] First, I will put the PHP code because it is the framework, but it also includes the SPIP templates. (spip11)
> Q: You say that the skeletons are part of the source code?
> Oh yes! (spip07)

These analyses of what is included in "source code" show the ambiguities of defining this notion. Whereas a commonsense definition of source code would lead one to define source code as a set of computer program instructions, a closer analysis shows that source code includes many other artifacts than text comprising computer instructions. Furthermore, depending on the programming language that is used, even files that match the definition of source code as a set of computer instructions are not necessarily identified as "real code." To summarize, an important ambiguity of the definition of source code relates to the *forms* a specific artifact takes, whether it is a text, an image, or a particular programming language.

The Core, the Periphery, and Other Pieces of Code: The Positionalities of Source Code

Definitional ambiguities also relate to the position of a particular piece of source code in relation to the broader project. Indeed, source code that is "available for downloading," as mentioned previously, does not constitute the whole of the artifacts considered as source code and that surround a software project. Building a website with SPIP or Symfony implies gathering several pieces of source code. This interview excerpt demonstrates well the blending of source codes of different forms and statuses: "There is the project's source code, and within, it includes Symfony's source code, the plugins' source code, and the application's source code. A project in itself, it's a blending of source code" (sf09). I call the source code that is published on the project website the "authorized source code," in reference to Bourdieu's "authorized language" concept (Bourdieu 1991) and in the sense that it is that part of the source code that was decreed at a specific moment and by certain people only, as the one that must figure on the download page. However, there are several other locations where artifacts that are described by study participants as source code may be encountered. Indeed, open source software development is often organized around a limited staff team that determines the project's orientations, and sanctions modifications of the source code. This team is often referred to as the core team in the case of SPIP and Symfony. In order to allow a broader participation to the collective production of source code, different methods of participation are put in place. *Plugins* are prime locations where source code is encountered and allow broader participation. Plugins are extension modules that allow adding functions to the software's core, even reconfiguring certain parts. They allow further decentralization of the source code's production by enclosing certain functions, and by being more liberal regarding the granting of authorizations of modification of these parts of the source code. These plugins are often created by developers outside of the "core team." On August 27, 2010, 225 plugins existed within the SPIP project and 1,103 plugins within the Symfony project. Just as is the case for the project's code source, it is possible to obtain the source code for the different plugins by downloading it through a .zip archive directly from its web page. Note that in the case of programs written in interpreted languages such as PHP, the plugin's source code is downloaded with the plugin itself. For study participants, while those files that compose the plugins are designated as "source code," they don't possess the same status as the project's source code, as explained by this participant: "So, I distinguish Symfony's source code, which is developed by the core team, from the plugins, which are sources that are developed by the community, by PHP developers mainly outside of the core team, developers that use Symfony in their enterprise. For me, there is this distinction all the same. One of Symfony's plugin is not part of Symfony's source code, it is part of the source code of an application that is in fact developed above it" (sf07). This quote highlights the way in which source code's position in the project hinges upon whether it is developed or not by the core team. It also introduces the existence of certain forms of authority and authorization in the writing of source code, to which I will return later.

Patches are another kind of artifact that may correspond to source code. Since the modification of source code is often restricted to a few people, participation in source code writing often consists in suggesting patches that will then be integrated within the source code. Patches are a kind of modification proposal, line by line, of the source code, as shown in figure 4. These patch proposals moreover are often the subject of debate and discussion.

Around these elements of source code mentioned above, "bits of code" or "pieces of code" may be found in several locations, such as within emails or on wiki pages. These bits of source code are sometimes shaped as "code snippets," and catalogued on a web page reminiscent of the "social network" form, where those pieces of code can be shared, commented on, and categorized by tags, and synchronized by RSS feeds. Of note, this repository of "code snippets" was one of the only places within both of the studied projects where the expression "source code" was explicitly used.

All these artifacts that circulate around a given project are sometimes referred to as "source code." However, these artifacts have changing statuses and identities de-

```
Index: lib/helper/JavascriptBaseHelper.php
===================================================================
--- lib/helper/JavascriptBaseHelper.php       (revision 19983)
+++ lib/helper/JavascriptBaseHelper.php       (working copy)
@@ -44,7 +44,7 @@
      if ( isset($html_options['confirm']) )
      {
        $confirm = escape_javascript($html_options['confirm']);
-       $html_options['onclick'] = "if(confirm('$confirm')){ $function;}; return false;";
+       $html_options['onclick'] = "if(window.confirm('$confirm')){ $function;}; return false;";
      }
      else
      {
```

FIGURE 4: Patch within Symfony. Source: Symfony project.

pending on their position or location in the project, whether they are, for instance, at the core or the periphery of the project. As I will show later, these positionalities of source code also have implications in the valuing of work in the project.

Defining Source Code and Its Consequences: The Politics of Definitions

The way in which definitions or categories are tied to political dynamics is well documented in the literature. Referring to Krippendorff (1993), Proulx (2007) argues that any metaphor and concept organizes some kind of reality and is also susceptible to becoming the object of political controversy. Feminist authors have also insisted upon the political character of definitions and categories. For example, they have shown that the categories used within computer-supported cooperative work (CSCW) projects are not trivial (Suchman 1993) and could organize users' experience and even "enact silence" when a reality does not fit into preestablished categories (Star and Bowker 2007). Closer to my argument, Wajcman (2004) emphasizes that the definition of technology changes across time and, more importantly, that these changes can be associated with dynamics of exclusion, especially tied to gender. Wajcman notes for instance that programming was first assigned to women because it was seen as tedious and clerical work. But when programming started to be more creative and intellectually demanding, it became increasingly seen as a male activity (Wajcman 2004).

In my case studies, the definition of source code also appears to entail certain similar political and gendered implications. These implications are tied to the valuing of certain activities according to their connection to what is considered source code. For instance, the participant who mentioned that RFCs could be the source code of the Internet also stated that his goal in doing these somewhat provocative statements was to propose a definition of source code that would be inclusive rather than restrictive. This person indeed noticed that some of the project's members (he named specific women, in particular) frequently devalue their own contributions because they don't think they are working on source code:

> Q: Do you think that it can have implications for the project, the way in which different people define "source code"?
>
> Yes, most likely. In any case, there are many people who minimize their contribution, or their participation, thinking that they are not coding, or who knows what. The discussions are always like this. (spip01)

In defining source code as the specification of the functioning of a software program or computer system (for instance, the RFCs being the source code of the Internet), my interlocutor tried to encompass several activities that are central to the making of software, such as designing interfaces or interaction processes, which are not usually considered as source code but are essential to the project. Both activities are mainly done by women in the SPIP project, as in many other projects.

The political and gendered implications of defining source code also appeared in my study at the time of recruiting participants for interviews: I would systematically be directed toward the men—never women—considered to be the most competent in computer programming. In particular, one women I approached for an interview, while consenting to participate in the research and happy to help me, didn't see what she could tell me since she, in her words, "didn't know anything" about source code. Despite her doubts, we ended up having a long and very stimulating discussion on the subject, during which she described what source code is, where it is stored, who contributes to its making, and so on. She also explained why she felt *illegitimate* to talk about source code: "Illegitimate, in the sense of a PHP coder. . . . Actually, illegitimate, it doesn't mean anything, but illegitimate, regarding the code. I know very well, and we are a few people in SPIP to know very well, that we are not capable of coding a line within *the core* of the software" (spip08). Thus, her feeling of illegitimacy is tied not so much to her actual competence, but rather to the position and status of the code to which she is contributing, in this case the periphery rather than the core. In another part of the interview, this participant also said that she didn't know how to program because she was only able to use "simplified language," referring to the SPIP language: "Yeah, but SPIP code, but which is completely different than the object language code, or than PHP language . . . *I really don't know how to program at all*. . . . I am self-taught. SPIP language, CSS and all that, I can manage, but that's because it's not language, it's actually simplified language" (spip08).

Another way to think about definitional politics is through the category of "the coder," which recurs often in study participants' discourse, as much in the interviews as in the exchanges on discussions lists. This category is often mobilized in a neutral way, simply to refer to someone who makes or produces code. In other instances, however, this category relates to the status of some actors and to the statuses and positionalities of source code. For example, a SPIP participant (a man this time) whom I interviewed used the category of "coder" in order to characterize the status of certain people who code within the "engine" of the project, a part of the source code that could be characterized as the "core of the core." However, despite being trained and getting most of his income as a programmer, he did not consider himself as a "coder" because he works at the "periphery" rather than at the core of the project.

> Q: Do you consider yourself to be a coder? You mentioned that J and D are coders?
>
> Huh. I am a coder, but I am not a SPIP expert. I am not a coder on the same level as them. Indeed, I code because it's my trade. But I don't have an adequate mastery of how SPIP works at the level of the core's engine, that I will rather intervene on the periphery, or I will not intervene on this at all. (spip10)

These comments show a relationship between what I refer to as the "positionalities of code" and the status of the person contributing to it: what counts as "real" code changes depending on the status of the artifact, in this case whether it is located at the core or at the periphery of the software. However, one should beware of the temptation of reification in attributing an exclusive relationship between the position of code and the status of people. In this interview, I tried to explore a bit more this category of "coders," and my interlocutor had a really hard time describing it. At some point, he decided to distinguish the "Coders," with a capital "c," from the "coders," with a lowercase "c." The "Coders," he said, are those who have a global vision of the code, and who are interested in how everything is assembled. Furthermore, it is worth noting that my interlocutor didn't consider the "Coders" as the most influential people in the project, although he did think of them as having a high status. The leadership (although not explicitly formalized) is rather attributed to the oldest active member of the project, who has the moral authority to settle disputes. This person is not a "Coder," who knows how everything is assembled, but rather someone whose interest relates to "how everything works and interacts with humans" (spip10).[7]

While being careful not to attribute an exclusive relationship between the position or form of code and the status of people, previously quoted interviewees nevertheless emphasize this relationship. I argue that these questions concerning what counts as programming and source code are particularly important when analyzing the contribution of women, who are generally invisible from the field of computer programming in general and open source in particular (Ghosh et al. 2002).[8] In fact, several feminist authors have emphasized the necessity of valuing activities other than those that are strictly tied to the production of code (Chun 2008; Lin 2006; Haralanova 2010). Haralanova, for instance, writes that "coding is neither the only nor the most important innovative activity in free software development" (Haralanova 2010, 43, free translation) and suggests an exploration of the different activities around open source software development, whether they relate or not to the "contribution to source code." I follow the same line, while adding that the very definitions of coding, code, and source code have implications for the valorization of certain activities and the recognition and visibility of the contributions of certain groups of people.

Visibility and Authority in the Making of Source Code

Another concrete way to understand the relationship between the valorization of work and the definitions and status of source code is through the metrics used by actors themselves (and sometimes by researchers who study them) to recognize contributions. In an article that analyzes the relationships between visible and invisible work, in particular within the context of CSCW devices, Star and Strauss (1999, 19) note that work is never completely visible or invisible, but is always "seen" through a selection of indicators. Citing several authors in CSCW (Bannon 1995; Suchman 1995; Blomberg et al. 1997), they insist that work is often inscribed in the neutral language of metrics that are actually quite political, in the sense that they are used to define what counts as valued work. Firms and organizations try to measure and better represent highly valued creative "knowledge work" so that it can be better billed or traded. In the case of open source development, where source code writing is not always directly "sold," measuring contributions can however enhance some people's reputation and prestige within the project, and even contribute to their employability or "marketable value." The ecology of visibility and visibility is narrowly tied to the definition of what is recognized as a contribution in a given project.

The act of *committing* and its associated metrics are important means to recognize contribution to source code in many free and open source software projects. The commit is a computer command that consists of publishing changes made to source code, in order to make them accessible to other members of the project or even to the general public in the case of free and open source software. For many software development projects, the collective making of source code is based on software devices known as *version control systems*. These devices allow the storing of files that make up the source code by preserving the chronology of the different modifications made to them. A person will thus work on the source code on her personal computer, in her local version, then will transfer her modifications into the common (or public) repository through the commit command. As figure 5 shows, each of the commits describes each modification, and is connected to a revision number, the name of the commit's author (redacted in figure 5), the moment when the commit was executed, and an optional message describing the modification made to the source code.

The commit is a highly valued indicator for measuring contributions to open source software development. The website openhub.net (ohloh.net at the time of the study), for instance, places the commit as a central indicator of contribution to source code. Openhub.net is a website that indexes a great number of open source software projects, and that proposes several metrics to analyze these projects and draw up a profile of their contributors. The metrics presented on openhub.net are based on an analysis of source code repositories, which are freely accessible in the case of open source software projects. Among these metrics, the *number* of commits has an important place and ranks the "contributors" of a given project. During my own research, several participants directed me to that website. As part of a public conference, one of the actors even mentioned that more and more employers use that website in order to recruit developers. Figure 6 shows a screenshot of the site where a project's contributors are ranked by order of commits.

What is the implication of this for the study of software source code and its making? It is important to understand that commit rights are associated with the capacity to modify only a restricted set of what is considered source code: what I previously called the "authorized source code," also called the "core" source code by actors of the projects. Contributing to the making of some of the peripheral parts of source code, such as plugins, or "skeletons" in the case of SPIP, is often granted less visibility by these metrics than contributions made to the "core." Producing formal specifications or functional prototypes—something that one participant tried to define as source code as an effort toward inclusiveness—is also not measured through the commit metrics. Moreover, as I showed in a previous article (Couture 2012), the concrete work involved in committing is sometimes limited to validating or topping off a broader chain of work: modifying a part of source code usually involves many discussions, tryouts of solutions, and production of "patches" (as previously described) that would then be applied by somebody who has the rights and authorization to do commits. All participants in this chain of work remain somewhat excluded by measurements of commits, and the recognition of contribution is rather attributed to the person who accomplishes the final act of commit.

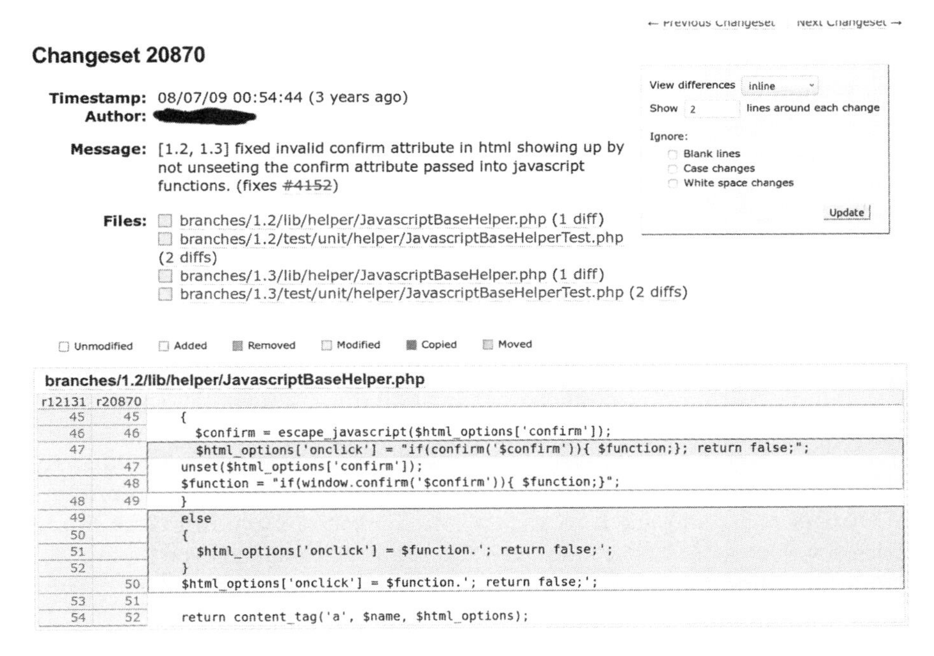

FIGURE 5: A commit in Symfony. Source: Symfony project.

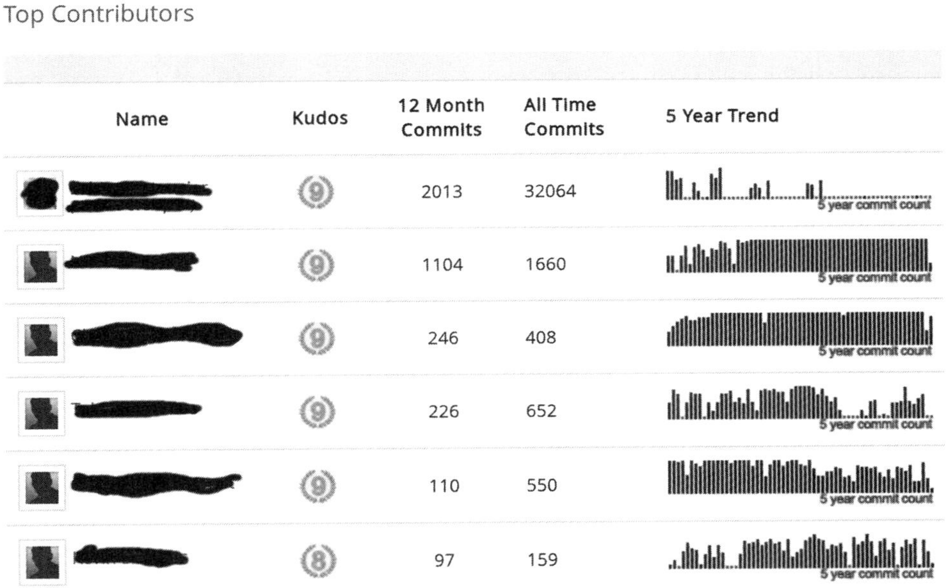

Top Contributors

Name	Kudos	12 Month Commits	All Time Commits	5 Year Trend
	9	2013	32064	5 year commit count
	9	1104	1660	5 year commit count
	9	246	408	5 year commit count
	9	226	652	5 year commit count
	9	110	550	5 year commit count
	8	97	159	5 year commit count

FIGURE 6: Contributors and commits on the website openhub.net.

The act of commit is moreover connected to authority, and to certain rights and privileges within the projects. Roberts et al. (2006) as well as Auray (2007) notice the existence of the status of "committers"[9] within certain open source software projects. Auray even writes, in the case he has studied, that a majority vote from these "committers" may cause the ousting of a member of the project's core team. Demazière, Horn, and Zune (2007b), who, as I mentioned, have also analyzed the SPIP project, note the differentiated access to commit rights that cause hierarchical relations between the project's actors. One of the SPIP project's particularities that I myself noticed is moreover that the granting of those commit rights is not necessarily the consequence of committer activity. Indeed, within SPIP, commit rights are sometimes granted to people who do not, in practice, produce commits or else very few. These people are rather chosen because their role of facilitator, moderator, or administrator is thought to be essential to the community's cohesion or to the software's development. Here commit rights have more to do with a symbolic recognition of authority, than with a pragmatic coordination role.

Interviews and observation also allow the grasping of other forms of work, which are invisible or less visible solely through the analysis of commits and online traces. For instance, in-person meetings, such as informal SPIP assistance sessions, and training in the case of Symfony, are also essential to the project. Several of the people active within SPIP are constantly present on the IRC chat rooms to answer questions or to simply take part in community life.[10] For Symfony, the business itself that sponsors the project employs several people, who may therefore meet in daily life and participate in the project in different ways. In some ways, these two types of work, "gathering a community" and "committing source code," could correspond to the distinction made by several authors and put forth by Star and Strauss (1999), between articulation work and cooperation work. Citing Schmidt and Simone (1996), the authors note that cooperative work is of a more rationalized nature and consists of interleaving distributed tasks, while articulation work deals with the consequences of this distributed nature of work (Star and Strauss 1999, 10). Articulation work, in other words, is the "behind-the-scenes" work and, for this reason, remains largely invisible to rationalized models of work.

Conclusion

My intent in this chapter was to show that "source code" is not a stabilized category, and that the ways actors define it have political consequences. As I demonstrated, a quick look at some recognized definitions of source code shows that it refers to a very diverse range of artifacts. While some definitions are quite specific and refer to source code as a "set of computer instructions written in a high level programming language," other definitions view it in a broader way as "the preferred form of the work to make modification to it." Interviews with actors and empirical analysis of what is considered as "source code" in two free and open source projects also echo these ambiguities. While many actors I interviewed did consider source code as a "text" written in a human readable language, others preferred to refer to it more broadly as the "specification" or the "reference code" of a software or digital application. Empirical analysis of the artifacts referred to as "source code" shows that multiple artifacts such as images, configurations, or comments are also included in source code, thus echoing findings by Robles and Gonzalez-Barahona's (2004) that "source code" encompasses more than the classical view as text written in a programming language. This ambivalence in the definition of source code is even more evident in the case of configuration files, such as those written with the SPIP language, which are considered as source code by some and not "real code" by others. The position of source code within a project—for instance if it's part of the "core" or the "periphery"—also seems to affect its status as "real code."

Most importantly, I showed that definitions of source code have political and gendered consequences in the sense that they may express the valorization of certain types of work and computerized activities over others. As one research participant pointed out to me, many actors in the projects—he had one woman in mind specifically—tend to minimize their contributions because they think that their work is not coding: "The discussions are always like this," he insisted (spip01). Another participant I met told me she felt "illegitimate" to talk about source code because she could write only in the SPIP language and is "not capable of coding a line within **the core** of the software," which clearly shows a valorization of work in relation to the form (the type of language used) and position (the "core" of the software) of source code.

This valuing of the code is also expressed in the interviews through the mention of the category of "coder," which shows a certain ranking of coding activity, according to its proximity to (source) code. The closer one gets to what is considered "true" code, in particular to "core" source code, the more the activity appears to be valued, especially on the quantitative level (for instance, in counting the number of commits). The opposite could also be argued: the higher one's status, the closer one is expected to be to "true" code. This is particularly the case in SPIP, where the leaders of the projects (members of the "core" team) are also given "commit" access to the core source code, even if they do not in practice modify this part of the code. While SPIP is doing this to be inclusive, in a way they also reify access to "true" code as the marker of authority. Finally, I also showed that many metrics used to recognize contribution to a software project are centered around the act of "committing," which involves modification to only a restricted set of artifacts circulating around a project, thus rendering invisible a wide range of contributions to software development.

Two perspectives could be put forth to make the diversity of the contributions to free and open source software development more visible. The first perspective, which I previously mentioned, has already been suggested by feminist authors who have focused on open source software practices. It consists of valuing activities other than those directly related to contribution to source code, like fixing bugs, writing documentation, and teaching how to use a program (Lin 2006, 1149; Haralanova 2010). My own perspective is symmetrical to this first one: it consists of looking at how "coding" and "contributions to source code" are defined and, ultimately, to question these definitions. From a feminist (or critical) point of view, it could also be interesting to take up the definition of source code given by the GNU General Public License (GPL) as the "privileged form of the work done to produce modifications," think about the "forms" that are privileged, and ask who is doing the privileging.

In terms of design, this latter perspective also leads us back to previous studies of *end-user programming* in human-computer interaction. In her work, Nardi (1993) looked at some interactive devices, like statistical packages, visual programming, and spreadsheets, that help give greater computational power to end users for configuring or even developing customized applications. For Nardi, an end-user programming language is defined not so much by its textual or written character but rather that its goal is to enable end users to accomplish their tasks. In a sense, a homemade programming language such as the "SPIP language" in my study could be characterized as an end-user programming system, as could any system that could empower people who do not necessarily identify themselves as coders to develop or configure customized applications.

As source code is a trope that is getting more attention—especially with the rise of free and open source software and increased interest in "coding classes" in the educational curriculum—the question is then to understand how artifacts and activities that relate to the creation, customization, and reconfiguration of digital media and technologies are valued and legitimized differently.

Acknowledgments

This research was supported financially by the Social Sciences and Humanities Research Council (SSHRC) and the Fonds de recherche du Québec—Société et culture (FRQSC). I would like to thank members of the SPIP and Symfony projects for their collaboration in this research, as well as the digitalSTS participants for their help in improving this chapter. Finally, many thanks to my partner Geneviève Szczepanik who reviewed, commented on, and discussed the different iterations of this work, and whose support is always invaluable.

Notes

1. The *10 PRINT* book is a good example of this, as a complete book that is dedicated to a popular one-line code at the time of the Commodore 64 (Montfort et al. 2012).
2. www.gnu.org/licenses/gpl-3.0.html.
3. My study differs from the one conducted by Demazière et al. (2007a) by the focus I put on the category of "source code." While these authors produce effective accounts of the rules and governing processes of SPIP, they do not address the artifact or the notion of source code.
4. Two women out of 20 people might seem too few, but in relative terms, this ratio is still superior to that found by Ghosh et al. (2002) of about 1.5% women active in free and open source projects (a number that is regularly reconfirmed by other less formal studies, albeit with some improvements; see http://geekfeminism.wikia.com/wiki/FLOSS). It was indeed very difficult to reach women in my study, especially in the context of Symfony, where I counted about 3 women out of 300 people participating in one conference.
5. Interview segments are identified by a code: "spip11" means the eleventh person I interviewed in SPIP, while "sf06" means the sixth person I interviewed in Symfony. All interview quotes were freely translated by the author.
6. The SPIP language is controversial, both within SPIP and outside of it, first because it is a language based on French, cutting SPIP from potential contributions from outside the francophone world and, second, because the language of skeletons does not match any recognized programming standard. See Couture (2017) for an analysis of this controversy.
7. The same could probably be said for any software project (including propriety software development), contrary to the idea that code writers "make laws" (Lessig 2006). While "coders," especially those working on central and complex parts of the projects, certainly have a high status, the real leadership of a project is probably most often delegated to a coordinator or CEO who has much less familiarity with code.
8. The Geek Feminism Wiki also reports other more recent surveys and evaluations of women in free and open source software: http://geekfeminism.wikia.com/wiki/FLOSS.
9. The different statuses mentioned by Roberts et al. (2006) are in order of importance: *developer, committer, member of the project's version control committee,* and *member of the Apache foundation.*
10. IRC means Internet Relay Chat, a system used to have synchronous "chat" discussions. IRC was one of the first systems invented for this purpose (see Latzko-Toth 2010), but it is still widely used within hacker and open source communities.

Works Cited

Auray, Nicolas. 2007. "Le Modèle Souverainiste Des Communautés En Ligne: Impératif Participatif et Désacralisation Du Vote." *Hermès*, no. 47: 137–44. doi:10.4267/2042/24086.

Bannon, Liam. J. 1995. "The Politics of Design: Representing Work." *Communications of the ACM* 38 (9): 66–68.

Blomberg, Jeannette, Lucy Suchman, and Randall Trigg. 1997. "Reflections on a Work-Oriented Design Project." In *Social Science, Technical Systems and Cooperative Work: Beyond the Great Divide*, edited by Geoffrey C. Bowker, Susan L. Star, William Turner, and Les Gasser, 188–215. Hillsdale, NJ: Erlbaum.

Bourdieu, Pierre. 1991. "Authorized Language. The Social Conditions for the Effectiveness of Ritual Discourse." In *Language and Symbolic Power*, 107–16. Cambridge, MA: Harvard University Press.

Byfield, Bruce. 2006. "FSF Reaches out to Social Activists." *NewsForge*. http://software.newsforge.com/article.pl?sid=06/08/31/158231.

Chun, Wendy Hui Kyong. 2008. "On 'Sourcery,' or Code as Fetish." *Configurations* 16 (3): 299–324. doi:10.1353/con.0.0064.

Coleman, Gabriella. 2009. "Code Is Speech: Legal Tinkering, Expertise, and Protest among Free and Open Source Software Developers." *Cultural Anthropology* 24 (3): 420–54. doi:10.1111/j.1548-1360.2009.01036.x.

——. 2012. *Coding Freedom: The Ethics and Aesthetics of Hacking*. Princeton, NJ: Princeton University Press.

Couture, Stéphane. 2012. "L'criture Collective du Code Source Informatique. Le Cas du Commit Comme Acte d'écriture Du Code Source." *Revue d'anthropologie des connaissances* 6 (1): 21–42.

——. 2017. "Values and Configuration of Users in the Design of Software Source Code. " *International Journal of Communication* 11:1112–21.

Demazière, Didier, François Horn, and Marc Zune. 2007a. "Des Relations de Travail sans Règles?" *Sociétés Contemporaines* 66 (2): 101. doi:10.3917/soco.066.0101.

——. 2007b. "The Functioning of a Free Software Community: Entanglement of Three Regulation Modes—Control, Autonomous and Distributed." *Science Studies* 20 (2): 34–54.

DeNardis, Laura. 2014. *Protocol Politics: The Globalization of Internet Governance*. Cambridge, MA: MIT Press.

Fuller, Matthew. 2008. *Software Studies: A Lexicon*. Cambridge, MA: MIT Press.

Ghosh, Rishab Aiyer, Gregorio Robles, and Ruediger Glott. 2002. *Floss Final Report. Software Source Code Survey*. V. Maastricht, Netherlands: University of Maastricht, International Institute of Infonomics..

Haralanova, Kristina. 2010. "L'apport des femmes dans le développement du logiciel libre." Master's thesis, Université du Québec à Montréal.

Krippendorff, Klaus. 1993. "Major Metaphors of Communication and Some Constructivist Reflections on Their Use." *Cybernetics & Human Knowing* 2 (1): 3–25.

Krysia, Josia, and Sedek Grzesiek. 2008. "Source Code." In *Software Studies: A Lexicon*, edited by Matthew Fuller, 236–43. Cambridge, MA: MIT Press.

Latour, Bruno. 2005. *Reassembling the Social: An Introduction to Actor-Network-Theory*. Oxford: Oxford University Press.

Latzko-Toth, Guillaume. 2010. "La Co-Construction d'un Dispositif Sociotechnique de Communication: le Cas de l'Internet Relay Chat." Doctoral thesis, Université du Québec à Montréal.

Lessig, Lawrence. 1999. *Code: And Other Laws of Cyberspace*. New York: Basic Books.

——. 2006. *Code: Version 2.0*. New York: Basic Books.

Lin, Yuwei. 2006. "Techno-Feminist View on the Open Source Software Development." In *Encyclopedia of Gender and Information Technology*, edited by Eileen Moore Trauth, 1148–54. Hershey, PA: Idea Group.

Mackenzie, Adrian. 2006. *Cutting Code: Software and Sociality*. New York: Peter Lang.

Marino, Mark C. 2010. "Critical Code Studies and the *Electronic Book Review*: An Introduction." *Electronic Book Review*. www.electronicbookreview.com/thread/firstperson/ningislanded.

Montfort, Nick, Patsy Baudoin, John Bell, Ian Bogost, Jeremy Douglass, Mark C. Marino, Michael Mateas, Casey Reas, Mark Sample, and Noah Vawter. 2012. *10 PRINT CHR$(205.5+RND(1));: GOTO 10*. Cambridge, MA: MIT Press.

Nardi, Bonnie A. 1993. *A Small Matter of Programming: Perspectives on End User Computing*. Cambridge, MA: MIT Press.

Proulx, Serge. 2007. "Interroger La Métaphore D'une Société de L'information: Horizon et Limites D'une Utopie." *Communication et Langages*, no. 152: 107–24.

Roberts, Jeffrey A., Il-Horn Hann, and Sandra A. Slaughter. 2006. "Understanding the Motivations, Participation, and Performance of Open Source Software Developers: A Longitudinal Study of the Apache Projects." *Management Science* 52 (7): 984–99.

Robles, Gregorio, and Jesus M. Gonzalez-Barahona. 2004. "Executable Source Code and Non-Executable Source Code: Analysis and Relationships." In *Proceedings of the Fourth IEEE International Workshop on Source Code Analysis and Manipulation (SCAM 04)*, 149–57. New York: ACM.

Schmidt, Kjeld, and Carla Simone. 1996. "Coordination Mechanisms: Towards a Conceptual Foundation of CSCW Systems Design." *Computer Supported Cooperative Work*, 5, 155–200.

Star, Susan, and Geoffrey Bowker. 2007. "Enacting Silence: Residual Categories as a Challenge for Ethics, Information Systems, and Communication." *Ethics and Information Technology* 9 (4): 273–80. doi:10.1007/s10676-007-9141-7.

Star, Susan Leigh, and Anselm Strauss. 1999. "Layers of Silence, Arenas of Voice: The Ecology of Visible and Invisible Work." *Computer Supported Cooperative Work* 8 (1–2): 9–30.

Suchman, Lucy. 1993. "Do Categories Have Politics? The Language/Action Perspective Reconsidered." In *Proceedings of the Third Conference on European Conference on Computer-Supported Cooperative Work*, 1–14. Milan: Kluwer. http://portal.acm.org/citation.cfm?id=1241935.

——. 1995. "Making Work Visible." *Communications of the ACM* 38:56–68. doi:10.1145/223248.223263.

——. 2007. *Human-Machine Reconfigurations: Plans and Situated Actions*. 2nd ed. Cambridge: Cambridge University Press.

Wajcman, Judy. 2004. *TechnoFeminism*. Cambridge: Polity.

Winner, Langdon. 1980. "Do Artifacts Have Politics?" *Daedalus* 109 (1): 121–36.

Introduction
Global Inequalities

Steven J. Jackson

STS concerns around the relationship between digital technologies and global inequalities are in many ways as old as the field itself. From photoelectric lighting kits (Akrich 1992) to Zimbabwean bush-pumps (de Laet and Mol 2000), STS scholars have long sought to untangle the complex linkages between technological growth and change, and the global flows and forms by which places, regions, and actors are drawn into new (and old) dependencies, both advantageous and otherwise. Against fantasies of a beneficent and frictionless globalism, or the march of happy capitals—Democracy, Development, Progress, etc.—work based in STS calls out the complex and ambivalent dynamics by which technology and "the global" meet. Studies that draw upon the rich heritage of STS theory and empirics are therefore especially well placed to engage with a now-growing set of concerns around the spread and differential impact of digital and human flows of all kinds, both within and across national and transnational spaces, proprietary platforms, and public spaces.

The chapters that follow deepen and extend this tradition. They provide new insight into the intersection between emerging digital forms and the (growing?) structures of inequality and differentiation to be found at the global scale. They speak to the mutual production of difference, inequality, and infrastructure: a relationship that leaves no side untouched. They underscore the distance between frequently utopian technological claims as framed by industry advocates and policy champions, and the messy social realities they are called upon to adjudicate and support. They suggest the frequent limits or brittleness of policy and design in relation to the complexities of social interaction. And they cast new and welcome light on sites and forms of technological labor and agency all too often obscured under prevailing accounts of technology.

Anita Say Chan charts the once and future promise of "edtech"—educational technologies from student laptops to massively open online courses (MOOCs)—and considers the recent explosion of public and private investments in educational technologies against the backdrop of prior histories and achieved (or not) results in countries of the Global South. Her chapter compares the contemporary "cycles of hype and hope" that characterize technology-centered aspirations for social change—here, around educational projects like One Laptop Per Child—against the obdurate realities of complex social and institutional environments, showing how the claims of tech evangelists and other champions of "venture education" work to efface both the realities of educational change (in the process often disempowering actors most needed for such transformations to succeed—local teachers, school leaders, and learners themselves) and the histories of past failed efforts that might themselves provide a more "productive starting point for imagining the present and future as otherwise." Not very good at achieving results on their own terms, such interventions turn out to be very good indeed at reproducing education as a space not of learning but of *data*—and therefore subject to the forms of extraction, oversight, analysis, and control that have come to characterize technology- and data-driven initiatives elsewhere in the economy.

Camilla Hawthorne explores the troubling intersections between Internet security, public safety, and racialized systems of differential control that emerge as liberal democracies (here, Italy) enter into states of exception in response to real and perceived threats of public violence. Tracing the genesis and differential impacts of the 2005 Pisanu Decree, her analysis reveals how racialized imaginaries of violence can lead in turn to racialized regimes of surveillance, practiced against immigrant bodies and places of gathering (notably, public Internet cafés). Her findings give the lie to both any lingering notions of the "placelessness" of online life and the idea of an open and undifferentiated Internet, with freedoms of movement, expression, and association available to all. Locating these trends against longer histories of race and identity stemming from Italy's transformation from emigration to immigration as a dominant imaginary of territorial control, the chapter traces the various "interiorizations" of the border achieved through the regulation of online space. Hawthorne shows how these interiorizations are then made *productive* of forms of racialization and differentiation that characterize the ongoing project of postwar (and post-Fascist) Italian social and political identity.

Carla Ilten and Paul-Brian McInerney explore the as yet underdeveloped juncture between activist adoption of new digital tools, theories of collective behavior and social movements (CBSMs) originating in sociology and media studies, and STS theory and scholarship. As argued by the authors, each of the sociological and media studies traditions around these questions demonstrates important blind spots. CBSM theory lacks adequate accounts of the mechanisms, infrastructures, and networks by which contemporary social movements and activist networks are increasingly constituted. Work in communication and media studies in turn has tended to be platform-centric, missing the important connections across and beyond platforms through which movement identity and communication are sustained. In response, the authors point to older and newer STS work—from social construction of technology (SCOT) and actor-network theory (ANT) to more recent efforts (for example, the Gillespie et al. *Media Technologies* volume)—as providing promising leads for how these worlds might be put back together. The result is a notably co-constitutive approach, built around forms of equivalence, symmetry, and methodological pluralism.

David Nemer and Padma Chirumamilla's moving account of precariousness and repair in a Brazilian favela speaks to growing STS concerns around maintenance and repair as modalities of technological life and engagement, and the complex and irreducible materiality of many objects classed (but too easily) as digital. Turning presumptions of stable infrastructure and predictable function on their head, they explore instead the ongoing labor—of fixing and of living—by which breakdown, failure, and uncertainty are recuperated in the service of a livable(-enough) life, under conditions in which violence and precarity stand as existential and ever-present threats. This sense of fragility—whether expressed through the ongoing battle to sustain LAN house and telecenter connectivity against the constant threat of brownouts and disconnection, or the equally fraught work of sustaining mobile phones and networks against the uncertain environments of favela living—opens up widely neglected domains of technological work and experience, and helps us toward richer and more satisfying accounts of how objects, infrastructures, and communities are sustained, evolved, and made durable through time.

Winifred Poster maps the emerging forms of communicative labor that subtend the global tech and service industries, with special focus on the nature and tensions of transnational customer service call centers. She tracks how sound and voice—and the increasingly sophisticated management of voice—operate as tools of affect, identity, and the negotiation of geographic and cultural difference. She details the role of accent management and synthetic forms of digital voice in constructing imaginary geographies and "façades of humanness" that mask and obscure the increasingly global organization of the service industries, while sustaining markers of difference and expectation that separate "good" cultural and geographical locations from "bad." The complex interactions of human and machine labor in this story belie any simple and wholesale movement toward automation in the service industries (as has been periodically predicted); rather, Poster's work shows how it is in the interplay of these forces—human, but not too human—that new globalized modes and transits of labor are produced.

Taken together, these chapters demonstrate the many rich possibilities for STS-inspired inquiry to cast new and needed light on "the global digital"—both in its inevitable and irreducible particularities, and in its common properties and gathering lines of force. Eschewing easy meta-stories and monochromatic moralities, the chapters speak instead to the messy entanglements of technology, practice, and power, the multiple forms of loss and violence they give rise to, and the artful (if rarely equal) ways in which actors navigate the shifting terrain of the digital as it moves across and constitutes global space.

Works Cited

Akrich, Madeleine. 1992. "The De-scription of Technological Objects." In *Shaping Technology/Building Society*, edited by Wiebe E. Bijker and John Law, 205–24. Cambridge, MA: MIT Press.

de Laet, Marianne, and Annemarie Mol. 2000. "The Zimbabwe Bush Pump: Mechanics of a Fluid Technology." *Social Studies of Science* 30 (2): 225–63.

Gillespie, Tarleton, Pablo Boczkowski, and Kirsten Foot, eds. 2014. *Media Technologies: Essays on Communication, Materiality, and Society*. Cambridge, MA: MIT Press.

Venture Ed

Recycling Hype, Fixing Futures, and the Temporal Order of Edtech

Anita Say Chan

The headlines rang clear: "Teaching Machines—Blessing or Curse?" (Gilmore 1961); "Do Teaching Machines Really Teach?" (Margolis 1963); "Which Is It? New World of Teaching Machines or Brave New Teaching Machines?" (Morello 1965); "Can People Be Taught Like Pigeons?" (Boehm 1960). In the 1960s, various articles in the popular US press underscored the public's growing concerns around what cultural historians characterized as a "boom" in the invention and promotion of teaching machines. Although the development of mechanical teaching devices can be traced back to at least the late 1800s, when patented devices to be used for education began to appear, it wasn't until the mid-20th century that historians note innovations in "teaching machines" had managed to generate a broader "movement" in actualizing wider applications, with national and international conferences held dedicated to new teaching technologies, and both popular and academic publications generating coverage of new research and applications (Benjamin 1988; Ferster 2014; Watters 2014a, 2015).

The National Education Association of the United Stated even published its own volume on *Teaching Machines and Programmed Learning* in 1960—reviewed as "indispensable for any one in the field [and that] will certainly be the major source book for many years to come" (Day 1961)—whose 47 chapters and 724 pages covered a range of commercially available machines (Lumsdaine and Glaser 1960). As the co-editor of the volume, A. A. Lumsdaine wrote in 1961, "The extension and inevitable partial replacement of human functions by the digital computer and similar electronic devices have already begun . . . to revolutionize such diverse aspects of our lives as industrial production and research techniques. This is only a bare beginning. Within a matter of decades it seems likely that . . . the extension of man's intellect through electronics will bring about profound changes in . . . our education system" (1961, 272). And even then, forecasts around the need to "increase greatly the resources and capabilities" of "good teachers" and the "accelerated requirements for a technically capable citizenry" (1961, 272) were the primary grounds on which teaching machines were advocated.

Within general publics, however, among the most frequently repeated preoccupation was the question of how machines would impact teacher retentions and workload: how, that is, a future of mechanized pedagogy might render educators to be a disposable, increasingly replaceable workforce. "Can Machines Replace Teachers?" (Luce 1960) and "Will Robots Teach Your Children?" (Bell 1961) ran through the headlines, pressing teaching machine advocates to respond. Attempting to assuage concerns in the years preceding the completion of his own famed "teaching machine" in 1953, the Harvard psychologist and inventor B. F. Skinner issued his own, brightened forecast: "Will machines replace teachers? On the contrary, they are capital equipment to be used by teachers to save time and labor. In assigning certain mechanizable functions to machines, the teacher emerges in his proper role as an indispensable human being. He may teach more students than heretofore—this is probably inevitable if the world-wide demand for education is to be satisfied—but he will do so in fewer hours and with fewer burdensome chores" (1958, 976).

Such publicly visible debates, and past prognostications of potential futures, are keen reminders that the current expansion of public and private sector investments in education to promote interactive technologies—from one-to-one student laptop and tablet programs to virtual platforms hosting massive open online courses (MOOCS)—as *the* key to preparing populations for a 21st-century information economy is far from the first revolutionary wave of its kind. Historians, indeed, have noted how various media technologies and "auto-instructional" media (Lumsdaine 1961)—from radio and film to educational television—newly promised to radically transform education when they were introduced into public schools throughout the 20th century (Cuban 1986, 2010; Ferster 2014). Initially presented as means to enhance textbook-based pedagogy with broadcastable instructional content—well suited for "banking" models of instruction that saw students as vessels that needed to be filled and formed (Freire 2006)—over time, they would be hailed for affording new capacities for student input and machine feedback, and for extending never-before-realized interactive capacities that could increase potentials for "customizable" instruction tailored to the personalized needs of individual learners. The promise of such new affordances spurred, too, deep concerns that the fervor for such technologically enhanced capacities could actually realize a displacement of existing teaching professionals. Time and time again, however, the repeated promise of revolution that accompanied the introduction of new teaching technologies would be followed by a later realization of the inability (disappointing to some, relieving for others) to meet elevated expectations.

Perhaps less than seeding a revolution in education, that is, teaching technologies over the course of the 20th century might be better said to accompany performances of a revolution in a perpetual cycle of reboot and delay. As important as the reminder that teaching machines have been among us before is the reminder that the revolutionary futures they assured publics they would bring about—what science and technology studies (STS) scholars have pointed to as the futuristic discourse of hype (Miyazaki 2013; Strathern 2011; Sunder Rajan 2006)—have been with us before too. Such attention to the temporal cycles of repetition and recycled process of hype in relation to teaching technologies shifts attention away from contemporary popular conceptions of hype as accompanying a sudden burst of high-impact innovation and disruptive activity. Yet even as rising concerns *today* around the projected impacts of digital education technologies on teaching and learning practice and labor worldwide echo those of past decades, there remains a curious lack of any reference to the cycles of project development, elevated expectations, failed futures, and lessons from those not-so-distant past experiments. Indeed, for as expansive as both state and commercial investments worldwide have been in visibilizing new, now often globally expansive contemporary digital education programs—through international conferences, government-launched programs, or elaborately staged education-oriented product launches and campaigns—the recent history surrounding past decades of deployments of educational technologies (or edtech) remains glaringly absent from contemporary accounts, much as the memory of the promise of revolution—and failure to realize projected visions for the creation of expanded value, even when global impacts were foretold—remains curiously absent from the official record as well.

But if the revolution-making promises of teaching technologies have manifested before over the decades[1]—and have done so accompanied time and again by memorable crests of elevated hopes and expanded anxieties—how is it that today they appear to be so easily wiped from memory? How, in other words, have the cycles of pronounced faith and eventual disappointment in teaching technologies and their unrealized revolution been so repeatedly forgettable? So much so that contemporary hype around edtech solutions can manifest now across broad, globally expansive contexts, as if such technologies were the first that had ever proposed to radically transform teaching and learning? And so much so that contemporary popular expressions hype are typically projected onto products of innovation, as if they resulted from a sudden burst of high-impact disruptive activity, rather than being part of a longer trajectory of cyclical fits and starts. Even the vaguest memory of past hyped (and arguably failed) teaching technologies, for instance, would presumably invite a healthy dose of skepticism around—or perhaps even outright rejection of—the asserted promise of contemporary edtech solutions. Yet the logic of hype—or what STS scholars have explored as a "strategic promissory grammar" of hype (Strathern 2011; Sunder Rajan 2006)—that frames edtech's increasingly global solutions today somehow manages to conjure and sustain a pronounced sense of belief, despite the evidence of past records demonstrating that doubt would perhaps be better warranted.

In this chapter, teaching technologies as such sticky objects to cycles of hype and hope—and the historical silences and omissions of memory around past records and evident cases of unrealized promise—are explored as symptomatic of a growing digital age orientation to education that's explored here as one of venture education. Such an orientation, I stress, places explicit value on education as a primary engine for spurring global innovation—even promoting tolerance to high-risk experimentation in educational contexts in the name of optimizing the economic competitivity of diverse populations in the digital age. Taking into ethnographic account the various devices—material, infrastructural, and discursive—constructed to promote global edtech initiatives, this chapter explores how they come to serve as especially productive digital age objects for inviting new engagements from knowledge workers targeted by international educational reform plans. Linked explicitly to high-tech imaginaries and the growing investments of corporate actors in edtech sectors, venture ed relies on shared investments from public and private sector actors. It thus deploys techniques designed to mobilize its imaginary in ways that can register—and resonate—for diversely situated global knowledge workers.

Speculative Futures of Global Edtech

Over the last decade, various studies have underscored how Latin America and the Caribbean have proven to be especially productive terrain for undertaking large-scale experiments with new edtech initiatives (Ames, forthcoming; Andersen 2013; Chan 2014; Medina et al. 2014; Warschauer and Ames 2010), with a World Bank report indicating that some 18 countries in the region had launched one-to-one laptop programs by 2012 alone (Severin and Capota 2011). Promoted explicitly within international policy arenas as global development tools under information and communication technology (ICT) for development (ICT4D) programs, teaching devices demonstrate a utility in enabling contemporary market logics to both *globally* extend and *locally* anchor among diverse knowledge sectors and publics. Much like the market devices that STS scholars and the social studies of finance (Callon 1998; Callon et al. 2007; Mackenzie 2006; 2009; Mackenzie et al. 2007; Mitchell 2005, 2008) studied as "material and discursive assemblages that intervene in the construction of markets" (Muniesa et al. 2007, 2), contemporary *teaching* technologies as devices developed in edtech sectors seek to intervene in the construction of contemporary educational practice and future outcomes.

Indeed, one such site where the interventions of contemporary teaching technologies unfold are the international events and forums, such as the Virtual Educa Conference, that aim to explicitly foster multilateral cooperation in education and innovation. Established in 2000 as an international initiative of the Organization of American States (OAS), the annual Virtual Educa Conference was created to represent an investment in education futures and "A New Educ@tion for a new era." As the secretary general of the OAS, César Gaviria, declared in 2001, shortly after the founding of the initiative, "Virtual Educa is called upon to create new forms of accessing and exchanging information about the technological revolution we're living in now, to establish relationships of cooperation between the educational institutions and private and public sectors." Some seven years later, various regional government heads could be seen affirming the annual conference meeting in its mission, with government representatives including Argentina's minister of education Juan Carlos Tedesco declaring at the opening of Virtual Educa Zaragoza 2008 that "we are aware that this forum of Virtual Educa has become the Iberoamerican event of greatest importance on this topic in discussions and articulations around Information and Communication Technologies and Education."

Indeed, the immediate entrance space for the 2014 Virtual Educa Conference, organized in collaboration with Peru's Ministry of Education and held that year in the typically austere museum space of the nation's Ministry of Culture, provided an immediate visual showcase of new educational technologies that channeled visions of "A New Educ@tion for a new era." An array of corporate sponsors for that year's event—most prominently Intel, Google, HP, Oracle, Telefonica, and Microsoft—filled the open lobby of the Ministry of Culture with elaborate demo spaces, converting the ministry lobby into a buzzing maze of interlinked educational technology product stages. Weaving through the space, one would encounter multiple demo areas, aglow in their stagings of new educational laptops and tablets, instructional software packages, robot kits, digital encyclopedias, and wall upon wall of interactive, flat-screen blackboards. Other sponsors simply erected a mock lounge area, dressed minimalistically with silver-trimmed white leather couches and bar stools that invited attendees to sample their "cloud-based" solutions for schools (including those, apparently, that—as in much of Peru—were located in rural provinces in the Andes and the rain forest that operated without Internet connectivity). As Peruvian vice president Marisol Espinosa Cruz stated in her opening address to the thousands of attendees for the event, "During [this year's] Virtual Educa, you will have the opportunity to access different plenary sessions and fora themed around education, innovation and competitivity, where the development of new goals and opportunities that technologies offer education, exchanges on practical experiences in education and innovation, and exhibits of the latest generation of information and communication technologies [will be shared]."

While fellow government officials, regional planners from across the region, as well as an array of corporate sponsors were in attendance, it was the public school teachers who had traveled to Lima for the conference, representing hundreds of Peru's urban and rural schools, including from the remote mountain and jungle provinces, who made up the largest sector of the audience—and to whom Virtua Educa's clear theater and staged "performances" of the "latest generation" of ICTs offered for the education sector primarily targeted. And while much work and energy had evidently been invested to construct such an elaborate and extensive display of new educational technologies, what was consistently absent from the staging of such devices, representing the "latest" in education-based innovations, were any reports of past studies that had been completed to indicate findings on degrees of efficacy of *actual* technological deployments and real-world performances in specific learning contexts and local communities. Showcasing an array of the latest innovations presumably too new to have yet been actually tested, Virtual Educa's conference space seemed less concerned with offering information on how such technologies actually performed in context than with creating an occasion to enable such performances to come into being. As science studies scholars Donald Mackenzie and Yuval Millo put it in describing the pointed efficacy of economic models to bring about the market conditions they are meant to only depict, "Economics does not describe an existing external 'economy,' but brings that economy into being: economics performs the economy, creating the phenomena it describes" (Mackenzie and Millo 2003, 108).

Indeed, some of global edtech's most prominent stages were used to platform the message that there was not any need for any studies, evaluations, or even basic information gathering of deployed edtech solutions. Speaking in a September 2009 Interamerican Development Bank forum, Nicholas Negroponte, the founder of MIT's Media Lab and the One Laptop Per Child (OLPC) project alike, insisted that it was so plainly self-evident that laptops would only enhance a child's natural capacity for self-learning that calls to study the early deployments of OLPC in local sites would be a waste of resources. Even as nations, including Peru, were investing millions of dollars in nationally scaled deployments (that in Peru alone would grow to some $300 million), Negroponte flatly dismissed the idea that studies of OLPC's impacts would be revealing: "That somebody in the room would say the impact [of the XO] is unclear is to me amazing—unbelievably amazing. . . . There's only one question on the table and that's how to afford it. . . . There is no other question" (Negroponte 2009; Warschauer and Ames 2010).

While social scientists of finance underscore the persuasive power of highly rationalized market devices and mathematical formulations in economics—such as option-pricing theory and the Black-Scholes-Merton model—to "perform" and actualize market processes in the real world via their ability to affect the perceptions of other legal, political, or financial experts, anthropologist of science Kaushik Sunder Rajan suggests that in contemporary innovation sectors it may be the performance of market devices within the specific context of speculative markets—where companies face uncertainty and may still be years away from making a tangible product but can drive stock prices "by virtue of promise alone" (2006, 122)—that makes the key difference. Describing the means by which the fevered conditions of hype are fundamental to shaping contemporary innovation sectors, he writes that one can understand emergent technologies under high-tech capitalism only by simultaneously analyzing "the market frameworks" (2006, 33)—that is, public relations strategies, investor relations discourse, and hype within which they emerge. For Sunder Rajan, hype's discursive marketing techniques and apparatuses are not incidental, but are indeed designed, like market devices, to concretely reorder the temporal and material orders and chrono-logics in which they operate: "[Hype is] a game constantly played in the future—to generate the present that enables the future" (2006, 34). Insisting on the indispensable value of such future build-outs, hype operates "as a discursive mode of calling on the future to account for the present" (Sunder Rajan 2006, 116). However distant or out of step the actual present might seem from the eventual future promised, hype urges focused investments of present work and labor in the name of achieving such promise.

Indeed, like a magical electric current, hype and its affect could be felt tangibly coursing through the various glowing demo spaces of Virtual Educa. It was such spaces, excessively outfitted with new technologies, as much as the conference panels and talks themselves that were designed to draw attention. And it was such spaces that undoubtedly drew large, buzzing crowds, filling demo stages to capacity so that the floor—lit up with display after display of new educational laptops and tablets, instructional software packages, robot kits, and interactive, flat-screen blackboards—seemed to literally vibrate with the palpable faith in the suite of edtech solutions and their fantastically rendered techno-futuristic projections of things to come. However out of step such proposed edtech solutions might have been with the real known and lived conditions of the present tense of classroom and teaching contexts across Peru, it was an uncontained optimism that was animated across the various product stages weaving through Virtual Educa, and the public responses it invited. It was as if, for the crowds that swarmed around such displays, even more important than the possibility of being able to achieve such future projections was the ability to stake a claims and place bets on that *particular* vision of the future as worthy of such investments. Or as Sunder Rajan writes, "Excess, expenditure, exuberance, risk, and gambling can be generative because they can create that which is unanticipated, perhaps even unimagined." After all, "[a] vision of that future has to be sold, *even if it is a vision that will never be realized*" (2006, 115).

Edtech's Disremembered Pasts

Beyond the glowing demo spaces of Virtual Educa, in the conference rooms where corporate sponsors and government officials spoke as panelists in presentations to the audience, a distinctly different kind of affect was being summoned—one that was explicitly bleaker, and that relied as much on a temporal projection of heightened risk, disruption, and even destruction as proximate outlooks, as on the glowing promise of edtech as solutions to avert such impending outlooks. Speaking to a literally standing room only audience of Peruvian public school teachers, for instance, Matias Matias of Silicon Valley's Hewlett-Packard explained the company's view on the contemporary challenges of global educators under conditions of the digital innovation in economy: "We've worked in many countries, we've collaborated in many counties . . . and one of the challenges we've seen is that the education system in which we all participate today, in our opinion is a system that's now obsolete. It is not a system that generates talent, nor is it a system for personalized education." He continued by providing disheartening example after another of allegedly failed, ineffective teaching techniques still in use by public educators in classrooms, but then turned pointedly brighter. "But here's the news I am here to deliver: we are here to help you. We are here to support you in this process of technological adoption and practice of teaching. . . . HP turns 75 this year, as one of the world's leaders in technology, and all this experience we bring and make available to you, to change lives and today, to prepare for a better future. So count on us in this transformation!"

It was, indeed, a unique approach to public relations from corporate IT sector actors—who in this case not only seemed untroubled by their direct pronouncements to audiences of working professionals that they were obsolete and working against generating a society of future talent but actually seemed certain there was a virtue in their doing so. When Matias did switch over to a more conventional form of public relations address, to reassure and assuage his audience, he did so by telling them that even if the current education system they participate in is one that's become obsolete, there is at least one resource educators can count on to avert risk: HP and its 75-year history and "experience" as one of Silicon Valley's oldest corporate exemplars. Here, Silicon Valley is summoned as an existing material model whose own presumably self-evident success record is assumed to require little explanation. But Matias's framing of HP turns less on high modernist framings of the company as manifesting a preserved system of high order and rationality, or even about its ability to have leveraged such valued toward new wealth creation. Rather, this narrative turns more on the notion of HP having managed to cultivate a record of growth for nearly a century in the Valley's highly competitive, uncertain, and rapidly changing ecology of technologically paced, high-stakes survival, risk, and extinction. Here, in other words, Matias places his emphasis on the company's long (and unparalleled among Silicon Valley's technology companies) record of proven responsive to conditions of rapid market change and uncertainty. Its ability to become and remain a market leader throughout the sudden technological changes of the 20th century, and its ability to manage the uncertainties of the computing industry for nearly a century, it's suggested, is what makes it an unparalleled ally to confront dynamic change already manifesting in the 21st century's digital economy.

One only has to try to replace HP with a *Fortune* Global 500 company from another sector—from perhaps banking or energy or consumer goods and services—to appreciate the distinctiveness of HP's reception. For whatever symbolic power and global capital is surely shored up in the unique market dominance and recognizability of any *Fortune* 500 player, it's hard to reasonably imagine the same invective being issued from corporate actors of another sector, who could speak on behalf of another clear industry giant whose growth over the past century had made them household names worldwide, without at once expecting that they could be met with at least some degree of indignation or distrust from audiences. IT sector actors today somehow prove the exception, and HP, even following Matias's declarations of the teaching profession's contemporary obsolescence, was met instead with a roomful of applause.

Matias's chronology was, of course, quite selective about the relevant details it related as 20th-century developments in technology sectors. He left out any mention of how decades earlier, in the mid-20th century, new developments in teaching machines and innovations applied in educational contexts were indeed coming to greater public attention. While historians underscore that development in mechanical teaching devices can be traced back to at least the late 19th century, when patented devices to be used for education began to appear, it was in the mid-20th century that innovations in "teaching machines" began to garner wider public notoriety (Benjamin 1988; Watters 2014a, 2015). Education historian Ludy Benjamin notes that it was in the 1960s that broader public debates could be seen across diverse forums: "National and international conferences were held to discuss the new technology, and popular magazines and scientific journals published news of the emerging research and applications. . . . Interest was shown not only by the educational establishment but also by industry and the military who were especially interested in training application" (Benjamin 1988, 709).[2] Commentary in the popular press, he writes, was frequently cautionary, with articles stressing concerns on the dehumanization of education ranging from the moral and cultural implications of having children taught by machines (Bell 1961; Gilmore 1961) to the question of having people "taught like pigeons" (Boehm 1960) and the question of whether machines could really teach (Margolis 1963). And indeed, among the most frequently repeated preoccupations that teaching machine advocates and innovators were pressed to respond to was the question of how machines would impact teacher retentions and workload—that is, how a future of mechanized education might render a modularized educational workforce—with headlines querying "Can Machines Replace Teachers?" (Luce 1960) and "Will Robots Teach Your Children?" (Bell 1961).

Notwithstanding such concerns around labor and the mechanization of teaching, the promise of such devices on amplifying education's potential had clearly generated a new "movement" for promoting the uptake and invention of teaching machines. Educational organizations including the American Educational Research Association, National Education Association, and American Psychological Association were aware they had to intervene, and collaborated on a joint statement on "Self-Instructional Materials and Devices" in 1961 that stressed the importance of at least providing guidelines and materials to ensure quality teaching and assessment in machine-engaged learning programs in their institutional deployments (American Psychological Association 1961; Benjamin 1988).

A decade later, the "boom" in teaching machines had calmed significantly, as institutions began to encounter barriers to computerizing classrooms. Foremost among these were the persisting lack of guidelines and materials to support programs that ultimately were not "user-friendly" and high costs. Accounts of schools investing thousands with few pedagogical results began to accrue, such as one report of a school district that spent $5,000 on machines and discovered there were no supporting programs available for them (Benjamin 1988). Others machines, lacking curricular support for teaching, came to be seen as little more than toys with limited educational benefits (Weisenberg 1961). And indeed, the decreasing number of publications on teaching machines further reflected the waning interest. While the *Readers' Guide to Periodical Literature* claims the number of publications in popular literature between 1960 and 1964 at 65, a decade later, from 1970 to 1974, they were down to just 5. Psychological Abstracts followed a similar pattern, where there were 101 citations for 1960–64 and only 15 for 1970–74 (Benjamin 1988). It would take new developments in microprocessors and other components in the late 1970s that would allow fully assembled micro- and personal computers to become more largely available to the general public to overcome such sobering outcomes.

Such historical traces are keen reminders that developments—as well as debates—surrounding teaching machines have histories that long predate the emergence of today's hypervisibilized edtech solutions. The ambitious, explicitly globally ranging digital education initiatives that leverage the growing global accessibility to both personal and mobile computing devices—from the MIT-launched One Laptop Per Child program (that's often been credited itself with spurring global movements for one-to-one-based student laptop plans) to the numerous massive open online course (MOOC) platforms—are among the most prominently narrated and visible inheritors of such layered histories that surround the prior generation's teaching machines. Yet contemporary edtech initiatives make no reference to even the once-celebrated versions of past examples of educational technologies. Now designed to be scaled for deployments that can cover entire national territories, such contemporary teaching technology initiatives tend to stress how their virtual learning environments come ready made for massification, and can overcome the constraints—whether spatial, geographic, economic, or technical—of standard classroom and institutional accommodations. Consistently left unmentioned, however, in their framings and stagings is their connection to and frequent reanimation of prior debates regarding the costs in either labor or financing—to students, teachers, and existing institutions and organizational infrastructures—of such transformations. It was as if such debates had never existed, or as if they had emerged anew without prior histories, records of development, or disappointing outcomes.

Such omitted pasts demonstrate that part of the function of hype in relation to edtech solution is not just about projecting a promissory future, but about seeing to a temporal ordering that can render former failures forgettable, and disguise the recycled performances of hype. Such cleansed pasts around the reality of disappointing outcomes of deployed edtech solutions facilitate the recycling hype as a discursive ground, and the mobilizing of frameworks of innovation and novelty as fundamental narrative elements. But indeed, the work of such devices becomes apparent only when seen or witnessed outside the temporal orderings and architectural stagings of hype. As STS scholars have observed before, as well, such chronological orderings are necessary in order for hype to operate effectively as a discursive apparatus that depends more on effecting credibility than on either promising the truth or attempting to hide an outright lie (Strathern 2011; Sunder Rajan 2006). For all the casual associations of hype with a lie, hype's promoters prove to be less concerned with their proximity to proven truths than with actively graining a certain credibility in the marketplace and confidence among investors and relevant publics. Or as Sunder Rajan writes, "The enterprises that produce fact, evidence, and PR are completely intertwined. This is why dismissing hype as 'simply cynical' . . . [and] not a fruitful way of understanding the mechanisms of its operation. Attributions of cynicism serve to erect a simple binary between the truth and the lie (hype always being somehow associated, not just typologically but normatively with the lie), a binary that just does not serve to understand the ways in which the truth and the lie are co-constituted as different types of ~~truth~~" (2006, 135).

It's in part because of venture education efforts' need to cultivate faith in the possibility of "different kinds" of credibility among diversely situated publics and investor classes alike that its coordinations between distinct institutions spanning public and private sectors and knowledge- and research-based and industry-centered work become so valuable. Each can work to capture and perform credibility distinctly for their relevant audiences. To the degree that venture ed efforts work toward building and proving "facts," their promoters care less about proving the fact of the efficacy of edtech solutions than about the simple fact of having been able to *visibly and evidently* "capture credibility." As such, captured faith, belief, and credibility among diverse publics perform the realness of edtech and its array of promises.

The varieties of promotional language and strategies that extend from edtech hype thus reflect the targeting of global urban managerial, elite governing classes, and IT industry leaders who might lend financial backing to such projects, as much as educational professionals and broader publics whose faith and enthusiasm for such projects impacts the shape of markets and market valuations around for the edtech sector. Promotional stunts for the OLPC thus infamously included Nicholas Negroponte's promise to, and eventual execution of, air droppings of OLPC tablets to rural villages in Rwanda to demonstrate—he argued—how even without teachers or classrooms, and with only access to the "right" technology, digital literacy could be achieved. The varied private- and public-sector allies to OLPC that bridged tech and global development worlds alike—having been designed by MIT engineers, led by Media Lab founder Nicholas Negroponte, and supported by the likes of Google, Red Hat, and the chip maker AMD—indeed fanned much of the anticipation and excitement surrounding the project. Particularly in Latin America, the project found ready uptake, with nearly 70% of all OLPC's global deployments extending across the continent by 2010, and where states like Peru and Uruguay became rapid partners in ambitious, nationally scaled programs.

Indeed, a similar promise of what OLPC's multisectoral, global private-public partnerships could bring to such international deployment sites as means of transforming national education systems and economies alike arguably helped fuel the popularity of the project among diverse cosmopolitan audiences. In the past several years, multiple OLPC representatives have been regular speakers in the international circuit of TED (Technology Entertainment Design) conferences that are renown for drawing crowds of tech-savvy global professionals, planners, and entrepreneurs. Rodrigo Arboleda, the Colombian-born chairman and CEO for the One Laptop Per Child Foundation, opened his presentation at the 2011 TED conference in Brussels with a short video segment on the Uruguayan Program—and a reminder to his audience of OLPC's commitment to universal inclusion: "The challenge is to not leave people out—and to have every child of primary school age in third world countries have the same opportunities [and] access to knowledge in [the same] quantity and quality as the most privileged child in New York, Tokyo, Berlin, or Brussels. That is the challenge!"

Neither yet the truth nor exactly a "lie," hype stretches toward the ambition of a reality it still can only claim to work toward constituting—a reality that exists in other words as much as fantasy as actual possibility. It thus acts all the while to preempt the potential for failure in the present by fortifying the discursive grounds on which reality in the future will unfold, using a range of devices and techniques—from educational to tech industry conferences, reports and publications, new channels, and the good will of the audiences and publics they each hail—in order to do so. All this in an effort to register credibility and truth effects for such varied publics. However immaterial they might be, collective faith and belief turn out to be quite bankable investments in venture ed's ecologies of hype, and the work of securing the futures they seek to bring about.

Cleansing Chrono-logics of Hype Cycles

Perhaps one of the clearest indictors of the concrete value placed on hype is the considerable investment business analysts and technology industry actors have made on studying and tracking hype. Probably the most widely cited, watched, and followed studies within tech sectors is the Hype Cycle report annually released by the US research, advisory, and information technology firm Gartner, whose reports have been "evaluating market promotion and perceptions" for over 20 years. Its 2014 report, for instance, included an evaluation of some 2,000 technologies, and came accompanied by an optional document—which Garner charges $495 for—as a guide to help interpret it. As Gartner explained, "The Hype Cycle for Emerging Technologies . . . featur[es] technologies that are the focus of attention because of particularly high levels of hype, or those that Gartner believes have the potential for significant impact. . . . Enterprises should use this Hype Cycle to identify which technologies are emerging. . . . Understanding where your enterprise is on this journey and where you need to go will not only determine the amount of change expected for your enterprise, but also map out which combination of technologies support your progression" (Gartner Newsroom 2014). A flurry of business news outlets, all waiting in anticipation for the report's publication, would distill the report's findings to punchy headlines. When it was released in late 2014, articles from *Forbes* and the *Wall Street Journal*, among others, pronounced in bold headlines that year's takeaway message: "It's Official: The Internet of Things Takes over Big Data as the Most Hyped Technology" (Press 2014).

Arguably, what the report is most famous for—and what many look to the report to see—is Gartner's Hype Cycle Graph: a curious visual element that represents the life cycle stages projected for emerging technologies (see figure 1). It thus charts five essential stages for the more than 2,000 new technologies it now evaluates annually: (1) the Technology Trigger Stage (visualized as a sharply upwardly inclining line), (2) the Peak of Inflated Expectations (visualized as the top of the slope, before a sudden drop), (3) the Trough of Disillusionment (visualized as a valley at the bottom of the drop), (4) the Slope of Enlightenment (visualized as a gently rising slope from the valley), and (5) the Plateau of Productivity (visualized as a flat plane that extends from the rise out of the valley). All its evaluated technologies find a place within the stages of the graph, which—although shaped as more of a sloped line pointing forward than a circle—is nonetheless referenced as "a cycle" by Gartner. The visual emphasis on the graph, however, is undoubtedly on the progressive forward movement and evolution of each technology as it matures toward an eventual Plateau of Productivity over a period that might take five to ten years. As Gartner says, explaining the graph and the work of its report around it, "This report tells a story of how technologies, services and strategies *evolve* from market hype and excitement of value to becoming a mainstream part of business and IT."

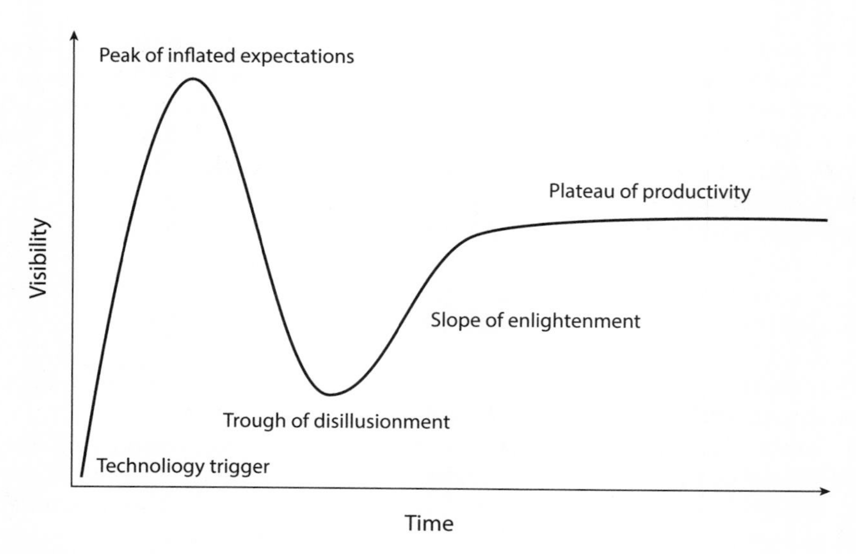

FIGURE 1: Hype Cycle's five stages.

But what's perhaps most interesting about the graph's narrative is its inclusion of a stage of overpromise at all—in the peak of inflated expectations—that's acknowledged as producing an inevitable deflation and a fall into the Trough of Disillusionment. Over time, however, a process of evolution is supposed to occur for each technology that is promised to eventually pass through a Slope of Enlightenment and later enter a Plateau of Productivity. STS scholars indeed have underscored the temporal function of hype and its game-like character: that hype has to overtly oversell the future in order to generate the present to be readied for its making. And yet for all the crucial work of visualization that Gartner's Hype Cycle Graph executes, one other crucial turn appears to occur that's less about visual generativity than about erasure. Once the question of what's missing from hype promoters' narrative becomes introduced, it becomes clear that hype depends not just on an overselling of the future to gain the credibility but also on calls for an explicit devisibilization, or wiping of the past. And once again, such omissions become apparent only when seen or witnessed beyond the border of hype's devices, temporal orderings, and architectural stagings.

Gartner's graph, after all, offers no moment of recollection or reflection. In its only forward-indicating movement, there is no folding back or movement to indicate a reconsideration of a particular technology's relationships to other past projects. And while the discursive and visual technologies the outlined in the Hype Cycle Report—much like the architectural stagings and discursive performances of conferences like Virtual Educa—may offer many projections of *the future*, what's explicitly omitted from all these elaborate investments and staging is any representation of history, whether around particular media and technology projects, their previous deployments, or histories of labor around such projects.

Among the elaborate stagings with their excessively displayed technology models and panels organized for the 2014 Virtual Educa Conference in Lima there was, indeed, no reference to the long history of experiments in teaching machines—whether dating back to 20th-century cases that spanned the "teaching and testing aids" patented in the United States by the Harvard psychologist B. F. Skinner or the globally deployed edtech initiatives of cases like the PLATO digital computing project that seeded international collaborations for the University of Illinois in the 1960s (Bitzer 1981).[3] Even more surprising was the absence of any direct address to a variety of recent cases of edtech deployments in Peru itself that had immediate relevance to the interdisciplinary, cross-sectoral audiences now routinely present at Virtual Educa's conferences. There was no mention of the fact that Peru, for instance, had a fairly long record of national deployments of digital education projects that included the prominent Plan Huascaran Initiative, launched in 2001, that promised nearly two decades earlier to network rural and urban classrooms alike. And there wasn't even any direct address of OLPC, the $300 million experiment that had been running since 2007 and had distributed nearly a million laptops across the nation, including into many of the classrooms of the hundreds of teachers present at Virtual Educa (Chan 2014).

Instead, the conference was used to announce that a new national edtech project had been designed and would shortly be deployed that would be structured around the Intel Corporation's Classmate computer. With speakers from Intel featured in the conference program—and not just one but multiple demo areas that displayed Classmate computers available for use and testing—Intel's corporate representatives and commercial products alike could appear at Virtual Educa's carefully curated demo spaces as ever-visible and ever-accessible objects. All this for the public school teachers, government officials, and corporate representatives present, whose distinct versions of faith could be hailed as important elements in realizing the nation's future of edtech. And with such an endless sea of new technologies on display, all framed as expressive of the new revolution in education available here for conference goers to get to interact with, what space or time was there to reflect on such mundane considerations as the past?

Conclusion

A consideration of the past waves of teaching technologies and the expansive promises that accompanied them, however, underscores the cyclical form of edtech revolutions that are consistently left out of contemporary venture ed discursive and material performances. As important as the reminder that teaching machines and their related broken and unrealized promises have been among us before, indeed, is the reminder that hype around the revolutionary futures they assured publics they would bring about have been with us before. Omissions of any references to past cycles thus are not merely casual lapses but are key operations to the strategic temporal ordering that hype seeks to effect. Minimizing attention to the temporal cycles of repetition and recycled process of hype in relation to teaching technologies, that is, enables a reinforcement of the perception of hyped technologies as products of a sudden burst of high-impact innovation and uniquely disruptive activity. Such conceptions in turn aid the justification of the expansion of new state and commercial investments in edtech initiatives promoted across a range of international conferences, government-launched national programs, and elaborately staged education-oriented product launches and campaigns that aim to enroll investments of faith and confidence among diversely situated relevant audience.

Much like the forms of 21st-century biocapital logics that anthropologist of science Kaushik Sunder Rajan investigates in global biotechnology industries, venture ed thus proves to be speculative in multiple capacities: partly as an outcome "of innovative experiments [that] are by definition unknowable, . . . [and where] the market inputs into these experiments are by unknowable," and partly too as an outcome of an "economic regime, overdetermined by the market" that channels the promise of commercial capitalism "almost to the exclusion of commodity capitalism" (2006, 111). That is, faith can be sustained and recaptured around edtech's proposed solutions—even at global scales—despite explicit failures to develop new and sustained value, or the absence of a working product of tested commercial value. Indeed, part of what makes teaching technologies associated with venture ed such curious objects of study has been their apparent stickiness as objects of expanded hope and hype across time and global terrains—as objects that have evidenced a certain resistance to cultivating disbelief, and that instead prove to be especially friendly (perhaps uniquely so) to resurrected cycles of internationally expansive hype. And without doubt, part of what appears to have enabled such notable capacities for recycling hype involves a dually oriented temporal ordering of teaching technologies: not only their ability to reanimate spectacular, promise-filled futures (Strathern 2011; Sunder Rajan 2006), but their ability to continually cleanse, expunge, and forget—rather than perhaps repurpose—disappointment in the actual past.

STS scholarship, however, has also gestured toward how mechanisms of hype—perhaps especially given its cyclical form—might be recycled and repurposed to other ends. Under such means, the past—even past records of failure and disappointment—could be framed as something other than obstacles to the presumed productivity and generativity of the "new." Instead, the past could be seen as a productive resource for imagining the present and future as otherwise. Even if hope—and hype—might be posed as oppositional frames to the work of critical scholarship, an "endpoint"—which indicates the moment when a project at last becomes retrospectively considered—opens the possibility of creating a moment too when the failure to realize projected promises could be made known and apprehendable (Miyazaki and Riles 2005). Marilyn Strathern thus underscores how in academia explicitly interdisciplinary projects frequently come to a "closure" when new partnerships fail, but that such closure is one that's realized as "only patchy." Thus, "as much as at its formal end, an identifiable shortfall [in a project] may turn into a tool for public evaluation, becoming a mark of internal criticism to the benefit of the rest of a project. . . . Disappointment in some elements of the project can work to reinforce hope in other elements" (Strathern 2011, 273).

This would require, however, a willingness of multiple sectors invested in venture ed to step outside the discursive terrain of hype and adopt an explicitly "retrospective" stance, characteristic of "contemplative knowledge" (Miyazaki 2004, 10). Such a retrospective orientation indeed becomes especially challenging in high-tech innovation fields that operate according to the temporal demands of continual high-stakes production, risk, and speculation. And the temporal conditions and demands for acceleration appear to produce further challenges for the work of critique. For while such retrospective orientations are continually necessary to press for, in the meantime it's become clear that a new generation of edtech tools and techniques is already drawing heated critique from educational practitioners concerned with the expanded ethical complications already evident in their early deployments. Of explicit concern for critics is how digital interactions in newly deployed edtech programs are being treated as means to generate new forms of information goods and digital commodities. These result in the collection and mining of data around teacher and student performances, and the quantification of learning assessments that massified digital interactions enable. Such calculations in contemporary education markets today enable powerful tools to allow micro-targeting decisions to be made on precisely where investments should be made, and where they should be withdrawn.

Aware that such new trends have begun to define educational technology markets even before they've begun to be generally perceived by the general public—and especially the vast majority of students and teachers they directly impact—education historians such as Audrey Watters have worked to underscore how increasingly today "education technology has become about control, surveillance, and data extraction" on participating actors shaped by the logics of venture ed. Human capital cultivated with the affordances of contemporary educational technology environments, that is, can be optimized as much via strategies of mining value from future human workers—produced *over the long term*—through contemporary educational technology investments and architectures, as through *immediate* value extractions strategies enacted via instantaneous data collections and calculations on current students' (and future workers') interactions within such digital architectures. As Watters writes, "Increasingly, education technology works in concert with efforts—in part, demanded by education policies—for more data. We hear these assertions that more data, more analytics will crack open the 'black box' of learning" (Watters 2014c). She cites among those making these claims most loudly as heads of educational data mining companies such as Knewton, a company that partners with educational institutions and textbook publishers to make content delivery "adaptive" and "personalizable" for individual students, gathering millions of (as it claims) "actionable" data points on millions of students worldwide each day. CEO Jose Ferreira calls education "the world's most data-mineable industry by far." As he boasted at the 2012 "Datapalooza" event, organized by the US Department of Education, "We have five orders of magnitude more data about you than Google has. . . . We literally have more data about our students than any company has about anybody else about anything, and it's not even close. . . . We literally know everything about what you know and how you learn best, everything" (Hill 2014; Ohanian 2014).

Ferreira, indeed, was only one among 150 leading entrepreneurs, software developers, policy makers, and education experts—key players in the growing educational technology software segment whose sales last year rose to nearly $8 billion, according to the Software and Information Industry Association—brought together for Datapalooza. And although messages like Ferreira's (and likely those of other product vendors at the event) stressed the new flood of "actionable" data the future promises to unleash, that prospect was framed against another, arguably more foundational message—one that began the event itself—on the sobering "reality" faced by educators today. As then secretary of education Arne Duncan stated after describing the quality of education as desperately subpar, "The factory model of education is the wrong model for the 21st century. . . . Our schools must . . . do far more to personalize instruction." And he minced few words in telling his audience that current methods and resources, including textbooks, "should become obsolete . . . as fast as we can" make it (Harber 2012).

Watters highlights that in such conditions, assessment and valuation can become rapidly economized—and turned into data calculations enabled and accelerated by new educational technology interactions. And although still largely unmentioned, the question of labor and of human capital in formation—both that of educators and now too of students—becomes key. As she writes, "Students' labor—students' test results, students' content, students' data—feeds the measurements used to reward or punish teachers. Students' labor feeds the algorithms—algorithms that further this larger narrative about teacher inadequacies, sure, and that serve to financially benefit technology, testing, and textbook companies, the makers of today's 'teaching machines'" (Watters 2014b). Which leads one to suspect that perhaps the most effective market devices are the ones you never could have remembered were worthy of suspicion to begin with.

Notes

1. The mid-20th-century excitement around teaching machines eventually calmed by the mid-1970s, as institutions began to encounter both high costs and the persistent lack of guidelines and materials to support programs that were neither "user-friendly" nor effective in producing clear outcomes as major barriers to computerizing classrooms (Benjamin 1988).
2. A survey of the *Readers' Guide to Periodical Literature* shows two citations for teaching machines prior to 1959; both were articles authored by Skinner. For 1959–60, there were 20 articles in popular magazines and another 31 in 1961–62. A similar trend can be seen in the scientific literature where "teaching machine" was initially used as an index term in the 1960 volume of *Psychological Abstracts*, which listed 12 entries. In 1961, that number increased to 20, and the following year to 24 entries.
3. The international collaborations were in Germany and South Africa.

Works Cited

American Psychological Association. 1961. "Self-Instructional Materials and Devices." *American Psychologist* 16 (8): 512–13.

Ames, Morgan G. Forthcoming. *The Charisma Machine: The Life, Death, and Legacy of One Laptop Per Child.* Cambridge, MA: MIT Press.

Andersen, Lars Bo. 2013. "A Travelogue of 100 Laptops." Doctoral thesis, Aarhus University.

Bell, J. N. 1961. "Will Robots Teach Your Children?" *Popular Mechanics*, October, 153–57, 246.

Benjamin, Ludy T. 1988. "A History of Teaching Machines." *American Psychologist* 43 (9): 703–12.

Bitzer, Donald. 1981. Interview with the Charles Babbage Institute, November 2.

Boehm, G. A. 1960. "Can People Be Taught Like Pigeons?" *Fortune*, October, 176–79, 259, 265–66.

Callon, Michel, ed. 1998. *The Laws of the Markets*. London: Blackwell.

Callon, Michel, Yuval Millo, and Fabian Muniesa, eds. 2007. *Market Devices*. Malden, MA: Blackwell.

Chan, Anita Say. 2014. *Networking Peripheries: Technologies Futures and the Myth of Digital Universalism.* Cambridge, MA: MIT Press.

Cuban, Larry. 1986. *Teachers and Machines: The Classroom Use of Technologies since 1920*. New York: Teachers College Press.

———. 2010. *Oversold and Underused: Computers in the Classroom.* Cambridge, MA: Harvard University Press.

Day, Jesse H. 1961. "Teaching Machines and Programmed Learning." *Journal of Chemical Education* 38 (5): 275.

Ferster, Bill. 2014. *Teaching Machines: Learning from the Intersection of Education and Technology.* Baltimore: Johns Hopkins University Press.

Freire, Paulo. 2006. *Pedagogy of the Oppressed.* 30th anniversary ed. New York: Continuum.

Gartner Newsroom. 2014. "Gartner's 2014 Hype Cycle for Emerging Technologies Maps the Journey to Digital Business." August 11. www.gartner.com/newsroom/id/2819918.

Gilmore, K. 1961. "Teaching Machines—Blessing or Curse?" *Science Digest*, February, 76–80.

Harber, Jonathan. 2012. "Datapalooza: A Game Changer for Education?" Pearson Research & Innovation Network. http://researchnetwork.pearson.com/digital-data-analytics-and-adaptive-learning/datapalooza-a-game-changer-for-education.

Hill, Adriene. 2014. "A Day in the Life of a Datamined Kid." *Marketplace*, September 15. www.marketplace.org/topics/education/learningcurve/day-life-data-mined-kid.

Luce, G. G. 1960. "Can Machines Replace Teachers?" *Saturday Evening Post* 36–37:102, 104–6.

Lumsdaine, A. A. 1961. "Teaching Machines and Auto-Instructional Programs: A Rationale of the Media and Analysis of the Programs." *Educational Leadership* 271–77:314.

Lumsdaine, A. A., and Robert Glaser, eds. 1960. *Teaching Machines and Programmed Learning.* Washington, DC: National Education Association.

Mackenzie, Donald. 2006. *An Engine, Not a Camera: How Financial Models Shape Markets.* Cambridge, MA: MIT Press.

——. 2009. *Material Markets: How Economic Agents Are Constructed.* Oxford: Oxford University Press.

Mackenzie, Donald, and Yuval Millo. 2003. "Negotiating a Market, Performing Theory: The Historical Sociology of a Financial Derivatives Exchange." *American Journal of Sociology* 109:107–45.

Mackenzie, Donald, Fabian Muniesa, and Lucia Sui. 2007. *Do Economists Make Markets? On the Performativity of Economics.* Princeton, NJ: Princeton University Press.

Margolis, R. J. 1963. "Do Teaching Machines Really Teach?" *Redbook* 51:98–99.

Medina, Eden, Ivan da Costa Marques, and Christina Holmes, eds. 2014. *Beyond Imported Magic: Essays on Science, Technology, and Society in Latin America.* Cambridge, MA: MIT Press.

Mitchell, Timothy. 2005. "The Work of Economics: How a Discipline Makes Its World." *European Journal of Sociology* 46:297–320.

——. 2008. "Rethinking Economy." *Geoforum* 39:1116–21.

Miyazaki, Hirokazu. 2004. *Search Results The Method of Hope: Anthropology, Philosophy, and Fijian Knowledge.* Stanford, CA: Stanford University Press.

——. 2013. *Arbitraging Japan: Dreams of Capitalism at the End of Finance.* Berkeley: University of California Press.

Miyazaki, Hirokazu, and Annelise Riles. 2005. "Failure as an Endpoint." In *Global Assemblages: Technology, Politics, and Ethics as Anthropological Problems*, edited by A. Ong, S. Collier, 320–31. Oxford: Blackwell.

Morello, T. 1965. "Which Is It? New World of Teaching Machines or Brave New Teaching Machines?" *UNESCO Courier*, March, 10–16.

Muniesa, Fabian, Yuval Millo, and Michel Callon. 2007. "An Introduction to Market Devices." In *Market Devices*, edited by Fabian Muniesa, Yuval Millo, and Michel Callon, 1–12. Malden, MA: Blackwell.

Negroponte, Nicolas. 2009. "Lessons Learned and Future Challenges." Speech at Reinventing the Classroom: Social and Educational Impact of Information and Communication Technologies in Education Forum, Washington, DC, September 15.

Ohanian, Susan. 2014. "An Update on Datapalooza at the U.S. Department of Education." *Substance News*, August 3. www.substancenews.net/articles.php?page=5135.

Press, Gil. 2014. "It's Official: The Internet of Things Takes over Big Data as the Most Hyped Technology." *Forbes*, August 18. https://www.forbes.com/sites/gilpress/2014/08/18/its-official-the-internet-of-things-takes-over-big-data-as-the-most-hyped-technology/#3fd1f83f3aca.

Severin, Eugenio, and Christine Capota. 2011. "One-to-One Laptop Programs in Latin America and the Caribbean: Panorama and Perspectives." Inter-American Development Bank Publications. http://publications.iadb.org/handle/11319/4919?locale-attribute=en.

Skinner, B. F. 1958. "Teaching Machines." *Science* 128:969–77.

Strathern, Marilyn. 2011. "An Experiment in Interdisciplinarity: Proposals and Promises." In *Social Knowledge in the Making*, edited by Charles Camic, Neil Gross, and Michelle Lamont, 257–84. Chicago: University of Chicago Press.

Sunder Rajan, Kaushik. 2006. *Biocapital: The Constitution of Postgenomic Life.* Durham, NC: Duke University Press.

Warschauer, Mark, and Morgan Ames. 2010. "Can One Laptop Per Child Save the World's Poor?" *Journal of International Affairs* 64 (1): 33–51.

Watters, Audrey. 2014a. "The History of Personalization and Teaching Machines." *Hack Education*, July 2. http://hackeducation.com/2014/07/02/personalization-teaching-machines/.

——. 2014b. "Teacher Wars and Teaching Machines." *Beacon Reader*, September 21. www.beaconreader.com/audrey-watters/teacher-wars-and-teaching-machines.

——. 2014c. "Convivial Tools in an Age of Surveillance." Talk delivered at NYU Steinhardt Educational Communication and Technology Colloquium Series, November 13. http://hackeducation.com/2014/11/13/convivial-tools-in-an-age-of-surveillance/.

——. 2015. "The First Teaching Machines." *Hack Education*, February 3. http://hackeducation.com/2015/02/03/the-first-teaching-machines/.

Weisenberg, C. M. 1961. "Teaching Machines." *Commonwealth*, January 27, 454–56.

Dangerous Networks
Internet Regulations as Racial Border Control in Italy

Camilla A. Hawthorne

How does a particular technology enable certain power-laden practices of "racializing surveillance" (Browne 2012), practices that also coincide with national borders?[1] Following the 2004 train bombings in Madrid and the 2005 attacks on London's public transportation system, the Italian government quickly enacted the strictest Internet regulations in the European Union. Law No. 155/2005, popularly known as the Pisanu Decree, was intended as an antiterrorism package that would protect Italian citizens from the threat of "Internet users who have been sensitized by Al Qaeda" ("Informativa Del Governo" 2005). The Pisanu Decree included a slate of new surveillance procedures targeting Internet cafés—businesses that in Italy are largely immigrant owned, operated, and frequented. The Pisanu regulations required the mandatory collection of identification documents for all Internet café users, as well as their web browsing histories. At the same time, the physical space of the Internet café emerged as a regular target for police inspections and raids against immigrants of color and Muslims. This chapter explores the lifespan of the Pisanu Decree's Internet café regulations (2005–12) to argue that sociotechnical imaginaries (Jasanoff and Kim 2009) and governmental practices of immigration control are always co-constructed and, in the case of the Italy, generated new practices of border drawing and racial classification that worked through the *placefulness* of Internet access. This networked surveillance was profoundly spatial, filtering the "metadata" of an individual's movements, associations, and situated technological habits through stereotypical understandings of Muslim culture in order to constitute a new, identifiable racial subject—the radicalized, male Muslim Internet user lurking within the nation's borders.

Early theorists of cybercultures portrayed the Internet as a deterritorialized "space of flows" (Castells 2000; cf. Shklovski et al. 2013, 12) with the power to simultaneously render borders insignificant, challenge the authority of centralized nation-states, and eliminate both identity politics and body-based forms of prejudice (Turner 2006). The idea of the "borderless" Internet was dealt a significant blow by the rise of Web 2.0 around 2005, as well as the 2013 revelations of NSA telecommunication surveillance through PRISM and MUSCULAR (the latter in collaboration with the British Government Communications Headquarters). While extensive literature exists on the use of biometric technologies for border securitization (Magnet 2011), however, the deployment of Internet regulations and surveillance to reinforce borders and control immigrant populations remains an underexplored topic. In the post-9/11 era, governments around the world have increasingly enrolled the Internet into their border control efforts through surveillance strategies that are folded into national security and antiterrorism programs that disproportionately target communities of color. The notion of the "splinternet" (Thompson 2010; Ananthaswamy 2011; Malcomson 2016), a term coined to describe the global instantiation of multiple, stratified Internets as a result of paid "walled gardens" and government firewalls, can thus be applied here to refer to the way in which the Internet is experienced as freedom for some and surveillance for others.

Just as the Internet is not frictionless, the Internet is not raceless; we must therefore continue to heed Donna Haraway's provocation in the "Cyborg Manifesto" (1991, 165) to take seriously the "rearrangements of race, sex, and class rooted in high-tech facilitated social relations." Central to the argument of this chapter is an understanding of race itself as a sociotechnical system—an "arrangement of humans, technologies, architectures, spaces, and policy regimes" that encompass "the biological, the sociological, the political, and the technical" (Forlano and Jungnickel 2015). Thus, race is not a stable category but rather articulates with human and nonhuman actors, policy regimes, and historical processes to generate new and often contradictory meanings in the present. And just as the theoretical tools of STS can help to open the "black box" of race, they can also uncover the embedded assumptions, imaginaries, practices, infrastructure, and relationships that are too often glossed as "the digital." If race is a sociotechnical rather than a biological phenomenon, then in the digital age technologically mediated social networks are the raw materials for the construction of racial subjects.

In the case of Italy, the government's sociotechnical imaginary of the Internet as an unruly, dangerous space of cyber-networking facilitated new forms of capillary control, surveillance, and territorialization. The Pisanu Decree operated through the temporal and spatial expansion of the border through networked surveillance practices, as well as the aggregation, analysis, and reassembly of web traffic data linked to nonvirtual sites such as urban Internet cafés in immigrant enclaves. By reworking older forms of racial boundary drawing in Italy, these everyday "border controls" made it possible for the state to identify threatening subjects through the isolation of specific, suspicious technological spaces, practices, and associations.

This chapter begins with a brief overview of Italy's transition from a country of emigration to a country of immigration, with a focus on the politics of difference and the criminalization of migration. I then turn to a review of theories pertaining to race and the Internet, as well as insights from postcolonial STS about the relationship between race, technology, and Western ideas of civilizational progress. The third section of the chapter analyzes the enactment, enforcement, and limitations of the Pisanu Decree through legislative reports, government documents, and popular media. I conclude with an analysis of the politics of digital circulation in Italy and the distinctiveness of racial profiling through Internet surveillance as opposed to systems such as biometric bordering.

Context: Immigration and Racism in Contemporary Italy

Historically, Italy was a country of mass emigration. Analysts have attributed Italy's transformation into a country of immigration in the 1980s to a variety of factors, with the result that by the end of the 1990s Italy's population began to represent a new plurality of national groups (Merrill 2006). The 2005 Pisanu Decree must be situated within the context of a much broader shift from laissez-faire to highly restrictive Italian immigration policies that also involved Italy's participation in the Schengen Agreement and the border regimes of Fortress Europe, Silvio Berlusconi's alliance with the far-right Northern League, and a post-9/11 emphasis on securitization. At the heart of the debates surrounding immigration to Italy is a profound ambivalence about Italy's position as a site of postcolonial immigration. Despite the fact that Muslims represent less than two percent of Italy's population, Islam in particular—as the second most widely practiced religion in Italy—has catalyzed a widespread moral panic about religious and cultural difference. Thus, immigration law, citizenship policies, the interiorization of the border through surveillance, and the proliferation of informal multiculturalisms in Italy can all be understood as efforts to manage internal difference in Italy.

It is also important to note that despite a long and troubled history of race thinking and the use of race as an organizing principle for the Italian nation, "race" as a descriptive or analytical category was disavowed in post–World War II Italy (Mellino 2012). Use of the word *razza* (race) remains taboo due to its association with fascism; instead, *etnia* (ethnicity) is commonly deployed to mark human groups. The discursive vacuum surrounding race in Italy has not repressed discussions about difference and national belonging, but instead has incited multiple state and non-state sources to engage different genealogies of Italianness and generate alternative modes of essentialization. These include, but are not limited to, geography, language, religion, and cultural inheritance. Indeed, as Stoler ([2002] 2010, 144) contends (via Foucault's refusal to conflate "race" with the biological), racisms gain their force from the "internal malleability assigned to the changing features of racial essence." In the context of digital surveillance, categories such as religion are fused to associations between technological practices and technologically mediated social relations to fix individuals as classifiable racial subjects. An STS-inflected understanding of race as a sociotechnical practice can therefore reveal the ways in which racial meanings are rearticulated with the digital even when the word "race" is not uttered.

STS and the Politics of Race on/and the Internet

The enactment of the Pisanu Decree's antiterrorism package suggests that deployments of technology cannot be analyzed separately from questions of difference, and that the meanings of "race" are themselves actively transformed by technologies. An understanding of Internet surveillance as co-constitutive of racial categorization requires a different theoretical genealogy of the relationship between race and the Internet. In other words, it requires shifting from an understanding of the Internet as merely a tool that can be used for racist purposes (for instance, by controlling circulation and access to information for specific groups, or spreading harmful stereotypes) to one in which the Internet, as a sociotechnical space of practices, is entangled with ever-evolving articulations of racism and race (for instance, through menu designs that unintentionally reify bounded racial groups, or surveillance algorithms that essentialize and criminalize individuals based on the geographies of their digital webs of association).[2] This is of course *not* to suggest that the Internet, as a sort of "sentient" agent, deterministically produces race out of whole cloth, creating racisms where they did not previously exist.

It is important to note that scholars of postcolonial and feminist science and technology studies have long pointed to the inextricable relationship between science, technology, racism, and modernity (Haraway 1989; Prakash 1999; Drayton 2000; Mitchell 2002; Phillip 2003). Their insights are relevant to any discussion of "racial formation" (Omi and Winant 1986) and the Internet. Within the context of European colonial expansion, technological difference was coded as racial difference and inferiority, and thus served as a justification—vis-à-vis notions of "civilizing missions"—for globe-spanning European imperial entanglements. Colonized natives, for failing to rise above the natural world with the help of domesticating technologies, were bounded as part of nature, which itself was constructed as a separate object for technological intervention by rational, detached experts.

Following the rupture of the dotcom bubble in 2000, however, a new body of scholarship emerged from the fields of new media and science and technology studies that critically reexamined some of the initial emancipatory promises and "founding fictions" (Nelson 2002, 1) of the Internet. The gleeful utopianism of early Silicon Valley cyberculture, as chronicled by Fred Turner (2006), was characterized by a conviction that virtual communities had the power to render categories such as race and gender obsolete and foster the elaboration of associational forms that were no longer bounded by spatial proximity. One important line of inquiry in the post-2000 literature focused specifically on challenging the claim that race and racism would cease to exist online. Key theorists of race and the Internet (Nakamura 2002, 2007; Nelson 2002; Chun 2006; Landzelius 2006; McGahan 2008; McLelland 2008; Daniels 2009; Everett 2009) argued that, in fact, race had actually *proliferated* on the Internet (Chun 2006) in the forms of racial identity tourism, visual and textual representations of race in community forums and games, and online white supremacist outposts. Prior to this work, most discussions of race on the Internet focused on "digital divides," a concept that scholars such as Nakamura (2002, 2007), Chun (2006), and Everett (2009) have critiqued in part for failing to capture the creative cultural productivity of minorities on the Internet. Attempts to revise this portrayal of minority groups as technologically lacking have taken the form of counternarratives that emphasize Black "technolust" (Everett 2009), emancipatory Afrofutures (Dery 1994), or digital innovation. Such redemptive stories of minoritarian cybercultures, however, are now being superseded by cautionary narratives of racist stereotyping, bullying, and surveillance online.

Key to this literature is the suggestion that digital disembodiment does not necessarily bring with it liberation from oppressive categories of race, gender, sex, or nation (Nguyen 2003). Instead, the Internet actually "propagates, disseminates, and commodifies" images of race (Nakamura 2002, 3). Reductive representations of race on the Internet serve to anchor or stabilize cyberspace precisely because it is imagined as a medium that severs the link between intelligible material bodies and representation; as Nakamura (2002, 5) writes, these "cybertypes" (online racial or ethnic stereotypes) "both stem from a common cultural logic and seek to redress anxieties about the ways that computer-enabled communication can challenge these old logics." The paranoia produced by the invisibility of technology is expressed as discourses of regulation and control that, in the post-9/11 period, now focus on dangerous people—e.g., terrorists—as opposed to dangerous content—e.g., pornography (Chun 2006). In other words, the Internet can be simultaneously reified as a postspatial, postracial utopia of unrestricted freedom and as a dangerous, unruly space that must be secured through the mobilization and reinscription of race and racial categories—a phenomenon Chun (2006) calls "control-freedom."

Implicit in these arguments is that centering the materiality and spatiality of the Internet can also reveal the reproduction of racisms and race online. Contrary to outdated predictions of "global villages" (McLuhan ([1962] 2011) and interactive communities based "not on common location but *common interest*" (Licklider and Taylor [1968] 1990, 3), Internet use is in fact characterized by experiences of friction (Tsing 2005), unevenness (Kraemer 2013), and boundaries (Shklovski et al. 2013)—including national boundaries. After all, the persistent influence of quotidian "offline" factors such as time zones (Boellstorff et al. 2012), along with political and infrastructural factors such as Internet censorship, surveillance, and bandwidth restrictions, serve as an important reminder that the Internet is not an abstract, intangible medium but is instead shaped by material realities and physical, embodied practices (see Nemer and Chirumamilla, this volume).

Race on the Internet can be thus understood as both restrictive—performing boundary work to militate against forms of cultural hybridity that do not conform to narrow, cosmetic multiculturalisms (Nakamura 2002, 20–21)—and generative of new modes of representing racial bodies (Nakamura 2007, 13). In addition, race itself is constitutive of the Internet. For example, race and gender are integral to hardware production and communication service provision (see Poster, this volume)—the individuals who assemble circuit boards and other electronic components, or who work in call centers, are largely women and/or of Asian heritage (Chun 2006, 72–73). In addition, race (or its disavowal) was central to the conception of cyberspace as utopian (Chun 2006, 129) and populated by unmarked "virtual homesteaders" (Rheingold 1993).

To conclude, the relationship between race and technology is structured in dominance, and can be expressed in three ways. First, technologies enable new forms of racialization, or racial categorization and boundary drawing. Second, the concept of technology is itself racial, fundamentally intertwined with Enlightenment-era scientific racisms that link the technological subjugation of the natural world with modernity, civilization, and progress (Adas 1990). Third, race can also be understood as a technology, a "levered mechanism" (Coleman 2009, 178) that "creates parallel social universes and premature death" (Benjamin 2016). But how are we to think specifically about the interplay of race and technology in relation to emerging border regimes? The case of the Pisanu Decree suggests that sociotechnical imaginaries of the Internet and Internet users can render technology as both *marker of* and *tool for* racial classification. The identification of certain forms and physical spaces of Internet activity can be used to isolate Others who are believed to threaten the integrity of the national body.

Securing Cyberspace: The Pisanu Decree and Internet Surveillance

Following a string of terrorist attacks in Madrid (2004) and London (2005), the Italian legislature moved quickly to enact a new set of national antiterrorism policies. The speed and near-unanimity with which this legislation was approved were remarkable, given the notoriously glacial pace of Italian government proceedings.

Law No. 155/2005 ("Urgent Measures for Combatting International Terrorism," also known as the Pisanu Decree) passed on July 31, 2005 and contained a series of new procedures governing residency permits, immigrant expulsion, telecommunication surveillance, and flight school administration (Conversione in Legge 2005). Article 7 of the law focused exclusively on public Internet use and included the following key provisions:[3]

1. Anyone who opens a new public Wi-Fi hotspot or public space with terminals for electronic communication (i.e., an Internet café) must first apply for a license from the local police headquarters. A license is not necessary for operators of public pay phones with only voice telephony services. This license is dependent upon the establishment having put in place appropriate data monitoring and retention systems; license applicants are also required to submit detailed information about their businesses, including floor plans (Hooper 2005).
2. The owner or operator of an Internet café must monitor the activities of customers and archive their data for at least six months—this includes documenting what computers they use, recording their log-in and log-out times as well as when they enter and exit the premises, and purchasing tracking software that saves a list of all sites visited (Celeste 2005). The web browsing logs must be submitted periodically to local police headquarters.
3. The owner or operator of an Internet café must record customers' personal data by photocopying an identity document such as a passport.

At this time, Italy already followed the 2002 amendment to the European Union Directive on Privacy. Article 15 permits Internet service providers to temporarily retain records of user activity, which can then be made available to law enforcement for the purpose of safeguarding "national security . . . defence, public security, and the prevention, investigation, detection, and prosecution of criminal offenses" (OpenNet Initiative 2010).[4] While the earlier 1997 EU privacy directive had required ISPs to erase customers' communication traffic data (Levi and Wall 2004, 203), the 2002 amendment allowed European Union member states to selectively restrict Internet users' right to privacy in the context of state security (European Parliament and Council 2002)—somewhat like the infamous USA PATRIOT Act 2001. In Italy, ISPs were required to cooperate with police and courts during investigations, but did not retain any online activity data except for the details of Internet payments for up to six months (Reporters without Borders 2004).

Although several European countries introduced new security measures after the London bombings (BBC 2005a), the Pisanu Decree established Italy as the only country in the European Union to require the presentation of identity documents at Internet cafés (Switzerland, which is not an EU member, did require Internet café customers to show ID). Some left-leaning politicians concerned with the suspension of individual rights to privacy contested these provisions; however, their reservations were ultimately outweighed by a more dominant preoccupation with terrorism and national security. The Pisanu Decree ultimately passed both chambers of the Italian Parliament with sweeping majorities.

Accordingly, the Pisanu Decree also marked the beginning of a period of intensified Internet surveillance in Italy, under the auspices of national security. Internet privacy advocates have previously suggested that the heavy-handedness of the Pisanu Decree and other attempts at Internet regulations in Italy can be attributed to the Italian government's lack of understanding of the Internet and privacy issues (Pavis 2000; OpenNet Initiative 2010); indeed, Internet penetration rates in Italy have consistently lagged behind those of most other European countries. The post-2005 move toward greater surveillance of the Internet and Internet access points in Italy can be more accurately described, however, as an imperfect attempt to resolve the inherent tensions between the ideal of Internet freedom and a concern with online terrorist networks. By focusing on specific *sites* of Internet use, which were in turn connected to specific *types* of Internet users, the Italian state could respond to growing popular and international calls for antiterrorist Internet surveillance in the wake of Madrid and London. A preoccupation in Italy with the dangerous connections that could be formed through the Internet, however, can be traced at least to 2004—in the wake of global post-9/11 surveillance and securitization efforts (Levi and Wall 2004). That year, the Ministry of the Interior reported to the Italian Parliament that "the use of telecommunication networks by fundamentalist groups represents an aspect of intense interest, as the Internet has now taken on the form of an interactive mass medium, whereas before the network was used as a means of internal communication among small groups with strictly operational needs" (Terrorismo Ed Eversione 2004, 17).[5] In this report, the Internet was figured as a sprawling network that facilitates unlimited communication and interaction among individuals who are not necessarily known to each other—a departure from a supposed earlier, less threatening iteration of Internet networking for benign purposes.

A 2010 Ministry of the Interior document titled "Security, Immigration, and Asylum" reflected the operationalization of this concern with jihadist networks on the Internet. According to the document, "Specific attention has been dedicated to the fight against radicalization and recruitment, starting with the monitoring of the Internet" (Ministero dell'Interno 2010, 11). In 2012, the Ministry of Justice published a report about Islamic radicalization in the penitentiary system. Although the report ostensibly addressed proselytization in Italian prisons, the authors devoted significant space to the dangers of the Internet:

> The development of the information society, in fact, has not escaped the Islamic world and the potential offered by new technologies (especially the Internet) constitutes one of the principle vehicles for the diffusion of ideologies, allowing something born at the local level to transform into something global. . . . The shared function of these various sites is that they sustain the jihadist infrastructure through the distribution of communications, secret messages, and propaganda materials, and we cannot forget the important role that they play in the recruitment of potential jihadist candidates. In fact, the main concern resides in the fact that the Internet has become a virtual training field. (Ministero della Giustizia 2012, 27)

As a result, the authors recommended the monitoring of "Internet networks" with a focus on jihadist sites and, in particular, discussion forms, along with the control of Internet cafés and other sites "frequented by radical elements" (Ministero della Giustizia 2012, 31). By incorporating a discussion of the Internet into a report about the Italian penitentiary system, the authors suggested that an unmonitored Internet has the potential to penetrate and undermine one of the most robust symbols of the state's power to discipline the population within its borders: the prison.

The concept of sociotechnical imaginaries is useful for understanding how the Internet was directly enrolled into Italy's national security and border control efforts in 2005. According to Jasanoff and Kim (2009, 120), sociotechnical imaginaries are "collectively imagined forms of social life and social order reflected in the design and fulfillment of nation-specific science and/or technological projects." Sociotechnical imaginaries prescribe "futures that ought to be attained" (120), but also warn against "risks or hazards" (123). Unlike popular media tropes, however, sociotechnical imaginaries are closely associated with the flexing of state power and the enactment of national policies (123).

Sociotechnical imaginaries move discussions of technology away from technologically determinist framings in which a technical system is locked in to a predetermined set of politics, policies, and social arrangements. Unlike Winner (1989), who suggests that particular technological artifacts are compatible with certain social and political orders, Jasanoff and Kim suggest that sociotechnical imaginaries and social orders or policies produce one another in a dynamic relationship that varies over both space and time. This insight is particularly valuable when discussing a technology such as the Internet. The openness of the Internet, as it is popularly understood, is simultaneously perceived as both liberating and threatening. In addition, the Internet as a decentralized network (admittedly, an oversimplification of its material infrastructure) is conducive both to freedom and to new forms of surveillance or "networked authoritarianism" (Pearce and Kendzior 2012). Italian policy makers' engagements with the Internet as a space of dangerous, unrestrained transnational networking that cannot be contained by national borders—influenced not only by post-9/11 terrorist attacks but also by Italy's own experiences with domestic terrorism by the Red Brigades and far-right groups during the Years of Lead and the vast criminal networks of the mafia—can thus be understood as a form of sociotechnical imagination.

Yet, as Benjamin (2016) notes, Jasanoff and Kim's formulation also acknowledges the coexistence of multiple imaginaries of a technology within a particular national space. These imaginaries can be deployed toward different ends and in relation to different populations. The Italian sociotechnical imaginary of the Internet is not unified, but rather incorporates racial distinctions between different kinds of Internet use: public versus private, individual versus communal, European versus non-European, cosmopolitan versus ethnically particularistic.

The debates in the Italian legislature surrounding the passage of the Pisanu Decree reflected a conflation of concerns about security and radical Islam, fears that were refracted through popular imaginaries of the Internet. Indeed, it was quite common during this time to see salacious news stories circulating in the Italian media that explicitly linked Internet use, the threat of terrorism, and undocumented immigration. The passage of the security package was justified with references to the London bombings and the fact that Internet surveillance and the policing of Internet cafés were used by police forces in both Italy and the United Kingdom to identify and locate suspected terrorists. In particular, the arrest in Rome of Hussein Oman (also known as Hamdi Isaac)—one of the men behind the London bombings who also supposedly communicated with his relatives in Italy over the Internet (BBC 2005b)—framed discussions about the securitization of the Internet and Internet cafés. Still, despite several vague terrorist threats against Italy due to the country's involvement in the Iraq War, Beppe Pisanu (then minister of the interior and namesake of Law No. 155/2005) admitted during discussions of the antiterrorism package that there was no specific evidence of an impending attack (BBC 2005a). According to Carlo Taormina, a member of the Chamber of Deputies with Silvio Berlusconi's conservative Forza Italia party, however, "The tools of investigation, which this decree will strengthen, enabled the effective identification of the person responsible for the failed attack on the London Underground on 21 July 2005. . . . It was possible to achieve the identification of the person responsible through the utilization of telephone tracking and through the identification of the people who provided the terrorist with logistical support—the operator of an Internet café, which will be subject to more restrictive control thanks to the application of the measures in the decree under discussion" (Decreto-legge 144/05 2005). During the debate, other legislators expressed concern that the Internet makes a wide spectrum of information about the preparation and use of explosive materials, firearms, and other types of weapons easily accessible. These anxieties about the information that can be accessed through the unrestricted networks of the Internet resurfaced in a December 2005 review about the first months of the Pisanu Decree's implementation. During this special parliamentary session, several legislators and government officials suggested that jihadist websites hosted in the Middle East could pose serious, material threats to the Italian social and political order. Pisanu reminded parliamentarians about online jihadist media outlets such as the Global Islamic Media Front:

> I remember as well that last November the Global Islamic Media Front, a promoter of what is considered "news" from Al Qaeda, distributed over the Internet a video reaffirming the strategy of the terrorist organization: recruitment, training, and encouragement of the mujahedeen over the web; promotion of jihadist media to intimidate the "crusaders"; and celebration of Al Qaeda in Mesopotamia as an example for all of the other armed movements. Subsequently, on 24 November, the same organization distributed a threat against the President of the Council of Ministers and the Italian people. In general, we can consider these various threats against our country as the work of Internet users who have been sensitized by Al Qaeda, who pose a very high risk, and play on their knowledge of current Italian politics. ("Informativa Del Governo" 2005)

Pisanu describes the Internet as a dangerous, decentralized transnational network for the dissemination of both propaganda and violence. In his statement, the primary danger to the Italian people is the Internet as an "*open* university in terrorism" (Chun 2010, 343), not terrorism itself. Notice, for instance, that he characterizes these threats against the Italian people as the work of "Internet users," *not* terrorists, who have been radicalized by Al Qaeda through their online encounters with jihadist websites. These Internet users, whose physical location is unspecified and uncertain, are also able to infiltrate the Italian political system and then use this information to generate threats to the country's national security.

The Internet Café as Zone of Alterity

In the implementation of the Pisanu Decree, Internet cafés became material spaces of intervention into the supposedly dangerous, immaterial networks of the Internet. In order to grasp the significance of the Internet café, it is important to first situate these spaces within the broader context of immigration to Italy. Internet cafés are popularly imagined as immigrant spaces, even though they also cater to tourists. They are usually clustered in areas with large immigrant populations, such as the neighborhood behind Rome's main Termini train station, and often serve as meeting places for "a community of newcomers" (Carter 2013, 203). Over the past 20 years, a growing need for money transfer centers (which are usually colocated with Internet cafés) for diasporic remittances, along with restrictive business and employment regulations that also favor Italian citizens, have created a niche for migrant self-employment through Internet cafés. A 2012 report found, for instance, that almost 94% of the Internet cafés in Rome are operated by foreigners (*Yalla Italia* 2012). While Internet cafés are themselves not highly lucrative enterprises, a factor that explains the dearth of Italian-owned Internet cafés, immigrant families often manage other businesses such as restaurants or shops that help to distribute costs, profits, and risk among different family ventures.

The association of Internet cafés with immigrants in Italy has been a major source of tension and fear for many "native" Italians. In the Reggio Emilia province, for instance, Forza Italia leader Claudio Guidetti remarked that Internet cafés and call centers are frequented mostly by "undocumented or irregular immigrants or those dedicated to terrorism" (Provincia Di Reggio Emilia 2006). In 2009, journalist Tom Kington interviewed an anti-immigrant protester in Tuscany who lamented that not just the kebab shops but also the call centers and Internet cafés were all "managed by foreigners" (Kington 2009).

The association of Internet cafés with immigrants is also closely linked to the relationship between software and modernity (Chun 2010). The computer and Internet access, like the plow for an earlier generation of anthropologists (see Goody [1971] 1980), are commonly used as markers of civilization and modernity. It is for this reason, as Chun (2006) argues, that "digital divide" rhetorics have troublingly colonial undertones. A conception of "bridging the digital divide" solely in terms of Internet access or laptop usage, for instance, would create "'junior users' not unlike 'colonized' subjects who were structurally dependent on knowledge from the motherland" (Chun 2006, 152; see also Burrell 2012). Importantly, however, computer usage must be linked to *personal* ownership (tied to Lockean understandings of individual property rights) if one is to be situated on the "correct" side of this digital divide (Chun 2010; see also Chan, this volume).

Viewed in this context, the Internet café stands as a marker of an uncivilized, premodern form of communal computer use. The Internet café is not just an "immigrant space," then; it is also a zone of technological Otherness and backwardness. This distinction between personal and communal ownership of technology articulates with Italian parliamentarians' association of Islam with collectivism rather than liberal individualism. In a Chamber of Deputies hearing, Umberto Ranieri of the social democratic Democratici di Sinistra party observed,

> This is a complex question: immigrants in the West express a religious question that serves to reinforce an identity put in crisis by the disorienting experience of immigration and that directs them to the network of mosques and Islamic associations. However, even on this point it is important to be careful: the association of Islam with a total ideology, a system that does not accept distinctions between religion and politics, is not shared by the majority of Muslim immigrants. All of the experts say it, the move to the West changes the relationship of Muslims with their religion and signals a progressive individualization. (Disegno di Legge di Conversione 2005)

Islam in this statement is associated with strong cultural ties—and potentially dangerous religious networks—that subsume the individual beneath a larger set of community obligations. This is, of course, a classically Orientalist portrayal of the Muslim world (Said [1978] 2014). According to Ranieri's "experts," however, migration to the West leads to a teleological process of individualization in which the individual Muslim immigrant is able to extricate herself or himself from these overbearing, non-Western cultural networks. It is no surprise, then, that Internet cafés, which allow Muslim immigrants to potentially access jihadist websites, are associated with this sort of dangerous collectivism. Internet cafés, as spaces of communal technology use, can reverse the modernization process catalyzed by the experience of migration to Italy and immersion in "Western" or "European" society. In other words, Internet cafés are spaces that enable new kinds of social relations that Italian policy makers perceive as highly threatening to Italian national security and social order.

Once enacted, the Pisanu Decree had a disproportionate impact on immigrant communities in Italy, and particularly on undocumented immigrants. A Bangladeshi immigrant and Internet café owner interviewed shortly after the law passed lamented, for instance, that he had no clients left (Sanminiatelli 2005). This reduction in clientele has been largely attributed to the law's ID registration requirement. Several cafés whose customers included undocumented immigrants turned a blind eye to this provision; in fact, one sociologist described the law as "useless" because customers intent on using Internet cafés could simply present false identification documents in order to skirt the regulations (Sanminiatelli 2005). Still, the ID requirement had a significant chilling effect on Internet café use among those immigrants, who chose to employ analog "privacy enhancing strategies" (Levi and Wall 2004, 210) to displace the impact of surveillance, such as avoiding Internet cafés altogether. In addition, the labyrinthine requirements for Internet café registration discouraged many entrepreneurially minded immigrants from opening new businesses.

In addition to the Pisanu Decree's requirement of ID collection and data tracking in Internet cafés, the Italian police also regularly targeted Internet cafés as sites for inspections and immigration raids. At the height of the decree's enforcement, many cafés experienced weekly—and in some cases even daily—police inspections. In many cities, municipal governments attempted to shut down immigrant-owned Internet cafés under the auspices of "terrorism," though in most cases they were actually closed for neglecting to collect customers' identification data. In just two days of August 2005 alone, police sweeps were carried out across Italy against 7,318 call centers, Internet cafés, money transfer points, and halal butchers—spaces targeted as meeting points in Muslim communities (Camera dei Deputati 2005). A total of 32,703 people were identified, 341 were arrested, and 426 were charged with various crimes (Camera dei Deputati 2005). In addition, 701 expulsion procedures were initiated and 325 fines were levied against call centers, Internet cafés, and money transfers for "administrative irregularities" (Camera dei Deputati 2005). While butcher shops represent material incursions of cultural difference into the boundaries of Italian territory, Internet cafés and related businesses signal the formation and elaboration of communication and financial linkages with potentially dangerous elements *outside* of the Italian territory. Both types of establishments threaten Italian national security, but in markedly different ways.

While the Italian police used Internet cafés as a way to target "undesirable," "dangerous," or "marginal" populations (i.e., undocumented immigrants of color and Muslims), there was more to the focus on Internet cafés than simply efficient policing. As seen in the parliamentary discussions of the Pisanu Decree described earlier, Internet cafés also became shorthand for the sorts of dangerous transnational networking that are facilitated by the technology of the Internet. In other words, the Italian government's targeting of Internet cafés represented an effort to simultaneously fix mobile immigrant populations *and* the unruly networks of the Internet. Muslim immigrants were conflated with the "dark side" of the Internet, and this racial categorization took material form in the space of the Internet café. The Internet, after all, is both spatial *and* material, and this is perhaps most obvious in the Internet café. Individuals must negotiate the particularities of place—urban geographies, zoning laws, material infrastructure—that coalesce in the Internet café in order to gain access to the transnational flows of the Internet. The Internet café can therefore be understood as a material node in the unruly, spatially extended Internet, and because of its stability, its fixedness in place, it became a natural target for immigration control and antiterrorism policing efforts seeking to contain radical transnational communication and intervene in dangerous networks.

Circulation, Control, and Racialized Networks

In post-2005 Italy, an overarching sociotechnical imaginary of the Internet as a space of freedom was simultaneously linked to techno-utopian visions of innovation *and* to alarmist fears of terrorist networking that cannot be contained by the borders of the nation-state. For this reason, debates about the Pisanu Decree encompassed concerns about both the restriction of entrepreneurship and invasions of user privacy (i.e., the maintenance of an open and free Internet) and the need to control suspicious immigrants (i.e., the dangers of unrestricted online communication)—what Foucault ([2004] 2007, 18) described in his "Security, Territory, Population" lectures as the division between good and bad circulation. The official imaginary of the Internet in Italy as both *constructive* circulation and *dangerous* circulation is therefore deeply racialized, with the latter deployed against people of color, and Muslims in particular. This tension between freedom and control (Chun 2006) can also be read as an extension of an opposition that has been central to the structure of the Internet since its development—namely, what Galloway (2004, 8) describes as the contradiction between anarchic distribution, represented by the TCP/IP protocols, and rigidly controlled hierarchies, represented by the DNS protocol.

If the idea of race is central to the formation of the modern European nation-state (Gilroy 1987; Balibar 1991; Goldberg 2002; Foucault [1997] 2003), then the border regimes of Fortress Europe and the technologies of digital and biometric surveillance they employ are also about race and the management of difference. Indeed, as Browne (2012, 2015) argues, modern practices of surveillance have their roots in the regulation of the movement of enslaved and colonized people. These bordering practices, while not limited to territorial borders, have material and violent consequences (Pugliese 2013a) even as they come to encompass increasingly mundane and quotidian activities such as frequenting an Internet café or responding to emails from family members in distant countries. These technologies allow for the policing of racial difference within the national body, and also constitute new types of racial subjects that can be categorized as Others on the basis of their social associations and technology use, as opposed to just their phenotype or biological makeup.

Studies of biometric borders as a mechanism for regulating circulation often emphasize a visual logic—the scanning of faces to reveal the "foreign terrorist in the nation" (Chun 2009, 25). Pugliese (2013b, 571), for instance, argues that biopolitical technologies of extraterritorialization constitute "regimes of statist visuality." Following this logic, the visual, face-centered cultures of social networking on the Internet (González 2009) provide new opportunities for panoptic facial scanning. As critical race theorists argue, however, race and classification are about much more than the visual—even Fanon ([1952] 2008; see Browne 2009) acknowledged that epidermalization implicated not only the gaze, but also interlinked psychological, structural, and spatial processes in the production of racial subjects. Thus, it is possible to think about Internet surveillance without privileging scopic metaphors, but rather by considering vision as one among many techniques of perception (Crary 2001).

In Italy, for instance, many Internet café owners installed video cameras to monitor their computer stations after the passage of the Pisanu Decree; in addition, businesses were required to register comprehensive floor plans with local police and track exactly which computer terminals their customers used and for how long. This optic surveillance, however, was inoperable if separated from its relationship to other information such as photocopies of passports and logs of Internet browsing and communication data (collected both through software installed on individual Internet café computers and at the level of the ISP). Surveillance, therefore, is not simply a matter of making threatening subjects visible, but is instead about forming *new* subjects from the assembly of discrete data points such as one's web traffic histories, online communications, patterns of Internet café use, and movement through space. This sort of "dataveillance" (Clarke 1994; Levi and Wall 2004; Amoore and de Goede 2005) works by compiling and processing raw data as inputs (van Dijk 2014) that, via algorithmic logics, can produce certain bodies and networks as "risky" (Epstein 2008, 179). Algorithmic war, Amoore (2009, 49) argues, is powerful precisely because of this invisibility. It creates "association rules" between people, places, objects, and events, using the prosaic and the everyday to make preemptive security decisions in a Foucauldian continuation of war by other means.

Unlike biometric surveillance, which attempts to make bodies visible as faces or organic molecules, Internet surveillance is (perhaps counterintuitively) also bound up with questions of placefulness: location, mobility (understood as movement through physical and virtual spaces), and spatially situated social associations. In an age of ubiquitous cybersurveillance, therefore, it is not only faces that must be scanned in order to unveil racial threats hidden within the body of the nation. The surveillance of Internet browsing and other technological practices involve the scanning of spatially extended *networks* as well.

Within the context of this sort of networked surveillance, it is not merely what one *is* that discloses a person as a threatening "raced Other," it is also that person's activities, movements, occupation of particular spaces, relations, and technological habits. As Noble (2018) argues, algorithms are not neutral; they are thoroughly suffused with the prejudices of their makers and in turn help to reinforce and reproduce structural racism and inequality. In the case of Italy, surveillance "metadata" are filtered through stereotypical understandings of Muslim culture as inherently illiberal and premodern in order to produce a new form of identifiable racial subject (see Puar 2007)—the male, Muslim immigrant Internet café patron who has been radicalized by jihadist websites and is now a potential terrorist lurking within Italy's borders. In this way, traces of movement through physical (Internet café terminal) and virtual (transnational Internet networking) spaces become the (t)races, or what Harrell (2013) calls "phantasms," for algorithmically assembling categorizable, raced subjects.

Of course, this is a process always replete with contradictions and slippages. Surveillance, algorithmic and otherwise, may have as its goal comprehensive predictive power, but in practice it generates "actual asymmetries and uncertainties" (Crampton and Miller 2017, 5). The Pisanu Decree, for instance, was a blunt-edged tool that sweepingly conflated Muslims with undocumented immigrants and potential terrorists, regardless of their origins; in addition, Pisanu also ensnared tourists (Hooper 2005), who would arguably constitute an example of "good" (orderly, nonthreatening) circulation in the eyes of the Italian state. This was not a problem of misidentification—as Hacking (2006, 23) argues, categories actively transform the people being classified in a recursive "looping effect." In the Italian state's efforts to fix a particular kind of threatening racial subject on the basis of recognizable technological characteristics, however, the communities under surveillance consistently exceeded or evaded categorization. Tactics as simple as presenting false identification documents could effectively disrupt the associational logics of the decree. In addition, the racial distinction between economically entrepreneurial Internet circulation and dangerous religious Internet circulation faltered on the figure of the (Muslim) immigrant Internet café owner, a contradiction that would ultimately lead to Pisanu's repeal.

Conclusion: Beyond the Pisanu Decree to New Terrains of Control

> *Digital connectedness does not come as a utopian alternative to histories of dislocation, rejection and expulsion. . . . Furthermore, the use of digital technologies has created new forms of surveillance, bordering and monitoring access to Europe. Fortress Europe becomes a highly virtualized concept, whose paradox is being poised on embracing a project of expansion and inclusion versus digital and physical re-walling and refencing.*
>
> **—Ponzanesi and Leurs (2016)**

In the last decade, the Internet has emerged as an important mechanism for the regulation and enforcement of borders in Fortress Europe. Indeed, the spatial potentialities of the Internet do not make nation-states obsolete but instead offer new terrains for control. As the case of the Pisanu Decree reveals, imaginaries of technology are intimately linked to race, and produce new kinds of racial subjectivities through technologically specific modes of profiling. While the surveillance of online communications (email, social networking sites, forums, etc.), web browsing, money transfers, and charitable donations has emerged as an important field of scholarly research, however, this new wave of Internet surveillance and control has for the most part not been integrated with broader conversations about race on the Internet, nor has it been connected to discussions about transnational flows, immigration, and border control. These lacunae are problematic because, as the case of Italy demonstrates, antiterrorism programs targeting the Internet are closely articulated with immigration control, border securitization, and racial profiling.

By neglecting to situate Internet surveillance and regulation within the context of broader debates about race and immigration, scholars risk overlooking the disproportionate impact of state-sanctioned Internet surveillance on marginalized communities of color. An unintended consequence of Edward Snowden's 2013 leaks of classified NSA documents has been a "whitewashing" of cybersurveillance—as Mohamad Tabbaa (2013) sardonically quipped in an editorial for *Salon*, "Suddenly, white people care about privacy incursions." Telecommunication surveillance and nonconsensual privacy incursions are indeed frighteningly pervasive in the contemporary moment; however, racializing digital surveillance has been a relatively unacknowledged yet profoundly troubling reality for people of color—and Muslims in particular—since 9/11. Politically, an STS-based understanding of race as a sociotechnical system allows us to consider the implications of Internet surveillance for both the transformation of contemporary racial and nationalist ideologies, and for the development of viable antiracist and antixenophobic practices.

It is important to note, however, that the techniques of surveillance are dynamic and shifting. Arguably, the time of the Internet café is passing, and the Pisanu Decree represented a snapshot in time of a particular effort in Italy to secure the Internet and monitor the people who use it. A 2012 report found that 70% of immigrants in Italy use the Internet, and of these, 65% percent browse the web from their own homes (Micheli 2012). The increasing ubiquity and affordability of personal Internet-connected devices is slowly shifting technological practices away from Internet cafés. While Internet cafés are no longer isolated by the Italian state as strategic sites of intervention into dangerous online networks, however, questions of space and placefulness are still important to any analysis of racializing surveillance. Perhaps signaling this transition, a 2010 report about terrorism in Italy described a suspect as having turned the space of his personal *living room* into a "virtual madrassa" (Bjorkman 2010, 241) linked to jihadist web forums and online resources.

As of January 1, 2012, Article 7 of the Pisanu Decree is no longer in force, following a lengthy and labyrinthine repeal process (*Punto Informatico* 2010; *Apogeonline* 2011; Scialdone 2011; Zambardino 2011; *ASAT* 2013). The eventual repeal of regulations on Internet cafés and Wi-Fi hotspots was justified due to concerns about privacy, business growth, and the difficulty of establishing and accessing public wireless Internet networks in Italy—questions of racism and immigrants' rights were never part of the public debate about the law. Rather, objections to the Pisanu Decree associated the "liberalization" of Internet cafés and Wi-Fi with political freedom, business development, and technological advancement—in other words, the techno-utopian belief that the Internet inherently "wants to be free."

But while identity documents are no longer required at Italian Internet cafés, most public wireless networks still require a lengthy user registration process (Monti 2013). And despite the repeal of the Pisanu Decree, Internet cafés are still sometimes subject to police surveillance. Similarly, Internet surveillance and monitoring in Italy continue, including increased attention to sex trafficking and online ISIS recruitment and radicalization rings. In addition, prompted by the EU, local governments and nongovernmental organizations have increasingly embraced information and communication technologies for immigrant integration (Borkert et al. 2009; Boccagni and Pasquinelli 2010; European Commission Joint Research Centre 2012), emblematic of a broader shift in Italian discourse and policy-making concerning immigration toward the goal of "integration." Finally, these developments must also be situated within the context of the intensified patrolling of the Mediterranean for refugees arriving to Italy by sea, including the absorption of Italy's Mare Nostrum search-and-rescue program into the EU-Frontex border management operation Triton (European Council on Refugees and Exiles 2014). This array of new border management and surveillance strategies targeting refugees and migrants—intervening in transnational digital communication networks and transnational maritime travel routes (Stierl 2015), as well as overland paths to and within Europe—represents diverse efforts to control circulation through the telescoping of borders both within and beyond the boundaries of the nation-state.

Notes

1. Following Browne (2012, 72), "racializing surveillance" refers to "moments when enactments of surveillance reify boundaries and borders along racial lines, and where the outcome is often discriminatory treatment."
2. See Ilten and McInerney (this volume) for a discussion about the significance of studying the social construction of ICTs and new media, rather than approaching these technologies simply as "tools."
3. While the Pisanu Decree covered many areas, I focus on Internet cafés in this chapter because they were significant objects of state concern during the law's implementation, as seen in legislative hearings and policing records.
4. A clause in a 2003 Italian bill that would have required ISPs to monitor Internet activity and retain data (including email data) for five years, which could then be turned over to the courts, was removed after fierce opposition by cyber-freedom activists, opposition parties, and the Italian Office for the Protection of Personal Data (Reporters without Borders 2004).
5. All translations by the author.

Works Cited

Adas, Michael. 1990. *Machines as the Measure of Man: Science, Technology, and Ideologies of Western Dominance*. Ithaca, NY: Cornell University Press.

Amoore, Louisa. 2009. "Algorithmic War: Everyday Geographies of the War on Terror." *Antipode* 41 (1): 49–69.

Amoore, Louisa, and Marieke de Goede. 2005. "Governance, Risk and Dataveillance in the War on Terror." *Crime, Law and Social Change* 34:149–73.

Ananthaswamy, Anil. 2011. "Welcome to the Age of the Splinternet." *New Scientist* 211 (2821): 42–45.

Angel-Ajani, Asale. 2000. "Italy's Racial Cauldron: Immigration, Criminalization, and the Cultural Politics of Race." *Cultural Dynamics* 12 (3): 331–52.

Apogeonline. 2011. "A Che Punto Sono le 'Leggi' di Internet?" January 10. www.apogeonline.com/filirossi/leggi-internet.

ASAT. 2013. "Internet e WiFi—Decreto Del Fare." July 24. www.asat.it/internet-e-wifi---decreto-del-fare/53-4558/.

Balibar, Etienne. 1991. "Is There a 'Neo-Racism'?" In *Race, Nation, and Class: Ambiguous Identities*, edited by Etienne Balibar and Immanuel Wallerstein, 17–28. London: Verso.

BBC. 2005a. "Italy Approves Anti-terror Steps." July 29. http://news.bbc.co.uk/2/hi/europe/4728873.stm.

———. 2005b. "Bombings Suspect Charged in Italy." August 1. http://news.bbc.co.uk/2/hi/4733867.stm.

Benjamin, Ruha. 2016. "Catching Our Breath: Critical Race STS and the Carceral Imagination." *Engaging Science, Technology, and Society* 2:145–56.

Bjorkman, Carl. 2010. "Salafi-Jihadi Terrorism in Italy." In *Understanding Violent Radicalisation: Terrorist and Jihadi Movements in Europe*, edited by Magnus Ranstorm, 231–55. New York: Routledge.

Boccagni, Paolo, and Sergio Pasquinelli. 2010. "The Potential of ICT in Supporting Immigrants in Domiciliary Care in Italy." Luxembourg: European Union Joint Research Centre, Institute for Prospective Technological Studies.

Boellstorff, Tom, Bonnie Nardie, Cecilia Pearce, and T. L. Taylor. 2012. *Ethnography and Virtual Worlds: A Handbook of Method*. Princeton, NJ: Princeton University Press.

Borkert, Maren, Pietro Cingolani, and Viviana Premazzi. 2009. "The State of the Art of Research in the EU on the Uptake and Use of ICT by Immigrants and Ethnic Minorities." 23991 EN. Seville, Spain: European Commission Joint Research Centre, Institute for Prospective Technological Studies. http://ipts.jrc.ec.europa.eu/publications/pub.cfm?id=2560.

Browne, Simone. 2009. "Digital Epidermalization: Race, Identity, and Biometrics." *Critical Sociology* 36 (1): 131–50.

——. 2012. "Race and Surveillance." In *Routledge Handbook of Surveillance Studies*, edited by Kirstie Ball, Kevin Haggerty, and David Lyon, 72–79. New York: Routledge.

——. 2015. *Dark Matters: On the Surveillance of Blackness*. Durham, NC: Duke University Press.

Burrell, Jenna. 2012. *Invisible Users: Youth in the Internet Cafés of Urban Ghana*. Cambridge, MA: MIT Press.

Camera dei Deputati. 2005. "Misure per prevenire il radicamento del fondamentalismo islamico sul territorio italiano - n. 2-01633." Seduta n. 676 del 22/9/2005. Dibatti svolti alla Camera nella XIV Legislatura. http://legislature.camera.it/_dati/leg14/lavori/stenografici/framevar.asp?sedpag=Sed676/s050.htm|STitolo9%2016.

Carter, Donald Martin. 2013. "Blackness over Europe: Meditations on Cultural Belonging." In *Africa in Europe: Studies in Transnational Practice in the Long Twentieth Century*, edited by Robbie Aitken and Eve Rosenhaft, 201–13. Liverpool: University of Liverpool Press.

Castells, Manuel. 2000. *The Rise of the Network Society*. 2nd ed. Malden, MA: Blackwell.

Celeste, Sofia. 2005. "Want to Check Your E-mail in Italy? Bring Your Passport." *Christian Science Monitor*, October 4. www.csmonitor.com/2005/1004/p07s01-woeu.html.

Chun, Wendy Hui Kyong. 2006. *Control and Freedom: Power and Paranoia in the Age of Fiber Optics*. Cambridge, MA: MIT Press.

——. 2009. "Introduction: Race and/as Technology; or, How to Do Things to Race." *Camera Obscura* 24 (1): 7–35.

——. 2010. "Imaged Networks: Digital Media, Race, and the University." In *Universities in Translation the Mental Labor of Globalization*, edited by Brett de Bary, 341–54. Hong Kong: Hong Kong University Press.

Clarke, Roger. 1994. "Dataveillance: Delivering 1984." In *Framing Technology: Society, Choice, and Change*, edited by Leila Green and Roger Guinery, 117–30. London: Routledge.

Cole, Jeffrey. 1997. *The New Racism in Europe: A Sicilian Ethnography*. Cambridge: Cambridge University Press.

Coleman, Beth. 2009. "Race as Technology." *Camera Obscura* 24 (1): 176–206.

Conversione in Legge, con Modificazioni, del Decreto-Legge 27 Luglio 2005, n. 144, Recante Misure Urgenti per il Contrasto del Terrorismo Internazionale. 2005. Legge 31 Luglio 2005, n. 155. Accessed January 6, 2015. www.camera.it/parlam/leggi/05155l.htm.

Crampton, Jeremy, and Andrea Miller. 2017. "Introduction: Intervention Symposium—'Algorithmic Governance.'" *Antipode*, May 19. https://antipodefoundation.org/2017/05/19/algorithmic-governance/.

Crary, Jonathan. 2001. *Suspensions of Perception: Attention, Spectacle, and Modern Culture*. Cambridge, MA: MIT Press.

Daniels, Jessie. 2009. *Cyber Racism: White Supremacy Online and the New Attack on Civil Rights*. Lanham, MD: Rowman & Littlefield.

Decreto-legge 144/05: Misure Urgenti Per il Contrasto del Terrorismo Internazionale. C. 6045 Governo, Approvato dal Senato C. 6045 Governo (Esame e Conclusione). 2005. Seduta di Venerdi 29 Luglio 2005. Dibatti Svolti alla Camera nella XIV Legislatura. http://legxiv.camera.it/_dati/leg14/lavori/bollet/200507/0729/HTML/frontesp.htm.

Dery, Mark. 1994. *Flame Wars: The Discourse of Cyberculture*. Albany: State University of New York Press.

Disegno di Legge di Conversione, con Modificazioni, del Decreto-Legge n. 144 del 2005: Misure Urgenti per il Contrasto del Terrorismo Internazionale (Approvato dal Senato) (A.C. 6045) (Discussione ed approvazione). 2005. Seduta n. 666 di Sabato 30 Luglio 2005. Dibatti Svolti alla Camera nella XIV Legislatura. http://legxiv.camera.it/_dati/leg14/lavori/stenografici/sed666/s000r.htm.

Drayton, Richard. 2000. *Nature's Government: Science, Imperial Britain, and the "Improvement" of the World.* New Haven, CT: Yale University Press.

Epstein, Charlotte. 2008. "Embodying Risk. Using Biometrics to Protect the Borders." In *Risk and the War on Terror*, edited by Louise Amoore and Marieke De Goede, 178–93. New York: Routledge.

European Commission Joint Research Centre. 2012. "ICT for Integration of Immigrants & Ethnic Minorities (IEM)." Brussels, Belgium: European Commission Joint Research Centre, Information Society Unit. http://is.jrc.ec.europa.eu/pages/EAP/eInclusion/IEM.html.

European Council on Refugees and Exiles. 2014. "Mare Nostrum to End—New Frontex Operation Will Not Ensure Rescue of Migrants in International Waters." October 10. http://ecre.org/component/content/article/70-weekly-bulletin-articles/855-operation-mare-nostrum-to-end-frontex-triton-operation-will-not-ensure-rescue-at-sea-of-migrants-in-international-waters.html.

European Parliament and Council. 2002. "Directive 2002/58/EC of the European Parliament and of the Council of 12 July 2002 Concerning the Processing of Personal Data and the Protection of Privacy in the Electronic Communications Sector (Directive on Privacy and Electronic Communications)." http://eur-lex.europa.eu/LexUriServ/LexUriServ.do?uri=CELEX:32002L0058:EN:HTML.

Everett, Anna. 2009. *Digital Diaspora: A Race for Cyberspace.* New York: State University of New York Press.

Fanon, Frantz. [1952] 2008. *Black Skin, White Masks.* New York: Grove Press.

Forlano, Laura, and Kat Jungnickel. 2015. "Hacking Binaries/Hacking Hybrids: Understanding the Black/White Binary as a Socio-technical Practices." *Ada* 6. http://adanewmedia.org/2015/01/issue6-forlano-jungnickel/.

Foucault, Michel. [1997] 2003. *"Society Must Be Defended: Lectures at the Collège de France, 1975–1976.* Edited by Mauro Bertani and Alessandro Fontana. Translated by David Macey. New York: Picador.

———. [2004] 2007. *Security, Territory, Population: Lectures at the Collège de France, 1977–1978.* Edited by Michel Senellart. Translated by Graham Burchell. New York: Picador.

Galloway, Alexander R. 2004. *Protocol: How Control Exists after Decentralization.* Cambridge, MA: MIT Press.

Gilroy, Paul. 1987. *There Ain't No Black in the Union Jack: The Cultural Politics of Race and Nation.* Chicago: University of Chicago Press.

Goldberg, David Theo. 2002. *The Racial State.* Oxford: Wiley-Blackwell.

González, Jennifer. 2009. "The Face and the Public: Race, Secrecy, and Digital Art Practice." *Camera Obscura* 24 (1): 37–65.

Goody, Jack. [1971] 1980. *Technology, Tradition, and the State in Africa.* Cambridge: Cambridge University Press.

Hacking, Ian. 2006. "Making Up People." *New York Review of Books* 28 (16): 23–26.

Haraway, Donna. 1989. *Primate Visions: Gender, Race, and Nature in the World of Modern Science.* New York: Routledge.

———. 1991. *Simians, Cyborgs and Women: The Reinvention of Nature.* New York: Routledge.

Harrell, D. Fox. 2013. *Phantasmal Media: An Approach to Imagination, Computation, and Expression.* Cambridge, MA: MIT Press.

Hooper, John. 2005. "Passport to Surf." *Guardian*, September 29. www.theguardian.com/news/blog/2005/sep/29/passporttosur.

Informativa del Governo Concernente la Prima Applicazione della Recente Normativa sul Contrasto del Terrorismo Internazionale. 2005. Seduta n. 716 di Venerdi 2 Dicembre 2005. Dibatti Svolti alla Camera nella XIV Legislatura. http://legxiv.camera.it/_dati/leg14/lavori/stenografici/sed716/s000r.htm.

Jasanoff, Sheila, and Sang-Hyun Kim. 2009. "Containing the Atom: Sociotechnical Imaginaries and Nuclear Power in the United States and South Korea." *Minerva* 47 (2): 119–46.

Kington, Tom. 2009. "Anti-immigrant Italians Find a New Foe: Food from Abroad." *Guardian*, November 15. www.theguardian.com/world/2009/nov/15/italys-kebab-war-hots-up.

Kraemer, Jordan. 2013. "Friend or *Freund*: Social Media and Transnational Connections in Berlin." *Human-Computer Interaction* 29 (1): 53–77.

Landzelius, Kyra, ed. 2006. *Native on the Net: Indigenous and Diasporic Peoples in the Virtual Age.* New York: Routledge.

Levi, Michael, and David S. Wall. 2004. "Technologies, Security, and Privacy in the Post-9/11 European Information Society." *Journal of Law and Society* 31 (2): 194–220.

Licklider, J.C.R., and Robert Taylor. [1968] 1990. "The Computer as a Communication Device." Reprinted in *In Memoriam: J.C.R. Licklider 1915–1990*, Research Report 61, Digital Equipment Corporation Systems Research Center (August), 21-41. http://memex.org/licklider.pdf.

Magnet, Shoshana. 2011. *When Biometrics Fail: Gender, Race, and the Technology of Identity.* Durham, NC: Duke University Press.

Malcomson, Scott. 2016. *Splinternet: How Geopolitics and Commerce Are Fragmenting the World Wide Web.* New York: OR Books.

McGahan, Christopher L. 2008. *Racing Cyberculture: Minoritarian Art and Cultural Politics on the Internet.* New York: Routledge.

McLelland, Mark J. 2008. "'Race' on the Japanese Internet: Discussing Korea and Koreans on '2-Channeru.'" *New Media & Society* 10 (6): 811–29.

McLuhan, Marshall. [1962] 2011. *The Gutenberg Galaxy: The Making of Typographic Man.* Toronto: University of Toronto Press.

Mellino, Miguel. 2012. "De-provincializing Italy: Notes on Race, Racialization, and Italy's Coloniality." In *Postcolonial Italy: Challenging National Homogeneity,* edited by Cristina Lombardi-Diop and Caterina Romeo, 83–99. New York: Palgrave Macmillan.

Merrill, Heather. 2006. *An Alliance of Women: Immigration and the Politics of Race.* Minneapolis: University of Minnesota Press.

Micheli, Massimo. 2012. "Il 70% degli Immigrati Naviga Sul Web." *Italiani nel Mondo,* February 1. www.italianitalianinelmondo.com/2010/notizie.php?id=640&s=4.

Ministero della Giustizia. 2012. "La Radicalizzazione Del Terrorismo Islamico: Elementi Per Uno Studio Del Fenomeno Di Proselitismo in Carcere." Numero 9. Quaderni ISSP. Rome: Ministero della Giustizia, Dipartimento dell'Amministrazione Penitenziaria.

——. 2010. "Iniziative dell'Italia: Sicurezza, Immigrazione e Asilo." Rome: Ministero dell'Interno. www.cnel.it/application/xmanager/projects/cnel/attachments/shadow_documentazioni_attachment/file_allegatos/000/142/460/0843_Opuscolo_ITA.pdf.

Mitchell, Timothy. 2002. *Rule of Experts: Egypt, Techno-Politics, Modernity.* Berkeley: University of California Press.

Monti, Andrea. 2013. "Il Decreto Pisanu è morto, I suoi obblighi, no." *Ictlex,* April 30. www.ictlex.net/?p=1475.

Nakamura, Lisa. 2002. *Cybertypes: Race, Ethnicity, and Identity on the Internet.* New York: Routledge.

——. 2007. *Digitizing Race: Visual Cultures of the Internet.* Minneapolis: University of Minnesota Press.

Nelson, Alondra. 2002. "Introduction: Future Texts." *Social Text* 20 (2): 1–15.

Nguyen, Mimi. 2003. "Queer Cyborgs and New Mutants: Race, Sexuality, and Prosthetic Sociality in Digital Space." In *AsianAmerica.net: Ethnicity, Nationalism, and Cyberspace,* edited by Rachel Lee and Sau-Ling Wong, 281–305. New York: Routledge.

Noble, Safiya Umoja. 2018. *Algorithms of Oppression: How Search Engines Reinforce Racism.* New York: New York University Press.

Omi, Michael, and Howard Winant. 1986. *Racial Formation in the United States: From the 1960s to the 1980s.* New York: Routledge.

OpenNet Initiative. 2010. "Italy." December 15. https://opennet.net/research/profiles/italy

Pavis, Theta. 2000. "Euros Catching up with Net." *Wired,* April 7. www.wired.com/2000/04/euros-catching-up-with-net/?currentPage=1.

Pearce, Katy E., and Sarah Kendzior. 2012. "Networked Authoritarianism and Social Media in Azerbaijan." *Journal of Communication* 62 (2): 283–98.

Phillip, Kavita. 2003. *Civilizing Natures: Race, Resources, and Modernity in Colonial South India.* New Brunswick, NJ: Rutgers University Press.

Ponzanesi, Sandra, and Koen Leurs. 2016. "On Digital Crossings in Europe." *Crossings* 5 (1): 3–22.

Prakash, Gayan. 1999. *Another Reason: Science and the Imagination of Modern India.* Princeton, NJ: Princeton University Press.

Provincia Di Reggio Emilia. 2006. "Terrorismo Islamico, a Reggio Servono Più Controlli." September 1. www.provincia.re.it/page.asp?IDCategoria=703&IDSezione=5244&ID=93292.

Puar, Jasbir. 2007. *Terrorist Assemblage: Homonationalism in Queer Times.* Durham, NC: Duke University Press.

Pugliese, Joseph. 2013a. *State Violence and the Execution of Law: Biopolitical Caesurae of Torture, Black Sites, and Drones.* New York: Routledge.

——. 2013b. "Technologies of Extraterritorialization, Statist Visuality, and Irregular Migrants and Refugees." *Griffith Law Review* 22 (3): 571–97.

Punto Informatico. 2010. "Decreto Pisanu, Pronto il Cestino?" October 6. http://punto-informatico.it/3004602/PI/News/decreto-pisanu-pronto-cestino.aspx.

Reporters without Borders. 2004. "Internet under Surveillance 2004—Italy." www.refworld.org/docid/46e6918b21.html.
Rheingold, Howard. 1993. *The Virtual Community: Homesteading on the Electronic Frontier*. New York: Addison-Wesley.
Said, Edward. [1978] 2014. *Orientalism*. New York: Vintage.
Sandoval, Chela. 2000. *Methodology of the Oppressed*. Minneapolis: University of Minnesota Press.
Sanminiatelli, Maria. 2005. "Anti-terror Law Forces Cybercafé Owners to Take Names." *USA Today*, December 8. http://usatoday30.usatoday.com/tech/news/computersecurity/2005-12-08-cybercafe-law_x.htm.
Scialdone, Mario. 2011. "Decreto Pisanu, l'Addio Definitivo?" *In tutta sincerità . . .*, December 30. http://scialdone.blogspot.com/search/label/decreto%20milleproroghe.
Shklovski, Irina, Janet Vertesi, and Silvia Lindtner. 2013. "Introduction to This Special Issue on Transnational HCI." *Human-Computer Interaction* 29 (1): 1–21.
Stierl, Maurice. 2015. "The WatchTheMed Alarm Phone. A Disobedient Border-Intervention." *Movements* 1 (2). http://movements-journal.org/issues/02.kaempfe/13.stierl--watchthemed-alarmphone.html.
Stoler, Ann Laura. [2002] 2010. *Carnal Knowledge and Imperial Power: Race and the Intimate in Colonial Rule*. Berkeley: University of California Press.
Tabbaa, Mohamad. 2013. "Suddenly, White People Care about Privacy Incursions." *Salon*, June 13. www.salon.com/2013/06/13/suddenly_white_people_care_about_privacy_incursions/.
Terrorismo Ed Eversione. 2004. "Relazione Al Parlamento—Anno 2004." Rome: Ministero dell'Interno.
Thompson, Derek. 2010. "The Fall of the Internet and the Rise of the 'Splinternet.'" *Atlantic*, March 8. www.theatlantic.com/business/archive/2010/03/the-fall-of-the-internet-and-the-rise-of-the-splinternet/37181/.
Tsing, Anna L. 2005. *Friction: An Ethnography of Global Connection*. Princeton, NJ: Princeton University Press.
Turner, Fred. 2006. *From Counterculture to Cyberculture: Stewart Brand, the Whole Earth Network, and the Rise of Digital Utopianism*. Chicago: University of Chicago Press.
van Dijk, José. 2014. "Datafication, Dataism and Dataveillance: Big Data between Scientific Paradigm and Ideology." *Surveillance & Society* 12 (2): 197–208.
Winner, Langdon, ed. 1989. *The Whale and the Reactor: A Search for Limits in an Age of High Technology*. Chicago: University of Chicago Press.
Yalla Italia. 2012. "Sempre Più Italiani Lavorano per gli Immigrati." April 26. www.yallaitalia.it/2012/04/sempre-piu-italiani-lavorano-per-gli-immigrati/.
Zambardino, Vittorio. 2011. "Ve lo Ricordate il Decreto Pisanu? Solo Adesso, Forse, Se Ne Va Davvero." *La Reppublica*, December 31. http://archive.is/G3AWU.

Social Movements and Digital Technology
A Research Agenda

Carla Ilten and Paul-Brian McInerney

Digital technologies have entered civic engagement and political participation in ways that scholars are only beginning to analyze. From the Zapatista Army of National Liberation movement of the early 1990s to the 2011 Arab Spring, activists have found myriad ways to employ digital technologies to advance their causes. In that time, the use of digital technology has evolved from broadcasting grievances worldwide on websites to tactics such as denial of service attacks and "doxing." As digital technologies have grown more complex, so have the ways that activists have employed them.

In this chapter we review the literatures from sociology and media studies that have attempted to understand how digital technologies have changed and continue to change civic engagement and political participation. Our review focuses on the sociology of collective behavior and social movements to work out in detail how the sunsetting of its theoretical era of studying "new social movements" and social movement organizations coincides with the rise of information and communication technologies (ICTs) on its horizon. The scholarly study of collective behavior and social movements has begun taking ICTs seriously, a trend that has great potential for fruitful conversation with science and technology studies (STS), especially if we renew and activate links with economic sociology and organization studies. Rather than excluding other social sciences from the conversation, this in-depth review hopes to highlight pathways of interdisciplinary pollination.

In our reviews, we find that the sociological literature tends to be movement-centric, focusing on specific social movements and highlighting how their use of digital technologies changes mobilization and tactics. Conversely, the media studies literature tends to be platform-centric, meaning that it analyzes how specific digital technologies afford new types of activism. Both approaches have provided scholars with useful ways to think about digital technologies and activism. However, their movement- or platform-centric focuses lead each to miss important elements of how activists deploy digital technologies to engage in the political process.

Based on insights from science and technology studies, we argue that digital technologies and activism are co-constituted: meaning that in the contemporary era of social movements, it is no longer possible to talk about one without the other. A discussion of social movements necessitates a focus on the new digital technologies its adherents use to communicate, mobilize, organize, and act. Further, we assert that any study of digital technology platforms necessitates a discussion of the forms of activism that it enables (or conversely, the modes of surveillance it affords to governments seeking to stem challengers). A co-constitutive approach focuses on digital technology use in situ: that is, how such technologies act as extensions of activism. We call for a co-constitutive approach to technology and social movements, one that reconciles conceptual differences and is substantiated by studies of newer social movements for which activism and digital technology use are coterminous. Such an approach begins by examining technological platforms as they are used in activist contexts, including tactics, mobilizing techniques, and organizational forms. Some of the most innovative platforms that organize for social change are not clearly part of social movements or even of civil society. We therefore also question extant theoretical categories of what constitutes a movement or legitimate cause.

The chapter is organized as follows. We begin by introducing the topic of digitally enabled activism. Next, we outline major perspectives within the sociological study of collective behavior and social movements, showing how these perspectives have changed as they incorporate analyses of digital technology use among activists. We then critique the sociological literature for its movement-centric focus. Next, we outline the major perspectives within the communications field as they relate to digital technologies and activism, showing how scholars have conceived of technology as activists in different settings have employed it. We then critique the communications literature for its platform-centric focus. Finally, we draw on science and technology studies to offer what we call a co-constitutive approach to the study of digitally enabled activism, which we believe bridges the gap between sociology and communications and advances the study of collective behavior and social movements.

The Conceptual Gap I: Collective Behavior and Social Movement Studies and ICTs

Within sociology, the study of collective behavior and social movements (CBSM) is a relatively mature subfield. Created in 1980, it has grown to be one of the largest sections of the American Sociological Association, with over 800 members in 2014. Yet, as the subfield has grown, it has become increasingly narrow in focus. Walder (2009) explains how the concerns of the subfield have shifted from social structure and political behavior to mobilization. McAdam and Boudet (2012) concur, adding that most of the research in the area "selects on the dependent variable." In other words, the sociological study of CBSM has come to mean the study of mobilization within or across particular movements.

The focus on mobilization has curiously little to say about the role of ICTs. With few exceptions, which will be discussed below, any focus on the role of ICTs in recruiting, mobilizing, and organizing activists remains at the periphery of the subfield. For example, perusing the table of contents of the major edited volumes on CBSM yields no mention of ICTs. With the ubiquity of social media, especially among youth, we feel this is an oversight that needs remedy. We argue that the movement-centric focus of sociological studies of CBSM leads scholars to overlook the role ICTs play in various aspects of mobilization and civic engagement more broadly.

Theoretical Perspectives within CBSM

There are several theoretical perspectives within the sociological scholarship of CBSM. Among them, the dominant perspectives are resource mobilization, political opportunity theory, and new social movement theory (under which we include theories of collective identity, social psychological perspectives such as framing theory, and sociology of emotions as it relates to the topic). Each perspective focuses on different facets of social movement activity and collective behavior.

For instance, resource mobilization theory is concerned with how activists acquire and deploy various resources, such as money, volunteer time, materials, and legitimacy, toward achieving their goals (McCarthy and Zald 1977). Resource mobilization theory generally focuses on the organizational aspects of social movements, for example, how movements form organizations as a way to collect and distribute resources, such as money and activists' time (Clemens and Minkoff 2004; Fisher et al. 2005; Minkoff and Agnone 2010; Minkoff and McCarthy 2005). According to the theory, organizations play a crucial role in accumulating resources and sustaining social movements over time (Staggenborg 1991; Taylor 1989). Alternatively, political opportunity theory focuses on how social movements target and exploit vulnerabilities in state organization and continuity (Tarrow 2011; Tilly and Wood 2009). Here, the focus is on political opportunity structures, such as changes and fissures in state leadership. This literature has also contributed to understanding how social movements organize political parties to influence change within extant state governance systems (Goldstone 2003).

New social movement theory is a term used to describe various perspectives that emerged to explain features of collective action that were not well explained by resource mobilization and political opportunity theories. Such features include the role of collective action frames, collective identity, networks, and emotion in mobilizing activists (Larana et al. 1994). New social movement theories focus on the micro- and meso-level facets of mobilization. Collective action frames are the cognitive schema that activists use to identify social problems and make claims about how best to solve them (Benford and Snow 2000). These frames also help activists enroll and mobilize other activists (Hunt et al. 1994; Snow and Benford 1992). New social movement theories also focus on the role of social networks in enrolling and mobilizing activists (Melucci 1989; Snow et al. 1980). Within new social movement theory, scholars have shown how networks operate to enroll and mobilize activists (Diani 2003). For example, Munson (2008) shows how activists are enrolled through the influence of friends and family members, contrary to accounts that assume activists seek out opportunities to mobilize. In contrast, Fisher (2006) and Fisher and McInerney (2008) show how the social networks responsible for recruiting young people into canvassing organizations may also pull them out of those organizations and mobilize them in other activist opportunities.

CBSM: Communications Technologies and ICTs

The importance of communications technologies has not been lost on sociologists of CBSM. On the contrary, scholars have long examined how activists interact with traditional media (Amenta et al. 2009; Andrews and Biggs 2006; Andrews and Caren 2010; Gamson 1992; Gamson and Wolfsfeld 1993; Gitlin 2003; Rucht 2004) as well as alternative media outlets (Brinson 2006). However, Myers (1994) makes one of the earliest calls for social movement scholars to pay attention to ICTs, highlighting the speed and cost, accuracy, and interactivity of these technologies as regards collective action.

Several events in the late 1990s drew CBSM scholars' attention to activists' uses of Internet technologies. The Zapatista Movement provided a case study for scholars to examine how activists can leverage Internet technologies to broadcast their grievances globally (Castells 1997; Garrido and Halavais 2003; Martinez-Torres 2001; Schulz 1998). The protests surrounding the 1999 World Trade Organization meetings in Seattle brought to light how activists can use Internet technologies to mobilize resources and gain new adherents (Eagleton-Pierce 2001; Smith 2001). These studies demonstrated the possibilities ICTs provided to garner a larger audience and grow public support worldwide (Ayres 1999; Fisher 1998; van Aelst and Walgrave 2004). In doing so, they drew further attention to ICTs as technologies of mobilization.

Since these early studies, the literature linking CBSM and ICTs has grown theoretically sophisticated. For instance, Jennifer Earl and colleagues (Earl et al. 2013; Earl and Kimport 2009; Earl et al. 2010) build on early work outlining the different ways activists use ICTs (Earl et al. 2010) to argue that scholars of CBSM should focus not on specific ICTs, but rather on how they are used in context (Earl and Kimport 2011). The new forms of protests afforded by ICTs, especially those taking place predominantly or exclusively online, may lessen the need for formal social movement organizations (Earl 2015). With the lower cost of mobilization and participation combined with the superseding of organizations, resource mobilization theories may have become less relevant for explaining these new forms of activism. Furthermore, certain ICTs, such as social media platforms, are designed for making and maintaining connections, necessitating theorizing about the role of identity and networks. Contemporary ICTs also modify existing forms of activism, such as online petitioning, and afford entirely new forms of protest, such as denial of service attacks (Coleman 2015; Phillips 2016; Tufekci 2017). The continual emergence of new ICTs and communications platforms necessitates revisiting and updating existing theories of CBSM and may ultimately require entirely new ways of thinking about activism. We will discuss the prospects for theory in further detail below.

Collective Identity Online

These studies showed how activists were able to leverage the web as a broadcast platform. Subsequent studies built on a growing body of literature on virtual communities to show how activists communicated with one another using these new technologies. In particular, they focus on how activists use virtual environments to facilitate collective identities. As social movements scholars Polletta and Jasper articulate it, collective identity is

> an individual's cognitive, moral, and emotional connections with a broader community, category, practice, or institution. It is a perception of a shared status or relation, which may be imagined rather than experienced directly, and it is distinct from personal identities, although it may form part of a personal identity. A collective identity may have been first constructed by outsiders (for example, as in the case of "Hispanics" in the U.S.), who may still enforce it, but it

depends on some acceptance by those to whom it is applied. Collective identities are expressed in cultural materials—names, narratives, symbols, verbal styles, rituals, clothing, and so on—but not all cultural materials express collective identities. (Polletta and Jasper 2001, 284)

Collective identity formation is a form of micromobilization, meaning that it occurs among activists at the level of small-scale interaction (Hunt and Benford 2004). Because they lack a face-to-face component, virtual environments may undermine what CBSM scholars traditionally consider markers of collective identity among movement members (Calhoun 1998; Diani 2000). For example, Wall (2007) argues that while ICTs are useful for activists to achieve concrete goals, certain technologies, like email, are not well suited for symbolic goals, such as collective identity formation.

However, research shows how activists negotiate collective identity on a wider range of Internet platforms. Ayers (2003) shows how the National Organization for Women struggled, but succeeded in developing a collective identity for members who were not colocated through their use of websites. Further research finds that ICTs provide strong support for collective identities when movement members are spread across time and space. Haenfler (2004) studied the straight edge movement to show how ICTs provide platforms for social movement members to demonstrate their commitment to a collective identity, by sharing symbols and meanings despite never meeting other groups of members. Online platforms may be especially effective for facilitating collective identity formation when members feel their identity is stigmatized, as in the cases of White Supremacists studied by Adams and Roscigno (2005) as well as Simi and Futrell (2006).

Protest Online

Social movement organizations that deploy ICTs often embrace new tactical repertoires (Chadwick 2007). Otherwise put, ICTs change what we mean by "mobilization" in the context of social movement behavior (Shumante and Pike 2006). New interactive technologies engage users in a variety of different ways, as research shows that ICT-enabled participation takes various forms. For some activists, participation simply means communicating in chatrooms and on electronic bulletin boards (Nip 2004). For others, it means online support coupled with real-world interaction (Simi and Futrell 2006). For still others, participation means coordinating real-world activities, like protest marches, online (Bennett 2005).

Earl and Kimport (2011) provide the most extensive treatment of online protest to date. Employing Gibson's (1979) theory of affordances, Earl and Kimport detail how activists leverage certain Internet tools in ways that offer qualitative differences to existing protest methods. Examining cases such as the growth of online petitions, the authors show how these new technologies allow smaller groups of activists to mobilize in new ways and have potentially greater impacts. According to Earl and Kimport's approach, ICTs present activists with two main affordances: cost and copresence. The "copresence affordance" reflects the ability of activists to coordinate their activities across time and space. The "cost affordance" describes how ICTs allow activists to organize and mobilize more people with fewer resources. Some suggest that the lower costs of mobilizing will lead to the trivialization of protests and therefore lessen their impact (van de Donk et al. 2004, 18).

However, Earl and Kimport (2009) examine the case of fan activism to show how seemingly trivial forms of protest may provide testing grounds for online tactics, which can then diffuse to other movements.

Taking advantage of these affordances produces supersize effects. Low costs mean ICTs can help to mobilize more protesters or more people to sign a petition. Copresence means ICTs can allow activists to plan protest actions without having to meet. However, creatively leveraging these affordances transforms activism in fundamental ways. In the hands of technologically savvy activists, low costs allow groups to reach previously unreachable audiences or to deploy large-scale outsider tactics online, such as denial of service attacks on target servers (Coleman 2011). Similarly, copresence allows activists to leverage distributed activities for outsized impacts (Carty 2002) and even organize without organizations (Earl 2015).

Despite the potential for global impact, movements from the Global South do not engage in online forms of protest as often as their northern counterparts. In a study commissioned by the Social Science Research Council, McInerney and Berman (2003) systematically collected data from social movement organizations around the world, finding a large discrepancy in the number and level of sophistication of online protest activities between first world and third world countries, due in large part to technical capacities and access to ICTs. Furthermore, as technologies have become more complex, the abilities of states to monitor the activities of civil society organizations and activists have become more sophisticated (Wong 2001; Yang 2003). For instance, Qiang (2011) provides a compelling account of the interplay between activists and the state in contemporary China, as the former learn new ways of expressing critique and the latter develop new methods of quashing such expressions.

Gaps and Prospects

Several key exceptions notwithstanding, the CBSM literature has not accounted for ICT use among activists in its theories. This is an outcome of the movement-centric focus of the literature on social movements and collective behavior. In other words, scholars of CBSM tend to focus exclusively on specific movements, either tracing some aspect of them or studying their historical arc. Recent work by Blee (2012) and McAdam and Boudet (2012) breaks with this trend by studying the formation of movements independent of topic areas. However, neither focuses on ICT use among these activists in ways that contribute to our understanding of them. Earl and Kimport's (2011) contribution represents a generative encounter between CBSM and studies of ICTs. Their work conceptualizes ICT-borne activism in terms of a spectrum from "low-leveraging" to "high-leveraging" tactics. However, as recent studies have shown, activists use digital tools seamlessly across platforms and from online to offline activities (Carty 2011; Milan 2013).

Paying attention to ICTs presents key challenges for scholars of CBSM. ICTs change rapidly. New ICTs change the costs of mobilizing and may afford entirely new tactics. Twitter was unheard of in 2002 when many of the early studies of ICTs and activism were conducted. Activists struggled with how best to use Facebook through the early 2000s. Even Causes.com, a Facebook spinoff site dedicated to raising awareness (and money) for movements, has struggled to gain and maintain relevance. Comparing the two volumes edited by Martha McCaughey on cyberactivism is telling. The chapters in McCaughey and Ayers (2003) are generally about how extant social movements are adopting new technologies, such as websites and message boards, and adapting to the broader new media landscape. In contrast, the chapters in McCaughey (2014) demonstrate the relatively seamless integration of movements and ICTs, drawing empirical content from various social media technologies.

Furthermore, technologically enabled activism requires scholars to update existing theories of CBSM. Below, we will explain what a co-constitutive approach to the study of CBSM and ICTs might look like. For now, we point to key examples. In her groundbreaking study of open source programmers, Coleman (2013) shows how programmers are simultaneously technicians and activists. They develop new technologies that reflect a particular ethical code. Similarly, McInerney (2014) explains how the Circuit Rider movement depended on ICTs to grow while promoting certain technologies that reflected their political ideologies. In both instances, ICTs were co-constitutive of the movements and their activities and not simply a new set of tools. Furthermore, the ICTs in question were (re)constructed in use.

The Conceptual Gap II: Media Studies

It is safe to say that media and communication scholars have outpaced social movement scholars when it comes to the volume of publications on the topic of digital activism. The journal *New Media & Society* has evolved into the primary platform for US media scholars publishing on social change activity. The number of contributions concerned with "activism" in particular has increased exponentially from 1999 on, picking up speed once more with the highly visible political movements of the Arab Spring. Early contributions investigated blogging as a new form of democratic participation in the public sphere (Kahn and Kellner 2004). With the widespread use of social media, the focus has shifted toward studies of the most popular platforms. An exploding volume of studies analyzing specific instances of mobilization via the biggest social networking sites is being published in the other pertinent journals in the field, *Information, Communication & Society* and *Journal of Communication*. This section is an attempt to synthesize the most important achievements as well as blind spots of recent media and communication scholarship on social movements and activism. While media studies have been providing the most coverage on media and movements, so to speak, they still cover certain grounds in-depth and leave others uncharted.

From Mass Media to "New Media"

The most consequential conceptual move has been the paradigm shift from a "transmission model" of mass communication to more interactive and varied models of communication (Lievrouw 2009; Livingstone 2009; Mattelart and Mattelart 1998). The classic "television and . . ." format, which unilaterally emphasized media *effects* on social life, became untenable with the introduction and diffusion of digital technologies and infrastructures such as the Internet, where users seemed to be communicators themselves rather than passive audience. This shift from audience to participants constitutes the defining moment of "new" media and is generally associated with *digital* media, due to the low cost of horizontal interactivity. The advent of "new media" was met with " 'improvisational' conceptual strateg[ies]" that drew on a wide range of disciplines, resulting in subfields such as "computer-mediated communication (CMC)" (Lievrouw 2009, 310).

Yet, the traditional focus on an established medium and its "effects" is still well and alive, but with a new set of main actors: "Twitter and . . ." (Weller et al. 2013) has become a staple title format for media studies of specific (activist) practices that make use of a particular platform. Social media, in a sense, have become the new mass media for communication and media studies and dominate current research agendas: "With some recent exceptions, however, studies on media and social movements tend to focus on specific types of media outlets (e.g., mainstream media), technologies (e.g., print media) and mediation processes (e.g., journalistic reporting)" (Mattoni 2013, 42). This focus on media outlets is strong in the voluminous literature on social networking sites, which are analyzed in a fashion quite reminiscent of the mass communication paradigm, except that the focus has shifted toward user activity, networks, and interaction rather than on pure medium effects. Yet, the "effects" question remains a thread that is woven into the fabric of media studies.

Social Networking Sites and "Big Data"

Popularly associated with the coordination of a number of uprisings and revolutions, the use and impact of Twitter has received particular attention in the past couple years (Earl et al. 2013; Guo and Saxton 2014; Lindgren and Lundström 2011; Thorson et al. 2013; Wojcieszak and Smith 2014). Facebook and YouTube are the other two platforms that have taken center stage in media studies (Nitschke et al. 2014; Harlow 2012; Thorson et al. 2013; Postigo 2008; Caers et al. 2013). Scholars of these platforms tend to focus on user practice and the facilitation of mobilization and widespread discourse.

A common methodological choice is to harvest a sample of speech acts or relational data from social networking sites. While this is promising in terms of access to large amounts of data, which can be analyzed using "big data" techniques, there remain important caveats. In particular, spatial analysis and social network analysis of data such as networks of tweets or "likes" result in understanding activities taking place only online. These approaches therefore reinforce our theoretical focus on and reassert the theoretical importance of online media. "Cyber-archaeology" retrieves online artifacts, not the offline meanings that actors associate with them (Zimbra et al. 2010).

The widely debated question in Internet studies of whether we are falling into a quantitative trap through the lure of "big data" applies to the study of activism in social networking sites as well. Social media research labs have sprung up to provide the infrastructure for developing methods appropriate to research digital infrastructures. As media studies in general, social media labs combine a wide range of disciplines including geography, information sciences, and social sciences. While multidisciplinarity has certainly contributed to innovation in methods and research questions, it has not helped bring forth more integrative theory (Lovink 2011, 77). Mattoni still finds "a lack of common concepts and integrative middle-range theories from which to develop comprehensive analysis of communication flows in grassroots political communication" (Mattoni 2013, 42).

Everything Is Mediated?

More often than not, media studies still take the medium as the starting point of analysis—similarly to social movement studies, where movements provide the analytic starting point and use of ICT comes as an afterthought. Both disciplines exhibit strong institutions with regard to what constitutes legitimate objects of analysis.

Livingstone's important discussion of the new paradigm of *mediation* is a case in point. The concept of mediation is the currently most powerful attempt at moving beyond "improvisational" theorizing (Lievrouw 2011). The question that many authors in the field of media studies have attempted to clarify is, what kinds of mediated relationships are there, and how do they differ? (Ruben and Lievrouw 1990). The question, though, has been heavily disciplined by the contours of the emerging discipline of media studies, where only relationships mediated by "the media" or its offspring "new media" are proper subjects of analysis. Livingstone makes this point when she argues that in a world where "everything is mediated," the core business of media studies, so to speak, must be "to understand how the media mediate" (Livingstone 2009, 4).

In response to this call, scholars have offered typologies of mediated relationships that take specific, historic media as their starting point and can usually be read on a timeline, but also as an analytic table—the prominent shift here is from the transmission model of communication to new, more diverse models of mediation (Rasmussen 2000). While Livingstone's job description for media studies implies that there are other mediating agents, such as language, money, literature, and material goods, it also once again confirms that there is such a thing as "the media." So while Livingstone's and Lievrouw's model of "media infrastructure" builds on Star and Bowker's work on infrastructure to include three elements (artifacts and devices, activities and practices, and arrangements and organizational form), it is meant to be used only in the realm of what has come to be defined as "the media" in media studies. Even Lievrouw's discussion and theorizing of "activist and alternative media" builds on the conventional definition of "new information and communication technologies" that are employed in alternative/activist ways (Lievrouw 2011, 19). This is a limitation from the perspective of STS: "the media" is a blackboxing move par excellence that clouds both technological and economic relations that make up quite divergent configurations.

Types of Media Matter

Mattoni (2013, 43) delineates "four clusters of literature dealing with media and social movements": (1) nondigital mainstream media, (2) digital mainstream media, (3) nondigital alternative media, and (4) digital alternative media. Such a categorical scheme supports our analysis that media studies remain guided by the distinction between the media (mainstream) and alternatives. While the popular social networking sites are grouped under digital mainstream, alternative media are considered as challengers to institutionalized "media power," and provide alternative channels for movements' communication with the public (Couldry and Curran 2003; Lievrouw 2011). Similarly to Earl and Kimport's resource mobilization perspective on ICT, media scholars see alternative media as means for information, communication, and mobilization that circumvent powerful mainstream media. In *Alternative and Activist New Media*, Lievrouw (2011) takes a specific communication theoretical perspective by identifying a number of genres of alternative and activist media (and avoiding the trap of ahistorical abstraction). The impetus of listing and distinguishing types of media is widespread in media studies, but is complemented by typologies of mediated relations (Rasmussen 2000; Calhoun 1992) as well as media rituals (Couldry 2003) and media practice (Cohen 2012; Couldry 2012).

The STS Concept of Affordances

One analytical lens that has been picked up widely by media scholars is the concept of affordances. It is used to describe the structural quality of sociotechnical environments that "afford" certain possibilities for (inter)action and disallow others. Media scholars have used it extensively to describe online infrastructures and their features (Ahy 2014; boyd 2011; Earl and Kimport 2011; Graves 2007; Postigo 2014; Wellman et al. 2003). Technical and design features such as "like buttons" and "rating systems" are analyzed with a view to the actions that they facilitate (or constrain). Yet, comparative studies of disparate uses are rare, given media studies' focus on *the* medium (see, though, a study of military versus movement use of social media by Gray and Gordo 2014). In STS, however, the concept has been critiqued for its technological-deterministic and essentialist tendencies (see Vertesi, this volume).

A Media Environment

The paradigm shift toward mediation, a more abstract relational concept, has helped media scholars see beyond individual media outlets: "A major innovation in research on social movements and the media is the conceptualization of a media environment (similar to Bourdieu's field) in which different spokespersons intervene and different types of media interact" (della Porta 2013, 31). Beyond the focus on individual platforms or media, scholars have investigated the interaction between media in social movement activity (Wolover 2014), as well as how "new" and "old" media have been integrated by activists (Dunbar-Hester 2009) in a "convergence culture" (Jenkins 2006). Especially with the tangibly place-based revolutionary action in Egypt and Occupy Wall Street, the online-offline nexus in movement action has started receiving more attention (Fernandez-Planells et al. 2014; Harlow 2012; Rucht 2013; Thorson et al. 2013; Tufekci and Wilson 2012). In line with other media research, scholars find that certain "offline" or face-to-face activities not only remain important to movement building, but are vital to producing movement outcomes. Rucht (2013, 261) observes that "organizations such as MoveOn in the United States and Campact.de in Germany, who at first enthusiastically and almost exclusively used the Internet as a mobilizing tool, have now gradually shifted to a strategy of combining offline and online activism, especially when focusing on a campaign that they perceive as crucially important." The media environment perspective has the potential to bridge media scholarship with (especially institutionalist) approaches from social movement and organization studies, where ecological concepts have a long-standing tradition.

A Networked Self

Media and communication scholars have been much more likely than social movement scholars to pay attention to the individual media user. While collective identity has been a theme in the study of social movements and collective behavior, changes in activist identities have received little attention beyond the "slacktivism versus high risk (read: real) activism" debate. Media scholars have investigated changes in identity and everyday practice in terms such as "subactivism"—the mediated everyday politics of somewhat activist people (Bakardjieva 2012)—or the "networked self," which emerges as the combination of a multiplicity of identities that can be played out on various distinct platforms (Papacharissi 2011). The perspective on the individual activist can also enable a critical analysis of the conditions for participation in certain forms of activism (Svensson 2014). Bringing Foucault into the picture, Bakardjieva and Gaden (2012) discuss the "Web 2.0 Technologies of the Self" to start thinking about the empowering but also rationalizing and disciplining aspects of learning how to navigate media as the activist subject.

Integrating Perspectives

The recently published edited volume *Mediation and Protest Movements* (Cammaerts et al. 2013) is a promising work that integrates media, communication, and social movement perspectives and offers new conceptual frameworks. The authors connect scholarship on democracy, social movements, and communication (della Porta 2013), bring in the notion of Internet cultures (Kavada 2013), and provide some historical context for understanding movements' media usages (Rucht 2013). A historical perspective on technology in social movements is still largely missing—a gap that makes it harder to ground statements about the changes brought about by *digital* technology in data. One contribution presents a case study using "group history telling" as a method to establish the mix of media use and communications in organizing protests—a refreshing approach that provides rich insights about groups' decision-making processes, timelines of media usage, and the combination of multiple media, ranging from social media outreach and coordination to printing stickers and organizing luncheon events (Ryan et al. 2013). Having organizers discuss their media repertoires is particularly enlightening when it shows how actors' perceptions of media use and effect diverge, and how media use is grounded in local context as well as in group knowledge and deliberation. The call for "more flexibility in thinking about the scales at which we approach technological life" (Nemer and Chirumamilla, this volume) applies to the study of activism as well: the analysis of systems should not eclipse the study of actually occurring use and bricolage among actors.

Bridging the Gaps: Toward a Co-constitutive Approach

Media Studies and STS

While media scholars have outright adopted the affordances concept—which has mostly fallen out of favor with STS scholars otherwise—a gap persists between STS knowledge and media studies. The STS community has seemed to spin off the topic of digital media, as the listing of pertinent journals on the Society for the Social Study of Science's website shows. ICT has—once more in the logic of following the medium—its own venues for publication now. Exceptions exist, such as STS scholars Ruppert et al.'s (2013) discussion of the methodological challenge in analyzing the interplay of digital *devices* and older media. They write on the impetus of unpacking the black box, offering a first outline of an approach.

Yet, digital technologies are being reintroduced to 4S through the efforts in organizing a digitalSTS and media studies community. Pushing in the same direction, the recent edited volume *Media Technologies* by Gillespie et al. (2014) begins to seriously engage media studies with an STS perspective on materiality. Such concerted efforts will provide one avenue for the fruitful theorizing of social movement-technology relations as well.

Some recent economic sociology has also begun to theorize the role of media in establishing and operating for example financial markets. Concepts such as "scopic media" or performativity (Knorr Cetina and Bruegger 2002; Millo and MacKenzie 2009; MacKenzie 2009) have not found their way into research on movements and activism yet, but are theoretically promising. The edited volume *Living in a Material World* (Pinch and Swedberg 2008) bridges economic sociology and STS and provides a compelling template for setting an agenda connecting social movement scholarship with materiality studies. We articulate two classic STS approaches that offer opportunities for the study of media and activism, then turn to other ways in which the co-constitutive approach may be generative for both fields.

STS Classics Waiting for a Media Sequel I: Social Construction of Technology

First and foremost, STS has the theoretical history and power to help media scholarship unpack the blackboxed media concept (Hughes et al. 1989; Kline and Pinch 1996; Pinch and Bijker 1987; Bijker 1997). Much research on activists' use of ICT focuses on technologies as *tools* rather than including broader perspectives on the social construction of media and their economic and organizational location (Lievrouw and Livingstone 2006; Postigo 2011; Tatarchevskiy 2011; Hawthorne, this volume). There is little systematic attention to the industries that develop and provide media infrastructures that movements use (Sandoval 2014; Fuchs and Sandoval 2014; Fuchs 2014). As Gillespie et al. put it in their 2014 introduction to *Media Technologies*, "There has been no STS-based analysis of the Internet or the World Wide Web on par with Latour's (1996) experimental French train systems, Winner's (1980) bridges, Vaughan's (1997) space shuttle disaster, or Pinch and Bijker's (1984) bicycles" (4).

STS Classics Waiting for a Media Sequel II: Artifacts Have Politics—So Do Networks

Social movement studies have also been largely silent on the politics and economics of access to infrastructure that affect social movements' options. In the debate around net neutrality, it is an important question how corporate and state control of large-scale telecommunications infrastructures impacts civil society actors' agency (Lovink 2011). This issue goes well beyond the digital divide: the very ownership structures of the most important social networking sites today make users highly dependent on providers' policies and potentially deliver them to various kinds of surveillance (Albrechtslund 2018; Bauman and Lyon 2013; Graham and Wood 2003; Papacharissi 2010). This should be of the utmost relevance to scholars concerned with contentious movements that face adverse state action: corporate-owned communications infrastructures—while not necessarily "mainstream media"—are easily accessed by state agencies, as has finally become widely understood with the uncovering of NSA surveillance programs.

An important field of research, therefore, not only for activists themselves but also for movement scholars, is the alternative technology infrastructure and associated movements that advocate for public noncommercial infrastructures (Guagnin and Ilten 2011; Youmans and York 2012). Platform cooperatives have emerged in several domains to challenge the for-profit drive of the so-called "sharing economy." For example, FairBNB is a platform designed to create more equitable opportunities for home sharing. Recent theorizing on "platforms" and their politics points the direction that this research can take (Gillespie 2010; Gillespie et al. 2014). From a different theoretical angle, Fuchs has been the most vocal developer of a critical theory of media, analyzing media and information economies from a Marxist perspective (Fuchs 2011).

Technology-Oriented Movements

Technology-oriented movements have received surprisingly little attention, even within STS (with the notable exception of Hess 2005, 2007; Hess et al. 2008; Postigo 2008; McInerney 2009, 2014). The net neutrality movement, community wireless and radio movements, digital rights movements, and of course the free software movement are important fields of study for both social movement scholars and media scholars (Atton 2002; Benkler 2001; Couldry and Curran 2003; Dickson 1974; Dunbar-Hester 2014; Flickenger 2003; Forlano et al. 2011; Lievrouw 2011). In media studies, movements targeting "the media" and power structures of mediation have been analyzed as "media activism" (Jansen et al. 2011).

While movement scholars agree that technologies play a tremendous role in social and political processes today, the politics that target those technologies and technology policies remain somewhat opaque. If power is increasingly leveraged through online and mobile infrastructures—both on the part of movements and on the part of states—then some of the most important (and radical) movements will emerge around the use of those powerful technologies in societies. While hacking and hacktivism have become topics in media studies (Coleman 2014; Jordan and Taylor 2004; Taylor 2005), social movement scholars have tended to ignore technology movements. Notable exceptions are David Hess's (2007) systematic work on science- and technology-oriented movements with a focus on environmental movements that intervene in industrial structures. McInerney's (2014) study of the Circuit Rider movement captures both moments of the social movement-ICT nexus: this technology-oriented movement promoted IT products for nonprofits that were in line with their political economic values, and the movement used ICTs to mobilize Circuit Riders and reach out. Beyer's (2014) study of four important online communities also provides some compelling insight into how technology-oriented political organization can originate in nonpolitical spaces, especially in architectures built around anonymity. Related to technology-oriented activism, the politics of nonuse and resistance to media imperatives constitute another uncharted field for studies of (anti-?) digital activism (Portwood-Stacer 2013).

Finally, scholars are also themselves involved in technology-oriented movements and sometimes act as their intellectual vanguard, as the high profile of Lawrence Lessig, a proponent of "free culture" and founder of Creative Commons, illustrates. Technology-oriented movements create not only alternative infrastructures, but also legal objects, as in the case of the Free Software GNU Public License and the Creative Commons license system. "Commonism" enabled by digital technologies is a theme that both US and European scholars discuss, albeit with somewhat different political emphases (Hands 2011; Dyer-Witheford 2013).

A Tentative Map

The scholarly communities currently invested in studying a combination of social movements and media are not only theoretically divided, there are also real boundaries that limit intellectual exchange. Much of the literature reviewed here can be subsumed under "US media and communication community," and very little European (let alone non-Western) research is on the radar of this academic field (China is an exception and receives much attention in US research on media—this seems largely accounted for by expatriates [Yang 2009]). The same can be said for social movement studies, which revolve around a strong US scholarly community and a somewhat separate European community. Movements on other continents have made it onto the map as a result of media use, not because social movement studies routinely turn to non-Western sites for analysis (Tufekci and Wilson 2012).

Within sociology, disciplinary fragmentation seemed to intensify when a Media Sociology section was proposed in the American Sociological Association. Potential competition between communities that research "media" with those that research "communication and information technologies" has been averted by adding "media" to the existing Communication and Information Technologies section (CITASA is now CITAMS).

A Moving, but Consolidating, Target

Part of the challenge of understanding the conceptual gaps between social movement studies and STS/media studies is that the latter are in the process of combining forces as we write. The edited volume *Media Technologies* (Gillespie et al. 2014) addresses many of the conceptual gaps identified in this review. Another attempt to bridge "communication studies and science and technology studies" is forthcoming in the next International encyclopedia *Encyclopedia of Communication Theory and Philosophy*.

Much responsibility falls upon social movement scholars to engage with this fast-moving conceptual field and community. If scholars remain stuck in disciplinary patterns of claiming authority over specific social phenomena due to a priori definitions we have employed for decades, we will lose theoretical ground. While the variety of literature and contributions around "media" can be dizzying, to say the least, sociologists cannot ignore the perspectives of media and technology scholars. The growing body of literature on movements that does not build on social movement theory whatsoever (or is largely nontheoretical to begin with [Lovink 2011, chap. 8]) should be a wake-up call to social movement scholars.

While some core theoretical tools of the traditional CBSM are losing purchase in changing conditions, others remain valuable to an integrated co-constitutive perspective on mobilization and technology: sociological perspectives on institutions, institutionalization, organizations, and power relations are building blocks that can situate both movements and new technologies within larger societal structures and processes. An initial contribution by CBSM to understanding movements in an era of digital technology is to reassess how social movement organizations are evolving empirically. Next, its institutional perspective on mobilization can help embed the study of digital activism in relevant contexts beyond the medium. What is at stake is the development of balanced theoretical perspectives that account both for social and institutional dimensions of mobilization and for the effects and uses of technologies, digital and otherwise.

A Co-constitutive Approach: Neither Movement nor Technology Takes Precedence

The current urgency of analyzing new media and mobilization only emphasizes long-standing gaps in social movement studies, where technology has not figured prior to widespread web use. We call for a co-constitutive approach to technology and social movements, one that reconciles conceptual differences and is substantiated by studies of newer social movements for which activism and digital technology use are coterminous. A co-constitutive approach to technology and social movements begins by examining new technological platforms for activism and their associated tactics, mobilizing techniques, and organizational forms—but it needs to move beyond the *movement–tool–outcomes* causality employed in movement-centric scholarship.

Rethinking Social Movement Theory's Movement Focus . . .

The imperative to incorporate a whole new set of questions into social movement scholarship should be stimulating in a number of ways. The first obvious limitation to especially US social movement scholarship is the traditional focus on *a movement* as the unit of analysis, an intellectual legacy from the widely recognized "new social movements" of the 20th century, which is strongly oriented toward movements represented by social movement organizations. As media studies have picked up on, new decentralized forms of organizing do not fit this hierarchical and historical model of movement emergence, growth, and maturation. We perceive a need to broaden movement scholarship's scope beyond well-defined social movements as they have been identified by "new social movements" scholarship.

Causes and *tactics* must be analytically distinguished in order to understand how they interact. Social media scholarship still exhibits a tendency to select for "appropriate" causes (read: social justice) rather than agnostically distinguishing movement causes from the tactics employed. While "rich people's movements" (Martin 2013) and right-wing activism have made their way into movement scholarship, less evidently political movements are largely excluded. This is problematic not only since the tactics of fan activism might be quite similar to social justice activism but also because there may be a good deal of tactic learning across spheres that we need to understand. This is a central element of a co-constitutive approach: to make room for discovering movement activity where we were not looking, and to use an analytical lens to see tactics where they emerge, rather than searching for them in established movements.

... While Not Falling into the Medium Focus Trap

Digital activism scholarship needs to move away from an overly instrumental perspective on technology (the "tool" mediating variable), yet it must also avoid media studies' problem with granting *the medium* precedence. It is clear that both limitations arise from disciplinary intellectual boundaries. The medium focus in media studies can be critiqued fruitfully with a rich body of STS knowledge—so there is no need to reinvent the wheel. Yet, the tendency to fall into technologically deterministic (now: platform deterministic) accounts is widespread. Bringing the different uses, users, and nonusers back in seems one helpful way of avoiding determinism—also a tried and tested theoretical move in STS. Some of the research discussed above successfully demonstrates this approach, for example by investigating activists' deliberation about media uses rather than observing uses only.

Again, no a priori seems the best strategy for overcoming this limitation: much research is already moving from single-medium units of analysis to understanding media ecologies, or repertoires that actors draw on. This implies combining online and "new" media with all other technologies that become mobilized for movements. It also begs the question of more spatial, relational, and temporal analyses of when which media and technologies are used for what. For example, we have reviewed evidence that online mobilizing frequently leads to offline place-based protest with continued online reporting. A map or typology of combinations of media use and nonuse is conceivable.

A Renewed Perspective on Organizations

At the level of organizations, we are also facing challenges of (disciplinary) definition: some of the most innovative platforms that organize for social change are not clearly part of social movements or even of civil society—this supports the case to question extant theoretical categories of what constitutes a movement or legitimate cause. The corporate actors that provide platforms used for activism are part of the "assemblage" and cannot be bracketed out of the equation. What is the relationship between social movement organizations, activists, media, and their designers and owners? What is the state's role in this field?

Social movement scholarship could benefit tremendously from a new round of cross-pollination with current organization theory (Davis 2005) and its relational cousins in economic sociology. After the hype around "organizing without organizations," we need to turn our attention to the new forms of organizations that facilitate campaigns, network membership, and participation (Karpf 2012). Much activism takes place in social spaces that have characteristics of markets or become marketized (McInerney 2014). One does not need to side with actor-network theory in order to use more *flattening* conceptual tools that can help scholars unlearn the convention of starting their research with individual movements. Again, it is clear that certain tactics are shared with many other "mobilizers" in nonmovement spheres, for example in marketing. Movement theory can draw on the dynamic literatures on these phenomena to take a step back and see a bigger picture including digital technology producers and regulators.

Methodological Variety

All of the above theoretical goals of a co-constitutive approach imply that we diversify our methodological toolbox. While current movement scholarship on technology is building the first large datasets on movements and online mobilization, media studies continue to favor case studies. A third fast-growing approach is social network analysis, where large collections of ties or speech acts are measured quantitatively and spatially. We can develop a most useful variety of methodologies when we reconsider our research questions and theorizing goals as suggested above. The current division of labor can turn into fruitful collaboration with an overarching goal of integrating theoretical frameworks as well. Practice-oriented methodologies can complement the quantitative character of social movement studies. While we need to work on creating large datasets that can represent technologically mediated activism online, the tasks ahead in developing theoretical tools require different methodologies.

Recent attempts to integrate media scholarship with STS provide a fortunate starting point for social movement scholars to become involved in integrating the rich literatures of STS, and now media studies, with the sociological body of knowledge on social movements. The title could read: "Mobilizing in a Digital/Material World."

Works Cited

Adams, Josh, and Vincent Roscigno. 2005. "White Supremacists, Oppositional Culture, and the World Wide Web." *Social Forces* 84 (2): 759–79.

Ahy, Maximillian Hänska. 2014. "Networked Communication and the Arab Spring: Linking Broadcast and Social Media." *New Media & Society* 18 (1): 99–116.

Albrechtslund, Anders. 2008. "Online Social Networking as Participatory Surveillance." *First Monday* 13 (3). http://firstmonday.org/article/view/2142/1949.

Amenta, Edwin, Neal Caren, Sheera Joy Olasky, and James E. Stobaugh. 2009. "All the Movements Fit to Print: Who, What, When, Where, and Why SMOs Appeared in the New York Times in the Twentieth Century." *American Sociological Review* 74:636–56.

Andrews, Kenneth T., and Mary Biggs. 2006. "The Dynamics of Protest Diffusion: Movement Organizations, Social Networks, and News Media in the 1960 Sit-Ins." *American Sociological Review* 71:752–77.

Andrews, Kenneth, and Neal Caren 2010. "Making the News: Movement Organizations, Media Attention, and the Public Agenda." *American Sociological Review* 75:841–66.

Atton, Chris. 2002. *Alternative Media*. Thousand Oaks, CA: Sage.

Ayers, Michael. 2003. "Comparing Collective Identity in Online and Offline Feminist Activities." In *Cyberactivism: Online Activism in Theory and Practice*, edited by M. McCaughey and M. Ayers, 145–64. New York: Routledge.

Ayres, Jeffrey. 1999. "From the Streets to the Internet: The Cyber-Diffusion of Contention." *Annals of the American Academy of Political and Social Science* 566:132–43.

Bakardjieva, Maria. 2012. "Subactivism: Lifeworld and Politics in the Age of the Internet." In *(Re)inventing the Internet: Critical Case Studies*, edited by A. Feenberg and N. Friesen, 85–108. Rotterdam, Netherlands: Sense.

Bakardjieva, Maria, and Georgia Gaden. 2012. "Web 2.0 Technologies of the Self." *Philosophy & Technology* 25 (3): 399–413.

Bauman, Zygmunt, and David Lyon. 2013. *Liquid Surveillance: A Conversation*. Cambridge: Polity.
Benford, Robert D., and David A. Snow. 2000. "Framing Processes and Social Movements: An Overview and Assessment." *Annual Review of Sociology* 26:611–39.
Benkler, Yochai. 2001. "The Battle over the Institutional Ecosystem in the Digital Environment." *Communications of the ACM* 44 (2): 84–90.
Bennett, W. Lance. 2005. "Social Movements beyond Borders: Understanding Two Eras of Transnational Activism." In *Transnational Protest and Global Activism*, edited by D. della Porta and S. G. Tarrow, 203–26. Lanham, MD: Rowman & Littlefield.
Beyer, Jessica. 2014. *Expect Us: Online Communities and Political Mobilization*. New York: Oxford University Press.
Bijker, Wiebe E. 1997. *Of Bicycles, Bakelites, and Bulbs: Toward a Theory of Sociotechnical Change*. Cambridge, MA: MIT Press.
Blee, Kathleen. 2012. *Making Democracy Work: How Activist Groups Form*. New York: Oxford University Press.
boyd, danah. 2011. "Social Network Sites as Networked Publics. Affordances, Dynamics, and Implications." In *A Networked Self: Identity, Community and Culture on Social Network Sites*, edited by Zizi Papacharissi, 39–58. New York: Routledge.
Brinson, Peter. 2006. "Liberation Frequency: The Free Radio Movement and Alternative Strategies of Media Relations." *Sociological Quarterly* 47 (4): 543–68.
Caers, Ralf, Tim De Feyter, Marjike De Couck, Talia Stough, Claudia Vigna, and Cind Du Bois. 2013. "Facebook: A Literature Review." *New Media & Society* 15 (6): 982–1002.
Calhoun, Craig. 1992. "The Infrastructure of Modernity. Indirect Social Relationships, Information Technology, and Social Integration." In *Social Change and Modernity*, edited by H. Haferkamp and N. J. Smelser, 205–36. Berkeley: University of California Press.
——. 1998. "Community without Propinquity Revisited: Communications Technology and the Transformation of the Public Sphere." *Sociological Inquiry* 68 (3): 373–97.
Cammaerts, Bart, Alice Mattoni, and Patrick McCurdy, eds. 2013. *Mediation and Protest Movements*. Bristol: Intellect.
Carty, Victoria. 2002. "Technology and Counter-Hegemonic Movements: The Case of Nike Corporation." *Social Movement Studies* 1 (2): 129–46.
——. 2011. *Wired and Mobilizing: Social Movements, New Technology, and Electoral Politics*. New York: Routledge.
Castells, Manuel. 1997. *The Power of Identity*. Malden, MA: Blackwell.
Chadwick, Andrew. 2007. "Digital Network Repertoires and Organizational Hybridity." *Political Communication* 24:283–301.
Clemens, Elisabeth S., and Debra Minkoff. 2004. "Beyond the Iron Law: Rethinking the Place of Organizations in Social Movement Research." In *The Blackwell Companion to Social Movements*, edited by D. A. Snow, S. A. Soule and H. Kriesi, 155–70. Malden, MA: Blackwell.
Cohen, Julie E. 2012. *Configuring the Networked Self: Law, Code, and the Play of Everyday Practice*. New Haven, CT: Yale University Press.
Coleman, E. Gabriella. 2011. "Anonymous: From the Lulz to Collective Action." The New Everyday: A Media Commons Project. http://mediacommons.futureofthebook.org/tne/pieces/anonymous-lulz-collective-action.
——. 2013. *Coding Freedom: The Ethics and Aesthetics of Hacking*. Princeton, NJ: Princeton University Press.
——. 2014. *Hacker, Hoaxer, Whistleblower, Spy: The Many Faces of Anonymous*. New York: Verso.
——. 2015. *Hacker, Hoaxer, Whistleblower, Spy: The Many Faces of Anonymous*. London: Verso.
Conover, Michael D., Clayton Davis, Emilio Ferrara, Karissa McKelvey, Filippo Menczer, and Alessandro Flammini. 2013. "The Geospatial Characteristics of a Social Movement Communication Network." *PLOS ONE* 8 (3): e55957.
Couldry, Nick. 2003. *Media Rituals: A Critical Approach*. London: Routledge.
——. 2012. *Media, Society, World: Social Theory and Digital Media Practice*. Cambridge: Polity.
Couldry, Nick, and James Curran. 2003. *Contesting Media Power: Alternative Media in a Networked World*. Lanham, MD: Rowman & Littlefield.
Davis, Gerald. 2005. *Social Movements and Organization Theory*. New York: Cambridge University Press.
della Porta, Donatella. 2013. "Bridging Research on Democracy, Social Movements and Communication." In *Mediation and Protest Movements*, edited by Bart Cammaerts, Alice Mattoni, and Patrick McCurdy, 21–38. Bristol: Intellect.

Diani, Mario. 2000. "Social Movement Networks Virtual and Real." *Information, Communication & Society* 3 (3): 386–401.
——, ed. 2003. *Social Movements and Networks: Relational Approaches to Collective Action*. New York: Oxford University Press.
Dickson, David. 1974. *Alternative Technology and the Politics of Technical Change*. Glasgow: Fontana.
Dunbar-Hester, Christina. 2009. "'Free the Spectrum!' Activist Encounters with Old and New Media Technology." *New Media & Society* 11 (1–2): 221–40.
——. 2014. *Low Power to the People: Pirates, Protest, and Politics in FM Radio Activism*. Cambridge, MA: MIT Press.
Dyer-Witheford, Nick. 2013. "Red Plenty Platforms." *Culture Machine* 14. www.culturemachine.net/index.php/cm/article/view/511/526.
Eagleton-Pierce, Matthew. 2001. "The Internet and the Seattle WTO Protests." *Peace Review* 13 (3): 331–37.
Earl, Jennifer. 2015. "The Future of Social Movements: The Waning Dominance of SMOs Online." *American Behavioral Scientist* 59 (1): 35–52.
Earl, Jennifer, Heather McKee Hurwitz, Analicia Mesinas, Margaret Tolan, and Ashley Arlotti. 2013. "This Protest Will be Tweeted: Twitter and Protest Policing during the Pittsburgh G20." *Information, Communication & Society* 16:459–78.
Earl, Jennifer, and Katrina Kimport. 2009. "Movement Societies and Digital Protest: Fan Activism and Other Non-political Protest Online." *Sociological Theory* 23:220–43.
——. 2011. *Digitally Enabled Social Change: Activism in the Internet Age*. Cambridge, MA: MIT Press.
Earl, Jennifer, Katrina Kimport, Greg Prieto, Carly Rush, and Kimberly Reynoso. 2010. "Changing the World One Webpage at a Time: Conceptualizing and Explaining Internet Activism." *Mobilization* 15:425–46.
Fernandez-Planells, A., M. Figueras-Maz, and C. F. Pàmpols. 2014. "Communication among Young People in the #spanishrevolution: Uses of Online-Offline Tools to Obtain Information about the #acampadabcn." *New Media & Society* 16 (8): 1287–1308.
Fisher, Dana. 1998. "Rumoring Theory and the Internet: A Framework for Analyzing the Grass Roots." *Social Science Computer Review* 16 (2): 158–68.
——. 2006. *Activism, Inc: How the Outsourcing of Grassroots Campaigns Is Strangling Progressive Politics in America*. Stanford, CA: Stanford University Press.
Fisher, Dana, and Paul-Brian McInerney. 2008. "The Limits of Networks for Organizational Mobilization." Paper presented at the American Sociological Association annual meeting, Boston.
Fisher, Dana, Kevin Stanley, David Berman, and Gina Neff. 2005. "How Do Organizations Matter? Mobilization and Support for Participants at Five Globalization Protests." *Social Problems* 52 (1): 102–21.
Flickenger, Rob. 2003. *Building Wireless Community Networks. 2. Auflage*. Beijing: O'Reilly Media.
Forlano, Laura, Alison Powell, Gwen Shaffer, and Benjamin Lennett. 2011. "From the Digital Divide to Digital Excellence. Global Best Practices to Aid Development of Municipal and Community Wireless Networks in the United States." http://eprints.lse.ac.uk/29461/1/__lse.ac.uk_storage_LIBRARY_Secondary_libfile_shared_repository_Content_Powell,%20A_From%20digital%20devide_Powell_From%20the%20digital%20divide_2014.pdf.
Fuchs, Christian. 2011. *Foundations of Critical Media and Information Studies*. Abingdon: Routledge.
——. 2014. *Social Media: A Critical Introduction*. Thousand Oaks, CA: Sage.
Fuchs, Christian, and Marisol Sandoval. 2014. *Critique, Social Media and the Information Society*. New York: Routledge.
Gamson, William. 1992. *Talking Politics*. New York: Cambridge University Press.
Gamson, William, and Gadi Wolfsfeld. 1993. "Movements and Media as Interacting Systems." *Annals of the American Academy of Political and Social Science* 528:114–25.
Garrido, Maria, and Alexander Halavais. 2003. "Mapping Networks of Support for the Zapatista Movement: Applying Social-Network Analysis to Study Contemporary Social Movements." In *Cyberactivism: Online Activism in Theory and Practice*, edited by M. McCaughey and M. D. Ayers, 165–84. New York: Routledge.
Gibson, James. 1979. *The Ecological Approach to Visual Perception*. New York: Houghton Mifflin.
Gillespie, Tarleton. 2010. "The Politics of 'Platforms.'" *New Media & Society* 12 (3): 347–64.
Gillespie, Tarleton, Pablo J. Boczkowski, and Kirsten A. Foot. 2014. *Media Technologies: Essays on Communication, Materiality, and Society*. Cambridge, MA: MIT Press.

Gitlin, Todd. 2003. *The Whole World Is Watching: Mass Media in the Making and Unmaking of the New Left*. Berkeley: University of California Press.

Goldstone, Jack, ed. 2003. *States, Parties, and Social Movements: Protest and the Dynamics of Institutional Change*. New York: Cambridge University Press.

Graham, Stephen, and David Wood. 2003. "Digitizing Surveillance: Categorization, Space, Inequality." *Critical Social Policy* 23 (2): 227–48.

Graves, Lucas. 2007. "The Affordances of Blogging: A Case Study in Culture and Technological Effects." *Journal of Communication Inquiry* 31 (4): 331–46.

Gray, Chris Hables, and Ángel J. Gordo. 2014. "Social Media in Conflict: Comparing Military and Social-Movement Technocultures." *Cultural Politics* 10 (3): 251–61.

Guagnin, Daniel, and C. Ilten. 2011. "Self-Governed Socio-technical Infrastructures. Autonomy and Co-operation through Free Software and Community Wireless Networks." In *Net Neutrality and Other Challenges for the Future of the Internet: Proceedings of the 7th International Conference on Internet, Law & Politics*, edited by A. Cerrillo-i-Martínez et al., 497–512. Barcelona: Open University of Catalonia, UOC and Huygens Editorial.

Guo, Chao, and Gregory D. Saxton. 2014. "Tweeting Social Change: How Social Media Are Changing Nonprofit Advocacy." *Nonprofit and Voluntary Sector Quarterly* 43 (1): 57–79.

Haenfler, Ross. 2004. "Collective Identity in the Straight Edge Movement: How Diffuse Movements Foster Commitment, Encourage Individualized Participation, and Promote Cultural Change." *Sociological Quarterly* 45 (4): 785–805.

Hands, Joss. 2011. *@ Is for Activism. Dissent, Resistance and Rebellion in a Digital Culture*. London: Pluto.

Harlow, Summer. 2012. "Social Media and Social Movements: Facebook and an Online Guatemalan Justice Movement That Moved Offline." *New Media & Society* 14 (2): 225–43.

Hess, David. 2005. "Technology- and Product-Oriented Movements: Approximating Social Movement Studies and Science and Technology Studies." *Science, Technology, & Human Values* 30 (4): 515–35.

———. 2007. *Alternative Pathways in Science and Industry. Activism, Innovation, and the Environment in an Era of Globalization*. Cambridge, MA: MIT Press.

Hess, David, Steve Breyman, Nancy Campbell, and Brian Martin. 2008. "Science, Technology, and Social Movements." In *The Handbook of Science and Technology Studies*, vol. 3, edited by E. J. Hackett, O. Amsterdamska, M. Lynch, and J. Wajcman, 473–98. Cambridge, MA: MIT Press.

Hughes, Thomas P., Wiebe E. Bijker, and Trevor Pinch, eds. 1989. *The Social Construction of Technological Systems: New Directions in the Sociology and History of Technology*. Cambridge, MA: MIT Press.

Hunt, Scott, and Robert Benford. 2004. "Collective Identity, Solidarity, and Commitment." In *The Blackwell Companion to Social Movements*, edited by D. A. Snow, S. A. Soule, and H. Kriesi, 433–57. Malden, MA: Blackwell.

Hunt, Scott, Robert Benford, and David Snow. 1994. "Identity Fields: Framing Processes and the Social Construction of Movement Identities." In *New Social Movements: From Ideology to Identity*, edited by E. Larana, H. Johnston, and J. R. Gusfield, 185–208. Philadelphia: Temple University Press.

Jansen, Sue Curry, Jefferson Pooley, and Lora Taub-Pervizpour. 2011. *Media and Social Justice*. New York: Palgrave Macmillan.

Jenkins, Henry. 2006. *Convergence Culture: Where Old and New Media Collide*. New York: New York University Press.

Jordan, Tim, and Paul A. Taylor. 2004. *Hacktivism and Cyberwars*. London: Routledge.

Kahn, Richard, and Douglas Kellner. 2004. "New Media and Internet Activism: From the "Battle of Seattle" to Blogging." *New Media & Society* 6 (1): 87–95.

Karpf, David. 2012. *The MoveOn Effect: The Unexpected Transformation of American Political Advocacy*. New York: Oxford University Press.

Kavada, Anastasia. 2013. "Internet Cultures and Protest Movements: The Cultural links Between Strategy, Organizing and Online Communication." In *Mediation and Protest Movements*, edited by Bart Cammaerts, Alice Mattoni, and Patrick McCurdy, 75–95. Bristol: Intellect.

Kline, R., and T. Pinch. 1996. "Users as Agents of Technological Change: The Social Construction of the Automobile in the Rural United States." *Technology and Culture* 37 (4): 763–95.

Knorr Cetina, Karin, and Urs Bruegger. 2002. "Global Microstructures: The Virtual Societies of Financial Markets." *American Journal of Sociology* 107 (4): 905–50.

Larana, Enrique, Johnston, Hank, and Joseph Gusfield, eds. 1994. *New Social Movements*. Philadelphia: Temple University Press.

Lievrouw Leah. 2009. "New Media, Mediation, and Communication Study." *Information, Communication & Society* 12 (3): 303–25.
———. 2011. *Alternative and Activist New Media*. Cambridge: Polity.
Lievrouw, Leah, and Sonia Livingstone, eds. 2006. *Handbook of New Media: Social Shaping and Social Consequences of ICTs*. Thousand Oaks, CA: Sage.
Lindgren, Simon, and Ragnar Lundström. 2011. "Pirate Culture and Hacktivist Mobilization: The Cultural and Social Protocols of #WikiLeaks on Twitter." *New Media & Society* 13 (6): 999–1018.
Livingstone, Sarah. 2009. "On the Mediation of Everything: ICA Presidential Address 2008." *JCOM Journal of Communication* 59 (1): 1–18.
Lovink, Geert. 2011. *Networks Without a Cause: A Critique of Social Media*. Cambridge: Polity.
MacKenzie, Donald. 2009. *Material Markets: How Economic Agents Are Constructed*. Oxford: Oxford University Press.
Martin, Isaac William. 2013. *Rich People's Movements: Grassroots Campaigns to Untax the One Percent*. Oxford: Oxford University Press.
Martinez-Torres, Maria Elena. 2001. "Civil Society, the Internet, and the Zapatistas." *Peace Review* 13 (3): 347–55.
Mattelart, Armand, and Michele Mattelart. 1998. *Theories of Communication: A Short Introduction*. Thousand Oaks, CA: Sage.
Mattoni, Alice. 2013. "Repertoires of Communication in Social Movement Processes." In *Mediation and Protest Movements*, edited by Bart Cammaerts, Alice Mattoni, and Patrick McCurdy, 39–56. Bristol: Intellect.
McAdam, Doug, and Hilary Boudet. 2012. *Putting Movements in Their Place: Explaining Variation in Community Response to the Siting of Proposed Energy Projects*. New York: Cambridge University Press.
McCarthy, John D., and Mayer N. Zald. 1977. "Resource Mobilization and Social Movements: A Partial Theory." *American Journal of Sociology* 82 (6): 1212–41.
McCaughey, Martha, ed. 2014. *Cyberactivism on the Participatory Web*. New York: Routledge.
McCaughey, Martha, and Michael D. Ayers, eds. 2003. *Cyberactivism: Online Activism in Theory and Practice*. New York: Routledge.
McInerney, Paul-Brian. 2009. "Technology Movements and the Politics of Free/Open Source Software." *Science, Technology, & Human Values* 34 (2): 206–33.
———. 2014. *From Social Movement to Moral Market: How the Circuit Riders Sparked an IT Revolution and Created a Technology Market*. Palo Alto, CA: Stanford University Press.
McInerney, Paul-Brian, and David Berman. 2003. *IT and Activism around the War in Iraq*. New York: Social Science Research Council.
Melucci, Alberto. 1989. *Nomads of the Present: Social Movements and Individual Needs in Contemporary Society*. Philadelphia: Temple University Press.
Milan, Stefania. 2013. *Social Movements and Their Technologies: Wiring Social Change*. London: Palgrave Macmillan.
Millo, Yuval, and Donald MacKenzie. 2009. "The Usefulness of Inaccurate Models: Towards an Understanding of the Emergence of Financial Risk Management." *Accounting, Organizations and Society* 34 (5): 638–53.
Minkoff, Debra, and Jon Agnone. 2010. "Consolidating Social Change: The Consequences of Foundation Funding for Developing Social Movement Infrastructures." In *American Foundations: Roles and Contributions*, edited by D. C. Hammack and H. K. Anheier, 347–68. Washington, DC: Brookings Institution Press.
Minkoff, Debra, and John McCarthy. 2005. "Reinvigorating the Study of Organizational Processes in Social Movements." *Mobilization* 10 (2): 289–308.
Munson, Ziad. 2008. *The Making of Pro-life Activists: How Social Movement Mobilization Works*. Chicago: University of Chicago Press.
Myers, Daniel. 1994. "Communication Technology and Social Movements: Contributions of Computer Networks to Activism." *Social Science Computer Review* 12 (2): 250–60.
Nip, Joyce Y. M. 2004. "The Queer Sisters and Its Electronic Bulletin Board: A Study of the Internet for Social Movement Mobilization." In *Cyberprotest: New Media, Citizens and Social Movements*, edited by W. van de Donk, B. Loader, P. G. Nixon, and D. Rucht, 233–58. New York: Routledge.
Nitschke, Paula, Patrick Donges, and Henriette Schade. 2014. "Political Organizations' Use of Websites and Facebook." *New Media & Society* 18 (5): 744–64.
Papacharissi, Zizi. 2010. *A Private Sphere: Democracy in a Digital Age*. Cambridge: Polity.

———. 2011. *A Networked Self: Identity, Community and Culture on Social Network Sites*. New York: Routledge.
Phillips, Whitney. 2016. *This Is Why We Can't Have Nice Things: Mapping the Relationship between Online Trolling and Mainstream Culture*. Cambridge, MA: MIT Press.
Pinch, Trevor J., and Wiebe E. Bijker. 1987. "The Social Construction of Facts and Artifacts: Or How the Sociology of Science and the Sociology of Technology Might Benefit Each Other." In *The Social Construction of Technological Systems: New Directions in the Sociology and History of Technology*, edited by Thomas P. Hughes, Wiebe E. Bijker, and Trevor Pinch, 17–50. Cambridge, MA: MIT Press.
Pinch, Trevor J., and Richard Swedberg. 2008. *Living in a Material World: Economic Sociology Meets Science and Technology Studies*. Cambridge; MA: MIT Press.
Polletta, Francesca, and James Jasper. 2001. "Collective Identity and Social Movements." *Annual Review of Sociology* 27:283–305.
Portwood-Stacer, Laura. 2013. "Media Refusal and Conspicuous Non-Consumption: The Performative and Political Dimensions of Facebook Abstention." *New Media & Society* 15 (7): 1041–57.
Postigo, Hector. 2008. "Capturing Fair Use for the YouTube Generation: The Digital Rights Movement, the Electronic Frontier Foundation and the User-Centered Framing of Fair Use." *Information, Communication and Society* 11:1008–27.
———. 2011. "Questioning the Web 2.0 Discourse: Social Roles, Production, Values, and the Case of the Human Rights Portal." *Information Society* 27 (3): 181–93.
———. 2014. "The Socio-technical Architecture of Digital Labor: Converting Play into YouTube Money." *New Media & Society* 18:332–49.
Qiang, Xiao. 2011. "Liberation Technology: The Battle for the Chinese Internet." *Journal of Democracy* 22 (2): 47–61.
Rasmussen, Terje. 2000. *Social Theory and Communication Technology*. Aldershot: Ashgate.
Ruben, Brent D., and Leah Lievrouw. 1990. *Mediation, Information, and Communication*. New Brunswick, NJ: Transaction.
Rucht, Dieter. 2004. "The Quadruple 'A': Media Strategies of Protest Movements since the 1960s." In *Cyberprotest: New Media, Citizens and Social Movements*, edited by W. van de Donk, B. D. Loader, P. G. Nixon, and D. Rucht, 29–56. New York: Routledge.
———. 2013. "Protest Movements and Their Media Usages." In *Mediation and Protest Movements*, edited by Bart Cammaerts, Alice Mattoni, and Patrick McCurdy, 249–69. Bristol: Intellect.
Ruppert, Evelyn, John Law, and Mike Savage. 2013. "Reassembling Social Science Methods: The Challenge of Digital Devices." *Theory, Culture & Society* 30 (4): 22–46.
Ryan, Charlotte, Karen Jeffreys, Taylor Ellowitz, and Jim Ryczek. 2013. "Walk, Talk, Fax or Tweet: Reconstructing Media-Movement Interactions through Group History Telling." In *Mediation and Protest Movements*, edited by Bart Cammaerts, Alice Mattoni, and Patrick McCurdy, 133–58. Bristol: Intellect.
Sandoval, Marisol. 2014. *From Corporate to Social Media: Critical Perspectives on Corporate Social Responsibility in Media and Communication Industries*. New York: Routledge.
Schulz, Markus. 1998. "Collective Action across Borders: Opportunity Structures, Network Capacities, and Communicative Praxis in the Age of Advanced Globalizations." *Sociological Perspectives* 41 (3): 587–616.
Shumante, Michelle, and Jon Pike. 2006. "Trouble in Geographically-Distributed Virtual Network Organization: Organizing Tensions in Continental Direct Action Network." *Journal of Computer Mediated Communication* 11:802–24.
Simi, Pete, and Robert Futrell. 2006. "Cyberculture and the Endurance of White Power Activism." *Journal of Political and Military Sociology* 34 (1): 115–42.
Smith, Jackie. 2001. "Globalizing Resistance: The Battle of Seattle and the Future of Social Movements." *Mobilization* 6 (1): 1–19.
Snow, David, and Robert Benford. 1992. "Master Frames and Cycles of Protest." In *Frontiers in Social Movement Theory*, edited by A. D. Morris and C. M. Meuller, 156–73. New Haven, CT: Yale University Press.
Snow, David A., Louis A. Zurcher Jr., and Sheldon Ekland-Olson. 1980. "Social Networks and Social Movements: A Microstructural Approach to Differential Recruitment." *American Sociological Review* 45:787–801.
Staggenborg, Suzanne. 1991. *The Pro-choice Movement: Organization and Activism in the Abortion Conflict*. New York: Oxford University Press.

Svensson, Jakob. 2014. "Activist Capitals in Network Societies: Towards a Typology for Studying Networking Power within Contemporary Activist Demands." *First Monday* 19 (8). http://firstmonday.org/ojs/index.php/fm/article/view/5207/4104.

Tarrow, Sidney. 2011. *Power in Movement: Social Movements and Contentious Politics*. 3rd ed. New York: Cambridge University Press.

Tatarchevskiy, Tatiana. 2011. "The 'Popular' Culture of Internet Activism." *New Media & Society* 13 (2): 297–313.

Taylor, Paul. 2005. "From Hackers to Hacktivists: Speed Bumps on the Global Superhighway?" *New Media & Society* 7 (5): 625–46.

Taylor, Verta. 1989. "Social Movement Continuity: The Women's Movement in Abeyance." *American Sociological Review* 54 (5): 761–75.

Thorson, Kjerstin, Kevin Driscoll, Brian Ekdale, Stephanie Edgerly, Liana Gamber Thompson, Andrew Schrock, and Chris Wells. 2013. "YouTube, Twitter and the Occupy Movement. Information." *Communication & Society* 16 (3): 421–51.

Tilly, Charles, and Lesley Wood. 2009. *Social Movements 1768–2008*. 2nd ed. Boulder, CO: Paradigm.

Tufekci, Zeynep. 2017. *Twitter and Tear Gas: The Power and Fragility of Networked Protest*. New Haven, CT: Yale University Press.

Tufekci, Zeynep, and Christopher Wilson. 2012. "Social Media and the Decision to Participate in Political Protest: Observations from Tahrir Square." *Journal of Communication* 62 (2): 363–79.

van Aelst, Peter, and Stefaan Walgrave. 2004. "New Media, New Movement? The Role of the Internet in Shaping the 'Anti-globalization' Movement." In *Cyberprotest: New Media, Citizens and Social Movements*, edited by W. van de Donk, B. Loader, P. G. Nixon, and D. Rucht, 97–122. New York: Routledge.

van de Donk, Wim, Brian Loader, Paul G. Nixon, and Dieter Rucht. 2004. "Introduction: Social Movements and ICTs." In *Cyberprotest: New Media, Citizens and Social Movements*, edited by W. van de Donk, B. Loader, P. G. Nixon, and D. Rucht, 1–25. New York: Routledge.

Walder, Andrew. 2009. "Political Sociology and Social Movements." *Annual Review of Sociology* 35:393–412.

Wall, Melissa. 2007. "Social Movements and Email: Expressions of Online Identity in the Globalization Protests." *New Media & Society* 9 (2): 258–77.

Weller, Katrin, Axel Bruns, Jean Burgess, and Merja Mahrt. 2013. *Twitter and Society*. New York: Peter Lang.

Wellman, Barry, et al. 2003. "The Social Affordances of the Internet for Networked Individualism." *JCC4 Journal of Computer-Mediated Communication* 8 (3): 0.

Wojcieszak, Magdalena, and Briar Smith. 2014. "Will Politics Be Tweeted? New Media Use by Iranian Youth in 2011." *New Media & Society* 16 (1): 91–109.

Wolover, David. 2014. "An Issue of Attribution: The Tunisian Revolution, Media Interaction, and Agency." *New Media & Society* 18 (2): 185–200.

Wong, Loong. 2001. "The Internet and Social Change in Asia." *Peace Review* 13 (3): 381–87.

Yang, Guobin. 2003. "The Internet and Civil Society in China: A Preliminary Assessment." *Journal of Contemporary China* 12 (36): 453–75.

———. 2009. *The Power of the Internet in China: Citizen Activism Online*. New York: Columbia University Press.

Youmans, William L., and Jillian York. 2012. "Social Media and the Activist Toolkit: User Agreements, Corporate Interests, and the Information Infrastructure of Modern Movements." *Journal of Communication* 62 (2): 315–29.

Zimbra, D., A. Abbasi, and H. Chen. 2010. "A Cyber-Archaeology Approach to Social Movement Research: Framework and Case Study." *Journal of Computer-Mediated Communication* 16 (1): 48–70.

Living in the Broken City
Infrastructural Inequity, Uncertainty, and the Materiality of the Digital in Brazil

David Nemer and Padma Chirumamilla

In this chapter, we theorize the work of repair as a site—or perhaps more accurately a kind of sustained encounter—through which conditions of pervasive infrastructural neglect and decay are contended with and managed on a contingent, ongoing basis. If the future of the city—as Mike Davis (2009) notes in *Planet of Slums*—lies in the Global South, and is a future characterized primarily by the severity and continuity of legal, economic, and social precarity, then it falls within our remit to theorize and bring to light those ways of living (of *making do*, even if *getting ahead* is an impossibly constrained task) that can render these conditions of ongoing uncertainty somewhat more manageable for those who live within the cities of our future.

Drawing from David's ethnographic fieldwork conducted in the favelas—urban slums—of a Brazilian capital city, we examine the way in which repair sustains the everyday technological lives of the favela's inhabitants. How do favela dwellers—living as they do in a zone of extreme uncertainty, marked by violences writ large and small—maintain a sense of stability in their technological lives? Breakdown and failure are the forms, we argue, in which technology is most commonly encountered within an everyday life soaked through with uncertainty. Instead of technological normalcy being marked by uninterrupted functionality (as it would be in rich neighborhoods or developed countries), technological normalcy here is the normalcy of the continually dropping signal, the easily scuffed and broken "off-brand" phone.

Infrastructural breakdown is not, then, an exception to a "normal" state of continuity, but is rather the backdrop against which the rhythms of everyday life must be forged. If one has to live with breakdown—and in zones of informality such as the favela and Davis's city of the future it seems inevitable that one must—then studying the work by which breakdowns are managed will give us some insight into the kinds of work, people, and spaces that allow for everyday life to maintain a semblance of continuity against a background of built and experiential uncertainty.

What we wish to emphasize through studying people and places considered "out of the way" by corporations or states is the necessity of more flexibility in thinking about the scales at which we approach technological life. The kinds of interactions that we encounter at the level of the neighborhood and its individual residents give us affective and intimate insight that systems-level thinking can obscure. These insights, we will show, are crucial to understanding how technologies—especially the kinds of communicative technologies marked as newfangled and "digital"—are taken up and received in areas far from the centers of technological and political power.

The Favela and Repair as Sustenance

The city we focus on is Vitória, an island and the capital of the Brazilian state of Espírito Santo. In focusing on cities outside those traditionally considered part of the global ambit, such as New Delhi or Rio de Janeiro, we wish to draw attention to the fact that these in-between regions—regions such as China's Pearl River delta, or mid-tier Indian cities such as Kanpur and Ludhiana—are where most of the world's population is expected to live in the coming years (Davis 2006).

More specifically, the sites of this study are the neighboring favelas of Gurigica, São Benedito, Bairro da Penha, and Itararé, all of which are located in the center of the island of Vitória. Informal settlements such as favelas—or even regional centers such as Vitória—are sometimes considered the "wrong" kind of site for studying technology (Takhteyev 2012), given their distance (infrastructural and otherwise) from the seats of power and capital. We believe, however, that concentrating our attention on these peripheral locations allows us to think more critically about the continuing importance of location to success in today's "knowledge economy."

In an environment like the favela, where a pervasive uncertainty governs the small works and acts that constitute an ordinary life, it is even more critical that we understand the kinds of work and acts of care that manage to hold together a semblance of continuity, and allow for everyday life to go on. Repair is one of these ongoing works of sustenance. As Steve Jackson (2014, 222) notes, it is through repair that "order and meaning in complex sociotechnical systems are maintained and transformed, human value is preserved and extended, and the complicated work of fitting to the varied circumstances of organizations, systems, and lives is accomplished."

Thus, we move away from the progressive narratives that have been embedded in discourse surrounding technology in zones of informality, such as Negroponte's One Laptop Per Child (OLPC) (Negroponte 2005), and instead move toward thinking through the ways in which ICTs slot into the continuance and ongoing maintenance of everyday life. This approach builds on the studies of repair and maintenance work by a diverse group of scholars, including Daniela Rosner and Morgan Ames (2014), David Edgerton (2011), and Douglas Harper (1987). Repair is a lens that highlights both the systemic instability and the individual creativity that constitute the effort to create workable technological systems in the favela. If not moving forward, then making do—this, above all, is an acknowledgment that while acts of creativity and of small-scale works of repair knit together zones of informality and neglect, they too are subject to the pervasive disruptions and disparities that long-term infrastructural neglect and abandonment bring about in their wake.

In the favela, these larger infrastructural breakdowns are highlighted by ongoing struggles such as the continuing fight over legal recognition of land titles, and the reliable provision of systems-level infrastructural services by the private sector and the government, which is closely tied to this legal recognition.[1] Instead of focusing upon the broad sweep of legal recognition and its infrastructural consequences, or on the difficulties in implementing large-scale technological projects within zones of uncertainty such as the favela, we chose instead to focus upon how technology works (and breaks) at a smaller, more personal level. This approach, we believe, is more suitable to understanding the intersecting scales of failure and reconstitution at work in the daily life of the favela.

The ICTs of interest in this chapter—ranging from mobile phones to the wired Internet system—are precisely those small-scale technologies that saturate the work and goings-on of everyday life. They have also been designed to fail over a short period of time and be replaced, instead of fixed (Rosner and Ames 2014). Whereas previous works on infrastructural systems focused on their large-scale constitution (Hughes 1987), the affective and symbolic power of these grand technological systems (Nye 1996), or the consequences of the partition and breakup of large-scale technological systems (Graham and Marvin 2001), we choose here to instead focus on those kinds of technologies the historian David Arnold (2013)—thinking of devices such as bicycles and sewing machines—has called "everyday technologies."

Why look at these small-scale technologies, and why characterize repair here as an encounter? We contend, drawing some inspiration here from Arnold, that it is small-scale technologies such as a keyboard or a mobile phone that figure most prominently in the intimacies of everyday life—that is, they are scattered throughout the background of daily work and domestic life, and as such are more tightly entwined with everyday practices than large-scale systems. This isn't to say that infrastructural systems have no place in a discussion such as ours. Clearly such large-scale systems as reliable water, sanitation, and electricity matter tremendously to the experience of one's everyday life. But in looking at and through everyday technologies, it is hoped that we can get a better sense of the affective and material intimacy and ordinariness that characterizes the uses of these devices and the fragile lives and systems that they inhabit.

A History of Making Do

Growing up in Vitória, a mile's walk away from Itararé, David encountered the favela primarily through stories of difference and fear: through reports in police blotters, through media reports and programming that assumed *criminality* to be the default state of the favela and its inhabitants, through derogatory remarks in conversations among private-school-educated friends and family members. Yet the history of these favelas—and of their relationship to the recognized city of Vitória and the state of Espírito Santo—was far more complicated than these portrayals allowed for.

Gurigica, São Benedito, Itararé, and Bairro da Penha are located east of Vitória, on the São Benedito hill, in between Av. Marechal Campos and Av. Leitão da Silva, as shown in figure 1. Unrecognized occupation of the area now part of the favelas

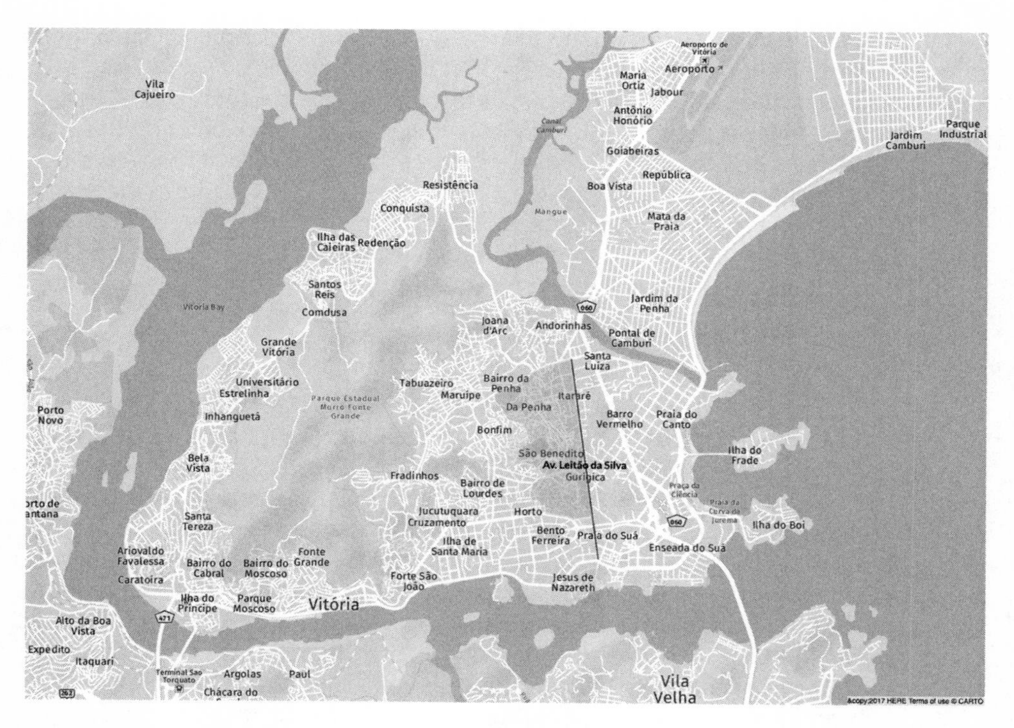

FIGURE 1: Map of the city of Vitória. The shaded area is where the favelas are located, and the solid line marks the Leitão da Silva Avenue, which divides wealthy neighborhoods from the poor ones. Source: Carto.

began in earnest in the 1940s, when the city of Vitória began building into the old mangrove swamps that once covered the area. The expansion of the earthworks and the advancement of urbanization in Vitória forced the poorer segments of the city's population to relocate to the unoccupied hillsides of the Farm Baixada da Égua and Farm Maruípe. The earliest occupations took place in the flatlands at the bottom of the hills, due to their proximity to the city center. Migrant workers from the neighboring states of Minas Gerais and Bahia, as well as rural Espírito Santo, also began to occupy the hill as they arrived in Vitória seeking employment. As people continued to relocate, they pushed higher into the hills.

The owners of the Farm Baixada da Égua and Farm Maruípe, the Hilal and Monjardim families, appealed to the police to protect their lands from the continuing occupation. The police responded to their appeals with a violent crackdown against the occupiers. Arcendino Fagundes de Aguiar, a retired military man from Rio de Janeiro whom the locals nicknamed "Sergeant Carioca," then helped the occupiers develop a strategy to take over the lands on which they intended to live, having nowhere else to go.

These strategies included "night raids" to divert police action. During the day, residents acted as "lookouts," blocking police and organizing protests. They tried to occupy as many public spaces as possible, including roads of access. Occupiers themselves, with machetes and hoes, built the first streets and alleys on the hill—a not-insignificant precursor to the self-reliant strategies that continue to color residents' technological lives.

Following a successful occupation attempt in 1954, and continuing struggles between police and protesters, the government of Espírito Santo intervened and made the land public property. In 2003, nearly 50 years after people had begun to settle the area, the city officially defined the boundaries and neighborhoods of Gurigica, São Benedito, Itararé, and Bairro da Penha. These boundaries were not necessarily recognized or acknowledged by the residents themselves, and could sometimes lead to mundane frustrations, like missing or misdelivered mail.

Given the historical and continuing absence of formal state institutions in favelas in general, non-state actors—such as cartels of heavily armed drug dealers—emerged to fill the vacuum left by the state (Ferraz and Ottoni 2014). The traffickers dictated and violently maintained a semblance of order within the favelas, and were respected by the residents because they created an environment in which critical segments of the local population felt safe, despite high levels of violence overall.

Since the residents settled at the bottom and middle part of the hill, the drug lords found the perfect location for their base at the very top of the hill. They worked as watchtowers, since they could see when cops or rival gangs were coming up the hill and prepare for battles. Traffickers' presence in Gurigica also turned the favela into a veritable war zone. Drug lords, originally from Rio de Janeiro, teamed up with a rival cartel in another favela, Bairro da Penha, and were trying to take over Gurigica and São Benedito at the time of research (Nemer 2016b).

Technology and the Everyday

Given this history of struggle and abandonment, capturing the tenor of everyday life—of the *ordinary*—in these areas was thus quite challenging. David found himself in the middle of three shootouts between rival gangs during the course of his research. It was the goodwill and help of the community leader, Serginho, and the telecenter manager, Christina, which helped him gain access to the favelas and their residents. The LAN house owners and inclusion agents, who were locals, helped him to learn his way around and gain trust within the community.[2] David—male and upper class—was constantly aware of the differences of power and status that he embodied. In order to alleviate the barriers that such differences may have caused, he first approached participants as conversational partners (Rubin and Rubin 2011). David's ethnographic approach (2015) centered on listening to participants with an open heart and mind, and to then truly consider what they expressed. His motivation resided not in judgment, but in understanding.

David's presence inside the favela was controversial—for varying reasons—among different groups of people, both in his own life as well as within his interlocutors' lives. David's family was worried and not pleased with the timing of his fieldwork, due to the intense drug war that was happening then. His friends thought he was insane for risking his life "to teach poor people how to use computers," and some participants did not want an outsider doing research within the favelas. The word "research" stirred up silence, conjured up bad memories, and caused distrust. Favela residents mentioned being approached by previous researchers as nothing more than guinea pigs. The project of research was implicated, for the favela residents, within an excess of classism, rude questions, inequality, and condescension. Residents were of the opinion that research, instead of benefitting them, instead classified and bracketed their life experiences into whatever framework the researcher demanded of them.

Favela residents thus did not identify with how their stories were being told and exploited. Throughout the project, David believed that critical and postcolonial researchers have the ethical obligation to represent marginalized communities according to their terms, respecting their history, values, and beliefs, as advocated by Hall (1997): when culture is a set of practices, representations are influential forces that shape how people are treated. He "resisted domestication" by using the resources, skills, and privileges available to him to make residents' voices accessible. He strove to provide a fair and empowering account of favela residents whose stories were otherwise suppressed (Madison 2012). In order to address this, David, along with favela residents, started a participatory project called "Favela Digital" where they attempted to provide an empowering account of the everyday life in the favelas. The outcome of the project was the bilingual (Portuguese and English) book called *Favela Digital: The Other Side of Technology/O Outro Lado Da Tecnologia* (Nemer 2013). The book was aimed to create awareness by showing how favela residents appropriate technology in innovative and meaningful ways, as well as to provide a channel to bring different networks into the same conversation. Such networks were policy makers and favela residents, and the "conversation" was about the issues that favela residents face with regard to accessing digital technology. Telling the "untold side of the story," as the one told here in this chapter, through the lenses of those who have been suffering the consequences of marginalization and exploitation would promote their recognition as human beings who deserve respect and recognition for their values and beliefs.

As is the case with most favelas, most of Gurigica, São Benedito, Bairro da Penha, and Itararé's population relies on LAN houses and telecenters to gain access to computers and the Internet. David's fieldwork focused on two of these LAN houses:[3] Life Games, a LAN house in Bairro da Penha, and Guetto LAN house in Gurigica; and on two telecenters in the area: one in Itararé and another one in São Benedito. Drawing from these studies, in this chapter we examine two communication technologies, both of which are crucial to the everyday technological life of the favela, though in subtly different ways. First, we examine the (usually questionably legal) systems of wired Internet connections that supply the LAN houses, and second, we take a close look at the mobile phone of favela residents.[4]

While these technologies are both tightly bound up with the rhythms of everyday life in the favela, the forms by which their functionality and breakdown manifest themselves within daily practice are quite different, and the demands they expect and pleasures that they afford their users are also quite varied. In looking at this divergent set of small-scale technologies and patched-together systems, it is our objective to offer a richer picture of both the technologies present in the favela and the practices, tricks, and acts of care that are needed to manage them and render them livable and perhaps even enjoyable.

Internet

The urbanization of favelas in Brazil was recent and inefficient (Perlman 2010), and the makeshift character of the most fundamental infrastructural services, such as electricity and water connections, reflects this disparity. While the government did not forcibly remove favela residents, it deliberately did not acknowledge inhab-

FIGURE 2: Cables and "gatos" running around the favelas. Photo: Leandro Recoba.

itants' existence in an infrastructural sense, as services such as water, electricity, and gas connections were never formally implemented. Forced to turn to their own devices in order to ameliorate this institutionalized neglect, favela inhabitants frequently acquired utilities illegally through homemade or makeshift wire and pipe taps, called *gatos* ("cats") (figure 2).[5]

The LAN houses Gyga Point and Guetto acquired their utilities through legal means, but even so the *gatos* all around the favela directly affected them. The irregular wire taps, for example, affected the voltage running through the power lines to the LAN houses (figure 3), damaging their computers, as Leonardo, the owner of Gyga Point, explained: "Changing light bulbs here is a frequent activity, but they are cheap, what really concerns me is how often the power supply units fry. Most of the time I don't have the money to buy a new one right away, so I have to put the computers away until I can buy new power supply units." Lisa, the owner of the Guetto LAN house, echoed these sentiments. She, like Leonardo, blamed cheap power supply units rather than fluctuating voltage in the power lines: "These power unit supplies are bad and fry all the time, they really hurt my business. I guess the ones I can afford are not good. I wish there were stronger and cheaper units."

This irregular and makeshift infrastructure also impacted the ability (or willingness) of companies to provide services to customers located in the favela. Internet providers were not able (or unwilling) to provide a reliable broadband connection to customers located in the favela. Lisa and Leonardo contracted a 3 Mbps Internet plan for their LAN houses, the fastest available to them. This connection, however, had to be shared with at least five computers. Lisa noted that the speed of their connection was not a trivial matter, given that Windows updates and security patches were *only* available online: "The users don't complain too much

FIGURE 3: Storage of PSU (power supply units) at Life Games LAN house. Photo: David Nemer.

because this is the only Internet they can access . . . the problem is when I have to make a security or Windows update. It takes forever to update every computer I have. It is dangerous because I have to stay in for the whole night and expensive since I have to pay for electricity." As Lisa mentioned, it was dangerous to have any business running after the curfew set by the drug cartel. Although the Internet providers were responsible for maintaining their infrastructure in the favelas, they were not keen on improving it and making it more accessible. Such disregard also affected the favela dwellers who weren't running LAN houses. Paula, an older woman from Gurigica, told David: "I called GVT [Internet provider] and they told me that the outdoor Internet box for Gurigica has been completely 'taken,' thus, they can't offer me an Internet connection . . . they suggested me to find a neighbor who has Internet and share the connection with him because they won't expand their box here." To overcome the arbitrary limitations imposed by the ISPs and reinforced by decades of institutional neglect, Internet connectivity also began to acquire the makeshift character common to other resources in the favela. Like the electrical taps or strung-out cable television connections, the Internet too had to be somehow acquired and reliably maintained in the face of continued institutional neglect. It was not the deeply embedded, easily available resource that, say, teams who developed Windows updates imagined their users to possess.

Rafael, a young man from São Benedito, sadly noted: "[Internet providers] say they won't improve their Internet infrastructures because there aren't enough costumers for them in the morro [hill], but it's not true . . . if you look out there every light pole you will see tons of blue cables going to every direction and every house . . . we need more and better Internet." Gyga Point LAN house owner Leonardo ended up searching out information on computer networking himself, noting pointedly that it was a task that challenged the well-worn institutionalized neglect the favela had always been subject to: "I can't stay here and wait around. . . . The

FIGURE 4: Bringing wired Internet to the top of the São Benedito hill. Photo: Brenda Shade.

government is not interested in us, so I might as well do something about it [Internet]. The people here don't have the time to learn about technology and Internet, and since this is what I do, I decided to look for articles on Google and YouTube that could teach me how to do this [bring the Internet to his community]. This is indeed another source of income for me but I also feel I'm doing some good for my community." Following his self-imposed crash course on computer networking, Leonardo proceeded to subscribe to a faster Internet connection through his uncle's house, which was located on the border of the favela and a legally recognized, richer neighborhood. Gustavo used 15 Linksys routers (which were placed inside plastic boxes on the light poles) and 500 meters of Ethernet cable to connect both his LAN house and subscribers in the neighboring community, as seen in figure 4. He charged R$35.00/month (approximately US$10.00) per subscription, and was working at maximum capacity. The price was still inaccessible to several residents—but it was much cheaper than a R$160.00/month subscription (average cost) from one of the Internet providers.

The LAN houses had, in their centralization of technological availability, become a source of help and technological knowledge for the favela residents. The increasing affordability of technology led to an increasing number of first-time users in the favela. The LAN houses capitalized on this trend by providing a base from which residents could acquire the knowledge and help to maintain their purchases: nowadays the government made it easier so most people could buy a computer, especially because they could buy it in installments and pay for it in 48 months, as Life Games LAN house owner Alberto described:[6] "The problem is that they don't know how to use it properly. . . . People would come and ask me if I could fix their computers since I maintain the computer at my LAN house. I saw it as an opportunity to broaden my business . . . now I get computers with a thousand viruses, fried boards . . . and if it wasn't for me they wouldn't be able to fix their computers since

I charge them a fair price and usually recycle boards." The LAN house operators themselves scraped together their desirable technological knowledge through a combination of hands-on interaction and online videos and articles, rather than any formalized certification or training. The knowledge of the computer's proper use, much like the knowledge of computer repair or the provision of cable television, was cobbled together, piecemeal, with little to no emphasis on broader technical skills or theoretical acumen.

The threads of favela infrastructure were thus severely strained in terms of durability and resilience: two characteristics that define the classic "large technical system" of Thomas P. Hughes (1987). While the abstractions used in high-level code (such as Windows updates) may treat resources as if they are inexhaustible (Blanchette 2011), it is clear this particular imagining of the world and potential users of the system falls far short of the reality of technological life in the favela, where connective cables, personal security, and reliable electricity are all in short supply, and are all things that need to be *struggled for*, pieced together out of what is immediately available.

The assumptions made in the delivery of critical system updates (that one has access to a reliable Internet connection, that one can stay with one's computer throughout the long process without risk to one's life, that the power will remain on throughout the download time) are all assumptions that hinder the easy working of computers in the favela. The system, it seems, never imagined such users (or such places where they might be), despite all its pretensions to global coverage. In many ways, then, the LAN house owners in the favelas fought an uphill battle not only against the most obvious and visible signs of infrastructural neglect and degradation (the failing power, the tangles of wire that are the *gatos*), but also against these more latent assumptions within technological systems—assumptions whose intent was not obviously harmful, and assumptions whose limitations appear only in those places and among those people deemed peripheral.

Mobile Phones: The Networks

Infrastructural neglect in the favela manifested itself not only in the physical network of wired Internet connections, but also within the seemingly intangible and untouchable wireless communications networks. Mobile phone carriers did not provide satisfactory signal coverage of the favelas, which led to constant complaints from the residents, especially because the innocuous act of walking around, searching for a signal to complete a call, was in actuality an extremely dangerous activity. This was due to the shootouts from the intense drug war that was happening during the period of David's fieldwork (2016a). One of David's informants, a teenage girl named Fernanda, said: "My smartphone has no bars up in here [at the top of the hill], my calls are never completed and it is really hard to communicate with people from here. I don't even know [why] I pay for this thing. When I need to make urgent calls, I try to go to Bairro da Penha, forcing me to walk through Av. Hermínio Blackman. You know that avenue is known as the Gaza Strip of Vitória, right?" Ironically, the hill where the favelas were located was known as the Morro da Antena (Hill of the Tower; figure 5) because of the eponymous cell phone tower located at its peak. Due to its lack of utility for the favela residents, some did not even know what the tower was for, as Rodrigo, a young man, noted: "I come up here on the hill almost every week. I guess that's one way to move up in life. I never went

FIGURE 5: The cell phone tower on top of São Benedito hill. Photo: Thais Gobbo.

up this crazy thing [cell phone tower] but I look at it and see that there's still more to achieve. It gives me hope." During fieldwork, the major cell phone carriers, Vivo (owned by Spain's Telefonica) and TIM (the Brazilian subsidiary of Italy's Telecom Italia Mobile), were both under investigation by state prosecutors in Espírito Santo for enforcing a sort of social segregation within their networks. The carriers' customers who were in the peripheral neighborhoods of Vitória, or in the legally unrecognized favela, had significantly more difficulty completing calls than users who were in richer locations.

Anatel, Brazil's national telecommunications agency, noted that these phone carriers had a blocking rate (the percentage of calls not allowed into the system) in the favelas that was significantly higher than 5%, the maximum allowable rate set by the agency (Campos 2012).[7] In São Benedito, for example, the average blocking rate was of 15%, three times the maximum, whereas in Bairro da Penha, Gurigica, and Itararé, it was 10%. PROCON, Brazil's consumer protection agency, would eventually fine Vivo R$7.5 million and ban sales of new SIM cards in the state of Espírito Santo for three months. Anatel noted that over 40% of calls made on the carrier during peak hours had not been completed in certain neighborhoods, a rate well over the allowable rate of 33%:[8] "Users . . . were being discriminated against in relation to the enjoyment of the carriers' network service, i.e., the blocking rate was much higher in some peripheral neighborhoods within Vitória, while in others this rate was negligible." This is a condition of institutionally enforced infrastructural discrimination that LAN house owners, however industrious or entrepreneurial, cannot hope to fix by themselves. It is not our suggestion that the LAN house owners' clever fixes and ongoing small-scale repair works and projects formed an enduring solution to this much larger problem of infrastructural neglect.

Rather, the informal patch jobs of the LAN house owners served as a means by which favela residents could begin to forge some kind of pleasurable, useful technological life *within* a zone of institutionalized infrastructural neglect. The LAN houses were hardly a curative for greater ills, but they did provide some means by which residents could begin to fashion small, ordinary pleasures within the anxiety and uncertainty that defined their daily lives: the pleasure of talking to friends on their phones, for example, or looking up sports scores on the Internet. Small-scale pleasures, to be sure, but no less meaningful for that.

Mobile Phones, the Phones Themselves

The most widely used smartphones in the favelas were called *xinglings*, though these too were only really "smart" when Wi-Fi was available, since the carriers' infrastructural choices segregated the favela users from consistent wireless service and data packages were relatively unaffordable.[9] The *xinglings* were smuggled into the favelas by people related to the drug cartel and sold in the black market. The cartel had a deal with the sellers, who gave 30% of the sales to the traffickers in exchange for protection. The sellers were secretive about the origins of the smartphones, but Rafael, a former black market seller, mentioned that the *xinglings* were smuggled in from China through Paraguay (figure 6).

The *xinglings* only came with a charger and did not have a warranty or any guarantee of working condition. The constant power outages in the favelas often damaged the chargers, which were of bad quality, and sometimes the smartphones themselves. The favela residents felt neglected since they did not have the money to keep buying new chargers. Hence, sharing cables and power cords was an activity that affected group formations and power relations, as mentioned by Beto, a teenager from Bairro da Penha: "Here, we purchase *xinglings* in the back alleys or in the neighborhood market. If you're lucky it comes with a charger and that's it . . . the charger lasts a week. I bought the USB cable separately and now everyone wants to go to the telecenter with me so they can transfer the photos to the computer and upload them on Face [Facebook]. I've got tons of friends and respect now. I'm even picked first to play soccer." As observed in the community technology centers (CTCs), the *xinglings* were also shared among groups of three or four friends since not everyone could afford to buy one. Usually, each person of the group would contribute to the *xingling* experience: one would bring the smartphone, one would bring the USB cable, and another would bring a charger. When offline, favela residents used the *xingling* mainly as media devices, utilizing the camera, music, and video players, rather than the phone itself.

The CTCs served not only as the primary wired ISPs of the favelas, but also as their wireless hotspot. In the telecenters, the residents were able to connect to *Vitória OnLine*, an open and free wireless network maintained by the City of Vitória, which was accessible in several public places such as parks, city buildings, and telecenters. In the LAN houses, the users had to pay a fee of R$2.00 (approximately U$0.75) per hour.

Favela residents perceived the *xinglings* as an extension to the CTCs. Although the devices were mobile, the access to the Internet was still bounded to such centers. Thus, CTCs provided a place for their users to not only lend their cables and cords, but also to promote other social dynamics such as becoming a hangout spot

FIGURE 6: Informal market in Itararé where *xinglings* are sold. Photo: Jeferson Louis.

for young people. Teenage girls, for example, went in groups to the bathroom to take selfies so they could share them later on Facebook.

When online, the users mostly chatted on Facebook messenger and played Facebook games. The content, such as photos, was not directly uploaded to Facebook from their *xinglings*. For example, the smartphone used by the female teenagers had several photos of different people, thus they preferred to first upload their photos to the CTCs' computers so they could choose the best photos and distribute them in an easier and faster manner—instead of signing into each teenager's account and upload a photo through the *xingling*. One of Nemer's young interviewees, Mariana, mentioned that she preferred to use Facebook on the computer since it offers a better experience than on her mobile phone. "I can't use it [xingling] the way I want. Like on the screen, most of the websites turn into English in the mobile version. I like to use the computer because on the phone it doesn't work quite right. It is not easy to use the phone . . . all these terms that I don't understand. I have lots of difficulties in downloading stuff from the Internet: music, photos, videos."

Downloading content from the Internet also worked in this manner: they first downloaded music or videos to the CTCs' computer, to check if the files were not corrupted, and then they transferred them to their smartphones through a USB cable, as mentioned by Roni, a young adult from Itararé. Since walking around in the favelas was risky, most users tried to download as much content as they could, for example episodes of television series. "I come here [telecenter] to transfer songs to my smartphone. Music is everything in my life. It sets me free, like when I read a book. The music goes well according to my mood, but everything in life is music. Car noise is music, tin banging is music. . . . Music is like a world where there's no prejudice and judgment, and the smartphone is like the spaceship that takes me

FIGURE 7: Women using the telecenter Wi-Fi in the back alleys of Bairro da Penha. Photo: Jeferson Louis.

there." As mentioned by Roni, his *xingling* allowed him to "be" in a place where he felt comfortable; Joana, a young teenager from São Benedito, found in her smartphone a safer place to express her individuality, feelings, and emotions (figure 7): "The Life Games [LAN house] is right next to my house and I can use their Wi-Fi connection. . . . I have the opportunity to have access most of the time. . . . I put my hand in my pocket and just to know that it is in there [xingling with Facebook], I feel safer. With it I can get out of this crazy reality anytime I want. I can cry, scream about my pain because I know that someone will be here, online, listening to me. . . . I can for a moment be myself."

The smartphones were seen as precious materials in the favelas. The users were afforded with bargaining power and had the possibility to exchange their *xinglings* for pretty much any good they desired, said Janine, a woman from Itararé: "Cellphones are the most democratic kind of money here in the favela; they're worth a lot and everyone needs one. I can buy one in the back alley . . . talk to everyone . . . and then if I want to buy something else, I just trade it for something else. The other day I was crazy about a bike I saw. What did I do? I didn't think twice and offered my cell phone . . . the trade was fair. This cell phone will still come back to me." Smartphones gave favela residents a sense of being included socially. They felt more courageous to cross social boundaries when they possessed the device, as proudly mentioned by Marcos, a teenager from Bairro da Penha: "I got the phone from my mom. This smartphone makes me empowered, because I can just go around to Praia do Canto or Jardim da Penha [rich neighborhoods] and not worrying being judged as poor or favelado. When I went to the mall the other day, I had my cell phone in my hands the whole time, it felt like the it worked as a key and was opening every door I was walking thought."

Discussion and Conclusions

What can we learn from these snapshots of ordinary technological use within a zone of continual infrastructural neglect? Some preliminary suggestions are presented below, with the caveat that further research, as always, is necessary—particularly in trying to think about what forms of difference a backdrop of legal informality and persistent precarity lends to the notion of "infrastructure." If not always functional, if not steadily reliable, what (and *how*) do infrastructural objects and systems gain meaning in people's everyday lives and practices? What kinds of character can be attached to an infrastructure that is blatantly visible (because it is so often nonfunctional to varying degrees), to systems and technologies that cannot be rendered an invisible part of the landscape, as it so often is in the developed world?

The mobile phones provide a particularly striking illustration in regard to the multiple meanings that may lie within the differing scales of technological systems. While the larger, systems-level picture is one of segregation by the major wireless carriers and dysfunction, with areas of the city delimited by their (in)ability to receive calls, at a more intimate level the smuggled *xingling* smartphones are clearly an integral part of everyday life, affectionately described by their users as "doors" leading to better (or at least, *other*) places.

How do we understand these two competing sentiments? The presence of affective attachment at this more intimate level does not negate the breakdowns and failures occurring within the system at a wider level. One could think of these differing sentiments—the sharp awareness of the broader picture of technological neglect residing alongside the deep appreciation for those devices that do embed themselves within everyday life—as a strategy of sorts, a way of rendering livable the extremely inequitable and uncertain technological environment in which one finds oneself.

The shared *xingling*, unlike the dystopian and individualistic smartphone experience that Hollywood movies such as *Her* envision, fosters a communal sociality that emerges out of conditions of technological unevenness and lack: friends gather at the CTCs with one bringing a charger, another the data cable, and the third the *xingling* itself, in order to load data onto and off the phone. In the interlocutors' accounts, USB cords—those *wired*, material connectors—were a bridge to a more joyous and richer social life, and *xinglings* were keepers of exchange value within the favela (tradeable for other goods) and sources of confidence in spaces outside of the favela.

None of these observations about the mobile phone are directly connected to the kinds of informal knowledge and labor practices that we most commonly associate with technological repair. But, in these stories surrounding the mobile phone, we can perhaps begin to see the outlines of what an "act of care" (to return to Jackson's conception of repair) could look like. It is a deeply affective (and perhaps, affectionate) relationship with what one has, with those things that *do*, despite everything, manage to find themselves embedded within the circle of one's everyday and ongoing existence.

Looking at the CTCs, including LAN houses, and the way in which they function as a node drawing together patched-together and piecemeal communications infrastructure, gives us another kind of technological negotiation to think through—negotiations that are certain to become the bedrock upon which Davis's envisioned cities of the future will be built. The LAN houses in the favelas are dependent upon illegal taps of electricity and telephone lines in order to maintain their piecemeal technological connections—connections that are acknowledged by the authorities and utility providers only as a drain on resources, rather than indicative of an unmet need.

The owners of the LAN houses utilize a mix of personal relations (such as relatives in legally recognized areas who can be convinced to buy an Internet connection), informally acquired and tested knowledge (such as learning repairs off of YouTube), and cheap parts (the constantly frying power units) in order to maintain some semblance of continuity and stability. The infrastructure, in this instance, is not *invisible* when rendered normal, as older understandings of infrastructure (Star 1999) would lead us to believe. Within normal practice in the favela, the Internet and electrical infrastructure remains constantly in sight, constantly in need of care in order to successfully undergird the goings-on of everyday use and practice. This constant attention—both to the deficiencies of the built environment and to those fixes that can be acquired through means outside of legal or traditional avenues—could be seen, arguably, as a kind of stabilizing force.

This stability, we think, is always contingent, always in flux—dependent on the whims of relatives, on the (not always sensible) will of the utility companies to enforce what rights they can and on the skill of the LAN house owners to put in what fixes they can. The infrastructural stability of the LAN houses' Internet and electrical connectivity is constantly, visibly *produced.* In some sense, this is qualitatively different from works of maintenance as such—instead of working to keep a system of technologies functioning at an acceptable level or a standardized ideal, here instead is the visible, constant struggle to *bring forth*, to ensure such things as relatively reliable power or stable connections exist at all to begin with.

This is not to say that providing systemic infrastructural intervention would "solve the problem" of the favelas. These communities have been historically marginalized, following decades of the social segregation and neglect from the state government and private sector. This is not a neglect that lends itself to easy solutions. Infrastructural improvements, after all, are the bread and butter of politicians who show up in the favelas seeking the vote during election season, promising things like cable cars to facilitate the transportation of local residents.[10] Instead of waiting for the promise of "decent infrastructure" to be delivered from without, we must understand infrastructure, maintenance, and repair more broadly; scenes such as the favela's must be made empirically and conceptually familiar, even normal.

While repair has been framed as a process of promoting political action and improving environmental awareness in repair clinics in the Bay Area, repair was approached in favelas as a critically necessary process, an act perpetuated in order to survive and perhaps thrive. Though Jackson (2014) states that repair fills in the moment of hope in which bridges from old worlds to new worlds are built, in the favelas this hope is a transient one.

It is the hope that the beginnings of a desirable life, a life that can begin to approximate the comfort and stability that marks the upper class, can be snatched out of what the favela has to offer. Favela residents must develop and rely on their own repair work since what infrastructure is offered to them is in constant breakdown, another forsaken promise. Hence, instead of thinking of such repair work as merely *gatos*, we instead imagine it as a quiet kind of *caring*, in some contrast to those overtly political understandings of acts of hacking and the zone of the hackerspace.

What does it mean to *care* for technology, as Jackson asks us to envision the work of repair doing? What does it mean to *suffer* and become enraged with it, in ways that aren't entirely comprehensible to one's self? Users' thoughts on the possible motivations of the technology's designers (surely, they cannot be more evil than the government)[11] may seem fanciful, but they do offer an entirely new way of thinking about how technological failure forces itself to the forefront of imagination and comprehension. Large-scale breakdown may inspire a kind of fear and helplessness (any number of disaster movies will attest to this), but it is this small-scale frustration—this daily compounding of anger toward an incomprehensible *thing* that invisibly works for richer, better *others*—that provides us with a more poignant and meaningful window into thinking about what technology does to us, and what we can do with it.

Affect and affection—how one *feels* for those technological things that surround oneself—permeate the stories that we present here, though perhaps not in the same ways that scholars such as Sherry Turkle have theorized (see Turkle 2011). Here in the LAN houses of the favela, there is less sublimation of the self, and a stronger awareness of just how closely intertwined the wider breakdown within the environment is with one's own experience of a particular technology. Any account of what technological life looks like amid precarity must consider these emotions as central to its experience. So too with repair, we think. If we are to take the metaphor of an "act of care" out to some kind of conclusion, then thinking about what acts of work and sorts of feeling are necessary to manage these sensations of frustration and incomprehension—sensations that are embedded within the uncertainty that defines everyday life in the favela—is a good place to start.

Notes

1. For more on informality (and the infrastructural breakdowns that characterize and color it) as a conceptual approach to the city, see Ananya Roy (2005). For more on the legal struggles concerning recognition and regularization of Brazilian favelas, see Macedo (2008), Fernandes (2002), and Durand-Lasserve and Royston (2002).
2. Inclusion agents are the people responsible for taking care of each telecenter, promoting computer-related workshops and classes and helping users.
3. LAN houses are privately owned establishments where, like in a cybercafé, people can pay to use a computer with a local area network and Internet access LAN. As opposed to LAN houses, telecenters are facilities supported by the state and NGOs, ones where the general public can access computers for free. LAN houses and telecenters, along with other technology access establishments, are considered community technology centers (CTCs) (Nemer and Reed 2013).
4. The types of CTCs studied for this chapter were state-supported telecenters and locally owned LAN houses.
5. *Gato* was also used to describe makeshift Internet and cable television connections.
6. It should be noted that even this seemingly mundane ability—to pay for goods in monthly installments—is itself a fairly recent development, following decades of inflation and the introduction of a new currency; see Chana Joffe-Walt (2010).
7. The block rate is the percentage of calls offered that are not allowed into the system, generally the percentage busy, but may also include messages and forced disconnects.
8. See Anatel's (2011) report for more details.
9. "Xingling" is a term used to refer to Chinese imitation and pirated brands, such as HiPhone, Galaxia, and Lumiax.
10. The project of cable cars in the favelas was announced to the local residents on October 2012, but up until February 2016 nothing was done. For a news article about the project to build cable cars in the favelas of Vitória, see Gildo Loyola (2012).

11. One of David's informants complained about the layout of the QWERTY keyboard, "Even the Government said that the keys on the Electronic Ballot are arranged like on the phones to make our lives easier, so why this [keyboard] is arranged this way? The technologists can't be more evil than the Government (laughing) (Teresa, 25)" (Nemer et al. 2013).

Works Cited

Anatel. 2011. "Relatório Anual 2011." www.anatel.gov.br/Portal/verificaDocumentos/documento.asp?null&filtro=1&documentoPath=278637.pdf.

Arnold, David. 2013. *Everyday Technology: Machines and the Making of India's Modernity*. Chicago: University of Chicago Press.

Blanchette, Jean-François. 2011. "A Material History of Bits." *Journal of the American Society for Information Science and Technology* 62 (6): 1042–57. doi:10.1002/asi.21542.

Campos, Mikaella. 2012. "Operadora Vivo É Acusada Pela Anatel de Discriminar Bairros." *A Gazeta*, November 22. http://gazetaonline.globo.com/_conteudo/2012/11/noticias/dinheiro/1375775-operadora-vivo-e-acusada-pela-anatel-de-discriminar-bairros.html.

Davis, Mike. 2006. *Planet of Slums*. Verso Books.

——. 2009. "Planet of Slums: Urban Involution and the Informal Proletariat." *New Left Review* 26. https://newleftreview.org/II/26/mike-davis-planet-of-slums.

Durand-Lasserve, Alain, and Lauren Royston, eds. 2002. *Holding Their Ground: Secure Land Tenure for the Urban Poor in Developing Countries*. New York: Earthscan.

Edgerton, David. 2011. *The Shock of the Old: Technology and Global History since 1900*. London: Profile Books.

Fernandes, Edesio. 2002. "Providing Security of Land Tenure for the Urban Poor: The Brazilian Experience." In *Holding Their Ground: Secure Land Tenure for the Urban Poor in Developing Countries*, edited by Alain Durand-Lasserve and Lauren Royston, 101–26. New York: Earthscan.

Ferraz, C., and B. Ottoni. 2014. "State Presence and Urban Violence: Evidence from the Pacification of Rio's Favelas." Paper presented at the 36th Meeting of the Brazilian Econometric Society, Natal.

Graham, Stephen, and Simon Marvin. 2001. *Splintering Urbanism: Networked Infrastructures, Technological Mobilities and the Urban Condition*. New York: Routledge.

Hall, Stuart. 1997. *Representation: Cultural Representations and Signifying Practices*. Thousand Oaks, CA: Sage.

Harper, Douglas A. 1987. *Working Knowledge: Skill and Community in a Small Shop*. Chicago: University of Chicago Press.

Hughes, Thomas P. 1987. "The Evolution of Large Technological Systems." In *The Social Construction of Technological Systems: New Directions in the Sociology and History of Technology*, edited by Wiebe E. Bijker, Thomas P. Hughes, and Trevor Pinch, 51–82. Cambridge, MA: MIT Press.

Jackson, Steven J. 2014. "Rethinking Repair." In *Media Technologies: Essays on Communication, Materiality and Society*, edited by Tarleton Gillespie, Pablo J. Boczkowski, and Kirsten A. Foot, 221–39. Cambridge, MA: MIT Press.

Joffe-Walt, Chana. 2010. "How Fake Money Saved Brazil: Planet Money: NPR." *NPR*, October 4. www.npr.org/sections/money/2010/10/04/130329523/how-fake-money-saved-brazil.

Loyola, Gildo. 2012. "Morros de Vitória Terão Teleférico." *A Gazeta*, October 27. www.gazetaonline.com.br/_conteudo/2013/07/noticias/cidades/1452879-morros-de-vitoria-terao-teleferico.html.

Macedo, Joseli. 2008. "Urban Land Policy and New Land Tenure Paradigms: Legitimacy vs. Legality in Brazilian Cities." *Land Use Policy* 25 (2): 259–70.

Madison, D. Soyini. 2012. *Critical Ethnography: Method, Ethics, and Performance*. 2nd ed. Thousand Oaks, CA: Sage.

Negroponte, Nicholas. 2005. "The Hundred Dollar Laptop-Computing for Developing Nations." Paper presented at the American Technologists Conference, Massachusetts Institute of Technology.

Nemer, David. 2013. *Favela Digital: The Other Side of Technology/O Outro Lado Da Tecnologia*. Vitória: GSA Grafica e Editora.

——. 2015. "Rethinking Digital Inequalities: The Experience of the Marginalized in Community Technology Centers." Bloomington: Indiana University.

——. 2016a. "Online Favela: The Use of Social Media by the Marginalized in Brazil." *Information Technology for Development* 22 (3): 364–79. doi:10.1080/02681102.2015.1011598.

——. 2016b. "Rethinking Social Change: The Promises of Web 2.0 for the Marginalized." *First Monday* 21 (6). doi:10.5210/fm.v21i6.6786.
Nemer, David, Shad Gross, and Nic True. 2013. "Materializing Digital Inequalities." In *Proceedings of the Sixth International Conference on Information and Communications Technologies and Development Notes—ICTD '13—Volume 2*, 108–11. New York: ACM. doi:10.1145/2517899.2517915.
Nemer, David, and P. J. Reed. 2013. "Can a Community Technology Center Be For-Profit? A Case Study of LAN Houses in Brazil." In *CIRN 2013 Community Informatics Conference*, 1–9. Prato, Italy. http://ccnr.infotech.monash.edu/assets/docs/prato2013_papers/nemerreedprato2013.pdf.
Nye, David E. 1996. *American Technological Sublime*. Cambridge, MA: MIT Press.
Perlman, J. E. 2010. *Favela: Four Decades of Living on the Edge in Rio de Janeiro*. Oxford: Oxford University Press.
Rosner, Daniela K., and Morgan Ames. 2014. "Designing for Repair?" In *Proceedings of the 17th ACM Conference on Computer Supported Cooperative Work & Social Computing—CSCW '14*, 319–31. New York: ACM. doi:10.1145/2531602.2531692.
Roy, Ananya. 2005. "Urban Informality: Toward an Epistemology of Planning." *Journal of the American Planning Association* 71 (2): 147–58. doi:10.1080/01944360508976689.
Rubin, Herbert J., and Irene S. Rubin. 2011. *Qualitative Interviewing: The Art of Hearing Data*. Thousand Oaks, CA: Sage.
Star, Susan Leigh. 1999. "The Ethnography of Infrastructure." *American Behavioral Scientist* 43 (3): 377–91.
Takhteyev, Yuri. 2012. *Coding Places: Software Practice in a South American City*. Cambridge, MA: MIT Press.
Turkle, Sherry. 2011. *Evocative Objects: Things We Think With*. Cambridge, MA: MIT Press.

Sound Bites, Sentiments, and Accents
Digitizing Communicative Labor in the Era of Global Outsourcing

Winifred R. Poster

Firms are relying upon the communicative labors of automated bots and electronically mediated live workers as a means of connecting to their consumers. In the process, they are digitizing sound. One might presume these sounds are reflective of "objective" technologies, design and business concerns, and thus socially neutral. Yet I show how such communications are embedded in sociopolitical groundings and tensions of ethnicity, citizenship, and geography. Because of transnational dynamics that stretch communicative labor across state borders (particularly through business process outsourcing), the sounds of service are nationalized as they are digitized.

This chapter examines what globalization does to the ICT design of service labor. Social agents and bots (as well as live employees on computers) are fashioned to be intelligent, deferent, and, very often, feminine (Suchman 2007). But in addition, they are also designed to mirror the nationality of the consumer and to mask that of the worker. This happens through linguistic accents and vocalized emotions within communication software and organizational labor processes. Sites in the United States, India, and the Philippines are examples of such transnational relations, and the focus of this study. Through ethnographic analysis, I draw upon original research conducted in the Indian outsourcing industry, and web research on technology design firms from around the world.

I illustrate these trends through the case of customer service call centers. Call centers are organizations, or parts of organizations, that handle customer relations, telemarketing, collections, and other telephone-based functions. These work processes—which operate transnationally through phone lines, satellite connections, and the Internet—signify the rise of communicative capitalism, and how sound has become the focal point for measuring productivity in the service economy. As managers attempt to assess the quality of worker-customer conversations, they deploy many kinds of technologies to intervene on the labor process and track the mundane sounds of talk. Their aim is to monitor, manage, and, to varying degrees, replace that sound digitally. In particular, several elements of worker vocalization are under contention in this context—the conveyance of affect, the humanness of the voice, and the nationality of the accent.

The analysis proceeds first by unraveling the ways that voice is harnessed and shaped in call centers. Managers use a variety of tools from affective computing and artificial intelligence in an attempt to cue broader cognitive mappings of emotion, identity, and location. These include emotion detection software, bots with identities, and online databases of human vocalizations.

Subsequently, I document the various forays into nationalizing sound by the call center industry. Vocal sounds are signifiers for a host of national stereotypes, through dynamics of service enthnocentrism and accent discrimination—even for bots. Some technical and scientific fields, accordingly, have been moving toward the global, using algorithms to account for accent and emotion cross-nationally.

The customer service industry applies such strategies to nationalize (and renationalize) communication. Call centers use a range of strategies to produce the "right" accents: from the training of live workers in "national identity management" to the creation of multilingual avatar workers and the design of a digital mix board that plays desired accents for (muted) human workers. While some of these techniques involve full automation, others integrate the sounds of live workers into avatars, and still others implant digital sound bites into the communications that humans do.

Methods

Outsourcing is rapidly spreading around the globe. For a close-up analysis of how it operates, I focus specifically on contracts from the United States and United Kingdom, to English-speaking countries such as India and the Philippines. My ethnographic research of this industry has been ongoing. From 2002 to 2004, I did fieldwork in northern India, in the cities of New Delhi (National Capital Region), Noida (state of Uttar Pradesh), and Gurgaon (state of Haryana). This triadic region is where the call center industry began, and still has one of the largest concentrations of organizations.

Three call centers were the sites for fieldwork, representing various size, ownership, and global positioning within the industry: a multinational firm, with about 3,000 employees; a joint venture firm with a US company and about 200 employees; and an Indian-owned firm, with 40 employees. Methods involved interviews and observations. The majority of interviews were with employees, but also with HR managers, quality control personnel, recruiters, trainers, on-site nurses, and others. Outside these firms, I conducted interviews in the community with representatives of industry associations, government offices, and employee associations. To get a feel for the experience of call center work, I observed the "production floor," attended training seminars, joined agents for dinner in the cafeteria, and so forth.

Between 2009 and 2014, I did further research to explore actors in the United States who participate in this story—the vendors, clients, consumers, and so on. This involved analyzing websites of call center companies and technology vendors, and watching the "webinars" or online videos about their products and programs (see the companies listed in the Works Cited). I examined consumer organizations and conducted interviews with consumer advocates and experts in the call center industry.

By 2015–16, the industry had begun to change in a number of ways, first geographically. The Philippines became a prominent actor in global outsourcing, in fact surpassing India as a destination for call center work. In addition, artificial intelligence started taking a more prominent role in automating outsourced customer service. So I examined a number of firms for their operations, flyers, products, and especially their technology design for customer service.

Digitizing the Sounds of Service

Voice has become important in the context of what scholars call "communicative capitalism" (Dean 2009). Rising economic sectors are now in the *infrastructure* of communications (cell phones, Internet providers, etc.), as well as the *content* of those communications (analysis of public texts, tweets, posts, etc.). Call centers have a primary role in a third feature of communicative capitalism—the industries that *facilitate the exchanges* of firms with other firms and more importantly with their customers (Brophy 2010).

Within these one-on-one conversations over the phone, voice is a key tool of interactive service labor. It conveys three components: emotion, identity, and location. To shape these dynamics, designers, vendors, and call center firms are developing technologies to capture digitally the sounds of service. They turn to fields such as affective computing and human-computer interaction as guides. With the examples below, we see how call center actors are not simply automating the worker as a whole, but rather computationally harnessing aspects of the human worker—like his or her voice—for a variety of purposes.

Affect

Voice is important, first of all in conveying to customers the emotional quality of service. Many features of the human sound—pitch, tone, pacing, phrasing, word choice, etc.—underpin the subtle meanings of the conversation, and relate to customers a crucial aspect of the service economy: care. Exchanged as part of the service is the feeling of being cared for by the organization. A call center employee's voice communicates that message.

As a foundation, the sociological literature gave us the concept of "emotion work," revealing how employers often ask workers to invoke, perform, and deliver particular emotions as part of the job. In her seminal study of the airline industry, Hochschild (1983) showed how flight attendants use smiles, polite gestures, and soothing words of comfort to enhance the status of the customer. The bill collectors, alternatively, perform the opposite type of emotion work within the same larger company. Through their conversations on the phone in airline call centers, they use their tone and language to insult and coerce the customer, ultimately to deflate his or her status. One is the emotional "heel" and the other the emotional "toe" of the service industry.

My research on Indian outsourcing reveals how call centers do much more with voice in the pursuit of affect. They create, display, perform—and manipulate—emotion in the service of credit and debt (Poster 2013b). They utilize emotions to get consumers to enter into, stay on, and pay back debt. While on the phone, employees do emotional investigative work to figure out consumers' personal sensitivities, and exploit their emotional motivations for paying. They tap into consumer ethics concerning debt, and lean upon their sense of honor, status, and respectability. In short, they use intricate emotional strategies to target consumer intimacies and moralities.

Many computerized programs have been designed to track, monitor, and analyze the emotion in call center speech. Straddling the fields of affective computing, communications studies, and psychology, researchers are well at work investigating the affective state of the customer. Some are curious what makes a customer angry (Schmitt et al. 2010). They use "acoustic, linguistic, and interaction parameter-based information for anger detection" (Neustein 2010, xiv). Others are curious what makes a customer happy (Gavalda and Schlueter 2010). They use data mining techniques of recorded and live calls to "search for words, phrases, jargon, slang, and other terminology" for evidence of customer satisfaction with their service interaction.

This "emotion detection" software is used in the workplace itself as well, by call center managers who seek to evaluate their employees (Poster 2011). It enables managers to technologically surveille the most "human" part of the service relation—the emotional engagement between customer and worker. They use the wave frequencies of a person's voice to detect a wide range of human emotions—irritation, duplicity, delight, or sexual arousal. Words themselves (such as "frustrated" and "angry") are evaluated for emotional content. The software also assesses features of the conversation (such as pitch, tone, cadence, and speed) for more subtle indicators of emotion. Rapid speech or rising tone can signal excitement. Slower speech or moments of silence can indicate distress, discontent, or unwillingness of a consumer, for instance, to sign up for a health insurance plan.

These systems help firms evaluate the effectiveness of the worker's voice in communicating the appropriate emotions of service. Accordingly, alarms can be sent to supervisors on the shop floor the moment that inappropriate emotions are expressed by an employee. Thus, it's notable how affective computing scholars use call centers as a test case for learning how to analyze emotions through algorithms (Neustein 2010). The service economy is a convenient site for combining the interests of communicative capital and HCI research.

Identity

Voice is important for a second reason within call centers: in communicating identity. This refers in a direct sense to the identity of the worker herself or himself. But by extension, it also refers to what the presentation of the worker's identity means for the organization he or she works for. Labor scholars have considered for some time how worker identity has value for firms: to elicit loyalty to the job, to achieve consent for various kinds of organizational controls, or to project the firm's corporate image (Poster 2013a). Toward this end, employers commit many kinds of formal and informal resources into shaping workers' identities.

This includes bots as well. AI designers are creating identities for website social agents, the V-reps, since many firms consider it a critical part of their public display of service. At one point, firms were hiring specialized professionals in Silicon Valley to create worker bot personalities, which have several components. One is a name and physique. United airlines had Julie; Unilever had Katie; Sprint PCS had Claire; Pepsi used Lisa. Sometimes the V-reps have a catchphrase, such as "Okay, let's get started" and "Got it!" They may have interests, hobbies, and jobs. Cal North was constructed for a California transit system to be a "retired cop who likes football and kids and hates decaf coffee" (Perry 2003). Yahoo's Jenni, who dictates your email over the phone, has an entire background including a resume (Wong 2005). In a 700+ word biography, Jenni is described with fake job references, university degrees, and boyfriends. This biography details her physical features as well: 5 feet 5 inches, 108 pounds, blue eyes, brown hair.

It is often the voice, however, that transmits this identity to the consumer, especially for live employees on the phone. Voice indicates status features of the speaker—his or her gender, age, background, and so on. Accordingly, the status of the worker becomes a reflection of the broader company, and for this reason managers reshape worker features to conform to idealized identities. For instance, I heard often from managers in my research that the voices of female workers are more suited for customer care, such as on helpdesks, in terms of soothing angry callers who are phoning in their problems. Male workers, alternatively, are said to be more suited for telemarketing, given that their voices convey the authority and aggression needed for sales.

Voice also indicates class. Call centers sometimes prefer higher-class sounds from their workers, in order to upgrade the status of the firm. This becomes problematic for some employers in the United Kingdom, who are distasteful of particular accents that are associated with low-class status. In turn, they have developed accent training programs for these workers to sound more middle class in their service interactions (Warhurst 2016).

Recent trends in the technologies of service are revealing how critical voice is for conveying "identity." VocalID is an online organization that collects and preserves human sound bites in virtual storage. Donors log onto the VocalID website and speak 3,000 scripted words or phrases that a person might say in a typical conversation. The organization then records and keeps them in a voice databank. Its goal is explicitly "Connecting voices to identities: Synthetic voices, as unique as fingerprints" (VocalID 2015). The banner on VocalID's homepage reads, "Say goodbye to uniform voices. Voices are not identical. They are our identities."

Originally, this system was designed for the "tens of millions worldwide" who cannot speak and therefore rely on synthetic voices to communicate (think of renowned physicist Stephen Hawking). For them, VocalID serves as an alternative to the limited options of most automated voice synthesizers, in which every user sounds the same regardless of age, gender, and so forth. Many feel that "Perfect Paul," the most efficient and common of these synthesizers, does not align with their image of themselves, such as 17-year-old female Samantha interviewed by National Public Radio (Spiegel 2013). Instead, donor sounds are electronically blended with a recipient's original base pitch, breathiness, and other characteristics (Spiegel 2013).

However, VocalID has an additional—and much broader—applicability in the digitized service economy. It will be used to imbue automated service workers with human-like sound. VocalID systems will be integrated in the software of cell phones and computers, so that the users of this technology will be expanded to the "hundreds of millions" who have text-to-speech technologies on their mobile devices. In the future, everyday consumers will likely have a range of options among real and modified human voices from the database of their own choosing—even their own—that speak to them. Virtual assistants, in other words, can be programmed to appear audibly more human, with more personalized voices and thus identities. This means that the "social agents" who serve the public in Suchman's (2007) scenario will be one step further from the generic and impersonal bots of the early days of HCI and AI.

VocalID, in this way, reveals the premium placed on the sound of a human voice. Furthermore, it reveals the technological strategies developed to preserve and integrate it into everyday platforms for communication. This will be become relevant again when we discuss the sounds of nationality below.

Geography

Finally, voice indicates place. Call center employees communicate aspects of location, space, and time—not directly through words and statements, but subtly through their accents. These accents can signify the worker's personal place of origin, as well as the location of the firm she or he works for, and its site of operations.

For call centers, there are many wrong kinds of accents (and accordingly wrong places) with which firms do not like to be associated. Call center firms in the United States are known to choose their locations in part based on the desirability of the accent of the workers (Bain 2001). Some setup operations in states such as Nebraska and Arizona on the premise that the accents of employees are most "neutral." This contrasts to states such as New York, Texas, Alabama, and Minnesota, which have accents with undesirable sociocultural signifiers—too tough, too urban, too cosmopolitan, and even too "dumb." Relocating based on these criteria then saves the firm investments in "accent training." As human resources are the main expenditure for these firms, such training can represent a large share of their labor costs.

Voice also conveys the foreignness (and/or foreign location) of a worker. In some cases, that foreignness is desired by the firm and enhances their service (Hill and Tombs 2011). Take the example of a French restaurant in the United States. A waiter's French accent supposedly improves the experience for the customer by legitimizing the coveted foreignness of the product. Other times, however, that foreignness is not desired by the firm. This point leads us to the next section: how and why firms are investing resources into transforming the sounds of service.

Outsourcing and the Global Problems of Voice

Employee voices carry meanings of nation, along with the factors above. For the call center industry, this becomes apparent in the context of outsourcing. *Outsourcing* is the contracting out of particular functions to a secondary firm that is specialized in those services and provides them more inexpensively (Poster and Yolmo 2016). The offshoring of services began to proliferate around 2000, when Internet connections, fiber optic cables, and satellite communications systems began to enable data and voice transfer easily and cheaply across distances.

Yet sending service contracts abroad means that workers and consumers are interacting directly across Global North and South. For firms from the United States and United Kingdom in particular, the most common destinations have been India and the Philippines. One of the primary reasons for moving to former colonies, in fact, is language—to make use of English-speaking capabilities of the workforce.

What may seem like a cost-saving endeavor in language parity, however, can backfire. Call center firms become troubled by local accents, which reflect different varieties of English from those used in the United States and United Kingdom—a nuanced phenomenon that the sociolinguistic field of World Englishes has illuminated (Sridhar 2008). Even within India, there are many Englishes. As Cowie (2014) describes, the more distinctive version may be marked by some of the following: a trilled "r" sound, a retroflex consonant (e.g., the pronounced "t" at the end of a word), and a British-style, long "a" sound (e.g., in words such as "class" and "chance").

In this context of outsourced call centers, then, a worker's voice (indirectly) relays her or his location in a global distribution of labor. It conveys not only the immediate "place," but a large grab bag of transnational codes and flashpoints: citizenship, nationality, meanings of service, politics of outsourcing, and others.

Nationalized Sound Bites

Voice over the phone itself carries nationalized meanings that customers, workers, and employees regularly evaluate. For consumers in the United States and United Kingdom who are on the phone with Indian and Filipino workers, nation is present in both positive and negative ways—as a trivial concern, a benign curiosity, or a point of virulent contention. In a parallel trend, the fields of affective computing, AI, and HCI have been moving toward the global. A preoccupation of these scholars has been dissecting particular patterns of talk along national lines using computerized techniques. As designers create automated workers for the service economy, these tendencies are spreading to bots and V-reps in the service economy as well.

Accent Discrimination

Sound is nationalized, to begin with, through accent. Psychologists show us that people make predictions and assumptions about the nationality of others during interactions. Moreover, they often do so based on just the sight and sound of a person—what psychologists call their "nonverbal accents" (Marsh et al. 2007). Nonverbal accents include a range of supra-linguistic tools that humans use to communicate—some of which are visual on the body (e.g., shrugging one's shoulders, raising an open hand, etc.), but others of which are vocal. This includes obvious audible cues such as words and pronunciation, but may include subtle markers as well: a tendency to laugh, certain vocal intonations, and so forth (Elfenbein 2007).

In fact, scholars predict that it is easier for people to identify national identity based on voice (e.g., a vocal recording) than vision (e.g., a photograph). In experimental studies, informants use such nonverbal accents to identify Americans versus Australians. Moreover, they attribute qualities to each based on these factors: Americans as more leader-like and dominant, Australians as more likable and friendly. Given that participants in these studies were given minimal information on which to interpret the cues, Marsh and colleagues (2007) conclude that participants were using stereotypes to attribute particular nationalities to accents.

In the employment context, such stereotypes are applied to accent on a regular basis. Many studies have shown that *employers* make decisions about a worker's career and earning potential based on his or her accent (at least in the United States). In the interview process, for instance, Asian, Latino, and African American applicants with minimal accents are rated as more employable than those with maximal accents (Carlson and McHenry 2006). Speakers of nonstandard English are seen as lazy, incompetent, unprofessional, uncreative, and so forth (Atkins 2000). Within that group, moreover, the nonstandard speech of blacks is rated more negatively than that of whites (e.g., who speak Appalachian English). Transnationally, employers privilege French and American accents over Japanese accents (Hosoda and Stone-Romero 2010).

Accent matters even more than foreign names. Job candidates with foreign sounding names and no accent were viewed favorably by recruiters, while those who had an accent along with the foreign name were viewed unfavorably (Segrest Purkiss et al. 2006). Critically, accents matter when wages are assessed. Earnings penalties are higher among workers of foreign ancestry who have lower proficiency in English (Hamilton et al. 2008).

In the service industry, *consumers* are found to react to accents this way as well. Of US consumers, 32% report negative responses to Asian-sounding call center workers just based on their accent (Sridhar 2008). Customers in Australia report negative responses to hearing Indian or Filipino accents. They report reduced tolerance, and beliefs that the worker can neither understand nor assist them (Hill and Tombs 2011). Another study found that people are more likely to stereotype and thus negatively rate the call center performance of workers with Indian versus British or American accents (Wang et al. 2009).

Human-computer interaction scholars are showing that the same is true for bots. As some experiments show, American consumers prefer listening to bots and other social agents who have their same accent, and even find them to be more knowledgeable than similar voices with foreign accents (Dahlback et al. 2007; Nass and Brave 2005). This was found comparing white American participants to Koreans, in one study, and Americans to Swedes in another. Informants preferred hearing online consumer information (descriptions of products) from online agents with their own accent.

Sound is nationalized in a second way, through affect. Until recently, much of the research on emotion assumed a geographically shared set of meanings. Yet parallel to the rise of the global economy and network society, scholars have turned their attention to the way affect is broken down by nation. Their focus is on the way emotions are localized rather than universal. This prompts a connection of psychology and linguistics to affective computing.

Research in affective computing has begun dissecting emotions for particular national settings—and what they mean within the consumer mind-set. With the tools mentioned earlier, scholars have used "machine learning" to classify affective sounds in five countries (Laukka et al. 2014): Australia, India, Kenya, Singapore, and the United States. They find with their algorithmic emotion analysis that certain affective sounds are more likely to be nation-specific than others: "anger, contempt, fear, interest, neutral, pride, and sadness" versus "happiness, lust, relief, or shame" (447).

Service workers, in particular, are being evaluated for the geographic contours of their emotional expression—largely through the sound of their voices on the phone. For instance, French workers are found to be less emotionally controlled versus those in the United States, who hide real feelings of negativity while putting on a (proverbial) smile (Grandey et al. 2005). In the Philippines, call center workers are found to be reluctant to handle confrontation. This leads to "the CSR [customer service representative] retreating into silence or resorting to formulaic responses to arrest the anger" (Hood and Forey 2008; Lockwood et al. 2008, 237). According to the studies, these workers are accustomed to implied expressions of discomfort rather than those that are direct. They also reportedly lack sociocultural training in problem solving. In turn, these linguistic barriers to the use of (American) English are interpreted by customers as emotional failings (e.g., in using words such as "would").

Given these varying emotional expressions, specific national pairings of worker and customer become problematic for communication. In one study, Filipino agents were "too polite" for American customers. Americans are reported to shout aggressive things during the call: "don't apologize, just fix it" (Friginal 2009, 59). Filipino agents would respond with apologies and deference, yet this was not received by American customers as friendly. Rather, they interpreted such talk as "ineptitude or condescension," which ultimately "exacerbated the communication breakdown, resulting in an unsuccessful transaction" (59). Alternatively, Chinese customers are found to be emotionally and expressively reserved (Xu et al. 2010). In turn, researchers suggest structuring the labor force as an affective complement to that, such as hiring emotionally assertive employees who are capable of applying "more interactional steps . . . to work out this customer's real intention" (466).

Within affective computing, scholars are also applying such frameworks in the design of emotional service worker bots. Take, for instance, a study from Northwestern and Harvard Business Schools on attitudes toward "botsourcing" and "outsourcing" (Waytz and Norton 2014). Findings indicate that people in the United States prefer to use robots for thinking versus feeling jobs. However, they are more comfortable giving robots feeling jobs if they are more "humanlike." Respondents also prefer to outsource "emotional" jobs to particular countries like Spain and Australia, instead of Germany and China, which they perceive to be generally more "robotic" as nations. Here, the narrative of the cyborg is extended to nationally defined workforces.

Thus, literature is showing us how consumers read emotions within conversations in the same way that they do accents—that is, through a nationalized prism. There is evidence that the two are interactive, moreover. In a study by Wang et al. (2013), accent bias tended to increase with a state of anger. Consumers who are upset while receiving calls from telemarketers and debt collectors, the main activities of call centers, are less likely to suppress their biases about accents. In these ways, then, accents are interwoven with emotions in the context of global services.

Service Ethnocentrism

All of this points to a common core dynamic: the sounds of workers in global call centers (as accents and affective displays) are codes or flashpoints for underlying tensions within the political economy of service. Thelen and colleagues conceptualize this phenomenon as "service ethnocentrism" (Thelen, Yoo, and Magnini 2010; Thelen, Honeycutt, and Murphy 2010). They have been studying American consumers regionally and nationally, and find that over 70% of their informants oppose outsourcing. Their work describes how, and explores why, many US customers prefer to talk on the phone with service workers of their own nationality.

Their research has pinpointed several reasons. Some are practical, such as concern for security (e.g., privacy protection and safeguarding of information) and ease of communication (e.g., understanding accents). Some have to do with protecting the national economy (e.g., expressing loyalty to American firms and jobs). Other reasons, however, are more subjective and reflect nationalist hostilities: such as "foreign enmity beliefs" that offshore workers are not familiar enough with American culture to provide effective services, and "nativist beliefs" that local workers are generally superior (smarter, more helpful, etc.).

When charted against types of service, these feelings are associated more closely with financial-related activities (such as taxes) and less so with problem-solving activities (such as computer help desks). In other words, service encounters that involve money heighten the ethnic/national unease among US consumers. And significantly, service ethnocentrism varies by geographic destination of the work. Customers differentiate the desirability of outsourced employees by country; for instance, they prefer Canada ahead of China, India, the Philippines, and Mexico. Global North countries are ranked over those of the Global South.

This suggests that accent and affect in call centers may reflect many levels of global politics for the consumers. They may be reacting to a deeper set of meanings and conflicts, rather than simply being put off by the sound itself (i.e., the immediate encounter with the worker's voice). In fact, their tensions may have less to do with an Indian worker per se than with actors and organizations in their own setting: the US firm that the Indian employee works for, the US government that has failed to regulate the labor practices of outsourcing firms, and so forth.

Especially troubling is the offensive language that some consumers use on the phone, in the form of hostile and explicitly nationalized abuse. It may include refusals to buy things from foreigners, demands to be transferred to an American, and the shouting of racist slurs. Such cases have been documented in call center research in India (Das and Brandes 2008; Mirchandani 2008; Noronha and D'Cruz 2007), the Philippines (Friginal 2009), and other countries. While these extreme cases tend to be infrequent relative to total call volumes (Poster 2007), they do represent an important segment of consumer reaction to globalized customers service.

In the digital era, and with expanding technologies of call center communication, customer service ethnocentrism has found new kinds of outlets. There is mounting evidence of hate talk circulated through online media, confirming that electronic communication can be fertile ground for racist, sexist, and xenophobic sentiment (Citron 2014). Within call centers in particular, employees are now experiencing this verbal abuse, not only directly from consumers' voices on the phone, but through many other media and algorithmic sources as well.

My research has examined how consumers are expressing nationalized emotions in a range of places (Poster 2011). They create websites to post complaints about overseas call centers and their workers, they input negative sentiments on customer satisfaction ratings and software, and they generate databases to log companies that have too many foreign workers. Given the growing role of technological platforms as sites through which this anger appears, D'Cruz and Noronha refer to such emotional outbursts and nationalized talk as "customer cyberbullying" (D'Cruz and Noronha 2014; D'Cruz 2014).

In fact, they argue that the anonymity of call center technology facilitates such virulent xenophobia. In line with my earlier discussion, they find in their interviews with call center workers that audio phone communications can decrease social presence and insert anonymity to service interactions. Customers, accordingly, may experience this anonymity as freedom to express their service ethnocentrism:

> In participants' view, the invisibility and partial anonymity of the interaction, aided by its one-time and perceived one-on-one occurrence, lowered customer inhibitions. . . . That customers neither could see participants (and their reactions) nor knew them . . . brought in an element of personal and social dissociation that diluted customers' regard for politeness and restraint about incivility. . . . [The] limited cue capacity . . . triggers misbehaviour. That is, . . . restrictions on the scope of observation due to the mode of communication lead customers to greater degrees of detachment and lower levels of propriety which, along with their sense of customer sovereignty, give rise to bullying behavior. (D'Cruz and Noronha 2014, 187, 190)

Anonymity of the service interaction, therefore, does not necessarily benefit the worker. However, as we'll see next, it may benefit the call center and its corporate clients. Service ethnocentrism then provides a context for understanding the backlash against employee voices and the "wrong" kinds of sounds.

Reconstructing "Appropriate" Accents, Digitally and Organizationally

With these transnational dynamics of accent and affect, the value of sound for communicative capitalism comes under threat. Features of the worker's voice (its transmission of care, humanness, etc.) that are helpful for firms above are now undercut by the troubles of global politics and nationalism within consumer economy.

Accordingly, many firms do not seek to address this issue head-on (by opening a dialog with consumers, for instance), but rather indirectly and deceptively (by hiding). They mask their identities within the customer exchange, often to obscure the process of outsourcing (Poster 2007). Significant for this analysis, they do it through sound. The voice of the worker has the potential to "give away" the location of the firm (as well as its identity) and to invite backlash, as we saw above. Thus, altering the voice can protect the firm in terms of maintaining its anonymity. Consumers will assume a synchronicity of nation, and the firm's outsourcing practices will remain hidden.

Notable is how many of the current managerial trends for handling the dilemmas of nation involve using AI and affective computing. Call center managers and technology entrepreneurs have responded with attempts to renationalize voices. Their solution for smoothing globalized tension is to create—through careful labor processes and employment digitization—the "right' kinds of voices, accents, and affects. Their strategies range in the extent and use of automation. Some are highly integrative of technology, to the point of full automation; others are only partially

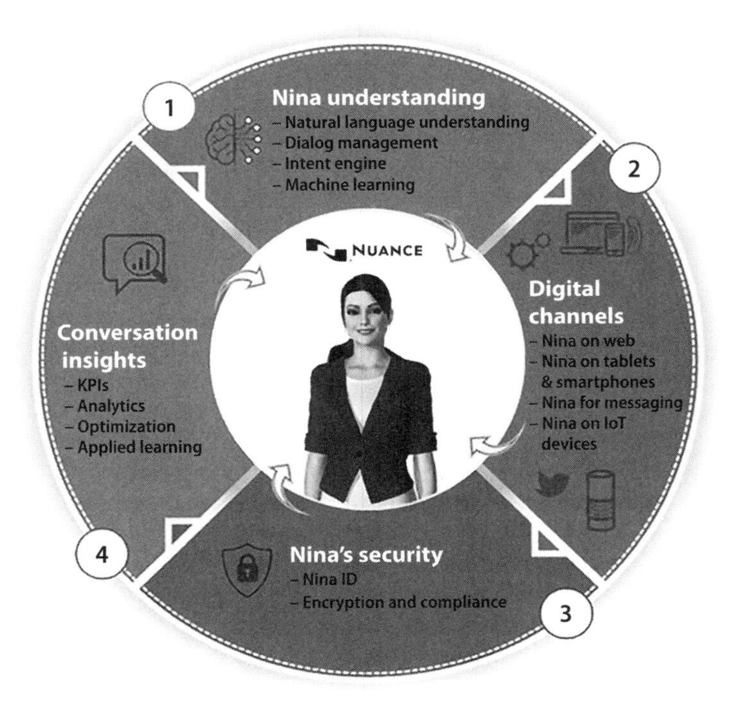

FIGURE 1: Nina, the multilingual V-rep. Source: Nuance Communications (2018).

so; and some not at all. In each, however, firms are manipulating and/or replacing human sound with preferred accents. Thus, firms are digitally capturing not only *voice* (as shown above), but *nationalized* voices.

The Multilingual Bot

The *fully automated* solution is to create bots that perform many accents and speak multiple languages. V-reps, appearing as online avatars on company websites, have become global. Just a few years ago when I researched the V-reps, they were plainly American (or subtextually "neutral" in nationality). But now, the bots are explicitly transnationalized. Nina from Nuance (figure 1), used by Coca Cola, represents a new age for the V-reps. As an early promotion on the website announced, she "speaks 38 languages" and "lives in the cloud." These V-reps are meant to be untethered to geography. Visually, they may display the national identity of the home country of the firm (i.e., the Global North hegemonic ideal of whiteness), yet vocally they are flexible for communicating across countries (and for doing so convincingly).

The value of the worldly bot is in its linguistic range. This automated employee is set up for breadth: she is a storage facility of global speech. Some of these bots also perform affective labor algorithmically. Amelia from Isoft speaks 20 languages and "understands language and emotion" (Isoft 2015).

Thus, the AI of service work is moving into the transnational economy. The design of V-reps takes into account nation and language. The social agents that

Suchman describes are now capable of interacting globally, or acting as global-functioning citizens, who acknowledge and communicate with multiple nationalities of consumers.

National Identity Management

The *fully human* option, on the other hand, is to alter live workers' behaviors within the labor process to match desired vocalizations. This option became prevalent among Indian call centers when the industry first took off in the early 2000s. The idea was to train (and retrain) employees in a variety of communicative and behavioral skills, so that they can hide their locations and instead convey to American (and British) consumers that they are in fact *in the United States*. I refer to this process as "national identity management" or NIM (Poster 2007).

Compared to the bots, the Indian employee has a different value—she or he produces one nationalized sound (the American accent) with incredible depth and detail. The call center worker can embellish her or his sound with other vocal capacities, such as geographically appropriate dialog, conversation imbued with localized meanings, and so forth.

NIM involves several components. Through induction sessions and ongoing human resource department activities, call centers train workers in a variety of communicative skills and resources: (1) *voice and accent* to reproduce American diction, voice modulation, rhythm (number of beats per second), and grammar; (2) an *alias* to announce American identity to the customers through their name; and (3) *conversational skills* to convey through small talk that they are in the United States. This includes extensive knowledge of American consumer items, retail outlets, restaurants, and so on. It also includes lingo, current events, sports, weather, and time zones for the locations they are calling. And finally, (4) the worker learns a *script* to repeat when customers test the boundaries of that façade and pose the looming question: "Where are you calling from?" The predefined responses range from the opaque: "an outbound call center," to the semispecific (and somewhat truthful) "in Asia," to the less honest "a US office of the client firm." And "if they ask again, then we change the subject." As the HR trainer summarized, "It's a marketing strategy—if you cannot convince, confuse."

These four practices lie on a continuum of layers of locational masking, ranging from the lesser forms that are indirect and more suggestive, to the more extreme forms that involve direct, outright lying. They may be applied individually or in combination in routine conversation. Workers vary in how much they actually practice it. Employers vary in how extensively they are committed to the endeavor, and to what lengths they go in promoting it. In some call centers, workers can be fired for failing to carry out elements of this process effectively.

NIM has broad reach across the Global South, as my colleague Kiran Mirchandani and I are observing. While much of the original research on locational masking focused on India, our book *Borders in Service* (Mirchandani and Poster 2016) collects cases from Morocco, Mauritius, the Philippines, and others, in which workers are often asked to participate in linguistic and conversational obfuscations of their nationality.

At the same time, it is important to note that some of these NIM strategies (especially the most devious ones) are on the decline (Mirchandani 2012; Nadeem 2011). The outright lying (step 4 above) has become less common in the second decade of

FIGURE 2: The accent soundboard. Source: Avatar and Echo Live Agents (2018).

the outsourcing industry in India. As consumers have protested the use of these strategies by Indian call centers, a few US firms have even revoked their contracts and pulled their work processes back to the United States.

Indeed, wary of retraining workers as a strategy for anonymizing, firms have turned to other options. We see next how the managerial practices of deception with voice are now appearing elsewhere—digitally.

The Accent Mix Board

A third solution for fixing accents in global centers (and a midpoint between the two poles of human versus robotic workers above) is *partial automation.* It involves integrating live labor with algorithmic labor, through call center "soundboards" (figure 2). The idea is to manufacture appropriately nationalized sounds through digital means. Then, the worker invokes those recordings to "talk" with customers instead of using his or her own voice.

The soundboard is a database of prerecorded phrases, questions, and answers appearing on a computer screen. After the worker makes the call and listens to what the customer says, he or she then presses a button to play the corresponding response. This may be an opening greeting to pitch a product, a follow-up to a question, or a statement transferring the call to a supervisor. If customers don't understand the first time a sound bite is played, the board will have several additional responses in slightly different wordings or intonation. The board also displays options for nonworded sounds, such as laughs and affirmative interjections: "exactly," "uh-huh," and "great" (figure 3). The purpose is to fill out the emotional contours of the conversation so that it feels like a "natural interaction," as one firm describes it (Madrigal 2013).

The point is that, instead of using their own verbal communications, workers use these prerecorded voices—with desired accents—as a stand-in. The industry hook for this service is "outsourcing without the accent." Avatar and Echo Live

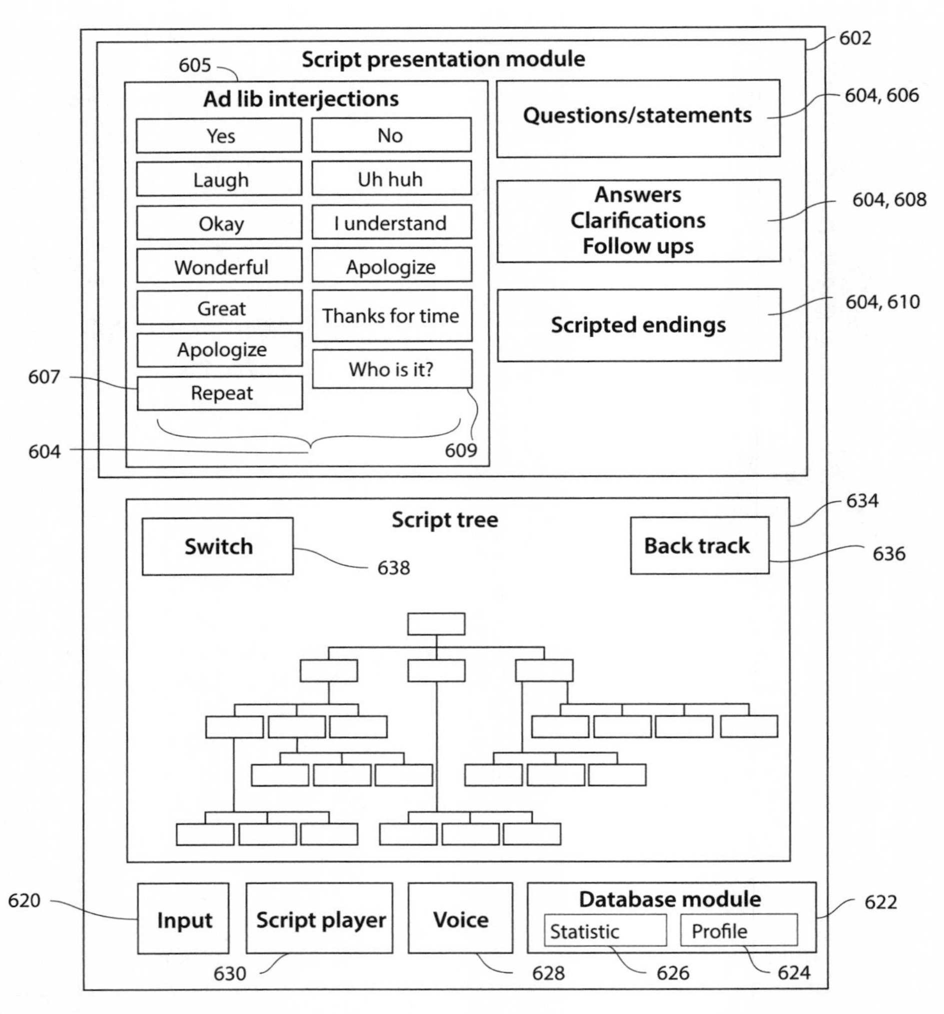

FIGURE 3: Patent for computerized sequence of scripts of compressed audio files. Source: Sirstins et al. (2013).

Agents, for instance, runs out of the Philippines and provides services for the United States, the United Kingdom, Ireland, and Australia. Their website offers sample buttons so that potential clients can hear the same phrase avatared in all four of these accents.

Corporate justifications lay bare the neoliberal motivations for this labor arbitrage. "International accents," they claim, translate into reduced profits. Avatar technology in turn solves this problem by removing those accents from the service encounter. Without any veiled language whatsoever, the firm achieves this goal by tapping into "cheap off shore labor" (as posted on their website in 2015): "Our patented Avatar™ Program is specifically designed to mask our foreign agents accents so that they can prospect for sales leads without any productivity loss. It is no secret that prospects do not like to hear a sales pitch delivered by a foreigner. When lead generation is performed with an international accent, there is an 80% conversion loss that negates the benefit of using cheap off shore labor. Avatar™ software allows our agents to convert prospects into potential sales while giving our clients the benefit of inexpensive labor." From the point of view of the call center, the vocal capacity of the Filipino worker is burdened by its grounding in national contexts and its display of accented sounds. His or her voice inhibits the social congruence of talk with the customer that is desired by the firms, and is therefore problematic to the transnational interests of communicative capital.

Enter the soundboard. Digitizing sound ameliorates the transnational nuisances of verbal labor by partially automating the worker. As such, it represents a hybrid of human-machine communicative labor. Some companies call their workers "cyborg telemarketers" or "avatar agents." One journalist labels it "ventriloquist telemarketing" (Madrigal 2013). The soundboard technologizes customer service labor to such an extent that workers can just listen and click buttons—without talking at all. Furthermore, it is also streamlined to the point that workers can respond to multiple customers at one time. With this software, workers can punch buttons, engage in conversations, and become different avatar workers for two to three calls simultaneously.

In fact, this changes (and/or adds to) the type of identity labor that workers have to do on the call. *National* identity management transforms into *cyborg* identity management. Much the same as the call center workers above, these soundboard workers *hide their identities* through a variety of conversational tactics. In this case, they disguise the fact that they are using technologies to semiautomate their conversations.

Yet, similar to the case of national identity management, soundboard workers have a script for when the customer asks "Am I talking to a robot?" Some firms (such as PerfectPitch) "proactively tell them that we are using prerecorded audio." Others (such as KomBea) however, have a more nuanced and complicated strategy. Workers state to customers (either live or recorded), "You are talking to a live person, but to ensure the information is accurate, I'm using prerecorded audio messages." Ironically, they announce their humanity through a computerized voice.

An adapted form of the mix board attempts to overcome the leaked identity of the avatars. They create their own sound bites—in-house. Instead of using the stock of prerecorded sounds (prepared elsewhere), the call center produces its own mini-audio clips on site (in its own offshore facility), by one of its own (better speaking) employees. In this case, he or she still may have an accent, but one that is "in between"—not too unbelievable as a robot fabrication, but not too off-putting as a foreigner. Most importantly, it is an accent that is navigable, so that the employee can switch back and forth, from the soundboard to interjections of his or her own voice.

With this system, managers are quite forthright with their intention to fool the customer. The CEO of KomBea (based in Utah) says, "I can promise you that 99 percent of the people do not know that the agent just shifted from pre-recorded to a live voice and back to pre-recorded audio" (Madrigal 2013, 4).

What we are seeing is a digitization of identity management. Even while some accounts of Indian call centers suggest a decline in the more egregious forms of national identity management (e.g., outright lying, as described above), these cases reveal how identity management is being adapted and shifted. It is now transferred into the digital realm through the soundboard equipment and the practices of cyborg service workers.

Heard and Not Seen: The Value of Digital Voice in the Global Economy

The questions remain, in an era of such varied and diffuse ICTs (including video, etc.), why has the call center industry been so focused on digital voice by itself? And why are workers in India, the Philippines, and other areas of the Global South being heard and not seen?

I argue that, quite in contrast to corporate narratives, the global service industry is not interested in technologies exclusively for "better communication" (in the sense of shared understandings, identities, participants, etc.). Rather, firms are seeking and adopting technology that will *filter* the social presence of outsourcing firms and their workers for specific purposes. Digital voice does this in three ways.

First, digital voice reveals enough to the consumer to expose the humanness of the worker, but not too much to reveal the electronic mediations of that service interaction. For call center executives, *digital voice helps to ameliorate opposition to labor automation in services—especially backlash from consumers about not being able to talk to a real person.* Some customers don't mind interfacing with machines, of course. Young people may be more accustomed than older people to using technology for retail and sales purposes. However, many consumers do mind talking to bots and are raising public objections in the forms of consumer campaigns, online social movements, and legal actions (Poster 2011).

For these consumers who oppose automated services, then, hearing a voice enables a verification of humanness for the worker. One of the markers for this is affect. As call center labor becomes more routinized and standardized (e.g., reading a script, typing in details, etc.), firms are emphasizing how human employees contribute to service by expressing emotion. The empathetic words and intonations of the live worker are a means to do this.

Alternatively, synthetic voices in the transnational outsourcing industry provide a façade of humanness. For tech entrepreneurs and call center firms, these automated reproductions of live workers help to enhance customer service. They may not be as well received by consumers as human workers, but they have value as "human-like" workers (Suchman 2007). This has been the argument of HCI scholar Cliff Nass and his colleagues. They argue that humans treat machines like they are people (Nass and Brave 2005). Furthermore, our brains don't distinguish between speech that is human versus machine: "Because humans will respond socially to voice interfaces, designers can tap into the automatic and powerful responses elicited by all voices, whether of human or machine origin, to increase liking, trust, efficiency, learning and even buying" (Nass and Brave 2005, 4). Call center administrators subscribe to this notion under the premise that, even if consumers are aware they are talking to bots, they prefer bots with human-like features relative to those that sound like machines. Whether or not we agree, this paradigm helps to illuminate why firms continue to research and produce "conversational" agents and "likable" talking bots so prodigiously (Markoff 2016).

Digital voice acts as a filter in a second way: by removing or hiding the undesirable elements of human sound in global call centers. In particular, it erases what the consuming public in the Global North may perceive as the wrong nationalities, accents, and affects. Such sentiments about accent are brought out and become evident as firms turn to Global South workforces as service providers for Global North customers. Accordingly, by manipulating sound technologically and organizationally, outsourcing companies can more easily create *nationalized symmetries—*

fictitious or otherwise—between consumers, workers, and firms across borders. More so than the images that would be transferred via video communications, then, voice by itself—through accent, tone, and substance of talk—can easily be altered and renationalized. Instead of realizing they are talking to a Filipino or an Indian, customers in the United States and United Kingdom can believe they are talking to another American or Brit. In these capacities, *digital voice provides a quick fix for the interactional dilemmas of "nation" for global customer service firms.*

This adds another layer to our understanding of bots, avatars, and social agents. Critical race scholars have shown us how online avatars are not racially neutral—but instead imbued with "cybertypes" (Nakamura 2002, 2008). Adapting the idea of the "stereotype," this concept "describes the distinctive ways that the Internet propagates, disseminates, and commodifies images of race and racism" (Nakamura 2002, 3). Online game avatars, buddy avatars, and digital signature icons are examples of cybertypes in action, and the way race is embedded in "digital technologies as a form of code, as well as a visual representation of a raced body" (Nakamura and Chow-White 2012, 8). I would add that citizenship and nationality are equally significant mediators of how social agents and other digital workers are represented. This is true of multilingual bots such as Nina, many of whom look and sound like white Americans, while they represent global firms or cater to international consumer bases.

Digital voice acts as a filter, in a third way, by obscuring the national location of the firm. It is not uncommon for organizations to hide themselves completely, or various aspects of themselves (Scott 2013). Increasingly, they use ICTs to carry this out. "Cloaked sites," for instance, involve the presentation of fictitious or misleading Internet homepages to hide political, social, or corporate agendas (Daniels 2009a, 2009b). They are created by many different kinds of groups, ranging from corporations (e.g., the retail giant Walmart) to hate groups (e.g., the KKK). If Daniels reveals how digital deception is racialized, this chapter points us to a parallel process in global contexts—how it is nationalized as well.

With the case of outsourced sound, we see how national identities and locations are technologically concealed for the fluid operations of global capitalism. *By manipulating workers' voice, firms can mask their geography to customers and proceed undisturbed in their transnational outsourcing.* Examples in this chapter show how voice by itself—through accent, tone, and substance of talk—can easily be used to change the connotations of place and citizenship of the speaking employee. The consuming public, in turn, takes comfort in the idea that their service interaction has "never left home."

Concealing the Global South workforce is very much embedded in this sociotechnical system. Suchman recounts how the "dream of technology innovators in the service economy" erases much of the stuff behind the scenes that enables it to happen. It rests on a narrative of humans as the masters (i.e., employers, managers, designers) and robots as the servants (i.e., the automated workers), which in turn "further obscure[s] the specific sociomaterial infrastructures—including growing numbers of human workers—on which smooth interactions at the interface continue to depend" (2007, 224–25). If, locally within the United States, women and people of color who do that work are often erased as Suchman notes, so are workers in Global South countries such as India and the Philippines (especially to the view of consumers in the Global North).

All this suggests, then, that workers in global call centers are heard and not seen because their voice has utility for deceptability. Voice can be conveniently altered (by firms and their technologies) to filter cues that are relayed to the consumer. Video, in contrast, would likely reveal too much information—the citizenship of the worker, the location of the firm, the automation or semiautomation of the labor process, and so on. Yet, digital sound enables an intricate dance between communication methods. It can heighten the social presence of human labor while reducing the social presence of nation. In the process, it obscures automation and geography, while enhancing the (perceived) quality of service.

Conclusion

The purpose of this analysis has been to show the ways that affect and nation are technologically inscribed in the work of interactive service. Sound, as a form of embodied labor, matters in the contemporary ICT economy. For call centers, voice has the capacity to impart a number of markers and symbols, including the emotion, identity, and location of the worker and firm.

The global call center industry has been making use of fields of affective computing, human-computer interaction, and artificial intelligence to capture and manipulate those voices in a number of ways. Voice is a medium to convey (however subtly) the requirements for service and to cue its humanness. But in addition, voice is also a tool to help firms avoid transnational tensions. Through audio communications, speech can be manipulated by the worker so that it masks location. And through digital recordings and software, speech can be reproduced digitally so that it matches the desired requirements of the consumer base even better (from the perspective of employers). Technologies of sound enhance the anonymizing practices by firms vis-à-vis consumers.

Digitizing sound in this way has several implications. It signals, first, the rise of communicative labor. With the increasing commodification of communications, and the rise of industries to harness and manage them, the burdens of performing those communications in desired ways are placed on workers. The communicative body (Lan 2001, 2003) of call center employees may be called upon for distinct tasks at various points in the labor process: for the mind in analyzing and inputting data from the consumer, for ears to hear and interpret the consumer's talk, and for the voice to speak. These functions may happen in combinations, but significantly, not necessarily as a connected set. In fact, as examples here show, the communicative body may be valued for its capacity *not to communicate* (that is, when the Filipino worker is told to mute his or her voice).

Second, this case underscores the transnational dynamics of such industries. The search for inexpensive labor may send call centers to the Global South on economic grounds, but the search for common linguistic resources sends firms specifically to former colonies. These combined practices—seeking out Global South workforces, renationalizing their talk, and digitally reshaping their accents and affect to suit those of the Global North—provides no better illustration of "postcolonial computing" (Irani et al. 2010; Philip et al. 2010).[1] Instead of reflecting "neutral" design, these new technologies of the service industry reflect and reify historic geopolitical relations.

Finally, important is what these trends represent in terms of automation. In the service industry, we are *not* seeing a unidirectional leap toward roboticization, as many have prophesized. Rather, we see highly complex and varying patterns of how technology is integrated into these new ICT-based forms of work. This means new constellations of the cyborg worker. In some cases, firms put human voices inside of bot workers and cell phone aps. In other cases, live workers use digitally recorded voices as a stand-in for their own (e.g., in the case of the soundboard).

Most of these forms, in fact, represent an intermediate stage of automation. Audio technologies—as opposed to video—help firms toward this endeavor. They filter the social presence of the worker *in just the right amount*: enough to show humanness, but not too much to compromise the (national) anonymity of the outsourced firm. Voice, in short, has utility in its deceptability. I have argued that this partial automation solves the dilemmas of nation in service work for global firms. Alternatively, for workers, it creates new demands. They perform *national* identity management to obscure their geographies and citizenship (through accent, affect, and sound). And now they perform *cyborg* identity management to obscure how digital or human they are (by covering up how much technology they are using to mediate the conversations). Communicative labor for the digital service economy, it appears, is a complex process with contradictory dynamics.

Acknowledgments

This research was conducted with support from the National Science Foundation (Awards SES-0240575 and SES-1535218). I'm indebted to the Labor Tech Reading Group for providing me with more inspirational ideas for this chapter than I'm able to outline in a small note. Thanks to Marion Crain for passing on key materials for the analysis, to Ernesto Vasquez del Aguila for encouraging me to fold in prior ideas with new ones, and to Miriam Cherry for being a general source of support. My team at the digitalSTS workshop in Denver 2015 (Luke Stark, Alexandre Camus, and Dominique Vinck) offered excellent comments, as did volume editors Janet Vertesi, David Ribes, and Steve Jackson, and three anonymous reviewers. Responsibility for the content is my own.

Note

1. According to Irani et al., "Postcolonial computing . . . is a project of understanding how all design research and practice is culturally located and power laden, even if considered fairly general" (2010, 1312). It is both a "shift in perspective" for understanding transnational forms of technoscience (particularly for transfers of technological knowledge and systems from global north to south), but also a "bag of tools" (Philip et al. 2010) for critiquing assumptions of Western technoscience and providing alternative lenses on computational practices.

Works Cited

Atkins, Carolyn Peluso. 2000. "Do Employment Recruiters Discriminate on the Basis of Nonstandard Dialect?" *Journal of Employment Counseling* 30:108–19.

Avatar and Echo Live Agents. 2018. "Avatar-Powered Telemarketer." www.avatarandecholiveagents.com.

Bain, Peter. 2001. "Some Sectoral and Locational Factors in the Development of Call Centres in the USA and the Netherlands." Occasional Paper 11, Department of Human Resource Management, University of Strathclyde.

Brophy, Enda. 2010. "The Subterranean Stream: Communicative Capitalism and Call Centre Labour." *Ephemera* 10 (3/4): 470–83.

Carlson, Holly K., and Monica A. McHenry. 2006. "Effect of Accent and Dialect on Employability." *Journal of Employment Counseling* 43:70–84.

Citron, Danielle Keats. 2014. *Hate Crimes in Cyberspace*. Cambridge, MA: Harvard University Press.

Cowie, Claire. 2014. "Oh, Narendra Modi, It's High Time to Accept English as India's Lingua Franca." *Quarz India*, July 1, 1–6.

Dahlback, Nils, Wang Qian Ying, Cliff Nass, and Jenny Alwin. 2007. "Similarity Is More Important Than Expertise: Accent Effects in Speech Interfaces." In *Proceedings of the Computer Human Interaction Conference*, 1–4. New York: ACM.

Daniels, Jessie. 2009a. "Cloaked Websites: Propaganda, Cyber-racism and Epistemology in the Digital Era." *New Media & Society* 11 (5): 659–83.

———. 2009b. *Cyber Racism: White Supremacy Online and the New Attack on Civil Rights*. Lanham, MD: Rowman & Littlefield.

Das, Diya, and Pamela Brandes. 2008. "The Importance of Being 'Indian': Identity Centrality and Work Outcomes in an Off-Shored Call Center in India." *Human Relations* 61 (11): 1499–1530.

D'Cruz, Premilla. 2014. *Workplace Bullying in India*. New Delhi: Routledge.

D'Cruz, Premilla, and Ernesto Noronha. 2014. "The Interface between Technology and Customer Cyberbullying: Evidence from India." *Information and Organization* 24 (3): 176–93.

Dean, Jodi. 2009. *Democracy and Other Neoliberal Fantasies: Communicative Capitalism and Left Politics*. Durham, NC: Duke University Press.

Elfenbein, Hillary Anger. 2007. "Emotion in Organizations: A Review and Theoretical Integration." *Academy of Management Annals* 1 (1): 315–86.

Friginal, Eric. 2009. "Threats to the Sustainability of the Outsourced Call Center Industry in the Philippines: Implications for Language Policy." *Language Policy* 8:51–86.

Gavalda, Marsal, and Jeff Schlueter. 2010. "'The Truth Is Out There': Using Advanced Speech Analytics to Learn Why Customers Call Help-Line Desks and How Effectively They Are Being Served by the Call Center Agent." In *Advances in Speech Recognition: Mobile Environments, Call Centers and Clinics*, edited by Amy Neustein, 221–46. New York: Springer.

Grandey, Alicia A., Glenda M. Fisk, and Dirk D. Steiner. 2005. "Must 'Service with a Smile' Be Stressful? The Moderating Role of Personal Control for American and French Employees." *Journal of Applied Psychology* 90 (5): 893–904.

Hamilton, D., A. H. Goldsmith, and W. Darity. 2008. "Measuring the Wage Costs of Limited English: Issues with Using Interviewer versus Self-Reports in Determining Latino Wages." *Hispanic Journal of Behavioral Sciences* 30 (3): 257–79.

Hill, Ally Rao, and Alastair Tombs. 2011. "The Effect of Accent of Service Employee on Customer Service Evaluation." *Managing Service Quality* 21 (6): 649–66.

Hochschild, Arlie Russell. 1983. *The Managed Heart: Commercialization of Human Feeling*. Berkeley: University of California Press.

Hood, Susan, and Gail Forey. 2008. "The Interpersonal Dynamics of Call-Centre Interactions: Co-constructing the Rise and Fall of Emotion." *Discourse & Communication* 2 (4): 389–409.

Hosoda, Megumi, and Eugene Stone-Romero. 2010. "The Effects of Foreign Accents on Employment-Related Decisions." *Journal of Managerial Psychology* 25 (2): 113–32.

Irani, Lilly, Janet Vertesi, Paul Dourish, Kavita Philip, and Rebecca E. Grinter. 2010. "Postcolonial Computing: A Lens on Design and Development." In *Proceedings of the SIGCHI Conference on Human Factors in Computing Systems*, 1311–20. New York: ACM.

Isoft. 2015. "Amelia." www.ipsoft.com/what-we-do/amelia/.

Lan, Pei-Chia. 2001. "The Body as a Contested Terrain for Labor Control: Cosmetics Retailers in Department Stores and Direct Selling." In *The Critical Study of Work*, edited by Rick Baldoz, Charles Koeber, and Philip Kraft, 83–105. Philadelphia: Temple University Press.

———. 2003. "Working in a Neon Cage: Bodily Labor of Cosmetics Saleswomen in Taiwan." *Feminist Studies* 29 (1): 21–45.

Laukka, Petri, Daniel Neiberg, and Hillary Anger Elfenbein. 2014. "Evidence for Cultural Dialects in Vocal Emotion Expression: Acoustic Classification Within and Across Five Nations." *Emotion* 14 (3): 445–49.

Lockwood, Jane, Gail Forey, and Helen Price. 2008. "English in Philippine Call Centers and BPO Operations: Issues, Opportunities Research." In *Philippine English*, edited by M. A. Lourdes, S. Bautista, and Kingsley Bolton, 219–42. Hong Kong: Hong Kong University Press.

Madrigal, Alexis C. 2013. "Almost Human: The Surreal, Cyborg Future of Telemarketing." *Atlantic*, December 20. www.theatlantic.com.

Markoff, John. 2016. "An Artificial, Likeable Voice." *New York Times*, February 1, B1, B4.
Marsh, Abigail A., Hillary Anger Elfenbein, and Nalini Ambady. 2007. "Separated by a Common Language: Nonverbal Accents and Cultural Stereotypes about Americans and Australians." *Journal of Cross-Cultural Psychology* 38 (3): 284–301.
Mirchandani, Kiran. 2008. "The Call Center: Enactments of Class and Nationality in Transnational Call Centers." In *The Emotional Organization*, edited by Stephen Fineman, 88–101. Malden, MA: Blackwell.
——. 2012. *Phone Clones*. Ithaca, NY: Cornell University Press.
Mirchandani, Kiran, and Winifred R. Poster, eds. 2016. *Borders in Service: Enactments of Nationhood in Transnational Call Centers*. Toronto: University of Toronto Press.
Nadeem, Shehzad. 2011. *Dead Ringers*. Princeton, NJ: Princeton University Press.
Nakamura, Lisa. 2002. *Cybertypes*. New York: Routledge.
——. 2008. *Digitizing Race*. Minneapolis: University of Minnesota Press.
Nakamura, Lisa, and Peter A. Chow-White, eds. 2012. *Race after the Internet*. New York: Routledge.
Nass, Cliff, and Scott Brave. 2005. *Wired for Speech*. Cambridge, MA: MIT Press.
Neustein, Amy, ed. 2010. *Advances in Speech Recognition: Mobile Environments, Call Centers and Clinics*. New York: Springer.
Noronha, Ernesto, and Premilla D'Cruz. 2007. "Reconciling Dichotomous Demands: Telemarketing Agents in Bangalore and Mumbia, India." *Qualitative Report* 12 (2): 255–80.
Nuance Communications. 2018. "Nina—The Intelligent Virtual Assistant." www.nuance.com.
Perry, Joellen. 2003. "Voice of Experience." *U.S. News & World Report*. www.usnews.com.
Philip, Kavita, Lilly Irani, and Paul Dourish. 2010. "Postcolonial Computing: A Tactical Survey." *Science, Technology, & Human Values* 37 (1): 3–29.
Poster, Winifred R. 2007. "Who's On the Line? Indian Call Center Agents Pose as Americans for U.S.-Outsourced Firms." *Industrial Relations* 46 (2): 271–304.
——. 2011. "Emotion Detectors, Answering Machines and E-Unions: Multisurveillances in the Global Interactive Services Industry." *American Behavioral Scientist* 55 (7): 868–901.
——. 2013a. "Identity at Work." In *Sociology of Work: An Encyclopedia*, 406–11. Thousand Oaks, CA: Sage.
——. 2013b. "Hidden Sides of the Credit Economy: Emotions, Outsourcing, and Indian Call Centers." *International Journal of Comparative Sociology* 54 (3): 205–27.
Poster, Winifred R., and Nima Lamu Yolmo. 2016. "Globalization and Outsourcing." In *The Sage Handbook of the Sociology of Work and Employment*, edited by Stephen Edgell, Heidi Gottfried, and Edward Granter, 576–96. Thousand Oaks, CA: Sage.
Schmitt, Alexander, Roberto Pieraccini, and Tim Polzehl. 2010. "'For Heaven's Sake, Gimme a Live Person!' Using Advanced Speech Analytics to Learn Why Customers Call Help-Line Desks and How Effectively They Are Being Served by the Call Center Agent." In *Advances in Speech Recognition: Mobile Environments, Call Centers and Clinics*, edited by Amy Neustein, 191–220. New York: Springer.
Scott, Craig. 2013. *Anonymous Agencies, Backstreet Businesses, and Covert Collectives: Rethinking Organizations in the 21st Century*. Palo Alto, CA: Stanford University Press.
Segrest Purkiss, Sharon L., Pamela L. Perrewé, Treena L. Gillespie, Bronston T. Mayes, and Gerald R. Ferris. 2006. "Implicit Sources of Bias in Employment Interview Judgments and Decisions." *Organizational Behavior and Human Decision Processes* 101 (2): 152–67.
Sirstins, John, Forest Baker III, and Forest Baker IV. 2013. "Apparatus, System, and Method for Designing Scripts and Script Segments for Presentation to a Contact by an Agent." US Patent 8,438,494, May 7.
Spiegel, Alix. 2013. "New Voices for the Voiceless: Synthetic Speech Gets an Upgrade." *NPR*, March 11. www.npr.org.
Sridhar, Kamal K. 2008. "Customer Attitudes to Non-native Accented Outsource Service." Unpublished presentation, Stony Brook University.
Suchman, Lucy A. 2007. *Human-Machine Reconfigurations*. New York: Cambridge University Press.
Thelen, Shawn T., Earl D. Honeycutt Jr., and Thomas P. Murphy. 2010. "Services Offshoring: Does Perceived Service Quality Affect Country-of-Service-Origin Preference?" *Managing Service Quality* 20 (3): 196–212.
Thelen, Shawn T., Boonghoo Yoo, and Vincent P. Magnini. 2010. "An Examination of Consumer Sentiment toward Offshored Services." *Journal of the Academy of Marketing Science* 39 (2): 270–89.
VocalID. 2015. www.vocalid.co.
Wang, Ze, Aaron Arndt, Surendra Singh, and Monica Biernat. 2009. "The Impact of Accent Stereotypes on Service Outcomes and Its Boundary Conditions." *Advances in Consumer Research* 36:940–41.

Wang, Ze, Aaron D. Arndt, Surendra N. Singh, Monica Biernat, and Fan Liu. 2013. "'You Lost Me at Hello': How and When Accent-Based Biases Are Expressed and Suppressed." *International Journal of Research in Marketing* 30 (2): 185–96.

Warhurst, Chris. 2016. "From Invisible Work to Invisible Workers: The Impact of Service Employers' Speech Demands on the Working Class." In *Invisible Labor*, edited by Marion G. Crain, Winifred R. Poster, and Miriam A. Cherry, 214–36. Berkeley: University of California Press.

Waytz, Adam, and Michael I. Norton. 2014. "Botsourcing and Outsourcing: Robot, British, Chinese, and German Workers Are for Thinking—Not Feeling—Jobs." *Emotion* 14 (2): 434–44.

Wong, Nicole C. 2005. "Voice Recognition: They Speak Thereby They Brand." *San Jose Mercury News*, March 21.

Xu, Xunfeng, Yan Wang, Gail Forey, and Lan Li. 2010. "Analyzing the Genre Structure of Chinese Call-Center Communication." *Journal of Business and Technical Communication* 24 (4): 445–75.

Introduction
Infrastructure

Janet Vertesi

The study of infrastructural systems, their assumptions, and their exclusions has been a staple of science and technology studies for over 20 years. Foundational work in the 1980s and 1990s by the late Susan Leigh Star set the scene for how to examine these otherwise submerged systems (Star and Ruhleder 1996; Star 1999). This work brought historian Thomas Hughes's work on sociotechnical systems into new terrain (Hughes 1983), with especial attention to how such large-scale, ubiquitous, and submerged systems could be productively analyzed as sites of contestation, of practice, and of power. Geof Bowker's concept of the "infrastructural inversion" (Bowker 1994) and later their coauthored book *Sorting Things Out* (Bowker and Star 1999) demonstrated this analytical technique for a new generation of STS scholars already well versed in the politics of technology and actor-network theory.

In many ways, the study of infrastructure is a perfect starting point for analyzing aspects of the Internet and digital technologies, which rely upon a sprawling, networked backbone of hardware, software, and invisible technicians. This sense that infrastructure must matter to digital systems is itself a product of the intellectual and institutional overlaps between Star and her colleagues, based in computer science departments and information schools in the United States in the 1990s. From this location, infrastructural scholars continue to speak frequently and clearly alongside system designers and builders, contributing to computer science conferences like computer-supported cooperative work (CSCW) and establishing parameters for design (for more on this history, see Vertesi et al. 2016). The recent success of "cyberinfrastructure" studies in STS and in CSCW speaks to this enduring crossover among generations of Bowker and Star's students (see, for instance, Bietz and Lee 2009; Ribes 2014; Ribes and Finholt 2009; Jackson et al. 2011; Millerand and Baker 2010). Given the extant influence of such theories and framings in existing studies of Internet technologies, it is no surprise that many chapters in this volume draw on this STS concern with infrastructure and its discontents.

The chapters in this section all address infrastructural systems in different ways, participating in sustained STS attention to the often-invisible sociotechnical systems that influence scientific and daily life. Since tracing the enormity of cybersystems is a difficult task, they instead attend to points of encounter and practice wherein individuals confront, contest, or otherwise deploy networked systems. For instance, seeking to move from discussing infrastructure to understanding *infrastructures* in use, Sawyer and colleagues advance the concept of "infrastructural competency." They examine how mobile workers such as real estate agents and architects must develop fluency with multiple infrastructural nodes, materials, hardware, and software systems in order to achieve their very mobility. Competency is locally performed as a "goal-oriented practice" as individuals thread together text messages, mobile Internet access, car seat file cabinets, even portable fax machines to produce seamless client interactions and accomplish their work. Centered in infrastructural studies, this essay speaks eloquently from the continuing, fruitful dialog between STS and CSCW at the information schools.

In contrast to instances of infrastructural competence, Singh et al.'s chapter on GPS systems in use focuses on instances of problem solving, getting lost, and situational repair. The authors examine how individuals manage digital "instructed actions" as practical, situated achievement. This draws on a different yet synergistic stream of analysis in science studies from that of infrastructures: ethnomethodology (Garfinkel 1967). The authors use this approach, advanced in STS by scholars like Lucy Suchman and Michael Lynch (Suchman 2006; Lynch 1991), to demonstrate how digital media "reconfigure" the relationship between maps and journeys through the simple use of a GPS-enabled smartphone.

These examples treat everyday digital systems, but what of systems in the sciences and industry? Addressing the "politics of digital infrastructure," Parmiggiani and Monteiro analyze a Norwegian web portal designed to carefully balance various stakeholder needs and identities in an oil-drilling region. Situated amid environmental advocates, scientists, and the NorthOil company, the "in-the-making" status of the web portal–cum–sensor network keeps the status of debates about climate change, science, and business interests open to enable conversation between constitutive groups. Doris Allhutter, like Singh et al., departs from the classic infrastructural lens, deploying the analytical framework of ontologies in her case. Examining a computational system that itself categorizes computational ontologies, Allhutter uses changes in the sociotechnical system to show how system engineers perform ontological work in the world, then deftly uses the example to speak back to STS theorizing about ontology itself (i.e., Mol 2002; Thompson 2007; Woolgar and Lezaun 2013).

Analyzing infrastructures in case studies is one way to understand their power and potency for sense making, sensing and confronting them in the wild is another. In their "techno-imaginative setup" of the Energy Walk, Winthereik and colleagues explore the "imaginative dimensions of infrastructures" by building and installing a digital system that guides individuals on a walking tour through a northern Danish village. They describe the system as an emergent infrastructure "for making relations within and between" elements of the environment, its publics, and the researchers themselves. As a result of this makerly approach, the authors suggest that STS scholars and practitioners come to view digital systems as "unruly research participants" in their sites. With such a range of theoretical and practical ways of attending to this classic topic in science and technology studies, as well as a wide range of contemporary examples, these chapters on infrastructure offer productive pathways for STS scholars wherever their theoretical and empirical sympathies may lie.

Works Cited

Bietz, Matthew J., and Charlotte P. Lee. 2009. "Collaboration in Metagenomics: Sequence Databases and the Organization of Scientific Work." In *Proceedings of the 11th European Conference on Computer Supported Cooperative Work, 7–11 September 2009, Vienna, Austria*, 243–62. London: Springer.

Bowker, Geoffrey C. 1994. *Science on the Run: Information Management and Industrial Geophysics at Schlumberger, 1920–1940*. Inside Technology. Cambridge, MA: MIT Press.

Bowker, Geoffrey C., and Susan Leigh Star. 1999. *Sorting Things Out: Classification and Its Consequences*. Cambridge, MA: MIT Press.

Garfinkel, Harold. 1967. *Studies in Ethnomethodology*. Englewood Cliffs, NJ: Prentice Hall.

Hughes, Thomas Parke. 1983. *Networks of Power: Electrification in Western Society, 1880–1930*. Baltimore: Johns Hopkins University Press.

Jackson, Steven, David Ribes, Ayse Buyuktur, and Geoffrey Bowker. 2011. "Collaborative Rhythm: Temporal Dissonance and Alignment in Collaborative Scientific Work. In *Proceedings of the ACM 2011 Conference on Computer Supported Cooperative Work*, 245–54. New York: ACM.

Lynch, Michael. 1991. "Laboratory Space and the Technological Complex: An Investigation of Topical Contextures." *Science in Context* 4 (1): 51–78.

Millerand, Florence, and Karen S. Baker. 2010. "Who Are the Users? Who Are the Developers? Webs of Users and Developers in the Development Process of a Technical Standard." *Information Systems Journal* 20 (2): 137–61. doi:10.1111/j.1365-2575.2009.00338.x.

Mol, Annemarie. 2002. *The Body Multiple: Ontology in Medical Practice*. Durham, NC: Duke University Press.

Ribes, David. 2014. "Ethnography of Scaling; or, How to a Fit a National Research Infrastructure in the Room." In *Proceedings of the 17th ACM Conference on Computer Supported Cooperative Work & Social Computing*, 158–70. New York: ACM. doi:10.1145/2531602.2531624.

Ribes, David, and Thomas A. Finholt. 2009. "The Long Now of Technology Infrastructure: Articulating Tensions in Development." *Journal of the Association for Information Systems* 10 (5): 375–98.

Star, Susan Leigh. 1999. "The Ethnography of Infrastructure." *American Behavioral Scientist* 43 (3): 377–91. doi:10.1177/00027649921955326.

Star, Susan Leigh, and Karen Ruhleder. 1996. "Steps toward an Ecology of Infrastructure: Design and Access for Large Information Spaces." *Information Systems Research* 7 (1): 111–34.

Suchman, Lucy. 2006. *Human–Machine Reconfigurations: Plans and Situated Actions*. 2nd ed. Cambridge: Cambridge University Press.

Thompson, Charis. 2007. *Making Parents: The Ontological Choreography of Reproductive Technologies*. Cambridge, MA: MIT Press.

Vertesi, Janet, David Ribes, Laura Forlano, Yanni A. Loukissas, and Marisa Cohn. 2016. "Engaging, Making, and Designing Digital Systems." In *Handbook of Science and Technology Studies*, 4th ed., edited by Ulrike Felt, Rayvon Fouché, Clark A. Miller, and Laurel Smith-Doerr, 169–94. Cambridge, MA: MIT Press.

Woolgar, Steve, and Javier Lezaun. 2013. "The Wrong Bin Bag: A Turn to Ontology in Science and Technology Studies?" *Social Studies of Science* 43 (3): 321–40. doi:10.1177/0306312713488820.

Infrastructural Competence

Steve Sawyer, Ingrid Erickson,
and Mohammad Hossein Jarrahi

We focus here on relationships workers have with information infrastructures (Erickson and Jarrahi 2016; Henningsson and Hanseth 2011; Jackson et al. 2007), particularly as they move about between locations and institutions as part of their working. In doing so, these workers rely on stores of knowledge that allow them to continue moving forward—both figuratively and literally. These knowledge workers exist within a professional landscape that is increasingly expansive (i.e., globalized, interorganizational) and decomposed (i.e., expertise driven, project based), and we assert that this knowledge not only is becoming highly individualized, but also is progressively more elemental to what it means to be a knowledge worker (National Academies of Science 2017; Spinuzzi 2015; Barley and Kunda 2006; Davenport 2005).[1]

We articulate the stores of knowledge that workers draw on as a form of "infrastructural competence." This idea builds on Star and Ruhleder's (1996) contention that infrastructures are sociotechnical entities, comprising not only a set of interdependent technical elements but also the social layers of norms and knowledge that make these technical elements function in situ. It also builds from the work of the late Claudio Ciborra (2000) and his concept of bricolage, or the making-do practices people use to ply resources at hand toward desired goals.

These conceptual foundations reflect both the scholarly influence of science and technology studies (STS) and the contributions of organizational scholarship (where the authors situate our work, straddling the two intellectual communities). Within organization studies, there has been a constant focus on the roles of information and communication technologies (ICT) as means and mechanism for replacing human effort, automating work, and reducing both errors and tedium. The STS influence has led many organizational scholars to contest this naïve utopianism with empirical understandings of the conditionality of computing's roles (e.g., Zuboff 1985). In the case of certain kinds of work, such as clerical and administrative tasks, for example, the introduction of ICTs served to "informate" rather than liberate, simultaneously creating a class of worker whose agency was stripped by the demands of the database or the data entry field. In this chapter we argue from this analytical position, building out the idea of "infrastructural competence" with both critical and empirical insight.

Given this chapter's focus, it is useful to contextualize some of the complex conditions that help to shape contemporary forms of knowledge work. At the technological level, the rising prevalence of (and reliance on) mobile devices in professional practice has increased communication flexibility while also expanding expectations of availability and the speed of response (e.g., Mazmanian et al. 2013; Mazmanian and Erickson 2014; Wajcman et al. 2009; Wajcman 2014). Mobile computing technologies have also helped to hasten the dissolution of work/life boundaries and contributed to the belief that life appears to be speeding up (Wajcman 2014). Cloud-based computing infrastructures make it possible for workers to decouple device-application pairings (i.e., Excel spreadsheets can run on a Windows PC and on any number of smartphones, accessing the file stored in some cloud-based storage service) and reassemble individuated ensembles that they find personally effective. As such, cloud services like Google Docs and Dropbox have become infrastructural satellites around which workers now orbit, connecting and reconnecting in different assemblage patterns as each specific situation demands.

At the socioeconomic level, knowledge work is also being reshaped. The globalization of work, a shift that began many decades ago, has resulted in an expanded network of actors and institutions upon which knowledge work must be executed today (e.g., Leonard et al. 2014). The normalcy of these international, interorganizational collaborations accords a diverse array of boundaries that must be negotiated—organizational, cultural, temporal, linguistic, legal, and so on. At the same time, these arrangements have instantiated a new logic for teamwork built ever more on specialized, rather than localized, forms of expertise. Coupled with this demand for expertise is the rising standardization of a project-based economy, an organizational structure in which specialists can be efficiently leveraged.

The ways of working have also been evolving, and the current primacy of project-based work not only has increased the shift to specialization among workers, but also is one of the forces underpinning today's "gig economy" and its related dependence on freelance or contract workers. Global platforms such as Mechanical Turk and Upwork reify the identity of knowledge workers as itinerant experts who move from one project to the next as they amalgamate a career. In some ways, the rising recognition of expertise in knowledge work has been the undoing of work itself, as workers are now more valued for their skills than they are for their humanity.

In all, these shifts in the sociotechnical and socioeconomic landscape of knowledge work amount to an unsated need for workers to be productive anytime, anywhere irrespective of context and location (e.g., Czarniawska 2014; Kleinrock 1996; Pittinsky and Shih 2004). For many workers, this may mean an opportunity to exercise agency, but it also implies the parallel imperative to manage all of the components of our productivity individually. What are the connotations of this demand?

A Model Modern Worker: Kaylie, the Realtor

Consider one example of today's mobile knowledge worker (and increasingly, doing so provides us a window into tomorrow). She is an agent of her own destiny as her goal is to excel at her job—whatever its form (e.g., contract worker, independent worker, full-time employee). In order to do this, she has to manage herself to achieve maximum performance across all of the contexts and situations in which (or through which) her work takes her. She might look something like Kaylie, a successful (by which we mean well paid) real estate agent.

We begin with Kaylie's car, a Subaru, that is clean, inside and out, which is rare enough in snowy, road-mucky Syracuse that it deserves notice. In the 22 years that she's been a Realtor, Kaylie's car choices have changed. It used to be that the back seat was the most critical feature because clients were always getting in and out. Now, her focus is on the car's front seat: she needs to see her laptop, printer, and chargers on the swiveling work desk she had installed after market. Clients no longer get in her car so that she can drive them around; instead, she meets them at the specific houses they have identified via online searches. More broadly and for the purposes of this chapter, the car is not the focus, but the means for mobility.

Kaylie is active on several local media spaces. Indeed, Kaylie is active enough online that she now outsources the work of maintaining her substantial web presence—posting pictures, videos, and other details of her realty listings along with a set of news feeds culled from various sources like the county tax office and a mortgage rate source, among other things—to someone else. A two-person company that specializes in listing houses for sale posts and updates Kaylie's residential links. Kaylie also works with three assistants: two help her to show houses, one focuses on helping clients ready homes for closing.

Kaylie is always on the phone, talking. She texts using Siri and takes voice notes. She is also always taking photos, sharing them with her network of lawyers, clients, contractors, mortgage brokers, the two-person team staging her listings, and her web people. Kaylie controls her own calendar, but it also remains visible to some of her most trusted staff and professional associates; access seems to be granted depending on trust—most of her staff do not has access, but very few contractors do.

Kaylie is always on the move. She always arranges to meet in places that are on her way from one location to another. Sometimes we simply talk in the car, though I am in the back seat—as noted, the front passenger seat is for her laptop and printer, and the foot well is where she stacks her files and papers. Kaylie puts about 25,000 miles on her car per year, but never leaves the county and rarely takes a trip of more than 10 miles.

From her car/office, Kaylie can copy, print, scan, email, and carry out document preparation through a common program that keeps all real estate forms preloaded, ready to be populated, printed, and shared digitally. Faxing signed papers is still required for some legal realty matters, and Kaylie is able to do that from her car as well. She keeps two old phones in her glove compartment and uses one or the other when her usual phone dies (something I've never seen happen). She also keeps backup batteries, phone chargers, and the like stuffed into several "cubbies" and pockets in her car, home, purse, and gym bag. Power matters.

During one of our recent meetings, Kaylie fielded 14 calls, sent dozens of text messages and emails, enlisted me to scan and send some papers for her, finished paperwork while I bought us coffee, and bumped into one of her contractors. We also were able to swing by her dry cleaner and run an errand to the drugstore. She dropped me at my home as she headed off to Curves for a workout class she never misses. Kaylie is always talking, noting that it is pretty quiet in her home in the evenings as by that point she is "talked out." Kaylie is an early riser and uses the 4:30 to 6:30 AM timeslot, before her spouse wakes up, "most days" to write her blog posts. For Realtors like Kaylie, "most days" include weekends, which tend to be busier. Wednesday and Thursday are her quieter days.

Kaylie's case provides us the means to unpack a concept that we are calling *infrastructural competence*. We advance infrastructural competence as an individual's use-oriented relationship with infrastructures that combines social abilities, goal orientedness, and leveraging of digital and material resources in a way that enables one to generate a functional, operable, and patterned or routinized (while also being personalized) set of sociotechnical practices to accomplish a necessary task or set of tasks.

With the rest of this chapter, we situate infrastructural competence as a frame that empirically illuminates new types of knowledge work practices but, in doing so, also questions the implications of this competence as a new imperative for workers of the future. In doing this, we draw on interviews with more than 50 knowledge workers located in New York (and primarily in New York City) and the Research Triangle Park area of North Carolina. These interviewees represent a variety of occupations such as architects, consultants, web designers, and salespeople, yet what bonds them as a group is that they are all engaged in work that requires them to be highly mobile. We have chosen these "mobile knowledge workers" to build our case for the idea of infrastructural competence because these professionals largely represent an extreme version of the modern knowledge worker—embedded in expertise-driven, project-based work and heavy users of cloud-based mobile infrastructures. In short, we posit that the case of mobile knowledge workers will help not only to explicate our ideas of infrastructural competence, but likewise to uncover how deft infrastructural use may be morphing into a required element of all types of work in the near future.

From Infrastructure Use to Infrastructural Competence

The mobile workers with whom we spoke all leverage interconnected digital ecosystems, increasingly cloud-based, to accomplish their work. Analytically, we refer to these digital ecosystems as "knowledge" or "information" infrastructures (e.g., Edwards et al. 2007; Pipek and Wulf 2009). However, unlike the large-scale cyberinfrastructures used in the sciences, the knowledge infrastructures in use by the knowledge workers in our study rarely have a recognizable installed base (Monteiro et al. 2014). This is primarily due to the fact that the demands of these workers are broad and vary from person to person. Moreover, this lack of an installed base derives from the fact that there are so many possible tools available to the modern knowledge worker. The key in understanding these infrastructural patterns, thus, lies less in the way a specific installed base learns to accommodate communal changes and more in the patterned ways that workers develop functional, personalized instantiations of infrastructure for immediate use.

Infrastructures, when considered as vernacular assemblages of software, hardware, and related technological artifacts (Hanseth et al. 1996), belie their configurational natures—configurations that either "move with the worker or are found in the places in which the worker moves" (Davis 2002, 69). Indeed, for mobile workers, infrastructure is often highly visible (e.g., Rossitto et al. 2014). As they move between offices or from building to sidewalk, the digital assemblages they create are at once scripted and dynamic: specific parts are uniquely geared toward narrower task(s), but the whole is able to shift and change over time as professional needs continue to arise and dissipate (e.g., Sawyer et al. 2014; Ribes and Polk 2014). When these infrastructural scripts are tested, infrastructural seams (Vertesi 2014) are revealed. Evident seams or boundaries can be physical, in terms of geographies and companies, as well as virtual, in terms of suites and platforms of preferred ICTs.

Mobile workers relate to infrastructure in a different way than workers in more static environments (Erickson et al. 2014; Ciolfi and de Carvalho 2014; see also Plantin et al. 2016). In traditional workplaces, infrastructure, if working effectively and as designed, tends to be invisible (Pipek and Wulf 2009). For mobile workers, alternately, infrastructures are often starkly visible. As they move from one place to another or from building to street, for example, the limits of their vernacular infrastructures begin to show—one piece may work well in one context and not another or infrastructure can fail altogether in other environments. It is not that mobility forces infrastructural breakdown, rather that it raises illuminates infrastructural seams (Vertesi 2014) or contextual boundaries (Cousins and Robey 2005). Evident seams or boundaries can be physical, in terms of geographies and companies, as well as virtual, in terms of suites and platforms of preferred ICTs.

Consider the case of Monica, an architect in New York City who is building her own consultancy. Because she is engaged in a lot of business development, Monica must traverse the city quite extensively on a daily basis. It is not uncommon for her to meet clients in their offices (as the demands of her expertise are often such that being "on site" is a critical part of her problem-solving process), attend and organize professional lectures and meet-ups for the architectural community, and work with a revolving set of interns at a coworking place where she shares a small office. There is nothing necessarily unusual about this scenario upon hearing it, but look a little more closely at the multiple contexts in which Monica must perform on a daily basis and you begin to see that she is regularly traversing multiple boundaries—temporal, spatial, social, institutional, and digital (Cousins and Robey 2005; Kietzmann et al. 2013; Koroma et al. 2014). She has a strong incentive to maintain connected to her clients and coworkers as she traverses the city; she also has a strong incentive to do this as smoothly as possible because any gaps or breakdowns will reflect poorly on the professional reputation she is working hard to build. Despite this suite of sociotechnical challenges, she manages all of these traversals quite agilely. Is this merely infrastructural use or something more? We would suggest that her agility is made possible by her infrastructural competence.

Infrastructural Competence: Artful Uses of Infrastructure

We define *infrastructural competence* as an individual's use-oriented relationship with infrastructure that enables him or her to generate a functional, operable, personalized, patterned, or routinized set of sociotechnical practices that accomplish a necessary task or set of tasks. The knowledge puts infrastructure into action in such a way that draws together social norms, goal directionalities, and the particularities of digital and material resources at hand.

In keeping with recent discussions of infrastructural generativity, Monica's agility with infrastructure is not just (or even mostly) about her ability to use ICTs well. And we are not suggesting that workers simply need "better technical skills." Rather, we advance infrastructural competence as a perspective, a way of framing and seeing the ways in which digital and material infrastructures provide actors possibilities (Hanseth and Nielsen 2013; Johnson et al. 2014). We see infrastructural competence as generative for those who possess it. This understanding of generativity asserts that actors who possess the infrastructural competence to recognize where productive infrastructural adjustments or interventions might be made can leverage this knowledge endlessly. They can use the seams and boundaries articulated above to achieve a goal again and again. Seen this way, Monica's agility is not just (or even mostly) about her ability to use ICTs well.

Developing the notion of infrastructural competence shifts the focus necessarily to the behaviors and practices of people. Workers actively draw together a set of digital components, colleagues, business partners, and working arrangements to support an intentional agenda—even if these elements are not located in one place or resist control. These efforts make the invisible visible—identifying, adopting, and configuring infrastructural arrangements. These actions reflect Vertesi's (2014) discussion of "seamfulness"—an idea developed in relation to distributed teams of scientists who navigate, negotiate, and integrate many disparate infrastructures to accomplish their work. Referring to these navigational, negotiated, and integrative alignments as "multi-infrastructural work practice" (Vertesi 2014, 278), Vertesi further suggests that individuals range over a spectrum of greater or lesser "ability"—or, possibly, circumstantial opportunity—to bring often conflicted sociotechnical systems (or device ecologies) into "artfully integrated" union.

Vertesi goes on to contend that the ability to achieve artful integration with multi-infrastructural environments connotes a degree of local knowledge and membership (Vertesi 2014, 270–71). Likewise, we suggest that someone with infrastructural competence must also be able to grasp the reality that one person's needed infrastructure is another's obstacle—what count as affordances or constraints in infrastructural terms are constructed out of individuals' own personal backgrounds, goals (Star and Ruhleder 1996), and moment-to-moment needs.

For example, a knowledge worker's complex use of calendaring (with different rights of access) reflects both the sociality of who she wants to share with and the technicalities of sharing (relative to the constantly evolving vagaries of Apple's Calendar, Google Calendar, and Microsoft Outlook). As such, any enacted infrastructural arrangement not only must work technically, but also must conform to social conventions, routines, and norms (as Kaylie's case helps to make clear). This aspect of infrastructural competence is made evident in the awareness of and respect for the shared expectations of work outcomes (if not necessarily processes). This awareness forces one's own "infrastructured actions" (Vertesi 2014, 267) to continually acknowledge the constellation of players who may be involved in shaping and maintaining infrastructures in use.

Infrastructural competence is visible through individual workers' patterns of action that showcase their ability to bridge and adjust to local sociotechnical and sociomaterial conditions. It can be identified by the set of routines that a person uses to address these conditions. Like all routines, these practices encode a set of recognizable "best practices," a template or genre known to both the individual and his or her collaborators that allows them to evoke a script of sociotechnical actions that helps "bound" (in the spirit of "bounded rationality") the particular details of a situation while still acting effectively and efficiently. Routines enable rapid action while lessening the need for perpetual sense making within interactions, but they lose their usefulness if they don't adapt as needed. Doing so aligns with organizational scholars like Feldman and Pentland (2003) who assert that even established patterns of action, like routines, are dynamic, constantly evolving to align with changing conditions while simultaneously maintaining a coherent wholeness that is both externally and internally recognizable. We develop each of these points in greater detail below using an example from our ongoing fieldwork.

Infrastructure Competence Is a Use-Centered View

Vertesi's argument that scientists engage in individual, artful ways to use the multiple infrastructures that define their work suggests both that the work-arounds they produce to bridge infrastructural gaps and pair otherwise disparate parts together as a functional whole are expected and useful. In the large cyberinfrastructure projects Vertesi studies, these artful bridges and work-arounds occur continuously as infrastructures grow and evolve. In the case of mobile knowledge work, these artful interventions are more frequent, but also likely more lightweight. This occurs because the need to configure—or reconfigure—infrastructure(s) arises both in direct proportion to a person's physical mobility as well as in relation to the social arrangements in which a particular work task is situated. In most knowledge work situations, these are shifting on a daily, if not hourly, rate because of the distributed, project-based nature of the work. The dynamism of these disparate, yet oft sequential, situations forces individuals to focus on tasks as triggers for unique, in situ problem solving. In slight contrast to other conversations about infrastructural use, the individuated work-arounds and improvisations—this infrastructure-in-use focus—rely on individual technological response. What may work best for me in achieving the presenting task's goals may not be the same configured arrangement as my colleague.

Building on the use-centered and practice- or routines-oriented perspective on using infrastructure, we identify five characteristics or attributes of infrastructure competence: goal orientation, reliance on digital assemblages, enacted and operationally resilient, situated and relational, and an expectation based on professional identity. We discuss each of these attributes in the rest of this section.

Goal Oriented

Mobile knowledge workers develop the skills and abilities to assemble and leverage a digital assemblage to pursue an outcome. Our subjects are not interested in the ICTs they are using for any intrinsic or computational goals; these are resources being marshalled to the needs, ends, or goals of their work. This goal orientation leads to practicing, even routinizing, the skills learned through repeated uses. The use- or task-related orientation of infrastructural competence may connote, falsely, that infrastructural practices are merely solution-driven engineering feats. This is not at all the case.

In order for an infrastructural practice to be successful, it must not only work technically, but be socially recognized as legitimate and also be accountable to shared needs. Workers may arrange their working environment in whatever way they would like, for instance, or deploy a particular set of tools to address a problem, but at its core an active technological assemblage—like any infrastructure—must adhere to the social guidelines that define it. In reality, this often means that one cannot stray out of the bounds of a shared toolset, a common digital platform, or some other socially determined sociotechnical baseline that anchors infrastructural actors to one another.

Reliant on Digital Assemblages

Knowledge workers are dependent on the collections of digital resources they assemble, but they are neither building these resources as hackers nor thinking of themselves as computing experts. We observe that most mobile knowledge workers are in a knowledge-based profession, are well educated, possess unique skills and experiences, and are busy. The collections of commodity technologies (laptops, phones, public Wi-Fi, cloud storage, available apps) they pull together and the skills they develop to assemble this dynamic collection create a digital infrastructure to support their work. We have called these collections "digital assemblages" (Sawyer et al. 2014) in prior work. Others have called these sets "digital" or "artifact ecologies" (Bødker and Klokmose 2012; Forlano 2010), "digital kits" (Mainwaring et al. 2005), "individual information systems" (Baskerville 2011), or "constellations of technologies" (Rossitto et al. 2013).

Such collections are not "systems" in that they are neither designed nor controlled by one person, reflecting instead a multiparty, distributed, and often commodity-based set of arrangements. They are at once purposeful and ad hoc. A digital assemblage is the patterned collection of digital resources that a person brings to bear on a task or problem in order to pursue a goal or solution. A digital assemblage comprises the devices, information resources, applications and platforms, connectivity options, software and computational resources, and other systems that a person brings together for a particular use. A digital assemblage is neither an infrastructure (which is shared by many and owned by many) or a specific system (which is owned by a few even if used by many), it is a personal collection.

Our ideas here build on some of the foundations in infrastructure studies (e.g., Hanseth and Monteiro 1997; Hanseth et al. 1996; Hanseth and Lyytinen 2010; David and Bunn 1988). Without going fully into the resemblances between the digital assemblages of mobile knowledge workers and infrastructural examples in prior studies, it is important to underscore the parallelisms in our ideas. The digital assemblages contrived by the workers in our study are personal, but they are built on, shaped by, and constrained by their relationship to a preinstalled base. This pattern is consistent with the relational nature of infrastructures; different individuals are positioned differently in relation to the infrastructure based on the goals they want to achieve, and they use gateway technologies to facilitate information sharing and communication among fragmented systems and interconnect them into a single integrated system (Jackson et al. 2007). Another way to say this is that the creation of a technological assemblage is an example of "installed base cultivation" (Hanseth et al. 1996), meaning that an important aspect of infrastructural competence is learning "how to wrestle with the inertia of the installed base" (Bowker and Star 1999, 382; see also Edwards et al. 2009), or a set of given constraints, to achieve what you want to achieve.

Enacted and Operationally Resilient

This use in doing is reflected in the efforts directed at keeping all the elements of a digital assemblage working together. Doing so requires substantial micro-innovation to learn, problem solve, adapt, and be productive. People develop redundancies and work-arounds through trial and error, and they measure new devices, platforms, and other digital resources relative to their operational usefulness and reliability. It is a pragmatic and evidence-based use, visible in the stories of Monica above as well as those of many of our other interview subjects. In essence, enacting operational resilience is what brings the digital assemblage and goals together.

Situated and Relational

This characteristic of infrastructural competence is reflected in these workers' skills and abilities to balance goals, dependencies, resources (such as what the digital assemblage can enable), and priorities. We have observed that mobile knowledge workers are able to maintain situational awareness across several tasks and to leverage their digital resources to support multithreading. Such activities reflect an expertise, an ability to "riff" on routines (adapting) and balance competing goals and issues, that appears as a distinct set of skills beyond their particular and specific expertise and professional knowledge. In this way, nomadic workers have a common set of skills related to their mobility, even as they are not professional colleagues, in the same profession (or even aware of one another's existence). They share the experiences and skills of being mobile.

A Professional Expectation

More broadly, infrastructural competence reflects an expectation by others that the mobile workers can perform their role (as professionals, in this location, as needed). As such, infrastructure competence reflects social accountability: what others expect of you regarding acceptable professional behavior, norms of connection and availability, and knowledge base. This relates to the way that one learns to enact infrastructure as a member of a community (Sandvig 2013). In this way, the social environment (e.g., a particular professional, social, or cultural community) articulates and perpetuates a common set of understandings about how certain infrastructural practices should unfold or take shape. Within the roles of actors such as real estate agents, for instance, these that can often be identified as professional routines (Pentland and Feldman 2005). In more mixed environments, these routines are solidified at the course of a collaboration engagement or as the result of certain technical requirements (e.g., privacy concerns). It is these routines that enable different people to play the same song with different instruments, at different tempos and in different keys, and sometimes with new riffs added. Difference is allowed in the expression of the routine in situ, but nevertheless the pattern is socially translucent enough among colleagues that a shared convention is identifiable and, thus, trusted. At the same time, it is imperative to note that routines are always flexible and evolving (Feldman 2000), so these patterns that serve to order sociotechnical actions are likewise ever in flux and open to updating.

Conclusion: Expectations of Infrastructural Competence

To possess infrastructural competence, then, is to recognize the goal-oriented practices that rely on smartly tuned and constantly evolving digital assemblages. These digital assemblages must be operationally resilient and are enacted in very situated and relational ways. There is an expectation by others that professional work requires such a set of skills and resources. Seen this way, infrastructural competence reflects the stream of research that illuminates the ways in which humans (and in our particular case, workers) take up and make uses of ICT.

Some readers will see the link between these insights and the broader scholarship of STS that builds on concepts of bricolage (as noted). Some scholars may also see the partial overlap of artful doing that is at the core of infrastructural competence with the concept of articulation, drawn from the sociology of work (e.g., Strauss 1985; Star 1991). Organizational scholars might also point to Pickering's (1995) mangle of practice or the Edinburgh school's long use of "configurational technologies and learning by doing" (e.g., Fleck 1994). The value of infrastructural competence, set against these, is its attention to the artfulness of doing, the ongoing and embedded nature of this set of skills, and the inherent and evolving flexibilities of the digital and material resources, the specific situations, and the goals.

A second implication of modern workers' need to create smartly tuned digital assemblages is that doing is also a means to showcase infrastructural competence. The two are bound up: each a window to the other. To understand how digital resources are developed, deployed, and exploited is also to understand the goals, situations, and practices of the person who assembled the digital kit. Infrastructural competence is what gives rise to digital assemblages. The uses and value of digital assemblages are visible in the practices of infrastructural competence.

There is a need for more empirical work and conceptual development of this nascent concept: both to provide more clarity and precision regarding the characteristics and processes of infrastructural competence and to identify what makes a person better or worse at doing or performing competently. Both require describing and mapping local practices and the competence that emerge as these serve as bridges among various scales of infrastructure. In the specifics of our study of mobile work, what becomes clear is that more attention is needed on the intertwining of infrastructural competence and the work lives of organizations and individuals. This need to see both the work and the digital and material elements that are bound up in that effort, within the frame created by realities of large-scale infrastructures and their constitution as platforms, demands extended attention (Edwards et al. 2009; Plantin et al. 2016; Fenwick 2004).

We are acutely aware of the empirical challenges of studying infrastructural competence and see the need for methodological developments in order to better understand infrastructural competence. For example, we have valued digital images and worked to get trace data from workers' devices even as we interviewed participants, observed them, and pursued other data. Linking all of these sources of insight together and doing more to connect the digital and material aspects of mobility, while also getting more data on the multiple goals and trade-offs that mobile knowledge workers seem to pursue, are both rich and demanding empirical spaces.

Third, programmatic attention to infrastructural competence demands increased attention regarding the roles of public and organizational policies, institutional guidance and norms, training, and so forth on individuals and groups. Currently we rely on individuals (or small groups) to develop infrastructural competence on their own. If this is indeed a core element of more and more forms of work, more attention is needed to the relational nature of the conception and to the development of technologies that may advance the cause.

Fourth, we need to go beyond the observations from the outset regarding the changing nature of work to connect with ongoing research streams regarding changes to labor force structures, the forms of organizing, and the work of macro-scale scholars such as Manuel Castells (2000, and his concept of networked societies), Peter Drucker (1969, and his prescient insights on the knowledge economy), Richard Florida (2001, with his articulation of the "creative class"), and others. These theorists might be seen as extolling knowledge work and elevating this form of work and mobility as models of the future. Harkening Neil Postman (1993), we also know that technological arrangements are only new to the generation in which they emerge. This leaves us wondering if, in a relatively short period of time, infrastructural competence will move from being a desired and differentiating characteristic of workers to being an expected and basic skill demanded from workers.

Beyond the open questions of specifics, methods, and expectations, we contribute here a new construct to the digital STS community as well as those in the burgeoning area of infrastructure studies. Our work progresses and deepens the discourse on seams within sociotechnical systems, and further contributes to our understanding of the skills and competencies that may be coming to define 21st-century work.

Note

1. "Knowledge work" and "knowledge workers" are contested terms, and we use them fully aware of this debate, as we discuss later in the chapter.

Works Cited

Barley, Stephen R., and Gideon Kunda. 2006. *Gurus, Hired Guns, and Warm Bodies: Itinerant Experts in a Knowledge Economy.* Princeton, NJ: Princeton University Press.

Baskerville, Richard. 2011. "Individual Information Systems as a Research Arena." *European Journal of Information Systems* 20 (3): 251–53.

Bødker, S., and C. Klokmose. 2012. "Dynamics in Artifact Ecologies." Paper presented at the 7th Nordic Conference on Human-Computer Interaction, Copenhagen.

Bowker, Geoffrey. and Susan Leigh Star. 1999. *Sorting Things Out: Classification and its Consequences.* Cambridge, MA: MIT Press.

Castells, Manuel. 2000. *The Rise of the Network Society.* New York: John Wiley.

Ciborra, Claudio, ed. 2000. *From Control to Drift: The Dynamics of Corporate Information Infrastructures.* Oxford: Oxford University Press.

Ciolfi, Luigina, and Aparecido de Carvalho. 2014. "Work Practices, Nomadicity and the Mediational Role of Technology." *Computer Supported Cooperative Work* 23 (2): 119–36.

Cousins, Karlene, and Daniel Robey. 2005. "Human Agency in a Wireless World: Patterns of Technology Use in Nomadic Computing Environments." *Weekly Bulletin* 15 (2): 151–80.

Czarniawska, Barbara. 2014. "Nomadic Work as Life-Story Plot." *Computer Supported Cooperative Work* 23 (2): 205–21.

Davenport, Thomas. 2005. *Thinking for a Living: How to Get Better Performances and Results from Knowledge Workers.* Cambridge, MA: Harvard Business Review Press.

David, Paul, and Julie Bunn. 1988. "The Economics of Gateway Technologies and Network Evolution: Lessons from Electricity Supply History." *Information Economics and Policy* 3:165–202.

Davis, Gordon. 2002. "Anytime/Anyplace Computing and the Future of Knowledge Work." *Communications of the ACM* 45:67–73.

Drucker, Peter. 1969. *The Age of Discontinuity: Guidelines to Our Changing Society.* New York: Harper & Row.

Edwards, Paul, Geoffrey Bowker, Steven Jackson, and Robin Williams. 2009. "Introduction: An Agenda for Infrastructure Studies." *Journal of the Association for Information Systems* 10 (1): 374–84.

Edwards, Paul, Steven Jackson, Geoffrey Bowker, and Cory Knobel. 2007. "Understanding Infrastructure: Dynamics, Tensions, and Design." Report of the NSF Workshop on History and Theory of Infrastructure: Lessons for New Scientific Cyberinfrastructures, Ann Arbor, MI.

Erickson, Ingrid, and Mohammad Jarrahi. 2016. "Infrastructuring as Appropriating in the Context of Mobile Knowledge Work." In *Proceedings of the 2016 Conference on Computer-Supported Collaborative Work*, 1323–36. New York: ACM.

Erickson, Ingrid, Mohammad Jarrahi, Leslie Thomson, and Steve Sawyer. 2014. "More Than Nomads: Mobility, Knowledge Work and Infrastructure." Paper presented at the EGOS Colloquium, Rotterdam.

Feldman, Martha. 2000. "Organizational Routines as a Source of Continuous Change." *Organization Science* 11 (6): 611–29.

Feldman, Martha S., and Brian T. Pentland. 2003. "Reconceptualizing Organizational Routines as a Source of Flexibility and Change." *Administrative Science Quarterly* 48 (1): 94–118.

Fenwick, Tara. 2004. "Learning in Portfolio Work: Anchored Innovation and Mobile Identity." *Studies in Continuing Education* 26 (2): 229–45.

Fleck, James. 1994. "Learning by Trying: The Implementation of Configurational Technology." *Research Policy* 23:637–51.

Florida, Richard. 2001. *The Rise of the Creative Class: And How It's Transforming Work, Leisure and Everyday Life*. New York: Basic Books.

Forlano, Laura. 2010. "WiFi Geographies: When Code Meets Place." *Information Society* 25:1–9.

Hanseth, Ole, and Kalle Lyytinen. 2010. "Design Theory for Dynamic Complexity in Information Infrastructures: The Case of Building the Internet." *Journal of Information Technology* 25 (1): 1–19.

Hanseth, Ole, and Eric Monteiro. 1997. "Inscribing Behaviour in Information Infrastructure Standards." *Accounting, Management and Information Technologies* 7 (4): 183–211.

Hanseth, Ole, Eric Monteiro, and Morten Hatling. 1996. "Developing Information Infrastructure: The Tension between Standardization and Flexibility." *Science, Technology, & Human Values* 21 (4): 407–26.

Hanseth, Ole, and Petter Nielsen. 2013. "Infrastructural Innovation: Flexibility, Generativity and the Mobile Internet." *International Journal of IT Standards and Standardization Research* 11 (1): 27–45.

Henningsson, Sven, and Ole Hanseth. 2011. "The Essential Dynamics of Information Infrastructures." In *ICIS 2011 Proceedings*, 1–19. New York: IEEE.

Jackson, Steven J., Paul N. Edwards, Geoffrey C. Bowker, and Cory P. Knobel. 2007. "Understanding Infrastructure: History, Heuristics and Cyberinfrastructure Policy." *First Monday* 12. https://journals.uic.edu/ojs/index.php/fm/article/view/1904/1786.

Johnson, Michael, Hajah Mozaffar, Gregoria Campagnolo, S. Hyysalo, Neil Pollock and Robin Williams. 2014. "The Managed Prosumer: Evolving Knowledge Strategies in the Design of Information Infrastructures." *Information, Communication & Society* 17 (7): 795–813.

Kietzmann, Jan, Kirk Plangger, Ben Eaton, Kerstin Heilgenberg, Leyland Pitt, and Pierre Berthon. 2013. "Mobility at Work: A Typology of Mobile Communities of Practice and Contextual Ambidexterity." *Journal of Strategic Information Systems* 22 (4): 282–97.

Kleinrock, Leonard. 1996. "Nomadicity: Anytime, Anywhere in a Disconnected World." *Mobile Networks and Applications* 1 (4): 351–57.

Koroma, Johanna, Ursula Hyrkkänen, and Matti Vartiainen. 2014. "Looking for People, Places and Connections: Hindrances When Working in Multiple Locations: A Review." *New Technology, Work and Employment* 29 (2): 139–59.

Leonard, Dorothy, Walter Swap, and Gavin Barton. 2014. *Critical Knowledge Transfer: Tools for Managing Your Company's Deep Smarts*. Cambridge, MA: Harvard Business Review Press.

Mainwaring, Scott, Ken Anderson, and Michele F. Chang. 2005. "Living for the Global City: Mobile Kits, Urban Interfaces, and Ubicomp." Paper presented at the International Conference on Ubiquitous Computing, Tokyo.

Mazmanian, Melissa, and Ingrid Erickson. 2014. "The Product of Availability: Understanding the Economic Underpinnings of Constant Connectivity." In *Proceedings of the SIGCHI Conference on Human Factors in Computing Systems*, 763–72. New York: ACM.

Mazmanian, Melissa, Wanda J. Orlikowski, and Joanne Yates. 2013. "The Autonomy Paradox: The Implications of Mobile Email Devices for Knowledge Professionals." *Organization Science* 24 (5): 1337–57.

Monteiro, Eric, Neil Pollock, and Robin Williams. 2014. "Innovation in Information Infrastructures: Introduction to the Special Issue." *Journal of the Association for Information Systems* 15 (4): article 4. https://aisel.aisnet.org/jais/vol15/iss4/4.

National Academies of Science. 2017. *Information Technology and the U.S. Workforce: Where Are We and Where Do We Go from Here?* Washington, DC: National Academies of Science Press.

Pentland, Brian T., and Martha S. Feldman. 2005. "Organizational Routines as a Unit of Analysis." *Industrial and Corporate Change* 14 (5): 793–815.

Pickering, Andrew. 1995. *The Mangle of Practice: Time, Agency, and Science.* Chicago: University of Chicago Press.

Pipek, Volkmar, and Volker Wulf. 2009. "Infrastructuring: Toward an Integrated Perspective on the Design and Use of Information Technology." *Journal of the Association for Information Systems* 10 (5): 306–32.

Pittinsky, Todd L., and Margaret J. Shih. 2004. "Knowledge Nomads Organizational Commitment and Worker Mobility in Positive Perspective." *American Behavioral Scientist* 47 (6): 791–807.

Plantin, J., C. Lagoze, P. N. Edwards, and C. Sandvig. 2016. "Infrastructure Studies Meet Platform Studies in the Age of Google and Facebook." *New Media & Society* 10:1–18.

Postman, Neil. 1993. *Technopoly: The Surrender of Culture to Technology.* New York: Knopf.

Ribes, D., and J. B. Polk. 2014. "Flexibility Relative to What? Change to Research Infrastructure." *Journal of the Association for Information Systems.* http://aisel.aisnet.org/cgi/viewcontent.cgi?article=1675&context=jais.

Rossitto, Chiara, Cristian Bogdan, and Kerstin Severinson-Eklundh. 2014. "Understanding Constellations of Technologies in Use in a Collaborative Nomadic Setting." *Computer Supported Cooperative Work* 23 (2): 137–61.

Sandvig, Christian. 2013. "The Internet as an Infrastructure." In *The Oxford Handbook of Internet Studies,* edited by W. H. Dutton, 86–108. Oxford: Oxford University Press.

Sawyer, Steve, Rolf Wigand, and Kevin Crowston. 2014. "Digital Assemblages: Evidence and Theorizing from a Study of Residential Real Estate." *New Technology, Work, and Employment* 29 (1): 40–54.

Spinuzzi, Clay. 2015. *All Edge: Inside the New Workplace Networks.* Chicago: University of Chicago Press.

Star, Susan Leigh. 1991. "The Sociology of the Invisible: The Primacy of Work in the Writings of Anselm Strauss." In *Social Organization and Social Process: Essays in Honor of Anselm Strauss,* ed. David Maines, 265–83. Hawthorne, NY: Aldine De Gruyter.

Star, Susan Leigh, and Karen Ruhleder. 1996. "Steps toward an Ecology of Infrastructure: Design and Access for Large Information Spaces." *INFOR. Information Systems and Operational Research* 7 (1): 111–34.

Strauss, Anselm. 1985. "Work and the Division of Labor." *Sociological Quarterly* 26 (1): 1–19.

Vertesi, Janet. 2014. "Seamful Spaces: Heterogeneous Infrastructures in Interaction." *Science, Technology, & Human Values* 39 (2): 264–84.

Wajcman, Judy. 2014. *Pressed for Time: The Acceleration of Life in Digital Capitalism.* Chicago: University of Chicago Press.

Wajcman, Judy, Michael Bittman, and Judith E. Brown. 2009. "Intimate Connections: The Impact of the Mobile Phone on Work/Life Boundaries." In *Mobile Technologies: From Telecommunications to Media,* edited by Gerard Goggin and Larissa Hjorth, 9–22. London: Routledge.

Zuboff, Shoshana. 1985. *In the Age of the Smart Machine.* New York: Basic Books.

Getting "There" from the Ever-Changing "Here"

Following Digital Directions

Ranjit Singh, Chris Hesselbein, Jessica Price, and Michael Lynch

> *Having just come out of a jungle, I can't promise you that in leading you in to show you what I've found that I won't lose the way for all of us.*
>
> **—Garfinkel (2008, 101)**

"Lost" in the Ever-Changing "Here"

The research for this chapter began with an exercise for a seminar on visualization in science and technology studies (STS), and each of the authors performed variations of it.[1] The exercise was designed to bring into relief the situated practices of following rules, plans, and other formal instructions.[2] Such practices can be subsumed under a theme that ethnomethodologist Harold Garfinkel dubbed "instructed actions" (2002, 197–218), a theme that also became a fixture in STS and information studies largely through the influence of Harry Collins's investigations of the tacit knowledge involved in efforts to replicate scientific methods and instruments (1974, 1975) and Lucy Suchman's research on the situated actions of enacting technical protocols in workplaces ([1987] 2007).[3]

The exercise we discuss was designed to explore the emergent contingencies, practical problems, and repairs that arise during efforts to follow different versions of maps and directions in the course of particular journeys. A number of variations were possible, depending upon the kind of "map" used: a standard printed map, a line or two of written directions, a hand-drawn "occasion map" that directs the user to a particular address or place, or a GPS (Global Positioning System) device. Before, during, and after our journeys, we took detailed notes about what happened along the way. We paid attention to the particular map-reading practices, and took notes on how maps informed and confused our wayfinding. We also took notes on the difficulties and contingencies we encountered and how we contended with them. In our notes and discussions we paid special attention to what is distinctive about different kinds of maps and directions, as visual/textual devices for organizing practical actions.

In addition to performing one or another variant of the exercise, we read selections on the design of maps and their epistemic and practical implications (Brown and Laurier 2005; Vertesi 2008; November et al. 2010).[4] We wrote up brief reports on the exercises, often focusing on wayfinding difficulties and experiences of being lost. A low-tech example will help set the stage for identifying themes that arose in the GPS-aided examples to follow. In this instance, one of us attempted to use instructions written by a friend on how to get from a residence in Brooklyn to a museum in Manhattan. The field notes begin with a quote from the directions:

> "Exit at 77th. Met is the big white building at 82nd and 5th Avenue."
>
> It was my first time in New York City. I alighted the 6 at 77th and walked up the subway steps into the street. The sidewalk was crammed with people, the buildings were tall, and I could hear yellow taxicabs honking at each other. It was exactly what I had hoped Manhattan would be like and I was delighted. Sadly, this initial rush gave way to a less pleasant feeling: I realized I was lost.
>
> Until I exited at 77th I had successfully followed my friend's written instructions. Entering the subway in Brooklyn I had changed at Union Square and taken the 6 to the Upper East Side without any problems. Allegedly, I had prepared myself for this part of the journey by looking at an online map of the area. In my mind it had all seemed so simple. The museum was only a few blocks across, then a few blocks up. The problem in that moment outside the subway steps on 77th was that I didn't know which direction was across and which was up. I grew up in Great Britain, which doesn't have any grid systems and it hadn't occurred to me that I might find the layout of Manhattan, in so many ways so logical, so utterly discombobulating. I looked around trying to overlay the map I had in my head of the route against what I could see in front of me and just kept failing. I decided that my chance of going the correct way to the park was 50% and that it would be worth walking a few blocks to see whether I could get there without having to ask for directions. Knowing Central Park was a large tourist attraction, I judged the direction most people were going and followed the crowd. My gamble paid off and once I had arrived at the park I was able to see the museum, according to my written instructions, "the big white building," very easily.

Even in this case, online resources came into play, as the written instructions were supplemented by consulting online maps for the layout of subway lines and their identification with street numbers, and these directional resources were crucial for coordinating instructions with the signs and directional indicators placed in subway stations and on street corners. And yet, our navigator still found that she was *lost* after exiting the subway stop, as the spatial direction "up" on a map had no clear correlation with her immediate embodied experience of the cityscape. After finding herself lost, our navigator improvised by reading the legible infrastructure of the city, and following pedestrian traffic flow until the description of the destination *came true* in immediate experience. We shall return to these themes of improvisation, and the coordination of various instructions with the legibility provided by the local infrastructure. But first we shall discuss how the exercise critically relates to navigational tools, especially the GPS.

We became intrigued by the GPS because, unlike a map or a set of written or verbal directions given prior to a journey, it adapts its instructions during the journey. Accordingly, the apparent gap between formal instructions and situated actions collapses. However, as we shall see, the GPS turns out to be less of an earth-shattering, paradigm-shifting device than it seemed at first glance, and its use requires some older and trusted forms of supplementation and repair. Specifically, as suggested by the above vignette about navigating in Manhattan, various backup systems and legible environments often supplement or override what initial instructions show and tell us to do. Although the GPS continually adapts to our position as we go along, the salience and authority of its directions can be, and frequently are, overridden.

Before turning to our exercises with the GPS, we shall discuss specific modalities of tacit knowledge that are tied to different sets of navigational instruction. We focus on the experience of "being lost" because, however disconcerting it can be at the time, it is an excellent "tool" for gaining insight into wayfinding practices. Our chapter draws upon numerous instances of journeys performed by the authors with different forms of maps and directions in different situations, but the focal point will be a series of exercises involving navigation through familiar and unfamiliar terrain using GPS directions. Therefore, after engaging with the literature, we will provide insight into how the relationship between maps and journeys is reconfigured with the use of digital media. In the spirit of the present book, we also encourage readers to perform similar observational exercises.

Ways of Getting from A to B

The wayfinding exercise provided a simple way to explore the limits of rational action. A journey from point A to point B is a prototype of rational action, in the classic sense in which an actor in a situation selects an efficient means for achieving an end.[5] Point B is the "end" while point A is the origin or starting point of such an ideal-typical act, and the act is performed under the jurisdiction of an aim to reach the end with minimal difficulty. In this case the "actor" is what Goffman (1972) defines as vehicular unit: "a shell of some kind controlled (usually from within) by a human pilot or navigator." This definition can be expanded beyond mechanical vehicles to include pedestrians navigating with their bodies (though the Cartesian "ghost in the machine" implications are disconcerting, and we wouldn't want to assume that the navigator simply parks the body and walks away at the end of the journey). Importantly, for Goffman, the navigator's choice of means (direction, path, available type of vehicular unit, etc.) is deeply attentive to moral, social, and physical barriers, rather than being an exercise in what we could call "Euclidean geography." Among other things, it is enabled and constrained by a *traffic code*: "a set of rules whose maintenance allows vehicular units independent use of a set of thoroughfares for the purpose of moving from one point to another" (6). Accordingly, the route taken as a means to the end follows standardized and normative (socially available and positively sanctioned) pathways.

As a sociologist, Goffman emphasizes traffic *rules* that prohibit what Robert Merton (1938) once called "innovative" actions: actions that pursue normatively approved ends by means that transgress legalities and social conventions. Imagine an idealized Euclidean navigator, attempting to follow the most direct geometrical path from A to B—a straight line drawn on a map—which would require cutting across streets, climbing over fences and scaling walls, trespassing on private property, and walking through gardens and homes.[6] Aside from encountering formidable physical barriers and hazards, such an actor would very likely provoke an extreme social and legal reaction about as quickly as if she or he were walking naked in public.[7] When walking from A to B, we normally do not even think about such possibilities, though when driving we occasionally do, for example, go against a directional arrow to enter a parking lot at its exit; or, more rarely, quickly cut across a divider strip or scoot for a short distance the wrong way up a one-way street. And we may do so by mistake. In brief, when points A and B are established as origin and destination, the rational actor does not choose the shortest distance between the two points. However, the traffic code and related rules are far from the only barriers the deter our traveler from navigating along an ideal geometrical pathway: rules are materialized, supplemented, and occasioned by traffic lights, pedestrian crossings, warning signs, and entrenched roadways. Rules and their rationalities are emplotted and inscribed into the territory, and the very surface of the terrain is underlaid by and overlaid with "mundane artifacts" (Latour 1992), making up a complex infrastructure of steel-reinforced concrete pathways that guide and block the movement of vehicular units.

Our wayfinding exercise was not designed to illustrate a conception of rational action with empirical examples; instead, it was designed to *make trouble* (Garfinkel 1967, 37) for any such conception by exposing the potential complexities and contingencies that arise in relation to whatever counts (or not) as an optimal path from A to B.[8] Instances of such trouble would then provide leverage for opening up discussion of what is missing *both* from particular forms of instruction *and* from theories of rational action. In a critical discussion of "plans as programs," Phil Agre and David Chapman suggest a different conception of plans as communications in which actions are a matter of "following natural language instructions" where "the agent uses the plan as one resource among others in continually redeciding what to do. Using a plan requires figuring out how to make it relevant to the situation at hand . . . as participating in the world, not as controlling it" (Agre and Chapman 1990, 17). Consequently, an existing plan may be revised or aborted in light of the unique contingencies and opportunities that arise in a specific course of action. This conception of navigation is related to the resilient theme of *tacit knowledge*.

Tacit Knowledge

The theme of tacit knowledge has a long association with STS and the information sciences, dating back to Michael Polanyi's (1958) philosophical writings on scientific practice, and Hubert Dreyfus's (1979) critique of artificial intelligence. Tacit knowledge covers an array of phenomena, including embodied skills acquired through practice, covert understandings shared among members of a guild or profession, and ubiquitous ways of acting acquired through participation in a society (see Collins 2010 for a classification of domains of tacit knowledge). Tacit knowledge is one of a family of terms such as know-how, knack, *Fingerspitzenfühl*, improvisation, tinkering, and bricolage—all of which reference the practical judgment, interpretive flexibility, situated action, and embodied dexterity required for performing skilled work. One common way to highlight the role of tacit knowledge in the natural sciences, which became important for establishing STS as a field, was to point to a *gap* between formal accounts of scientific method and close observations of scientific and technical practices. Well before STS became established, Peter Medawar asked in the title of a popular magazine article, "Is the scientific paper fraudulent?" He answered in the subtitle, "Yes; it misrepresents scientific thought"

(1964). Medawar went on to explain that the charge of fraud was an exaggeration, and that what he was suggesting was that scientific papers provide systematically misleading narratives of scientific "thought" (and, as later work in STS documented, "thought" was a thin way of describing the complex assemblages of embodied work, technological virtuosity, craft, artistry, and routine interaction that compose scientific practices). Not only were such reports misleading as post facto accounts of scientific work, they also were virtually useless as sources of instruction on how to perform and replicate such work. This gap between formal, written prescriptions and the actual performance of technical actions also framed Michael Polanyi's writings about tacit knowledge—the embodied skills and know-how learned on the job through experience and apprenticeship, rather than through formal instruction: "Textbooks of diagnostics teach the medical student the several symptoms of different diseases, but this knowledge is useless, unless the student has learnt to apply it at the bedside. The identification of a species to which an animal or plant belongs, resembles the task of diagnosing a disease; it too can be learnt only by practicing under a teacher's guidance. . . . Thus, both the medical diagnostician and the taxonomist acquire much diagnostic knowledge that they could not learn from books" (Polanyi 1962, 603). Textbook instructions not only are written but also attempt to formulate abstract accounts of activity that specify what practitioners must do under highly variable conditions, and while such instructions might give helpful guidance for novices, they cannot possibly encompass any and every single episode. Empirical STS research, starting in the 1970s, addressed a gap between, on the one hand, formal accounts of practice and, on the other, the embodied and mechanical performance of practices in particular situations (Collins 1974, 1975; Suchman [1987] 2007; Garfinkel 2002). The critical import of the gap between formal instructions and practices also applied to technological efforts to incorporate human actions into "instructions" programmed into a computer.

It is tempting to think of tacit knowledge in terms of a dichotomy between a formal (written, programmed, decontextualized) part and an informal (situated, enacted, contextual) part, and to develop a critique of the formal part by elaborating upon its failure to specify the embodied practices and skills necessary to perform the informal part in particular circumstances. However, instructions conveyed by word of mouth, written recipe, nonverbal diagrammatic sequence, video demonstration, and interactive exchange are not all of a piece, and many of them incorporate or reconfigure activities that might otherwise be assigned to the "tacit" (systematically hidden, informal, situationally specific) side of the ledger. This lack of strict demarcation provides hope for behavioral engineers who suppose that mechanisms discovered by cognitive neuroscience and/or built into sophisticated programs will eventually encompass even the most recalcitrant aspects of human and nonhuman actions. Hubert Dreyfus (1992, 100) once remarked that such hopes are akin to an attempt to build a ladder from the earth to the moon—it is always possible to add another rung, but the project as a whole will literally never get off the ground. Whether or not Dreyfus was right about this is not germane to our interest in this chapter. Instead, the reconfiguration of tacit knowledge with different communicative devices itself calls for investigation (Lynch 2013). Without specifying a fixed point at which formal directions leave off and tacit knowledge takes over, we begin with the idea that the relations between instructed actions and tacit knowledge continually change, not only with the situations in which instructions are applied, but also with particular technologies through which instructions are embodied and conveyed. Consequently, for any existing device that somehow instructs or guides a journey, there will be an investigable domain of tacit knowledge in relation to *that* device and its uses.

Maps and Their Gaps

Directions, maps, and navigational devices are instructional technologies:[9] they do not simply convey information in accordance with standard designs. When used for a particular journey, these technologies convey navigational directions. They are materially, semiotically, and narratively configured, and their configuration is tied to their use in specific environments. Like other tools, they exhibit transparency or transitivity. Polanyi used the example of a stick or probe that a blind person uses to navigate, suggesting that the instrument user "dwells" within the instrument. It is as though the instrument becomes an extension of the body, as the person "feels" the texture of the sidewalk touched by the end of the probe, and the sidewalk acquires distinctive properties tied to the probe (as well as the soles of shoes) through which it is "felt" (also see Merleau-Ponty 1962). This also applies to directions and maps, as the landscape acquires properties in relation to their communicative interface. Written or spoken directions typically begin with a point of origin and end with a destination, while including reference points to what "you" should look for, and where "you" should turn (or turn back, if "you" have missed a turn) during the journey. These reference points embed "phenomenal field properties" (Garfinkel and Livingston 2010) within the material/semiotic form of the directions, implying a phenomenological situation *in the world* within an objectified account of a landscape. Google Earth and GPS navigational aids embed further phenomenal properties into a more dynamic and circumstantial display of the immediate field of action (November et al. 2010).

Drivers, walkers, and ocean navigators have access to an array of formal navigational devices, and the practices of using them have been studied for decades by anthropologists, ethnomethodologists, and communication scholars. Commonplace devices used for driving or walking along streets and thoroughfares range from mass-produced and standardized tools to specific ones tailored for a particular journey:

1. **Comprehensive maps**. A comprehensive map is designed to be used for an open-ended array of purposes. There are, of course, many different varieties of comprehensive map, and in this chapter we discuss only road and street maps designed to guide wayfinders within and through the mapped territory. There also are many varieties of these, varying in scale and resolution, and deploying different conventions and orientations (e.g., for tourists, marking sites of special interest, scenic routes, etc.). The design of such maps is suitable for a range of specific purposes, but they do not include singular points of origin and destination, though a particular route can be inscribed by hand on them prior to a journey.[10] Trouble can occur during a journey when you presume to be "on" the map at a moment when you are actually "off" of it, or when you presume to be "off" the map and you are "on" it, or when you think you are "on" the map in one place, but you are actually "on" it in another place.

2. **Placed maps.** These are "immutable *im*mobiles" that are fixed in place, and preserved against the elements.[11] As noted above, they can be found on campuses, in city centers, and at highway rest stops. Typically, they are standardized maps, of "appropriate" scale, which may be identical in many respects to portable road and city maps distributed by a tourist office or found on a campus website. Typically, they include a "you are here" mark (for discussion of such placed maps, see Latour 1987, 162; Laurier et al. 2016), and they may include special features of interest to tourists or visitors to a campus. In this case, the fixed locale of the map enables its virtual-phenomenological "here" to be inscribed. It is a peculiar "here" that usually leaves unspecified where else "you" might go from "here," though it might provide helpful indications of where and how "you" *might* reach any of an indefinite number of different destinations via available roads, underground lines, or walking routes.
3. **Verbal directions**. These are delivered orally and/or in writing and denote a linear sequence with conditional features. George Psathas (1990, 183) likens them to "stories," because the sequential organization of their telling narrates a temporal progression. This analogy links the Christian/Western notion of *telos* in storytelling to the way in which navigation involves moving from "here" to "there." The directions are ordered not only as a story, but also as a series of commands for a series of actions to be taken. Conventional sequential features include such instructions as "Take Route 17 East to Binghamton, and several miles past Binghamton, take US Rt. 81 South." Of course, the story is highly compressed, and the journey will include long intervals between directions, sometimes denoted by distance, time, or other measures ("after turning left, go five blocks"; "stay on 81 for around 20 miles, until . . ."; "in five minutes or so, you should see . . ."). Conditional features often are included in instructions, such as, "if you pass under a railroad bridge, you've missed the turn." There is a significant difference between written and oral instructions, as the written instructions can repeatedly be consulted, while the oral instructions are subject to the vagaries of recall.
4. **Occasion maps.** Like verbal directions, and often combined with them, these maps are drawn for a particular occasion and typically discarded afterward, though in some cases they may be retained and reused on other occasions. Psathas (1979, 204) observes that these maps are non-topographical, in the sense that they do not include standard coordinates and scales, and provide little or no mapping of terrain beyond the linear route to be taken.[12] Consequently, it is easy to get "off" the map and hard to find one's way back "on" to it. Written directions often include a variable amount of sketched detail, and more elaborate sketch maps typically include written directions and labels. Occasion maps typically include a starting point and destination (which can be reversed for a return journey), lines for routes, denotation of cross streets, traffic lights, names for key streets, and landmarks, and a few other features. Although occasion maps are drawn for a specific journey, there is no less of a "gap" between their features and those that arise in the course of a journey than there is for standardized road maps (Liberman 2013).
5. **Navigational instruments**. This category is worth exploring further, though we shall not do so here, except in the case of GPS. Historically, navigational instruments (sextant, compass, etc.) were crucial for reckoning direction and distance. In an automobile, the most commonplace navigational instru-

ments are the speedometer and odometer, and less commonly a compass. A built-in or portable clock or wristwatch also may be included among the instruments. Increasingly, digital aids are built into the vehicle, including GPS.

6. **GPS**. There is a rapidly proliferating variety of GPS and related navigational tools. A GPS is, in a sense, a navigational Turing machine,[13] in that it combines and affords variable combinations of maps, directions, and instruments. It can be adjusted to simulate voiced directions, inscribed directions, road maps, or placed maps. It also can simulate a vantage point akin to that of a driver or passenger viewing an unfolding road scene, but it's "here" differs from the scene available through the windshield. When fixed to the windshield, a GPS is a second, miniaturized window that delivers a distinctively organized vantage point from a virtual position "above" a schematized version of the vehicle and the space surrounding it that is more characteristic of a map than the phenomenal field of driving. Using it while driving is a matter of seeing double. In the case of walking with a smartphone, the schematic virtual window differs from the environment that surrounds the walker.[14] In the instances of driving described in this chapter, a Garmin device attached by suction cup to the windshield was used; in cases of walking, the navigation was with Google Maps on a smartphone.

A Turn-Taking Machine

The GPS does more than incorporate and combine aspects of road maps, placed maps, occasion maps, navigational instruments, and verbal directions. The most notable features are its voice, which all too literally (mis)pronounces street names while giving instructions about upcoming maneuvers, and displaying a continually updated, forward-looking street map that shows where "you" are at any point in the journey. The voicing of directions is akin to a companion with a limited repertoire consisting mainly of turn-by-turn instructions, such as "In about a quarter of a mile take a left exit onto University Avenue" (Brown and Laurier 2012, 1623). These instructions are organized differently from the turn-taking machinery of conversation (Sacks et al. 1974); instead of being tied to the speaking turns and turn-transition-relevance places of an ongoing conversation, they are programmed in relation to upcoming intersections at which turns should be taken or not in the course of the journey. Although drivers and passengers may respond to the voice, talk around it, yell insults at it, or shut it off, the order of the GPS's turns is more akin to a list of commands, not unlike verbal directions, except that the directions are adjusted and are recalibrated to the position of the vehicle on its path toward a programmed destination. In this way, following the GPS is akin to what Psathas (1986) and other ethnomethodologists (Psathas and Kozloff 1976; Garfinkel 2002, 179) describe in their studies of verbal direction giving and direction following.

Wayfinding Troubles and Repairs

As noted earlier, instructed action exercises are designed to elicit troubles—troubles that disrupt the transition from formal instructions to particular efforts to enact them. Such troubles have methodological significance as "aids to a sluggish imagination" (Garfinkel 1967, 38), as they allow for both critical attention to the limits of preset plans or programs of action, and articulation of how tacit knowledge or situated action comes into play in specific cases and domains of action. And, as we shall argue, such exercises encourage us to reconsider what we might possibly mean when we use catchall categories such as "tacit knowledge."

Wayfinding troubles often occur during efforts to bridge the "gap" between a plan for a route composed in advance and an ongoing journey that attempts to follow the plan. This gap might be said to arise from what Tim Ingold calls a "paradox at the heart of modern cartography": "The more it aims to furnish a precise and comprehensive representation of reality, the less true to life this representation appears" (2000, 242). The immobility of inscribed detail in maps (of any kind), so suitable for reproduction and dissemination (Latour 1990), contrasts with the mobile vantage point of the journey. Because of the emergent, site-specific, and idiosyncratic contingencies arising in the course of unique journeys, there is no possible way for a map to include enough detail to anticipate such contingencies. However, with a GPS one might suppose that Ingold's paradox is dissolved, as the gap between map and journey closes up, even though it never quite disappears. This is because the GPS continually updates its instructions, adjusting to changes in location and recalibrating the route when the vehicle goes off course, takes a side journey, or heads to a new destination. Although the GPS receives digital signals from a satellite or an array of cell phone towers and computes geographical coordinates, it translates these coordinates into a simulation of a mobile existential "here" rather than a static objective location.

However, while the GPS may seem to close up the gap between instructions and situations of action, it opens up another gap between the display on its screen and a more encompassing scene in which the screen is embedded.[15] The screen displays a small schematic field organized around "here and now" incorporated within the vehicular unit's *umwelt*: a field of action and instrumentation with its own "here and now" that extends from the vehicle dashboard, through the windshield and mirrors, reaching into the terrain along the line of vehicular movement. The practical alignment of the two fields is a task and occasional source of trouble for navigation (Brown and Laurier 2012). Various kinds of trouble arise when using a GPS, especially for novice users but also for others who have more practice with the device. In what follows, we focus upon an array of troubles and repair efforts arising from (apparent) misalignments between the field accessed through the GPS interface and the more encompassing environment of its immediate use.[16] Repairs in these instances are not a matter of fixing a faulty device but a matter of contending with momentary troubles that arise through "normal" interactions between the device and its user. Some of these troubles may be eased or eliminated through future improvements in the devices and their programming, but many are endemic to following directions. Repair is less a matter of fixing technology than of improvising ways to use it in particular situations—of relying upon as well as further developing what Sawyer et al. (this volume) call "infrastructural competence." As Jackson (2014, 226) points out, repair involves innovative work, though often of a kind that receives little notice.[17]

Two different situations were used in this study to explore the relationship between the GPS and the environment. One situation involved using a GPS to navigate from A to B in familiar environments (e.g., using the GPS while walking or driving from office to home), while the other involved using the GPS as a resource to guide journeys through unknown (or vaguely known) territory. Performing the exercise alerted us to the ambiguity of this distinction, as it challenged us to reconsider what it means to *know* where you are and how to get to somewhere else. When we performed these exercises (or when we described experiences that were not originally planned as exercises), the distinction between familiar and unfamiliar destinations became complicated.

Using GPS in Familiar Environments

When a GPS was used as a guide, familiar environments acquired elements of unfamiliarity, and following the GPS directions also created awkward interactional situations. One of us used a GPS program on a smartphone to guide a walk of approximately two miles from a campus location to home. As happened in most instances in which we attempted this variant of the exercise, the route suggested by the GPS differed from the "usual" route.

> As I walked on this route, my first observation was that interacting with the visual interface of the GPS and the sense of continuity that it offered [in comparison to sporadic consultation of a static map or written directions] had considerable impact on my awareness of the environment. My focus shifted from the roads and crossings to the visual interface of the GPS and its pointer indicating where to go next, which was continuously updated as I went along. I was walking with a friend who knew the route as well as I did, and our conversations became sparse as my focus was continuously on the GPS interface. This lack of communication became an issue as the exercise progressed, because I had not informed my friend in advance about this exercise. Ultimately, I had to change my performance of the exercise to some extent to preserve the texture of the everyday ritual of walking back home from campus.

At first, the companion on the walk did not take issue with the use of the phone, since walking while fixated to the screen of a phone is far from uncommon these days, but when the path guided by the phone diverged from the usual route, the companion began to raise questions:

> Dryden Road has a curve situated between Aladdin's Natural Eatery and Sangam Indian Cuisine. A staircase allows you to cut across the curve and is a known shortcut. While some of the stairways that can be taken as shortcuts are listed on Google Maps, this one is not recorded. Again as I tried to follow the curve (*absent-mindedly looking at the GPS interface on my phone*), my friend finally asked me what I was up to with my phone and why was I deviating from the normal route that we usually take. I explained to him the objective of my exercise and he suggested an alternative to continuously looking at the interface while I walk.
>
> Friend: Increase the volume of your phone, so that you can hear the instructions of the GPS, and then you can avoid having to look at the interface continuously!
> Me: Should we follow the instructions together?
> Friend: No, we can also just look at where the GPS route deviates from the route that we usually take.

Following the companion's advice led to a significant change in the exercise: instead of following the GPS instructions blindly, largely oblivious to the surroundings and the familiar routes through them, the two walkers treated the GPS directions as an option. When they ignored the GPS directions it recalibrated its instructions, and when they ignored the new instructions it recalibrated them again. The GPS consequently *followed* rather than *led* the route taken by adapting its directions continually with each divergence from its prior directions. This change in procedure opened up questions concerning "hardness" of instructions and also the relationship of dependency between human and machine. While some instructions are "hard" in the sense that if they are not followed correctly, the end result will be disastrous, not following GPS instructions in this case turned out to be easy and inconsequential for the ultimate success of the journey.[18] The *gap* (or, rather, series of gaps) between the instructions and the situated actions became gratuitous and was repeatedly ignored.

Being Lost with the GPS in a Familiar(?) Environment

On some occasions, following the directions given by the GPS led to being lost. There is something paradoxical about being lost with a GPS, since a major reason for using the instrument is to *avoid* getting lost. Moreover, a GPS system specifies with great accuracy where exactly you are on its map, even when you have no idea of where that is. However, from our experiences with using the GPS, both in specifically designed exercises for this project and in everyday use, getting lost with a GPS was far from uncommon. Though disconcerting at the time, such occasions provide some insight into what it means to *be* lost as an existential experience—a way of (not) being in the world; and, as Ingold (2000, 219) observes, being lost also allows for reflection on "what it actually *means* to know where one is, or the way to go." When it is functioning correctly (getting a signal and displaying it on the screen), a GPS should provide the user with a precisely located moving position on a map and a schematic display of local surroundings, as well as other verbal and numerical information about street names, intersections, and the distance and time to destination. However, experiences with getting lost involve a kind of double vision where the "view" provided by the instructions gets out of alignment with the other view of "where you are" provided by the vehicular *Umwelt*.

With verbal directions, an occasion map, or even a local road map, you may find (or suspect) that you are no longer "on" the described route, or even on the map, and finding your way back to the route and/or map (or otherwise finding your way to your destination) can become quite difficult as you search for landmarks—singular features that render the landscape *legible* (Lynch 1960). You may also find (or suspect) that you are still "on" a road map or topographic map, but are no longer where you previously thought you were. The immediate environment may provide few if any clues. For hikers lost in a desert landscape, the plants, sandy washes, rock formations, and horizons may seem indistinguishable from those found at any other locale for miles around. For drivers in another kind of desert—a seemingly endless series of franchised fast-food restaurants, gas stations, and strip malls—the landscape provides no clue about "where you are," even though everything in the surroundings seems all too familiar. With a GPS, however, even if you deliberately or inadvertently abandon the original route from A to B and have no idea of where you are, the device *should* provide you with a continuous display of where you are on a map, and it *should* instruct you on how to get to your destination from there. Sometimes, however, the GPS fails, or the user fails to program or follow it correctly (these two possibilities often are indistinguishable in the course of a journey).

When we performed the exercise, instances of getting lost with a GPS were not clear-cut. In many instances, it dawned on us gradually that we were lost, and we were never entirely sure that if we kept following the route the GPS prescribed, it would get us to our destination more quickly than if we abandoned that route. The following description of trouble that occurred during a drive from Ithaca to Boston is one of several cases in which we used the GPS in more or less familiar territory—*more or less* familiar because the GPS tended to "discover" routes that the user had not previously taken or even imagined.

> I hadn't used the GPS in several weeks, and it didn't set up too quickly or easily when I entered the destination address. At one point after we set out, it seemed to be showing where we were after a time delay of a minute or so, but we knew where we were, and could ignore it. Eventually, it seemed to catch up with us. As we drove through Whitney Point, went on Route 26 S., and headed toward the onramp for I-81 South, the GPS instructed us to go straight on Route 26 past the onramp, and I decided to go along with it on the chance that it "knew" a shortcut. As we followed the designated route, it became increasingly clear that we were not going a good way. Rt. 26 goes to Endicott, several miles west and south of where we would pick up I-88. And, it goes through several small towns. There didn't seem to be a good way to backtrack or to improvise another route, so we stayed with the GPS, and lost at least a half-hour of travel time. Eventually, we pulled the plug of the GPS from the cigarette lighter, and then re-inserted it, and after setting it up again it seemed to behave itself, and our trip was uneventful from there until we hit Boston.

When reflecting on events like this, we are left with questions about whether to blame the device for absurd directions or to acknowledge the possibility that we set up the trip wrong and/or misread the directions. Like other digital devices, the GPS does not provide transparent indications of some of the settings, and either you have to remember how it is set it up or consult the manual, both of which are difficult to do on the fly. One particularly annoying possibility is that the GPS may inadvertently have been set for surface roads, to avoid freeways, or set for the shortest distance rather than quickest journey, and the particular settings are not easy to find or change, especially when the journey is under way. Even if we take the blame for incompetence, our failures point to the requirement for distinct competences with following GPS instructions, and also anticipating such instructions, setting up the device correctly, and knowing when to disregard the GPS.

In this instance, the driver already knew how to get from A to B. But unlike in the prior example of using a GPS on a walk through familiar territory, there were junctions at which trust (or hope) invested in the GPS directions overrode the driver's sense of the best route to take. This relates to a point that Leshed et al. (2008) make that the GPS can alienate the user from the environment by offering an unquestioned source for the user to follow instructions blindly. Accordingly, while easing the burden of reading the landscape in an unfamiliar territory, the GPS obviates the necessity of taking note of the environment. When it dawns on us that the GPS is leading us astray, a dilemma begins to take shape: should we stay with the device in hopes that it "knows" the route better than we *thought* we did, or should we privilege our own judgment and abandon it? A variant of this dilemma is more playful and leisurely: should we follow the GPS to see if it will find a novel way to get to where we know we are going, or should we abandon such curiosity and revert to our habitual route? The possibility of abandoning reliance on the GPS, and relying upon other navigational resources, including, when available, our own familiarity, points to options and even forms of play that counteract the tendency to invest unquestioning trust in the machine.

Getting Off the Beaten Track: "Here Be Dogs"

When discussing their conception of plans as communicative resources rather than as self-sufficient programs that drive action, Agre and Chapman (1990, 17) observe that "the agent uses the plan as one resource among others in continually redeciding what to do." Although a GPS itself performs a calculative version of such "redeciding," effectively using this tool also requires treating it as one resource among others. Experiences of being lost with the GPS provide clear reminders of that necessity, and they also provide analytical insight into the alternative sources of directional instruction that may be at hand or found in the environment. A vivid example of how such backup systems can come into play occurred during the performance of the exercise in San Juan, Costa Rica.

> This trip occurred during a vacation, and involved a driver (me) with a rental car, and a local guide who did not drive but supplied a windshield-mounted GPS programmed for our destination. I spent a fair amount of time in the days before the trip reviewing and rehearsing with online resources how to get to the destination . . . a national park around 90 kilometers from the city. On the appointed morning, we set out confidently.[19] The initial problem was to get from the hotel to the Pan American Highway, which would take us to the national park. The guide was unfamiliar with the neighborhood around the hotel, and so we relied on the GPS to get us to the highway and out of the city. This is when the trouble began. At the start, the guide turned off the GPS "voice" because he could not tolerate the way it pronounced Spanish street names, but the visual display gave us clear directions and the lack of the voiced directions was not germane to the problems we encountered. Not long after we started driving, we became increasingly uneasy as the GPS led us through numerous local roads at the outskirts of the city, with no main highway in sight. We stayed with the directions in hope that the GPS was taking a shortcut. After about a half-hour of traveling through winding streets, the GPS directed us up a hill and the pothole studded paved road gave way to dirt road. As we wended our way uphill, we witnessed a dog threesome—two mating, one watching—in the middle of the road. The sight of the unruly dogs was enough to initiate an abrupt change of course, as we were like pre-modern seafarers turning back from uncharted territory marked by the mapmaker's convention "here be dragons."[20] The guide directed me to stop beside a pedestrian who was walking downhill. In Spanish, he asked him about where the road led and if it was reliable. The fellow answered affirmatively, pointing toward the way we had been going up the hill. However, the guide seemed resolute at this point, and said he did not like the way this route was turning out, and we headed back by the route

> we had come. Every kilometer or so, the guide indicated for me to pull over and he would ask directions from someone waiting for a bus, walking, or even jogging. As I understood it, with each stop for directions the guide was not only trying to verify the directions he had been given earlier, though I think that was an issue, he also was updating directions in a point-by-point way as we proceeded. It was both a way to check that we were still en-route and to "refresh" the directions from a new point along the way. Meanwhile, the GPS kept "insisting" that we revert to our previous, seemingly hopeless, route even after we joined the main highway. The guide shut off the GPS, and we made it to our destination without further mishap, though much later than we had initially planned.

It was clear in this case that the guide's trust in the machine was far from absolute, as he ended up completely disregarding the GPS and using other resources available in the environment, none of which was fully trusted either. The many permanent and impermanent features of the scene outside the vehicle, which were nowhere to be seen on the GPS screen, were overwhelmingly evident during the journey. These gratuitous details—gratuitous for the GPS instructions, but not for the driver and navigator—included pedestrians, unleashed dogs, potholes, and other hazardous features of the local road. The layout of streets in the hills made for confusing reading of the GPS, as it was difficult to see at a glance which of the network of roads was *the route*, even though the route to follow was clearly indicated by the magenta line on the GPS screen. Such streets were unlike a main highway, which is wider than the incoming streets, has clearly marked lanes and road edges, and is outfitted with multiple signed exits. The roads and traffic also were less "disciplined," as they included an indiscriminate mix of humans, animals, streets, intersections, and types of machines (cars, large trucks, motorbikes, bicycles). Such an environment, though far less congruent with the GPS's "world" than a modern cityscape and highway infrastructure (patches of which can be found in San Juan and Costa Rica, but not where we were at the time), also provided a resource—a legible backup system. Potholes and unruly dogs alerted us that we were off the beaten track, and the guide used the low-tech system of asking for local directions to repair the confusion and anomalous route sketched out for us by the GPS.

Discussion: Navigation as an Intertextual Achievement

The concept of tacit knowledge is often treated as a matter of *personal*, often nonverbal, knowledge (Polanyi 1958). The commonly used example of riding a bicycle—an embodied competence that is much easier to master than it is to articulate—suggests that tacit knowledge is embedded in the person, as a holistic assemblage of perceptual, motor, and cognitive skills that operates beneath language and is acquired through personal contact and example. As Collins (2010) emphasizes, tacit knowledge is not limited to personal somatic skill, since the cultivation of such skill arises from and is afforded by membership in the social groups (such as scientific disciplines) that sustain the relevant practices and establish the standards through which such skill is enacted and evaluated. Whether construed as personal or social, or both, tacit knowledge tends to be elucidated through contrasts to formal knowledge that takes the form either, or both, of written instructions for humans or as programmed instructions for automatons. Accordingly, tacit knowledge extends beyond the limits of formal instructions and automated actions. However, when we examine examples such as those elicited through our exercises, not only do we discover how tacit knowledge is featured differently in navigational work, in relation to the specific formats of instruction used, we also have the opportunity to investigate just how we contend with the contingencies that arise on particular occasions. Although personal skills, habits, and abilities to improvise certainly are important, when we performed variants of the exercise, much of what guided us with, without, or in spite of the instructions we had at hand, and/or on screen, was embedded in the environment through which we navigated. Our examples included such social resources as the flow of pedestrians and directions from passers-by. However, they also included legible aspects of the built environment.

In our example of the Euclidean navigator, networks of material infrastructure (roads and fences), semiotic notifications (road signs), legal frameworks (one-way streets), and so on, block or prohibit any effort to get from A to B through the shortest possible route. As we have seen, however, these same barriers also provide resources for navigation. Navigation—with or without the latest digital tools—calls into play an intertwining of infrastructures (material pathways and barriers, systems of road signs, rules of the road), to bring them into alignment through an infrastructural mashup of paper maps, occasion maps, GPS, and so on, to practically accomplish a journey from A to B. By *infrastructural mashup*, we mean the ability of combine several different sources of information into a coherent context and source of guidance for any next action. A GPS combines information about these different infrastructures (material, road signs, legal speed limits, and so forth) into a unified interface that gives ongoing directions, but such directions become intelligible (or not) in relation to the multiplicity of other instructional resources at hand and in the local environment.

In a precursor to what later became known as infrastructure studies, Bruno Latour describes how legibility is on *both sides* of the gap between a map and the landscape:

> When we use a map, we rarely compare what is written on the map with the landscape—to be capable of such a feat you would need to be yourselves a well-trained topographer, that is to be *closer* to the profession of geographer. No, we most often *compare* the readings of the map with the road *signs* written in the *same* language. The outside world is fit for an application of the map only when all its relevant features have themselves been written and marked by beacons, landmarks, boards, arrows, street names and so on. The easiest proof of this is to navigate with a very good map along an unmarked coast, or in a country where all the road boards have been torn off (as happened to the Russians invading Czechoslovakia in 1968). (Latour 1987, 254, emphasis original)[21]

In this sense, a map does not simply represent what is "out there" in the environment. Both the built environment and the map deploy a coordinated array of signs and markings. Navigation is thus an intertextual achievement. A GPS device enhances legibility, not only by including names and directional arrows that correspond to those printed on road signs, but also by making gratuitous details and forbidden pathways disappear. Here, the GPS device is a Latourian (1992) mundane technology in the way it obviates and supplements signs, rules, and injunctions with an algorithm. It does not replace such infrastructure, but is instead parasitic on it. Unlike infrastructures such as sewer systems, which are hidden beneath the surface of the urban landscape, and accessible only to specialists (or in the case of the Paris sewers, tourists), much of the infrastructure that affords street and sidewalk navigation is on the surface and out in the open. However, the GPS also systematically hides features in its doubled vision of that infrastructure. Blocked and forbidden paths simply do not exist in its display of possibilities, although as we learned when walking with a GPS along familiar pathways, like other digital rights management tools, the GPS also eliminates other paths and infrastructures that are available to "fair use" (Gillespie 2009).

Conclusion

So, what difference does this particular digital technology make for the familiar tasks of wayfinding? As we have discussed, using a GPS not only alters the gap between instructions and instructed actions in specific ways, but also alters the relationship users have with the environment through which they are navigating. A GPS device, in this context, draws together a distinctive set of relationships among constituents of the infrastructure for travel (roads, pathways, signs, and so on). Functionally, a GPS device used for navigating between A and B reconstructs the landscape to be traversed in specific algorithmic ways. It is not that other routes are inconceivable, it is rather that when the user chooses to take another route, the GPS prescribes a route for *that* journey, and it can do so for others ad infinitum. This mundane mode of prescription brings together pathways affording travel into an account of a singular, if ever-changing, route to be traversed during the journey. Considered in this way, a GPS is a distinctive kind of scalar device (Ribes 2014) whose legibility also requires a reading of the environment "out there," and both the device and the environment deploy established infrastructures composed of signs, signals, roads, walkways, and so forth. Occasions, such as those described above, when there is a mismatch between these two systems of legibility provide opportunities both to reflect upon the distinctive properties of the GPS *and* upon the legibility of the infrastructures in which its use is embedded. The GPS does not by itself close the gap between map and journey—indeed, as we have seen, it opens up as well as closes; it reproduces and reduces the legibility that is already present in the built environment.

Instances of getting lost instructed us that the infrastructural mashup is not a seamless web—there are many "infrastructural seams" (Vertesi 2014) to stumble over, and sometimes to exploit, in order to form local pathways and linkages in the course of a journey. And, now that it is far from novel, the GPS itself is a resource at the back end of a multiplicity of location services such as Uber, Foursquare, and so on. These services are parasitic on the GPS, just as it is itself parasitic on older forms of road maps, occasion maps, and verbal directions. GPS also is parasitic on orbiting satellites and cell phone towers that it taps into, even though they were not purpose-built for it.

While the word "parasite" has negative connotations (tapeworms and leeches, etc.), it also has broader theoretical relevance to constructive systems and relationships (Serres 1982; Brown 2013). Our reference to "parasitic infrastructures" thus means a borrowing of existing infrastructures for uses that differ from their dedicated purposes. Accordingly, we are treating infrastructures as manifestations of human ingenuity for instructed action in a built environment. Parasitic infrastructures are pervasive in this sense. In line with Winthereik and colleagues' conception of "third wave infrastructure studies" (this volume), which focus on "experimental" borrowings from existing infrastructures to exploit their unanticipated uses, we suspect that many other information infrastructures are parasitic in the way they create infrastructural mashups, which need to be continuously recombined and worked upon in the course of achieving their dedicated purpose. To pursue this and other suggestions we have made in this chapter, we encourage our readers to use this chapter as a field guide for further exploration.

Acknowledgments

The authors are grateful to Yanni Loukissas, Shreeharsh Kelkar, and two anonymous reviewers for their helpful comments on earlier drafts of this paper.

Notes

1. The course (Science & Technology Studies 6251, Visualization and Discourse in Science) was taught by Michael Lynch at Cornell University in spring 2014. Ranjit Singh, Jessica Price, and Christopher Hesselbein were enrolled in the course. The professor and students performed the exercises in preparation for and in some instances as part of each weekly seminar.
2. Other assigned exercises involved performing classic experiments from published descriptions, following recipes for preparing a dish or making beer, tying knots described in diagrams (such as a Windsor knot for formal men's attire), and many others.
3. Interpretations of Ludwig Wittgenstein's (1953) treatment of actions in accord with rules provided a philosophical starting point for these investigations.
4. Further readings also were recommended and were used in both the design and interpretation of the exercise and the experiences associated with it: Psathas (1979, 1986), Psathas and Kozloff (1976), Wallace (2004), Garfinkel (2002, 197–218), Hutchins (1993), and Liberman (2013). Many other sources were brought into play during our discussions and are cited in this chapter.
5. Such a journey from A to B appears at first glance to highlight the elements of the "unit act" formulated by Talcott Parsons (1937, 43–48) as a sociological parallel to the Newtonian conception of motion. The essential elements of the unit act include an *actor*; a clear-cut *end*; a choice of *means* to efficiently realize that end; and culturally prescribed *normative considerations*, which encourage or deter the choice of one or another possible means to that end.
6. See Ziewitz's (2017) account of the troubles occasioned by walking in a way guided by a simple algorithm.
7. Such barriers and prohibitions are by no means limited to modern cityscapes. One of the examples Bourdieu (1977, 37–38) uses of *habitus* is a worn pathway through a village that generations of villagers have used and beaten into the landscape with their feet. When following a footpath, one frequently finds normatively sanctioned and unsanctioned (or specifically prohibited) pathways. Even with footpaths through natural parkland, there are signs stating the imperative to keep to the path, and at strategic points there are fences, barriers, and warning signs (complete with threats of arrest), which are contravened by improvised paths that cross the barriers into the forbidden zones. In a town or city, a material infrastructure of paved roads, curbs and other barriers, road signs, pedestrian directions, kiosks with city and campus maps, and so on leads the way with an integrated set of possible pathways. The conditions of possibility for the journey depend on the ways in which instructions for getting from A to B are framed, and as discussed later they are also limited and afforded by an infrastructure of constructed pathways.
8. Garfinkel (1967) famously devised exercises to deliberately *induce* trouble by disrupting familiar interactional routines. The troubles solicited through the exercises discussed in this chapter use a less disruptive strategy (and one more compatible with IRB requirements) that Brown and Laurier (2012) effectively use in their study of "normal natural troubles" arising through GPS navigation.

9. Our treatment of older navigational aids, as well as contemporary gadgets and apps, as "technologies" reminds us of a remark by a student of Janet Vertesi's who was new to science and technology studies: "I never thought of the bicycle or the car as technology" (Janet Vertesi, personal communication).
10. Not too long ago, when customers of the American Automobile Association (AAA) sought advice on planning a journey, an employee would use a marking pen (often with a bright color such as magenta) to trace a route on a road map. Nowadays, online direction maps retain that convention by marking a route with a similar magenta (or other brightly colored) line.
11. This term is a transformation of Bruno Latour's (1990) notion of "immutable mobile"—a map (or other rendering)—that is inscribed in a fixed medium and can be transported and reproduced without (at least in principle) changing its features. In contrast, a map at a kiosk that includes a singular reference point to "here" cannot be moved from its place without requiring erasure or revision of the deictic expression "here."
12. Kevin Lynch (1960) includes numerous examples of such maps in his classic treatment of the city. The London Underground map discussed by Vertesi (2008) is an interesting hybrid of an occasion map and a road (or in this case rail line) map. The Underground map is like a road map in its layout of a complete array of possible routes, independent of any particular journey, but like an occasion map it displays underground lines in a stylized, non-topographical way.
13. We use the term "Turing machine" here in the contemporary popularized sense of a "universal" machine that can be configured (programmed) to perform a broad range of tasks once associated with specialized human jobs and technologies: calculators, typewriters, word processors, and so on.
14. McCormick (2013), in a discussion of "mobility and the city" with smartphones in different international contexts, observes, "What is baffling, often times across class divides, are the ways in which our actual physical location becomes rendered on digital interpretations of space: on a colored screen, with a pulsing blue dot representing ourselves. This logic, portrayed through the cartography of services such as Google Maps, can be incomprehensible to someone who lacks the necessary literacy to read, interact, and decipher maps. This can then recast the physical-spatial representations we all have in our minds with the visual and experiential images we come to interact with in the city." He goes on to observe that the difficulty some friends of his in Cairo had in following directions he gave them in terms of such features on the screen "foretells an altered way of learning, being, and moving in the city. These virtual representations of our physical environments are like an electronic guide, to be followed on our screens, as we step over curbs, through traffic, and around corners, all the while connected and existing in space in a different way." We are grateful to Shreeharsh Kelkar for alerting us to this article.
15. This is akin to the gap Vertesi (2008) identifies between the London Underground map and the above-ground experience of surface travel in that city (itself often guided by maps such as the London A-Z).
16. From a phenomenological point of view, the GPS screen presents the user with a "small" world embedded in a more comprehensive lifeworld. But from an engineering point of view, the system in which that screen is embedded is "global" in scope and the user's situation is but a "micro" node in the information network. In the case of GPS devices, the visible interface on the small screen is a product of an invisible infrastructure of cell phone towers and satellites, together with the labor force that constructs and maintains it.
17. Repair in conversation was first discussed by Sacks et al. (1974, 723) and subsequently developed in conversation analysis. For development of the theme in the context of STS, see Sims and Henke (2012).
18. This turn in the exercise was analogous to some of Garfinkel's (1967) "experiments" that involved violations of normal routines (such as bargaining for prices of goods purchased in a supermarket) that revealed that seemingly inflexible rules were more open to negotiation than had previously been imagined.
19. See Liberman (2013, 62) on the confident beginnings of trips guided by occasion maps. In such cases, the journey starts from a known place and proceeds to an unknown destination, and the confidence tends to wane as the terrain becomes unfamiliar and the map directions become more difficult to correlate with features of the terrain.
20. See Vertesi (2008, 18) on variations of this convention in hand-drawn maps of London.
21. Brown and Laurier (2005, 18) recite another story (from Holub 1977) that can be positioned as a counterpoint to what Latour mentions about Russians invading Czechoslovakia. In this story, a

detachment of Hungarian soldiers sent into the Alps encountered a snowstorm, but the soldiers managed to find their way back from the wasteland. When asked how they achieved this navigational feat, they pointed to a map one of the soldiers had in his pocket. Upon closer inspection, however, it turned out to be a map of the Pyrenees. Accordingly, the map projected its legibility onto the blank canvas of the Alpine wasteland, inspiring and enabling the troupe to go on with its journey.

Works Cited

Agre, Philip, and David Chapman. 1990. "What Are Plans For?" *Robotics and Autonomous Systems* 6:17–34.

Bourdieu, Pierre. 1977. *Outline of a Theory of Practice*. Cambridge: Cambridge University Press.

Brown, Barry, and Eric Laurier. 2005. "Maps and Journeys: An Ethno-methodological Investigation." *Cartographica* 4 (3): 17–33.

——. 2012. "The Normal-Natural Troubles of Driving with GPS." In *Proceedings of the SIGCHI Conference on Human Factors in Computing Systems*, 1621–30. New York: ACM.

Brown, Steven D. 2013. "In Praise of the Parasite: The Dark Organizational Theory of Michel Serres." *Porto Allegre* 18 (1): 83–100.

Collins, H. M. 1974. "The TEA Set: Tacit Knowledge and Scientific Networks." *Science Studies* 4:165–86.

——. 1975. "The Seven Sexes: A Study in the Sociology of a Phenomenon, or the Replication of Experiments in Physics." *Sociology* 9:205–24.

——. 2010. *Tacit and Explicit Knowledge*. Chicago: University of Chicago Press.

Dreyfus, Hubert. 1979. *What Computers Can't Do*. New York: Harper & Row.

——. 1992. *What Computers Still Can't Do: A Critique of Artificial Reason*. Cambridge, MA: MIT Press.

Garfinkel, Harold. 1967. *Studies in Ethnomethodology*. Englewood Cliffs, NJ: Prentice Hall.

——. 2002. *Ethnomethodology's Program: Working Out Durkheim's Aphorism*. Lanham, MD: Rowman & Littlefield.

——. 2008. *Toward a Sociological Theory of Information*. Boulder, CO: Paradigm.

Garfinkel, Harold, and Eric Livingston. 2010. "Phenomenal Field Properties of Order in Formatted Queues and Their Neglected Standing in the Current Situation of Inquiry." *Visual Studies* 18 (1): 21–28.

Gillespie, Tarleton. 2009. *Wired Shut: Copyright and the Shape of Digital Culture*. Cambridge, MA: MIT Press.

Goffman, Erving. 1972. *Relations in Public*. New York: Harper & Row.

Holub, M. 1977. "Brief Thoughts on Maps." *Times Literary Supplement*, February 4, 118.

Hutchins, Edwin. 1993. "Learning to Navigate." In *Understanding Practice: Perspectives on Activity in Context*, edited by Seth Chaiklin and Jean Lave, 35–63. Cambridge: Cambridge University Press.

Ingold, Tim. 2000. *The Perception of the Environment: Essays on Livelihood, Dwelling, and Skill*. New York: Routledge.

Jackson, Steven J. 2014. "Rethinking Repair." In *Media Technologies: Essays on Communication, Materiality and Society*, edited by Tarleton Gillespie, Pablo J. Boczkowski, and Kirsten A. Foot, 221–39. Cambridge, MA: MIT Press.

Latour, Bruno. 1987. *Science in Action*. Cambridge, MA: Harvard University Press.

——. 1990. "Drawing Things Together." In *Representation in Scientific Practice*, edited by M. Lynch and S. Woolgar, 19–68. Cambridge, MA: MIT Press.

——. 1992. "Where Are the Missing Masses? The Sociology of a Few Mundane Artefacts." In *Shaping Technology/Building Society: Studies in Sociotechnical Change*, edited by W. E. Bijker and J. Law, 225–58. Cambridge, MA: MIT Press.

Laurier, Eric, Barry Brown, and Moira McGregor. 2016. "Mediated Pedestrian Mobility: Walking the Map App." *Mobilities* 11:117–34.

Leshed, Gilly, Theresa Velden, Oya Rieger, Blazej Kot, and Phoebe Sengers. 2008. "In-Car GPS Navigation: Engagement with and Disengagement from the Environment." In *Proceedings of the SIGCHI Conference on Human Factors in Computing Systems*, 1675–84. New York: ACM.

Liberman, Kenneth. 2013. "Following Sketch Maps." In *More Studies in Ethnomethodology*, 45–82. Albany: State University of New York Press.

Lipartito, Kenneth. 2003. "Picturephone and the Information Age: The Social Meaning of Failure." *Technology and Culture* 44 (1): 50–81.

Lynch, Kevin. 1960. *The Image of the City*. Cambridge, MA: MIT Press.
Lynch, Michael. 2013. "At the Margins of Tacit Knowledge." *Philosophia Scientiæ* 17 (3): 55–73.
McCormick, Jared. 2013. "The Whispers of WhatsApp: Beyond Facebook and Twitter in the Middle East." *Jadaliyya*, December 9. Available at: www.jadaliyya.com/pages/index/15495/the-whispers-of-whatsapp_beyond-facebook-and-twitt.
Medawar, Peter. 1964. "Is the Scientific Paper Fraudulent? Yes; It Misrepresents Scientific Thought." *Saturday Review*, August 1, 42–44.
Merleau-Ponty, Maurice. 1962. *Phenomenology of Perception*. London: Routledge & Kegan Paul.
Merton, Robert K. 1938. "Social Structure and Anomie." *American Sociological Review* 3 (5): 672–82.
November, Valérie, Eduardo Camago-Hübner, and Bruno Latour. 2010. "Entering a Risky Territory: Space in the Age of Digital Navigation." *Environment and Planning D: Society and Space* 28:581–99.
Parsons, Talcott. 1937. *The Structure of Social Action*. Vol. 1. New York: Free Press.
Polanyi, Michael. 1958. *Personal Knowledge*. Chicago: University of Chicago Press.
——. 1962. "Tacit Knowing: Its Bearing on Some Problems of Philosophy." *Review of Modern Physics* 34 (4): 601–16.
Psathas, George. 1979. "Organizational Features of Direction Maps." In *Everyday Language: Studies in Ethnomethodology*, edited by George Psathas, 203–26. New York: Irvington.
——. 1986. "Some Sequential Structures in Direction-Giving." *Human Studies* 9 (2/3): 231–46.
——. 1990. "Direction-Giving in Interaction." *Réseaux* 8:183–98.
Psathas, George, and Martin Kozloff. 1976. "The Structure of Directions." *Semiotica* 17:111–30.
Ribes, David. 2014. "Ethnography of Scaling: Or, How to Fit a National Research Infrastructure in the Room." *CSCW*, February 15–19.
Sacks, Harvey, Emanuel Schegloff, and Gail Jefferson. 1974. "A Simplest Systematics for the Organization of Turn-Taking in Conversation." *Language* 50 (4): 696–735.
Serres, Michel. 1982. *The Parasite*. Baltimore: Johns Hopkins University Press.
Sims, Benjamin, and Christopher R. Henke. 2012. "Repairing Credibility: Repositioning Nuclear Weapons Knowledge after the Cold War." *Social Studies of Science* 42 (3): 324–47.
Suchman, Lucy. [1987] 2007. *Human Machine Reconfigurations: Plans and Situated Actions*. 2nd ed. New York: Cambridge University Press.
Vertesi, Janet. 2008. "Mind the Gap: The London Underground Map and Users' Representations of Urban Space." *Social Studies of Science* 38 (1): 7–33.
——. 2014. "Seamful Spaces: Heterogeneous Infrastructures in Interaction." *Science, Technology, & Human Values* 39 (2): 264–84.
Wallace, Anthony F. C. 2004. "Driving to Work." In *Modernity and Mind: Essays on Cultural Change*, Vol. 2, edited by A.F.C. Wallace, 17–33. Lincoln: University of Nebraska Press.
Wittgenstein, Ludwig. 1953. *Philosophical Investigations*. Oxford: Blackwell.
Ziewitz, Malte. 2017. "A Not Quite Random Walk: Experimenting with the Ethnomethods of the Algorithm." *Big Data & Society* 4 (2): 1–13. https://doi.org/10.1177/2053951717738105.

Digitized Coral Reefs

Elena Parmiggiani and Eric Monteiro

Lophelia, SpongeBob SquarePants, and Mr. Krabs

In November 2013, a workshop is arranged in a small village on the coast of North Norway. A group of IT and environmental advisors from a Scandinavian oil and gas company (NorthOil, a pseudonym) and experts in marine acoustics from a company selling submarine sensors and from the national Institute for Marine Research (IMR) are gathered to present a new website to the local community of fishermen.

The presenters are very excited, the fishermen are curious. Also a few journalists are in the room to report on the event. The village is small and it is not often the case that something makes the headlines.

The web portal displays in real time the data collected by an ocean observatory, a sensor network installed on the seafloor off North Norway (Venus for short) approximately 20 kilometers off the village, at a depth of 250 meters. The picture of the nearby coastline in the background, it has a very simple layout (see figure 1 for a mockup). It is divided into three columns, with embedded modules showing updated pictures of a portion of a coral reef, a video made with the latest pictures, two graphs (called chromatograms) representing in different colors the concentration of fish and other biomass in the water column, and the values of a few oceanographic parameters such as underwater pressure, amount of chlorophyll, temperature, and direction and variation of the water current—from the seafloor up toward the sea surface. The fishermen are familiar with these parameters: they are similar to those they use to detect the fish in the water. The fishermen, however, usually conduct the measurements from their ships in the opposite direction, from the sea surface toward the floor. Some of them are also pleased to recognize two chromatograms on the web page, which they also adopt to visualize their fish concentration measurements. Upon closer scrutiny, however, they remark that the chromatograms on the Venus web portal are a bit more difficult to understand. They represent a larger area than those they generally scan, making it difficult to identify what is fish, and what is not fish.

A good portion of the web page is used to display images and a video of one specific inhabitant of the seafloor: a coral reef of the *Lophelia pertusa* species, which is very dense in the Venus area. It is photographed every thirty minutes by a stationary camera and the pictures from the two last days are used to build a video to animate the scene. As the animation runs, it shows that there is quite some activity going on around the reef. A sponge lives in the same neighborhood, and NorthOil's IT advisor responsible for the web portal development named it "SpongeBob

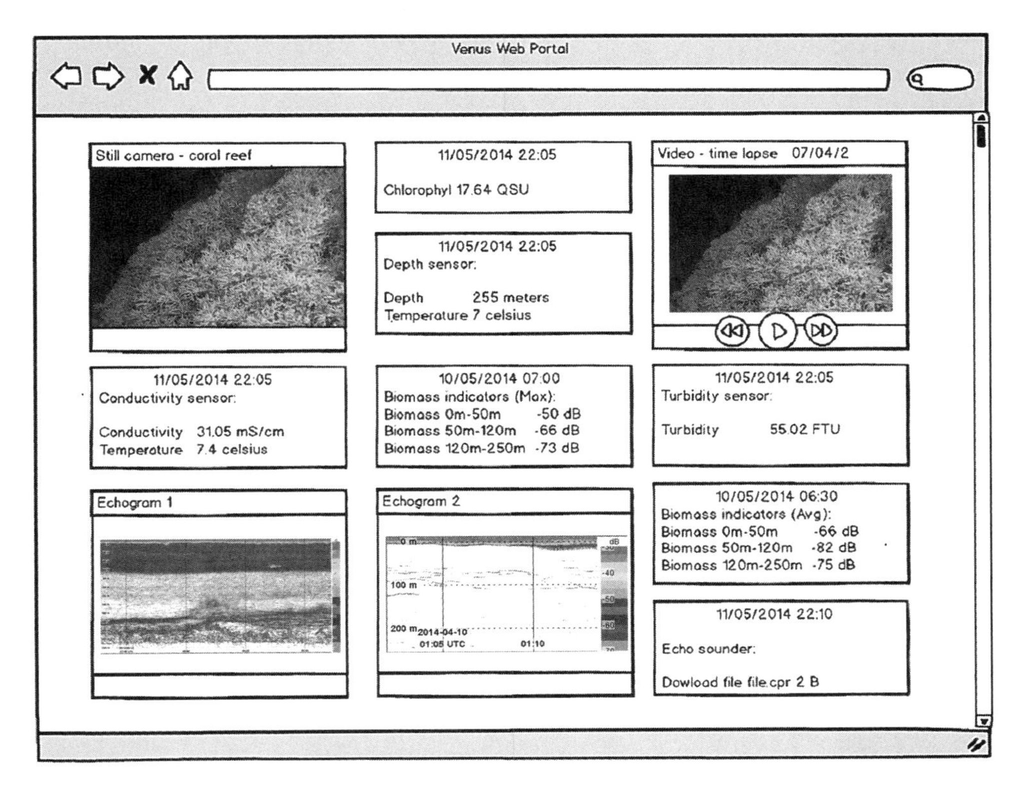

FIGURE 1: An illustrative mockup of the Venus web portal. Source: Authors' own drawing, with www.balsamiq.com. Credit for the coral photos: MAREANO/Institute of Marine Research, Norway.

SquarePants." There is also a crab that crawls all around on top of the reef. It goes without saying that its name is "Mr. Krabs." The reef is also home to a fish, a cod-like tusk. Its name is Bertil, one of the most common male names in Sweden. The IT advisor tells us that the developer team noticed that it comes out of the reef rather often. It gets close to the camera lens, stares at it for a while, and finally leaves. There is also a hydrophone (a marine microphone) in the ocean observatory. The interpretation of the recording has previously revealed that the fish actually speaks to the camera, something that to the human ear sounds like "*Shshshshshsh!*"

One of the journalists from a local newspaper later wrote that the web portal was becoming "more popular than the Disney Channel" and that "for the fishermen the observatory meant way more knowledge about the sea" (reference suppressed for anonymity). The article also reports—perhaps unexpected—very positive feedback from the fishermen community. One of them indeed commented that he wished there were more such observatories in the area.

What is happening in the small village in North Norway is an instance of the process to make a coral *digital*, which we unpack as an example of the *political work of infrastructuring.* Through the initiative and financial capacity of NorthOil, the mismatching agendas of multiple stakeholders of the Venus area are wrapped into the digital infrastructure that underlies the Venus web portal, aimed to generate new real-time "facts" about the unknown subsea life and, indirectly, the possible risks associated with oil operations in Venus—where oil operations are currently forbidden.

Those facts are, nonetheless, anything but neutral. They are digital representations of the subsea resources that enact the geography of the Venus region by maintaining the sociopolitical interests of several stakeholders (Edwards 2010; Barry 2013) while triggering a new order of relevant matters, actors, and interesting creatures to study in underwater environmental monitoring (cf. Latour 2004). This process of geography shaping is a form of *infrastructural inversion*, an "upside-down" perspective on infrastructure aimed to inscribe and support (nonoverlapping) agendas in new infrastructural relations (Bowker 1994).

Our main theoretical contribution is to characterize the concept of infrastructural inversion, specifically by joining the discourse on the politics of infrastructure (see also Winthereik et al., this volume). We do so by discussing how inversion emerges as a political tool that leverages the inherently in-the-making and open-ended status of the Venus infrastructure for different stakeholders with contrasting material interests (cf. Barry 2013; Jensen and Winthereik 2013). In so doing, it postpones quarrels on the environmental and business viability of oil and gas operations in the Venus area and consequently in the Arctic. The politics of environmental discourses in oil and gas is performed in three ways through the digital nature of the Venus technologies. First, the underspecification of the "organizing vision" around the need for technological innovation in the Venus region serves as an early stage tool that mobilizes actors' efforts around (nonoverlapping) broader social, economic, and technical concerns and reduces perceived uncertainties associated with operations in the Arctic (Swanson and Ramiller 1997).

Second, the infrastructuring work enacted in Venus foregrounds evolvability as a fundamental aspect of digital design. Rather than being a threat, the inherently incomplete and evolving nature of the Venus infrastructure is productive of uses and innovations (Garud et al. 2008; de Laet and Mol 2000).

Third, the making of the Venus infrastructure is illustrative of the politics entrenched with environmental data production. Environmental politics is currently subject of heated debates related to the melting of ice in the Arctic because of global warming. The strategy of openness adopted by NorthOil and its partners is an instantiation of the trend of developing quantified indicators of governance and tendency to "replace political debate with technical expertise" (Merry 2011, S83). Apparently paradoxically, openness legitimizes the trustworthiness of the oil company' role in recruiting stakeholders with the technical expertise to generate knowledge of the marine environment in the Arctic region.

Our analysis relates to and extends existing notions in infrastructure studies within science and technology studies and neighboring fields such as information systems focusing on the open-endedness of infrastructures and their generative potential for novel uses and functions (cf. Zittrain 2006). Our characterization of infrastructural inversion in Venus sits on the property of relationality of information infrastructure (cf. Star and Ruhleder 1996): the same infrastructure has very different meanings for different stakeholders. Our analysis also recognizes that changes (e.g., toward real-time subsea risk monitoring) are never solely due to new products (e.g., the Venus web portal), but always to the techno-political work to maintain the infrastructure that permits the development of these products (Star and Bowker 2002).

Our portrayal of the Venus case also highlights significant similitudes of the Venus infrastructure with the concept of boundary objects (Star 1989), or boundary infrastructures (Bowker and Star 1999)—workable solutions that remain ill structured (inconsistent, ambiguous) yet functional to resolve controversies despite lack of consensus. The Venus infrastructure is plastic enough to be domesticated differently in different contexts by means of material and rhetorical relations and for mismatching local purposes and agendas (Star and Griesemer 1989). Adding on to this concept, our analysis portrays a reality related to global discourse of Arctic politics (Mason 2016), where the infrastructure is prominently expansive and open-ended (cf. Jensen and Winthereik 2013), where the purposes of the actors involved are tied to an imagined future that has not materialized yet (the opening or not of the Venus area to oil and gas operations) (Swanson and Ramiller 1997), and where this very underspecification is used as a productive business strategy by powerful industrial players. Furthermore, the current large-scale and long-term evolution of digital technologies specific of the oil and gas sector provides new empirical turf to expand on long-established notions in the STS literature.

This article is based on a three-year ethnographic study (April 2012–December 2014) of a few NorthOil-led initiatives to design and develop an integrated ICT platform for real-time underwater environmental monitoring during offshore operations. The study was conducted as part of the first author's PhD research in information systems (2012–15) that focused on how tools, systems, and methods are developed or adapted and integrated to fit the existing systems and practices of NorthOil (Parmiggiani 2015). This article covers one research question asked in the thesis: how the adoption of specific digital technologies in the monitoring infrastructure reshuffles power/knowledge and order relations within environmental risk assessment in the oil and gas sector. The main data source consisted of participant observations (two to three days a week for three years on average) collected at NorthOil research and development department, during workshops, meetings, and seminars and during visits at the IMR and other industrial partners. In addition, 38 interviews were collected with environmental experts, computer scientists, and engineers from NorthOil, the IMR, and three partner companies. Observations and interviews allowed us to gather internal and public documentation about NorthOil, its partners, and their environmental monitoring initiatives, and norms and laws by authorities. We were also allowed to make use of the software that was used for environmental monitoring or was under development.

The remainder of this chapter is organized as follows. In the next section, we provide a description of the sequence of events that led NorthOil and its partners to invest in real-time environmental monitoring in Venus. We subsequently describe our theoretical basis by fleshing out the aspects of infrastructural inversion that are relevant to the Venus case to disclose the political work of infrastructuring. The chapter then continues by discussing aspects of underspecification, evolvability, and openness through which politics is performed in digital infrastructures. We then conclude by outlining some implications of our analysis for infrastructure studies.

The Politics of *Lophelia*

Lophelia pertusa is a species of cold-water coral that has its widest population on the Norwegian continental shelf (NCS) in at least 9,000 years, living at a depth range of 40 to 400 meters. Researchers estimated that 30% to 50% of the reefs in the NCS have been damaged since the 1980s by the extensive bottom trawling activity of the fishery industry, one of Norway's leading industrial sectors (Fosså et al. 2002). After the discovery of oil in the NCS in 1969, the offshore oil and gas sector became the country's main industrial force, populating the seafloor with an intricate network of rigs, platforms, subsea installations, pipelines, remotely operated vehicles, and vessels. The NCS spans from the North Sea in the south to the Barents Sea and the Arctic region in the north. It is divided into quadrants of one degree by one degree by the Norwegian Petroleum Directorate.[1] Each quadrant is licensed for a few years to a consortium of operators for exploration and—if resources are found—production. There is only one spot on the NCS where no quadrants can be assigned and oil and gas operations are forbidden: the Venus region, offshore North Norway, above the Arctic Circle. It also corresponds to the narrowest point of the NCS. There, not only is *Lophelia* very dense, but in the first half of the year its reefs offer shelter and food to a large flow of cod and herrings that migrate and spawn from south to north and are a vital resource to the fishery industry (see figures 2 and 3).

As a result, *Lophelia* reefs are of extreme value from a biological point of view (Costello et al. 2005). As the vignette with SpongeBob, Mr. Krabs, and Bertil exemplifies, they constitute the centers of varied ecosystems of marine animals, thus giving a good cross section of marine life in an area (figure 4). This point was soon very clear to researchers in marine biology and oceanography: by constantly monitoring one coral reef, access to other species that are more mobile and thus more difficult to track is also possible.

Researchers have adopted different types of measuring devices to study the corals (from biomass sampling to more or less powerful sensors), sometimes to discover that corals are more stubborn than we might think. Coral reefs have for instance grown on the pillars of oil and gas platforms in the North Sea (Gass and Roberts 2006). In the NCS, interest in the corals (in the guise of research funds and attention in the mass media) increased when the oil and gas industry developed its offshore operations enough to get very close to the reefs. As a result, knowledge about their genesis and behavior has evolved in parallel with the technical capacity and economic interest brought about by that industrial sector (cf. Mukerji 1989). Increased knowledge has also made areas of non-knowledge more visible (cf. Beck 1992): whereas a major oil spill would likely cause irreversible consequences to the marine ecosystem, it remains uncertain how routine operations might affect *Lophelia* and fish spawning over the long term (Blanchard et al. 2014).

Such uncertainty has left ample space for politically charged debates about the Venus region. It is still closed to operations, but the Norwegian political parties have recently been engaged in harsh discussions on the pros and cons of an opening. The area is in fact the crossroads of contrasting interests by several stakeholders, whose positions we attempt to sketch here.

First the oil and gas industry in Scandinavia, where NorthOil is one of the major operators, is aware of the possibility of finding resources to extract in the sub-Arctic area, as current calculations estimate that roughly 25% of the world's undiscovered fossil resources are buried in the Arctic region (Bird et al. 2008). Due to the 2015 drop in oil prices, some operators have been motivated to develop more cost- and time-efficient solutions to conduct their daily activities in harsh environments like the Arctic. NorthOil has been particularly involved in doing so since the mid-2000s, through initiatives to develop platforms for real-time subsea environmental monitoring in the Venus area.

FIGURES 2 AND 3: Left: map of the areas on the NCS licensed to oil companies, including announced licenses. Credit: Norwegian Petroleum Directorate. The Venus area is the only one where oil and gas operations are prohibited (approximately stretching the portion of sea between Bodø and Tromsø). Right: map of the *Lophelia pertusa* coral reef structures. Credit: MAREANO/Institute of Marine Research, Norway.

FIGURE 4: Fish and other species around a coral reef. Credit: MAREANO/Institute of Marine Research, Norway.

A second cardinal player in the Norwegian offshore industry is the fishery. Fishermen are concerned that a further expansion of oil and gas traffic on the NCS (especially with the extensive use of seismic signals to monitor subsurface reservoirs) might irreversibly subvert fish populations. Given this premise, one might expect harsh debates. Surprisingly, our study showed a smooth collaboration between the two players in the Venus area. On the one hand, especially in its early days, the oil and gas sector has undoubtedly benefited from the existence of knowledge and technology centers developed in the main Norwegian cities around fishing activities (Andersen 1997). Moreover, the fishermen have a (often unacknowledged) situated experience of the sea waters that might prove useful, for instance when deciding where to deploy a sensor-based environmental observatory or to get feedback on the quality of representations such as the chromatograms. On the other hand, a portion of the oil and gas revenues in Norway is invested by the authorities to support the development of the fishery sector.

A third important stakeholder on the NCS is the community of research institutions—represented by the aforementioned IMR—that are extremely concerned about the possible impact of an opening of the Venus area (Blanchard et al. 2014; Hauge et al. 2014). The IMR sits at the opposite end of the spectrum of stakeholders compared to NorthOil: it is involved in the Venus web portal to show that operations should *not* be allowed in the Venus area. Nevertheless, the IMR is renowned worldwide for its knowledge and expertise in marine biology. As an ally for NorthOil, it has the double effect of boosting the scientific work and the Venus initiative's credibility in the eyes of public opinion. Finally, the relationship between the oil industry and the public in Norway is a complex one. The Labor government in charge from 2005 to 2013 had a strong international environmental agenda promoting initiatives to decrease emissions of CO_2. For the international media, however, these initiatives clash with the fact that the Norwegian state has established the world's largest pension fund with the revenues and taxes generated by the oil and gas activities on the NCS.[2] In connection with the political elections in September 2013, harsh discussions emerged around the possibility of opening the Venus region to oil and gas activities. The public's environmental concerns were voiced by the IMR itself, stating that modern quantitative risk assessment procedures do not allow for a thorough understanding of the effects not only of a major accident, but also of routine operations (Blanchard et al. 2014). Researchers at the IMR have argued that sources of uncertainty depend on (possibly chronic) lack of knowledge about the subsea environment: "Uncertainty cannot fully be quantified when facing ignorance—what we do not know, and even further: what is beyond our conception of what is possible" (Hauge et al. 2014, 87). However, uncertainty is not only a matter of lacking knowledge (possibly to be solved with better methods and instruments), but is also a question of sociopolitical relations among the stakeholders listed above. In particular, issues of reputation—or, conversely, credibility—are at stake when knowledge about the seafloor is promoted by oil companies that are also investing billions of dollars to search for hydrocarbons to drill and produce. The mass media are well aware of this unstable equilibrium, especially around the theme of Arctic drilling. For example, in the United States, Royal Dutch Shell spent large sums of money to seek permission to operate offshore of Alaska, facing stark opposition from environmental organizations: "Shell is determined to drill for oil in the Arctic this summer [2015] if it can win the permits and overcome legal objections, although the energy company accepts *it will never win a battle with environmentalists over its reputation*" (Macalister and Carrington 2015, emphasis added). What we observed in Venus, however, was not public controversy inevitably ending with a loss of reputation for NorthOil. In contrast, the situation of general uncertainty, or "ignorance," turned out to be fertile terrain for both NorthOil and its stakeholders (IMR included) on both the material *and* social levels. We find that this uncertainty is carefully maintained by means of an ensemble of varied organizational techniques to generate a knowledge base about an area that is rather unknown to marine scientists and to create momentum in case of an opening of the Venus area to operations. These techniques are, as we shall see, what Bowker (1994) calls an infrastructural inversion. They consist of developing or repurposing instrumentations and information systems, approaches to gather and interpret the datasets, and, *simultaneously*, methods to weave them into the agenda of politically and scientifically strategic stakeholders by means of the digital infrastructure beneath the Venus web portal. The new information infrastructure for subsea environmental monitoring in Venus, therefore, unfolds as an ongoing process of building relationships among social and material elements, rather than as an innovative combination of artifacts and IT systems (Carse 2014, 5).

Subsea Environmental Monitoring in the NCS

What is the path that led to the current state of affairs in the Venus region? The oil and gas companies operating on the NCS have always been required by the Norwegian government and the Petroleum Safety Authority to carry out underwater environmental monitoring. Monitoring practices largely impinge on the expertise and experience of disciplines like marine biology, zoology, and oceanography, of which the IMR is a well-established knowledge custodian.

What parameters to measure has generally been a choice spurred by the authorities' requirements. *How* to measure varies greatly depending on the area and the operational phase (resource exploration, drilling, extraction). Whereas for some monitoring programs very detailed and established guidelines are issued, in some cases even the parameters are decided by third-party surveyors to fulfill generic demands. In sum, subsea environmental monitoring during the oil and gas operations has so far been a largely offline task, where experts from third-party risk assessment companies would go offshore with boats and, for example, collect samples of the benthic flora and fauna on the seafloor, track the amount of sediments generated by technical operations, and measure the concentration of particles in the water and the pH of the water. In the case of corals, the common practice is to sample small portions of a reef and take pictures and videos. The datasets resulting from these campaigns are stored on different hard disks, taken to onshore analysis centers, and managed by different personnel. Nine to twelve months generally separate the measurements from the publishing of the results. In the words of many environmental experts we interviewed, however, these fragmented and slow methods are inherited from the discipline of marine biology, suffering itself from problems of poorly standardized units of measure, lack of best practices, and fragmented data management and analysis procedures. One environmental chemist commented, "It is still a mess!"

In the early 2000s, the Norwegian Petroleum Safety Authority in collaboration with the Directorate for the Environment published new health, safety, and environment regulations for the oil and gas industry.[3] In particular, the companies operating offshore on the NCS are required to perform *remote* and *continuous* environmental monitoring. Those qualifiers, *remote* and *continuous*, were generic yet promising drivers of a change in the state of affairs. They have taken different forms, largely depending on the data transfer and storage technologies available in an offshore area. The technologies environmental monitoring must rely on are seldom task specific; their availability is rather driven by the need to perform more efficiently other operations, like drilling a well or producing oil or natural gas.

In the same period, governance also targeted technological innovation. In the aftermath of the Enron scandal, solutions were demanded to foster accountability and traceability of operations (Jarulaitis and Monteiro 2010). In addition, forecasts of a decrease in oil and gas production were worrying investors. Triggered by the need to operate deeper fields, elaborate more efficient work processes, and be accountable for its operations, the Norwegian Oil and Gas Association began to push toward a large technological innovation program for oil and gas operations in the NCS (Norwegian Oil and Gas Association 2005). Subsea fiber-optic cables to transfer real-time data from the offshore installations to the data centers onshore were no longer science fiction. Integrated videoconference and file sharing systems were also implemented for quicker communication between offshore and onshore personnel.

Despite these changes, environmental monitoring remains a largely slow and offline task. As a result, offshore environmental risk tends to be perceived as something to address ex post, a concern that is far in time and space from the general public.

The Venus Laboratory: Lophelia Goes Live

The Venus project represents a shift of paradigm. Through the web portal the seafloor is visible digitally, online, and is publicly accessible from a computer screen. The closer and quicker availability of environmental data goes hand in hand with epistemic questions, an attempt to let environmental monitoring evolve as a science within the oil and gas context. In other words, the Venus area acts as a laboratory where new science is constructed (Latour 1999a; Bowker 1994). Efforts are concentrated on a very situated natural and social reality (a portion of the sea offshore North Norway) to create the only infrastructure in the area and to let it have a global impact through a publicly accessible web portal (Parmiggiani et al. 2015).

Behind the Venus web portal is a long-term attempt by NorthOil to become infrastructural in doing scientific work in the Venus area. In fact, the story of the Venus laboratory does not begin with the workshop in North Norway. It goes back to at least the mid-2000s, when a group of NorthOil's environmental experts joined in an EU-funded project to study the wide but unknown populations of *Lophelia pertusa* of the Venus area. The project was headed by European research institutions and universities. NorthOil was the only oil and gas company involved. The initiative looked promising, so NorthOil's environmental experts organized to obtain the approval of the company to start a new project to study some aspects of the coral reefs that had become relevant during the EU project.

The way the Venus project performed the subsea environment was significant on the material and political levels. The Venus area was free from oil and gas activities, so it proved a good playground to study the behavior of *Lophelia* away from operations. As stated by one of the project managers whom we interviewed in 2013, the objective was to obtain a reference long-term trend of the environmental behavior in order learn if and to which extent the corals might be affected by activities. The reason for this focus was that the pipeline installations and well drilling operations were getting too close to *Lophelia* reefs: "The corals have started to be a concern maybe 10 years ago . . . it's been a concern not so much for drilling operations, but for pipeline operations and then like 10–12 years we started moving into areas where we had coral systems close to where we go to drill, and then it became a concern. Then we needed to do something to show . . . to find out whether these guys are sensitive or not for discharges." Since then, the corals have become the target of cyclical research and development projects and initiatives in NorthOil. The primary role that they acquired among the environmental experts in the company is strong in the personification (or anthropomorphization) implied by the expression "these guys"—a perspective that is also embedded in the adoption of the anthropomorphic features of the SpongeBob cartoon.

The monitoring campaigns consisted of deploying small sensor networks supported by a metallic structure (an *ocean observatory*) from a boat down to the seafloor. The sensors were off-the-shelf commercial devices, not very different from those currently sending online data to the Venus portal. The parameters measured were pressure, temperature, and salinity of the water, direction and speed of the current, and amount of biomass (fish and what else was detectable) in the water column. A camera was taking pictures of a nearby *Lophelia* structure.

There is a whole coral forest in the Venus region, but the sensor network available was only one. The choice of the one *Lophelia* reef was not simple, but rather emerged at the intersection of three main issues. First, the coral structure should enable the detection of as much fish traffic as possible. Second, due to its technical characteristics, the support to the sensor network needed a flat terrain. As recalled by one NorthOil environmental chemist, many "nicer reefs" had to be discarded because no place that was flat enough was found to position the sensors. Third, the sea is not silent. It is in fact a very noisy place. As one environmental monitoring expert from a company collaborating with NorthOil told us, in a "very quiet area [where] there are no ships. Also the plankton in the water column dampens the signal." Eventually, the seafloor had to be arranged in a way that fitted the digital technologies available.

In addition, the seafloor had to fit the Norwegian political context. The political elections in Norway were scheduled for 2013 against the backdrop of high uncertainty regarding which party would win. Neither of the two main parties (Labor and Right) could have a majority without forging a coalition with minor parties, all of which have strong opinions on the opening or not of the Venus region. An environmental advisor candidly explained how NorthOil had to develop the Venus infrastructure by postponing any bet on future outcomes:

> There is of course one reason why we are doing this: it is to gather background data for potential future operations. But this depends on a lot of things. It's very much dependent on the elections in Norway. . . . Because the situation is that in the government [Labor party] there is a clear majority for opening this area or at least to go for an impact assessment in the area for the potential opening. . . . But other two smaller parties are against. . . . And then this will also happen on the other side if they win the elections, depending on what is the majority, if they are to bring in those two smaller parties . . . they will go for a ban. . . . We don't know. But in the meantime, we have this observatory and we are going to use it for testing both software and also hardware technologies.

Although the Right party finally won the elections, as of 2017 and a few months away from another election round, the discussion on the opening of Venus is still subject to heated debates in the Norwegian parliament.

In addition, the seafloor was made to fit the social context in North Norway. None of these initiatives was ever carried out by NorthOil and its environmental experts alone. The company was always accompanied by a network of visible allies in the Norwegian context: not only the IMR and technology vendors for their expertise on subsea monitoring, but also groups of local fishermen with experience of those waters.

The IMR representatives whom we interviewed were strongly against opening the Venus area for oil and gas operations. Nevertheless, the institute plays a particularly significant role: on the one hand, it has a vast knowledge base of marine biology and oceanography; on the other, the scientific authority that the IMR embodies shifts the attention away from the political stakes involved in the Venus project. As a result, it is essential to NorthOil's agenda, as an environmental expert explained: "[The Venus project] is an initiative that is promoted strongly by the [IMR]. . . . And also because our ambition is that in order to be able to do a good job if it is opened, we need to know as much as possible about the natural conditions. . . . If it is opened it will give us a much better basis to operate safely in the area." The first occasion to enroll the fishermen came with the cruises to deploy the ocean observatory on suitable points of the seafloor. Renting a boat for an offshore campaign is extremely expensive: when fully equipped, it might cost around 1 million Norwegian kroner per day. The environmental monitoring project in Venus did not have sufficient funds at the time, so it was decided to ask the fishermen for help with the trip to the offshore location and the deployment of the sensor network.

Initially, data were stored on hard disks placed on the same support as the sensors and later retrieved and analyzed. To that point the monitoring activity was not so different from the traditional approaches, except for one detail: all this effort was made for monitoring an area where NorthOil was *not* operating. It could therefore rely on only limited funding. In connection with the escalation of the political debate around the opening of the Venus area, in 2011 an Arctic-oriented initiative promoted by NorthOil's production and development department decided to finance the Venus monitoring projects. The outlook on the business relevance of the project was now official, at least internally. The additional support served to pay for the installation of a fiber-optic cable from the Venus ocean observatory to a small data center on the coast of North Norway where online data would be stored. That step increased attention on the here-and-now relevance of the initiative. The fishermen were again enrolled for helping to position the cable on the seafloor and to connect it with the observatory on the one side and the data center on the other side. Moreover, the fast data transfer capacity needed to be supported by a fast Internet connection on the ground where the data center was located. NorthOil thus decided to also finance a fiber-optic Internet connection to the fishermen village nearby the data center. The local onshore communities were now part of NorthOil's infrastructure.

In the summer of 2013, the cable was connected and data began to flow in real time. What to do with these data remained partly unclear. In the same period, those same participants to the Venus project were also involved in a larger initiative conducted by NorthOil research and development department to design an integrated infrastructure for environmental monitoring in connection with all phases well operations, especially drilling. This latter initiative was however going slov than predicted, so the Venus participants obtained funds for developing a web to display the data from the Venus observatory for a few months—at least until tne larger initiative had delivered the expected results. The IT advisor leading the project explained in December 2013, after the portal went online, "The [Venus] observatory . . . especially the portal side of it, [was] a 'skunk works project,' which means that it's developed in secret by not following the normal procedures . . . that also means that at some stage we will throw it away. Perhaps." The Venus web portal was supposed to be a temporary, lighter tool without any GIS interface but with graphs and pictures to visualize the trend of real-time environmental parameters. Given these premises, developers opted for a Microsoft Azure cloud solution based on standardized technologies. Within three months, the Venus portal was published under a Creative Commons license and publicly accessible online. Its legal notice declares that all material on the website is owned by NorthOil, yet it is available for download for further use or for publishing in peer-reviewed outlets upon informing and acknowledging NorthOil and the IMR.

As soon as the portal went online, it attracted the attention of news media and research institutions. In November 2013, it was presented to the fishermen community. As a consequence, NorthOil and its business partners decided to continue improving it in parallel with the larger initiative. As of mid-2017 both the sensor station and the web portal are active and regularly maintained and improved with new features.

The Politics of Digital Infrastructures

> *To understand infrastructure, you have to invert it. You turn it upside down and look at the "bottom"—the parts you don't normally think about precisely because they have become standard, routine transparent, invisible.*
>
> **—Edwards (2010, 20)**

The Venus web portal, the fiber-optic cable, and the subsea sensors are useful prostheses to access previously invisible locations with short temporal gaps. In this respect, modern ICT has an unprecedented potential to generate information about unknown environments. However, new technologies per se do not solve problems—sometimes they cause new ones related to the management and interpretation of the new datasets. To be effective, they must be coupled with new organizational forms, encompassing new or adapted practices, computational systems, and expertise, which often entail a redistribution of labor. A powerful concept to address the intricacy of these elements is that of information infrastructure (Bowker and Star 1999; Monteiro et al. 2013; Edwards 2010). Within STS, infrastructure is a relational concept (Star and Ruhleder 1996) used to encompass always evolving sociotechnical "systems of systems" that stretch across space and time and are characterized by several users and stakeholders and by a bundle of multiple interconnected systems, tools, work practices, regulations, and agendas (Monteiro et al. 2013). Infrastructure is far from linear, but rather recursive, because local and global scales and perspectives constantly flicker (Jensen and

Winthereik 2013). Although infrastructure is by definition pervasive and distributed, it must be made part of the local context by being entangled with situated practices, which, in turn, determine the evolution of the infrastructure as a whole (Star and Ruhleder 1996; Almklov et al. 2014).

Following from this literature, infrastructure is "an expansive and open-ended but also fundamentally vague ontological entity" where "tools and technologies are never independent of the actors and organizations that promote, maintain, and use them" (Jensen and Winthereik 2013, 12–13). This perspective implies that the *open-endedness* of infrastructure is the very condition for the workability, or inversion, of infrastructure. As such, infrastructure shapes the world it represents in multiple ways at once that are necessarily never neutral. *Infrastructural inversion* is an analytical tool that acts as a gestalt switch to look at the strategies of both researchers and actors in the field to shift the attention from the users/artifacts relations toward the often unrewarded (Star and Ruhleder 1996) but politically meaningful work that sustains and maintains many interconnected, interacting, and distributed infrastructural relations and perform them in different ways in different contexts (Bowker and Star 1999; Ribes and Lee 2010; see also Monteiro et al. 2013). Scholars have proposed to rather speak of infrastructur*ing*, to better account for the continuity, variety, and flexibility that keep the infrastructure going (Pipek et al. 2017; Karasti et al. 2006). Harvey et al. (2016, 3) explain that infrastructural inversion involves a figure-ground reversal that highlights the simultaneous material and political natures of infrastructure: "Rather than assuming cultural, political, or knowledge-based explanation *of* infrastructures, Bowker thus suggested an analytical entry-point via a focus on materiality. . . . Bringing the infrastructural 'ground' up front in this way facilitated understanding of how complex chains of material relations reconfigure bodies, societies, and *also* knowledge and discourse in ways often unnoticed." An infrastructure inversion perspective therefore has implications for the way we intend the politics of infrastructure. Again, in the words of Harvey et al. (10), "If it is assumed that infrastructural development follows a linear, progressive, and rational path defined by governments or corporations, responsibility for the outcome is also placed squarely in the hands of these actors." On the contrary, Harvey and colleagues continue, the open-endedness of infrastructure lets us understand innovation as emergent over time from the interactions among multiple agents with competing and irreducible interests and capacities. According to this view and of relevance to this chapter, Mitchell (2013) tells how global oil infrastructures were not the result of one-way pressures by corporations onto the states, but were co-constructed over more than a century with political formations, democracy, and warfare that made oil the obligatory passage point of the modern lifestyle.

The constantly emergent and open-ended nature of infrastructure that follows from this conceptual framework poses challenges related to understanding how technological innovation unfolds and evolves. Nevertheless, *infrastructural inversion recognizes the productive potential of open-endedness*. As a result, we believe that it is worth considering the ways with which such inherent uncertainty is engaged from the earliest phases and framed in an *organizing vision* by different actors to legitimize infrastructural innovation and tie it to broader business and social concerns (Swanson and Ramiller 1997). The organizing vision is a sensitizing concept that we borrow from the information systems literature and adapt to describe a focal ideal used by the most powerful stakeholders involved in a technological innovation effort to motivate, legitimize, and organize this effort from its earliest stages and tie it to broader concerns (Swanson and Ramiller 1997). The organizing vision is by definition underspecified, and is supported by different—possibly opposite—discourses in different communities. A process of *mobilization* is what makes a vision productive, as it activates and shapes business and social forces that promote the material implementation of an innovation. For the sake of this chapter, we intend mobilization in the sense used by Latour (1999b): a process thanks to which things can be presented in terms that are directly useful in the arguments of different stakeholders. As a result, imagined futures and business strategies are constructed and the perception of uncertainty reduced.

Before we move on to characterizing the facts of open-endedness in the Venus infrastructure, we would like to point out that a perspective based on infrastructural inversion also has a methodological implication that is relevant to the study of the digital. The literature has often discussed infrastructural inversion as a "pair of glasses" worn by the researcher to conduct studies of infrastructuring (Parmiggiani 2015). In this chapter, we show how the same glasses are also worn by the actors observed—the designers, developers, and users of infrastructure. Even though they do not call it an "infrastructural inversion," our assumption is that these actors are asking themselves the same question that we pose (cf. Ribes 2014): how to support the work of innovating methods, instruments, and knowledge (the infrastructure) simultaneously and in the face of uncertainty? Digital infrastructures like the one underlying the Venus portal cannot be treated as big physical artifacts: they transfer bits, an impalpable material that can assume the most diverse meanings (Tilson et al. 2010)—thus making the answer to seemingly simple questions like "*Where* is the infrastructure?" nontrivial, but a simultaneously conceptual and empirical issue. A corollary to this observation is that watching how actors deal with it as part of their daily job may, in turn, sensitize scholars of STS to improve their methods to study the digital.

Taking this conceptual understanding as a point of departure, we characterize the development of the Venus infrastructure as a (ongoing) political negotiation that emerges from its (1) *underspecification*, which is used to mobilize networks of stakeholders to construct matters of concern and their spatial and temporal frame; (2) *evolvability*, which is enabled by digital technologies as they are productive of new uses and knowledge and thus new perceptions on phenomena (environmental risk); and (3) *openness*, which is translated into a discourse on open data and thus an effort to legitimize the role of an oil and gas company as a producer—and not only a consumer—of knowledge about the Arctic environment.

Open-Endedness and the Venus Digital Infrastructure

Underspecification

The legitimization of the need for technological innovation in the Venus area and the Arctic is framed and promoted through the staging of an organizing vision enacted by the most visible stakeholders such as NorthOil and the IMR. It is centered on the production of knowledge about the Arctic environment—which is largely considered unknown ("[The Venus project] is an initiative that is promoted strongly by the [IMR] . . . we need to know as much as possible about the natural conditions";

"There is of course one reason why we are doing this: it is to gather background data for potential future operations"). Therefore, the mobilization of different forces (not only NorthOil and the IMR, but also the fishermen, the local communities, political parties, and the oil and gas industry) happens by arguing in favor of conducting scientific work. Such work, we find, is made possible by organizational work, the ensemble of all the technical and collaborative efforts to create and organize a working infrastructure across different communities of stakeholders.[4]

NorthOil indeed saw the connection between organizational work and the acquisition of a solid knowledge base of the marine ecosystems to achieve its business goals. For the oil and gas company, it was a question of becoming infrastructural to *how* knowledge is built through political work, by letting the natural world of the seafloor and the Norwegian social world converge.

This feature of infrastructural inversion is evident in Bowker's (1994) account of the offstage work of the oil service company Schlumberger in the 1920s and 1930s to support the creation of a whole science of testing and measuring oil fields. In so doing, Schlumberger managed to become one of today's leading companies in its sector, and its methods are still infrastructural to petroleum science. Behind the seemingly value-free scientific measurements, the natural and the social contexts were seamlessly embedded. In Bowker's words, the heart of the infrastructural inversion performed by Schlumberger was that "the scientific work is rendered possible by the organizational work. . . . In the case of Schlumberger, the constant thread was organizational strategy, applied simultaneously to the social and the natural worlds so as to make them converge" (1994, 104). Similarly, Edwards (2010) makes use of infrastructural inversion to describe the evolution of climate science into what it is today. He details how climate scientists in the 1970s had to make global sense of inconsistent and poorly standardized datasets collected from surface stations spread worldwide, by reconstructing the history of the atmosphere and digitizing and interpolating data in many ways. Simultaneously, inversion also indicates the unofficial activity of "citizen science" websites to verify the results of official climate science.

As a consequence, infrastructural inversion does not simply reveal the social construction of facts (Pinch and Bijker 1984)—e.g., what counts as environmental risk in the Venus area?—but also the intrinsically infrastructural nature of these facts: "The difference between controversial claims and settled knowledge often lies in the degree to which the production process is submerged. Thus, *an established fact is one supported by an infrastructure*" (Edwards 2010, 22). The analysis of how the infrastructure is built and questioned (inverted) discloses how "an established fact" is constructed and turned into a public problem. This type of political work is another cardinal element of infrastructural inversion. As Allhutter (this volume) recognizes in her studies of the Semantic Web, infrastructure and epistemic practices coevolve and have economic and political implications. As an analytical lens, inversion helps us to unravel how infrastructure embodies and maintains controversies in this process. The environmental sustainability of oil and gas activities in the Venus region is still in the phase of controversial claims, to paraphrase Edwards's quote. NorthOil's mobilization strategy does not imply that controversies should necessarily be solved. Rather, they are generated and frontstaged by making hidden references more visible. This resonates with Edwards's account: "Inverting the weather and climate knowledge infrastructures and tracing their history reveal profound relationships, interdependences, and conflicts among their scientific, technological, social, and political elements" (Edwards 2010, 22). Ultimately, understanding such connections shows the underspecification of the Venus infrastructure is also linked to the underspecification of our knowledge of nature in the Arctic—as the IMR often reminded us on the Norwegian media. As Geirbo (2017, 23) compellingly illustrates, "attending to infrastructuring as the crafting of connections between people, things, and places may enable us to get 'the permanent imbrication of infrastructure in nature' (Edwards 2003, 9) into view." Following this argument, it is thus critical to take this aspect of underspecification seriously: "Infrastructures fail precisely because their developers approach nature as orderly, dependable, and separable from society and technology" (Edwards 2003, 195, as quoted in Geirbo 2017, 23). Carse (2014, 23) looks at how nature is made part of the infrastructure of the Panama Canal and its rural hinterland: "The boundaries between the technical, social, and environmental are always porous and in flux." From this perspective, the political dimension of infrastructure proves fundamental to understand the unfolding of environmental conflicts, because "infrastructure has also a poetics that shapes how people make sense of the past, the present, and their places in the world, thereby producing moral economies and expectations" (23).

In the case of Venus, the "moral economies and expectations" associated with its opening are not denied or blackboxed, but rather carefully staged and modulated. Conflicts and disagreement are turned into a resource and channeled into successful collaboration with fishermen, external environmental experts, and other stakeholders not around the final ends for environmental monitoring in Venus (operating vs. not operating) but around the need for better methods to represent and measure environmental risk in the area. Such an approach, moreover, acknowledges the need for a type of expertise that is not only distributed and sensitive to temporal concerns, but also politically aware and strategic: "If infrastructures of state and empire were once associated with large public and corporate laboratories, infrastructure expertise is now dispersed across a vast array of businesses, which deal with matters as diverse as risk assessment, quality control, community relations, corrosion, environmental protection, asset management, and finance. These forms of expertise are concerned not just with the current conditions of infrastructures, but also with their future stability. . . . Expertise in infrastructural futures has become a focal point for political opposition" (Barry 2015; see also Barry 2013).

In sum, bridging the underspecification of future use with the construction of infrastructural expertise, infrastructural inversion is a generative resource with which actors generate an organizing vision or conceive of a possible change to "reinterpret the status quo of infrastructure in light of potentiality, thus paving the way for embedding new tools in particular ways" (Kaltenbrunner 2014, 19). Staging the evolution of infrastructure around a vision relates to discourses on how infrastructure has the capacity to connect previously unrelated entities, as techno-imaginative dimensions are created for and by the stakeholders involved (Winthereik et al., this volume). However, the exact mechanisms by which infrastructural inversion acts as a generative resource by harnessing underspecification have been undertheorized. As an effort to fill this gap, we show how inversion *qua* political work is used to quantify environmental risk by leveraging the potential for evolvability of the digital technologies adopted in the Venus infrastructure.

Evolvability

A consequence of the Venus infrastructure being underspecified is its malleability for change based on emerging requirements and stakeholder groups. Openness to evolution is a cardinal characteristic to harness in all infrastructure design, especially against the backdrop of political and financial uncertainty (cf. Garud et al. 2008), as is the case for oil and gas operations in Venus. We observe that the same infrastructure in Venus emerges as different infrastructures for different purposes by different stakeholders, notably NorthOil, the fishermen, the IMR, and the public. The Venus web portal is the tip of the iceberg of this infrastructure and serves as a *fluid* object that is not bounded in use and scope, and is adaptable and flexible enough to serve different needs simultaneously (de Laet and Mol 2000).

As a result, as the infrastructure underlying the Venus portal is inverted and evolves, new infrastructural relations are established so that new knowledge and new risk re-presentations are developed to articulate environmental risk perception differently. What makes this process of infrastructuring environmental risk in the Venus project different in comparison to traditional approaches is that it happens *digitally*: the political work of infrastructuring is tightly embedded into how the digitalization process in the Norwegian oil and gas domain influences the generation of knowledge about underwater environmental monitoring.

We use the term "digitized" in the title to point to the transformation of the coral reefs into bits and then pixels on the operators' screen. However, the whole process of transformation is one of *digitalization*. While "to digitize" is defined by dictionaries as the process of converting analog information into a bit stream, "to digitalize" has a broader connotation and refers to the "sociotechnical process of applying digitizing techniques to broader social and institutional contexts that render digital technologies infrastructural" (Tilson et al. 2010, 749).

Scholars have addresses the performative role of digitalization in modern society in reshaping concepts such as materiality and place (Introna 2011; Kallinikos et al. 2013; Knorr Cetina 2009). Introna (2011) argues that digital encodings are normatively governed material enactments. Kallinikos et al. (2013) problematize the ontological status of digitality and argue that digital artifacts cannot be treated in the same way as physical tools because of their incompleteness, which is simultaneously a problem and an opportunity: "Digital artifacts are *intentionally incomplete and perpetually in the making* (Garud et al. 2008; Zittrain 2008). Incompleteness is both an opportunity and a problem. It is an opportunity insofar as it does not foreclose the range of tasks and operational links an artifact can or might accommodate. But it is a problem as well, in the sense of reducing control over the artifact and its use" (357–58, emphasis added). The evolvability of the Venus infrastructure—it being "*perpetually in the making*"—is epitomized by the way its digital nature changes the perception of environmental risk in Venus. Environmental risk is not a self-standing fact. It is a re-presentation shaped by the infrastructure that is used to measure it, the result of a process of construction that is imbued with (often contrasting) social and political meanings of diverse stakeholders and where scientific analysis and political deliberations entangle (Jasanoff 1999). The way these facts are constructed changes as the infrastructure to construct them is also changing: How do new digitally enabled infrastructures change—if they do—the people/things relationships in environmental monitoring? Do they mean that it is safe to drill because we have good visibility of the coral reefs, or rather that we should not drill? We conjecture that these two questions are interrelated.

The digitalization of environmental monitoring in Venus emerges as an occasion to reframe environmental risk as a different problem than it was in the past, especially focused on a rather unknown area like Venus. We find that NorthOil and its partners' strategies are grounded on the assumption that the construction of measures (facts) of environmental risk is primarily a problem of building a knowledge base. This is explicit in the regulations issued by the Norwegian Ministry of Climate and Environment that point to the need for a "reasonable" baseline for assessing the risks associated with human activity, including fishing, tourism, and oil and gas operations (Norwegian Ministry of Climate and Environment 2009, sec. 8): "Official decisions that affect biological, geological and landscape diversity shall, as far as is reasonable, be based on scientific knowledge of the population status of species, the range and ecological status of habitat types, and the impacts of environmental pressures." NorthOil speaks of this aspect in a positive meaning ("to gather background data for potential future operations"), and so do the fishermen in their accounts to the newspaper ("for the fishermen the observatory meant way more knowledge about the sea"). The IMR is more negative on the matter, speaking the impossibility of quantifying risk in Venus "when facing ignorance." Despite opposite rhetorical constructions, what surfaces is a conception of environmental risk not as a property of the outside world that waits to be discovered, but as the result of our (lack of) knowledge of it (Beck 1992). Quantification is used as a solution to decrease the perception of uncertainty and turn it into a risk that can be measured and, as such, managed (Porter 1996; Power 2007).

The way quantification of the environment is performed has, as we have mentioned above, changed in the last decades. Originally, a lot of human, manual work was involved in monitoring the seafloor during oil and gas activities. Since the early 2000s, however, technological evolution and, importantly, the investments by oil companies have triggered the availability of digital tools and platforms (e.g., web portals, fiber-optic cables), sometimes even in an off-the-shelf, rather inexpensive format (e.g., acoustic sensors). As a result, the agency of human work has been gradually passed to technological artifacts (Ribes et al. 2013). These technologies perform, namely acquire their materiality, *as* they are made part of specific sociomaterial practices over time through daily human organizational work—the very unromantic, unrewarded, and sometimes forgotten examples of infrastructural inversion. For instance, the web portal—seemingly impalpable—is rooted to the Venus infrastructure via both a fiber-optic cable and the fishing practices of the local fishermen.

Such space for experimentation to create knowledge constituted by the Venus project is productive of what counts as relevant risk for the environment. It thus impacts on the spatial and temporal order given to the perception of environmental risk. As recognized by Knorr Cetina (2009), digitalization is performative because it augments and temporalizes situations (e.g., the marine life around a *Lophelia* reef) through scopic components (e.g., the web, sensing devices) so that they become translocal and activities acquire a global order. In this process, digitalization amplifies the claims of new actors such as *Lophelia* that, through spokespersons like environmental experts, ask to become a relevant part of an ordered collective of humans and nonhumans (Latour 2004). Through the work of the environmental advisors first and their colleagues afterward, this due process for *Lophelia* to enter the realm of oil and gas has been delegated to the Venus web portal. Continuous organizational work to build a network to sustain it is conducted by deploying the first ocean observatories in externally funded research projects and later by obtaining funds from a core division in NorthOil to install the fiber-optic cable, networking with diverse and competing partners, arguing in favor of the implementation of a web portal, and implementing the portal *under the radar* (it was a "skunk works project . . . it's developed in secret"). The evolution of environmental monitoring practices has, in this way, become performative of what counts as environment in Venus. It raises the question, *why the corals?*

First of all, *Lophelia* literally sits in the area of interest of the largest national industry. Also, in addition to being very dense on the NCS, it is a good scaling tool: by spotting one coral structure with the subsea camera, you get the sponge, the crab, and the fish. In a sense, corals are natural infrastructures themselves. Moreover, they are beautiful. They are an example of Bowker's (2000) charismatic fauna, one of those species that are more likely to receive the attention of policy makers and research funding for studying their protection. The business side of infrastructural inversion also goes through leveraging the most appealing aspects of nature, which in turn shape the material properties of the end-user artifacts. In this respect, corals do well in front of the camera. The environment is thus reconstructed at the intersection of the technology available, the capabilities of that technology (*How can you measure?*), and the political and social interest in the Norwegian setting associated with what is close to the oil and gas fields (*What do you want to measure?*). How these portions become visible is a question that depends on the technological advances and the business interests of the oil and gas industry.

Our analysis also indicates that the specific materiality acquired by elements of the environment through digital technology has consequences for how we experience the world (cf. Carlile et al. 2013) as it becomes part and parcel of our infrastructure (Carse 2014). As the tip of the iceberg visible to everyone, the Venus web portal becomes the site for engagement with the environmental problems and produces new "facts" related to environmental risk in North Norway. The increasing attention to corals generates a network effect where more knowledge about the corals and the surrounding species is required and built. As a consequence, those elements are perceived as more at risk, whereas what is invisible is increasingly excluded (cf. Bowker 2000). An environmental expert comments, "We have started [to monitor] the corals because the authorities had a large focus on corals, but then afterwards the sponges have also become a type of organism that has received focus." SpongeBob SquarePants has become interesting as a result of hanging out with the right buddies.

The evolvability of the Venus infrastructure also has consequences for the perception of the temporal and spatial framing of environmental risk. As environmental monitoring becomes digital matter, it makes an online and public subject out of an offline and slow practice, and thus displays new temporal and spatial dimensions through which the subsea environment and the associated risks are perceived (cf. Knorr Cetina 2009).

First, the Venus web portal turns subsea environmental risk into something immediate, real time. As the temporal gap between the collection and publishing of the environmental data is diminished to zero, environmental risk is no longer a long-term effect, but something very visible and very close to potentially everyone. From NorthOil's end, this opportunity has business relevance: it is fundamental to demonstrate to the authorities that the company has a very timely overview of the state of the environment. The argument is straightforward: if permission to drill a well is granted to NorthOil, the company will be able to stop at the earliest sign of possible environmental damage or disturbance. The IMR has the opposite argument, which is also straightforward: the natural environmental in Venus should be preserved from oil and gas activities because "there is no exact science" behind the numbers that are used to measure environmental risk and more research is needed (Helgesen and Tunmo 2009). As a consequence, as obscure subsea life becomes so visible and so detailed, the very conception of what is "possible damage" is likely to change. The evolution of the science of marine biology and oceanography is ongoing. One environmental expert from the IMR told us that he is still surprised by the lack of precise reactive procedures in case of environmental hazard, even when a risk was correctly predicted. What if the assessment methodologies become faster as a consequence of the faster availability of data? Will reactive procedures grow more imprecise? The incompleteness of digitality can thus also be a problem (Kallinikos et al. 2013).

Second, environmental risk in Venus acquires a new spatial dimension. The more the Venus environmental monitoring project becomes situated around the portion of seafloor where the sensor network is installed, the more the web portal is visible globally. In this process, both the sea waters and the onshore space are shaped by the inversion operated by the Venus project.

For the web portal to become so easy to present to different audiences, it had to filter out the underlying messy infrastructure and synthetize the complexities of marine biology and the variety of the Venus region (its non-flat ground, its coral forests, its fish traffic) to fit a photograph of a portion of a *Lophelia* structure and the time/depth graph of the surrounding water column. For environmental risk to be assessed in real time, the decades of experience with marine science and the days lost looking for a right position for the subsea sensor network must sometimes be turned into a video that contains SpongeBob SquarePants, a crawling Mr. Krabs, and an angry Bertil. Leveraging popular culture to convey a message is also an art. As Latour has demonstrated (1999a), the complexities of the particular have to be reduced and inscribed into globally recognizable representations for knowledge to be amplified and become shareable with specific external audiences.

Openness and Knowledge Legitimization

The third aspect of infrastructuring work in Venus is enabled by its digital nature and relates to the way it is used to share open environmental datasets in real time. This adoption of openness, insofar as it allows for public scrutiny (Porter 1996), is a political tool in the Venus project to legitimize NorthOil and its partners' efforts to create knowledge bases in the area.

The re-presentations of environmental risk in the Venus project are molded by several technologies and strategies that are significant for being open—an aspect that does not resonate with the secretiveness and business sensitivity of the oil and gas realm. The Venus web portal materializes as a combination of an open architecture (it is based on nonproprietary modules) meant to accomplish a "skunk works project" and the availability of freely available datasets. It is exactly this openness that NorthOil leverages to face the still uncertain future possibilities in Venus: it responds to open-endedness with openness. This strategy, we find, is aimed for NorthOil to manage criticism, be central to the creation of knowledge in the unknown Arctic territory, and shift the competition of bigger oil companies away from financial capacity and toward the realm of knowledge.

First, upon a superficial scrutiny, a strategy of openness might seem at odds with NorthOil's agenda to obtain a permission to operate in the area. In fact, it is a move in the direction of legitimizing the company's position: anyone can see what NorthOil is seeing and doing down there. The Venus web portal thus acts by inviting and anticipating potential criticisms to NorthOil's approach of including everyone in the conversation. As a consequence, critique is not excluded but made visible, managed, and channeled (Barry 2013). The incompleteness of the digital infrastructure underlying the Venus web portal thus stimulates an emergent and cost-efficient way to avoid damage to reputation and produce new knowledge by inviting external parties to conduct research on NorthOil's data. By framing risk as a problem of (lack of) knowledge, rather than a danger associated with oil and gas activities, it elicits the contribution of external parties as stakeholders: the IMR, the fishermen, the general public, and the local communities close to the Venus observatory. For instance, the need to reduce this ignorance despite opposite agendas is what brought together NorthOil and the IMR. A strategy based on managing controversy (rather than achieving exact consensus) is also advocated by the IMR: since to decide whether to exploit or conserve the Venus area is a matter of incommensurable value between different stakeholders defending different socioeconomic benefits, controversy might be a useful instrument to explore heterogeneous views and let possible options surface (Olsen et al. 2016). In this context, a relevant choice in the direction of dampening the tensions toward NorthOil as an oil and gas company is the very choice of Venus as the place for a public environmental monitoring initiative. Venus is indeed a "clean slate." No oil and gas–related dirt is visible there, making it easier to emphasize the aforementioned epistemic problems and the charisma of nature.

Second, although the sharing of open datasets acts as an invitation for external parties (e.g., citizens) to participate in the creation of knowledge about the Arctic, NorthOil (and its partners, too) uses the Venus initiative to be a necessary player in the development of a science of the Arctic, encompassing suitable measurements and classification mechanisms. In this way, the methods to conduct scientific work (Bowker 1994) and produce theories and knowledge about the Arctic environment are centered on NorthOil and its partners, ultimately putting them at the center of the creation of imagined future possibilities for environmental management in the Arctic (Mason 2016). In the case of NorthOil and its partners, openness is, somewhat paradoxically, proving a useful strategy to concentrate the methods of knowledge creation in the hands of the earliest and most powerful incumbents, as observed by Sally Engle Merry (2011, S85) talking about the adoption of statistics in modern states:

> Statistical measures have embedded theories and values that shape apparently objective information and influence decisions. Despite the increase in democratic openness produced by the use of statistics in decision making, this is a technology that tends to consolidate power in the hands of those with expert knowledge. In many situations the turn to indicators as modes of governance does not eliminate the role of private knowledge and elite power in decision making but replaces it with technical, statistical expertise. Decisions that were carried out by political or judicial leaders are made by technical experts who construct the measures and develop the processes of classification and counting that produce the numbers.

Third and last, NorthOil's strategy of openness sheds light on the fact that the political work of infrastructuring is entwined with the relation between infrastructure and power. NorthOil's approach makes sense in the specific Scandinavian context because a tradition of openness is something that oil and gas operators are accustomed to in Norway. The Norwegian government requires that all oil companies share on a public database all the datasets resulting from the exploration for new reservoirs on the NCS. In addition, international competition is particularly harsh when it comes to obtaining permission to operate in new blocks leased by the states facing the Arctic region. NorthOil's efforts to promote itself as an open, "knowledgeable," and regulation-compliant company when it comes to not only adopting but also designing the best solutions possible to prevent risks to the Arctic environment should be understood in terms of its size. It is a big oil company in the European context (more than 30,000 employees), but it cannot financially compete with larger American, Russian, or Chinese competitors. Some of these companies have reportedly been in conversation with NorthOil, but proved to have a different perspective on the need to develop new environmental monitoring approaches: "We've been talking to some companies about this [project], and some companies, they do not want to touch this kind of concept because they fear that . . . the government, the regulatory bodies, will adapt it and make it a requirement. So that . . . bringing in new technologies and new ways of working like this might put new constraints to the operators" (environmental advisor, consultancy company, interview).

Conclusions and Implications for Infrastructure Studies

The purpose of this chapter has been to use the example of the Venus infrastructure to explore how oil and gas information infrastructures unfold with reference to environmental concerns. We adopted the notion of infrastructural inversion to showcase how a seemingly simple artifact such as the Venus web portal discloses a politically charged infrastructure in the making where knowledge about the environment is inextricably tied to industrial interests (Mukerji 1989).

Our study has three implications for the research of digital infrastructures in STS.

First, *the digital is political*. We illustrated in particular how digital re-presentations can contribute to perform a new infrastructural order. By recalling SpongeBob SquarePants, the Venus portal showcases the importance of images and metaphors in performing a non-neutral order that makes (different) business sense for the stakeholders involved (Monteiro and Hepsø 2002). NorthOil and its partners make extensive use of appealing representations to promote their goals. Corals are highlighted; other animals (e.g., marine mammals) are left out. The lens of infrastructural inversion might thus help researchers to look at the way metaphors and ultimately organizing visions are used to talk about technology to create a reality and to drive—or prevent—the design and development of infrastructures (Star and Bowker 2002). In addition, we have demonstrated how the inversion of the Venus infrastructure is enabled by open data sharing enabled by modern digital technologies, cloud computing, and open standard specifications. This application of openness is for NorthOil a strategy in and of itself in the background of the highly uncertain political future of the Venus area (Swanson and Ramiller 1997).

Second, our analysis also builds on and ultimately extends existing concepts in the infrastructure literature by tapping into arguments about infrastructural, digital technologies inherently being incomplete and thus underspecified (Garud et al. 2008). The underspecification, evolvability, and openness of the Venus infrastructure that we have discussed take to the foreground the potential for relationality of information infrastructures (Star and Ruhleder 1996). As infrastructure becomes real only in relation to given organized practices, these faces of the Venus infrastructure allow NorthOil to tie it to contrasting, even competing organized practices by establishing intra- and inter-infrastructural relations that are politically meaningful in different ways for the involved stakeholders. In this way, tensions between local and global concerns are resolved because Venus acts as a "global" infrastructure that also affords heterogeneous "local" practices (cf. Star and Ruhleder 1996). As a result, environmental risk becomes an emergent phenomenon that exists only as related to the underlying infrastructure: it is enacted in practice by the different stakeholders and left incoherent and made to temporarily coexist on different levels through ad hoc modes of political coordination (Mol 2002).

Finally, our analysis relates to the notion of generativity (Zittrain 2006). The political perspective has recently been fed into the logic of platforms (see, e.g., Tilson et al. 2010), that is, architectural arrangements that acknowledge open-endedness by accommodating multiple and never prespecified purposes. We believe that it is important for researchers in STS and neighboring fields to dig more into the way platforms are sustained by the political work of infrastructuring, in the sense that their components are artfully integrated through rhetorical strategies that hide whatever tensions exist among constituents (Gillespie 2010).

Notes

1. www.npd.no.
2. More than 8,000 billion NOK as of May 2017 (www.nbim.no/).
3. www.psa.no/framework-hse/category403.html#p48.
4. In this sense, infrastructural inversion can be considered as a generalization of the CSCW concept of articulation work (cf. Schmidt and Bannon 1992), but used to emphasize collaborative and epistemic work that sustains infrastructure and unfolds over time and space (Parmiggiani et al. 2015).

Works Cited

Almklov, Petter G., Thomas Østerlie, and Torgeir Haavik. 2014. "Situated with Infrastructures: Interactivity and Entanglement in Sensor Data Interpretation." *Journal of the Association for Information Systems* 15 (5): 263–86.

Andersen, Håkon. 1997. "Producing Producers: Shippers, Shipyards and the Cooperative Infrastructure of the Norwegian Maritime Complex since 1850." In *World of Possibilities: Flexibility and Mass Production in Western Industrialization*, edited by Charles F. Sabel and Jonathan Zeitlin, 461–500. Cambridge: Cambridge University Press.

Barry, Andrew. 2013. *Material Politics: Disputes along the Pipeline*. Chichester: Wiley-Blackwell.

——. 2015. "Discussion: Infrastructural Times." *Cultural Anthropology*, August 24. www.culanth.org/fieldsights/724-discussion-infrastructural-times.

Beck, Ulrich. 1992. *Risk Society: Towards a New Modernity*. London: Sage.

Bird, Kenneth J., Ronald R. Charpentier, Donald L. Gautier, David W. Houseknecht, Timothy R. Klett, Janet K. Pitman, Thomas E. Moore, Christopher J. Schenk, Marilyn E. Tennyson, and Craig J. Wandrey. 2008. "Circum-Arctic Resource Appraisal: Estimates of Undiscovered Oil and Gas North of the Arctic Circle." Fact Sheet 2008-3049. Menlo Park, CA: US Department of the Interior, US Geological Survey. http://pubs.usgs.gov/fs/2008/3049/.

Blanchard, Anne, Kjellrun Hiis Hauge, Gisle Andersen, Jan Helge Fosså, Bjørn Einar Grøsvik, Nils Olav Handegard, Matthias Kaiser, Sonnich Meier, Erik Olsen, and Frode Vikebø. 2014. "Harmful Routines? Uncertainty in Science and Conflicting Views on Routine Petroleum Operations in Norway." *Marine Policy* 43:313–20.

Bowker, Geoffrey. 1994. *Science on the Run: Information Management and Industrial Geophysics at Schlumberger, 1920–1940*. Cambridge, MA: MIT Press.

——. 2000. "Biodiversity Datadiversity." *Social Studies of Science* 30 (5): 643–83.

Bowker, Geoffrey C., and Susan Leigh Star. 1999. *Sorting Things Out: Classification and Its Consequences*. Cambridge, MA: MIT Press.

Carlile, Paul R., Ann Langley, and Haridimos Tsoukas. 2013. *How Matter Matters: Objects, Artifacts, and Materiality in Organization Studies*. Oxford: Oxford University Press.

Carse, Ashley. 2014. *Beyond the Big Ditch*. Infrastructures Series. Cambridge, MA: MIT Press.

Costello, Mark J., Mona McCrea, André Freiwald, Tomas Lundälv, Lisbeth Jonsson, Brian J. Bett, Tjeerd C. E. Van Weering, Henk de Haas, J. Murray Roberts, and Damian Allen. 2005. "Role of Cold-Water Lophelia Pertusa Coral Reefs as Fish Habitat in the NE Atlantic." In *Cold-Water Corals and Ecosystems*, edited by André Freiwald and J. Murray Roberts, 771–805. Erlangen Earth Conference Series. Berlin: Springer.

Edwards, Paul N. 2003. "Infrastructure and Modernity: Force, Time, and Social Organization in the History of Sociotechnical Systems." In *Modernity and Technology*, edited by Thomas J. Misa, Philip Brey, and Andrew Feenberg, 185–225. Cambridge, MA: MIT Press.

——. 2010. *A Vast Machine. Computer Models, Climate Data, and the Politics of Global Warming*. Cambridge, MA: MIT Press.

Fosså, Jan Helge, Paal Mortensen, and Dag Furevik. 2002. "The Deep-Water Coral Lophelia Pertusa in Norwegian Waters: Distribution and Fishery Impact." *Hydrobiologia* 471 (1): 1–12.

Garud, Raghu, Sanjay Jain, and Philipp Tuertscher. 2008. "Incomplete by Design and Designing for Incompleteness." *Organization Studies* 29 (3): 351–71.

Gass, Susan E., and J. Murray Roberts. 2006. "The Occurrence of the Cold-Water Coral Lophelia Pertusa (Scleratctinia) on Oil and Gas Platforms in the North Sea: Colony Growth, Recruitment and Environmental Controls on Distributions." *Marine Pollution Bullettin* 52:549–59.

Geirbo, Hanne Cecilie. 2017. "Crafting Connections—Practices of Infrastructuring. An Ethnographic Study of Developing a Village Electricity Grid in Bangladesh." Doctoral thesis, University of Oslo.

Gillespie, Tarleton. 2010. "The Politics of 'Platforms.'" *New Media & Society* 12 (3): 347–64.

Harvey, Penelope, Casper Bruun Jensen, and Atsuro Morita, eds. 2016. *Infrastructures and Social Complexity: A Companion*. London: Routledge.

Hauge, Kjellrun Hiis, Anne Blanchard, Gisle Andersen, Ragnhild Boland, Bjørn Einar Grøsvik, Daniel Howell, Sonnich Meier, Erik Olsen, and Frode Vikebø. 2014. "Inadequate Risk Assessments—A Study on Worst-Case Scenarios Related to Petroleum Exploitation in the Lofoten Area." *Marine Policy* 44:82–89.

Helgesen, Ole K., and Truls Tunmo. 2009. "Beskyldes for å Sabotere Oljedebatten." *TU*, April 24. www.tu.no/petroleum/2009/04/24/beskyldes-for-a-sabotere-oljedebatten.

Introna, Lucas D. 2011. "The Enframing of Code Agency, Originality and the Plagiarist." *Theory, Culture & Society* 28 (6): 113–41.

Jarulaitis, Gasparas, and Eric Monteiro. 2010. "Unity in Multiplicity: Towards Working Enterprise Systems." In *Proceedings of the 18th European Conference on Information Systems*. https://aisel.aisnet.org/ecis2010/107/.

Jasanoff, S. 1999. "The Songlines of Risk." *Environmental Values* 8 (2): 135–52.

Jensen, Casper Bruun, and Brit Ross Winthereik. 2013. *Monitoring Movements in Development Aid: Recursive Partnership and Infrastructures*. Infrastructures. Cambridge, MA: MIT Press.

Kallinikos, Jannis, Aleksi Aaltonen, and Attila Marton. 2013. "The Ambivalent Ontology of Digital Artifacts." *Management Information Systems Quarterly* 37 (2): 357–70.

Kaltenbrunner, Wolfgang. 2014. "Infrastructural Inversion as a Generative Resource in Digital Scholarship." *Science as Culture* 24 (1): 1–23.

Karasti, Helena, Karen S. Baker, and Eija Halkola. 2006. "Enriching the Notion of Data Curation in E-Science: Data Managing and Information Infrastructuring in the Long Term Ecological Research (LTER) Network." *Computer Supported Cooperative Work* 15 (4): 321–58.

Knorr Cetina, K. 2009. "The Synthetic Situation: Interactionism for a Global World." *Symbolic Interaction* 32 (1): 61–87.

Laet, Marianne de, and Annemarie Mol. 2000. "The Zimbabwe Bush Pump: Mechanics of a Fluid Technology." *Social Studies of Science* 30:225–63.

Latour, Bruno. 1999a. "Circulating Reference: Sampling the Soil in the Amazon Forest." In *Pandora's Hope: Essays on the Reality of Science Studies*, 24–79. Cambridge, MA: Harvard University Press.

———. 1999b. "Science's Blood Flow: An Example from Joliot's Scientific Intelligence." In *Pandora's Hope: Essays on the Reality of Science Studies*, 80–112. Cambridge, MA: Harvard University Press.

———. 2004. *The Politics of Nature*. Cambridge, MA: Harvard University Press.

Macalister, T., and D. Carrington. 2015. "Shell Determined to Start Arctic Oil Drilling This Summer." *Guardian*, January 29. www.theguardian.com/business/2015/jan/29/shell-determined-arctic-oil-drilling-summer.

Mason, Arthur. 2016. "Arctic Abstractive Industry—Cultural Anthropology." July 29. https://culanth.org/fieldsights/945-arctic-abstractive-industry.

Merry, Sally Engle. 2011. "Measuring the World: Indicators, Human Rights, and Global Governance: With CA Comment by John M. Conley." *Current Anthropology* 52 (S3): S83–95.

Mitchell, Timothy. 2013. *Carbon Democracy: Political Power in the Age of Oil*. London: Verso.

Mol, Annemarie. 2002. *The Body Multiple: Ontology in Medical Practice*. Science and Cultural Theory. Durham, UK: Duke University Press.

Monteiro, Eric, and Vidar Hepsø. 2002. "Purity and Danger of an Information Infrastructure." *Systemic Practice and Action Research* 15 (2): 145–67.

Monteiro, Eric, Neil Pollock, Ole Hanseth, and Robin Williams. 2013. "From Artefacts to Infrastructures." *Computer Supported Cooperative Work* 22 (4–6): 575–607.

Mukerji, Chandra. 1989. *A Fragile Power: Scientists and the State*. Princeton, NJ: Princeton University Press.

Norwegian Ministry of Climate and Environment. 2009. "The Royal Norwegian Ministry of Climate and Environment—Act Relating to the Management of Biological, Geological and Landscape Diversity (Nature Diversity Act)." LOV-2009-06-19-100, Norwegian Ministry of Climate and Environment. https://lovdata.no/dokument/NL/lov/2009-06-19-100.

Norwegian Oil and Gas Association. 2005. "Integrated Work Processes: Future Work Processes on the NCS." www.norskoljeoggass.no/PageFiles/14295/051101%20Integrerte%20arbeidsprosesser,%20rapport.pdf?epslanguage=no.

Olsen, Erik, Silje Holen, Alf Håkon Hoel, Lene Buhl-Mortensen, and Ingolf Røttingen. 2016. "How Integrated Ocean Governance in the Barents Sea Was Created by a Drive for Increased Oil Production." *Marine Policy* 71:293–300. doi:10.1016/j.marpol.2015.12.005.

Parmiggiani, Elena. 2015. "Integration by Infrastructuring: The Case of Subsea Environmental Monitoring in Oil and Gas Offshore Operations." Doctoral thesis, NTNU. http://hdl.handle.net/11250/2358470.

Parmiggiani, Elena, Eric Monteiro, and Vidar Hepsø. 2015. "The Digital Coral: Infrastructuring Environmental Monitoring." *Computer Supported Cooperative Work* 24 (5): 423–60.

Pinch, Trevor J., and Wiebe E. Bijker. 1984. "The Social Construction of Facts and Artifacts: Or How the Sociology of Science and the Sociology of Technology Might Benefit Each Other." *Social Studies of Science* 14 (3): 399–441.

Pipek, Volkmar, Helena Karasti, and Geoffrey C. Bowker. 2017. "A Preface to 'Infrastructuring and Collaborative Design.'" *Computer Supported Cooperative Work* 26 (1–2): 1–5. doi:10.1007/s10606-017-9271-3.

Porter, Theodore M. 1996. *Trust in Numbers: The Pursuit of Objectivity in Science and Public Life*. Princeton, NJ: Princeton University Press.

Power, Michael. 2007. *Organized Uncertainty: Designing a World of Risk Management*. Oxford: Oxford University Press.

Ribes, David. 2014. "Ethnography of Scaling—Or, How to Fit a National Research Infrastructure in the Room." In *Proceedings of the 17th ACM Conference on Computer Supported Cooperative Work & Social Computing*, 158–70. New York: ACM.

Ribes, David, Steven J. Jackson, Stuart Geiger, Matthew Burton, and Thomas Finholt. 2013. "Artifacts That Organize: Delegation in the Distributed Organization." *Information and Organization* 23:1–14.

Ribes, David, and Charlotte P. Lee. 2010. "Sociotechnical Studies of Cyberinfrastructure and E-Research: Current Themes and Future Trajectories." *Computer Supported Cooperative Work* 19 (3–4): 231–44.

Schmidt, Kjeld, and Liam Bannon. 1992. "Taking CSCW Seriously." *Computer Supported Cooperative Work* 1 (1–2): 7–40.
Star, Susan Leigh. 1989. "The Structure of Ill-Structured Solutions: Boundary Objects and Heterogeneous Distributed Problem Solving." In *Distributed Artificial Intelligence*, vol. 2, edited by M. Huhns, 37–54. San Francisco: Morgan Kaufmann. http://dl.acm.org/citation.cfm?id=94079.94081.
Star, Susan Leigh, and Geoffrey C. Bowker. 2002. "How to Infrastructure." In *Handbook of New Media. Social Shaping and Consequences of ICTs*, edited by Leah A. Lievrouw and Sonia Livingstone, 151–62. Thousand Oaks, CA: Sage.
Star, Susan Leigh, and James R. Griesemer. 1989. "Institutional Ecology, 'Translations' and Boundary Objects: Amateurs and Professionals in Berkeley's Museum of Vertebrate Zoology, 1907–39." *Social Studies of Science* 19:387–420. doi:10.1177/030631289019003001.
Star, Susan Leigh, and Karen Ruhleder. 1996. "Steps toward an Ecology of Infrastructure: Design and Access for Large Information Spaces." *Information System Research* 7 (1): 111–34.
Swanson, E. Burton, and Neil C. Ramiller. 1997. "The Organizing Vision in Information Systems Innovation." *Organization Science* 8 (5): 458–74.
Tilson, David, Kalle Lyytinen, and Carsten Sørensen. 2010. "Research Commentary—Digital Infrastructures: The Missing IS Research Agenda." *Information Systems Research* 21 (4): 748–59.
Zittrain, Jonathan L. 2006. "The Generative Internet." *Harvard Law Review* 119 (7): 1974–2040.

Of "Working Ontologists" and "High-Quality Human Components"

The Politics of Semantic Infrastructures

Doris Allhutter

> *A given infrastructure may become transparent, but a number of significant political, ethical and social choices have without doubt been folded into its development—and this background needs to be understood if we are to produce thoughtful analyses of the nature of infrastructural work.*
>
> **—Star and Bowker (2006, 233)**

Nearly twenty years ago, Tim Berners-Lee (2000)—the inventor of the World Wide Web—proclaimed his vision of an intelligent networked infrastructure that will equip computers with common sense, automated reasoning, and the capacity to communicate with a variety of other electronic devices.[1] At that time, Berners-Lee described it as a "web of data"—a "web of everything we know and use from day to day" (183)—equipped with logic that communicates the meaning of the data to the computer.[2] This meaning-centered "semantic" infrastructure "can assist the evolution of human knowledge as a whole" (Berners-Lee et al. 2001). And once again, just like in 1989, when he initiated the development of global hypertext at CERN, "it's going to be a grassroots thing" (Berners-Lee 2000, 195). The decentralized architecture of the emerging semantic infrastructure intends to allow for the coexistence of a plurality of perspectives and multiple worldviews. In contrast to earlier expert-driven knowledge representation systems, the inclusion of multiple sources and of peer-produced knowledge reflects the idea of a more inclusive and democratic process of knowledge generation.

Semantic technology is a key component of artificial intelligence (AI). Research that contributes to the field of semantic computing is conducted in different areas such as data mining, computational linguistics, and machine learning and overlaps with fields and applications such as linked data or context-aware devices. To endow machines "with human-like understanding of certain problem domains has been and still is a main goal of artificial intelligence research" (Cimiano 2014, v). Semantic computing has adapted the AI technologies of the 1980s to the vast web resources of the beginning of the 21st century (Sowa 2011). Early AI approaches mostly used deductive inference aimed at deriving theorems from axioms. The open, distributed, and inherently incomplete nature of the semantic infrastructure has entailed an increasing interest in machine learning techniques. These techniques use inductive inference that generalizes from a gradually increasing set of observed instances (Ławrynowicz and Tresp 2014, 47). Machine learning evolves based on its exposure to an ever-increasing amount of data. However, "a crucial question is how high the cost actually is for encoding all the relevant knowledge in such a way that it can be exploited by machines for automatic reasoning" (Cimiano 2014, v).

In the past ten years, there has been a massive increase in semantically structured data. Semantics is used everywhere from Facebook and Google to Linked Open Data and Apple's Siri. Given the huge amounts of open and private data that are "out there," the high performance of existing computer networks, and the significant increase in research funding in the field, semantic computing is moving from simple text mining toward deep text understanding.[3] As a consequence, representations of the knowledge available in databases and on the Internet can evolve more easily from simple classifications to "richly axiomatized top-level formalisations" (Lehmann and Völker 2014, ix). The engineering of such ordering systems or "ontologies" can be done from the ground up, particularly when a domain's conceptual relations are highly complex.[4] But researchers seek semiautomated processes to construct ontologies, notably to reduce the overhead costs for human participation in the overall process. Simply said, ontology engineering is a way of reworking (domain-specific) knowledge in informational terms (Ribes and Bowker 2009, 200). Examples include domain ontologies such as the "disease ontology" that integrates extensive vocabularies of specific knowledge domains, "terrorism ontologies," or the "friend-of-a-friend ontology" that supports social network analysis, and DBpedia, which extracts structured information from Wikipedia, classifies it in a consistent ontology, and, in this way, makes information across multiple resources available for semantic query. Ontologies of everyday commonsense knowledge seek to perform human-like reasoning based on a "formal representation of the universe of discourse" (Indiana University Bloomington 2016). The latter are particularly interesting due to their inherent power relevant dimensions: commonsense knowledge may not be subject to or backed by scientific findings or expert knowledge, but it represents widely uncontested knowledge that has become hegemonic in a particular geohistorical context. It implicitly informs our everyday practices and ad hoc decisions. In her political theory of the everyday, Brigitte Bargetz (2014, 2016) describes everyday practices as a crucial site of political contestation: they are how power structures are enacted (2016, 208). The everyday is "a mode of exercising power" (2016, 35) that, at the same time, carries a potential for agency and political resistance.

In this chapter, I argue that the semantic infrastructure coemerges with a set of epistemic and economic practices that require careful scrutiny of their multilayered and entangled political implications. Representing knowledge in ontologies in terms of axiomatic relationships between the objects and concepts in a domain and designing (semi)automated methods to reason about these relations reflects a particular way of conceptualizing the world. It is part of a sociomaterial process in which classifications, standards, (meta)data, and methods coemerge with processes of signification that reconstitute and/or shift hegemonic ecologies of knowledge (Bowker and Star 1999). These processes of signification materialize in hybrid practices that rely on algorithmic agencies and the human labor of various actors. Scholars in ontology engineering and related semantic techniques suggest integrating "the human in the loop" so as to overcome technological challenges posed by the use of natural language that includes irony, sarcasm and slang, the imprecision of human knowledge, and the problem of decoding nontextual data such as images, videos, and speech. In particular, they suggest a division of infrastructural work between ontology engineers, domain experts, and other "high-quality human components" (ISWC 2013). This chapter presents an ethnographic study into practices of infrastructuring through techniques of semantic computing. Based on the example of building commonsense ontologies, it focuses on different modalities of power that interlock in these practices and thus (re)constitute phenomena of difference and structural inequality.

Infrastructuring and the Apparatus of Semantic Computing

During the past two decades, scholars in science and technology studies (STS), information systems, and computer-supported cooperative work have examined how information systems and infrastructures emerge with a set of practices that relate different knowledge areas—and thus also their sociopolitical implications—to each other (e.g., Star and Ruhleder 1996; Edwards et al. 2009). The notion of information infrastructure succeeds that of information systems and more strongly emphasizes the emergent aspects of networked environments (Hanseth et al. 1996). Infrastructures are "pervasive enabling resources in network form" consisting of static and dynamic elements (Bowker et al. 2010, 98). They are designed as extensions and improvements of existing infrastructures and are evolving over time (Hanseth 2010). Therefore, the focus of study has moved from artifacts to the work of designing "infrastructure as a contextual 'relation' rather than a 'thing'" (Karasti et al. 2016, 4; see Monteiro et al. 2013). "By their nature knowledge infrastructures are often accrued/layered and dispersed rather than discrete identifiable objects" (Karasti et al. 2016, 7). Star and Bowker (2006) have used the notion of "infrastructuring" to emphasize that infrastructure relates to the people who design, maintain, and use it. Infrastructure is embedded: it "is sunk into, inside of, other structures, social arrangements and technologies" and "takes on transparency by plugging into other infrastructures and tools in a standardized fashion" (231). It is invisible "almost by definition, disappearing into the background along with the work and the workers that create or maintain them" (Karasti et al. 2016, 8).

These characteristics point to the multilayered apparatus that is in place for a semantic infrastructure to emerge. This apparatus comprises the existing web architecture turning into the new infrastructure, the software tools (e.g., ontology editors), the methods and practices that are explored and tailored (e.g., ontology learning and semantic annotation), the data stored in clouds and databases and collected in social media and by devices such as cameras, microphones, and sensors, and the standards and protocols orchestrating this data-driven restructuring of the web (e.g., RDF standards). Nonetheless, the apparatus includes narratives of an economically beneficial and smart future growing from epistemic traditions in computer science and AI research, but also socioeconomic and political structures and practices that embed and coemerge with these technological enactments. All this appears to be held together by affective investments into a spirit of collectivity and peer production (Allhutter and Bargetz 2017). These intertwined structures, standards, materialities, practices, and affects configure a sociomaterial apparatus that is historically contingent and performative and, thus, deeply political. The STS-inspired study of information systems and knowledge infrastructures has a long-standing tradition of questioning power relations. Two intersecting issues are pertinent in the study of infrastructure and also crucial for my research: The first problem involves the power of infrastructure itself, which has been analyzed by considering the ways that classification and standardization shape knowledge practices (e.g., Bowker and Star 1999; Bowker 2008). In particular, Parmiggiani and Monteiro (this volume) elaborate on the performativity of digital infrastructures by revealing how digital representation can perform a non-neutral order of knowledge that is tied to stakeholder interests. A second aspect involves the development of infrastructures that has been politicized in terms of the relations of labor it (re)produces. This research examines the invisible work of making, maintaining, and repairing infrastructures (Suchman 1995; Jackson 2014; Goëta and Davies 2016) and the question of which work is made visible or invisible by infrastructures (Star and Strauss 1999; Lin et al. 2016).

Tying in with this research, I also sharpen the concepts of power I use to grasp the *sociomaterial entanglements* that my notion of the apparatus of semantic computing alludes to. In this spirit, I refer to different modalities of power. Karen Barad's (2003) notion of *material-discursive performativity* conceptualizes the "doings" and agential capacities of intertwined discursive and material relations that come to bear in processes of "mattering" (i.e., entangled processes of signification, materialization, and embodiment). The political nature of infrastructure becomes visible by analyzing how information infrastructures or particular enactments of an infrastructure coemerge with societal structures, individuals, bodies, and their knowledge practices. In Barad's view, some things, views, practices matter—they are made possible—and others are excluded and do not materialize.[5] The *performativity* of infrastructure and its agentive capacities show in the way in which it accommodates some practices, people, and viewpoints more than others (Star 1990; Star and Ruhleder 1996).

However, Barad's concept does not analytically grasp how these processes of mattering are interwoven with social power relations that congeal in the "intermediate structures of political economy and broader macro-level systems" (Coole 2013, 453). When elaborating on the inseparability of processes of knowing and being (Barad 2007) and on questions of in/visible labor, I suggest that the notion of *ideological practices* and the concept of *hegemony* add important analytical perspectives to the study of sociomaterial practices. The concept of ideology refers to how social practices reproduce capitalist relations of production. According to Marxist theory, practices arise from and reflect the material conditions in which they are generated. These practices express social relations and the resulting contradictions between different social groups or classes. This means that in capitalism, practices are ideological and position subjects in a society-in-dominance. For instance, Astrid Mager (2014) has analyzed how corporate search engines manifest capitalist value systems in search technology and how they spread through algorithmic logics. The existence of ideology is material "because it is inscribed in practices" (Hall 1985, 100)—in work practices and everyday practices such as developing and using information infrastructures. The structure of social practices is therefore not random or immaterial. Practices are how a structure is actively produced. Antonio Gramsci's (1971) work on hegemony shows that capitalism is (re)produced not by force and economic principles but by people's consent, through their beliefs and everyday practices. If a discourse or practice is uncontested, it means that "subjects have come to forget the contingency of a particular articulation and have accepted it and its elements as necessary or natural" (West 2011, 418). I suggest that ideological practices and hegemony are useful concepts to identify the structural dynamics of historically contingent (post-Fordist) politico-economic systems and to zoom in on how these dynamics "intra-act" (Barad 2003, 815) with current practices of semantic infrastructuring (see also Fukushima 2016).[6] Barad's concept of material-discursive performativity, however, helps to grasp not only how human-machine practices are situated, embodied, and material, but also how they unfold discursively and materially to situate different bodies differently.

Ontology Engineering as a Hybrid Practice

In computer science, ontologies are "a formal, explicit specification of a shared conceptualization" (Studer et al. 1998, 184, building on Gruber 1995). They are a means to model the structure of a system, to identify "the objects, concepts, and other entities that are assumed to exist in some area of interest and the relationships that hold among them" (Gruber 1995, 907). In simple terms, building ontologies involves gathering and formalizing domain knowledge and encoding it into machine language (Ribes and Bowker 2009). The field of ontology engineering provides the methods and methodologies for formally representing concepts and relationships of a domain. As mentioned earlier, ontologies can be built from scratch by an ontology engineer and a domain expert or they can be *learned*. In the latter case, an ontology algorithm learns from a set of data (typically text) and the results are turned into an ontology (Cimiano 2014). The field of ontology learning is concerned with the development of methods that can induce relevant ontological knowledge from data (Mädche and Staab 2004). It integrates ontology engineering and machine learning and thus semiautomatically supports the ontology engineer.

In this respect, scholars in semantic computing have largely defined two important research directions. First, in order to design ontologies that *create not-yet-known knowledge, better learning algorithms* shall "assist a human in the process of modelling more complex axioms" by inducing "more complex and non-trivial relationships" from the vocabulary of large datasets (Cimiano 2014, vi). Second, "in order to exploit the capabilities of both humans and machines optimally" (vi), *methodologies for human-machine collaboration* are seen as an urgent research desideratum. Simperl et al. (2014, 225) emphasize that "despite constant progress in improving the performance of corresponding algorithms and the quality of their results, experiences show that human assistance is nevertheless required." Thus, infrastructural elements such as ontologies are fabricated in semiautomated processes by learning algorithms and humans. And they are envisioned to be designed in this hybrid way in a computationally rigorous and methodically sound manner. With a focus on these two aspects, the remainder of this chapter discusses the entanglement of sociomaterial practices, techno-epistemic frameworks, and socioeconomic structures.

Distributed Agencies in Ontology Learning

This section illustrates how ontologies are *learned* from various sources. In order to show some continuities and differences between older and newer systems, I refer to two projects: Cyc, which was started in 1984 by Douglas Lenat and is developed by the Cycorp company, and ConceptNet, which was launched in 2004 at the MIT Media Lab as part of the Open Mind Common Sense project and is developed as an open source project by Luminoso Technologies. Cyc is an expert-driven AI project assembling a comprehensive ontology and knowledgebase of everyday commonsense knowledge. ConceptNet is a semantic network organizing a crowdsourced knowledge base of common sense. According to the website of Cycorp, the domain of Cyc's core ontology is "all of human consensus reality." In her analysis of the intrinsic politics of the Cyc project, Alison Adam (1998, 82) explains how Lenat, "sees the route to common sense through lots and lots of knowledge" and not as fundamentally different than the knowledge represented in expert systems. Adam (86) and others (e.g., Forsythe 1993) have criticized Cyc's foundationalist epistemology that is based on the assumption that a "real world about which we will all agree" can be represented. Cyc has been set up as an expert-defined ontology, and all assertions and facts have been entered by knowledge engineers of Cycorp. To accommodate multiple points of view and contradictory assertions, Cyc is divided into thousands of microtheories or contexts. Microtheories are focused on such issues as a particular domain of knowledge, interval in time, and level of detail. However, "there is little evidence to suggest that individuals compartmentalize their common sense in such a manner," Adam (1998, 88) remarks critically: "all these myriad assumptions have to be made explicit" in Cyc because it "is not situated in the world." As Lucy Suchman (2007, 144) explains, projects such as Cyc "exemplify the assumption, endemic to AI projects, that the very particular domains of knowing familiar to AI practitioners comprise an adequate basis for imagining and implementing 'the human.'" Ultimately, for Adam (1998), the use of microtheories still represents the universality of a knowing subject that structures its own knowledge and beliefs but just in greater detail. Meanwhile, the knowledge base contains over 500,000 terms, including about 17,000 types of relations, and about 7 million assertions relating these terms (Cycorp 2017). Most entries into Cyc are in the form of semiautomated, natural language acquisition. It can access relational databases and semantic resources to supplement its own knowledge base.

ConceptNet is set up as a huge semantic network of commonsense knowledge that includes knowledge from expert-created resources but also uses crowdsourcing to retrieve data from the public. To get an idea of the epistemic nature of the common sense in ConceptNet, I tried a simple query searching for the concepts "man" and "woman." I performed this first query in January 2016 with ConceptNet's version 5.4 that was then accessible at the MIT website. ConceptNet infers everyday basic knowledge or cultural knowledge. It defines concepts and connects them by relations such as "IsA" or "RelatedTo." The output of a query is a list that represents relations between words as triples of their start node, relational label, and end node: the top result of my query for "man" is the assertion that "a man is a male person" and is expressed as "man–IsA→male person." In contrast, the top result for the concept "woman" is "woman–HasA→baby" referring to the assertion that "women can have babies." In summary, the search generates long lists of results defining what a man or a woman "Is," what s/he is "UsedFor," what s/he "Desires," is "CapableOf," and so forth. "Man" and "woman" both are defined by descriptions qualifying them as "adult persons" and the two concepts have some common characteristics that, at times, seem rather random (e.g., a wo/man can chair a committee). However, the overall results reconstitute asymmetric and heteronormative gender differences. They disproportionally refer to women's reproductive qualities (e.g., woman–CapableOf→getting pregnant), looks (e.g., woman–CapableOf→carrying purses), love interests (e.g., woman–CapableOf→hunting a husband), sexuality (e.g., woman–CapableOf→pleasing a man and→touching her cunt) and other stereotypical attributions (e.g., woman–CapableOf→spending money). The results for "man" do also contain sexually explicit language (e.g., man–CapableOf→eating pussy) and invoke stereotypes of masculinity (e.g., man–CapableOf→oiling a hinge and→hunting an animal). However, they show a wider variety of capabilities and interests (e.g., man–CapableOf→believing in god). The relation "Desires" specifies that "a woman wants to be loved and wants a man," while "a man wants a woman, respect, and honesty." Whereas a woman is "UsedFor having sex with, having fun, bearing children," there is no entry in this relation for the concept "man."

While expert-driven ontologies might represent what Adam referred to as TheWorldAsTheirBuildersBelieveItToBe,[7] the content generated by ConceptNet 5.4 cannot be attributed to a particular group of authors, to a particular database, or to a human or algorithmic practice. The results of these queries simply seem to reflect particular discourses represented on the Internet. In recent years, an emerging community of researchers in discrimination-aware data mining (DADAM) and fairness, accountability, and transparency in sociotechnical systems (FAT) has been working on removing structural bias from automated decision making and machine learning (e.g., Pedreshi et al. 2008; Kamiran and Calders 2009; Sweeney 2013; Berendt and Preibusch 2014, 2017; Barocas and Selbst 2016; Crawford et al. 2016; Veale and Binns 2017; Dobbe et al. 2018).[8] Caliskan-Islam et al. (2016, 1) acknowledge the fact that machine learning from ordinary language results in "human-like biases" and thus may "perpetuate the prejudice and unfairness that unfortunately characterizes many human institutions." They refer to meaning generated from text corpora and to so-called word embeddings, a widely used semantic technique that represents words as a vector in a vector space, based on the textual context in which the word is found. For example, if the word "programmer" frequently appears close to the word "man" or "male" in text corpora used for machine learning, the word embedding of "programmer" implies that this profession is held by men. Bolukbasi et al. (2016, 4349) show that "even word embeddings trained on Google News articles exhibit female/male gender stereotypes to a disturbing extent." They demonstrate in detail the way in which "word embeddings contain biases in their geometry" and warn that "their wide-spread usage . . . not only reflects such stereotypes but can also amplify them" (4350). For instance, due to the semantic proximity of "programmer" and "man," a decision support system may automatically prioritize job applications of male programmers over applications of female programmers. Using the judgment of crowdworkers from Amazon Mechanical Turk (see next section), Bolukbasi et al. claim that they have developed algorithms that "significantly" reduce gender bias.

In November 2016, ConceptNet 5.5 was released. It builds on the same data model as the previous version but includes new word-embedding features (Speer et al. 2017).[9] "We want to provide word vectors that are not just the technical best, but also morally good," Rob Speer (2017a) blogged when he announced the latest changes to the system.[10] To improve the system, it uses "lexical and world knowledge from many different sources in many languages" (Speer et al. 2017, 4444) and introduces the "semantic space ConceptNet Numberbatch," "which incorporates de-biasing to avoid harmful stereotypes being encoded in word representations"

(Speer 2017b). Extending Bolukbasi and colleagues' (2016) debiasing method, Speer et al. (2017) cover multiple types of prejudices, including biases based on gender, religion, and ethnic descent. From my query examples above, it seems obvious that the text corpora included pornographic sources when learning what is a "woman" or a "man." And indeed, as Speer (2017a) explains, a system "is going to end up learning associations for many kinds of words, such as 'girlfriend,' 'teen,' and 'Asian'" from the huge amount of porn on the web. In the case of gender bias, Speer uses Bolukbasi et al.'s "crowed-sourced data about what people consider appropriate" (e.g., the distinction between a female "mother" and a male "father") and then algebraically eliminates inappropriate distinctions between the genders (e.g., the assumption that a "homemaker" is female and a "programmer" is male). To see the changes that ConceptNet 5.5 entails, I returned to my earlier example and repeated my query for the concepts "woman" and "man." The results included lists of "synonyms," "related terms," "types of woman/man," and "context" or "derived terms." The web interface now also shows the provenance of the data. My query results include content from OpenCyc and lexical databases such as Open Multilingual WordNet, bot-generated content from Wiktionary, as well as content from crowdsourced platforms such as DBpedia and from Verbosity, a game that collects commonsense facts from its players. A comparison with the previous results shows that pornographic references have largely been eliminated. However, despite the debiasing efforts, the results still evoke hierarchical gender discourses. "Woman" is related to qualities such as "effeminate" and "womanish" and to objects such as "dress" (whereas "man" is not related to any type of clothes). Some relations do not make sense without further inquiry. For instance, the assertion that "woman" is related to "furnish" has been taken from Wiktionary which states that "to woman" has the meaning of "to furnish with a woman."[11] Some relations are not traceable.[12] The output for "woman" lists person descriptions such as "lady," "fiancée," "mother," "human," "servant," and "wife." A "man" is described as "alumnus," "crew," "boyfriend," and "hawk." In the category "types of woman" we find stereotypical prototypes such as "goddess," "man eater," "queen," "slut," "wife," and "Cinderella." "Types of man" are, for instance, "father figure," "Adonis," "black," "boyfriend," "checker," "dandy," "ejaculator," "eunuch," and "ex-husband." In fact, these definitions are misleading at first sight. When going back to the respective sources, I found that a "slut" is a "woman" or that a "eunuch" is a "man" rather than the other way around. This reveals a core problem when trying to define a "man" or a "woman." From the perspective of feminist and queer theory, the assignment of defining "man" and "woman" identifies these concepts as empty signifiers that can only be defined as a relation of difference to the other without being able to define an *ontological* essence of "man" or "woman." In line with this, the list of antonyms defines that a "woman" is not a "man,"[13] a "girl," "male," or a "fisherman" and that a "man" is not a "boy," "female," "woman," "animal," "baby," or "servant" (note that "a woman is related to servant" as indicated above).

Despite these findings, Speer (2017a) claims that after debiasing, "the inappropriate gender distinctions have been set to nearly zero." But with regard to racist bias and bias relating to religion, "we don't have nice crowd-sourced data for exactly what should be removed in those cases." Instead, Speer aims to "de-correlate them with words representing positive and negative sentiment." Since the concept of "black man" came up in my searches for "man," I followed that path and tried a query for "black man/woman" and "white man/woman." In my first search before the debiasing efforts, a query for "black man" defined this concept with relations such as "a black man is reaching into a clothes dryer," ". . . is cleaning a laundry machine," ". . . is related to evil," and ". . . has been shot." A query for "black woman" delivered no entries. After the latest changes, the query for "black man" generates a short list simply stating that the concept is derived from "black" and from "man" and relating it to "demon" and "evil."[14] Interestingly, ConceptNet picked up the figurative meaning of "black man" as an "evil spirit or demon" from English Wiktionary, whereas references such as "male member of an ethic group" or "sub-Saharan African decent" did not find their way into ConceptNet 5.5. In contrast, "white man" is related to "Caucasian," "European," "good old boy," or "white trash." A "white woman" is simply a "white person." The concept "black woman" still does not exist, it "is not a node in ConceptNet."[15] My query results reveal asymmetries between the representation of "black" and "white" man as well as between "black man" and "black woman" and between "white man" and "white woman." Drawing the line between positive and negative sentiments requires context, as Speer (2017c) points out, and clearly these simple queries cannot capture the way in which "algorithmic racism" becomes performative. Indeed, Speer's own examples provide a better understanding of how indirect discrimination is enacted through word embeddings. In order to explore which sentiments the system associates with names of different origins, he uses lists of names that reflect different ethnic backgrounds "mostly from a United States perspective." His research shows that "the sentiment is generally more positive for stereotypically-white names, and more negative for stereotypically-black names." Another example reveals how a racist bias that relates "Mexican" to "illegal" calculates negative sentiment for Mexican food and, as a consequence, may rate Mexican restaurants lower than others. Clearly, consequences are even more severe in areas such as education, employment, health, counterterrorism, and policing (see, e.g., Browne 2015; Campolo et al. 2017; Buolamwini and Gebru 2018).

From a technological perspective, debiasing is not a trivial task. The research that is done to avoid "representing harmful stereotypes" (ConceptNet5 2017) and "amplifying human-like biases" (Caliskan-Islam et al. 2016, 1) seems to be a driver to improve infrastructural components of machine learning and AI overall. However, as some of the above quotes imply, the problem is largely diagnosed as existing in society; the technical solutions seek to adjust human shortcomings. Apparently, practices of debiasing rely on the idea that gender stereotypes can be removed easier than racist bias. Practices of solving gender bias aim to remove the hierarchy of two gendered terms. They do not question the gender binary itself but attempt to create symmetry between binary terms. Acknowledging that "gender associations vary by culture and person," Bolukbasi et al. (2016, 4354) legitimize their method by using the judgment of US-based crowdworkers to draw a line between "appropriate" (e.g., sister/brother, matriarch/patriarch) and "inappropriate" (e.g., nurse/surgeon, midwife/doctor) gender differences.[16] In comparison, racism and religious prejudice are seen to be multilayered and more complex (Speer 2017a). However, using "positive" and "negative" sentiment to identify racism in the meaning of language still follows a binary logic.

As Winthereik, Maguire, and Watts (this volume) point out, infrastructures are sites of experimentation and methodological challenge. Developers indeed experiment with and learn about bias in language when working to improve methods and systems (see also Schmidt 2015). Based on these experiments, they define biases and find methods for debiasing. Thus, they also negotiate differences based on constructions of gender, race, ethnicity, and religion. Speer et al. (2017, 4444) describe ConceptNet as "designed to represent the general knowledge involved in understanding language, improving natural language applications by allowing the application to better understand the meanings behind the words people use." *To understand meaning*, as the quote states, involves general knowledge and context. However, I want to add, it is not only about the way that people use words but about how language is entrenched in discourse and in situated and embodied knowledge.[17] Creating a debiasing method is a material-discursive practice of negotiating boundaries. It reactivates gender discourses and embodied knowledge that are deeply rooted in history, society, science, and technology. This is also the case for experimenting with sentiment analysis when trying to interpret the affective ecologies of racist structures and practices in society. Following Judith Butler, the reaffirmation of difference and of thinking in binaries is a practice that becomes performative through acts of reiteration.[18] As Barad explains, performativity is not merely a discursive but also a material process. Practices of boundary drawing and of (re)constituting difference as exclusionary practices are effects of and, at the same time, (re)constitutive of power relations. The methods deployed by developers reiterate/shift meaning; they are material practices with infrastructural consequences.

Boundaries Unfold

Alison Adam (1998) has raised the issues of how different worldviews are represented in a system like Cyc and what happens to perspectives and sense making that "do not have enough epistemic status to be assigned to a model" (86). "Cyc will have to decide whether to be a Marxist or a capitalist and presumably decisions will have to be made as to whether CYC is Christian or Jewish, male or female, old, young or middle-aged. Cyc's models of the world are hegemonic models—unconsciously reflecting the views of those in powerful, privileged positions" (86). The commonsense knowledge represented in ConceptNet is not expert-defined (in the narrower sense of the word) but is induced from various sources including crowdsourced platforms and applications. However, the information collected from these sources and the methods used for information gathering do not represent multiple worldviews. The very aim of understanding "common sense" narrows the "universe of discourse" down to hegemonic perspectives. They are generated and structured through the (boundary-drawing) practices of developers and the agentive capacities of methods and other infrastructural components. These distributed agencies are situated, embodied, and material in (at least) two ways. First, practices and methods of debiasing reflect a particular perspective. For instance, they include discourses that are specific to the (present-day) US-American context in which they are performed and exclude discourses that a different perspective might imply. They reflect ethical claims that are aligned with professional knowledge that has been described as being shaped by white, middle-class, male engineering practices (e.g., Adam 1998; Faulkner 2007; Noble 2018).[19] Second, ConceptNet is an infrastructural resource that plugs into other systems and practices, which require commonsense reasoning. It can be used for tasks such as data exploration, emotion analysis, trend monitoring, autolabeling, and autoclustering or simply "as part of making an AI that understands the meanings of words people use" (ConceptNet5 2017). For example, Aditya et al. (2018, 1) suggest using "background ontological knowledge" from ConceptNet to improve

systems that answers questions about images in natural language. These systems combine image understanding, natural language processing, and commonsense reasoning.[20] This exemplifies that ConceptNet's hybrid practices and distributed agencies unfold discursively and materially in manifold contexts and in multilayered ways.[21]

Capitalizing on the Power of Collective Intelligence

The first of two current research directions in semantic computing has been covered in the previous section: the aspiration to create better learning algorithms to induce more complex and nontrivial relationships from large datasets. The following sections consider the political implications of the second research direction: developing methodologies for human-machine collaboration. Ontology learning has different approaches that conceptualize this collaboration. (1) *Expert-driven approaches* create ontologies by learning axioms from domain experts or existing ontologies (Lehmann and Völker 2014). According to the literature, interaction with experts needs to be kept to a minimum to reduce their workload (e.g., Rudolph and Sertkaya 2014). Consequently, a user-assisted ontology learning approach suggests that interactive knowledge acquisition algorithms gather knowledge from domain experts by asking them questions. (2) *Ontology learning from databases* uses extraction algorithms to automatically build ontology models (Cerbah and Lammari 2014, 207). Yet, low-level tasks, such as data cleaning and removing inconsistencies in databases, form the foundation for ontology learning algorithms (Maynard and Bontcheva 2014). To improve the quality, these tasks are done manually. (3) *Capturing the semantics from social tagging systems* exploits the emergent semantics within the large bodies of human-annotated content in social media (Benz and Hotho 2014). Tagging or social annotation is a form of creating metadata.[22] "The fact that large user populations are effectively involved in the creation of metadata opened up new possibilities and challenges to the field of ontology learning" (Benz and Hotho 2014, 176). Collaborative tagging is seen as "grassroots semantics" and thus an alternative to the top-down paradigm of expert-created ontologies. From this, the roles of human actors in human-machine collaboration become clearer.

Ontologists are knowledge engineers who cooperate with domain experts or specialists in a particular field of knowledge. Tool developers build semantic artefacts conforming to user requirements. Considering the previously mentioned lower-level tasks, moreover, recent methodological developments suggest that "specific Semantic Web problems such as entity extraction and linking, ontology mapping, semantic annotation, conceptual modeling, or query resolution and processing could be approached by *assemblies of scalable, automatic and high-quality human components*" (ISWC 2013, emphasis added). With the emergence of labor marketplaces such as Amazon Mechanical Turk and CrowdFlower, the idea of the "wisdom of crowds" or "collective intelligence" also started to pop up in semantic computing. In 2013, workshops and conference papers at the major semantic computing conferences began to explore the potential of "harnessing human computation" (Simperl et al. 2014, 230) and of "capitalizing on the power of collective intelligence" (226). Meanwhile, the research community widely agrees that some core techniques in building a semantic infrastructure fall short of the vision of automation. Also, they are too time-consuming and costly to be solved by experts and ontologists and thus need to be solved by "the crowd." While there is an ongoing debate on crowdwork from a workers' rights perspective (see Cherry 2009; Scholz 2013; Platform Cooperativism Consortium 2018), I will examine how crowdwork is being discussed and practiced in technical terms. I base this on an analysis of technical papers and my own observations at crowdsourcing tutorials and workshops.[23]

The Crowd as Part of the Apparatus of Semantic Computing

Generally speaking, crowdwork has different genres (Simperl et al. 2013): (1) *Merchandised labor or microwork* is performed by unskilled workers supposedly with only basic numerical and literacy skills. Workers are paid a small amount of money for completing low-complexity to medium-complexity microtasks via the Internet. (2) *Games with a purpose*, such as Verbosity (see above), are small games during which players describe images, text, or videos and may not necessarily be aware that they contribute work or conceptualizations to a system. (3) *Altruistic crowdsourcing* is done for free, for example, in citizen science projects. These three genres are described as highly complementary and can be integrated into hybrid-genre workflows.

In semantic computing, tasks that can be performed by crowdworkers are, for example, specifying term relatedness (e.g., for ontology creation), verifying relation correctness (e.g., for ontology evaluation), verifying domain relevance (for automatic ontology learning), and specifying relation types such as equivalence, subsumption, disjointness, or instanceOF (Hanika et al. 2014, 186). Figure 1 shows examples of respective microtasks. In regard to the quality of crowdsourcing results, several studies have shown that the crowd performs better in commonsense knowledge than in domain-specific knowledge, but in aggregate, the workers perform on par with experts (6).

To create and manipulate ontologies, engineers use ontology editors that assist them in structuring and relating concepts to each other. Stanford's Protégé has become the leading ontology editor and earned its developers the SWSA Ten-Year Award at the International Semantic Web Conference 2014. WebProtégé, an extension of the editor, allows a larger distributed group of knowledge experts to contribute to the process of engineering large scale ontologies. Considering the good results of crowdwork, several research groups have also suggested facilitating the integration of crowdsourcing into ontology engineering (see Noy et al. 2013; Bontcheva et al. 2014; Hanika et al. 2014). For instance, the uComp Protégé plugin helps ontology engineers to crowdsource tasks directly from within the popular ontology-editing environment (see figure 2; Wohlgenannt et al. 2016). As the authors explain, a hybrid-genre crowdsourcing platform flexibly allocates the received tasks to games with a purpose and/or microwork labor platforms. The interface selects the appropriate crowdsourcing genre and creates the relevant crowd jobs. By default, all created jobs are assigned to "level 3" CrowdFlower workers who are assumed to be the contributors delivering the highest quality work. They have a high confidence value and accuracy rate. For ontologies in English, the authors suggest restricting the jobs to workers in Australia, the United Kingdom, and the United States. Completing a microtask, such as the ones depicted in figure 1, pays an average of $0.05. As I learned at the Tutorial@ISWC2014, the reward scheme not only considers the level of fair pay for particular classes of workers but also

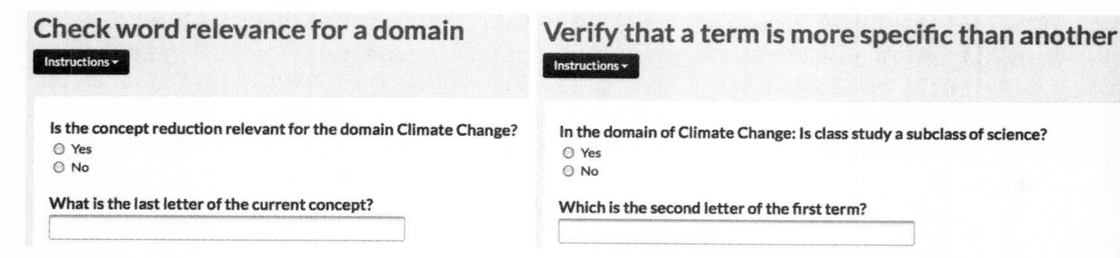

FIGURE 1: CrowdFlower job for (1) the verification of domain relevance and (2) the verification of relation correctness. Source: Hanika et al. (2014, 8).

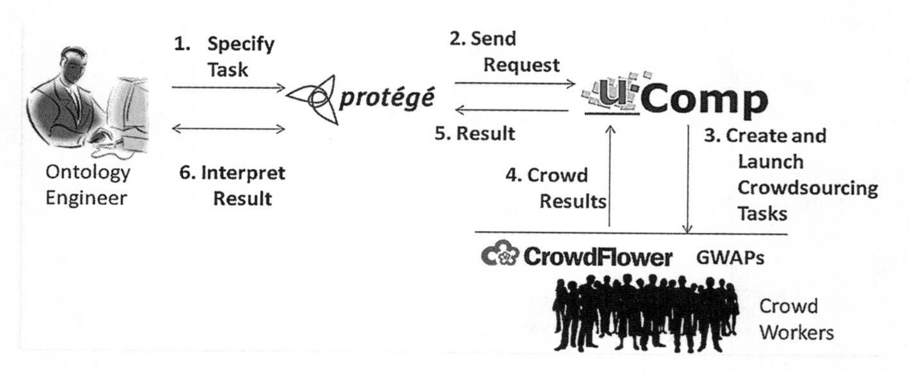

FIGURE 2: Ontology engineering workflow using the uCopm plugin. Source: Hanika et al. (2014, 5).

seeks to identify the pay level that attracts enough good workers but not too many spammers. Higher pay will get a job done faster but not necessarily with better quality. Even higher pay lowers the quality of data gathered because it increases the number of spammers. The definition of high and low pay differs widely from country to country. With the option to restrict the tasks to workers from a particular country, systems such as the uComp plugin also allow, for example, a different pay rate for workers from the United States than for workers from India. Hence, due to its embeddedness in digital marketplaces and tools such as uComp, low pay is inherent to microwork. However, many researchers in this field argue that it is not always just about the money but about intrinsic motivation "driven by the interest or enjoyment in the task itself" (Simperl et al. 2013, 25). The uComp platform finally collects and combines crowdwork harvested by various genres and presents the results to the ontology engineer.

The digitization of labor and the post-Fordist formation of work relations have been theorized under the labels of cognitive capitalism, communicative capitalism, and immaterial labor. Scholars such as Yann Moulier Boutang, Jodi Dean, Michael Hardt, and Antonio Negri or Paolo Virno analyze how knowledge work has become a main source of capital accumulation in present-day capitalism. The research fields of semantic computing and ontology engineering strongly resort to epistemic narratives of automation, while at the same time, in practice, they rely on "assemblies of scalable, automatic and high-quality human components" and thus the cognitive and embodied labor of crowdworkers. Crowdworkers edit the basic data infrastructure underlying semantic technologies and machine learning (see Irani and Silberman 2013). *They* are supposed to be the reasoning machines that solve what algorithms cannot. This ideological practice relates the notion of scalability to "human components" and discursively dehumanizes crowdworkers. The material equivalent of this sort of "othering" is a form of economic precarization that has become an integral part of the semantic infrastructure that is being built. Even monitoring the quality of microwork "takes place largely automatically through algorithms" (Kuek et al. 2015, 7) that will eventually decide if a worker gets paid at all.

Building an Infrastructure for the Global North

Whether or not researchers suggest restricting microwork to English-speaking countries, the statistics on crowdworkers provide a better insight into who helps build the foundation of the evolving semantic infrastructure. According to a report on "The Global Opportunity in Online Outsourcing" by the World Bank Group, almost two-thirds of workers in the global online-outsourcing industry (including crowdwork and online freelancing) come from the United States, India, and the Philippines, followed by Pakistan and the United Kingdom (Kuek et al. 2015, 29). Serbia and Romania contribute a high number of workers in Europe. Measured in the percentage of crowdworkers in a country's total labor force, the Philippines leads Serbia, Romania, Canada, and the United States. Demand for online outsourcing is driven by private-sector clients in Australia, the United Kingdom, and the United States. Whereas 38% of all microworkers have a bachelor's degree or higher, 80% of Indian workers for Amazon Mechanical Turk have a bachelor's degree or higher (32). Globalization also has a peculiar dynamic with gendered participation in crowdwork: Amazon Mechanical Turk in the United States employs a large number of women (65%), while in India they use mostly men (70%) (31). Because microtasks in ontology engineering require some understanding of data structures and classifications, workers need more than basic numerical and literacy skills (see Erickson, Sawyer, and Jarrahi, this volume). The field benefits massively from the technically educated, cheap workers available in the Global South who also perform extremely low-paid work such as annotating visual content. That means that the researchers and consumers of the Global North presume that automated, nonhuman processes deliver results. In fact the "human components" of the South deliver invisible and, moreover, badly paid work (see also Aytes 2013 and Crain et al. 2016).

The microwork market is dominated by two big players, Amazon Mechanical Turk and CrowdFlower. Most platforms do not have a minimum wage; workers typically earn $2 to $3 per hour.[24] "Online workers in most countries do not receive the benefits of unionization, collective bargaining, social benefits, or legal protection such as minimum wage laws" (Kuek et al. 2015, 4). While governments of the Global North largely ignore this silent undermining of labor rights and despite an emerging resistance from US-based crowdworkers (see Scholz 2016), the World Bank sees good opportunities in digital labor for the developing world. In particular, because of businesses' growing demand for data gathering, cleaning, mining, and packaging, the World Bank estimated that the market for digital labor would reach "USD 32 billion by 2017—about six time the growth rate of the overall information and communication (ICT) market" (16). New opportunities "that would not feasibly be performed" without digital marketplaces include large-scale image tagging and categorization, driven by the rapid growth in data (22). Therefore, the report suggests removing legislative and regulatory barriers that inhibit the growth of the online outsourcing industry. With this, it refers to workers' rights and taxes that "can create asymmetries in the market" (60). "Governments in developing countries need to make an explicit decision between the social good of ensuring a fair minimum wage for workers and the potential economic impact that wage laws may have on the growth of the industry," the World Bank notes (59). Instead of advocating fair wages, it emphasizes the importance of ensuring a supply of suitable workers, for example, by raising English-language literacy and IT proficiency (53).

When analyzing crowdwork's role in expanding the reach of the neoliberal economy in cognitive capitalism, Ayhan Aytes (2013, 90) points to the distinct conditions for the workers of the South, which "could be described as the gray zones of international laws that are designed by neoliberal policies to take advantage of stark regional differences in labor costs." Referring to Aihwa Ong (2006), who conceptualizes this as a "system of exception," he explains that the deterritorialization of labor is linked to a labor arbitrage that "breaks apart the traditional relationships between the national labor legislations and the worker as citizen" (Aytes 2013, 91). In her analysis of the gendered and racialized dimensions of migrant household and care work, Encarnación Gutiérrez-Rodríguez (2010) elaborates on the persistence of the "coloniality of labor." Migrant household and care work "is codified by a historical moment of appropriation as 'available' and 'disposable' labor due to its cultural predication. . . . Feminized and racialized subjects are targeted as 'raw material' as their labor is codified as 'natural' [referring to 'natural female' faculties], not in need of capital investment or pursuing a strategy of capital accumulation" (44). In the words of Gutiérrez-Rodríguez, the societal devaluation of domestic work and of crowdwork results from its cultural codification: "Neither the concrete labor-time nor the concrete labor-power defines its value" (92). "The availability of a critical mass of semantic content . . . is generally accepted as being one of the core ingredients for the wide adoption of semantic ontologies," Bürger et al. (2014, 209) emphasize. Nevertheless, crowdwork is claimed to be unskilled and driven by intrinsic motivation and thus not by a need to generate a fair income. Gutiérrez-Rodríguez (2010, 92) goes on to explain, "If the labor produced by the worker is socially characterized as 'unskilled labor,' then its value character is socially considered inferior to 'skilled labor.'"[25] The devaluation of some sort of labor depends on the quality attributed to it by society. The value of domestic work and crowdwork is "preset by a cultural system of meaning production based on historical and sociopolitical systems of gender difference and racialized hierarchies" (92). Digital marketplaces built to organize work as distributed microtasks exploit cheap female labor and even cheaper labor from the South for the economic interests of the Global North.[26] Value is "a historical, social and cultural outcome" (92). Gutiérrez-Rodríguez describes a phenomenon like labor arbitrage not merely in market terms, but as a colonial legacy and a social process of hegemonic struggles. As Aytes (92) suggests, the consequences of labor arbitrage are further accentuated by the temporality of the digital network that creates another effect—the time arbitrage: "The crowds of the global South are materially configured within the machinic always-on time of the networks through their immaterial labor in order to fuel the linear material progress that characterizes Western temporality."[27]

As the similarities between the two fields of labor show, (undocumented) domestic work and crowdwork "engender a place of 'exteriority' or 'colonial difference'" (Gutiérrez-Rodríguez 2010, 44). Both domestic workers and crowdworkers suffer vulnerable working conditions, unregulated work times, and a high dependency on the employer or an opaque technical evaluation system. The unprotected state of both of these groups of workers is maintained by the circumvention of national labor laws. Yet, while the global care chain relies on the migration of female workers to the Global North, the deterritorialization of human microwork keeps national border regimes in place.

Performing Difference and Structural Inequality

This chapter has analyzed how the apparatus of semantic computing and its infrastructuring practices are materially and discursively performative in their co-emergence with techno-epistemic discourses and politico-economic structures. It has focused on practices of ontology learning and has traced two research trajectories that shape the field: (1) the aim to create not-yet-known knowledge by inducing more complex and nontrivial relationships from large datasets and (2) doing this by combining human and machine intelligence. The availability of huge amounts of different data has inspired new, semiautomated ontology engineering and machine learning practices. In combination with practices of crowdsourcing and the use of a global labor force, this dynamic seems to integrate multiple perspectives and to prompt a democratization of knowledge creation. It appears to shift ecologies of knowledge that were relying on the deductive, expert-driven systems of early AI toward decentralized and more open processes of reasoning within the inductive, data-driven semantic infrastructure. Building infrastructural elements from everyday language and commonsense knowledge is a particularly valuable resource for training intelligent systems.

The intransparency and distributed agencies of ontology learning processes do not allow for a general empirical statement on the kind of hegemonic struggles enacted by the apparatus of semantic computing; nevertheless, my ethnography of practices of infrastructuring strongly points to the problematic nature of epistemic claims raised by techno-scientific expert discourses. With its roots in AI and machine learning, the epistemic narratives of automation and machine intelligence have provided a foundation for the field of semantic computing. However, "human computation" is becoming an integral part of its methodologies. This contradiction obscures the way in which the knowledge represented through commonsense ontologies and algorithmic performances of "the everyday" (Bargetz 2014) is entrenched in power relations (see also Allhutter, forthcoming): (1) When creating ontologies, the hybrid practices of boundary-drawing essentially try to define concepts and their relations by excluding other material-discursive configurations of the world, and, thus, perform difference. I suggest that acknowledging that semantic infrastructuring (or making meaning infrastructural) affects and is affected by power relations allows us to analytically grasp the hegemonic ways in which difference becomes performative. In reference to Lucas Introna (2016, 27) and John Law, I want to suggest that ontology engineering and machine learning are "*empirical practices with ontological contours. Their actions are not just in the world, they make worlds*" (emphasis original). And, I want to add, they affect different subjects, bodies, and the objects related to them in different ways. (2) Crowdsourced data do not reflect a free competition of opinions and views but reflect particular discourses. Shaped by restrictions in global access and other aspects such as censoring practices of platforms or the anonymity of much (e.g., offensive) web content, this discourse represents communication and information specific to web culture(s). The view that online content merely reflects society (e.g., biases that characterize many human institutions) ignores that this content does not represent real-world diversity. It ignores that the data that are used to train systems are already a product of the sociomaterial transformation of language and discourse and of intransparent practices of data processing. (3) I have argued that human microwork plays a substantial role in building the foundations of a semantic infrastructure and in supporting epistemic claims of automatic reasoning and machine intelligence. As I have shown, methodologies and practices of human-machine collaboration configure human-machine relations differentially with regard to automation and scalability. While the workload of "working ontologists" and experts is to be reduced through machine-learning techniques, "the crowd" is subject to ideologies building on the unquestioned presumption that a global cheap work force is available around the clock. Crowdwork includes knowledge providers from the Global South, but the nature of microtasks does not allow thinking outside predefined options. It is hardly possible to articulate alternative ways of knowing. Moreover, the system measures workers' performance against a "gold standard," and workers with high "error rates" have less or no pay. While crowdworkers perform infrastructure work that generates economic value and promises economic prosperity for the future, the perspectives of these knowledge providers will hardly shape the infrastructural components that they help build.

These hybrid practices for creating a semantic infrastructure are enacting phenomena of difference and structural inequality. They are ideologically invested and contingent on a transnational division of labor. They emerge in relation to colonial pasts and presents and enact, continue, and transform global economic processes. The digitization of knowledge work and the implementation of crowdwork into the apparatus of semantic computing are part of the post-Fordist transformation of a transnational division of labor. We have to consider this phenomenon also in light of ongoing changes in other sectors.[28] Looking at infrastructure work from this broader perspective shows how a meaning-centered "semantic" infrastructure for machine learning and AI co-emerges with economic and political practices that adapt the regime of capital accumulation by establishing a renewed form of transnational exploitation.

Acknowledgments

This research was funded by the Austrian Science Fund (FWF): V273-G15. I want to thank Brigitte Bargetz, Bettina Berendt, Cecile Crutzen, Robert Kahlert, Harold Otto, Rob Speer, and two anonymous reviewers for their feedback on earlier versions of this chapter.

Notes

1. The notion of "working ontologists" in the title of this chapter was inspired by a practice guide to semantic web technologies by Allemang and Hendler (2011).
2. Computers "understand" a machine-processible relative form of "meaning" by linking terms and concepts.
3. Researchers in this field have access to user data collected by free apps that they provide; users consent (often unwittingly) to loss of privacy when agreeing to the terms and conditions. "We have no

privacy anymore. Our information is somewhere and I'd like to be able to use it" (computer scientist Yolanda Gill in Q&A after her keynote at ISWC2014).

4. In computer science, the term "ontology" roughly refers to a formal representation of a domain of knowledge. In philosophy, ontology is used "to talk of what is in the world, or what reality out there is made of" (Law 2017, 31). Rather than attempting to define a reality or several contested realities, in STS "ontologies are relational effects that arise in practices . . . and . . . since practices vary, so too do objects" (43). STS scholars usually use "ontologies in the plural, and accordingly the focus of analysis is on the always fluid and unstable processes by which ontologies come into being" (Van Heur et al. 2013, 342). The notion of ontology is used "to signify the centrality of objects in constituting socio-technical relations" (355). This chapter examines the relational effects emerging with practices of ontology engineering as applied in computer science. In this text, I use the term "ontology" and its compositions such as "ontology engineering" predominantly with reference to computer science. To differentiate, I use the notion *ontological* in italics when referring to the meaning prevalent in STS.
5. For instance, Bowker (2008, 109) has described how the entangled technical, formal, and social practices that surround information infrastructures are "implicated with the core of our being and of our understanding of the world."
6. Despite fundamental conceptual differences between new materialist and historical materialist thought—especially as to whether agency is exclusive to the realm of the human subject, authors such as Diana Coole (2013) and Hanna Meißner (2016) have pointed to the common threats of these traditions of thought.
7. I borrow this from Adam's (1998, 88) "TheWorldAsTheBuildersOfCycBelieveItToBe," which she describes as informed by middle-class, male, professional knowledge. Ribes and Bowker (2009, 211) describe ontology engineering as a process of negotiating knowledge in which "tacitly held differences should be articulated for the purpose of overcoming them in a single unifying ontology."
8. Feminist scholars in STS, HCI, CSCW, and IS have been arguing for the existence of biases in computer systems for many years and have called for computer engineers to take responsibility for the social orderings that their technologies produce (see, e.g., Floyd 1992; Oudshoorn et al. 2004; Trauth 2006; Suchman 2008; Crutzen and Hein 2009; Bardzell and Churchill 2011; Rommes et al. 2012; Fox and Rosner 2016; Breslin and Wadhwa 2018). While this research has largely been ignored in most computing fields, some of the critique has found resonance in AI research (Suchman 2007, 230).
9. ConceptNet "word embeddings represent words as dense unit vectors of real numbers, where vectors that are close together are semantically related. . . . It represents meaning as a continuous space, where similarity and relatedness can be treated as a metric" (Speer et al. 2017, 4445).
10. With the release of version 5.5, the website has been relaunched at http://conceptnet.io and ConceptNet is not linked to MIT anymore.
11. A Shakespeare quote from 1603 is given as an example for this meaning.
12. For example, the assertion that "man" is related to "unanimously" is stated to come from Wiktionary. The only link between the words is that the term "unanimously" means "in a unanimous *man*ner."
13. Note that the assertion that a "woman is a type of man" in the query results refers to "man" meaning "human."
14. Moreover, a relation labeled "black man is capable of" is linked to "do this better." It is not comprehensible if this means a Black man "can do something better" or if he is supposed to "do something better."
15. Joy Buolamwini and Timnit Gebru (2018) reveal the nonrepresentation of women of color in facial recognition systems. Safiya Umoja Noble (2018) analyzes how search engines misrepresent women of color.
16. The analogy between sister and brother reflects a family relationship between siblings and is ontologically correct. However, the notion of patriarchy refers to a system of male domination over women, while the notion of matriarchy refers to a matrilinear society and does not imply female domination over men. Thus, this analogy is not correct.
17. As Barad (2003, 819) reminds us, "Discourse does not refer to linguistic or signifying systems, grammars, speech acts, or conversations. . . . Discourse is not what is said; it is that which constrains and enables what can be said."
18. Butler emphasizes that reiteration carries a potential for shifts and change since a reiteration is never exactly the same as what it reiterates.

19. See also "Black in AI," a platform that fosters collaboration and initiatives to increase the presence of Black people in AI (https://blackinai.github.io/).
20. Aditya et al. (2018) combine ontological knowledge and distributional knowledge from ConceptNet 5.5 and word2vec. For example, a picture of a horse in front of a building may raise the question "what is the building in the background?"; the reasoning engine can figure out that it is a barn, because a barn is a building (ontological) and barn relates more closely (distributional) to horse than to, e.g., church.
21. For example, Allhutter and Hofmann (2014) show how affective gender-technology relations unfold in subjective, collective, and structural ways.
22. Metadata are structured data about an object. They are created by experts or the public or are generated automatically (Lytras et al. 2014, 46). Nontextual resources are mostly annotated manually, which, despite the routine character of this task, means significant costs (Bürger et al. 2014, 199). The objective of any metadata mechanism is semantic interoperability: the ability of systems to create meaning-based relations between entities.
23. As part of my fieldwork, I attended the following events: Semantic Annotation of Social Media: A Hands-on Tutorial at the 13th International Semantic Web Conference (indicated in the text as Tutorial@ISWC2014), the tutorial on Practical Annotation and Processing of Social Media with Gate at the 12th European Semantic Web Conference (Tutorial@ESWC2015), and the 2nd International Workshop on Crowdsourcing in Software Engineering at the 37th International Conference on Software Engineering (Workshop@ICSE2015).
24. The World Bank report does not mention differences in wages between workers of different countries, even though they are an inherent aspect of the digital marketplace. This was mentioned in one way or the other at all events that I attended as part of my fieldwork. In all cases, discussions on the fairness of unequal pay were dismissed quickly by concluding that this is a problem of the political realm that researchers in semantic computing cannot solve (with the exception of Dustdar 2015 at Workshop@ICSE2015).
25. As Aytes (2013, 91) notes, some US-based Amazon Mechanical Turk workers state they were indeed motivated by the novelty of the experience, while workers from India or China mostly sell their labor for a basic income (see also Kuek et al. 2015, 31).
26. The weekly contributor profiles on CrowdFlower (2016) show a series of usually one Indian worker, one Vietnamese worker, and one American homemaker and mother who enjoys the freedom of microtasking while her baby is napping.
27. Another indication of this is that the World Bank report's list of benefits for online workers includes acquiring soft skills such as the ability to manage their own time and to work to a deadline (Kuek et al. 2015, 4).
28. The ongoing changes include the growing service industry entailing a digitization of service work but also a feminization of migration and a care drain from low-income countries, or the digital networking of industrial value chains (the "Industrial Internet").

Works Cited

Adam, Alison. 1998. *Artificial Knowing. Gender and the Thinking Machine.* London: Routledge.

Aditya, Somak, Yezhou Yang, and Chitta Baral. 2018. "Explicit Reasoning over End-to-End Neural Architectures for Visual Question Answering." arXiv:1803.08896v1.

Allemang, Dean, and James Hendler. 2011. *Semantic Web for the Working Ontologist. Effective Modeling in RDFS and OWL.* Amsterdam: Elsevier.

Allhutter, Doris. Forthcoming. "Infrastructural Power." *Feminist Theory.*

Allhutter, Doris, and Brigitte Bargetz. 2017. "Emergent Society-Technology-Formations in Affective Capitalism." Presentation at the Annual Meeting of the Society for Social Studies of Science, Boston.

Allhutter, Doris, and Roswitha Hofmann. 2014. "Affektive Materialitäten in Geschlechter-Technikverhältnissen. Handlungs- und theorie-politische Implikationen einer antikategorialen Geschlechteranalyse." *Freiburger Zeitschrift für GeschlechterStudien* 20 (2): 59–78.

Aytes, Ayhan. 2013. "Return of the Crowds. Mechanical Turk and Neoliberal States of Exception." In *Digital Labor: The Internet as Playground and Factory,* edited by Trebor Scholz, 79–97. New York: Routledge.

Barad, Karen. 2003. "Posthumanist Performativity. Toward an Understanding of How Matter Comes to Matter." *Signs* 28 (3): 801–31.
——. 2007. *Meeting the Universe Halfway: Quantum Physics and the Entanglement of Matter and Meaning*. Durham, NC: Duke University Press.
Bardzell, Shaowen, and Elizabeth F. Churchill. 2011. "Feminism and HCI: New Perspectives. Special Issue Editors' Introduction." *Interacting with Computers* 23 (5): iii–xi.
Bargetz, Brigitte. 2014. "Figuring Ambivalence, Capturing the Political. An Everyday Perspective." In *Multistable Figures. On the Critical Potentials of Ir/reversible Aspect-Seeing*, edited by Christoph F. E. Holzhey, 191–214. Wien: Turia + Kant.
——. 2016. *Ambivalenzen des Alltags. Neuorientierungen für eine Theorie des Politischen*. Bielefeld: Transcript Verlag.
Barocas, Solon, and Andrew D. Selbst. 2016. "Big Data's Disparate Impact." *California Law Review* 104:671–732.
Benz, Dominik, and Andreas Hotho. 2014. "Capturing Emergent Semantics from Social Tagging Systems." In *Perspectives on Ontology Learning*, edited by Jens Lehmann and Johanna Völker, 175–88. Berlin: IOS Press.
Berendt, Bettina, and Sören Preibusch. 2014. "Better Decision Support through Exploratory Discrimination-Aware Data Mining: Foundations and Empirical Evidence." *Artificial Intelligence and Law* 22 (2): 175–209.
——. 2017. "Toward Accountable Discrimination-Aware Data Mining: The Importance of Keeping the Human in the Loop—and Under the Looking Glass." *Big Data* 5 (2): 135–52.
Berners-Lee, Tim. 2000. *Weaving the Web: The Original Design and Ultimate Destiny of the World Wide Web*. New York: HarperCollins.
Berners-Lee, Tim, James Hendler, and Ora Lassila. 2001. "The Semantic Web." *Scientific American* 284 (5): 28–37.
Bolukbasi, Tolga, Kai-Wei Chang, James Zou, Venkatesh Saligrama, and Adam Kalai. 2016. "Man Is to Computer Programmer as Woman Is to Homemaker? Debiasing Word Embeddings." In *NIPS '16 Proceedings of the 30th International Conference on Neural Information Processing Systems*, 4349–57. New York: ACM.
Bontcheva, Kalina, Ian Roberts, Leon Derczynski, and Dominic Rout. 2014. "The GATE Crowdsourcing Plugin: Crowdsourcing Annotated Corpora Made Easy." In *Proceedings of the 14th Conference of the European Chapter of the Association for Computational Linguistics*, 97–100. Stroudsburg, PA: ACL.
Bowker, Geoffrey C. 2008. *Memory Practices in the Sciences*. Cambridge, MA: MIT Press.
Bowker, Geoffrey, Karen Baker, Florence Millerand, and David Ribes. 2010. "Toward Information Infrastructure Studies: Ways of Knowing in a Networked Environment." In *International Handbook of Internet Research*, edited by Jeremy Hunsinger, Lisbeth Klastrup, and Matthew Allen, 97–117. Berlin: Springer.
Bowker, Geoffrey, and Susan Leigh Star. 1999. *Sorting Things Out: Classification and Its Consequences*. Cambridge, MA: MIT Press.
Breslin, Samantha, and Bimlesh Wadhwa. 2018. "Gender and Human-Computer Interaction." In *The Wiley Handbook of Human Computer Interaction*, edited by Kent L. Norman and Jurek Kirakowski, 71–87. Oxford: Wiley-Blackwell.
Browne, Simone. 2015. *Dark Matters: On the Surveillance of Blackness*. Durham, DC: Duke University Press.
Buolamwini, Joy, and Timnit Gebru. 2018. "Gender Shades: Intersectional Accuracy Disparities in Commercial Gender Classification." *Proceedings of Machine Learning Research* 81 (1): 77–91.
Bürger, Tobias, Elena Simperl, and Christoph Tempich. 2014. "Methodologies for the Creation of Semantic Data." In *Handbook of Metadata, Semantics and Ontologies*, edited by Miguel-Angel Sicilia, 185–215. Hackensack, NJ: World Scientific.
Caliskan-Islam, Aylin, Joanna J. Bryson, and Arvind Narayanan. 2016. "Semantics Derived Automatically from Language Corpora Necessarily Contain Human Biases." arXiv:1608.07187.
Campolo, Alex, Madelyn Sanfilippo, Meredith Whittaker, and Kate Crawford. 2017. "AI Now 2017 Report." https://ainowinstitute.org/AI_Now_2017_Report.pdf.
Cerbah, Farid, and Nadira Lammari. 2014. "Ontology Learning from Databases: Some Efficient Methods to Discover Semantic Patterns in Data." In *Perspectives on Ontology Learning*, edited by Jens Lehmann and Johanna Völker, 207–22. Berlin: IOS Press.
Cherry, Miriam A. 2009. "Working for (Virtually) Minimum Wage: Applying the Fair Labor Standards Act in Cyberspace." *Alabama Law Review* 60 (5): 1077–1110.

Cimiano, Phillip 2014. "Foreword." In *Perspectives on Ontology Learning*, edited by Jens Lehmann and Johanna Völker, v–viii. Berlin: IOS Press.

ConceptNet. 2017. "An Open, Multilingual Knowledge Graph." http://conceptnet.io/.

ConceptNet5. 2017. "FAQ." July 11. https://github.com/commonsense/conceptnet5/wiki/FAQ.

Coole, Diana. 2013. "Agentic Capacities and Capacious Historical Materialism: Thinking with New Materialisms in the Political Sciences." *Millennium-Journal of International Studies* 41 (3): 451–69.

Crain, Marion, Winifred R. Poster, and Miriam A. Cherry. 2016. *Invisible Labor: Hidden Work in the Contemporary World*. Berkeley: University of California Press.

Crawford, Kate, Meredith Whittaker, Madeleine Clare Elish, Solon Barocas, Aaron Plasek, and Kadija Ferryman. 2016. "The AI Now Report: The Social and Economic Implications of Artificial Intelligence Technologies in the Near-Term." https://ainowinstitute.org/AI_Now_2016_Report.pdf.

CrowdFlower. 2016. "Contributors." www.crowdflower.com/blog/topic/contributors.

Crutzen, Cecile K. M., and Hans-Werner Hein. 2009. "Invisibility and Visibility: The Shadows of Artificial Intelligence." In *Handbook of Research on Synthetic Emotions and Sociable Robotics: New Applications in Affective Computing and Artificial Intelligence*, edited by Jordt Vallverdú and David Casacuberta, 472–500. Hershey, PA: IGI Global.

Cycorp. 2017. "Knowledge Base." www.cyc.com/kb/.

Dobbe, Roel, Sarah Dean, Thomas Gilbert, and Nitin Kohli. 2018. "A Broader View on Bias in Automated Decision-Making: Reflecting on Epistemology and Dynamics." Paper presented at the Workshop on Fairness, Accountability, and Transparency in Machine Learning, ICML, Stockholm.

Dustdar, Schahram. 2015. "Challenges in Engineering Social Machines." Invited talk at the 2nd International Workshop on Crowdsourcing in Software Engineering, Florence.

Edwards, Paul, Geoffrey Bowker, Steven Jackson, and Robin Williams 2009. "Introduction: An Agenda for Infrastructure Studies." *Journal of the Association for Information Systems* 10 (5): 364–74.

Faulkner, Wendy. 2007. "Nuts and Bolts and People: Gender-Troubled Engineering Identities." *Social Studies of Science* 37 (3): 331–56.

Floyd, Christiane. 1992. "Software Development as Reality Construction." In *Software Development and Reality Construction*, edited by Christiane Floyd, Heinz Züllighoven, Reinhard Budde, and Reinhard Keil-Slawik, 86–100. Berlin: Springer.

Forsythe, Diana 1993. "The Construction of Work in Artificial Intelligence." *Science, Technology, & Human Values* 18 (4): 460–79.

Fox, Sarah, and Daniela Rosner. 2016. "Inversions of Design: Examining the Limits of Human-Centered Perspectives in a Feminist Design Workshop Image." *Journal of Peer Production* 8. http://peerproduction.net/issues/issue-8-feminism-and-unhacking/peer-reviewed-papers/inquiry-through-inversion-collisions-of-feminism-and-design-in-two-workshops/.

Fukushima, Masato. 2016. "Value Oscillation in Knowledge Infrastructure: Observing Its Dynamic in Japan's Drug Discovery Pipeline." *Science & Technology Studies* 29 (2): 7–25.

Goëta, Samuel, and Tim Davies. 2016. "The Daily Shaping of State Transparency: Standards, Machine-Readability and the Configuration of Open Government Data Policies." *Science & Technology Studies* 29 (4): 10–30.

Gramsci, Antonio. 1971. *Selections from the Prison Notebooks*. New York: International.

Gruber, Thomas 1995. "Toward Principles for the Design of Ontologies Used for Knowledge Sharing?" *International Journal of Human-Computer Studies* 43 (5): 907–28.

Gutiérrez-Rodríguez, Encarnación. 2010. *Migration, Domestic Work and Affect. A Deconolonial Approach on Value and the Feminization of Labor*. New York: Routledge.

Hall, Stuart. 1985. "Signification, Representation, Ideology: Althusser and the Post-structuralist Debates." *Critical Studies in Mass Communication* 2 (2): 91–114.

Hanika, Florian, Gerhard Wohlgenannt, and Marta Sabou. 2014. "The uComp Protégé Plugin: Crowdsourcing Enabled Ontology Engineering." In *Knowledge Engineering and Knowledge Management*, edited by Krzysztof Janowicz, Stefan Schlobach, Patrick Lambrix, and Eero Hyvönen, 181–96. Berlin: Springer.

Hanseth, Ole. 2010. "From Systems and Tools to Networks and Infrastructures—from Design to Cultivation: Towards a Design Theory of Information Infrastructures." In *Industrial Informatics Design, Use and Innovation*, edited by Jonny Holmstrøm, 122–56. Hershey, PA: IGI Global.

Hanseth, Ole, Eric Monteiro, and Morten Hatling. 1996. "Developing Information Infrastructure: The Tension between Standardization and Flexibility." *Science, Technology and Human Values* 21 (4): 407–26.

Indiana University Bloomington. 2016. "Ontology List." http://info.slis.indiana.edu/~dingying/Teaching/S604/OntologyList.html.

Introna, Lucas. 2016. "Algorithms, Governance, and Governmentality: On Governing Academic Writing." *Science, Technology and Human Values* 41 (1): 17–49.

Irani, Lilly C., and M. Six Silberman. 2013. "Turkopticon: Interrupting Worker Invisibility in Amazon Mechanical Turk." In *Proceedings of the SIGCHI Conference on Human Factors in Computing Systems*, 611–19. New York: ACM.

ISWC. 2013. "Microtask Crowdsourcing to Solve Semantic Web Problems." https://sites.google.com/site/microtasktutorial/.

Jackson, Steven J. 2014. "Rethinking Repair." In *Media Technologies: Essays on Communication, Materiality, and Society*, edited by Tarleton Gillespie, Pablo J. Boczkowski, and Kirsten A. Foot, 221–39. Cambridge, MA: MIT Press.

Kamiran, Faisal, and Toon Calders. 2009. "Classifying without Discriminating." In *2nd International Conference on Computer, Control and Communication*, 1–6. New York: IEEE.

Karasti, Helena, Florence Millerand, Christine Hine, and Geoffrey Bowker. 2016. "Knowledge Infrastructures: Part I." *Science and Technology Studies* 29 (1): 2–12.

Kuek, Siou Chew, Cecilia Paradi-Guilford, Toks Fayomi, Saori Imaizumi, Panos Ipeirotis, Patricia Pina, and Manpreet Singh. 2015. "The Global Opportunity in Online Outsourcing." Washington, DC: World Bank.

Law, John. 2017. "STS as Method." In *The Handbook of Science and Technology*, edited by Ulrike Felt, Rayvon Fouché, Clark A. Miller, and Laurel Smith-Doerr, 31–57. Cambridge, MA: MIT Press.

Ławrynowicz, Agnieszka, and Volker Tresp. 2014. "Introducing Machine Learning." In *Perspectives on Ontology Learning*, edited by Jens Lehmann and Johanna Völker, 35–50. Berlin: IOS Press.

Lehmann, Jens, and Johanna Völker. 2014. "An Introduction to Ontology Leaning." In *Perspectives on Ontology Learning*, edited by Jens Lehmann and Johanna Völker, ix–xvi. Berlin: IOS Press.

Lin, Yu-Wei, Jo Bates, and Paula Goodale. 2016. "Co-observing the Weather, Co-predicting the Climate: Human Factors in Building Infrastructures for Crowdsourced Data." *Science & Technology Studies* 29 (3): 10–27.

Lytras, Miltiadis, Miguel-Ángel Sicilia, and Cristian Cechinel. 2014. "The Value and Cost of Metadata." In *Handbook of Metadata, Semantics and Ontologies*, edited by Miguel-Angel Sicilia, 41–62. Hackensack, NJ: World Scientific.

Mädche, Alexander, and Steffen Staab. 2004. "Ontology Learning." In *Handbook on Ontologies*, edited by Steffen Staab and Rudi Studer, 173–90. Berlin: Springer.

Mager, Astrid. 2014. "Defining Algorithmic Ideology: Using Ideology Critique to Scrutinize Corporate Search Engines." *tripleC: Communication, Capitalism & Critique* 12 (1): 28–39.

Maynard, Diana, and Kalina Bontcheva. 2014. "Natural Language Processing." In *Perspectives on Ontology Learning*, edited by Jens Lehmann and Johanna Völker, 51–67. Berlin: IOS Press.

Meißner, Hanna. 2016. "New Material Feminism and Historical Materialism. A Diffractive Reading of Two (Ostensibly) Unrelated Perspectives." In *Mattering: Feminism, Science, and Materialism*, edited by Victoria Pitts-Taylor, 43–57. New York: New York University Press.

Monteiro, Eric, Neil Pollock, Ole Hanseth, and Robin Williams. 2013. "From Artefacts to Infrastructures." *CSCW* 22 (4–6): 575–607.

Noble, Safiya Umoja. 2018. *Algorithms of Oppression. How Search Engines Reinforce Racism*. New York: New York University Press.

Noy, Natalya, Jonathan Mortensen, Mark Musen, and Paul Alexander. 2013. "Mechanical Turk as an Ontology Engineer? Using Microtasks as a Component of an Ontology Engineering Workflow." In *Proceedings of the 5th Annual ACM Web Science Conference*, 262–71. New York: ACM.

Ong, Aihwa. 2006. *Neoliberalism as Exception. Mutations in Citizenship and Sovereignty.* Durham, DC: Duke University Press.

Oudshoorn, Nelly, Els Rommes, and Marcelle Stienstra. 2004. "Configuring the User as Everybody: Gender and Design Cultures in Information and Communication Technologies." *Science, Technology, & Human Values* 29 (1): 30–63.

Pedreshi, Dino, Salvatore Ruggieri, and Franco Turini. 2008. "Discrimination-aware data mining." In *Proceedings of the 14th ACM SIGKDD International Conference on Knowledge Discovery and Data Mining*, 560–68. New York: ACM.

Platform Cooperativism Consortium. 2018. "Platform Cooperativism." https://platform.coop/.

Ribes, David, and Geoffrey Bowker. 2009. "Between Meaning and Machine: Learning to Represent the Knowledge of Communities." *Information and Organization* 19 (4): 199–217.

Rommes, Els, Corinna Bath, and Susanne Maass. 2012. "Methods for Intervention: Gender Analysis and Feminist Design of ICT." *Science, Technology, & Human Values* 37 (6): 653–62.

Rudolph, Sebastian, and Bariş Sertkaya. 2014. "Formal Concept Analysis Methods for Interactive Ontology Learning Methods." In *Perspectives on Ontology Learning*, edited by Jens Lehmann and Johanna Völker, 247–61. Berlin: IOS Press.

Schmidt, Ben. 2015. "Rejecting the Gender Binary: A Vector-Space Operation." October 30. http://bookworm.benschmidt.org/posts/2015-10-30-rejecting-the-gender-binary.html.

Scholz, Trebor, ed. 2013. *Digital Labor: The Internet as Playground and Factory.* New York: Routledge.

——, ed. 2016. *Platform Cooperativism. Challenging the Corporate Sharing Economy.* New York: Rosa Luxemburg Foundation.

Simperl, Elena, Gianluca Demartini, and Maribel Costa. 2013. "Crowdsourcing for the Semantic Web." Lecture slides. https://sites.google.com/site/crowdsourcingtutorial/.

Simperl, Elena, Stephan Wölger, Stefan Thaler, and Katharina Siorpaes. 2014. "Learning Ontologies via Games with a Purpose." In *Perspectives on Ontology Learning*, edited by Jens Lehmann and Johanna Völker, 225–46. Berlin: IOS Press.

Sowa, John 2011. "Future Directions for Semantic Systems." In *Intelligence-Based Systems Engineering*, edited by Andreas Tolk and Lakhmi Jain, 23–47. Berlin: Springer.

Speer, Rob. 2017a. "ConceptNet Numberbatch 17.04: Better, Less-Stereotyped Word Vectors." April 24. https://blog.conceptnet.io/2017/04/24/conceptnet-numberbatch-17-04-better-less-stereotyped-word-vectors/.

——. 2017b. "Changelog." July 6. http://github.com/commonsense/conceptnet5/wiki/Changelog.

——. 2017c. "How to Make a Racist AI Without Really Trying." July 13. https://blog.conceptnet.io/2017/07/13/how-to-make-a-racist-ai-without-really-trying/.

Speer, Robert, Joshua Chin, and Catherine Havasi. 2017. "ConceptNet 5.5: An Open Multilingual Graph of General Knowledge." In *Proceedings of the Thirty-First AAAI Conference on Artificial Intelligence*, 4444–51. Menlo Park, CA: AAAI.

Star, Susan Leigh. 1990. "Power, Technology and the Phenomenology of Conventions. On Being Allergic to Onions." *Sociological Review* 38 (1): 26–56.

Star, Susan Leigh, and Geoffrey Bowker. 2006. "How to Infrastructure." In *The Handbook of New Media: Social Shaping and Social Consequences of ICTs*, edited by Leah A. Lievrouw and Sonja Livingstone, 230–45. London: Sage.

Star, Susan Leigh, and Karen Ruhleder. 1996. "Steps toward an Ecology of Infrastructure: Design and Access for Large Information Spaces." *Information Systems Research* 7 (1): 111–34.

Star, Susan Leigh, and Anselm Strauss. 1999. "Layers of Silence, Arenas of Voice: The Ecology of Visible and Invisible Work." *Computer Supported Cooperative Work* 8:9–30.

Studer, Rudi, Richard Benjamins, and Dieter Fensel. 1998. "Knowledge Engineering: Principles and Methods." *Data & Knowledge Engineering* 25 (1): 161–97.

Suchman, Lucy. 1995. "Making Work Visible." *Communications of the ACM* 38 (9): 56–64.

——. 2007. *Human-Machine Reconfigurations: Plans and Situated Actions*. 2nd ed. New York: Cambridge University Press.

——. 2008. "Feminist STS and the Sciences of the Artificial." In *The Handbook of Science and Technology Studies*, edited by Edward J. Hackett, Olga Amsterdamska, Michael E. Lynch, and Judy Wajcman, 139–63. Cambridge, MA: MIT Press.

Sweeney, Latanya. 2013. "Discrimination in Online Ad Delivery." *Communications of the ACM* 56 (5): 44–54.

Trauth, Eileen M. ed. 2006. *Encyclopedia of Gender and Information Technology.* Hershey, PA: IGI Global.

Van Heur, Bas, Loet Leydesdorff, and Sally Wyatt. 2013. "Turning to Ontology in STS? Turning to STS through 'Ontology.'" *Social Studies of Science* 43 (3): 341–62.

Veale, Michael, and Reuben Binns. 2017. "Fairer Machine Learning in the Real World: Mitigating Discrimination without Collecting Sensitive Data." *Big Data & Society* 4 (2): 1–17.

West, Karen. 2011. "Articulating Discursive and Materialist Conceptions of Practice in the Logics Approach to Critical Policy Analysis." *Critical Policy Studies* 5 (4): 414–33.

Wohlgenannt, Gerhard, Marta Sabou, and Florian Hanika. 2016. "Crowd-Based Ontology Engineering with the uComp Protégé Plugin." *Semantic Web* 7 (4): 379–98.

The Energy Walk

Infrastructuring the Imagination

Brit Ross Winthereik, James Maguire, and Laura Watts

We need stories (and theories) that are just big enough to gather up the complexities and keep the edges open and greedy for surprising new and old connections.

—Haraway (2015)

Imagine a place at the windswept edge of Denmark, where the salty waves batter the sandy shores. A harbor place, where boats once fished aplenty and fishermen lorded their catch. Today, the sight of a fishing boat in the large industrial harbor is rare and the fishing that takes place is increasingly regulated by onboard monitoring technology fitted to the trawlers. This harbor, Denmark's second largest measured by catch, fails to secure jobs for local inhabitants as the trawler industry becomes ever increasingly globalized, attracting workers from around the world. New ideas for countering this development in order to secure the next generation's future are welcome, says the director for the association of local businesses, adding that a solid focus on renewable energy may be a way forward in the effort to create more jobs in the region.

There are initiatives to counter falling local employment, but the future, as always in small places, is uncertain. In 2007, a national park, Nationalpark Thy, was established to protect the dunes, heather, and coastal lakes of this region on the edge of Denmark. In addition to the tourism related to the national park, *energy tourism* is becoming a locus of interest for the local government. The municipality received the European solar prize in 2007 and since then has been working to brand itself as *the* "climate municipality of Denmark." In addition to this, it collaborates with local businesses regarding sustainable solutions that benefit the local community, as well as organizing trips to sites where climate-friendly technologies are being demonstrated. With the area being self-sufficient in terms of green energy—notably wind, but biomass and geothermal power also play a role—there are indeed many such sites for the public to visit. But the question still remains whether energy tourism can leverage an increase in new jobs to replace those lost as a consequence of the changes occurring in the fishing industry.

How might energy tourism contribute to this? In the dune plantation Østerild, large-scale wind turbines are being tested. Transforming the forest into a test site for three 200-meter-tall wind turbines was highly controversial at first, as a plantation had to be cut and the landscape transformed. Not everyone thought it a great idea that Siemens and potential Chinese companies put such a huge mark on the landscape. The turbines attract tourists. Not in large numbers, but enough to have engaged the municipality as tour organizer. A large device for the conversion of ocean waves into electricity is included as one of the demonstration sites that tourists can visit when they go to this edgy place. The idea is to offer firsthand experiences with clean tech engineering and renewable energy to the public. According to a consultant working in wave energy, such initiatives are crucial for innovation in renewable energy technology. If people have not felt the "forces of nature on their own bodies," if they have not witnessed that natural forces can indeed be transformed, they will be unable to "get" and trust renewable sources. Note that this consultant does not use the word "understand"; he is speaking about "getting" renewability in an embodied way.

While the public must learn how to trust renewables as a stable form of energy, they must also learn to trust the technology developer's dreams that something viable, and lasting, is afoot. A consultant from the Danish Wave Energy Center (DanWEC), which is an organization for wave energy stakeholders in Denmark, told us that in Denmark wave energy projects are often considered "utopian." He said that the biggest threat to leveraging public and private investment in the wave energy sector is the reputation that wave energy developers have of being on some sort of "unrealistic" or "idiotic" quest.

Being conjured up here is a catch-22 for the emerging Danish wave energy sector and for "getting" renewables in an embodied way: A public that does not get wave energy is unlikely to elect politicians who support experimentation with new technology and develop funding schemes that are likely to bridge "the valley of death" between technology development and private investment. At the same time, without experimentation, there simply is nothing to showcase and the public will continue to "not get" renewability and wave power.

At stake are questions of how technology, sociality/community, nature, and politics are related, and related well (Latour 2003; Ang 2011), and of how nature becomes kin (Haraway 2015). How can science and technology studies participate in authoring fictions that help address the environmental urgencies we are facing, and that connect us to green and sustainable energy futures? In this chapter we propose engaging the digital in fabricating an embodied connection to that which is not quite seen.

Alien Energy

Our ethnographic studies of water-based energy forms (marine energy and geothermal) had led us to the Danish northwest coast, where DanWEC is located. Here we learned about the activities around energy tourism, and found that experimentation with ethnography through the design of a digitally mediated infrastructure for public involvement might be a generative way of engaging with the place and its energies.

The Energy Walk invites the public to see the landscape around DanWEC as a place of energy. Walking with an interactive, digital walking stick and headphones, visitors experience the wind-swept geographical edge augmented by a poetic narrative on the relation between energy and landscape. The Energy Walk engages the public in an imaginative comparison of energy landscapes in Scotland, Iceland, and Denmark, and their respective renewable energies. As we show in the following, the Energy Walk is also a site for exploring the role of the imaginative and the digital in science and technology studies.

Prior to designing and installing the Energy Walk in the town that is known for its strong wind as well as "Denmark's best waves," we had interviewed developers of wave energy devices, observed their meetings, studied their reports, and followed their mail correspondence as well as the public debate on renewability. However, it was not before we had thrown ourselves into the task of designing the Energy Walk that the town emerged as a place where "energy tourism" formed a tentative link between a growing clean tech industry and a public learning to trust alien energy forms.[1]

At the time of making the Energy Walk we were two senior scholars, two junior scholars, and a research assistant affiliated with the research project.[2] One was conducting fieldwork in Denmark around the wave energy industry, another was continuing fieldwork around wave and tide energy in Scotland. Our third field site was focused on geothermal energy in Iceland. Three different places, people, and industries, but with something common across their energetic and watery experience.

We all had substantive experience with ethnographic fieldwork, and were eager to pursue a more design-oriented and interventionist version of ethnography than the anthropology-inspired version that is routinely embraced in science and technology studies (Winthereik et al. 2002; Watts 2012; Winthereik and Verran 2012). Very often such STS ethnography is a mix of multisited observation, structured and semistructured interviews, and document analysis and relies on a separation of "home" and "the field" (despite much emphasis on this being otherwise; Gupta and Ferguson 1997; Strathern 1991). And while certainly a mode of intervention, this method of data generation usually does not hinge on the installation of designed objects.[3]

In the following, we describe some of the practicalities involved in designing and installing the Energy Walk. We then analyze the walk as an infrastructure of the imagination and an intervention that inspired methodological experimentation with futures not yet seen. We first offer a brief introduction to infrastructure studies and place our intervention in this theoretical context as we move toward literature dealing with the imaginative and its relation to technology.

Infrastructure Interventions

Studies of infrastructure have always been central to STS since early work on large technological systems, such as Thomas Hughes's study of electrification in the United States, United Kingdom, and Germany (Hughes 1993). Hughes argued that electrical systems are constituted by not only interconnected technological artifacts, but also national politics, policy, as well as geography. Electricity grids, Hughes argued, are not technically determined, but a result of constant negotiation among a host of heterogeneous actors—a classic STS move. Susan Leigh Star soon applied a more nuanced and ethnographic attention to the "invisible work" involved in making and maintaining infrastructure (Timmermans et al. 1998; Star 1999). She and her colleagues were interested in the success of infrastructure invisibility, in how infrastructure has a taken-for-granted banality (until it breaks down), which often means it is overlooked in social and cultural research. Operating infrastructures, from water to data to electricity grids, are mundane, unremarked, and unremarkable.

There is an ongoing commitment in STS to infrastructure, to making it visible, remarked upon. For example, Nicole Starosielski's work on the undersea cables that constitute the Internet shows how the invisible "cloud" connecting our devices and data is actually a rather centralized and very material set of fiber-optic cables, with implications for both the politics and imaginaries of what might otherwise be considered invisible data (Starosielski 2015). Ashley Carse has pursued the inseparability of what might be considered "natural" landscape and water-management infrastructure through a study of the Panama Canal system. The canal requires the surrounding forest water catchment to operate (Carse 2012). Both these examples were important to our intentions with the Energy Walk: we knew that landscape and energy infrastructure were entangled, beyond the usual markers such as a wind turbine or solar panel; and we knew that data infrastructures were another "invisible" energy infrastructure in the landscape. Simply, electricity and electrically operated telecommunications are classic invisible infrastructures in the landscape.

Studies of infrastructure do not end with making critical remarks, however. They are also a site for intervention, a site where entities and actors might spin out to make new relations (Jensen and Morita 2017), where different worlds might be brought into being. Infrastructures are *sites of experimentation* and a methodological challenge (Jensen and Winthereik 2013; Blok et al. 2016; Maguire and Winthereik 2016). In these cases, it is neither their social aspects nor their embedded politics alone, but how sociopolitical techno-worlds are made as entities (including the not yet seen) are assembled together, in continuous dynamic interaction (Allhutter, this volume).

Following this, our approach to the Energy Walk was to commit to making an intervention into the existing energy landscape. We wanted to open up the possibility for participants, on the walk, to be in a more dynamic interaction with the energy infrastructures around them. Indeed, as we shall later discuss, we struggled with how deeply entangled our assembly of Energy Walk and energy landscape became.

In our work, we considered ourselves as extending this long history of infrastructure studies and its recent interest in world making and mutual remaking of worlds and infrastructures. Researchers have long been involved in such worldly infrastructural composition. But we wanted to explore how this might be done with the landscape, not just at the desk, not just through the printed page. What other kinds of experiments might be helpful in making infrastructure interventions? In the next section we describe the Energy Walk in more detail as one such attempt.

The Energy Walk

Listen . . . to my voice. Just sit for a moment, breathe, listen.
Listen. There are seagulls in the air, following the fish from the sea.
Listen. Can you hear the engines of a ferry? [thum thum, thum thum, thum thum]
Those four white ticket booths, there, across the road in front of you, can hear.

I have been trying to listen to these "edge of the world" places and their energy.
Why is there wave energy here, in Hanstholm? Why in Orkney? Why geothermal energy in Iceland?
So I walk. I listen through the soles of my feet, listen through the walking stick.
Walk with me, along the pavement. Let the walking stick guide you.

—Extracts from the Energy Walk

In the Alien Energy research project, we were inspired by the propositions put forward by our interlocutors in the local community and its local industries. They told us that to "get" new renewable energy, such as wave, tide, and geothermal energy, people (including politicians and policy makers) must learn to be affected by it. The new renewable energy forms we were working with are invisible and unknown to most people. Keen to take up the challenge of making nature kin (see Haraway 2015) by means of affect rather than through educational or various rhetorical devices, we began working on the Energy Walk. Guided by the idea that on-site, firsthand experiences are effective for "getting" relations between landscapes and renewable energy, we studied site-specific art installations. In particular, we were taken in by the walk and sound installations of Janet Cardiff and George Bures Miller.[4] One of these installations, a walk through Munich Central Station, uses a mobile application to send the visitor back to World War II, when the station was used for the deportation of Jews. We were intrigued by the way Cardiff and Miller used walking as a means to explore human history, and wondered if we could make a design that would allow for exploring our common future. We discussed this as an experiment in method rather than art: a method for repatterning situated knowledges with that which does not yet or only partly exists (Hughes and Lury 2013). An invention.

Previously, as part of a video project, we had conceptualized the Alien Energy project as an attempt to "make the invisible visible."[5] In this video, the spectator is taken to a town where devices for marine renewable technology are being tested, and is invited inside the technology to see the rusty practicalities of future making. Similarly, the Energy Walk emphasizes the importance of energy as a *site*, its landscapes, technologies, politics, and humans. The walk takes place in the landscape of the town that hosts DanWEC, and since it is an installation designed for that particular place, people, architectures, and infrastructures, it cannot be moved or scaled, which is an important feature of the work.

The Energy Walk is an immersive experience of walking through a landscape accompanied by an interactive, digital walking stick, which augments the experience through sound. The walk takes about 40 minutes and can be done without a map as directions for walking are given by audio. In addition to walking with a digital object, the intimacy of the relationship between the visitor and the voice recording is something we adopted from Cardiff and Miller's artwork. The Energy Walk differs in several ways from the audio guides in use at most major touristic sites. At places of agreed tourist value, the audio is often installed to *teach* the visitor a comprehensive, fact-based account of the place. In comparison, while the Energy Walk does tell some "facts," it mostly seeks to defamiliarize the place and the landscape, and to do so in a way that affects the visitor emotionally. The voice in the headphones is one means to achieve this. This voice is not of a neutral kind, meant to enlighten, but is of a friendly companion sharing her knowledge of and experiences with renewable energy in the places she has traveled–our three field sites in the research project.

> If you do not live here, at the edge, then you may not know about wave energy.
>
> Its history is already long. Both Orkney and Denmark have been testing . . . and testing . . . prototypes and possibilities, since the 1970s. Out there, the North Sea has seen wave dragons, and wave stars; sea snakes, and salter's ducks.
>
> —Extracts from the Energy Walk

Standard tourist sites tend to produce a sense of being overwhelmed and are imagined to be, in some sense, speaking for themselves. These sites can be explored without audio. While the coastal landscape in northern Denmark is impressive with its beaches and cliffs overlooking a gray-blue ocean, the Energy Walk conjures up a world that is not, and cannot be, without the audio, a world not present and not seen without the sound (see figure 1). While one could argue that this, in principle, could also be the case at major tourist sites, a key difference is that if we failed to engage visitors emotionally and morally, there simply would be nothing for them to see. Therefore, we sought to place the visitor in the middle of energy pasts, presents, and futures using the specific landscape and the visitor's imagination to achieve this here-now.

We worked in close collaboration with an interaction design company, led by two experienced fabricators and engineers, who specialized in making innovative devices for public engagement,[6] They helped us develop the concept, and designed and constructed the walking sticks, including the electronics, audio controls, charging system, and RFID tagging. They also fitted the sticks with a wooden bulb to protect the hardware, and to which headphones could be inserted. They made four walking sticks for us, so even though the walk itself was intended to be an immersive and solitary experience, the walk could be done as a group.[7] Laura Watts, who is also a poet, authored the text for the Energy Walk audio with input from the other team members. Together they created the path for the walk through which the poetic text would move. The final soundscape was the result of a careful calibration with the landscape, and we made several test runs to ensure alignment between the sound and posts in the landscape, as well as to ensure a suitable walking pace. For example, a gravel path had to be visible when the audio suggested that the visitor walked toward one. To make the Energy Walk a personal experience that would allow for a pace befitting the physical condition of all possible visitors, the audio file was separated into six chapters. Each chapter was activated by touching the walking stick to posts in the landscape fitted with RFID tags.[8]

Six posts, adorned with the Alien Energy logo, created a recognizable physical infrastructure within the landscape (see figure 2). As we found out when applying for local permission to install these posts, a certain type of bureaucratic tension is embedded in the area; some parts are administered by the Nature Agency and other parts by the municipality. This indexes a long and complicated history of land ownership and governance. Interestingly, yet outside the scope of this chapter, ownership issues are also at stake at sea, and pertinent to know about when planning to install wave energy converters at sea.

To make manifest the link between the soundscape and the landscape, we used stripped birch branches from the nearby national park for the walking sticks. Thus, the makeup of the Energy Walk infrastructure was manifold, including the

FIGURE 1: View over Hanstholm harbor. © Laura Watts.

FIGURE 2: Energy Walk post with RFID tag. © IT University of Copenhagen.

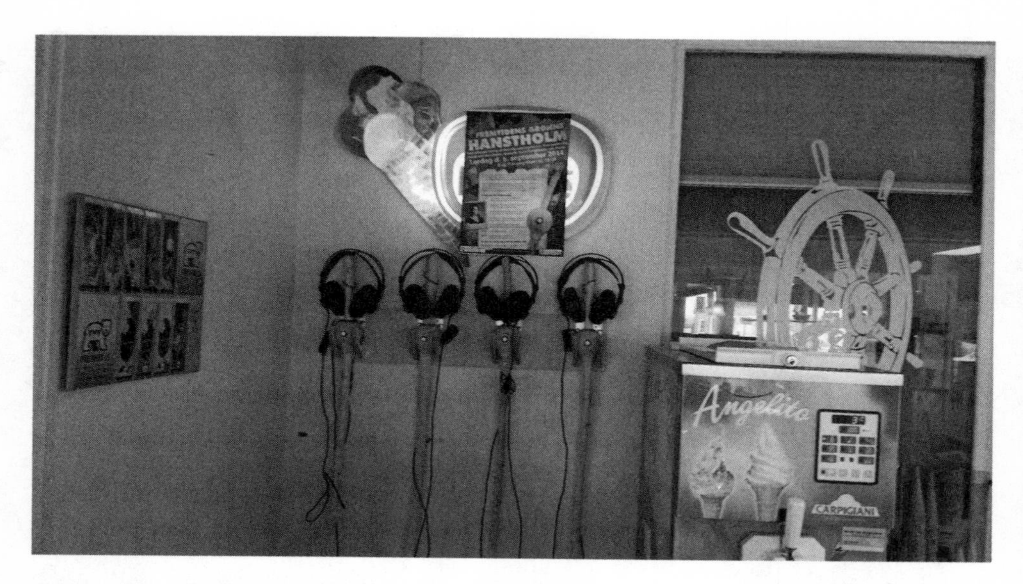

FIGURE 3: Walking sticks recharging on the wall at the Ferry Grill. © Brit Ross Winthereik.

posts in the landscape, ethnographic records, birch sticks, visitor bodies, local bureaucracy and its history, as well as antennas, monuments, gates, fences, and vistas in the landscape.

The owners of a harbor café, the Ferry Grill (Færgegrillen), agreed to act as keepers of the walking sticks (see figure 3). We contacted the grill when we discovered that the town had no public service organizations, such as a library that could hand out the walking sticks and inform visitors about their use. The café seemed the perfect place, as it already aspired to be the town's "cultural center," selling grilled fish as well as handicrafts. Moreover, it is locally renowned for entertaining a large number of regular customers, locals as well as tourists, and for sponsoring a local soccer team.

When showing the pamphlet that advertised the walk to a fellow anthropologist, Karen Waltorp, who had grown up in the area and moved to Copenhagen in her teens, she said, "The photos are nice, but it is strange to see a photoshopped version of [the town]." Interestingly, the photographs in the pamphlet had not been through any optimization or treatment prior to printing, but thinking about our colleague's reaction revealed how our rendering of the town is in stark contrast to the general opinion of it. When hearing about the location of the Energy Walk, many have expressed an opinion of the town as being remote, industrial, and not a place they would ever consider visiting, even if on holidays in the area. This points to our endeavors as part of a difficult "selling" or "branding" exercise, the Energy Walk being a participant in the efforts to explore and perform new relations on the geographical edge.

> It feels like we are standing on the edge of an island, below us, a sea of heather.
>
> And we are. Hanstholm was an island, surrounded by water, until tectonic plate forces, makers of geothermal energy, made the land rise out of the water.
>
> Icelanders are right to feel at home, for this is a geothermal landscape.
>
> —Extracts from the Energy Walk

The Energy Walk begins and ends at the Ferry Grill. The visitor walks from the pavement outside the grill and climbs up through the sand dunes on wooden stairs. As she steps onto the stairs, the narrator's voice invites her to think of wood as a historically significant source of energy. Now standing on a cliff, she is facing the harbor and the sea. She then moves along the cliff heading inland, eventually passing the historical lighthouse, the church, and the cemetery where fishermen and innovators and many others are buried. She is invited to think of the social relations that make up this small town and consider the notion of "a community" as making invisible conflicts and tensions that are also an intricate part of social relations. Walking past electricity generators, the visitor is nudged to think of invisible networks of power and the global construction of standards allowing for the travel of goods and utilities.

The walk returns the visitor to a viewpoint on the cliff from where a huge prototype construction for harnessing wave energy dominates the view. The narrator makes a proposal to the visitor: imagine thousands of machines for wave energy capture filling the sea in a not-too-distant future, and imagine that you are one of the people who thinks this is not a utopian dream, but a possible future.[9]

> I think, I hope, that what I dream may, in some small part, make a future that never was: wave energy, geothermal energy, all things that once never were, but may almost be.
>
> Listen. There are seagulls in the air, following the fish from the sea.
>
> Listen. Can you hear the engines of the ferry? Can you hear the waves, far off, feeding a hundred devices with energy? The waves are bone deep. [thum thum, thum thum, thum thum]
>
> I am going to leave you now, to your own dreams, your own futures.
>
> Stand here for a while, with your dreams, and then walk back down the wooden stairs, and go back to the cafe.
>
> —Extracts from the Energy Walk

Infrastructuring the Imagination

After this description of the Energy Walk, we now return to the notion of infrastructures as experimental sites and to the question of how ethnographers can participate in composing them. We bring into focus the importance of the imagination for understanding infrastructures' world-making capacities and consider how we as ethnographers relate to the compositional work done by infrastructures (Singh, Hesselbein, Price, and Lynch, this volume).

On launch day researchers from the Alien Energy project followed the first walkers around. Descending from the place where one looks out to the sea at the wave energy device one of the walkers, a woman in her late 50s exclaimed, "It is impressive how you have been able to grasp and communicate exactly what it feels like living here on the edge. It is spot-on!" She said it with tears in her eyes. The situation was moving for us because we felt we had made a situated intervention that had mattered to someone, but in that moment we had no way of knowing what made her cry. Was it because her relationship with the landscape, the wind, and the place was rendered in a poetic narrative form, one that articulated that which usually remains implicit and unspoken? Was it because walking linked the embodied experience of dwelling in a specific place with conceptual resources?

We noted with curiosity that the woman used the notion of the edge to characterize the place, which was a term that we had invented for the Energy Walk. Her use of it was noteworthy because in Danish "edge" (*kant*) is not commonly used for locations. We introduced "edge," "edge place" (*kantsted*), and "edge dweller" (*kantboer*) because we consider these terms as expressing a politics that runs counter to the standard way of presenting certain places. More specifically, the edge forms a significant contrast to the notion of outskirt (*udkant*). *Udkant* has been used a lot in the Danish national media since 2010 and is increasingly used to characterize regions that are relatively far removed from a city. Needless to say the notion of *udkant* is considered degrading by the people, who are supposedly living in *udkants*-Denmark. During the walk the woman might have experienced that she possessed *infrastructural competence*, meaning that her knowledge of the place emerged as valuable and relevant for thinking about the place as interconnected with other places and in the context of the place as part of sustainable futures (for an introduction to infrastructural competence, see Erikson, Sawyer, and Jarrahi, this volume).

We were taken by surprise by the woman's response, but have since come to entertain the possibility that we were, together, composing a world, one that was solid enough for her to enter into and become part of, yet flexible enough to spur an unanticipated affective response. In reflecting on imagination in anthropological thinking, it has been argued that imagination is an aspect of a human being's existential power. "It describes . . . the potentiality that we are able to see in the material world, the myriad ways in which we are able to make sense of it and to remake that sense." According to this approach, imagination describes the drive humans have to make decisions about "who and what they will be (their essence)" (Rapport and Harris 2015, 20). This drive often sits in tension with the specificity of particular lives that seem to ground the imagination.

To observe tensions between general human bodily capacities and individual bodies in the here-now, our methods must be imaginative. In the context of digitalSTS, we are interested in the technological makeup of imaginative methods. In relation to the example above, the woman-as-walker, the walking stick, infrastructure and technologies in the landscape, and the audio made "edge" emerge as a specific nature-culture relation.[10]

Going beyond the idea of the imagination as universal human capacity, an overarching template for thought and action or as a way of making sense of the world, anthropologists Sneath and colleagues (2009) argue that imagination is an effect. Following on from previous work on relations between technoscience and imagination, Sneath et al. argue that imagination is an effect of technological process, yet is thoroughly underdetermined by such processes (2009, 19). Thus, while imagination is an effect, it is not a direct effect of particular sets of technological processes. If we were to think of imagination as being determined by technology, imagination would lose its quality and no longer be imaginative. Imagination, in other words, is a result of certain (material, social) affordances, but also a play with such affordances (Marcus 1986; Watts 2012). To get a conceptual grip on this "mess" (Law 2004), we must explore the conditions under which unconditional outcomes happen. To Sneath et al. (2009), technologies of the imagination are technologies of the incidental.

If we consider imagination as an effect, and the walking sticks as technologies of the incidental, how does this fit with design as a structured and structuring endeavor? As we developed technology that would allow for new configurations of landscape and humans to emerge, we were occupied with avoiding anything incidental. Creating the walk was a highly controlled and structured process, as we needed to make an object that people would feel attracted to, an object that they would spend time using. It would have to be an object solid enough to be out in the wind and rain over a period of three months, and an object that we could "leave alone," hoping that others would adopt it. It would have to be a desirable object, but also one that visitors could not easily run away with. The infrastructure might hold boundary object quality, but could not be open to conceptual or technical vagueness (Parmaggiani and Monteiro, this volume).

You might say that we hardened the design of our infrastructure, including its conceptual components, such as the meaning of "edge." In retrospect, we were seeking to control the imaginative effects of the walk *and* allow for it to be a personal experience, one that would inspire emotional response and reflection. And so designing the walk was not a matter of creating infrastructural coherence to avoid emergence. Moreover, it was not about design phases (first we control, then we let go and see what comes out of it, then we record the imaginative effects). Rather, similar to Emilie Gomart and Antoine Hennion's analysis of music lovers and recreational drug users' expertise in creating apparatuses that allow for certain emotional and bodily effects (1999), the Energy Walk exemplifies a *dispositif.* Taken from Foucault, a dispositif is an arrangement that is both sufficiently constraining and sufficiently loose for it to have effects that might be considered generative. The efficacy of a dispositif lies in its capacity to *fait faire* (make do), that is, its nondetermining way of working as a structuring arrangement. In Gomart and Hennion's rendering, the generative power of dispositifs depends upon their capacity to form new subjectivities in the persons who pass through them. As such, the dispositif is an arrangement that is effective due to the generative constraints it offers to the passer-through.

Compared to Sneath and colleagues' (2009) telling of a space that is both (technologically) constrained and (imaginatively) open, in Gomart and Hennion's (1999) work emphasis is put on the work of designing the right kind of constraint. Their interlocutors were actively designing constraints that they anticipated would create the right kinds of effects. In a similar manner, the effects of the walking sticks did not just reveal themselves as effects after the infrastructure had been put into use. Instead, these effects hinged on the efforts we put into producing an enjoyable and effortless experience for the visitor that would nevertheless be constraining enough to be generative of several imaginative processes—the visitors' and our own.

Thinking through the digital walking stick has shown that technologies of the imagination can indeed be understood as "theoretical objects" as Sneath et al. argue (2009, 18). What this means is that such technologies are actively producing and distorting visions, beliefs, and knowledges as they open up a space for the imagination. However, an important aspect of them working as theories is that they are designed and crafted with much effort. In this particular instance, we, the ethnographers, undertook this effort. When ethnographers design the dispositif, it may or may not make a difference for the people who pass through this technologically enabled space. However, it makes a whole lot of difference for the ethnographers' imaginative capacities to not only observe working technologies of the imagination, but also participate in designing and installing them.

A final example from the launch day illustrates how. In preparing for the launch after infinite rounds of testing, we made one final test of the walking sticks. To our surprise, the audio behaved strangely, and it turned out there was interference between the trawlers' monitoring and communication devices and the mp3 players fitted in the sticks. While, in theory, this was an interesting case of technological unruliness, it also posed an immediate problem as we had invited the whole town to see the walking sticks in action and experience the Energy Walk. We called in one of the designers for help, and a work-around was established and the sticks were demonstrated more or less as planned.

This incident teaches us about the capacity of the digital to interrupt our technological setup and the dispositif that, in our imagination, would do the trick of passing visitors through a transformative collection of constraints. It was both the Energy Walk as a self-sufficient and bounded entity that was interrupted and our expectations that such a bounded entity could indeed be designed and controlled. The launch was a culmination of a strictly controlled process in which a researcher-designer-interlocutor collective had fabricated an object that could exist on its own. We saw the launch as a final deliberate hardening of the boundaries around the installation. As the walking sticks established relations with other actors in the environment, as they began communicating with other entities than the tags on the posts, the Energy Walk threatened to unravel as it became too entangled in its surrounding energy infrastructures. But more than that, it made the industrial fishing industry visible and reminded us that any intervention, including interventions that take place as a consequence of an emerging marine renewable energy industry, needs to both present itself as "new" and be able to coexist with whatever entities are already present. Emerging as semiautonomous actor, the sea was made available to us by the walking sticks in ways that made "edge issues" of disappearing jobs and hopes for energy tourism a relation already and always present in places where ocean and land meet.

Conclusion

What does the Energy Walk add to ideas about infrastructures as sites of experimentation? Our analysis of the Energy Walk suggests that there is an imagined dimension to infrastructure—for the visitors as well as for the researchers. Well-working infrastructures become invisible, not in the sense that they cannot be seen, but in the sense that there is a tendency to stop seeing them over time. Bringing them back into view requires imaginative capacities, but not in any universalized sense. Such capacities are activated within a techno-imaginative setup, as specific sets of effects are produced, yet remain contingent (underdetermined). In this ethnographic instance, the Energy Walk as dispositif activated the visibility of certain infrastructures in the landscape. In so doing, it brought to the fore the experience that all vision is situated and potentially a resource for political action (Haraway 1991, 193; Suchman 2011; Forlano 2013). We still do not know what specific *edge politics* around energy and renewability will look like, but energy tourism that highlights relations between place and knowledge (knowl-*edge*) might be a beginning.

Moreover, within this techno-imaginative space and of relevance for a potential counterpolitics, in our ethnographic example, one of the visitors became aware of herself as edge dweller and of situated knowledges, for example, differences between the edge and the center that do not conceptualize edge as remote (*udkant*). While it is hard to pass any final judgment over the effect of the Energy Walk dispositif with respect to the production and alignment of subjectivities, this surprising reconfiguration of the remote place as *edgy* created an imaginative shift. This shift indexes a situation in which our ideas and hers are put into experimental juxtaposition.

Of particular relevance to the digitalSTS community and anthropological infrastructure studies is how the digitally mediated walk elicited imaginative effects, for example by conjuring up a reality where sea-powered wave energy devices are brought out of the realm of "utopian devices" and placed in relation to a possible future, forming new relations in/with the environment. The role of "the digital" in the Energy Walk was that it distorted our, the researchers', perception of the landscape and afforded a reconfiguration of our thoughts about what can possibly enter into relation with what. The digital sticks recalcitrantly made visible how we had wrongly believed that the digital had blended into the background as infrastructure. As these beliefs were overturned, our imaginative processes were spurred and opened a space between sensing and understanding and between the seen and the not yet seen. We began imagining new things about the place and the industries based there; we imagined fish, we no longer saw the sea as a big empty space, but as an infrastructured space ordered by bureaucracy, and occupied by the military. In the middle of all this we made ourselves available to more of the issues that are at stake for an emerging wave energy sector needing to find a place for its devices amid all the stuff already present in the landscape. Our imaginative forces opened up to the idea that innovation is a local concept, and it must find a place among other conceptions of innovation, things, and ideas. Sometimes we need imaginative methods to create infrastructures that relate that which is planned, structured, controlled, and designed, and that which is incidental, ephemeral, and not yet seen.

Acknowledgments

This chapter is based on experiences generated from making and installing the Energy Walk, a collective undertaking of the Alien Energy research project. We would like to especially acknowledge Laura Watts's amazing poetry and Line Marie Thorsen's research assistance. The Energy Walk was funded by the Danish Independent Research Councils (grant 0602-02551B—Marine Renewable Energy as "Alien": Social Studies of an Emerging Industry). Thanks to Hans Jørgen Brodersen, Christian Grant, Per and Kirstine Gravgaard, all staff at Færgegrillen, and Jan Krogh for all kinds of generosity. Thanks to Peter Adolphsen, Vanessa Carpenter, and Dzl Møbius at GeekPhysical, and to Simon Carstensen and Rina Bjørn for co-laboring. Thanks to Marisa Cohn, Klara Benda, Casper Bruun Jensen, Lea Schick, and Helen Verran for textual inputs, and to Heather Swanson at the Department of Anthropology, Aarhus University, for hosting the AURA-Alien Energy seminar in November 2014.

Notes

1. By "alien" we mean energy forms whose existence among us is yet to be decided upon.
2. www.alienenergy.itu.dk.

3. This is changing and many believe, and have demonstrated, that there is more to learn from combining anthropological ethnography and digital design (for examples, see Otto and Smith 2013; Vertesi and Ribes, this volume).
4. www.cardiffmiller.com/artworks/walks/alterbahnhof_video.html.
5. http://tinyurl.com/y99yxacm.
6. Vanessa Carpenter and Dzl Møbius, GeekPhysical (www.geekphysical.com), are also members of Illutron, Collaborative Interactive Arts Studio, Copenhagen.
7. After our research period ended, a local entrepreneur adopted the sticks and included them in his catalogue of "energy tours." Using a QR code and a smartphone to access the sound files means that larger groups can do the walk together.
8. The audio is available in Danish and English. Watts's original English version was translated into Danish by Author Peter Adolphsen. You can listen to the audio via SoundCloud at https://soundcloud.com/alien-energy.
9. Britt Kramvig and Helen Verran, University of Tromsø, have created a video titled "Dreamland" (2016) that addresses questions of how the Arctic has been imagined and how these imaginations are linked to state control and colonial regime. Their intervention is similar in its attempt at putting the spectator in the middle of the landscape, while also framing what she or he might possibly see from that position.
10. See Watts and Winthereik (2018) on the concept of edge in a comparative exploration of ocean energy and its discontents.

Works Cited

Ang, Ien. 2011. "Navigating Complexity: From Cultural Critique to Cultural Intelligence." *Continuum* 25 (6): 779–94.

Appel, Hannah, Nikhil Anand, and Akhil Gupta. 2015. "The Infrastructure Toolbox. Theorizing the Contemporary." *Cultural Anthropology*, September 24. www.culanth.org/fieldsights/725-the-infrastructure-toolbox.

Blok, Anders, Moe Nakazura, and Brit Ross Winthereik. 2016. "Knowing and Organizing Nature-Cultures: Introduction to Special Issue on 'Infrastructuring Environments.'" *Science as Culture* 25:1–22.

Bowker, Geoffrey C., and Susan Leigh Star. 1999. *Sorting Things Out: Classification and Its Consequences*. Cambridge, MA: MIT Press.

Carse, Ashley. 2012. "Nature as Infrastructure: Making and Managing the Panama Canal Watershed." *Social Studies of Science* 42 (4): 539–63.

Forlano, Laura. 2013. "Making Waves: Urban Technology and the Co-production of Place." *First Monday* 18 (11). http://firstmonday.org/ojs/index.php/fm/article/view/4968/3797.

Gomart, Emilie, and Antoine Hennion. 1999. "A Sociology of Attachment: Music Amateurs, Drug Users." *Sociological Review* 47:220–47.

Gupta, A., and J. Ferguson, eds. 1997. *Anthropological Locations: Boundaries and Grounds of a Field Science*. Berkeley: University of California Press.

Haraway, Donna. 1988. "Situated Knowledges: The Science Question in Feminism and the Privilege of Partial Perspective." *Feminist Studies* 14 (3): 575–99.

——. 1991. *Simians, Cyborgs and Women: The Reinvention of Nature*. New York: Routledge.

——. 2015. "Anthropocene, Capitalocene, Plantationocene, Chthulucene: Making Kin." *Environmental Humanities* 6:159–65.

Hughes, Christina, and Celia Lury. 2013. "Re-turning Feminist Methodologies: From a Social to an Ecological Epistemology." *Gender and Education* 25 (6): 786–99.

Hughes, Thomas P. 1993. *Networks of Power: Electrification in Western Society, 1880–1930*. Baltimore: Johns Hopkins University Press.

Jensen, Casper Bruun, and Atsuro Morita. 2017. "Infrastructures as Ontological Experiments." *Ethnos* 82:615–26.

Jensen, Casper Bruun, and Brit Ross Winthereik. 2013. *Monitoring Movements in Development Aid: Recursive Partnerships and Infrastructures*. Infrastructures. Cambridge, MA: MIT Press ().

Larkin, B. 2013. "The Politics and Poetics of Infrastructure." *Annual Review of Anthropology* 42:327–43.

Latour, Bruno. 2003. "Why Has Critique Run Out of Steam? From Matters of Fact to Matters of Concern." *Critical Inquiry* 30:225–48.

Law, John. 2004. *After Method: Mess in Social Science Research.* London: Routledge.

Maguire, James, and Brit Ross Winthereik. 2016. "Protesting Infrastructures: More-Than-Human Ethnography in Seismic Landscapes." In *Infrastructures and Social Complexity: A Routledge Companion,* edited by P. Harvey, C. Bruun Jensen, and A. Morita, 161–73. London: Routledge.

Marcus, George E., ed. 1986. *Technoscientific Imaginaries: Conversations, Profiles, and Memoirs.* Chicago: University of Chicago Press.

Otto, Ton, and Rachel Smith. 2013. "Design Anthropology: A Distinct Way of Knowing." In *Design Anthropology: Theory and Practice,* edited by C. Gunn, T. Otto, and R. Smith, 1–27. New York: Bloomsbury.

Rapport, Nigel, and Mark Harris. 2015. *Reflections on Imagination: Human Capacity and Ethnographic Method.* Surrey: Ashgate.

Sneath, David, Martin Holbraad, and Morten Pedersen. 2009. "Technologies of the Imagination: An Introduction." *Ethnos* 74 (1): 5–30.

Star, Susan Leigh. 1999. "The Ethnography of Infrastructure." *American Behavioral Scientist* 43 (3): 377–91.

Starosielski, Nicole. 2015. *The Undersea Network.* Durham, NC: Duke University Press.

Strathern, Marilyn. 1991. *Partial Connections.* Association for Social Anthropology of Oceania Special Publications Series. Lanham MD: Rowman & Littlefield.

Suchman, Lucy. 2011. "Anthropological Relocations and the Limits of Design." *Annual Review of Anthropology* 40:1–18.

Timmermans, Stefan, Geoffrey Bowker, and Susan Leigh Star. 1998. "The Architecture of Difference: Visibility, Control and Comparability in Building a Nursing Intervention Classification." In *Differences in Medicine: Unraveling Practices, Techniques and Bodies,* edited by M. Berg and A. M. Mol, 202–25. Durham, NC: Duke University Press.

Tsing, Anna. 2008. "Alien vs. Predator." *Encounters* 1 (1). www.dasts.dk/wp-content/uploads/tsing-anna-2008-alien-vs-predator.pdf.

Watts, Laura. 2012. "OrkneyLab: An Archipelago Experiment in Futures." In *Imagining Landscapes: Past, Present and Future,* edited by T. Ingold and M. Janowski, 59–76. Oxford: Ashgate.

Watts, Laura, and Brit Ross Winthereik. 2018. "Ocean Energy at the Edge." In *Ocean Energy: Governance Challenges for Wave and Tidal Stream Technologies,* edited by Glen Wright, Sandy Kerr, and Kate Johnson. London & New York: Routledge, pp. 229–46.

Winthereik, Brit Ross, Antoinette de Bont, and Marc Berg. 2002. "Accessing the World of Doctors and Their Computers: 'Making Available' Objects of Study and the Research Site through Ethnographic Engagement." *Scandinavian Journal of Information Systems* 14 (2): 47–58.

Winthereik, Brit Ross, and Helen Verran. 2012. "Ethnographic Stories as Generalizations That Intervene." *Science Studies* 25 (1): 37–51.

Introduction
Software

Carl DiSalvo

Software is not a new object of inquiry for STS, but it is fair to say that our relationship with software, as both users and researchers, is changing. This handbook itself is a prime example. Software played a significant role in producing and organizing these chapters. We deployed php code for open online peer review to produce a collaborative space for our community's early proposals and essay drafts. Later, coeditor Yanni Loukissas, working together with Ben Sugar (then a graduate student at the Georgia Institute of Technology), analyzed the text of all of the essays using topic modeling algorithms to identify themes and form groupings, aiding in the decision of how to organize the essays in the volume. These software deployments were not rote processes, but involved experimentation, discussion, a bit of hesitancy, and finally some decisions. This is emblematic of our contemporary engagement with software as both commonplace and a social practice.

The ordinariness of software, its social construction, and its practices are themes that span the chapters in this section. These essays attend to the vernacular aspects of software. Although we live, work, and play with software, despite its many fantastic capacities, it is a rather mundane thing. That software is commonplace, however, does not mean it is not a worthy object of study. Rather, its ordinariness opens us up to new encounters and themes of inquiry. Building upon foundational work understanding how software shapes and is shaped by works practices in science and technology (Star 1995; Bowker and Star 2000), we now see an increasing number of investigations into the material characteristics of software (Rosner 2018; Dourish 2018), the roles of software throughout society, including financial markets (MacKenzie and Spears 2014) and governance (Introna 2016; Ziewitz 2016), and, increasingly, the political character of algorithms (Gillespie 2014; Crawford 2015).

The chapters in this section share a commitment to analyzing the practices that compose software: that is, how software works through associations, interactions, and performances between and among individuals, groups, organizations, and code. This line of inquiry connects to scholarship that examines the variety of ways that software is made and the cultures of its making, from historical inquiries (Turner 2010) to studies of free and open software (Kelty 2008; Coleman 2013) to global do-it-yourself and so-called "maker" communities (Lindtner 2015). It is perhaps the attention to the diverse practices of making, using, and maintaining software that distinguishes STS work in this area from other contemporary studies of software as cultural objects.

In "From Affordances to Accomplishments: PowerPoint and Excel at NASA," Janet Vertesi introduces readers to the use of everyday software packages in the context of space science. As Vertesi notes, although there is significant work examining the custom and expert tools of science, less attention has been directed to studying the use of the run-of-the-mill suites of software that permeate much of the work of science and technology research and development. In particular, Vertesi analyzes the concept of affordances in software design. She argues that discussions of affordances are "logically problematic" and their "explanatory capabilities are thin" because, in fact, affordances are not universal but specific to and differentiated among varied communities of users and contexts. Vertesi suggests a renewed commitment to attending to the situatedness of working with software, interrogating how software is used to order work, without giving precedence to any particular functionality or pattern of use.

"The Role of 'Misuses' and 'Misusers' in Digital Communication Technologies," by Guillaume Latzko-Toth, Johan Söderberg, Florence Millerand, and Steve Jones, returns to a critical theme in STS: how the actual use of technologies often extends the normative notions of its "proper use." The authors trace misuse and misusers through the three historical cases of PLATO (an educational platform), Bitnet (an international network for liberal arts faculty), and Internet Relay Chat (IRC). From across these cases, the authors proceed to identify the character of a "misuser" as an active, performative figure in the innovations process. In addition, the misuser can operate as an analytic category. Situating this figure and category of the "misuser" in STS literature, the authors go on to describe how the actions of a misusers are relational to what they term as the "plasticity" of a technology. The result of this inquiry, then, is a new perspective on innovation, situated in regard to digital communications technologies, that highlights misuses and misusers who resist the normative framings of "proper use."

Nick Seaver's chapter addresses a fundamental component of software: the algorithm. He begins "Knowing Algorithms" with the insight that algorithms should be (but often are not) treated as social constructions. Those who study algorithms regularly claim that lack of "access" and "expertise" hinders our understanding and evaluation algorithms, but these claims are dubious, "where facts are simple objects to be retrieved from obscurity and analyzed by ever more sophisticated experts." Through the chapter, Seaver carefully builds an argument for an ever more nuanced appreciation of algorithms, addressing key themes of experimentation and transparency. This leads Seaver to the conclusion that, from an ethnographic perspective, our focus should be on what he terms "algorithmic systems—intricate, dynamic arrangements of people and code." Pushing this idea further still, Seaver argues against the notion of a "cultural/technical distinction as a ground for analysis." In many ways, Seaver's insights into studying algorithms recall the features of a multisited ethnography (Marcus 1995), in which we are charged to follow and interpret the varied sites and practices of humans and nonhumans together in the construction and expression of algorithmic systems.

How do we study software as a thing that changes over time? This is the question that motivates Marisa Leavitt Cohn's chapter "Keeping Software Present: Software as a Timely Object for STS Studies of the Digital." Though we commonly think of software as new, as Cohn points out, much of the software that we rely upon for our most needed work are legacy systems, sometimes decades old. For Cohn, one implication of studying the temporalities of software is that we attend to "its forms of duration, the entangled lifetimes of careers, professional identities, programming languages and paradigms, all of which come and go." The phrase "keeping software present" speaks to this notion of duration, of "bringing old code together with new." Throughout the chapter Cohn describes the ways in which code is *lived*, the ways that engineers, managers, users, and others, in multiple and varied ways, endeavor to keep software working. For Cohn, these endeavors are part of the materiality of software, and like other materialities, possess distinctive temporalities: "its lifetimes, histories, and changes as drag, drift, and forgetting."

Taken together, these inquiries into software, its analysis, and its discontents offer opportunities for scholarship that examines the everydayness of software systems and their associated practices, while at the same time troubling our everyday distinctions between use and misuse, culture and technology, social and material, the novel and the aged.

Works Cited

Bowker, G. C., and S. L. Star. 2000. *Sorting Things Out: Classification and Its Consequences*. Cambridge, MA: MIT Press.

Coleman, E. G. 2013. *Coding Freedom: The Ethics and Aesthetics of Hacking*. Princeton, NJ: Princeton University Press.

Crawford, K. 2015. "Can an Algorithm Be Agonistic? Ten Scenes from Life in Calculated Publics." *Science, Technology, & Human Values* 41:77–92.

Dourish, P. 2018. *The Stuff of Bits: An Essay on the Materialities of Information*. Cambridge, MA: MIT Press.

Gillespie, Tarleton. 2014. "The Relevance of Algorithms." In *Media Technologies: Essays on Communication, Materiality, and Society*, edited by Tarleton Gillespie, Pablo Boczkowski, and Kirsten Foot, 167–95. Cambridge, MA: MIT Press.

Introna, L. D. 2016. "Algorithms, Governance, and Governmentality: On Governing Academic Writing." *Science, Technology, & Human Values* 41 (1): 17–49.

Kelty, C. M. 2008. *Two Bits: The Cultural Significance of Free Software*. Durham, NC: Duke University Press.

Lindtner, S. 2015. "Hacking with Chinese Characteristics: The Promises of the Maker Movement against China's Manufacturing Culture." *Science, Technology, & Human Values* 40 (5): 854–79.

MacKenzie, D., and T. Spears. 2014. "'The Formula That Killed Wall Street': The Gaussian Copula and Modelling Practices in Investment Banking." *Social Studies of Science* 44 (3): 393–417.

Marcus, G. E. 1995. "Ethnography In/Of the World System: The Emergence of Multi-sited Ethnography." *Annual Review of Anthropology* 24 (1): 95–117.

Rosner, D. 2018. *Critical Fabulations: Reworking the Methods and Margins of Design*. Cambridge, MA: MIT Press.

Star, Susan Leigh. 1995. "The Politics of Formal Representations: Wizards, Gurus, and Organizational Complexity." In *Ecologies of Knowledge*, edited by Susan Leigh Star, 88–118. Albany: State University of New York Press.

Turner, F. 2010. *From Counterculture to Cyberculture: Stewart Brand, the Whole Earth Network, and the Rise of Digital Utopianism*. Chicago: University of Chicago Press.

Ziewitz, Malte. 2016. "Governing Algorithms: Myth, Mess, and Methods." *Science, Technology, & Human Values* 41 (1): 3–16.

From Affordances to Accomplishments

PowerPoint and Excel at NASA

Janet Vertesi

A gray-haired scientist stands in front of the screen at the head of the room. Projected behind him is a colorful spreadsheet, the boxes in its rows and columns luminescent in shades of yellow, light green, orange, red, and purple. Forty of his colleagues are in the room, busily asking questions, referring to individual boxes, requesting changes. He dutifully edits the document as requested in real time so all can see. Their glances mutually entrained on the projected Excel spreadsheet, this group is hard at work trying to decide where their spacecraft should travel for the next seven years.

STS scholars have explored the politics and infrastructural assumptions of bespoke digital tools in the sciences, such as specialist databases and convergent cyberinfrastructures (Ribes and Bowker 2008; Lee et al. 2006). But STS studies of the off-the-shelf software suites are rare, despite the near ubiquity of smartphone apps, productivity suites, and social media platforms like Facebook or LinkedIn in the laboratories and communities we study (Vertesi 2014; Gillespie 2010). Technologies of this type are highly blackboxed for users, pre-prepared for particular conditions of use by the corporations that produce and update them. Yet they also facilitate a variety of usage patterns in order to secure the broadest possible user base. Hence the analysis of these commercial and widely distributed tools in specialist use contexts demands a vocabulary that makes sense of the technology's varied local contexts of meaning making. Such an analytical vocabulary should take into account existing STS tools for understanding commercial/user relationships, such as relevant social groups (Pinch and Bijker 1987) or market lock-in (Schwartz Cowan 1985), as well as frameworks for understanding varieties of usage contexts such as interpretive flexibility (Pinch and Bijker 1987) or de-scription (Akrich 1992). It must also note that, in mass commercial software, radical shifts in the understanding, design, or functioning of the artifact are much more limited than the framings of actor-network theory or the social construction of technology might imply, even if the social and networked constraints upon immediate technical use are relatively flexible.

Commercial software suites like Microsoft Office are an excellent site for this type of analysis given their extensive use across a variety of sectors. Conceived n a range of use cases and scenarios in mind, such systems are relatively open situated interpretation. Yet they are used by such a broad cross-section of the population that this delimits the ability for specific relevant social groups to demand sweeping changes in product design. As such, one might suppose that there are nearly as many ways of using Excel or PowerPoint as there are users: but this is not always the case.

In this essay, I show how the interactional norms that guide usage of such suites are not due to individual perception, widespread usage patterns, or particular "affordances" available for use. Instead, interactional norms are enrolled in the production and reproduction of the local organization and its membership, in practices that make available different opportunities for software interaction. Since the methods of working that I observed empirically in my field site required using the software suites in specific ways—not necessarily the primary use envisioned by their designers—I offer a way to address how the specific properties of the commercial software suite come to matter for its organizational use. I do so by inverting the question, asking instead how organizational usage settings come to matter for making software properties visible and accountable. This shifts our attention from taxonomizing built-in material properties or built-in "affordances" toward articulating *software accomplishments* and their deployment in the field.

Digital Properties and Affordance Theory

At stake in my account is how STS might approach the study of software suites in conversation with the concerns of organizational studies on the one hand, and communications and media technologies on the other. The former community has developed a notion of "sociomateriality" that draws on STS theories of hybridity (i.e., Latour 1991; Haraway 1991; Barad 2003) to describe the material and semiotic aspects of software in context (Orlikowski 2010; Leonardi 2011). In the latter community, the language of "affordances" aims to bridge the gap between social and technical determinisms. New media scholars frequently deploy this terminology, imported from Gibsonian psychology through human-computer interaction (Gibson 1979; Norman 2002), to delineate how material objects and digital systems alike both enable and constrain particular forms of social action.

In the classic sense, affordances are assumed to be relational: that is, only visible or perceptible to particular users in contexts of use. Norman's use of the term relies on the psychological qualities of the encounter between a user and an object with an eye toward design: how to make the usage properties of an object immediately perceptible to a novice user. This relational perspective offers a useful middle ground between object and subject, but it steers clear of familiar STS analytical perspectives on object hybridity and situated encounters. From an STS perspective, this runs the risk of neglecting the networked, political, and ontological work-in-the-world that makes such capabilities available for action in the first place. Even concepts such as "scripts" (Akrich 1992) do not taxonomize baked-in properties of technical objects, but instead seek to reveal the networked conditions that make particular use cases possible. Still, deployments of "affordances" in contemporary studies of digital media frequently elide those elements of a technology that are "found" in the object and guide its usage, and those elements that are explicitly designed in. In such work, material constraints are frequently considered to be "in" the object and available to perception, application, or subversion in various use contexts, whether these were designed in at the level of code or the result of the natural properties of digital materials (boyd 2010; Faraj and Azad 2012; Ilten 2015; Majchrzak et al. 2013; Neff et al. 2012).

Affordance theory has clear merits in the design sphere, where purposefully including certain capacities for action—and restricting others—can be an empowering way to approach wicked technical problems. But when it comes to analyzing digital artifacts, platforms, and tools in their sites of use, affordance theory renders users passive with respect to receiving, deploying, or reacting to elements that are simply "baked in" to the technology, ready-made for the perceiving.[1] Further, it makes an error in *post hoc propter hoc* reasoning by assuming that properties located by actors in contexts of use must somehow have been there prior to their use case, thereby making them perspicuous for action in the first place. Finally, it assumes that actors' categories that describe material conditions are in fact analytical categories: real, hard constraints in the world. In naturalizing these emic categories, affordance theory imbues objects with political qualities that reproduce instead of question the social relations at hand.

Assuming that "affordances" are the *explanandum*, not the *explanans*, I seek to reframe the conversation in terms more familiar to the STS scholar. I turn to microsociological approaches (along with Sharrock and Coulter 1998) to show how any such material "constraints" or properties are emergent, constructed, and made perspicuous at different times as members deploy quotidian software suites to participate in organizational activity. I argue, therefore, for shifting our analytical language from one of *software affordances* to *software accomplishments* (drawing on Garfinkel 1967). I mean *accomplishments* in two ways. Here, I address the *user's situated accomplishment* in producing or making perspicuous qualities of use relevant to their problem at hand. There is, of course, a parallel, no less situated accomplishment on behalf of the designer(s) that make such qualities *appear to be* natural, psychological, or decontextualized, and otherwise devoid of politics: this is beyond the scope of this essay. Instead, my aim for this chapter is to chart an analytical course for STS scholars that eschews post hoc taxonomies and assumptions about the material or political conditions of software usage, while at the same time remaining attuned to the creative, flexible, and sometimes unexpected things that actors *do* with such technologies on a regular basis.

To do so, I describe the use of popular Microsoft Office Suite software as it is deployed on two NASA spacecraft teams, each with different ways of collaborating and organizing work among their members. I first describe how the practices of such software use are both unusual with respect to the software's typical patterns of use, yet at the same time highly organizationally standardized. I then turn to the challenges that such practices pose to existing theories of software materialities. Finally, I demonstrate an alternative approach that emphasizes users' *situated accomplishments* to offer novel insights into the constructed-yet-obdurate nature of such software tools in practice.

Workplace Software in Space

The two missions arrived at their respective planets in the early years of the 21st century, carrying with them a suite of instruments to make observations in the different planetary environments. One is an orbiter, sweeping through a planetary system on a charted course over many years. As the spacecraft passes by moons and other features in the planetary environment it captures rich streams of data through 12 different instruments about the planet's physical environment, its moons' weathering and geological activity, and any observed changes over time. The other robots are landed, roving vehicles. They conduct scientific investigations on a planet by driving to new locations, observing rocks, dust, and atmospheric conditions, and protecting the robot overnight from the cold. At the time of my fieldwork there were almost three times as many mission personnel working on the orbiter as there were working on the ground-based team.

Although the missions enroll many of the same scientific and technical personnel and the same institutions, they are organized and operated quite differently. The orbiter mission coordinates its collaboration deploying a matrix organization. This features a bureaucratic hierarchy (Weber 1968) composed of 12 distinct instrument teams. A group of coinvestigators is selected along with team leaders as core team members, with affiliated researchers and students following in descending order. Additionally, scientists from across these teams gather in thematic groups responsible for planning segments of spacecraft activity depending on its location in the planetary system and when it flies past targets of interest such as various moons. The leaders of each instrument team and each thematic group meet as an executive committee to make critical decisions about the mission. The ground-based robot collaboration, by contrast, is structured laterally under a single principal investigator as a charismatic collective (Shils 1965; Weber 1968). Members describe themselves as a single team of scientists with many different skills and backgrounds, each of whom may request and collectively use observations from any robotic instrument and all of whom make decisions via consensus. While scientists have different official designations (PI, team leaders, co-investigators, graduate students), in practice there is no distinction between these roles in the daily planning practices of the collaboration.

Planning cycles for the matrix team's orbiter take several weeks and many rounds of interaction between scientists, science planners, and representatives of various engineering specialties before uploading to the spacecraft. Planning for the collective's ground-based mission takes place within a single day for the next day's events, requiring the construction of agreement on a rapid and highly compressed timescale. To manage these planning activities and their temporal rhythms, each group prefers different software with different visual metaphors and infrastructural capacities for capturing, representing, and collectively enacting various aspects of their work (Mazmanian et al. 2014). Certainly each mission has bespoke software systems that they use for the purposes of entering and managing observational requests (see Cohn's contribution to this volume). I focus here instead on the process of "mutual entrainment" (Collins 2004) in the teams' daily or weekly planning meetings. These center around shared PowerPoint presentations on the ground mission and shared Excel spreadsheets on the orbiter.

Attention to screens at these meetings is paramount. Members of each collaboration keep their software application of choice open on their desktop continuously, sometimes with several files loaded at once. They then use these spreadsheets or slide decks as the focus of their collective attention and the primary site of their decision-making work. As the postsocial milieu in which organizational interactions take place (Knorr Cetina and Bruegger 2002), these digital presentations serve to ground the conversation, guide the planning process, and produce spacecraft activity. In this way, plans are "enacted in practice" (Ziewitz 2011) through a continuous achievement of group-and-spacecraft coordination, producing activities to be implemented in space and documenting team agreement. Yet even as such software grounds the enactment of the planning process, each group departs from the software's built-in assumptions and templates, preferring to use their software in unusual ways that are consistent with the organizational orientation of each group. As such, they raise analytical possibilities for STS practitioners studying software in organizational and scientific contexts.

Roving with PowerPoint

Many organizational scholars have examined how PowerPoint presentations are both reflective of and constitutive of organizational cultures. Joanne Yates, Wanda Orlikowski, and Huburt Knobloch recommend analyzing PowerPoint presentations as a "genre" of business communication that plays a structurational role at the firm (Yates and Orlikowski 2007; Knoblauch 2013), while Sarah Kaplan focuses on the slide deck's material-discursive qualities as constitutive of the epistemic culture of the firm (Kaplan 2011). On the consensus team, PowerPoint and its presentation in ritual meetings also play a structurational and material-discursive role by means of which team members establish knowledge production priorities and (re)produce their lateral social order. But they do so by *departing from* the standard, built-in templates for the PowerPoint presentation used across the business world with a long history in business communication (see Robles-Anderson and Svensson 2016; Yates and Orlikowski 2007).

Even visualization expert Edward Tufte (n.d.) might be surprised by the PowerPoint slides on the collectivist team. Avoiding the templates that feature a title with bullet points, collaboration members import large images into their slide decks, frequently taking up the entire screen. These images are typically taken by the robot itself and marked up by a team member with annotations to indicate important features around the robot. This requires laboriously maneuvering text boxes, open circles, and arrows (not template bullet points) on screen until they sit over top of the image in just the right spot to direct attention to a feature on the planet's terrain. In this way, team members have their own slide templates that they adhere to day after day, frequently copying the prior day's presentation and updating its content for today's meeting.

I observed this style of PowerPoint usage at every mission team meeting over my two years of ethnographic immersion with the team. At each planning meeting, a daily ritual, the scientist in charge of the meeting used the transition between slides to pace conversation and get everyone on the same page, as well as to impart information. The opening slide always featured a full-screen image taken from the robot's current location with the date (both on the planet and on Earth) and the primary meeting-related roles for the day. The following slides set the scene for the immediate planning context with a series of images taken from the ground or from orbit, annotated with elements of the robot's current position and complemented with a list of activities to accomplish at or near this location. The rhythm of the meeting is established with, "Next slide . . . next slide" (figure 1).

Annotations on the slides such as colored dots, lines, and text might indicate prospective plans for investigation or identify areas around the robot; they might also cross-correlate elements visible from different vantage points or propose how to plan a drive (figures 2 and 3). Such slides provide an evolving snapshot of the team's planning process at a particular point in time. They also frequently carry

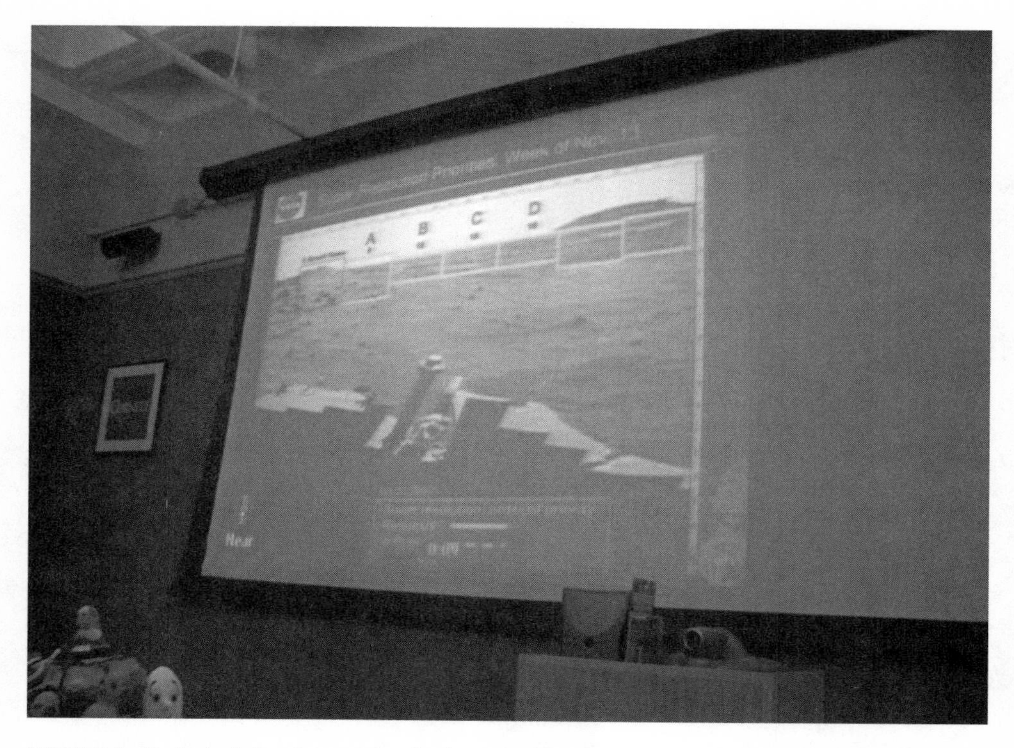

FIGURE 1: At a typical planning meeting for the ground-based team, PowerPoint slides deploy annotated images to get everyone on the same page.

forward from one meeting to the next, grounding collaboration members in prior decisions yet without constraining them too tightly to one course of action over another.

Annotation techniques are not impossible in PowerPoint, but the tools are clumsy. The lines, arrows, and built-in shapes were built for creating flowcharts for business and engineering presentations. Scientists struggled with the autoplace feature, resorted to default system fonts like Comic Sans and Arial to overlay text on the image, and frequently lost visuals in their presentations due to PC/Mac version incompatibly. One might wonder why they used PowerPoint in this way at all. However, this method of annotation aligns with a technique with a long heritage in planetary geology. Planetary scientists are taught, based on the principles of geomorphology, to draw their way through their work. They practice drawing geological contact points and identifying features on images taken from orbit or from the air.[2] Annotation also served an organizational purpose. As I have argued elsewhere, getting everyone on the same page toward consensus requires bringing far-flung collaboration members together into a singular position (Vertesi 2015). The PowerPoint slides on this team enable the process of arriving at this consensus by establishing place and shared conversational ground (Schegloff 1972), at the same time as they enable all team members to view the same elements on the terrain around the robot as the same type of thing ("seeing as"; Vertesi 2015). This encourages the conversational flow throughout the meeting as one that moves from a shared position at the meeting's outset to a consensus moment at its end.

Alongside this repetition of visual imagery, then, the mutual entrainment of the PowerPoint presentation with its recognizable local style was an essential part of the meeting's ritual practice (Collins 2004). The slide deck's relatively stable format and rhythmic presentation contributed to a sense of a ritual quality in each individual meeting, no matter the specific issues of the day. As such, the team's use of PowerPoint to enable a unified, shared perspective helped to reproduce the collaboration's organizational commitments toward collectivism as they planned a robot's actions on another planet.

How to Excel in Orbit

Just as PowerPoint is ubiquitous on the collectivist team, work with spreadsheets is a central part of life on the matrix collaboration. On the latter mission, an entire subgroup typically works on a spreadsheet at once during periods of copresence, projected onto a shared screen in the room and circulated by email for remote meeting participants to follow along during the conversation. As a site of mutual entrainment through which interactions are ordered, the spreadsheet coordinates work within the collaboration. It is also the site of recording those same interactions, as individuals input text into the cells to record the group's agreements. As such, work with spreadsheets on the orbiter mission is group work that also produces collective entrainment and paces the ritual elements of their meetings.

Much like PowerPoint is based on business presentation formats, the Excel program in the Microsoft Office productivity suite is based upon both digital and analog accounting and budgeting platforms, including Lotus software (see Dourish 2017). The software presents a grid organized in columns and rows, and is equipped with a variety of number-crunching tools to enable the quick and routine tabulation of money, percentages, and graphs. As part of producing these tables and graphs, Excel comes equipped with text functions (largely for titles and labels), and the ability to color spreadsheet cells or text to improve legibility in large tables of numbers. But while Excel spreadsheets are optimized for computation and tabulation of numbers, orbiter members rarely use its actuarial functions, numerical expressions, or graphic capabilities. Instead, they enter text and colors in rows and columns that suggest and produce an order to their conversation. The orbiter's scientists and engineers use these attributes of the Excel software to articulate their work and archive its achievement.[3]

There is no material element of the planets or exploration context that automatically recommends Excel to an orbital mission and PowerPoint to a land-based one. But the choice of software is consistent with each team's organizational context. Excel on the matrix mission provides an environment for text entry in a tabular, gridded environment to reflect the team's internal organization and to catalog priority assignments among what are often incommensurate observational requests from different subgroups. In a matrix organization where individuals may cluster by both research interest and instrument team membership, it is essential to locate the correct spokesperson for each subgroup as well as to visualize and seek coverage of all of the collaboration's priorities. The presence of the spreadsheet cell at the intersection between row and column speaks to the position of individual members of the collaboration: members responsible for atmospheric observations fill in the atmosphere boxes, while individuals on the infrared instrument team are responsible for those cells related to infrared observations.

This continues until all perspectives are accounted for and made visible across subgroup boundaries. In this way, the record of conversation, group convergence, and accountability is carried forward and made actionable as scientific priorities. These are ultimately transformed into spacecraft commands.

The spreadsheet therefore not only captures knowledge from different sides of the organization but also participates in the work of delineating and recording both ownership and prioritization. Like the collectivist team and PowerPoint, most scientists and engineers on the matrix collaboration had a number of colorful spreadsheets open on their desktops and splayed across communal screen. Through discussion at their meetings they either modified the spreadsheet or created a new one based on the values that came up through the discussion. That spreadsheet was then carried forward to the next task, where the process repeats. Like the many generations of PowerPoint slides described above, orbiter teammates typically use the spreadsheet as a guide for discussion, focusing attention, filling in or changing the spreadsheet subsequent to discussion, and then archiving the change in the spreadsheet. Thus these working spreadsheets also have a heritage and a trajectory, generating new spreadsheets or nestled amongst other spreadsheets.

For example, a meeting I attended in April 2009 was organized around and through spreadsheet work and annotation. The goal of the meeting was for the group to divide up over 100 flybys of a moon of interest in the planetary system, assigning observational time to different instrument teams. At the start of the three-day meeting, held at an engineering laboratory on the East Coast of the United States, a spreadsheet was projected at the screen at the front of the room and a hard copy was circulated in the room as well. Each row indicated a prospective flyby of the moon, assigned a number in sequence. Also listed were relevant details that located a particular pass in a particular year or period in the future such as start time, end time, operational modes, and any instruments that might observe as an additional "ride along" opportunity. Each flyby also had a tentative assignment of an observation requested by one or more instruments in advance. Some of these were noted to be in conflict or otherwise contested by more than one instrument: these were highlighted in yellow on the spreadsheet (figure 2).

At the outset of the meeting, the coordinator announced the process: "What we thought we would do is sort of go around the room and annotate the spreadsheet to give an idea of what observations have been allocated."[4] She then asked the group, "Is there any discussion we wanna have before we just jump into the spreadsheet?" There was some discussion over one pass, which was contested between two instruments and required intervention from a senior member of the collaboration to determine whether the pass should go to one instrument or another: he suggested the latter. As a result, a meeting coordinator expressed, "here's the plan": "[The senior scientist] is going to do a little more work on his spreadsheet, and then he's going to present what we have come to as a consensus . . . and then we thought what we'd do is present part of the answer as a suggestion . . . and see if that helps." Note that the spreadsheet was the location for working out a conflict between two of the matrix organization's subgroups. The team gravitated toward this spreadsheet row and its indicated conflict as a problem worth getting out of the way before they began the rest of their process. They then stepped away from the spreadsheet to resolve what should be placed in that cell. When the conflict was resolved, the spreadsheet was updated to represent the group's decision. The senior scientist would then confirm the agreed-upon solution to the conflict, indicated in the contested Excel cell. Following this, the group went through the spreadsheet line by

FIGURE 2: A subgroup for the orbiter team inputs their decisions into their spreadsheet, also printed out on paper (held by the woman on the left). Scientists sit in the rows behind, and one at the front desk on the right.

line, filling it in with details about each instrument's prospective observational request as they went. The spreadsheet thus served as a way of structuring discussion, highlighting conflict, and recording its resolution.

Organizational Consistencies, Technical Fluidities

These two examples establish a way for STS analysts to consider mundane software tools used in the context of scientific and technical work. Both presentation formats bring collaboration members together in a moment of collective focus, but each does so in a way that represents the organizational orientation of the collaboration. The collectivist team's PowerPoint slides bring individuals together into the same subject position from which there is a single, shared point of view, to address the challenges of decision making via consensus. The matrix team's Excel rows and columns make clear the responsibilities of different individuals for exerting ownership over different tasks and elements of exploration, with colorful boxes visually representing the plurality of perspectives that must be factored into any decision. Rows and columns account for the competing priorities across the matrix organization, while the annotated slides circulating in the collectivist mission serve as a site for alignment around a unified stance.

Thus both collaborations' members conscript these two ubiquitous software tools from the Microsoft Office Suite into the *coordination* of their collaborative activities, the *collective entrainment* of collaboration meetings, and the ultimate ritual reproduction of the *organizational orientation* on each mission. Both mission teams use software to structure their teammates' interactions in their ritual meetings to formulate plans for the spacecraft's activities. The experience of working with the same spreadsheet or observing the same presentation slides brings members of each collaboration into a moment of collective focus and mutual entrainment. In this interaction ritual (Collins 2004), each organization reproduces itself through its interactions, and representational forms. And the productivity software is an essential part of this interactional moment, with implications for the scientific data that are ultimately collected on each team. Each of these software conscriptions therefore represents a considerable *accomplishment* on behalf of the individuals using standardized tools for their specific, situated ends.

At first glance, this recalls early work in organizational studies of technology. Like Steve Barley's observations of CT scanners in two different radiology departments, we might read these contrasting use cases as examples of "structuration" and the development of local "scripts" that underline or undermine organizational hierarchies and work processes (Barley 1996). More recently, organizational scholars such as Joanne Yates, Wanda Orlikowski, Paul Leonardi, and others have explored how social norms and cues combine with material properties to shape and constrain activity in an organizational milieu. The material properties of these technologies acquire meaning by virtue of the context of organizational practices that imbue them with local significance: putting the "socio-" in "sociomateriality." Thus in a parallel study of a spacecraft collaboration, Mazmanian et al. (2014) argue that software is "figured and reconfigured" in contexts of practice that produce meaning and legibility on earth of matters in space. In another take on sociomateriality, Leonardi argues that material and social properties are "imbricated" in an organizational setting, stacked amid each other much like tiles on a rooftop (Leonardi 2012).

Yet the question remains as to which properties and which elements come to matter *in practice*, both in actors' and in analysts' views. What do actors (and analysts) assume is "written in" to the software? What do *they* think is the "material" with which "the social context" must contend? Science and technology studies has largely shied away from describing patterns of technology uptake in organizational context, so it is unclear how our present theoretical tools would help us to describe creative repurposing of commercial software. Current framings might instead reveal the commercial-governmental axis along which such software suites appear to seamlessly flow (Edwards 1997), presenting a lock-in situation where such tools are ready to hand even if they are imperfect or inefficient for the task at hand (Schwartz Cowan 1985) and perhaps suggesting an approach to the politics of platforms (Gillespie 2010). But this would not allow us to assess which properties come to matter in an organizational milieu, and why. They might suggest a form of interpretive flexibility deployed by a relevant social group (in this case, a scientific collaboration) with a problem that the technology does not quite solve, or a form of local tinkering to address a problem at hand (Kline and Pinch

1996). However, the properties of commercial, widely available software suites appear to be simply there for the taking: users cannot open the hood or the black box in order to alter inner workings (Gillespie 2006) or initiate interactions that disrupt patterns of usage (Woolgar 1990; Latzko-Toth et al., this volume). And while looking at how contexts of use influence design iterations is important for the SCOT framework (Kline and Pinch 1996; Oudshoorn and Pinch 2003), in the case of mundane

and ubiquitous technical tools not every set of users can be considered a "relevant social group" whose use may reshape the technology or influence design. Ubiquitous corporate software products typically ascribe limited agency to local users, making individual users ineffective writ large when it comes to shaping the attributes of the systems that they deploy.

This is also not a clear case of de-scripting a network. The networked conditions of use are very much present and still in place, from governmental-industrial partnerships to the institutional relations that make this software appear to be "ubiquitous" in the first place (Latour 2005; Kling 1991). Nor are these users "non-users" (Wyatt 2003) with a particular politics. And while their representational work of other planets may produce hybrid objects inscribed with earth-bound analytical work (Haraway 1991, 1997; Barad 2003; Myers 2014; Vertesi 2015), it is not clear if these mild diversions from expected use require meeting the material properties of the software halfway.

Enter the Gibsonian "affordance."

Affordance Theory and Its Discontents

The language of "affordances" might seem to safely occupy this middle ground between the social construction of technical tools and local patterns of use, between intentionally designed-in aspects of software and their potential creative repurposing in social context. It offers a lightweight material determinism that points to which properties of a technology people are working *with* when they do their (social) work. This notion of "affordances" is prominent in the sociotechnical analysis of digital systems, especially in the communications literature, to point to the role of the underlying architecture of systems in *influencing yet not entirely determining* digital interactions. In these accounts, technologies possess different capacities for action: "affordances" that are baked into hardware or software and can be uncovered, used, or resisted by users as they encounter or otherwise perceive them in the wild.

STS scholars moving into the study of software will encounter the phrase in widespread use. Social media sites are said to "have" the affordances of "visibility" and "persistence," affecting how these platforms are used in a range of contexts, from corporations to high schools to protest movements (boyd 2010; Treem and Leonardi 2013; Tufecki 2017). In a similar vein, the many affordances of paper over digital systems make the idea of a paperless office an obvious "myth" (Sellen and Harper 2001). These highly cited ethnographic accounts of user practices arise from taking the relational view of affordances, articulating how social media's designed-in qualities of persistence and visibility are managed *by coworkers*, *by teens*, or *by activists*, or how paper is preferable *for office workers*. But as their analysis travels, these affordances harden into essentialist, taxonomized properties of the system at hand: properties that analysts assume are then either taken up or resisted elsewhere in ways consistent with different cultural milieus (e.g., Vitak and Kim 2014; Raja-Yusof et al. 2016; Vaast and Kaganer 2013, etc.; see also critiques in Evans et al. 2017 and Costa 2018).

In line with prior work in this vein, the story of the present chapter could be one in which new affordances of Excel and PowerPoint are surfaced to add to a list of Microsoft Office properties, demonstrating how actors deploy these affordances and resist others to achieve their local tasks consistent with their cultures. But this would produce certain blind spots that trouble the STS scholar. After all, object properties are neither static or neutral. There was a time in the 1990s when the Internet represented ephemerality and anonymity, not persistence and visibility, a viewpoint that persists among librarians if not among commercial data-gathering platform engineers. Push a little further and each of these "affordances" appears to be baked into not the digital or technical product, but rather the eye of the beholding usage community. Again, although "affordances" initially arose from precisely this relational view, as affordance taxonomies travel they blackbox the social construction of the (software) system and focus too readily on the context of use as meaning determinative.

This sets up a peculiar logical fallacy. As a thought experiment, consider what it would take to list all the affordances of Excel as a software program up front. Not only would this be a very long list, but it could be populated only by particular attributes of the technology that are visible in particular times and places of situated practice. For instance, it is only in observing *the use of Excel by the matrix team members* that one might think to notice the use of color and text alongside formula management as an "affordance" of Excel in the first place. As such, any interest in an exhaustive list of technical affordances is a Sisyphean exercise, reminiscent of the Borgesian map that expands to represent a territory in its entirety. Because "affordances" are visible only in practice, they cannot necessarily be thoroughly cataloged in advance. Due to this mistake of *post hoc propter hoc*, their resulting explanatory capabilities are thin.

This visibility in practice leads to a frequent confusion between actors' categories and analytical categories. That is, it is only in local contexts of use, *in situated accomplishments*, that particular attributes of a technology become perspicuous as affordances at all—and then only to particular communities of users. This proposed vocabulary recalls the ethnomethodological concept of *accomplishment*, with its focus on how the intelligibility of social life and interaction cannot be taken for granted a priori, but must be achieved through practical, observable members' work (Garfinkel 1967). Casting work with software as a question of members' accomplishments therefore suggests a different orientation toward the problem of the material constraints of software and the relevance of its designed-in features in daily life. Software accomplishments remind the analyst that it is *members' work* that makes software systems legible in practice, and that casts certain features of a system as relevant or perspicuous to sense making or navigating local social worlds.[5]

A key element of orienting toward software use as situated accomplishment is observing members' own material-semiotic work in the world: the local dualities of nature-culture, human-machine, or object-agent that individuals routinely draw to make sense of software-facilitated action (cf. Suchman 2006). This is also part of members' work-in-the-world. But it also means that any post hoc account of the demarcation between "the material" from "the social" that assumes such categories arise from the properties of objects themselves is logically problematic. Because any attempted taxonomy of a technology's available affordances must, by definition, derive from a singular user group's perspective on a technology at hand—based on that group's local, material-semiotic, situated, practical accomplishments—elevating this emic definition to an etic one commits a second logical fallacy. Indeed, all "affordances" can only ever be what Nagy and Neff (2015) call "imagined affordances": whether imagined by users and projected into the context of design, imagined by designers and projected into context of use, or imagined by analysts and projected into the contexts of both use and design.

Because articulating "affordances" (and their discontents) requires us to take on a particular user (or designer) group's material-semiotic work as primary and to view alternatives as disruptive, attempts at defining a technology's affordances are a power-laden venture that writes certain users in and leaves others at the margins. To adopt a single group's local definition of a technology's "affordances" as canonical flies in the face of the "partial perspectives" (Haraway 1997) that are so essential to understanding the complex relationships between individuals and technologies, and contradicts many of the core principles that STS scholars espouse. We need instead some way to state the importance of the wide variety and modes of technology use to the production of group membership and identity, without assuming a priori categories that describe certain techniques as othering practices as opposed to practices of belonging.

Another Way Forward

One solution to this problem is to argue for a return to the original, relational concept of the affordance. But "affordances" even in this view are too often an *explanandum* masquerading as *explanans*. Instead of conceptualizing technical constraints as baked in and passively there for the using, or viewing material elements as available a priori and then selected for their virtues in structuring an organizational field site, I argue that an STS-centric approach to digital software systems in practice should instead focus on *software accomplishments*: how different opportunities for acting-with a technology are enacted or brought into focus *by each group*. It is not simply that a material property (an "affordance") becomes visible and can be catalogued through the relational processes of acting-with (or acting-against). Rather we must eschew the taxonomic and passive view suggested by contemporary deployments of "affordances" in exchange for a view of *how actors work with software in the world*: how individuals deploy various attributes, write novel scripts, or otherwise enroll software-enacted techniques as resources for producing and maintaining local social order.[6] After all, it is only by means of situated practice that certain elements available to software users, such as coloring or annotation tools, come to be seen as perspicuous properties to begin with. The language of software achievements captures this local, situated, organizational, and ultimately material-semiotic perspective.

Further, moving away from a passive, taxonomizing language of "affordances" allows us to better observe *where actors themselves draw the line* between software properties ("the material") and organizational expectations for software use ("the social"). Such a perspective shifts the sociomaterial view of constraint and enablement within technological systems. Rather than being etic categories—properties of an object in the world—locally achieved software properties and distinctions between the technical and the social should be analyzed as *emic* categories: local achievements of sense making and membership work. These local assumptions are ones that the analyst must endeavor to understand and ground in context of use.

We must therefore go a step further to identify what local actors consider to be "constraining" and what they consider to be "enabling" in the technologies they use. This assists the analyst in uncovering the politics at play in the local site (on a related example on "constraint," see Vertesi 2015, chap. 7). From this point of view, the technique of coloring in Excel cells or of importing large image files into PowerPoint does not become meaningful, perspicuous, or identifiable as an important attribute of the software until actors locate and engage that element of the software tool and narrate it as such. Should we take this work for granted, we would obscure the politics of the production of "the material" and "the social" in the software suite, casting these as something external to locally meaningful actors' work in the world.

This approach offers to shift present discussions of digital technologies in social context. For example, we might more readily frame "persistence" or "searchability" online not as natural qualities of the Internet or as designed-in properties of social media sites, but as the heterogeneous *accomplishments* of economic actors like Facebook, Twitter, and Google: companies that configure and stabilize activity online as an engine of capital, as opposed to alternative configurations of the online sphere as a place for anonymity (as in, for instance, Turner 2008 or Coleman 2014). This avoids elevating certain groups' understandings of digital interactions, made perspicuous in their practices and concerns at a particular time and place, as *the* matters of concern and objective realities of a digital tool itself, silencing other voices.[7] It also avoids assuming retrospectively that non-searchability was a pressing problem that needed solving (a Whig-historical approach), while revealing the invisible labor of those who work hard to stabilize such a view of the Internet as natural or possessing such objective properties. The goal of reframing software taxonomies as a question of *accomplishments* instead of as technological "affordances" is to surface these STS-relevant topics of concern for focused scholarly investigation.

Identifying Material/Social Properties as Members' Work

With this in mind, let us return to the spacecraft case for an example of what this perspective makes perspicuous to the STS scholar. I here draw attention to those moments when the stability of the software artifact and the social milieu are under negotiation: they must be *accomplished* in context. While certainly a process of stabilization would take place over the early days of a collaboration, even several years into the collaboration there are moments when such qualities are made visible. This is especially clear in jokes that question or draw attention to the standardized practices at hand, questioning their naturalness and wondering how it could be otherwise.

Above I described how the ground-based team's PowerPoint slides are said to "feed forward" into new presentations. It is tempting, perhaps, to label "feeding forward" or some archival property involved as an "affordance" of the file format or software. However, this is a property made perspicuous only in this particular context of use (consider how many slides are quickly forgotten elsewhere). This was evident when members of the team themselves made jokes or comments about this property of the slides, legible only in their social context. For instance, slide annotations were so critical to the collaboration's work that scientists often took each other to task if they felt the annotations did not align with their own observations, expectations, or discussion. When a senior team member was challenged over the interpretation they had written onto a slide, one of his former students explained to me that it was important that the image showed the right thing: "If you label the slide, it just snowballs . . . [team members] reuse the slide, and it becomes part of the lexicon" (personal conversation, August 25, 2011). Thus the "feeding forward" property of the slides was problematized not necessarily as a natural feature of the software, but as an element embedded in social practices associated with legitimacy and knowledge work.

This does not mean that "accuracy" was a taken-for-granted social norm either. In another case, one of the collaboration members on the mission used Photoshop to doctor an image that the spacecraft took to commemorate a distance milestone, embedding a fictional four-digit odometer at 9,997 on the robot's body, close to rolling over to 0,000. The image was imported into the PowerPoint deck as a joke and remained there for up to a week, generating chuckles from collaboration members each time it was displayed. This Y2K-style visual humor played on the archival sensibility that slides had acquired in this context, drawing attention to its import while at the same time subverting the importance of accuracy in the archival record as a social practice.

Like consensus team members' jokes about their slides, Excel is frequently the subject of humor on the matrix mission. In one mission meeting, scientists were debating the relative merits of a task they faced together when one of them in the room offered bluntly, "It could be a colorful spreadsheet, so we have to do it." This good-humored jab at the *work* of spreadsheet filling highlighted the activity's centrality to the collaboration's practices and processes. It also drew attention to how the embeddedness of spreadsheets—their heritage and trajectory within a wide variety of spreadsheets—was itself not a property of the spreadsheet itself, but an element of organizational practice that had become entangled with the spreadsheet as an obvious (to members) matter of fact.

Another example of this was the coordination work executed across instrument teams in the orbiting spacecraft's path selection process, the final stages of which I observed in 2009. The group had to rate seven possible pathway options according to how well these achieved local scientific goals, reporting back to the collaboration with a master spreadsheet. Each pathway had its own column, the scientific priorities were in rows, and the team had to fill in single color—red, yellow, or green—filling in the corresponding box to indicate the tour's satisfactory rating. Then, "everything" would be put into "one combined spreadsheet" (January 27, 2009). As one subgroup's meeting kicked off, the team's science planner described the purpose of the process, embedded in a history of spreadsheet work that had come before:

> The [group planning] approach was to take the prioritized spreadsheet that had been worked on very hard . . . priority that is well articulated, determine the seasonal changes in the methane—hydrocarbon. . . . Okay who contributes to this objective? They got their own row in this massive spreadsheet I've been talking about. . . . What is their piece of the pie, so to speak, in terms of how they're going to achieve this objective? Over to the right there are 20 rows that are, "what are the [instrument] priorities?" . . . Now we're at the point where each instrument is going to assess how well this part of the [path] [*gestures to columns*] achieves this objective [*gestures to row*]. (January 26, 2009)

The spreadsheets that the group faced on that day were themselves the latest in a chain of inscriptions (Latour 1995) that subsumed a long list of priorities and suggestions. As this science planner described its heritage, "The story so far is that we took the objectives that were drafted and the tour designers asked each group to draft the objectives . . . we have hundreds of pieces of input." Note how in this quote and above the spacecraft's path and the group's objectives were inextricably entangled and enacted with columns and rows, reflecting the wicked organizational problem of giving each group "their piece of the [observational] pie" through "their own row." This generated "hundreds of pieces of input" from across the heterogeneous organization.

As was typical in this team's collaboration, the spreadsheet became the site of their collective attention. It also provided the order for the conversation in the room as the group went through, coloring in one cell at a time. The spreadsheet was projected on a screen at the front of the room, "just so everyone's getting used to looking at this spreadsheet," as well as emailed out to members of the group (figure 3). The meeting proceeded by looking at the tours and priorities and assigning colors to the spreadsheet boxes. "For tour one, priority one . . . looking at that for this group, [tour number] seven is green, or yellow, or red?" The science planner called upon different subgroups to voice their opinions of the tour, filling in the spreadsheet accordingly. For example, one team said, "With our simulation . . . [that] our colleagues . . . , shows that number six is the worst." The planner then responded, "Is there any objections to turning this [box] to red?" The scientist confirmed, "It can turn into red."

It soon became clear that the three colors did not adequately express the range of opinions in the room. Scientists hedged their accounts by suggesting that a particular observation should be ranked below a green-ranked one, but not as far down as the yellow range. The science planner started suggesting intermediary colors, like "lemon-lime" and "orange" and the spreadsheet lit up in a rainbow of colors, with scientists eventually commenting along the lines of this request: "I want to change the color of that box. It was an orange but it should be a pink."

Excel provides built-in functions that add up, multiply, or otherwise compute the values listed in the cells. But text and color cannot be automatically tabulated. When the group reconvened later that afternoon, then, they were faced with the problem of aggregation. How to compute all these instrumental needs, represented on a rainbow scale, into a single recommendation for or against a tour option in one of only three colors? The scientist who led the group suggested assigning numbers to the boxes, which could then be computed as an average, giving a score that could produce a red, yellow, or green recommendation. He randomly values on a scale of 1 to 100 to correspond to each color, and began inputting them in the cells. This provoked consternation: How were these numbers chosen? What did it mean to give 85 to lemon-limes? Should all lemon-limes have 85s, or did the color represent a range of values? One scientist suggested using a continuous color scale. His colleague corrected, "we should use *numbers*, which are a continuous scale"—to which another piped up, to much laughter, "Can we use irrational numbers?" The question then arose as to whether or not the computed average should be weighted or not, and once the averages were revealed and a score of 83.6 was computed for one of the columns, one of the scientists asked puckishly if they shouldn't differentiate in the third significant figure. Someone else suggested consulting a different document, saying, "I think I have a spreadsheet somewhere. . . ." A senior scientist in the room jokingly described this process of assigning numbers to colors to his colleagues as an instance of "saving the hypothesis," and compared it to the epicycles in "the universe of Ptolemy."[8] Indeed, the numbers seemed to serve more of a gut check in the final analysis, as I noted the group started changing color assignments to end up with different averages for each column.

Certainly the team's humor pointed to the central role that Excel had come to play in their organizational life and decision-making process. Moving from colors

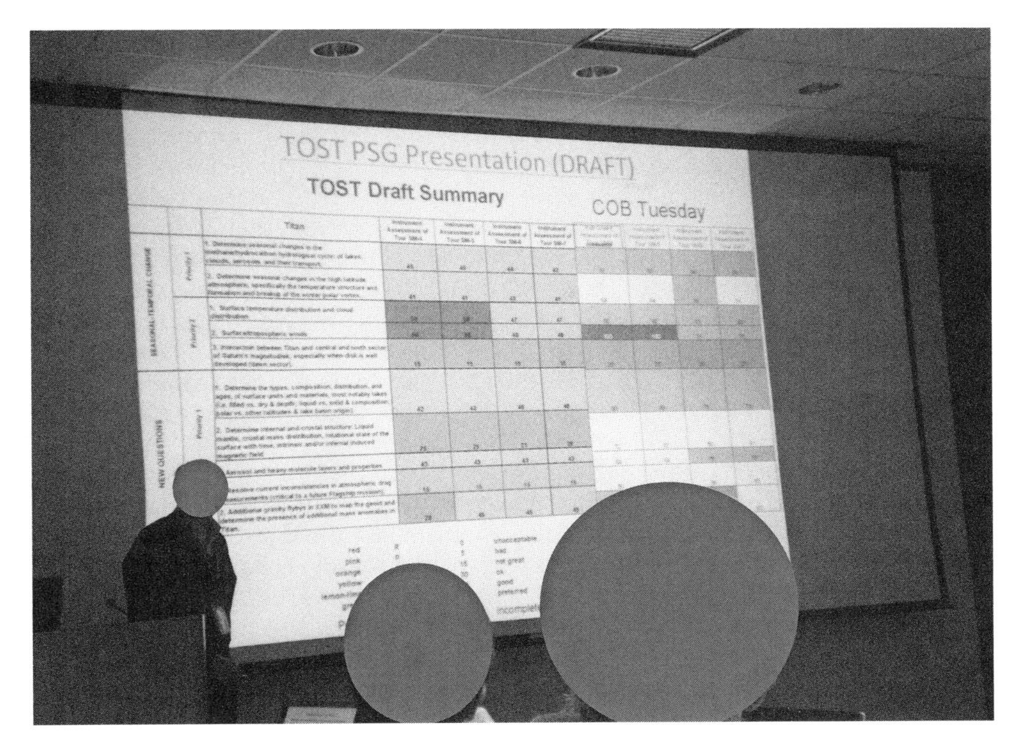

FIGURE 3: An orbiter planning subgroup works on a spreadsheet, coloring in cells with green, lemon-lime, yellow, orange, and pink.

to numbers and back again offered creative solutions to the local problem of deciding what the spacecraft should do by considering a plurality of inputs. However, this activity also represented a *software accomplishment*, at the same time opening up to analytical view the very processes of meaning-making assigned to the spreadsheet, its properties, and its suggested organizational outcomes. For a team well practiced in using color to depict preferences, this breakdown in their ability to use color to suggest an equitable and "objective" outcome of the practice laid bare their assumptions about spreadsheet work, especially those of its properties that were typically used to generate fair decisions. The dissatisfaction with numbers and the ultimate color scale that the team thought best preserved their preferences troubled the very question of *what Excel is good at doing*. Neither the spreadsheet's toolbar of potential options nor its organizational qualities of maintaining distinctions between groups were natural or obvious in this circumstance. Indeed, the team had to work to distinguish which decision-making elements were "of the spreadsheet" and which properties were "of the group"—the latter making the enterprise a house of cards "like the universe of Ptolemy."

Discussion

Because actors engage aspects of a mundane software tool as resources for structuring and ordering their local social world, calling attention to how a particular group uses software packages is a question of surfacing how software is enrolled in group membership and organizational practices. *Software is thus a ready-to-hand element in the situated achievement of work.* Attunement to the organizational context of this work can reveal patterned ways in which users deploy different attributes of software to different ends in ways that are nonconforming, yet not idiosyncratic: after all, these local techniques are ultimately both organizationally contingent and reproducing. Even when user groups identify certain practices as problematic or against the grain and others as mainstream, then (as in the "misusers" in Latzko-Toth et al., this volume), we would be remiss to elevate any single group's emic categorization to the level of an etic taxonomy, or anything other than an expression of group membership, participation, and local practices. When actors perform sense making and decision making with software, this is more frequently a question of ordering phenomena in the world than of sense making *about* software itself (on an ethnomethodological approach to that topic, see Singh et al., this volume).

The "here and now"–ness of these software forms—what Garfinkel would call their *quidditiy* or *haeccicity* (Garfinkel 1967)—enables them to become more than simply tools or a genre of organizational expression. While there are certainly elements of their local usage that beg the "genre" framing (Yates and Orlikowski 2007), they might also analytically be identified as the stuff of actors' work with which the social orders of each collaboration are performed, expressed, made, and remade. Hence examining how groups work with these digital tools does not require identifying passive attributes that can be taken up in social context; it is rather a question of revealing which situated *software accomplishments* groups turn to in order to simultaneously perform and reify their organizational and social milieu.

Software accomplishments are also the ways in which members of these collaborations perform membership and individual fluency with local norms and practices. Members' own ethnomethods (Garfinkel 1967) include facility with particular modes of software interaction and display, through which local sense making and management of team members with their varied interests can be accomplished. Membership on each team overlaps slightly, but members of both teams do not mix modes or presentational expectations. Thus the story of these tools and representational forms is not simply a story of how roving and orbiting team members have adjusted to or taken up these classic productivity software suites to local organizational ends. It is a story of how both collaborations have produced the local accomplishment of incorporating these tools, ways of visualizing, and projecting spacecraft planning into the very heart of their planning process, in ways consistent with their local organizational norms and structures. Ultimately, it is also a story about just how malleable and visible these competing elements are within the context of actors' work.

There are a few contrasts here to organizational studies of software at work. In a move popular in information systems study and design at the end of the 20th century, Kling and Scacchi (1982) proposed that isomorphism between organizational structures and software infrastructures is essential to tool uptake. This isomorphic tendency guided early computer-supported cooperative work and groupware project development. However, PowerPoint and Excel do not emerge from the (black-)box pre-prepared to align with *all* organizational structures. If we were to read these systems in such a way, the template architecture built into PowerPoint suggests a rather more bureaucratic form than the ground-based team enacts using that same software. Instead, these examples of technical practices demonstrate that contexts of use are more important than any structural tendencies embedded within the software itself.

This is equally the case in sociomaterial scholarship, which attends deeply to practice and to bottom-up theorizing as opposed to top-down categorization or structuration. The case of these two spacecraft teams reveals a paradox in which an even more micro perspective is necessary. "The social" and "the material" take shape through members' interaction, much as "affordances" were initially intended to arise from relational encounters. But it only through attenuation to practice, in the bottom-up work of collaborative planning and interaction, that individual software elements become analytically perspicuous, resolve into "social" or "material" categories, or are naturalized analytically as "sociomaterial" understandings. Arriving at the observational moment with categories assigned a priori—social and technical, normal and deviant, built in and uncovered—can too easily lead us to read such categories backward into the field, causing us to lose sight of the political or interinstitutional processes that shape what we, as analysts, might consider to be a digital property or an organizational imperative in the first place.

Conclusion

Thinking about mundane software tools in technoscientific context reveals several possibilities for STS analysis. Much existing work in STS has shown how categories and assumptions are built into information systems, and the work of wrestling with messy phenomena to produce data with no "residual categories" (Woolgar 1990; Ribes and Bowker 2008; Millerand et al. 2013; Bowker and Star 1999; Ribes and Jackson 2013). Studying software in scientific and technical organizations requires adopting an approach to representation and documentation in which the electronic documents in question are not only the result of work, but the sites of work as well (Ziewitz 2011; Vertesi 2014). As the center of collective attention, they are the place where work is done. It is therefore important to see them as evolving and live representational forms that both serve to structure a meeting and can be manipulated and changed as part of the meeting process. They are not passive sites for inscription, but an active part of meetings at which individuals are copresent virtually and physically (Beaulieu 2010). Whether projected on a shared screen or emailed out in advance, the documents participate in coordinating and focusing collaborative work. Even if the documents come to us "cold" from many years in an archive, analysts must work to see them as evolving documents with both a history and a trajectory, as well as a just-here and just-now-ness with which individuals actually do their work. The language of *software accomplishments* encompasses this active, ongoing, meaning-making process.

The ubiquity and general availability of these software tools speak to a broad circulation across laboratory environments and other contexts (for example, managerial contexts), perhaps on a scale not otherwise visible for other types of knowledge-making equipment. At the same time, it is clear that such one-size-fits-all solutions like office productivity software are used differently across local environments to suit quite different sorts of ends. We cannot truly call this a process of interpretive flexibility, stabilization, and closure, as such local groups cannot be said to be "relevant" in any way to the continuing development of the software.

Further, the software's use in these different contexts does not descript any networks or radically upend our understanding of how software should be used. Still, a standard software package's use in these different contexts suggests valuable ways of approaching software—like other tools—in and across technoscientific organizational milieus.

The first is a commitment to the organizational situatedness of knowledge work. In this case, the software is a representational tool that not only represents the organization to itself but is part of the active work of producing and reproducing that organization. Exactly *how* these software systems are put to use locally—not necessarily adapted, modified, or even appropriated per say, but quite simply and daily, *used* or *acted-with*—reveals organizational work of ordering the natural, social, and software worlds at the same time. In this sense, even the most mundane, ubiquitous, cookie-cutter tools can serve to produce and reproduce different organizations in different contexts. Further, neither the organization nor the tool come "preloaded" with organizationally enhancing ways of acting or interacting: both the tool as site of interaction and the organization as collaborative framework are enacted through the local use of the software.

Second, it is important to present an alternative way of examining the diverse ways in which technologies are put to local use without resorting to the language of affordances as a question of embedded material properties that permit certain patterns of use over others. The notion that technologies might in and of themselves suggest, prompt, or require different ways of using them from human bodies or interlocutors neglects the richness and complexity that occurs when different groups take up technological tools to achieve local ends: a richness and complexity that STS has always aimed to surface. It also risks reifying one social group's notion of patterns of use, due perhaps to their power in controlling the narrative, at the expense of another's, othering particular patterns of use without understanding local contingencies and expectations of group action. Instead of leaving "affordances" as an analytic black box, STS scholars should use it as a starting point for analysis: when something is glossed as an "affordance," whether by users, designers, or analysts, this is where our analytic process must begin. Continued attention to observable practice can help us to address how such software is put to use in different kinds of contexts as a tool for achieving local ends and making local distinctions, without resorting to taxonomies that unintentionally delimit particular groups of users.

Finally, there is value in studying and surfacing different forms of software as the invisible tools of knowledge work. There is much work in STS and computer-supported cooperative work that has examined the development of bespoke tools, drawing largely on the infrastructure or cyberinfrastructure perspective (Ribes and Bowker 2008; Lee et al. 2006). Microsoft products are perhaps less sexy than large-scale, government-funded e-science projects with their conflicting ontologies and interdisciplinary groups of computer scientists and domain experts. Yet without PowerPoint or Excel the work of these spacecraft collaborations would look quite different; without including these tools in the story of these scientists' work, any account of their practices would be severely limited. Returning to the shop floor to witness the everyday work of scientific practice can surface the most surprising of technologies and techniques that are essential to the conduct of scientific practice today.

Acknowledgments

The author thanks the spacecraft teams for permitting her observation of their work, the National Science Foundation for support of the empirical work in this project, and the Spaceteams research group Marisa Cohn, Matt Bietz, David Reinecke, Melissa Mazmanian, and especially Paul Dourish, in dialog with whom many of the early ideas in this piece developed. Helpful comments from danah boyd, Ingrid Erickson, Steve Saywer, Oliver Marsh, Michael Lynch, David Nemer, Malte Ziewitz, Carla Ilten, and the anonymous reviewers shaped and dramatically improved this piece. I am also thankful to Erika Robles-Anderson for circulating her award-winning paper about PowerPoint before its publication.

Notes

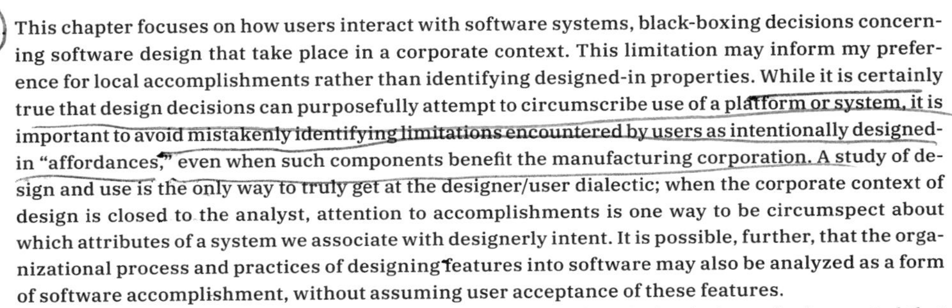

1. This chapter focuses on how users interact with software systems, black-boxing decisions concerning software design that take place in a corporate context. This limitation may inform my preference for local accomplishments rather than identifying designed-in properties. While it is certainly true that design decisions can purposefully attempt to circumscribe use of a platform or system, it is important to avoid mistakenly identifying limitations encountered by users as intentionally designed-in "affordances," even when such components benefit the manufacturing corporation. A study of design and use is the only way to truly get at the designer/user dialectic; when the corporate context of design is closed to the analyst, attention to accomplishments is one way to be circumspect about which attributes of a system we associate with designerly intent. It is possible, further, that the organizational process and practices of designing features into software may also be analyzed as a form of software accomplishment, without assuming user acceptance of these features.
2. The social scientists and computer scientists who studied the mission in its early days noted that science team members were more likely to print photos out and write all over them using colored pens than to use immersive, 3-D digital environments or even a purpose-built digital whiteboard to interpret the terrain and make decisions about robotic action. Eventually the social scientists therefore suggested bringing in large tables and printers so that scientists could pore over printed images together. This was more popular with the science team, who retired the digital whiteboards as "digital clocks." I am grateful to Roxana Wales, Bill Clancey, Wendy Ju, Jeff Norris, Alonso Rivera, and Mark Powell for sharing these recollections.
3. This analysis of the spreadsheet resonates with that of Dourish (2017); I am grateful for our many conversations about this phenomenon and its analysis.
4. These observations were acquired April 17–19, 2009.
5. See Reeves et al. (2017) for an application of the ethnomethodological "accomplishment" lens to software, in this case gaming systems.
6. This is not incompatible with the relational view of affordances that focuses on how actors confront technological limitations whether through daily practice, evasion, or creative repurposing. However, this view refrains from identifying as material properties of the system those elements that users encounter as limitations; and from assuming those properties to be the result of designerly or unintentional circumscription of user activity.
7. A synergistic argument is offered in Costa (2018), released as this essay went to press, critiquing communications scholars for elevating Anglo-American practices as stable, intrinsic social media "affordances" and arguing also for a renewed emphasis on practice.
8. For more on the Ptolemaic worldview, the epicycles as "saving the hypothesis," and the transition to Copernicanism, see Kuhn (1958).

Works Cited

Akrich, Madeleine. 1992. "The De-scription of Technological Objects." In *Shaping Technology/Building Society*, edited by Wiebe E. Bijker and John Law, 205–24. Cambridge, MA: MIT Press.

Barad, Karen. 2003. "Posthumanist Performativity: Toward an Understanding of How Matter Comes to Matter." *Signs* 28 (3): 801–31.

Barley, Stephen. 1996. "Technology as an Occasion for Structuring: Evidence from Observation of CT Scanners and the Social Order of Radiology Departments." *Administrative Science Quarterly* 31:78–108.

Beaulieu, Anne. 2010. "From Co-location to Co-presence." *Social Studies of Science* 40 (3): 453–70.

Bowker, Geof, and Susan Leigh Star. 1999. *Sorting Things Out: Classification and Its Consequences.* Cambridge, M.A.: MIT Press.

boyd, danah. 2010. "Social Network Sites as Networked Publics: Affordances, Dynamics, and Implications." In *A Networked Self: Identity, Community, and Culture on Social Network Sites*, edited by Zizi Papacharissi, 39–58. New York: Routledge.

Coleman, Gabriella. 2014. *Hacker, Hoaxer, Whistleblower, Spy: The Many Faces of Anonymous.* New York: Verso Press.

Collins, Randall. 2004. *Interaction Ritual Chains.* Princeton, NJ: Princeton University Press.

Costa, Elisabetta. 2018. "Affordances-in-Practice: An Ethnographic Critique of Social Media Logic and Context Collapse." *New Media & Society.* doi:10.1177/1461444818756290.

Dourish, Paul. 2017. *The Stuff of Bits: Essays on the Materialities of Information.* Cambridge, MA: MIT Press.

Edwards, Paul N. 1997. *The Closed World: Computers and the Politics of Discourse in Cold War America.* Cambridge, MA: MIT Press.

Evans, Sandra K., Katy E. Pearce, Jessica Vitak, and Jeffrey W. Treem. 2017. "Explicating Affordances: A Conceptual Framework for Understanding Affordances in Communication Research." *Journal of Computer-Mediated Communication* 22 (1): 35–52. doi:10.1111/jcc4.12180.

Faraj, Samer, and Bijan Azad. 2012. "The Materiality of Technology: An Affordance Perspective." In *Materiality and Organizing*, edited by Paul M. Leonardi, Bonnie A. Nardi, and Jannis Kallinikos, 237–58. Oxford: Oxford University Press. https://doi.org/10.1093/acprof:oso/9780199664054.003.0012.

Garfinkel, Harold. 1967. *Studies in Ethnomethodology.* Englewood Cliffs, NJ: Prentice Hall.

Gibson, James J. 1979. *The Ecological Approach to Visual Perception.* Boston: Houghton Mifflin.

Gillespie, Tarleton. 2006. "Designed to 'Effectively Frustrate': Copyright, Technology and the Agency of Users." *New Media & Society* 8 (4): 651–69. doi:10.1177/1461444806065662.

——. 2010. "The Politics of 'Platforms.'" *New Media & Society* 12 (3): 347–64. doi:10.1177/1461444809342738.

Haraway, Donna J. 1991. *Simians, Cyborgs, and Women.* New York: Routledge.

——. 1997. *Modest_Witness@Second Millenium. FemaleMan_Meets_OncoMouse: Feminism and Technoscience.* New York: Routledge.

Hutchby, Ian. 2001. "Technologies, Texts and Affordances." *Sociology* 35 (2): 441–56. doi:10.1177/S0038038501000219.

——. 2003. "Affordances and the Analysis of Technologically Mediated Interaction: A Response to Brian Rappert." *Sociology* 37 (3): 581–89. doi:10.1177/00380385030373011.

Ilten, Carla. 2015. "'Use Your Skills to Solve This Challenge!' The Platform Affordances and Politics of Digital Microvolunteering." *Social Media + Society* 1 (2). https://doi.org/10.1177/2056305115604175.

Kaplan, Sarah. 2011. "Strategy and PowerPoint: An Inquiry into the Epistemic Culture and Machinery of Strategy Making." *Organization Science* 22 (2): 320–46.

Kline, Ronald, and Trevor Pinch. 1996. "Users as Agents of Technological Change: The Social Con struction of the Automobile in the Rural United States." *Technology and Culture* 37 (4): 763–95. doi:10.2307/3107097.

Kling, Rob. 1991. "Computerization and Social Transformations." *Science, Technology, & Human Values* 16 (3): 342–67.

Kling, Rob, and Walt Scacchi. 1982. "The Web of Computing: Computer Technology as Social Organization." *Advances in Computers* 21:1–90.

Knoblauch, Hubert. 2013. *PowerPoint, Communication, and the Knowledge Society.* Cambridge: Cambridge University Press.

Knorr Cetina, Karen, and Urs Bruegger. 2000."The Market as an Object of Attachment: Exploring Post-social Relations in Financial Markets." *Canadian Journal of Sociology* 25 (2): 141–68.

Kuhn, Thomas S. 1958. *The Copernican Revolution: Planetary Astronomy in the Development of Western Thought.* Cambridge, MA: Harvard University Press.

Latour, Bruno. 1991. "Technology Is Society Made Durable." In *A Sociology of Monsters: Essays on Power, Technology and Domination*, Sociological Review Monograph 38, 103–32. New York: Routledge.

——. 1995. "The 'Pedofil' of Boa Vista: A Photo-Philosophical Montage." In *Common Knowledge* 4:143–87.

——. 2005. *Reassembling the Social: An Introduction to Actor-Network-Theory.* Oxford: Oxford University Press.

Lee, Charlotte, Paul Dourish, and Gloria Mark. 2006. "The Human Infrastructure of Cyberinfrastructure." In *Proceedings of the 2006 Conference on Computer Supported Cooperative Work*, 483–92. New York: ACM. doi:10.1145/1180875.1180950.

Leonardi, Paul M. 2011. "When Flexible Routines Meet Flexible Technologies: Affordance, Constraint, and the Imbrication of Human and Material Agencies." *MIS Quarterly* 35:147–67.

——. 2012. *Car Crashes without Cars: Lessons about Simulation Technology and Organizational Change from Automotive Design*. Cambridge, MA: MIT Press.

Majchrzak, Ann, Samer Faraj, Gerald C. Kane, and Bijan Azad. 2013. "The Contradictory Influence of Social Media Affordances on Online Communal Knowledge Sharing." *Journal of Computer-Mediated Communication* 19 (1): 38–55.

Mazmanian, Melissa, Marisa Cohn, and Paul Dourish. 2014. "Dynamic Reconfiguration in Planetary Exploration: A Sociomaterial Ethnography." *Management and Information Science Quarterly* 38 (3): 831–48.

Millerand, Florence, David Ribes, Karen Baker, and Geoffrey C. Bowker. 2013. "Making an Issue Out of a Standard: Storytelling Practices in a Scientific Community." *Science, Technology, & Human Values* 38 (1): 7–43.

Myers, Natasha. 2014. "Rendering Machinic Life." In *Representation in Scientific Practice Revisited*, edited by Catelijne Coopmans, Janet Vertesi, Michael Lynch, and Steve Woolgar, 153–76. Cambridge, MA: MIT Press.

X Nagy, Peter, and Gina Neff. 2015. "Imagined Affordance: Reconstructing a Keyword for Communication Theory." *Social Media + Society* 1 (2). doi:10.1177/2056305115603385.

Neff, Gina, Tim Jordan, Joshua McVeigh-Schultz, and Tarleton Gillespie. 2012. "Affordances, Technical Agency, and the Politics of Technologies of Cultural Production." *Journal of Broadcasting & Electronic Media* 56 (2): 299–313.

Norman, Donald A. 2002. *The Design of Everyday Things*. New York: Basic Books.

Orlikowski, Wanda. 2010. "The Sociomateriality of Organisational Life: Considering Technology in Management Research." *Cambridge Journal of Economics* 34 (1): 125–41. doi:10.1093/cje/bep058.

X Oudshoorn, Nelly, and T. J Pinch. 2003. *How Users Matter the Co-construction of Users and Technologies.* Cambridge, MA: MIT Press.

Pinch, Trevor, and Wiebe E. Bijker. 1987. "The Social Construction of Facts and Artifacts: Or How the Sociology of Science and the Sociology of Technology Might Benefit Each Other." In *The Social Construction of Technological Systems: New Directions in the Sociology and History of Technology*, edited by Trevor Pinch, Wiebe E. Bijker, and Thomas P. Hughes, 17–50. Cambridge, MA: MIT Press.

Raja-Yusof, Raja-Jamilah, Azah-Anir Norman, Siti-Soraya Abdul-Rahman, Nurul 'adilah Nazri, and Zulkifli Mohd-Yusoff. 2016. "Cyber-Volunteering: Social Media Affordances in Fulfilling NGO Social Missions." *Computers in Human Behavior* 57:388–97. doi:10.1016/j.chb.2015.12.029.

Reeves, S., Christian Greiffenhagen, and Eric Laurier. 2017. "Video Gaming as Practical Accomplishment: Ethnomethodology, Conversation Analysis, and Play." *Topics in Cognitive Science* 9:308–42. doi:10.1111/tops.12234

Ribes, David, and Geoffrey C. Bowker. 2008. "Organizing for Multidisciplinary Collaboration: The Case of the Geosciences Network." In *Scientific Collaboration on the Internet*, edited by Gary M. Olson, Ann Zimmerman, and Nathan Bos, 310–30. Cambridge, MA: MIT Press. http://mitpress.universitypressscholarship.com/view/10.7551/mitpress/9780262151207.001.0001/upso-9780262151207-chapter-18.

Ribes David, and Steve Jackson. 2013. "Data Bite Man: The Work of Sustaining Long-Term Study." In *'Raw Data' Is an Oxymoron*, edited by Lisa Gitelman, 147–66. Cambridge, MA: MIT Press.

Robles-Anderson, Erika, and Patrik Svensson. 2016. "'One Damn Slide after Another': PowerPoint at Every Occasion for Speech." *Computational Culture* 5. http://computationalculture.net/article/one-damn-slide-after-another-powerpoint-at-every-occasion-for-speech.

Schegloff, Emanuel. 1972. "Notes on a Conversational Practice: Formulating Place." In *Studies in Social Interaction*, edited by David Sudnow, 75–119. New York: Macmillan.

Schwartz Cowan, Ruth. 1985. "How the Refrigerator Got Its Hum." In *The Social Shaping of Technology*, edited by Donald MacKenzie and Judy Wajcman, 202–18. London: Open University Press.

Sellen, Abigail, and Richard Harper. 2001. *The Myth of the Paperless Office*. Cambridge, MA: MIT Press.

X Sharrock, Wes, and Jeff Coulter. 1998. "On What We Can See." *Theory and Psychology* 8(2):147–64.

Shils, Edward. 1965. "Charisma, Order, and Status." *American Sociological Review* 30 (2): 199–213. doi:10.2307/2091564.

Suchman, Lucy. 2006. *Human–Machine Reconfigurations: Plans and Situated Actions*. 2nd ed. Cambridge: Cambridge University Press. http://ebooks.cambridge.org/ref/id/CBO9780511808418.
Treem, Jeffrey W., and Paul M. Leonardi. 2013. "Social Media Use in Organizations: Exploring the Affordances of Visibility, Editability, Persistence, and Association." *Annals of the International Communication Association* 36 (1): 143–89. https://doi.org/10.1080/23808985.2013.11679130.
Tufecki, Zeynep. 2017. *Twitter and Tear Gas: The Power and Fragility of Networked Protest*. New Haven, CT: Yale University Press.
Tufte, Edward. N.d. "The Cognitive Style of PowerPoint: Pitching out Corrupts Within." http://edwardtufte.com/tufte/powerpoint.
Turner, Fred. 2008. *From Counterculture to Cyberculture: Stewart Brand, the Whole Earth Network, and the Rise of Digital Utopianism*. Chicago: University of Chicago Press.
Vaast, Emmanuelle, and Evgeny Kaganer. 2013. "Social Media Affordances and Governance in the Workplace: An Examination of Organizational Policies." *Journal of Computer-Mediated Communication* 19 (1): 78–101. doi:10.1111/jcc4.12032.
Vertesi, Janet. 2014. "Seamful Spaces: Heterogeneous Infrastructures in Interaction." *Science, Technology, & Human Values* 39 (2): 264–84. doi:10.1177/0162243913516012.

——. 2015. *Seeing Like a Rover: How Robots, Teams and Images Craft Knowledge of Mars*. Chicago: University of Chicago Press.
Vitak, Jessica, and Jinyoung Kim. 2014. "'You Can't Block People Offline': Examining How Facebook's Affordances Shape the Disclosure Process." In *Proceedings of the 17th ACM Conference on Computer Supported Cooperative Work & Social Computing*, 461–74. New York: ACM. doi:10.1145/2531602.2531672.
Weber, Max. 1968. *Economy and Society: An Outline of Interpretive Sociology*. Edited by Guenther Roth and Claus Wittich. Translated by Ephraim Fischoff et al. New York: Bedminster Press.
Woolgar, Steve. 1990. "Configuring the User: The Case of Usability Trials." *Sociological Review* 38 (S1): 58–99. doi:10.1111/j.1467-954X.1990.tb03349.x.
Wyatt, Sally. 2003. "Non-users Also Matter: The Construction of Users and Non-users of the Internet." In *How Users Matter*, edited by Trevor Pinch and Nelly Oudshoorn, 67–79. Cambridge, MA: MIT Press.
Yates, JoAnne, and Wanda Orlikowski. 2007. "The PowerPoint Presentation and Its Corollaries: How Genres Shape Communicative Action in Organizations." In *Communicative Practices in Workplaces and the Professions: Cultural Perspectives on the Regulation of Discourse and Organizations*, edited by Mark Zachry and Charlotte Thralls, 67–91. Amityville, NY: Baywood.
Ziewitz, Malte. 2011. "How to Attend to Screens? Technology, Ontology and Precarious Enactments." Special issue on screens, edited by Brit Ross Winthereik, Lucy Suchman, Peter Lutz, and Helen Verran. *Encounters* 4 (2): 203–28.

Misuser Innovations

The Role of "Misuses" and "Misusers" in Digital Communication Technologies

Guillaume Latzko-Toth, Johan Söderberg, Florence Millerand, and Steve Jones

Studying the digital is an opportunity to revisit and reconsider certain notions that may have been overlooked or did not receive full attention in STS scholarship based on nondigital technologies. It is old news that users of technology can twist and turn technological products to serve ends that were not foreseen from the outset. STS researchers have documented how unanticipated uses of technology and resistances to the designer's program of action have given way to new uses. Their work has examined moments of resistance (Levy 2015), tampering (Gillespie 2006), nonuse (Wyatt 2003), and de-scription of technical objects (Akrich 1992). Occasionally, such appropriations are invested with claims and values that are in conflict with prescribed uses of the technology (Oudshoorn and Pinch 2003). Engineers and designers too have learned the lesson by now: users often develop new functions for old products that had not been foreseen at the outset. This insight has even been incorporated in the design process itself and inscribed in the curriculum at engineering and design schools. What additional claims are then left to be made for the misuser, that has not already been extensively discussed in relation to the user?

In this chapter, we propose that reflections over misuses and misusers, while relevant for understanding the use and adoption of any kind of technology, become critical for digital communication technologies. The plasticity associated with them boils down to the difficulties of imposing any single norm on other parties and bring to closure the prescribed uses. Extending Flichy's concept of "frame of use" (2007), we suggest that sociotechnical practices are discursively and normatively framed by developers and users as being *proper uses* (desired, appropriate) or, conversely, *misuses* (deviant, disruptive). Then, what begins as a deviation from a preexisting normative frame of use may lead to innovations from which grows a new set of prescribed uses and misuses. Our reflection takes hold in a study of three controversies around alleged misuses that flourished at the dawn of networked computing: gaming on an educational system; recreational communication on a research infrastructure; and user robots in the context of IRC networks. In all of these cases, conflicts erupted over how to use the device in the proper way. As signs of conflict, we point to the many accusations and complaints users had about system misuse and misusers. The construction of use and misuse of digital devices is the central theme of our investigation. By misuse, we mean a category that is ascribed to some actors by other actors, all of whom are considered users of the device. This contrasts with hacking, file sharing, and other misuses that involve the law, where there are other actors involved (e.g., governments) that don't directly pertain to the device. We have limited our discussion to cases where the controversy is contained within the device. The advantage with illustrating this discussion with historical cases is that we can put in perspective the construction of the misuser. What at one point in time is framed by users as a distraction and a waste vis-à-vis the presumed, primary function of the device is, at a later point, the central activity that defines the purpose of the device. In the case of PLATO, Bitnet, and IRC, the conflict centered around users playing games and chatting, whereas the communication networks were initially meant for educational and informational purposes.

The observations that we draw from these cases resonate with previous findings in studies of users of technology: that users tend to act in unpredictable ways, that in doing so they sometimes stumble over functions that had not been anticipated by the designer or the manufacturer, and, finally, that this can become the starting point of new functionalities or markets (Von Hippel 1986, 2001, 2005). These are claims that have been made in regard to all kinds of technologies and with respect to all kinds of users. We make a more specific claim in proposing that there is a close connection between digital devices and conflicts over what counts as a proper or appropriate usage. This claim has already been made by Brunton (2013) in his thorough analysis of "spam" as a phenomenon that, he contends, is inextricably tied to the development of information technologies. As he puts it (2013, 203), since the early days of digital networks, "spam has always been there—as threat, as concept, or as fact. It provokes and demands the invention of governance and acts of self-definition on the part of those with whom it interferes. They must be clear about misbehavior, bad speech, and abuse of the system, which entails beginning to talk about what is worthwhile, good, and appropriate." It is this heightened level of conflict over how proper use is constructed that warrants our investigation into the notion of misuse. This in turn we attribute to how the digital has been constructed as flexible, plastic, in order for it to comply with the design ideal of horizontal, peer-to-peer cooperation.[1] Consequently, the system administrator, moderator/operator, or project leader cannot so easily bring any single interpretation of what constitutes the functionality of the device to closure. Lacking any decisive advantage over the other users, the only means for sanctioning misbehaving users, and, consequently, for stabilizing the prescribed function and use of the device, is through the mobilization of social norms and the active policing of uses. In other words, by wielding the "misuser" label. This strategy proved ineffective in the three cases discussed here, but the failure was a productive one, in the sense that out of the deviations sprang new markets and new industries, which in turn helped to redefine what the functionality of computers and communication networks is supposed to be. Whether or not a deviant practice will be reevaluated as an act of innovation by posterity depends on the success of this practice in establishing itself as the new norm for use.

Recreational Uses of Digital Resources: A Threefold Case Study

Our reflection is grounded in the study of three communication platforms that flourished at the dawn of home computing: PLATO, Bitnet, and Internet Relay Chat (IRC). These three cases are interesting because they prompted discussions around what is the proper use of a digital technology. Their historical character provides the necessary distance to put the misuser construct in perspective. A historical perspective is necessary because it is only in hindsight that one can tell, if misuse was just misuse or if it was part of the innovation process. A common denominator of these cases is that in each of these settings, misuse was associated with a recreational use of the device: playing games (PLATO), chatting (Bitnet), and bot programming (IRC). These activities could be subsumed under the overarching category of "play," as used by Danet et al. (1997).[2] They deviate from the convened frame of use for each of these platforms—respectively teaching, file transfer, and chat.

To conduct these historical case studies, we used an approach inspired from virtual ethnography (Hine 2000) that we call "asynchronous online ethnography" (Latzko-Toth 2013): the observation, description, and analysis of the past social life of an online group, based on archived traces of digitally mediated activities and interactions. Like "trace ethnography" (Geiger and Ribes 2011), this approach does not rely on trace data only but involves participant observation, interviews, and documentary research to get a deep understanding of sociotechnical practices under study. Latzko-Toth and Jones (2014) digitized a corpus of conversations that took place in the early online forum PLATO Notes (see figure 1). Following University of Illinois Archives nomenclature, the 68 files (plus one addendum) in our dataset are referred to as follows: LN01–LN19 (October 1972 to August 1973), ON01–ON09, ON18–ON41 (January 1974 to December 1975), PN01–PN16 (December 1975 to June 1976). The Bitnet corpus consists of a set of e-journal issues preserved on the Internet Archive website. Finally, the IRC corpus was gathered by Latzko-Toth from various sources, and mainly consists of archived posts from two mailing lists (Operlist and Wastelanders) and a Usenet newsgroup (alt.irc). First-hand experience with PLATO (Jones) and IRC (Latzko-Toth) as well as interviews with key actors of PLATO and IRC development provided triangulation in how we interpreted the data.

Our analysis focused on the rhetorical ways by which use and misuse were constructed and evolved through time, particularly in moments where conflicts erupted over accusations and complaints about misuse and misuser. A QDA software was used as a text processing and coding tool. This allowed for an inductive approach to the material, letting analytic categories and topics emerge through coding. Considering the amount of material, we turned to plain-text searches on keywords to focus our attention of relevant messages. We would then iterate the process once a new topic or significant term was identified.

PLATO Games

PLATO was an educational and social computer platform developed at the Computer-based Education Research Laboratory (CERL) at the University of Illinois at Urbana-Champaign in the 1960s and 1970s. Emerging in the context of a growing public interest for computerized teaching (see Chan, this volume), it is

```
FOR ME, SUPPORTS THE RECOMMENDATION FOR USE OF LAST NAMES.
↑CONSIDER↑; ↑I RECEIVED A FORM LETTER FROM YOU SOME TIME
AGO, SENT TO MANY PEOPLE, IN WHICH (OF COURSE) YOU SIGNED
YOUR PROPER NAME.  ↑NOW, IF YOU HAD NOT MADE A POINT OF
INCLUDING YOUR SIGNON IN THE LETTER, ↑I WOULD HAVE HAD
NO WAY TO CONTACT YOU ON LINE.  ↑I CERTAINLY HAD NO
INTUITIVE CONNECTION BETWEEN THE SIGNATURE ON THE LETTER
AND THE SIGNON ↑I HAD SEEN IN ↑PLATO NOTES.  ↑USE OF YOUR
FIRST NAME MAY BE PERFECTLY FAMILIAR TO THOSE IN YOUR
OWN PROJECT, BUT FOR COMMUNICATION WITH OTHER AUTHORS
(WHICH IS THE STATED PURPOSE OF THE RECOMMENDATIONS),
USE OF YOUR LAST NAME WOULD BE HELPFUL.

---------------------------------------------------------- RESPONSE 9
  03/20    17.43    WEASEL                 IU

↑HEY......IF ↑I HAD TO USE MY LAST NAME, NOT ONLY
WOULD MY FASCINATING AND INNOVATIVE SIGN-ON BE LOST
TO THE SYSTEM, BUT ↑I WOULD GO THROUGH THE HASSLE
OF HAVING TO TYPE ↑,RIESELBACH↑, IN ↑,IU↑, EACH TIME
↑I SIGNED ON↑,↑,↑,↑,↑,↑,

    WEASEL$$$$$$      $$$$$$↑4↑W↑E↑A↑S↑E↑L   $$$   $$$(↑AND DON↑7T YOU FORGET IT↑,)

--------------------------------------------------------------------------------------- NOTE 296  NICE GOING
  03/19    15.54    HODY                   MED

I AM REALLY SORRY TO SEE THE SYSTEM STAFF TAKING
SO MUCH ABUSE AND SARCASM FOR ONE OF THE SOUNDEST
DECISIONS MADE IN A LONG TIME.  THERE ARE CERTAINLY
CLEAR CASES OF USERS WHO TRY DELIBERATELY TO INTERFERE
WITH THE ENJOYMENT OF THE SYSTEM BY OTHERS AND IT
IS A WASTE OF RESOURCES AND TALENT TO DEMAND THAT THE
SYSTEM IMPOSE SOFTWARE LIMITS WHICH WILL MAKE THIS
ABUSE IMPOSSIBLE IN ALL CASES.  WHAT WOULD HELP WOULD
BE DELINEATION OF WHAT THE SYSTEM CONSIDERS IMPROPER
USE.

WITH RESPECT TO ABUSE, IT HAS BEEN OBSERVED THAT SOME USE
RESOURCES (DISC, CPU, CONDENSOR, TERMINALS) FOR
PURELY RECREATIONAL PURPOSES IN PRIME TIME WHEN
OTHER USERS WHO HAVE TO WORK ARE BEING DEPRIVED.  IF
THE SYSTEM ALLOCATED ALL RESOURCES IN SOME EQUITABLE
MANNER, THAT COULD BE DEFENDED. UNFORTUNATELY THE
ASSIGNMENT OF DISC SPACE IS ↑,ARBITRARY↑, TO A LARGE
DEGREE AND THE CONDENSOR AND CPU  (CONT)

---------------------------------------------------------- RESPONSE 1
  03/19    16.00    HODY                   MED

... CONT... THE CPU AND CONDENSOR ARE LIKEWISE NOT
ALLOCATED IN A FOOL PROOF MANNER.  THUS IT IS POSSIBLE
THAT IF RECREATIONAL USERS WERE NOT ON IN PRIME TIME,
THEN THE REMAINING RESOURCES WOULD SERVE THE REST BETTER.
I DO NOT ARGUE WITH THE MERITS OF ENCOURAGING PEOPLE
TO BE CREATIVE AND TO EXPERIENCE THE SYSTEM.  I DOUBT
THAT THAT FUNCTION IS SERVED BY INTERTERMINAL COMBATIVE
GAMES IN PRIME TIME.
```

FIGURE 1: A sample of PLATO Notes original printouts.

considered to be the first computer-assisted learning system to be widely disseminated (Van Meer 2003). A key element of PLATO plasticity was its powerful yet relatively simple programming language: TUTOR. It made possible a wide array of creative programs—called "lessons"—including purely recreational ones, but also communication and collaboration services that became essential to the system development.

Games and other "non-educational programs" were at the heart of an ongoing debate over proper allocation of resources. In fact, they were epitomizing a broader debate over the *purpose* of PLATO, highlighting the interpretive flexibility of the artifact. The following quotes are from three different PLATO users and developers, from the "general interest" Notes forum hosted on the system. They provide a good oversight of the conflict over this topic:

> I must point out that this is an experimental teaching computer, and its function is primarily that of an educational asset. During the day, there are so many authors and students, that the terminals and computer time are limited to those people who are using the computer *as it was designed, for educational purposes.* (File LN09, 3/28/1973, emphasis added)

> If recreational users were not on in prime time, then the remaining resources would serve the rest better. I do not argue with the merits of encouraging people to be creative and to experience the system. I doubt that that function is served by interterminal combative games in prime time. (File ON24, 3/19/1975)

> The machine should (given resources) be available to all students, not only those in specific courses, and for all purposes, not only lessons. *Just like a library*. The catch is the "given resources." If PLATO can allocate resources sensibly, . . . it should. (File ON26, 4/8/1975)

CERL authorities had a "freedom of authorship" policy (derived from notions of "academic freedom" common at US colleges and universities and early hacker ethos) that encouraged exploratory and creative practices on PLATO, including purely recreational ones. But that ethos had to deal with the scarcity of key digital resources as well as nondigital ones (e.g., lab operating hours, system staff labor, etc.). Thus, games and other programs from "unsupervised authors" were competing with other applications for resources, including terminals, memory space, and peak hour time slots:

> The objection is not to "game" lessons per se but to recreational use of terminals and site ecs [memory] when others have pressing needs for these resources. (File ON09, 5/13/1974)

> Why abuse [the system] with games during prime time, when games eat away at precious, valuable [memory] space, when this time is the ONLY time some authors can work? (File ON21, 2/4/1975)

The quotes illustrate the idea of disruptiveness that was associated with gaming, deemed as transforming the frame of use of the system:

> Room 165 CERL is a classroom, not an arcade. I can't believe you weren't aware of the crunch for terminals when you entered the games. (File PN11, 4/20/1976)

> Children of unspecified ages spend their days at cerl playing games. . . . Concomitant with playing is the issuing of loud, disgusting noises . . . when the noneducational use of the . . . system interferes with serious work, SOMETHING HAS TO BE DONE. (File ON34, 8/20/1975)

One participant in the discussion notes that "there are clear cases of users who try deliberately to interfere with the enjoyment of the system by others" and suggests that, instead of curbing them with material restrictions coded into the system, it "would be useful [to] delineat[e] . . . what the system considers improper use" (File ON24, 3/19/1975). This epitomizes how misusers are constructed as trespassing the line drawn by official designers (the "system people") around what constitutes the desired frame of use for the system. The following examples show how that norm is rhetorically constructed: what constitutes appropriate and inappropriate uses of PLATO, what is "of value," and what is not:

> We should be careful to know the difference between "serious" and "nonserious" usage. The usual game-playing that at times takes place should not constitute "serious" usage. (File ON02, 1/23/1974)

> Why would you want to play games when there are so many other neat things to do on Plato? (File LN12, 4/18/1973)

> How come you ********* keep stealing the games from this ****** computer. They're probably the only thing of value or interest on this machine. (File LN09, 3/27/1973)

The notion of "educational value"—a boundary object (Star and Griesemer 1989) between teachers, system developers, and students—served as a guiding principle and also as a rhetorical device, a claim to legitimize users' appeals vis-à-vis resources, to sort out what was and what was not a misuse of those resources. Thus, "theories" of education could be invoked either to ban games or, conversely, to justify their presence on the platform:

> The entire idea of PLATO is to show that learning need not be the rote methods in use before. Most "game" lessons . . . include some element of education in them. (File LN06, 1/22/1973)

> Systems people must realize the educational benefit of certain games both to those using them and to those (especially new) authors writing and coding them. If system capacity permits, certain game playing and games authoring should be considered educationally as vital as more discipline oriented lesson viewing and authoring. (File ON05, 2/28/1974)

Games were also seen by some as being beneficial to PLATO (and therefore to its users) because they were pushing the system to its limits and forcing PLATO developers to improve the design of the system:

> The question of redundant computation on the system is a more rational way of dealing with the REAL problem than making a neurotic attack on some symbolic organism called "games." It seems to be a matter of system engineering to *utilize the available resources profitably*. . . . Get rid of lousy lessons and programs that are less functional than the computer stuff they consume . . . entertainment and education are both functional, redundant computation and idle time are not. (File ON21, 2/5/1975, emphasis added)

Our case study of PLATO illustrates how the notions of proper use and device purpose are relative and constructed. It also sheds light on the discursive aspect of building a normative frame of use and negotiating its possible extension or transformation.

Bitnet Chats

Bitnet (Because It's Time network) is a computer networking initiative launched in 1981 to link liberal art faculties across the world, by interconnecting IBM computers using a proprietary protocol (RSCS). It was eventually superseded by the Internet.

Communication as Misuse

Reading through archives of Bitnet newsletters, one thing strikes the contemporary reader: the network was not originally conceived of as a communication infrastructure. According to its charter, written in October 1986, Bitnet is a computer network established with the "purpose of facilitating non-commercial exchange of information consistent with the academic purposes of its . . . members" (Bitnet Executive Committee 1986). The broad notion of "exchange of information" doesn't exclude communication, but the formulation stresses the academic value of the content transferred from an institution to another. The words chosen reflect a compromise and tolerance for communicational uses of Bitnet as long as they are deemed academically relevant. It results from years of sometimes heated debates about the primary purpose of the system. The crux of the matter lay in the design of the network protocol used in Bitnet: messages had priority over file transfers. File servers were the original dominant application on Bitnet (Condon 1990). In that sense, Bitnet was designed as a long-distance file exchange system. Communication applications (email, chat, bulletin boards) where secondary. Even more, real-time communication (chat) was a *détournement* of the system's message functionality. The more that feature was used, the less file serving could work properly, due to long delays (sometimes days!) generated by queues. In that sense, "recreational communication" appeared as a misuse insofar as it was competing with and preventing "legitimate uses" of the network:

> Since RSCS gives priority to message buffers ahead of file buffers, this has a tendency to slow file transfer to a standstill. . . . When I have a professor . . . who has thesis students sending their papers to him via the network for correcting and comes and complains that his files are not getting through and I check and see a few copies of MONOPOLY and ZORK and ADVENTURE going from Spain to Germany . . . and then I . . . see a very large number of message buffers (going to Chat systems in the United States) preempting the file transfer of legitimate computer work, I begin to see abuse with a capit[a]l A. (Nussbacher 1985)

A similar hijacking of the prioritized transmission of characters between PLATO terminals—which allowed for the development of chat programs (Dear 2017)—was not met with a similar opposition from PLATO administrators. This could be explained by the fact that this type of use had no detrimental consequences on other activities, nor drained them from resources.

The Disruptive Individual

Another interesting aspect of the Bitnet case is that it vividly illustrates the transformative power of a single individual within a digital infrastructure. Christopher Condon was both one of these "entrepreneur" individuals and a commentator of the phenomenon. While pursuing undergraduate studies at Sacred Heart University, he was involved in the Bitnet Network Information Center (BitNic) hosted at Yale University. Condon was the founder and editor of a series of periodical newsletters, including *BitList* and *NetMonth*. *BitList* started as a mere list of Bitnet servers providing information of new servers, "dead" servers, their network address, their type, and other relevant facts. It quickly became the essential link between Bitnet users. As Condon regularly recalls, it was considered the "official list" because it was the *only* list of its kind. The content of the newsletter diversified considerably, and it soon became a "hub" of information *and* communication between Bitnet users (and staff). It was also instrumental in promoting new programs and other periodicals (like *Class Four: The Relay Magazine*), hence giving traction to new initiatives. Overall, this electronic medium was transformative in turning a set of scattered servers and users into a reflexive online community. In one of his editorial notes, Condon cites two other individuals whose initiatives, just like his, had a transformative impact of the very nature of Bitnet: Eric Thomas, author of the ListServ mailing list program, and Jeffrey Kell, who wrote the distributed chat program "Relay":

> Most change is brought about not by work groups or committees, but by individuals (for whatever reason). Eric Thomas took it upon himself to improve the original BITNIC LISTSERV and propagate his version throughout the network. Jeff Kell wrote RELAY and did much the same thing. While these people received (and are receiving) assistance from others in the network to continue their efforts, the impetus for change [is] an individual one. Or rather, the effort in BITNET change is a group one, where the momentum behind that change, the groundwork, often comes from a single person. (Condon 1988)

Communication as the New Purpose

While it appears *positive* in hindsight, these transformative actions are in fact *disruptive*, in the sense that they profoundly reshape the distribution of resources in the network while redefining the artifact and its meaning to the point of a reversal of foreground and background, ends and means. As Grier and Campbell put it, Condon "was the individual most responsible for converting the back stage of the old Bitnet into the front stage of the new" (Grier and Campbell 2000, 36). Communication, which used to be a *means* to achieve the originally assigned tasks of the system, eventually became its very purpose: "BITNET was an end in itself. As the network has grown, this emphasis has changed. . . . The End is now to provide academic communications services, and BITNET has become the Means" (Condon 1988). Finally, the *identity* of the network is affected, in that a specific use or application of the system may be equated with the whole device: "I noticed something interesting. Many of the people I talked to were under the impression that Relay *was* BITNET. That is, they were not aware that BITNET existed for some other purpose" (Condon 1990).

Earning and Preserving Legitimacy

Toward the middle of the 1980s, chat programs started to bloom on Bitnet (Kell 1986). Due to limited buffer space and the priority on message transfer, they caused the saturation of the network. In a note sent in February 1985 to all Bitnet users and administrators, Henry Nussbacher, then head of the Network Information Center (BITNIC), condemned chat servers operating on Bitnet, stating that they were a serious threat to the network. Because public chat implies "rebroadcasting" of messages to a whole group of users, they imposed a heavy load on bandwidth. He recommended that such program should be banned from Bitnet. Along with technical reasons, Nussbacher justifies this veto by the frivolous content of messages, way beyond the scope of "academic" use of resources: "The bulk of data being transferred over TP [twisted pair] lines becomes a hackers CB world. High school students and college undergraduates discuss everything from dirty jokes to sex to crashing the VM system" (Nussbacher 1985). Kell recalls, "Nussbacher's infamous paper . . . was circulated. . . . And boom, the party was over. . . .

Management in general (contacts, administrators, technical people) developed a bad taste for 'chatting,' and most of it was put to a halt. Others went underground, and when discovered and exposed, only served to increase management distaste for the whole concept" (Kell 1986).

Enters Kell, who figures a way to alleviate the load on Bitnet links by using a decentralized infrastructure of chat servers *relaying* messages to one another—and broadcasting them locally—instead a broadcasting them to the entire network. The "Relay" program was born. Kell skillfully combined technical optimization and public relations to have his program viewed more positively than single-server "chat machines"—as they were called. It implied making concessions by establishing use policies and enforcing restrictions (on the time of the day, on user location, and even on which chat rooms—channels—were allowed). Because "bad programs" were always reappearing like the heads of the Hydra, Relay began to be seen by Bitnet managers as a way to police the practice of chatting: "The biggest 'boost' to Relay came when some management switched over to our side, although granted with considerable caution. . . . Thus the first set of statistics, limitations, and restrictions were imposed in an effort to make Relay 'legitimate.' . . . It would be of enormous value to have the Relay program finally reach a 'legitimate' status" (Kell 1986). Bitnet Relay also gained legitimacy because it was considered by users as the most efficient way to get technical help with the system, and for that reason, it had major advocates including Christopher Condon. But that legitimacy remained fragile and contested. For instance, flaws in the program were exploited by hackers from Computer Chaos Club in Germany to get into computers at Cornell, whose authorities incriminated the local Relay server and shut it down. But the largest threat on Relay came from "rogue" chat machines—the centralized, resource-consuming type. First, they were a bad publicity for online chat in general. But some of them were worse: they imitated the "look and feel" of Relay—and even allegedly took its name. Toward the end of 1986, a controversy on Relay "clones" revealed a struggle between Relay promoters who wanted to preserve its hardly acquired legitimacy and clone developers who, intentionally or not, took advantage of it to fool node administrators. From then on, Relay advocates would demonize and fight rogue chats, including modified versions of the Relay program: "Even though Relay is somewhat accepted by the network administration, a copy of Relay can be easily hacked to break the rules it is designed to enforce. . . . These 'dark-side hackers' would *misuse* anything" (Kell 1987, emphasis added). Building the Relay network was therefore also building a sociotechnical network, an *actor-network* strong enough to resist attacks from Bitnet management and to aggregate and absorb all other chatting devices; it is a struggle to establish a standard, a normative frame of use of online chat. If there were any hope that Relay would ever become "official"—legitimate—then, all chatting practices outside that frame had to be labeled as "misuse" and suppressed.

IRC Bots

IRC is an online chat protocol initially developed in Finland in the late 1980s. It started as a modest program with limited features and evolved into a large, complex technical infrastructure consisting of myriad independent networks of servers. Long before the arrival of ICQ, MSN Messenger, Skype, Twitter, and Snapchat, IRC emerged as a decentralized, always-up messaging infrastructure allowing millions of simultaneous users to have real-time, polyphonic, written conversations online (Latzko-Toth 2010). Just like its predecessors, IRC has struggled to be recognized as a proper use of network resources (particularly in academic institutions). In the 1990s, many universities banned IRC as it was considered a waste of bandwidth, an invitation to obnoxious hackers to attack their computers, and overall an Internet application with no educational value:

Sysadmins view it as a frivolity and pick-up mechanism, and the easiest thing to do when problems arise is to simply remove it—most sysadmins don't want to deal with IRC at all, and if it causes problems, removing it means they don't have to deal with it. (Operlist, 1/8/1992)

To quote [four] randomly asked administrators here at UC Berkeley today, of how the[y] viewed IRC, they all same something different, but all equalled the same thing. No educational value. (Operlist, 1/9/1992)

From the above, it can be seen that IRC was to the Internet what games had been to PLATO, or Relay to Bitnet, and this regular reference to educational value has to do with the academic settings within which these digital infrastructures had been developed as well as to rhetorical appeals to education as an intrinsically valuable purpose. But contested as the status of IRC was, in this section we will focus on the notion of misuse *within* the context of IRC, as it epitomizes the relationality of misuse. A robot or bot is an autonomous program capable of signing on to IRC by itself and interacting with other servers and clients. IRC bots serve various purposes, from policing use to mere fun. In addition to bots, there are less elaborated code objects known as scripts which, just like bots, can react to online events without any direct intervention from the user. Are programming a script and "running a bot" appropriate in the context of IRC? What is at stake here is the definition of what constitute a proper use of IRC.

The IRC device is a complex ecology of human and nonhuman actants, the latter consisting of various software entities including servers, services, user clients, and "user bots." Interestingly enough, the latter term emerged as a means to differentiate them from "official bots" endorsed by the administrators and operators of a specific IRC network. In other words, the very notion of user bot implies their potentially contested status and the fact that they can possibly fall outside the perimeter of what is considered to be part of the device (features) and what is deemed external and potentially disruptive to it (bugs). If one reads the discussions about bots that took place on the main online venues where IRC development was discussed in its early days (notably Operlist and alt.irc), it appears that the normative frame of use of the original IRC network (EFnet) tended to exclude running bots, as bluntly stated in the IRC FAQ (Rose 1994) and exemplified by the dialogue below, where R. R., an IRC operator, ironically rephrases a beginner's question to suggest that chatting has been superseded as the main purpose of IRC by "running bots":

> I was just wondering what is the current status of running bots on EFNet,
> is it permitted? and if so, what servers currently allow them? . . .

I was just wondering what is the current status of chatting on EFnet, is it permitted? and if so, what servers currently allow it? With all the clone/war/

takeover . . . and all the efforts of server admins to combat them, is there any chatting going on? Is it actually possible to chat without running a bot? (R.R., alt.irc, 5/13/1996)

As revealed by a thorough examination of IRC history, user bots played a key role in the dynamics of innovation within IRC networks (see Latzko-Toth 2014). The original IRC code was an "unfinished artifact" (Maxigas 2015) when it was released by Jarkko Oikarinen. It somehow came with an antidote to this incompleteness: the distribution package included a set of bot templates that constituted an invitation to add new features—called "services"—to a minimalist protocol by delegating these extra functionalities to bots. The term "bespoke code" was coined by Geiger (2014) to name this way for users of a software infrastructure to contribute to it by adding a nonofficial layer of code, notably in the form of bots.

Early on, user bots were controversial on IRC. Some were simply considered annoying and were treated like a particular and elaborated form of "spam," because they would make repetitive, frivolous comments on the public chat channel, send private messages to users, or duplicate (clone) themselves. Some IRC users would see them as an innocent form of entertainment. Other bots were frowned upon for very different reasons: they were interfering with the governance of IRC networks and channels. Who opens a conversation channel first rules it until he or she leaves the channel or nominates another "channel operator." Thus, the original IRC protocol did not provide any way to "register" a channel and to permanently establish oneself as its owner or manager, which resulted in constant fights over the control of a channel. Channel bots were developed to fix this problem. By continuously sitting on the channel they are taking care of, they serve a double function: they keep the channel open and keep control of it on behalf of their owner. Even more, since most bots feature access levels, they allow for a much more subtle scale of power than the binary structure inscribed in the IRC protocol (operators versus non-operators).

The "user bot" was disruptive to IRC in the sense that it established a new de facto regime of governance within the digital space of an IRC network, directly competing with the "official" regime of governance as reflected by the typology of IRC operators, channel operators, and so forth. On EFnet, this shift in channel governance was vigorously criticized by the self-appointed, co-opted official designers (server administrators, code developers) who established a *channels-are-not-owned* policy and who regarded channel bots as a violation of that policy—while setting a double standard when it came to keeping control on their own channels.

Interestingly, a common argument against the practice of connecting user bots was that they were "a waste of resources": they take away connection sockets—limited on every server—from human users and generate unnecessary data traffic inside the network. An impassioned debate around the status of user bots took place among the founders of another large IRC network (Undernet), which was set up precisely for the purpose of testing bots. There again, allocation of digital resources was at the core of discussions, as illustrated in the following excerpt of a "bot policy" draft, where the notion of useful bot is also outlined:

Idle bots are nothing more than a drain on the resources of the many servers which comprise the network. In an attempt to help keep the number of idle processes to a minimum, all processes on this server must display some activity at least once every eight hours. . . . The only exceptions to this clause are bots that

> provide useful services such as note services, database services, file services, or communication services, provided they are sanctioned by the opers of this server.[3]

To contenders who would insist on their frivolous nature, bot proponents would justify their existence by the educational value found in programming them, like in the following examples:

> When I first started to [use] IRC . . . I started to program bots :) At first that is ALL i did. . . . It hel[p]ed me to learn A LOT about irc and its calls!! (D. L., Wastelanders, 5/2/1993)

> Bots are educational. It's been criticized that IRC is nothing more then brainless chatting. However, anyone who ha[s] attempted to write a bot, figure[d] out how it works, etc. knows you can learn a little programming, logic, etc. (B., Wastelanders, 2/16/1995)

It may be useful to point out that soon after these discussions took place, Undernet operators decided to establish a channel registration procedure and set up an "official" bot (X/Cservice) to look after registered channels. It basically made the channel bot a formal component of the IRC network. Today, channel registration—or an equivalent function, ChanFix, allowing to recover ownership—has become a standard feature implemented on most IRC networks.

Like in previous cases, the controversy around the appropriateness of running bots on IRC as a way of using it revolves around what uses should be given priority in terms of resources directed to them, based on a normative frame of use that is constantly subject to redefinition attempts. But the case of the IRC user bot also exemplifies how "misusing" networked digital resources can lead to reconfiguring the whole device, with the consequence of profoundly transforming the way it operates and/or it is governed, thus exhibiting its plasticity.

Discussion: Steps toward a Theory of the Misuser in the Digital Age

The three historical case studies above allow us to outline what we call the figure of the misuser. This figure is rhetorically mobilized in discursive exchanges inside a community of innovation gathering designers, early users, and other stakeholders of a nascent digital innovation.[4] In analyzing the traces of these exchanges, we could find regular occurrences of such terms as "misuse/misuser," "disruptive use/user," and "abuse," and conversely the adjectives "proper," "legitimate," and "serious" associated with the word "use." This led us to making the hypothesis that the categories of "misuse" and "misuser," although they originate from the actors themselves, are performative and effectively contributing to shaping the innovation as they are deployed in boundary work (Gieryn 1999) by members of a community of innovation to circumscribe "proper" use versus improper use of a digital communication device. Furthermore, we propose that the importance of these categories in the dynamics of digital innovation is tied to the plasticity of these artifacts, which is why we refer to them as "misuser innovations"—to underline that innovative uses of digital communication technologies that contribute to define them are often stemming from marginalized users of an existing device established for other purposes.

In this final section, we articulate the notion of "misuser" as a useful analytical category to understand the dynamics of innovation at play in the construction of a networked, software-based communication device. To do so, we first have to explain why this category matters, and why it particularly matters in the digital context.

Some early instances of misusers are featured in Kline and Pinch's classic study of the reshaping of the automobile by recalcitrant farmers in the first half of 20th century (Kline and Pinch 1996). In terms of militancy and technical ingenuity, few contemporary users could measure up to the resistance of the farmers against the automobile. The case study was issued from a social construction of technology (SCOT) perspective, but it was written in response to criticism leveled at an earlier version of that theory. Initially, SCOT had put emphasis on how artifacts are shaped by relevant social groups up until the moment of a closure in the design and function of the artifact. However, to speak about a closure of an artifact implies a closure on who is counted as relevant, emphasizing the designer and manufacturer while leaving out the user and/or consumer from the equation. With the study of farmers and automobiles, the point on interpretive flexibility of products was reasserted but now extended beyond the singular point of closure. Kline and Pinch showed how parts of the farming community reacted with fierce resistance, in the extreme cases by digging traps and ambushing drivers. Later, other farmers begun to repurpose the automobile, for instance, by using the motor as a power source for various tasks. The attitude of manufacturers toward the interpretative flexibility exercised by farmers shifted with the seasons of the market. To start with, the inventiveness of the farmers expanded the market for general purpose-built cars. As the product range was diversified, and some models were developed that catered specifically to farmers' needs, such inventiveness became instead an infringement on the niche market. Consequently, manufacturers began condemning practices that they had initially condoned. Farmers, having had no say over the original development process of the automobile, exercised some influence on its subsequent developments. Although the interventions by the farmers were spectacular and entertaining, Kline and Pinch concede that, when all is said and done, "it is clear that one social group initially had more influence than any other in terms of giving a meaning to the artifact: the manufacturers" (1996, 774).

Our case for singling out the misuser takes foothold in the last remark, concerning the relative (un)importance of the farmer-user in the overall development process of the automobile. It is because of this power of disrupting—destabilizing—the whole system that the misuser stands out from misbehaving and unruly users in general. What is distinctive about the former is the capacity of a single (mis)user or a small subgroup of them to reconfigure the distribution of resources in the entire network and, with that, to performatively reinterpret how the device is supposed to work for everyone else. In the extreme case, every user will be equally affected by the exertion of interpretative flexibility by every other user. When the interpretative agency of all users approximates such a symmetry, we may say that the device demonstrates a high degree of plasticity. That is, the way the device functions is plastic and bendable to the interventions and interpretations of any of the users.

In the cases investigated in this chapter, PLATO, Bitnet, and IRC, this plasticity was mostly experienced in a negative way. Every user was dependent on the same and, at the time, very limited resource allocation of computation, bandwidth, memory, connection sockets, and so on. The link between this conflict of interests over resources and the application of the "misuser" label comes out from the case descriptions. The opposite situation is at hand when a single user or a small subgroup of users may deviate from prescribed uses and tinker with the device without this activity having much of a destabilizing (disruptive) effect on the network as a whole. Arguably, the automobile exemplifies a device of the latter kind. The process by which this device can be reinterpreted is relatively inflexible because it is sunk into the car manufacturing process, the transportation systems, traffic regulations, and so forth. A farmer who turned his automobile into a power generator did so without it having much of an effect on every other driver. Only indirectly, by demonstrating a demand in a niche market, did the farmer modify the product range of automobiles. As mentioned by Kline and Pinch in their article, manufacturers and others who felt that they ought to have an interpretative superiority over the farmers did nevertheless complain about the product being misused. Our proposition is, however, that the actors' need for distinguishing between proper uses and misuses of a device is proportionate to the plasticity of the device. Or, put differently, the need to distinguish between use and misuse is proportionate to users' perceived risk that a deviation from prescribed uses poses to the integrity and ascribed functions of the entire network. The felt need is greater for social sanctions against deviations that users believe put everybody's uses of the device at risk, by, for instance, causing system owners to restrict access or resources. It is the latter kind of situations that, we contend, warrant us to make a special case of misusers of digital communication technologies, provided we keep in mind that "misuse" is itself an actor's category, something we can trace in actors' accounts.

It is understood that the words "misuse" and "deviation" are loaded, derogatory labels. Their meanings derive from a taken-for-granted setup of the device at the time, corresponding to a dominant interpretation—within the group of designers or the larger user community—of what the function and the use of the device are supposed to be. This dominant interpretation is what we call the normative frame of use of the technology, drawing on Flichy's (2007) concept of "frame of use." The frame of use "describes the kind of social activities proposed by the technology, the integrated routines of daily life, sets of social practices, kinds of people, places and situations connected to the technical artifact" and "shows [what its] purpose" is to users and nonusers (Flichy 2007, 83). There is therefore a prescriptive, normative dimension inherent to such a frame. The term resonates with the notion of technological frame (Bijker 1987; Orlikowski and Gash 1994), which is a sociocognitive construct guiding how groups of actors "make sense" and interpret the purpose of a technology. However, while this concept suggests that different social groups may have different technological frames competing together, the concept of a normative frame of use rather stresses that there always tends to be a dominant interpretation of what proper use is for the technological device in question. This dominance results from an active policing process that Callon (1998), drawing on Goffman's theory of social interaction, calls "framing."

As noted by Callon, actors' framing efforts are constantly threatened by "overflows": activities and relations leaking outside the enclosure of the frame (Callon 1998). Or at least this is the perspective that we are invited to adopt from the point of view of an established normative frame of use. But what if, as Callon suggests, we adopt a reversed perspective and see overflows not as accidents but as the typical way by which creativity occurs? While adoption of established interpretations and norms leads on to the reproduction of past or current relations, a deviation points to something new and unanticipated. It is therefore conceivable that the same act that at one moment is labeled "misuse," at a later time will be spoken of as innovation. A new configuration of the device might arise from that act, transforming the expectations and the interpretations of the mass of users in the process.

The special connection between misuse and digital technology can be gleaned from the remarks above. The automobile, on the one hand, and the communication platforms discussed in this chapter, on the other hand, are located at opposite poles on a scale of plasticity. By this we do not say that digital devices are inherently more plastic than nondigital devices. On this ground, the high plasticity of a digital device should be understood not as an inherent quality of digital materiality but as a sociotechnical construct operating at two levels of users' perceptions. First, plasticity is constructed at the level of the *affordances* of the device that result from design decisions—for instance, leaving it "unfinished" and "incomplete" and providing interfaces to expand it, like authoring and scripting languages.[5] Indeed, plasticity was the sought-after quality when (reprogrammable) software was first separated from hardware in the subsequent developments of the ENIAC computer. At a secondary level, plasticity boils down to the perceived *fragility* of a normative frame of use vis-à-vis a single user's divergent interpretations and patterns of (mis)use, and how susceptible the device is to being repurposed and its functions transformed as a result of its interpretive flexibility.

This is to say that the high plasticity of some digital communication devices, while it may be facilitated by some design features (e.g., programmability, expandability, connectivity, recursivity), is constructed and can be reversed. As is demonstrated by digital rights management (DRM) devices, the plasticity of a digital artifact can be inhibited in the design process (Gillespie 2006). For a technical fix such as DRM to be effective in preventing users from tinkering with the device, however, requires the leverage of supplementary resources. This might include that the protocols for interoperability between industrialists are renegotiated, national laws are amended, the investigative powers of the police are boosted, and so on. In other words, it presupposes an asymmetry in (nontechnical) resources available to the various stakeholders and users of the device in question. Even then, as is amply shown by the filesharing controversy, technical fixes are always vulnerable to reverse-engineering attacks (Burkart and Andersson-Schwarz 2015). The labeling of filesharers as thieves and pirates (i.e., as misusers of the communication network) is part and parcel of the efforts to police behavior and enforce prescribed uses that could not be eliminated through the technical fix alone.

When investigating the concept of misuse, the most fruitful cases are likely to be those where no single party holds such an advantage that technical fixes can be imposed on the others. Well-behaving users of the device must bracket up their technical agency, given that it is more or less equally distributed to everyone in the network, with social norms. It is for this reason that the "Internet troll" takes on a heightened importance in the regulation of peer-to-peer collaborations on the Internet. Troll-labeling is the other side of the openness in such projects (Tkacz 2013). Labeling someone a "misuser" is a political act to deligitimize certain activities with, and interpretations of, the device, while promoting proper ones. Users who are in the risk zone are thereby discouraged from pulling the device further in a direction judged as inappropriate by the party applying the label (Söderberg 2010). The self-understanding of the one labeled a misuser, together with his or her reinterpretations of the technology in question, is defined in his or her relation with the owner or lawgiver or normsetter of the technology. In horizontal and informal projects, the lawgiver might be a sys-admin, a moderator in a discussion forum, or some other spokesperson of the silent majority. Of course, the accusations of being a misuser are sometimes adopted as a positive identity, around which like-minded (mis)users can be rallied, sometimes resulting in the balance of forces being overturned.

Conclusion

Our aim in this chapter was to shed light on a particular and underrated aspect of the dynamics of innovation in the field of digital communication technology: the key figure of the misuser and the tension between so-called "proper" uses and "misuses" in stabilizing a normative frame of use that will define the main purpose of the artifact. From the discussion about alleged misuses of PLATO (games), Bitnet (chat programs), and IRC (bots), we have seen how the notions of disruptiveness and misuse were frequently evoked as a rhetorical strategy by some actors to police the use of digital communication technologies and thus refrain the expression of their plasticity. At the same time, these cases demonstrate that such notions were temporary constructs, evolving together with a changing context—notably resource availability.

Initially, playing, idle chatting, and other activities were perceived as a waste of resources, taking place at the expense of proper uses. This standpoint within the affected communities drew on general, negative attitudes toward play and leisure, according to which the youth were distracted from spending their time more productively, getting an education and a job. In hindsight, however, we know that playful and social uses of early digital resources pushed technical development in new directions and gave chief impetus to two digital media industries: online games and social media. For instance, while PLATO eventually failed as an e-learning platform, its communication tools paved the way to social computing and groupware (Dear 2017). Lotus Notes is a direct offspring of PLATO Notes (Woolley 1994), while "Talkomatic" inspired other chat devices including Bitnet Relay and its Internet-based successor IRC. Microsoft's *Flight Simulator* was originally developed on PLATO, and *Empire* was seminal in the domain of multiuser online games (Dear 2017). Several key people in the software industry, including Ray Ozzie, former chief software architect at Microsoft, fulfilled course requirements by developing for PLATO.

Along the same lines, while Bitnet was originally intended as a file sharing infrastructure for the academic community, this functionality was eventually taken over by Internet protocols (Gopher, FTP, the Web). On the other hand, two applications stemming from communication practices once labeled as misuses (the ListServ mailing list manager and the Relay chat program) were in such demand that they were ported to the Internet (Grier and Campbell 2000; Latzko-Toth 2010). The IRC protocol in its turn came to serve as an underlying technical layer in numerous commercial "webchats" and served as the matchmaking infrastructure for major online games including Blizzard's through its Battle.net chat network. Furthermore, the introduction of new governance features within IRC networks through the use of bots has been established as standards not only for IRC networks—such as freenode, a key infrastructure for free software development nowadays—but also for chat platforms targeted at the general public.

Indeed, if judged by economic figures, computer games and online chat have eclipsed some of the activities that at the time were deemed to be more serious and important. The opposition between play and education dissolves when game design is inscribed in university curricula and regular career paths have been established in the field (Kucklich 2005; Scholz 2012). In a similar fashion, real-time text messaging has become a norm in social media and even in the workplace. Recreational uses of digital resources, at first marginalized, ultimately became the new "norm." Usage that is constructed as a transgression at one point in time might lead to changing the very coordinates by which that value judgment was made, thus presenting itself as the proper, prescribed usage of the device in question.

Notes

1. Digital technologies can sure be designed to be nonplastic too, as exemplified by digital rights management (DRM) devices, but then this is typically perceived as a betrayal of the ideal.
2. Using Caillois's (1961) typology, they convincingly framed IRC as a game and chat as play.
3. "General guidelines for bot usage on *.iastate.edu servers," document posted to Wastelanders (Undernet operators' mailing list), March 30, 1993.
4. Drawing on Von Hippel's concept of innovation community (Von Hippel 2005) and Tuomi's concept of "network of innovation" (Tuomi 2003), we call *community of innovation* the hybrid collective (Callon and Law 1997) gathering users, developers, and various actants, focused on the creation, improvement, and expansion of an artifact or a set of artifacts within an articulated ethical framework.
5. See Vertesi (this volume) for a discussion on affordances. On "unfinished artifacts," see Maxigas (2015); on "incompleteness by design," see Garud et al. (2008).

Works Cited

Akrich, Madeleine. 1992. "The De-scription of Technological Objects." In *Shaping Technology/Building Society*, edited by Wiebe E. Bijker and John Law, 205–24. Cambridge, MA: MIT Press.

Bijker, Wiebe E. 1987. "The Social Construction of Bakelite: Toward a Theory of Invention." In *The Social Construction of Technological Systems: New Directions in the Sociology and History of Technology*, edited by Wiebe E. Bijker, Thomas P. Hughes, and Trevor Pinch, 159–87. Cambridge, MA: MIT Press.

Bitnet Executive Committee. 1986. "The Bitnet Charter." https://archive.org/stream/bitnet_documents/bitchart.txt.

Brunton, Finn. 2013. *Spam: A Shadow History of the Internet*. Cambridge, MA: MIT Press.

Burkart, Patrick, and Jonas Andersson-Schwarz. 2015. "Piracy and Social Change: Revisiting Piracy Cultures." *International Journal of Communication* 9:792–97. http://ijoc.org/index.php/ijoc/article/view/3704.

Caillois, Roger. 1961. *Man, Play and Games*. Glencoe, IL: Free Press.

Callon, Michel. 1998. "An Essay on Framing and Overflowing: Economic Externalities Revisited by Sociology." *Sociological Review* 46 (S1): 244–69. doi:10.1111/j.1467-954X.1998.tb03477.x.

Callon, Michel, and John Law. 1997. "After the Individual in Society: Lessons on Collectivity from Science, Technology and Society." *Canadian Journal of Sociology* 22 (2): 165–82. doi:10.2307/3341747.

Condon, Chris. 1988. "Bitnotes." *NetMonth*, November. https://archive.org/stream/bitnet_documents/nm8811.txt.

——. 1990. "Bitnotes." *NetMonth*, September. https://archive.org/stream/bitnet_documents/nm9009.txt.

Danet, Brenda, Lucia Ruedenberg, and Yehudit Rosenbaum-Tamari. 1997. "'Hmmm . . . Where's That Smoke Coming From?' Writing, Play and Performance on Internet Relay Chat." *Journal of Computer-Mediated Communication* 2 (4). doi:10.1111/j.1083-6101.1997.tb00195.x.

Dear, Brian. 2017. *The Friendly Orange Glow: The Story of the PLATO System and the Dawn of Cyberculture*. New York: Pantheon.

Flichy, Patrice. 2007. *Understanding Technological Innovation: A Socio-technical Approach*. Northampton, MA: Edward Elgar.

Garud, Raghu, Sanjay Jain, and Philipp Tuertscher. 2008. "Incomplete by Design and Designing for Incompleteness." *Organization Studies* 29 (3): 351–71. doi:10.1177/0170840607088018.

Geiger, R. Stuart. 2014. "Bots, Bespoke Code, and the Materiality of Software Platforms." *Information, Communication & Society* 17 (3): 342–56. doi:10.1080/1369118X.2013.873069.

Geiger, R. Stuart, and David Ribes. 2011. "Trace Ethnography: Following Coordination through Documentary Practices." In *2011 44th Hawaii International Conference on System Sciences (HICSS)*, 1–10. doi:10.1109/HICSS.2011.455.

Gieryn, Thomas F. 1999. *Cultural Boundaries of Science: Credibility on the Line*. Chicago: University of Chicago Press.

Gillespie, Tarleton. 2006. "Designed to 'Effectively Frustrate': Copyright, Technology and the Agency of Users." *New Media & Society* 8 (4): 651–69. doi:10.1177/1461444806065662.

Grier, David A., and Mary Campbell. 2000. "A Social History of Bitnet and Listserv, 1985–1991." *IEEE Annals of the History of Computing* 22 (2): 32–41.

Hine, Christine. 2000. *Virtual Ethnography*. Thousand Oaks, CA: Sage.

Kell, Jeff. 1986. "Instant Insanity." *NetMonth*, August. https://ia800708.us.archive.org/28/items/bitnet_documents/nm8608.txt.

———. 1987. "Re: Your Relay Clone." Email originally sent on January 23, 1987, and published in *NetMonth*, February. https://ia800708.us.archive.org/28/items/bitnet_documents/nm8702.txt.

Kline, Ronald, and Trevor Pinch. 1996. "Users as Agents of Technological Change: The Social Construction of the Automobile in the Rural United States." *Technology and Culture* 37 (4): 763–95. doi:10.2307/3107097.

Kucklich, Julian. 2005. "Precarious Playbour. Modders and the Digital Games Industry." *Fibreculture Journal* 5. http://five.fibreculturejournal.org/fcj-025-precarious-playbour-modders-and-the-digital-games-industry/.

Latzko-Toth, Guillaume. 2010. "Metaphors of Synchrony: Emergence and Differentiation of Online Chat Devices." *Bulletin of Science, Technology & Society* 30 (5): 362–74.

———. 2013. "Asynchronous (Online) Ethnography: A Tale of Two IRC Networks." In *Qualitative Research: The Essential Guide to Theory and Practice*, edited by Maggi Savin-Baden and Claire H. Major, 209–10. London: Routledge.

———. 2014. "Users as Co-designers of Software-Based Media: The Co-construction of Internet Relay Chat." *Canadian Journal of Communication* 39 (4): 577–95.

Latzko-Toth, Guillaume, and Steve Jones. 2014. "Sharing Digital Resources: PLATO and the Emerging Ethics of Social Computing." Paper presented at the ETHICOMP 2014 Conference, Paris.

Levy, Karen E. C. 2015. "The Contexts of Control: Information, Power, and Truck-Driving Work." *Information Society* 31 (2): 160–74. doi:10.1080/01972243.2015.998105.

Maxigas. 2015. "Peer Production of Open Hardware: Unfinished Artefacts and Architectures in the Hackerspaces." Doctoral dissertation, Universitat Oberta de Catalunya.

Nussbacher, Henry. 1985. "Chat Server Machines." Note sent via Bitnet, February 26.

Orlikowski, Wanda J., and Debra C. Gash. 1994. "Technological Frames: Making Sense of Information Technology in Organizations." *ACM Transactions on Information Systems* 12 (2): 174–207. doi:10.1145/196734.196745.

Oudshoorn, Nelly, and Trevor J. Pinch. 2003. *How Users Matter: The Co-construction of Users and Technologies*. Cambridge, MA: MIT Press.

Rose, Helen. 1994. "IRC Frequently Asked Questions (FAQ)." Version 1.39, originally published on Usenet (alt.irc). www.ibiblio.org/pub/academic/communications/irc/help/alt-irc-faq.

Scholz, Trebor. 2012. *Digital Labor: The Internet as Playground and Factory*. New York: Routledge.

Söderberg, Johan. 2010. "Misuser Inventions and the Invention of the Misuser: Hackers, Crackers and Filesharers." *Science as Culture* 19 (2): 151–79.

Star, Susan Leigh, and James R. Griesemer. 1989. "Institutional Ecology, 'Translations' and Boundary Objects: Amateurs and Professionals in Berkeley's Museum of Vertebrate Zoology, 1907–39." *Social Studies of Science* 19 (3): 387–420.

Tkacz, Nathan. 2013. "Trolls, Peers and the Diagram of Collaboration." *Fibreculture Journal* 22:15–35. http://twentytwo.fibreculturejournal.org/fcj-154-trolls-peers-and-the-diagram-of-collaboration/.

Tuomi, Ilkka. 2003. *Networks of Innovation: Change and Meaning in the Age of the Internet*. Oxford: Oxford University Press.

Van Meer, Elisabeth. 2003. "PLATO: From Computer-Based Education to Corporate Social Responsibility." *Iterations* 2. www.cbi.umn.edu/iterations/vanmeer.pdf.

Von Hippel, Eric. 1986. "Lead Users: A Source of Novel Product Concepts." *Management Science* 32 (7): 791–805. doi:10.1287/mnsc.32.7.791.

——. 2001. "Innovation by User Communities: Learning from Open-Source Software." *Sloan Management Review* 42 (4): 82–86. http://sloanreview.mit.edu/article/innovation-by-user-communities-learning-from-opensource-software/.

——. 2005. *Democratizing Innovation.* Cambridge, MA: MIT Press.

Woolley, David R. 1994. "PLATO: The Emergence of Online Community." www.thinkofit.com/plato/dwplato.htm.

Wyatt, Sally. 2003. "Non-users Also Matter: The Construction of Users and Non-users of the Internet." In *How Users Matter*, edited by Trevor Pinch and Nelly Oudshoorn, 67–79. Cambridge, MA: MIT Press.

Knowing Algorithms

Nick Seaver

Definitions

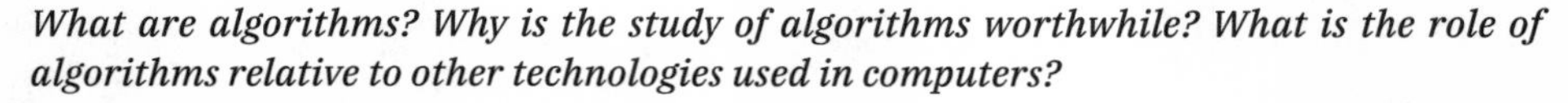

What are algorithms? Why is the study of algorithms worthwhile? What is the role of algorithms relative to other technologies used in computers?

Informally, an ***algorithm*** *is any well-defined computational procedure that takes some value, or set of values, as* ***input*** *and produces some value, or set of values, as* ***output****. An algorithm is thus a sequence of computational steps that transform the input into the output.*

We should consider algorithms, like computer hardware, as a ***technology****.*

—Cormen et al. (2009, 5, 13)

Algorithms have become objects of concern for people outside of computer science. As more and more of human life is conducted alongside and through computers, contoured by algorithmic selection both online and off, people who once had little interest in the workings of computers have growing concern about their effects. This concern has manifested in an explosion of popular and academic productions: books, articles, op-eds, blog posts, conferences, and research programs engage with algorithms in terms that never appear in the *Introduction to Algorithms* textbook from which I took the epigraph above.

For many undergraduates in computer science, Cormen et al.'s textbook is the definitive source for knowledge about algorithms. It is a massive book: 1,000+ pages containing descriptions of basic algorithms and the data structures on which they operate. It teaches how to design an algorithm, to determine its efficiency, and to prove its optimality. In this frame, knowledge about algorithms is a matter of mathematical proof, and "analyzing an algorithm has come to mean predicting the resources that the algorithm requires" (Cormen et al. 2009, 21). Other knowledge about algorithms—such as their applications, effects, and circulation—is strictly out of frame here, save for the obligatory mention in the introduction of genomics, network routing, cryptography, and resource allocation. Algorithms per se are supposed to be strictly rational concerns, marrying the certainties of mathematics with the objectivity of technology.

If one goal of a digital STS handbook like this one is to offer new exemplars—to bring the field up to date and to demonstrate its relevance beyond bridges, bicycles, and Bakelite—then algorithms seem like good candidates, having become "somewhat of a modern myth" (Barocas et al. 2013), representing advanced technology, creepy mathematical efficacy, and shadowy control. Popular and academic concern with algorithms has taken off over the last few years, generating a host of conferences and panels (e.g., "The Politics of Algorithms," 4S Copenhagen, 2012; "Governing Algorithms," NYU, 2013; "The Contours of Algorithmic Life," UC Davis, 2014; "Algorithms and Accountability," NYU, 2015) and popular-audience books (e.g., *Automate This*, Steiner 2012; *The Glass Cage*, Carr 2014; *The Formula*, Dormehl 2014; *Automating Inequality*, Eubanks 2018; *Algorithms of Oppression*, Noble 2018).

In academia, scholarship that I have taken to calling "critical algorithm studies" is transdisciplinary, spanning the humanities (Striphas 2010), critical theory (Galloway 2006; Parisi 2013), law (Pasquale 2015), communication (Cheney-Lippold 2011; Napoli 2013; Gillespie 2014; McKelvey 2014), media studies (Beer 2009; Nakamura 2009; Uricchio 2011; Mager 2012; Bucher 2012; Mahnke and Uprichard 2014), journalism (Anderson 2013; Diakopoulos 2014), history (Ensmenger 2012), geography (Lyon 2003; Graham 2005; Amoore 2011), sociology (Snider 2014; MacKenzie and Spears 2014), and anthropology (Helmreich 1998; Asher 2011; Seaver 2012; Kockelman 2013). The glut of literature is increased tenfold if we include algorithms' obverse: big data, or the widespread collection of transactional data like that of Target shoppers or web browsers, made tractable through the use of algorithmic systems.[1]

Work in critical algorithm studies tends to be animated by two primary concerns. First, algorithms produce or require the formalization of informal qualities: they take uncertain, personal, situational things like your taste in music or your political sensibility and render them as stark quantities. Second, the algorithms we care about are typically obscure: they require technical training to understand in detail, they are often so complex that they pose interpretive challenges even to their makers, and they are hidden behind veils of corporate secrecy. Both of these concerns have clear antecedents in STS, from the humanistic critique of rationalization, formalization, and quantification (e.g., Feyerabend 1975; Star 1995; Porter 1996) to the interest and worry about the details of complex, powerful, difficult-to-apprehend technical systems (e.g., Hughes 1983; Winner 1986). Critical algorithm studies carries these critiques on, pointing out formalism's embeddedness, pressing for algorithmic transparency, and noting that the ability to edit and impose algorithmic logics is striated by power.

But researchers interested in the empirical specificities of algorithms in action (rather than vague, broad statements about "the algorithmic") have a problem: how do we come to know things about algorithms? In this chapter, I argue that algorithms are tricky objects to know. This trickiness is not limited to their technical complexity and sophistication,[2] nor to their obscure locations in the recesses of trade secrecy. What makes knowing algorithms *really* tricky is that these are superficial problems: they suggest that the barriers to knowledge are simply access and expertise. If only we had access, and if only we had expertise, then we would know what was really going on, and we could begin to evaluate the social consequences accordingly.

From a long history of work in STS (particularly its constructivist strains), we should have learned to be incredulous of such accounts of knowledge, where facts are simply objects to be retrieved from obscurity and analyzed by ever more sophisticated experts. We need to be more critical of our own knowledge practices—to reflexively apply the tools of STS to the knowledge we produce.[3] "Algorithms," I propose, are social constructions that we, as outsiders and critics, contribute to. The point of declaring something a construction is to argue that it might be constructed differently, and that is also the point of this chapter.

In fixating on access and expertise (through calls for transparency or mystifications of algorithms' sophistication), work in critical algorithm studies reifies a deficient understanding of how algorithms work and exist in the world, beyond the theoretical definitions of computer science, the breathless spin of press releases, or the just-so stories that circulate through popular and academic discourse. In this chapter, I outline what I see as the dominant critical approach to knowing algorithms, describe the challenges that face this approach, and suggest an alternative way to construct "algorithms" that better captures the concerns of critics and the empirics of algorithmic systems in action.

Anagnorisis

In a TED Talk based on his book *The Filter Bubble* (2011a), Internet activist and entrepreneur Eli Pariser describes a moment of recognition—the realization that his experience online was being quietly mediated by algorithmic filters:

> I first noticed this in a place I spend a lot of time: my Facebook page. I'm progressive politically, but I've always gone out of my way to meet conservatives. I like hearing what they're thinking about; I like seeing what they're linking to; I like learning a thing or two. So I was kind of surprised, when I noticed one day that the conservatives had disappeared from my Facebook feed. And, what had turned out was going on was that Facebook was looking at which links I clicked on, and it was noticing that actually I was clicking more on my liberal friends' links than on my conservative friends' links. And without consulting me about it, it had edited them out. They disappeared. (2011b)

Pariser's story has a few remarkable features. First, it is likely a fiction. As director of MoveOn.org, a major online liberal activist organization, Pariser knew his way around the Internet, and it is unlikely that he was surprised that algorithmic filtering existed on a service like Facebook. But the truth of his story is beside the point: it is better understood as a kind of parable. As a parable it bears its second interesting feature: the entanglement of concerns about hiddenness and morality. Pariser poses Facebook's filtering as a moral problem in two ways: it hides divergent viewpoints, inhibiting debate and the establishment of a public sphere, and it hides itself, preventing users from realizing what has happened. The moral solution is transparency: filters and the content they hide should be made visible.

Pariser's moment of recognition is a didactic model for knowledge about algorithms. Where filters obscure, recognition brings sudden clarity. Such moments are common features of popular critical writing about algorithms. They are narrative devices in a literal sense: *anagnorisis* is Aristotle's term from the *Poetics,* a "change from ignorance to knowledge, producing love or hate" (1:XI). When Oedipus discovers his true parentage or King Lear discovers the treachery of his daughters, that is *anagnorisis. Anagnorisis* presupposes a particular definition of knowledge: the dramatic realization of a preexisting fact (with the drama often heightened by the fact that the audience has known it all along). By focusing attention narrowly on the moment of discovery, *anagnorisis* tempts us to take the facts discovered for granted—in the fictional worlds of Oedipus and Lear, there is no doubt of Jocasta's incest or Goneril and Regan's treachery.[4] Once recognized, knowledge is no longer in question.

In this model, knowing algorithms is a matter of revealing them. Once the basic facts of their functioning are revealed, then they can be critiqued. To know an algorithm is to get it out in the open where it can be seen. However, work in science studies, particularly its constructivist strains, has taught us to be suspicious of overly simple stories of revelation. These stories, we have learned, often elide important details about how knowledge is achieved and produced. Constructivist accounts of knowledge production emphasize the processes through which knowledge is achieved, not as the overcoming of barriers or pulling back of veils to reveal what is really going on, but as interactional work that produces local and contingent truths.

This may sound like a strange attitude to take toward algorithms, which are supposed to be human-designed paragons of rationality and objectivity. Surely algorithms cannot be mysterious and hard to specify in the same way that solar neutrinos or deep-sea thermophilic bacteria are, right? Taking my lead from science studies, I want to suggest that there is more going on in the making of knowledge about algorithms than the revelation of plain fact. What we recognize or "discover" when critically approaching algorithms from the outside is often partial, temporary, and contingent.

Experimentation

Faced with the corporate secrecy of a company like Facebook, some critical researchers turn to experimentation or reverse engineering to determine what is happening behind the curtain. In such studies, researchers augment their claims about algorithms with casual experiments conducted through the user interface: searching the same term through Google on multiple computers, systematically interacting with Facebook to see what content is surfaced, and so on. Within the usual constraints of social scientific and humanistic research, these experiments are typically run by one or two people, through ordinary user accounts. While these experiments can provide evidence that filtering has occurred (e.g., by receiving different search results from computers in different locations), they face serious limits when it comes to determining what algorithms are like with any specificity.

Given that personalization algorithms by definition alter one's experience according to interactions with the system, it is very difficult, if not impossible, for a lone researcher to abandon the subject position of "user" and get an unfiltered perspective. In the case of many recommender systems, "everything is a recommendation" (Amatriain and Basilico 2012), and all interactions with the system are tailored to specific user accounts.

To provide an example from my own fieldwork, I recently interviewed a scientist who works for an online radio company. When I asked him about the playlisting algorithm that picks which song to play next, he corrected me: there is not one playlisting algorithm, but five, and depending on how a user interacts with the system, her radio station is assigned to one of the five master algorithms, each of which uses a different logic to select music. What does this mean for the researcher who assumes she is experimenting on one "algorithm"? When not only the results but the logic by which results are chosen are adapting to the individual user, according to invisible meta-logics, single user experimentation can't begin to cover all the possible permutations. Some researchers have attempted to overcome this problem through the use of software bots, which, posing as users, can cover more ground than a single human; while promising, such approaches may be limited by the ability of the services in question to "sniff out" nonhuman patterns of interaction.

At a conference for designers of algorithmic recommender systems, I saw Ron Kohavi, a research scientist working on Microsoft's Bing search engine, present a keynote on "Online Controlled Experiments" (Kohavi 2012). These experiments, or "A/B tests," test modifications to Bing, from alternate designs of the "Search" button (should it have a magnifying glass or not?) to details of the ranking function (what are the best criteria for sorting the top five results?). At any moment, the engineers working on Bing are running about 50 of these experiments, sorting users into experimental groups and serving them with different versions of the site. In a given visit to Bing, a user will be part of around ten different experiments. Depending on how well these modifications perform according to various metrics (clicks, subsequent searches, etc.), they are incorporated into the system.

As a result of these experiments, there is no one "Bing" to speak of. At any moment, there are as many as *10 million* different permutations of Bing, varying from interface design to all sorts of algorithmic detail. A/B testing is now standard for large web companies, and there are even services that algorithmically assess the tests' results, automatically implementing successful modifications on the fly. While we try to experiment on algorithms, they are experimenting on us. This raises a number of issues for would-be experimentalist critics: If we want to talk about how Bing works, what are we talking about? Features shared across the 10 million Bings? One of 10 million possible combinations? How do we know our status relative to the test groups? Rather than thinking of algorithms-in-the-wild as singular objects to be experimented upon, perhaps we should start thinking of them as populations to be sampled.

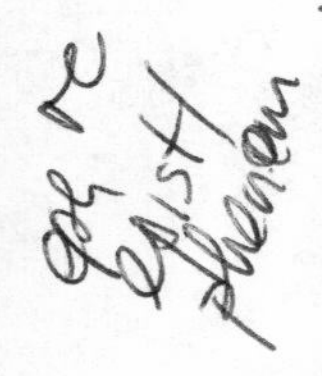

David Karpf (2012) has written about the problems that face social scientists working in "Internet time." The online landscape moves quickly, unlike social scientific research, publication timelines, or our traditional objects of study. What are we to make of claims about what "the Internet" is like that are recent (in academic time), but predate the introduction of Twitter, the opening of Facebook to nonacademic users, or the use of YouTube for US presidential debates? Many comparative claims are compromised by the fact that the Internet has changed beneath them. If research relies on ceteris paribus assumptions—that while we examine some variable, all else is held stable—then the quick pace of online change poses serious issues.

While Karpf is concerned with the broader Internet ecology, the ceteris paribus problem plagues even short-term experimental engagements with algorithms online. Following from the A/B testing described above, algorithmic systems are constantly changing. These changes may be minor or major, but significantly, they will often be invisible to users. For would-be experimenters, this rapid evolution is a serious validity threat: you can't log into the same Facebook twice.

Experimental methods are well suited to objects we have reason to believe are stable. While it makes sense to presume that gravity at the Earth's surface is more or less consistent, we have no such reason to believe that about the algorithmic operations under the hood of major search engines or social media services. Perhaps the persistence of experimentation belies a desire on the part of researchers: *if only* our objects of interest had a stable underlying logic, our jobs would be much easier.

Transparency

The knowledge problems faced by experimental approaches are fairly straightforward. As outsiders, "we" do not know for sure what is happening behind the curtain, but as I've described, we have reason to believe that it is constantly changing and, even at a single moment, multiple. If we were engineers at a high enough level inside Facebook, these problems in their most obvious form would go away: we could know what test group a user had been assigned to, we could know what changes had been made to the architecture, and depending on how much authority we had, we might even be able to hold algorithmic variables constant to achieve ceteris paribus conditions. In short, the problem for experimentation is that we are not insiders—our experiments are designed to get at knowledge that engineers already have and are confounded by hidden variables that engineers already know about and control.

Calls for transparency are aimed at resolving this situation—to let knowledge about how algorithms actually work out from behind the curtain. If these hidden details were better known, the argument goes, then we could engage in more effective critique and there would be incentive to design algorithms more justly. However, transparency is not as clear as it seems. As we know from anthropological research on governments, regulation, and "audit culture," transparency in practice is complicated and by no means a guarantee that the information provided will be complete, accurate, or understandable (Strathern 2000; Hetherington 2011; Hull 2012). Transparency has its limits (Ziewitz 2013). For the developers of algorithmic filtering systems themselves, there is a concern that revealing the details would render their algorithms useless, enabling bad-faith actors to game the system (not to mention aiding corporate competitors), thus negating transparency's benefits (Granka 2010).

At their most simple, calls for transparency assume that someone already knows what we want to know, and they just need to share their knowledge. If we are concerned about Google's ranking algorithm for its search results, presumably that knowledge exists inside of Google. The moral rhetoric accompanying calls for transparency assumes that Google does not or cannot have critical perspective, so its inside knowledge should be passed along to those who do, or that transparency is a moral, democratic end in itself. While transparency may provide a useful starting point for interventions, it does not solve the problem of knowing algorithms, because not everything we want to know is already known by someone on the inside. Here, I want to point to two issues that pose problems for transparency: complexity and inclusion.

In spite of what our formal definition might lead us to believe, algorithms in the wild are not simple objects. In a conference presentation, Mike Ananny described his encounter with an engineer working on Google's Android Marketplace, an online store for purchasing smartphone applications. Ananny had written an article about an offensive association made by the Android Marketplace's recommendation system: while looking at an application targeted at gay men, the recommender suggested an application for locating the residences of sex offenders (Ananny 2011). Several days after the article circulated online, he was contacted by an engineer at Google who offered to talk about the unfortunate association. While the "error" had quickly been manually corrected, it was still not clear why it had emerged in the first place. The operations of the recommender algorithm, incorporating machine learning techniques and changes made by numerous teams of numerous people, all accumulated over time, were so complex that there was no simple bit of code to point to as a cause.

The apparent simplicity and determinism of algorithms in theory does not account for their features in practice. In assuming that engineers have complete understanding of the systems they create, we make a mistake: first, these systems are works of collective authorship, made, maintained, and revised by many people with different goals at different times; second, once these systems reach a certain level of complexity, their outputs can be difficult to predict precisely, even for those with technical know-how. Simply making the system transparent does not resolve this basic knowledge problem, which afflicts even "insiders." A determined focus on revealing the operations of algorithms risks taking for granted that they operate clearly in the first place.

In presenting my own research on music recommendation algorithms, I am frequently asked questions about "inclusion": Does a given algorithm include lyrics when deciding that you might like a song? Does it include details about the music "itself" (tempo, key, etc.), or does it only include other listeners' ratings? The prevalence of these questions indicates a dominant attitude toward knowledge about algorithms: to know an algorithm is to know its decision-making criteria. Calls for transparency share this attitude: What is in there? We want to know!

However, the "what" of algorithmic inclusion is often less significant than the "how": to include something requires decisions about how to calculate and represent it. For example, the main variables included in Facebook's EdgeRank algorithm, which determines what content shows up on a user's news feed, are well publicized: it takes into account a user affinity score (a measure of affinity between the viewing user and the content posting user), a weight for different types of content (photos, for example, are reportedly weighted higher than text alone), and a time decay (older material is less likely to show up).[5] While this information is posted all over websites that claim to help brands promote themselves on Facebook, it tells us very little. How are user affinity scores calculated? How are content weights calculated? How do these calculations vary across different user types? None of these variables tell us more than we might have already guessed based on the fact that Facebook users share different kinds of material with each other over time. Add to this the fact that Facebook claims that EdgeRank contains other variables that are kept secret, and it becomes clear that even when we "know" the algorithm in broad strokes, we actually know precious little.

If we are interested in talking about algorithms' cultural effects, then it is not enough to know that something called "affinity" is included. Our questions should be more ambitious: What is affinity? How is it defined? What experiences do the engineers producing affinity scores draw on as they attempt to formalize it? How might these formalizations differ if we started from different assumptions and experiences? To assume that an algorithm can simply "include" something like affinity (or, for that matter, even such apparently objective features as tempo or key) is to miss the amount of interpretive, cultural work required to translate these features into computable form. It is to mistake maps for territories.

Redefinition: Toward the Ethnographic Study of Algorithmic Systems

The basic problem pointed to by these issues is this: the algorithms we are interested in are not really the same as the ones in *Introduction to Algorithms*. It is not the algorithm, narrowly defined, that has sociocultural effects, but *algorithmic systems*—intricate, dynamic arrangements of people and code. Outside of textbooks, "algorithms" are almost always "algorithmic systems." If, as I suggested earlier, critical algorithm studies scholars participate in the construction of algorithms as particular kinds of objects, an ethnographic approach allows us to construct algorithms as heterogeneous sociotechnical systems, influenced by cultural meanings and social structures. Taking an ethnographic eye to algorithmic systems allows us to see features that are typically elided or obscured.

We have inherited a number of technical distinctions from computer scientists for talking about algorithms: the difference between algorithms and data structures; the difference between an algorithm expressed theoretically and its implementation in a particular programming language; the difference between formally defined algorithms and ad hoc heuristics. These distinctions can be useful for building computational systems, and they are central elements of the emic understanding of computation among software engineers. However, they do not capture the dynamism and imprecision of algorithmic systems: we do not interact with algorithms and data structures in isolation from each other, there are no theoretical algorithms in actually existing software systems, and while a given algorithm may be well defined, the selection among algorithms may yet be ad hoc, guided by informal—not algorithmic—processes.

When we realize that we are not talking about algorithms in the technical sense, but rather algorithmic systems of which code in a strict sense is only a part, their defining features reverse: instead of formality, rigidity, and consistency, we find flux, revisability, and negotiation. The use of phrases like "the Google algorithm" or "the Facebook algorithm" should not fool us into thinking that our objects are simple, deterministic black boxes that need only to be opened. These algorithmic systems are not standalone little boxes, but massive, networked ones with hundreds of hands reaching into them, tweaking and tuning, swapping out parts and experimenting with new arrangements. If we care about the logic of these systems, we need to pay attention to more than the logic and control associated with singular algorithms. We need to examine the logic that guides the hands, picking certain algorithms rather than others, choosing particular representations of data, and translating ideas into code.[6] Unlike the technical details of algorithmic configuration at a given moment, these logics are more readily accessible and persistent over time, giving the critical researcher a firmer grasp.

This might be understood as a call to examine empirically the contexts—social, cultural, political, economic, legal, institutional, etc.—in which algorithms are developed, and that would be a welcome addition to the current state of algorithmic criticism. But thinking of these factors as merely context keeps algorithms themselves untouched, objective stones tossed about in a roily social stream. We should not abandon concern with technical details or "close readings" of algorithmic function; contextualization is not an explanatory panacea. My point is that when our object of interest is the algorithmic system, "cultural" details *are* technical details—the tendencies of an engineering team are as significant as the tendencies of a sorting algorithm. This is not so much an attempt to add the cultural back onto the technical as it is a refusal of the cultural/technical distinction as a ground for analysis.

This redefinition should be familiar to scholars of STS and the anthropology of technology, and it has significant consequences for how we examine the ethics of algorithms and their governability. Rather than starting our critiques from the premise that we already know what algorithms do in relation to culture—they reduce, quantize, sort, and control—we take that as our research question. How do algorithmic systems define and produce distinctions and relations between technology and culture?

With this as our guiding question, the relationship between "outsider" humanistic or social scientific critics and "insider" engineers changes. When all there is to know about an algorithm is its function and effect, then expertise is neatly split: engineers know about functions, social scientists about consequences. But, if we're willing to muddy the waters a little, a space opens up for engineers and social scientists to critically engage with algorithmic systems together. Algorithmic systems, with their blend of "technical" and "cultural" concerns, spread across institutional settings in broader social contexts, are choice objects for ethnographic study. You can visit and live within an algorithmic system, collaborating with the people who work there.[7] For critics and advocates alike, if we want to know algorithms, we may need to live with them.

Notes

1. The rate of publication in this area is accelerating dramatically; these citations are representative of the field as it was when this piece was first drafted in late 2012, with some periodic revisions over the intervening years. See Gillespie and Seaver (2015) for an updating bibliography, which aspired (as of early 2016) to comprehensiveness but demonstrates the challenge of keeping up with the astonishing breadth of work in the area.
2. Judging by popular news coverage, which tends to take its lead from press releases, all algorithms are "sophisticated."
3. This is not because reflexivity is an intrinsic virtue unique to STS scholars (Lynch 2000), but because it has specific benefits to the discussion at hand, which I outline through the rest of this chapter.
4. As these examples suggest, this way of thinking about knowledge has a sexual politics as well.
5. Since the original drafting of this piece, it has come out that "EdgeRank is dead" (McGee 2013), replaced with a more complex blend of variables. Internet time strikes again.
6. Of course, we do not always have the kind of access we might like to the systems we want to know about. In these cases we might have to resort to assessments of studies of "culture at a distance" or "reverse engineering" (Benedict 1946; Diakopoulos 2014), but as I have outlined here, these approaches face significant shortcomings that should not be ignored.
7. One promising, recent effort in the context of professional data science can be found in Neff et al. (2017).

Works Cited

Amatriain, Xavier, and Justin Basilico. 2012. "Netflix Recommendations: Beyond the 5 Stars (Part 1)." *Netflix Tech Blog*, April 6. http://techblog.netflix.com/2012/04/netflix-recommendations-beyond-5-stars.html.

Amoore, Louise. 2011. "Data Derivatives: On the Emergence of a Security Risk Calculus for Our Times." *Theory, Culture & Society* 28 (6): 24–43.

Ananny, Mike. 2011. "The Curious Connection between Apps for Gay Men and Sex Offenders." *Atlantic*, April 14. www.theatlantic.com/technology/archive/2011/04/the-curious-connection-between-apps-for-gay-men-and-sex-offenders/237340/.

Anderson, C. W. 2013. "Towards a Sociology of Computational and Algorithmic Journalism." *New Media & Society* 15 (7): 1005–21.

Asher, Andrew. 2011. "Search Magic: Discovering How Undergraduates Find Information." Paper presented at the 2011 American Anthropological Association meeting, Montreal. www.erialproject.org/wp-content/uploads/2011/11/Asher_AAA2011_Search-magic.pdf.

Barocas, Solon, Sophie Hood, and Malte Ziewitz. 2013. "Governing Algorithms: A Provocation Piece." Working paper. http://papers.ssrn.com/sol3/papers.cfm?abstract_id=2245322.

Beer, David. 2009. "Power through the Algorithm? Participatory Web Cultures and the Technological Unconscious." *New Media & Society* 11 (6): 985–1002.

Benedict, Ruth. 1946. *The Chrysanthemum and the Sword: Patterns of Japanese Culture*. Boston: Houghton Mifflin.

Bucher, Taina. 2012. "Want to Be on the Top? Algorithmic Power and the Threat of Invisibility on Facebook." *New Media & Society* 14 (7): 1164–80.

Carr, Nicholas G. 2014. *The Glass Cage: Automation and Us*. New York: Norton.

Cheney-Lippold, John. 2011. "A New Algorithmic Identity: Soft Biopolitics and the Modulation of Control." *Theory, Culture & Society* 28 (6): 164–81.

Cormen, Thomas H., Charles E. Leiserson, Ronald L. Rivest, and Clifford Stein. 2009. *Introduction to Algorithms*. 3rd ed. Cambridge, MA: MIT Press.

Diakopoulos, Nicholas. 2014. "Algorithmic Accountability Reporting: On the Investigation of Black Boxes." Working paper, Tow Center for Digital Journalism, Columbia University. http://towcenter.org/research/algorithmic-accountability-on-the-investigation-of-black-boxes-2/.

Dormehl, Luke. 2014. *The Formula: How Algorithms Solve All Our Problems . . . and Create More*. New York: Penguin.

Ensmenger, Nathan. 2012. "Is Chess the Drosophila of Artificial Intelligence? A Social History of an Algorithm." *Social Studies of Science* 42 (1): 5–30.

Eubanks, Virginia. 2018. *Automating Inequality: How High-Tech Tools Profile, Police, and Punish the Poor*. New York: St. Martin's.

Feyerabend, Paul. 1975. *Against Method*. New York: New Left Books.

Galloway, Alexander. 2006. *Gaming: Essays on Algorithmic Culture*. Minneapolis: University of Minnesota Press.

Gillespie, Tarleton. 2014. "The Relevance of Algorithms." In *Media Technologies: Essays on Communication, Materiality, and Society*, edited by Tarleton Gillespie, Pablo Boczkowski, and Kirsten Foot, 167–95. Cambridge, MA: MIT Press.

Gillespie, Tarleton, and Nick Seaver. 2015. "Critical Algorithm Studies: A Reading List." https://socialmediacollective.org/reading-lists/critical-algorithm-studies/.

Graham, Stephen. 2005. "Software-Sorted Geographies." *Progress in Human Geography* 29 (5): 562–80.

Granka, Laura. 2010. "The Politics of Search: A Decade Retrospective." *Information Society* 26:364–74.

Helmreich, Stefan. 1998. "Recombination, Rationality, Reductionism and Romantic Reactions: Culture, Computers, and the Genetic Algorithm." *Social Studies of Science* 28 (1): 39–71.

Hetherington, Kregg. 2011. *Guerrilla Auditors: The Politics of Transparency in Neoliberal Paraguay*. Durham, NC: Duke University Press.

Hughes, Thomas P. 1983. *Networks of Power: Electrification in Western Society, 1880–1930*. Baltimore: Johns Hopkins University Press.

Hull, Matthew. 2012. *Government of Paper: The Materiality of Bureaucracy in Urban Pakistan*. Berkeley, CA: UC Press.

Karpf, David. 2012. "Social Science Research Methods in Internet Time." *Information, Communication & Society* 15 (5): 639–61.

Kockelman, Paul. 2013. "The Anthropology of an Equation: Sieves, Spam Filters, Agentive Algorithms, and Ontologies of Transformation." *Hau: Journal of Ethnographic Theory* 3 (3): 33–61.

Kohavi, Ron. 2012. "Online Controlled Experiments: Introduction, Learnings, and Humbling Statistics." Keynote talk, RecSys conference, Dublin.

Lynch, Michael. 2000. "Against Reflexivity as an Academic Virtue and Source of Privileged Knowledge." *Theory, Culture & Society* 17 (3): 26–54.

Lyon, Stephen. 2003. *Surveillance as Social Sorting*. New York: Routledge.

MacKenzie, Donald, and Taylor Spears. 2014. " 'The Formula That Killed Wall Street': The Gaussian Copula and Modelling Practices in Investment Banking." *Social Studies of Science* 44 (3): 393–417.

Mager, Astrid. 2012. "Algorithmic Ideology: How Capitalist Society Shapes Search Engines." *Information, Communication & Society* 15 (5): 769–87.

Mahnke, Martina, and Emma Uprichard. 2014. "Algorithming the Algorithm." In *Society of the Query Reader: Reflections on Web Search*, edited by René König and Miriam Rasch, 256–70. Amsterdam: Institute of Network Cultures.

McGee, Matt. 2013. "EdgeRank Is Dead: Facebook's News Feed Algorithm Now Has Close to 100K Weight Factors." *Marketing Land*, August 16. Available at http://marketingland.com/edgerank-is-dead-facebooks-news-feed-algorithm-now-has-close-to-100k-weight-factors-55908

McKelvey, Fenwick. 2014. "Algorithmic Media Need Democratic Methods: Why Publics Matter." *Canadian Journal of Communication* 39:597–613.

Nakamura, Lisa. 2009. "The Socioalgorithmics of Race: Sorting It Out in Jihad Worlds." *In New Media and Surveillance*, edited by Kelly Gates and Shoshana Magnet, 149–61. New York: Routledge.

Napoli, Philip M. 2013. "The Algorithm as Institution: Toward a Theoretical Framework for Automated Media Production and Consumption." Scholarly Paper 2260923, Social Science Research Network. https://papers.ssrn.com/abstract=2260923.

Neff, Gina, Anissa Tanweer, Brittany Fiore-Gartland, and Laura Osburn. 2017. "Critique and Contribute: A Practice-Based Framework for Improving Critical Data Studies and Data Science." *Big Data* 5 (2): 85–97. doi:10.1089/big.2016.0050.

Noble, Safiya Umoja. 2018. *Algorithms of Oppression: How Search Engines Reinforce Racism*. New York: New York University Press.

Pariser, Eli. 2011a. *The Filter Bubble: What the Internet Is Hiding from You*. New York: Penguin.

——. 2011b. "Beware Online 'Filter Bubbles.'" *TED*. www.ted.com/talks/eli_pariser_beware_online_filter_bubbles.html

Parisi, Luciana. 2013. *Contagious Architecture: Computation, Aesthetics, and Space*. Cambridge, MA: MIT Press.

Pasquale, Frank. 2015. *The Black Box Society*. Cambridge, MA: Harvard University Press.

Porter, Theodore. 1996. *Trust in Numbers*. Princeton, NJ: Princeton University Press.

Seaver, Nick. 2012. "Algorithmic Recommendations and Synaptic Functions." *Limn* 2. http://limn.it/algorithmic-recommendations-and-synaptic-functions/.

Snider, Laureen. 2014. "Interrogating the Algorithm: Debt, Derivatives and the Social Reconstruction of Stock Market Trading." *Critical Sociology* 40 (5): 747–61.

Star, Susan Leigh. 1995. "The Politics of Formal Representations: Wizards, Gurus, and Organizational Complexity." In *Ecologies of Knowledge*, edited by Susan Leigh Star, 88–118. Albany: State University of New York Press.

Steiner, Christopher. 2012. *Automate This: How Algorithms Took over Our Markets, Our Jobs, and the World*. New York: Penguin.

Strathern, Marilyn. 2000. *Audit Cultures: Anthropological Studies in Accountability, Ethics and the Academy*. New York: Routledge.

Striphas, Ted. 2010. "How to Have Culture in an Algorithmic Age." *Late Age of Print*, June 14. www.thelateageofprint.org/2010/06/14/how-to-have-culture-in-an-algorithmic-age/.

Uricchio, William. 2011. "The Algorithmic Turn: Photosynth, Augmented Reality and the Changing Implications of the Image." *Visual Studies* 26 (1): 25–35.

Winner, Langdon, ed. 1986. *The Whale and the Reactor: A Search for Limits in an Age of High Technology*. Chicago: University of Chicago Press.

Ziewitz, Malte. 2013. "What Does Transparency Conceal?" Working paper, Privacy Research Group, New York University. http://ziewitz.org/files/Notes%20on%20transparency.pdf.

Keeping Software Present
Software as a Timely Object for STS Studies of the Digital

Marisa Leavitt Cohn

Lived Temporalities of Code

This chapter considers how we might study software as a timely object—that is, an object subject to continuous change and lived with over time as it evolves. As software engineers well know, code is lively and changing. Designs are criticized for being "frozen" if they cannot adapt and evolve alongside code as it changes. Engineers remind themselves of this aspect of their work with aphorisms about how documentation is out of date as soon as it is written. Yet at the same time, code persists. Legacy systems, while given little attention in social analyses of software, are the majority of the software that we rely upon, from the UNIX-based operating systems on our PCs to the COBOL-based systems supporting US social security.[1]

Much STS literature on software has emphasized code's promiscuous evolvability, how its mutability and dynamism led it to its inscrutability. As Adrian Mackenzie suggests, software "simply does not sit still long enough to be easily assigned to conventional explanatory categories" (Mackenzie 2006, 18). However, this difficulty to pin down code sufficiently in order to represent knowledge about it belongs not only to the STS analyst but also to engineers who work closely with code, particularly the software maintainers who deal with software as a continuously evolving object over the longue durée.[2]

STS has done much to unmask the presumed neutrality, immateriality, and automaticity of software, revealing instead the highly imbricated social relationality of software-in-the-making (Mackenzie 2006; Gillespie 2014; Crawford 2013; Introna 2016; Bucher 2012; Seaver, this volume). Less attention, however, has been placed on how coders working to maintain legacy systems grapple with this relationality as part and parcel of the nature of working with code.[3] These coders who occupy not the cutting edge but the dulled and fading expanse of ever evolving bodies of code are unlikely to deny this relationality, which increasingly manifests as a system ages. These are engineers who work to maintain obsolescent bodies of code under continuous pressures toward upgrade or keep track of the many different iterations of software that emerge over its lifetime or struggle to account for what software even exists within the organization.

It is this sensibility toward temporality that should guide STS approaches to software, its life cycles of change, its lifetimes within organizational practice,[4] and the lived experiences of its dynamism over time. I argue that the significance of software's materiality for the politics of computational work lies not only in its present performances—however hard these may be to trace—but also in its forms of duration, the entangled lifetimes of careers, professional identities, and programming languages and paradigms, all of which come and go. This of course still requires an STS understanding of code as a relational assemblage but also emphasizes that these entanglements have duration—that the relationality of code exists not only in the evocative present of *runtime* but also in the shared histories and anticipated futures that a body of code *binds together*.[5]

The examples presented in this essay are drawn from a case of an aging infrastructure and the work of engineers and software developers who continue to operate and maintain it. This infrastructure supports a large-scale outer planetary science mission to Saturn (henceforth referred to as the Mission) at a California NASA-funded laboratory (henceforth the Lab).[6] The Mission was developed in the 1980s and launched its spacecraft in 1997, reaching Saturn in 2004. Being a long-lived mission, its computational infrastructure is naturally quite old, including legacy software from earlier outer planetary missions at the Lab, some written in legacy programming languages like FORTRAN. During ethnographic fieldwork I conducted in 2010–11, I observed the work of engineers maintaining this aging and obsolescent infrastructure and planning for the end-of-mission in 2017.[7] In their work, engineers crisscrossed back and forth among pieces of code, scripts, and protocols written at different times in the life of the organization, weaving together old and new.

Taken together, these various pieces of software compose the Mission "ground system"[8] depicted in the architectural diagram in figure 1. This system includes all of the software needed to meaningfully command and control the spacecraft—from the applications used to communicate with the Deep Space Network antennas to those that allow engineers to translate plans for scientific data collection into executable commands and those needed to retrieve telemetry data and reassemble bits into meaningful data.

However, it is precisely this kind of depiction of software that we know to be suspect as it is incapable of representing software as an evolving system. In fact, this diagram was created years before I arrived at the Mission and has not been updated since. In its upper left corner, one discovers an inset with a time stamp noting that this, its 25th version, was saved at 4:30 PM on July 27, 2001. While this diagram demanded attention and work at that particular moment in the life of the organization, it has since become historical—the work of maintaining it deemed no longer relevant or economical.

As I shadowed the engineers and software developers working at the Mission, I found instead many partial diagrams of the software like the one in figure 2. These images, in contrast to that in figure 1, are quite lively—full of notes jotted down during ongoing work or at meetings where details about the system resurface from institutional memory to fill in gaps in current understanding. These diagrams—of which no two are exactly the same—reveal the horizons of any one individual's work within the broader tapestry of the organizational software. I even began to imagine that I could collect these partial images together and stitch them back into a new sort of whole, keeping the borders and seams intact in order to reveal contradictions and gaps. I briefly succumbed to an STS analyst's dream of using the

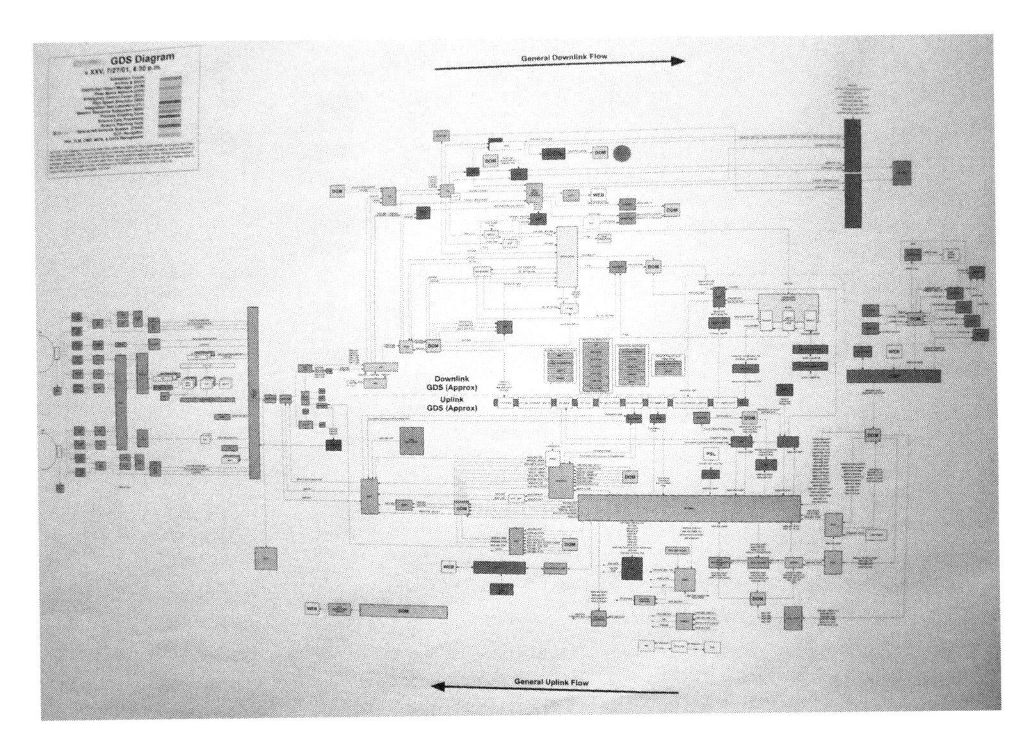

FIGURE 1: Mission infrastructure architecture.

software as a kind of cultural probe to uncover the lively and messy web of the organization as a relational network.

However alluring, the messy picture of software as a relational network is not necessarily where we might find a view of software as it is "lived." This presumes that the software itself (in its present performance), although difficult to retrieve or represent, could produce a map of the organizational relational assemblage. This mapping between the software and organization is a particular sociotechnical imaginary (Jasanoff and Kim 2009), and it is one that certainly lives within the organization, where relations between different teams are often figured by the software tools that sit between them, as I will discuss further. But this imaginary does not fully encompass the sense of software as *lived* that I want to address here. In fact, the "lived temporality" of software exists somewhere in the gray middle, somewhere between the excess of code's present performance, and the inevitable decay and obsolescence of software diagrams, documentation, and representations. Maintaining legacy software is to some extent always a matter of living between these two unrealities—namely, of the body of code and of its failures to be fully articulated and represented.

In the stories I collected and scenes I observed during ten months of ethnographic fieldwork, the work of keeping software present (maintaining its currency, knowability, relevance) continually brought software's lifetimes and durations to the foreground. It is not just software's performativity in the present that becomes troubling to those who work with it. It is also its *decay* over time—that is, its former selves and forms of forgetting, its once anticipated or still possible futures, what it might have once been had things been different, or what it could still be depending

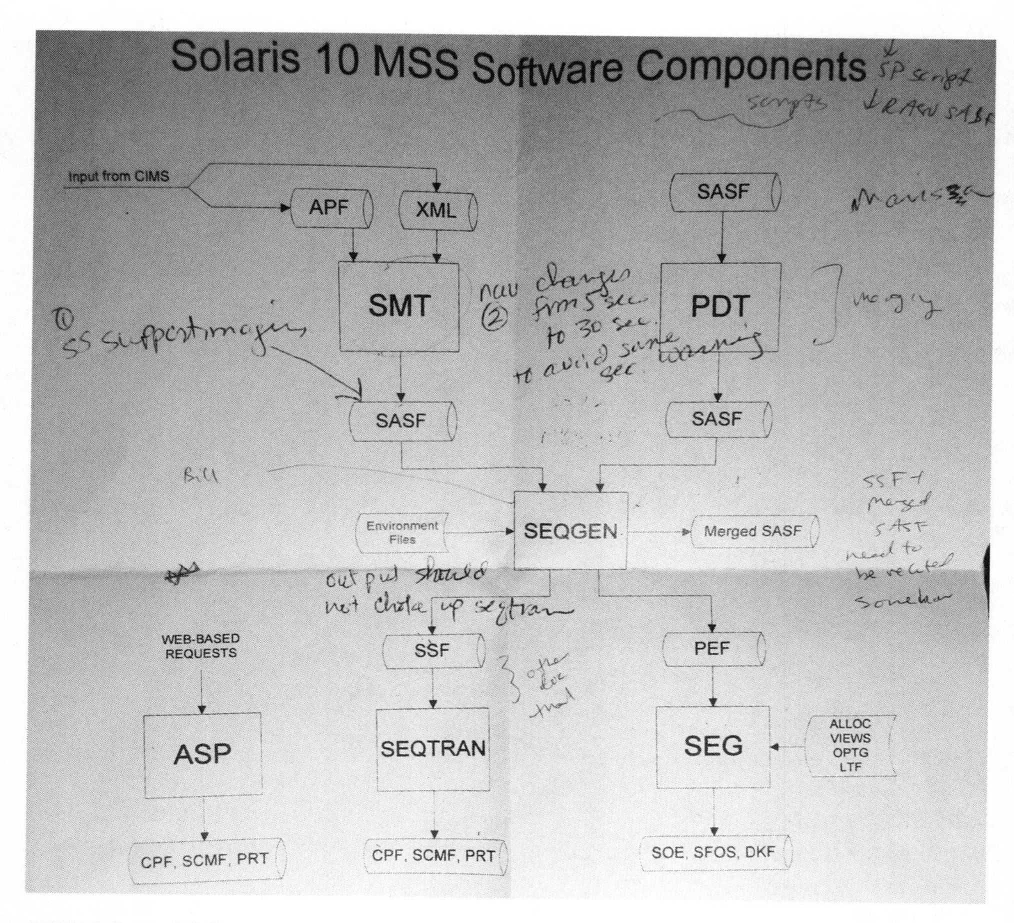

FIGURE 2: Partial diagram of software-in-maintenance.

on how things go—that must be confronted in practice. It is these temporalities of decay, forgetting, obsolescence, and legacy—of software's lived durations—to which this chapter turns.

One way to begin our investigation into these temporalities would be to look at particular moments in the work of software maintenance. We might place our quadrant somewhere in the midst of the organizational software ecology and see what forms of life live there. We might see people working with code, old and new, layering them together. If we do that, we might gain a sense of what it is like to work with legacy software as a material and a medium. We might, for example, find people sitting at terminals struggling to make sense of old code that others have written. But if we want to understand the *timeliness* of software, we might also need to see how the life stories of software are crafted and where these stories are made relevant.[9] We might have to take an approach inspired by Donna Haraway and, rather than use the techniques of the quadrant, see what happens when we grab hold of one thread in the tangle and mess and pull it.

Moments in Software Maintenance

The examples presented in the following sections are drawn from ethnographic fieldwork conducted at the Mission between 2010 and 2011 as the Mission was finishing its middle phase and preparing for its final one. This final phase, projected to end in 2017, came with an extension in funding from NASA, but at significantly reduced budget, leading to cuts in personnel and a winnowing of expectations of what the final years of the mission might comprise. It was also a time when some of the more senior engineers were anticipating retirement, and imagining how the Mission would run its final years without those who had been there from the start. They were also beginning to plan for the end-of-mission—the final operations and science activity to be performed before running out of fuel and plummeting the spacecraft to its death (via combustion) in the Saturnian atmosphere.

Such a case—of an aging and dying organizational infrastructure—is bound to reveal a different sort of temporality than that of a cutting-edge software system. The engineers whom I worked alongside often engaged in reminiscing about the Mission's earlier days and considered how they had come from where they started to where they are today. Their work also involved managing continual pressures toward upgrade coming from changing expectations at the Lab as well as from the software engineering field, which has professionalized significantly over the duration of the Mission. These pressures came in the form of programming languages reaching end-of-support and the need to reduce maintenance costs. They also came from the persistent difficulties that the Mission faced in adopting new tools developed by the Lab's centralized software development group; a group that has emerged in the intervening years while the Mission has been in operations. At times the challenge of these upgrades arises due to the critical nature of their work—how do you migrate the navigation software you use to fly a spacecraft mid-flight?! However, at the heart of this challenge was also an effort to know their own software, or make account of it, in ways that would mesh with the current software development managerial regime. To take this navigation software as an example, the navigation team was unable to provide proper unit tests for their existing code, to fully transition into an object-oriented paradigm, or to stay current with each version release of the new software without some other application breaking.

This work of bringing old code together with new, and managing software change in ways that interoperate with legacy systems, is the work that I am calling "keeping software present." It is the work of maintaining software's currency, its ability to perform in ways that are up to date with today's expectations of performance or disciplinary regimes of software management, or dealing with upgrades and changing demands. But I also refer to the second-order work that this requires; in order to discipline software into alignment with current expectations one must also know and articulate the software, and be able to make it accountable and available to managerial decisions.[10] These techniques of knowing software—knowing its extent, its limitations, knowing the software you do and do not have—is what prompts forms of documentation even as these forms of documentation fail.

Much of this work is quite bespoke and individual, as the partial diagram shown above reveals—almost no engineers share exactly the same view into the software systems that are relevant to their work. But, at the same time, software cannot be meaningfully used and maintained unless some shared articulation of the software is brought forth. This is the work of making software present, or copresent (Beaulieu 2010), to collective work within the organization. In order to maintain software, not only the code but also the software as relational organizational assemblage must be known. How software's maintainers do this is precisely what this essay is about; how they bring the software present and how this "presencing" of software is achieved, not only through knowledge of its "current performance," but also through the presencing of relations that have duration, shared histories, and shared futures.

In the following examples, we can see how this work of keeping software present, of making software knowable and making software matter, is work that repeatedly brings software's histories and durations to the foreground. In the first, we see Sarah, a systems engineer in charge of the Mission software infrastructure, working to make the body of code at the Mission known to the organization. When attempts to do this through visual forms of representation fail, she shifts toward tactics of making the software "felt"—of tugging on one thread in the relational mess of the software assemblage to give her colleagues an embodied sense of how the software in the organization touches all aspects of their work. The software that has grown over years to fill in the seams between different organizational teams and their tools connects them all together. In the second example, we see how Frank, a software developer at the Mission, in order to meet the needs of scientists, must grapple with a software tool in terms of its histories and anticipated futures, or even its obsolescent futures. These durations of the software present themselves simultaneously as limitations and as remembrances of what the software *might once have been able to do* had things evolved differently.

In both examples we see how software is indeed material and discursive, but presents troubles not only in the gaps between current needs and functions, but also in the forms of drift and drag, of software's creeping changes in which aspects of software are forgotten and remembered.[11]

Making Software Felt

It was Sarah, the lead engineer for the Mission ground system, who finally pointed out to me the architectural diagram shown earlier (figure 1), where it hung on a cubicle wall just outside her office, in a back corner of the software development "annex." The diagram had almost mythological status. I had heard during many interviews with Mission engineers that such a diagram had at one time existed. Now that I had seen it, my enthusiasm waned as Sarah explained to me that it was exceedingly out of date and there had been no efforts to maintain it since July 2001, when the organization had needed to capture an overview of the system for organizational accountability.[12]

It was the responsibility of Sarah, as a "level III" systems engineer in charge of software, to manage the maintenance and development of the computational infrastructure supporting the mission.[13] While other managers might be responsible for a particular software tool used by their team, it was solely her responsibility to consider how all these different tools worked together, and manage software changes and how they might affect different kinds of work. Since I came to the field with an interest in this relationality—in how software mediated interdisciplinary collaboration on the mission—I spent many hours shadowing Sarah and observing her work.

One of the first things I observed was Sarah working to update an Excel worksheet that listed all of the software tools at the Mission and their versions, last "change management review," status, and criticality.[14] Sarah explained to me that the software tools at the Lab fall into three categories, A, B, and C, according to their criticality.[15] A fourth quasi-category exists colloquially, referred to as "Cat D." Cat D is a category that serves as a placeholder for "all of the other software" that does not fit into A, B, or C, and remains undocumented. By definition this software is considered nonessential and therefore not requiring of documentation and rigorous change management. It is this category of undocumented software that Sarah was now working to document.

Cat D software includes what are referred to colloquially as *scripts* that individual engineers write to facilitate exchanges of files or to generate automatic alerts when a file is updated, *utilities* that improve quality of work by customizing the way data are visualized, or *hacks* that solve inconsistencies among different teams' naming conventions, just to give some examples. These pieces of software are not formally documented precisely because of the inordinate cost that would ensue from assessing every single change to every piece of software used by the organization. These tools are relegated to a category of "user tools" because they are defined as software that is not "in line to the craft"—a phrase used to denote the apparatus of tools that at some point touch commands that will be sent to and executed by the spacecraft. The assumption is that Cat D software primarily impinges on individual quality of work, and therefore is not of primary managerial concern.[16] These are tools that are deemed to be software that engineers should be able to "live without" if some more official change leads them to break down.

Yet, these are in fact tools that engineers have *lived with* for over a decade. They have proliferated into the thousands over the course of the Mission, written for needs as they arise and then shared and circulated. The engineers have lived with these tools for so long that they have become largely invisible to the organization. As Sarah explained to me (paraphrased in my field notes), "Cat D software is ostensibly the software we 'can live without,' but actually that isn't always the case. People might respond to a request to reduce Cat D software saying that they really need it and it is crucial to their work. There are a number of reasons that software ends up in Cat D. Sometimes it is that people just did not want to be formal about it. They will say that it is 'just a script' and then it turns out to be 45 pages of code with encoded knowledge about how the spacecraft operates and with no documentation and no sense of responsibility." Sarah was especially eager to convey the importance of these "Cat D" tools since, at the time of my fieldwork, she was nearing her own retirement in a few months and her role as head of the ground system would not be filled due to downsizing in the final phase of the mission. In making software present to the organization, Sarah seemed to struggle against the absurdity of documenting all the Cat D software, but in fact, it was this very absurdity that she told me she was attempting to show. Her Excel sheet of Cat D software already ran to 50 pages, and she estimated that a full list would be an order of magnitude more.

This software was what Sarah and others in the software development community sometimes refer to as "glueware." This is software that is developed to fill the gaps between official managerial software tools, software that covers seams or stitches together the work that is done by the Mission's primary software tools. She explained, "Without a systems perspective it can seem that software [at the Mission] is these separate and distinct tools with different requirements, but . . . when you think through a systems perspective you see that two software tools that are in different interlinking parts of the process have different ways that they may overlap. [Over time] you get some collisions, some overlaps, some gaps. And you get different kinds of patches or ways of smoothing the interface or covering the gap." What interested Sarah the most was not the fact that glueware exists, but what it covers over, what we might call the "seamfulness" of the infrastructure (Vertesi 2014) and the varying qualities of those seams—the multiple forms that they take. She related this to the historical reasons that these seams emerged in the first place, each representing a different "pain point" in the organization that had emerged at a particular moment or event in the Mission. The glueware exists as a kind of record or organizational memory of these various pain points, like scars from wounds that have mostly healed over.

As Sarah prepared for her retirement, she knew that a full documentation of all the glueware was implausible, but she wanted to impart some of her knowledge of these pain points to her boss, the project manager, and her team. She expressed some frustration that management had failed to understand the extent and "maturity" of the Mission software, and tended to discount the costs and risks of maintaining software, since software was largely considered "done at launch." "It does the job, and the job hasn't changed" is the attitude that she said she was pushing back against.

When her incomplete Excel tabulation failed to convey the criticality of the Cat D software, Sarah switched tactics. Rather than try to visually represent the burden of the quantity of software, she used a tactic that she called "making them say ouch" to make her colleagues *feel* the software even if they could not see it. In advance of a major file sharing system upgrade, Sarah asked each of the various teams (Navigation, Science Planning, and Spacecraft Engineering) to do some light "housecleaning" and create official shared repositories of their most relevant Cat D software. She asked them to create a list of these file repositories to prepare for a meeting about the migration, knowing that these lists would likely be incomplete as well. But she also let them know that she would go ahead and migrate a single file in the Mission workflow over from the old file-sharing platform to the new one for an upcoming science activity.

At the meeting, she went over the documentation she had received from each team. Everyone felt rather prepared for the move, presenting the various files and tools that would be impinged upon by the new system, until Sarah asked for a show of hands from those who noticed a disruption to their workflow due to the moved file (see figure 3).

At that point, every single hand in the room went up, followed immediately by an uneasy laughter and a few gasps of surprise, as navigators, science planners, and engineers from various teams looked around the room and saw that they were all impacted. I felt a sense of the uncanny realization that they were all so intimately connected by the software systems they use. Since files are exchanged through what appears to be a very linear workflow process, the so-called "upstream" and "downstream" of preparing files for upload to the craft, individuals felt largely insulated from the work of others beyond one degree removed from them in this workflow.

This tactic of Sarah's, which she repeated in subsequent meetings, seemed to work by pulling on one thread within the relational network of the Mission's software assemblage. Bypassing the representational mode, it worked by reminding everyone of the shared body of code that touches them all and binds them together. It also made clear that this relationality was something that had been largely submerged or forgotten.

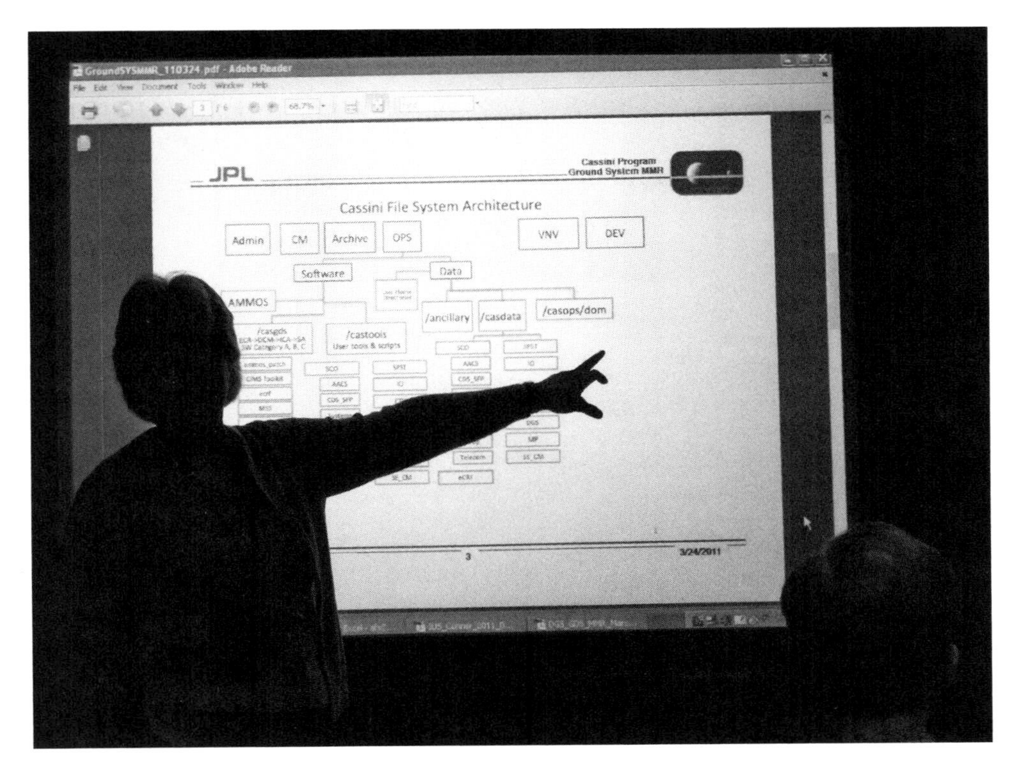

FIGURE 3: Category D house-cleaning meeting.

We can see this as a moment of deliberate infrastructural inversion (Bowker and Star 2000), a way to make the embedded and taken-for-granted parts of the software system visible. But at the same time Sarah's framing of this tactic, as "making them say ouch," makes the software known through a register of touch rather than vision,[17] recalling earlier pains that have been lived through but have healed. Effectively, she sought to artificially create a moment of shared pain in order to rebind the social ties of the organization,[18] which were originally forged through shared pain at earlier moments, for example in hotly debated decisions about what software change to prioritize or whether to absorb one software tool into another. Her tactic also followed a logic of anticipation and care,[19] to make such pains felt now, so that they might be better prepared for the inevitable breakdowns in the final years of the mission, the moments that might reopen these old forgotten wounds.

Software Occultations and Remembrances

Frank is an engineer at the Mission who has built a career working on a single piece of software, called the Science Observation Design Tool, or SDT, used by scientists and instrument operators to implement scientific observations into command files. SDT visualizes the fields of view of various Mission remote-sensing instruments and models the geometries of the various "targets" in the Saturn system from the perspective of the spacecraft based on its position along its orbital trajectory.[20]

Frank has been working at the Mission since 1978 and on other missions at the Lab before that. He specializes in the development of systems that support the implementation of science observation by pointing spacecraft instruments at targets and tracking those targets over time. This is particularly challenging on orbital spacecraft since both the instrument and the target are on the move.[21] Earlier in his career, he helped to develop a tool called POINTER that was used on previous missions and then formed the basis for SDT. For the planning of Mission science, "pointing" at targets is especially complicated since the instruments are mounted directly to the body of the spacecraft. This means that the commands used to point, or articulate, the spacecraft in space at targets are the same as those used for spacecraft operations to maneuver and navigate the spacecraft along its orbital trajectory (they rely upon the same engineering subsystem for controlling the spacecraft "attitude and articulation").

As mentioned above, software development at the Lab has professionalized over the years and now includes a dedicated organizational unit, the "Software Development Section,"[22] through which all programmers are employed. While Frank is now a member of this section, his career began before this unit was formed, and before software development was professionalized as its own subdiscipline of engineering work at the Lab (which lagged a bit behind professionalization of software in other engineering industries).

There was a time, not very long ago, when the very category of software did not demand its own professional unit, since there was no clear line between hardware and software work. Software development work on the Mission typically continues to follow this older paradigm in which software developers are embedded within the domains where their software tools are applied, close to those working with the hardware this software commands, and to those who use the software. Within this paradigm, software developers like Frank often continue to work with a piece of software, continually developing, maintaining, and evolving a software application over the duration of its lifetime.

As Sarah had explained to me, this older paradigm was one of "adopt and adapt"—picking up software from earlier missions and then adapting them to meet the next mission's needs. In the newer paradigm, software is managed and controlled centrally and each of the missions on the Lab is treated as a kind of customer for a suite of "multi-mission" software tools. Careers like Frank's, which align with the life cycle of a piece of software, have become obsolete.

So too is the software that Frank works to develop and maintain. SDT is written in one of the earliest visual user interface languages, tcl/tk (pronounced "tickle TK"), and at the time of this research, the third-party group that developed this language was about to close up shop. In response to the approaching end of support for tcl/tk, Frank had been working over the past five years on a new Java version of SDT. This new version was to be rolled out to the Mission's community of scientists for designing their scientific observations. However, these users, many of whom are at other institutions across the United States and Europe, were not keen on the new Java version and were not adopting it into their workflow.

At the time of my fieldwork, the fate of this particular software tool hung in the balance. At a meeting convened by the heads of the Software Development Section, Frank presented the new Java version of SDT, comparing the two systems, presenting the associated costs, risks, and benefits of each. Sitting in on the meeting, I felt the weight of this decision. It felt as if I was attending a trial, with the software as defendant and Frank as a proxy to voice its self-defense. It was ultimately a decision made by the Section whether the Lab would invest more of Frank's work effort into this tool, or whether its development would cease, thereby cutting their losses and requiring that they accept the associated risks of running a system on an obsolete programming language.

At stake, however, was not just the future of this one software tool, but also the future of Frank's career—the two being tightly coupled. The investments that Frank has made over the years to further evolve and develop SDT have given him refined knowledge of tcl/tk. Yet this has come at a cost to his own continued professionalization. His Java version failed to be taken up by the science community in part because most of these scientists had developed their own glueware—scripts that link into SDT or even entirely new programs that interface their work and blackbox SDT from view. Another reason was that the new version, while updated to Java, did not live up to changing trends in interface design.

This decision therefore assessed not only the value of five years' worth of programming work that Frank had put into the tool, but also his ability to align himself with a newer programming language, one that is more up to date and relevant to other software projects at the Lab. As one younger developer put it, in the face of various upgrades, many engineers face the following dilemma: Why would you want to switch to a new programming language when you are "killer at" and "ten years deep" in another one? At the same time, he said, there is a risk that these more senior engineers are "gripping the casket for fear of death."

This gives a stark view into the ways that software, skills, and expertise obsolesce together. This linked obsolescence is often talked about as if people too obsolesce in the context of the Mission, as natural and inescapable as the process of aging itself.[23] However, if we too adopt this naturalizing of obsolescence we risk essentializing the narrative of software's evolution as one of progression. When interviewing Frank about SDT, I came to understand software's evolution as much more convoluted.

Frank understands the evolution of the SDT system as a series of investments in some kinds of work over others during its lifetime—investments both in his own work and in the work of the collective organization. This software tool had grown over time in much the same way that Sarah described the growth of glueware. Frank spoke about the organization's earlier decisions about SDT, discussing how the "value of time at different parts of the mission" had varied, in terms of "when they can waste time," and when they had to be more cautious about the investments they made in updating software. There are therefore signs in the SDT code of both the "accruing of small changes" and of "big jumps and extreme customizations." In the beginning SDT provided an interface that allowed users to write commands to turn the spacecraft and animated a simulation to show how their instrument field of view might track a body over time. But over time, SDT also enabled the creation of "macros" that allowed a scientist to reuse commonly repeated commands, and later enabled the development of "templates" and "modules," each with increasing levels of abstraction away from the commands.

As one example of this evolution, the "rings guys" (the scientists studying Saturn's ring structure) had not provided requirements early on for the SDT tool and had come to feel shortchanged later on when realizing it did not support their needs. Subsequently, some effort was put into developing modules for their work. Frank described the development of SDT as a gradual series of investments made in certain kinds of observations over the years. If the rings science team had previously mapped the rings by performing "radial scans"—tracking a particular radial spoke

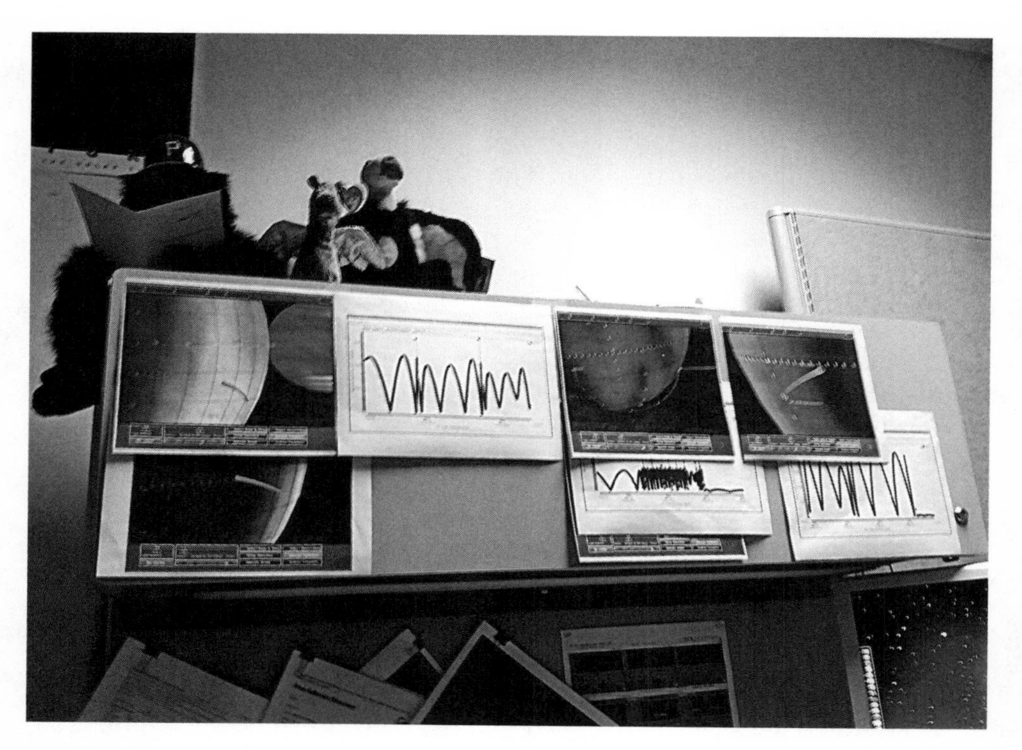

FIGURE 4: Frank's trophy shots.

of the rings from the center outward—then it continued to make sense to map radially. These radial scans became relatively routinized and embedded into SDT's modules. However, this development not only enabled certain ways of building commands together in blocks, it also made other ways of putting together those commands less visible or imaginable by instrument operators. Over time there emerged a tendency to forget how to work directly with the commands themselves.

Frank still receives special requests to craft custom pointing solutions "on the fly," to design observations for teams that they cannot achieve on their own. The file cabinets above Frank's desk were covered with screenshots from SDT showing the outcomes of this bespoke design work—a series of trophy shots for hard-won observations that required him to dip in and create customized strings of commands to achieve atypical observations with strange geometries (see figure 4).

These images evoke a metaphor for Frank's career. The needs of users have sedimented over time into routine observations and ways of putting commands together in such a way that articulates a bespoke trajectory for the evolution of SDT as a whole; such investments are simultaneously moves toward some ways of working and away from others. In a similar way, Frank's career has not stagnated but has continuously evolved toward niche knowledge work, at its own kind of bleeding edge, but in ways that have moved away from computing practices in the world around him. While his Java version worked exceedingly well for him—using a decision tree interface that allows him to plumb the depths of all prior observations ever designed to find that one time when the pointing was similar to a current request—it failed to work for the ways that other users had evolved their work practices.

For Frank, SDT is not separable from the history of its development alongside the history of the Mission and their mutual entanglement. As he responded to my question about the current task he was working on that day, he recalled stories about the tool, the various "wrinkles" they had encountered in the past or might encounter in the near future. He was working on one of his bespoke solutions, while explaining to me that some of the assumptions now embedded into SDT were soon to be turned inside out:

> So . . . let's see, other wrinkles that we came [to] along the way . . . not too much recently. Well, okay, we're going to have to go through [customization] for the ring guys because, for the proximal orbits and the F ring stuff, we're going to spend a lot of time where the spacecraft is basically inside the rings—[and because of] the orbit—[and the close] distance of the rings, the ansa target options don't work anymore.[24] That was just a limitation that we didn't recognize when the target option was defined. So I suspect we're going to have to go in and do some work on those to actually get a better algorithm.

In this quote, Frank is speaking about the final activities of the Mission, planned for 2016–17, when the spacecraft was to fly between the rings and the planet. Years of building up ways to point the spacecraft at the rings while flying outside of them would become completely obsolete, as the geometries of tracking the rings radially would be entirely different from the perspective of the spacecraft looking at the rings from within. Decisions that had essentially automated the process of designing observations from an external perspective would break when trying to design for a radically different perspective.

When looking toward this future, the ability for Frank to have a Java interface that would let him plumb the entire history of pointing designs on the Mission, even returning to the earliest days when they had not yet templatized and modularized the software, would be invaluable in these end days of the Mission, allowing him to think of potential solutions, treating almost any previous design as a template. Maybe the time that they tracked a moon as they traveled just inside of its orbit would be relevant? For Frank, this tool allowed him to think across different moments in the life of the Mission, the life of the software tool, and the trajectory of the spacecraft to recover specialized "paths" that SDT might still support. Its evolutionary trajectory as a software tool was not yet closed.

In this example we can see how Frank assesses the software and what it can do through a lens of its own histories, its past jumps, twists, and turns, and its potential future reconfigurations. Like the glueware that Sarah described covering over the seams between the various software programs in the organization, this tool has its own signs of earlier wrinkles in time, places where the weft of the fabric was pulled too tight or too loose, as the software tool was given more or less effort. We can see how Frank's work with code as a medium deals not only with the current performance of the code but also with its past and possible futures.

The Configuring of Code and Work

In both of these empirical stories we can see how working with legacy software invokes a relational perspective on the software for those who maintain it. Both stories deal with moments of software maintenance in the face of potential upgrades or migration to new systems—to manage change and loss. And in both cases it is not so much the code, but the relational assemblage of software and organizational work that these engineers must tweak and adapt in order to prepare for the future. These moments exemplify software work that is extra-algorithmic in that they go beyond making sense of the functioning of software or its proper execution or understanding what is inside the black box of software's operations. This includes work, for example, to understand what software an organization has in the first place or to identify what software might be able to do, or might once have been able to do, given particular organizational conditions.

These moments pertain to "software management," but I would also argue that a clear distinction between what software "does" and how it is "managed" is not easy to slice, since these are figured together. The software as a relational assemblage is not only what engineers and managers must wrestle with, but also a lens through which they understand their own organizational relations, as we see in the moment of Sarah invoking the "ouch" or the way that Frank figures the SDT tool itself as a kind of archive of Mission history. Software maintenance is thus made up not only of writing lines of code, but also of meetings, phone calls, laughter, and pain. All of these compose maintenance work, the work of keeping software present, knowable, up to date, and organizationally accountable. These empirical stories illustrate the work it takes to make software organizationally accountable so that software and organizational practice are open to revision and change.

This view into software maintenance reveals the ways that code and work are figured together—the relationality of work practices, processes, and people in the organization and the relationality between or within particular software applications in the infrastructure compose overlapping imaginaries. I saw this configuring—or figuring together[25]—of work and code throughout my fieldwork, from the ways that people negotiated epistemic and disciplinary boundaries through debates about software to the references to the power that comes with having a server inconveniently taking up room under one's desk.

Software, like other technologies explored in STS literature, is sociomaterially constituted. This case builds upon scholarship in STS, as well as STS-cross-cutting areas of software studies and human factors of computing, that has problematized software's assumed immateriality (Mackenzie 2006; Dourish 2014; Mazmanian et al. 2014) and explored the gap between idealizations of software code as abstraction and the embodied realities of computational work (e.g., Button and Dourish 1996; Galloway 2004; Star 2002; Kelty 2005; Coleman 2013; Bucher 2012). These authors show that, when we look closely at software work, we see that the materiality of software cannot be separated from its discursivity and performativity. Code is never just code, but exists within a "co-constitutive milieu of relations between humans and their technical supports" (Bucher 2012).

What emerges from this literature is a sense of software as inherently relational, hybrid, and dynamic. This is certainly the view on software that I encountered in the field, where the inability to distinguish between code that is running on a machine and actions performed by another human being in the organization frequently arose. I heard complaints about precisely this kind of error, of being treated like a robot by a colleague who did not realize that the information provided in a form would be read by a human. Even within one's own practice, the parts of code that do not run perfectly without a little help are often treated as part of the software. There is code that is just about to be written, but in the meantime can be finagled by running the code in parts and "holding its hand" in between. In the example of Frank's work, the close coupling of his work and the software can certainly be seen as a kind of hybridity. This is in part why I, along with other scholars of software, such as Webmoor and Mackenzie, have often referred not to code but to code work or codings as the object of study—drawing attention to the ongoing work of making any piece of code work (Webmoor 2014; Mackenzie 2006).

Much of the recent work in digital STS that addresses code as a material object of analysis has focused on the algorithm. Algorithms are seen to be both more than software, in that they can exist outside of it as "encoded procedures [that] transfor[m] input data into a desired output" (Gillespie 2014, 167), while at the same time fundamental to computational media that are deemed to be "algorithmic machines" (Gillespie 2014). These analyses of the algorithm make a similar point to the one developed here, arguing that algorithms are always part of larger assemblages, embedded within relations that require upkeep. Algorithms are "rarely stable and always in relation with people" (Neyland, cited in Crawford 2016, 81) and "must be understood in situated practices—as part of the heterogeneous sociomaterial assemblages within which they are embedded" (Introna 2016). All algorithms and code require upkeep, including the hygienic work users perform to groom their usage patterns in order to domesticate the algorithm (Gillespie 2014). An algorithm needs its suites of software tools and its maintainers in order to be enacted as "automatic and untarnished" (Gillespie 2014, 179).

It is thus precisely this embeddedness of the algorithm "into practice in the lived world" that brings us back to software (Gillespie 2014, 183). While my aim is not to draw a stark distinction between software and algorithm as concepts, what I find useful about software as a *unit of analysis* is that it tends to escape the gravitational pull of debates about code's abstract purity or the algorithm's deterministic agency. Instead of contending with that particular imaginary, software helps us begin from the relational. According to Ensmenger's historical analysis of the term, the definition of software came to encompass an understanding of code that embraces this relationality, expanding to include everything but hardware (Ensmenger 2016). "Software is where the technology of coding meets social relationships, organizational politics, and personal agendas" (Ensmenger 2011, 8) and remains "inescapably local and specific, [and] subject to the particular constraints imposed by corporate culture, informal industry standards" (Ensmenger 2011, 8). I agree with Ensmenger in suggesting that there is also *analytical value* in keeping the notion of software broad enough to encompass forms of work that go into making software work.

So, while I agree that the algorithm is a powerful unit of analysis that can be used to draw out concerns about power and economic relations in an era when algorithms are a key commodity, I think it is important not to lose sight of software as another kind of starting point for STS studies and theorization. These STS analyses succeed in establishing a healthy skepticism toward claims about the algorithm's "agency or power . . . to operate alone" (Neyland 2016, 52). Yet despite attempts to move away from a fetishized notion of the algorithm as generalized abstraction, the implicit presentism of the "in-the-making" or "in relation" analytical moves seems to bias the STS analyst toward conceptualizations of code as a performative object rather than as something that lingers within the organization and decays over time. It is precisely the way that software speaks to a body of code and its upkeep, to all the "soft" parts of computation, to that which is not commodified, that enables a perspective on the lived temporalities of code that are often missed in discussions of the algorithm.

Unruly Bodies of Code in Time

Software maintenance work not only deals with present instantiations of the relational assemblage but also involves the rebinding of social relations through both cuts and restitching of interconnections in the assemblage, pulling upon threads that connect the past and the future of these bodies of code. To make software work is to work with software's histories and anticipated futures, crisscrossing back and forth across different pieces of code written at different times. Working with code as a material is inseparable from working with the social relations within which it is embedded, and in many cases the histories of those social relations that are in turn embedded within the code.

In the empirical moments of maintenance work performed by Sarah and Frank, we can see how these longer time frames of lived temporal relations animate systems work. In both stories, there is first an attempt to disembed software from these lived temporalities—to treat code in a reified and ahistorical way—and in both cases this attempt fails. We see Sarah attempt to capture the extant software in database form, a catalog of wares that overflows, that cannot be contained by the Excel grid. And we see how Frank, in his attempt to modernize an old piece of software, attempts to place the two pieces of software side by side, to measure them according to performance, costs of upkeep, and risk. But the historicity of the SDT artifact turns out to exist not only in the obsolete programming language, but also in outdated aesthetics of interaction design and in the confounding ways that Frank's lived experiences of the tool's evolution diverge from those of its users. What surfaces in these attempts is a view into code as something that is related to not only its present state but also its histories, anticipated futures, and counterfactualities—what it was, might be, or might once have been able to become.

Software, as a body of code, binds the social relations of the organizational body together. As one young engineer, Toby, described it, the software at the Mission is a kind of "patchwork."

> I hear people say "the legacy code has been around for more than 40 years." And that is a really long time. When I hear that and they say "and it's been added to and modified as requirements have been constantly been changing." And it's just been more, in my mind, [a kind] of patchwork software that started out probably with certain goals and requirements in mind and then grew in a piecewise fashion to accommodate future needs. Whereas the architecture before is not, you know, 40 years later, the architecture that will still hold for future missions after that because so much has changed about what we're able to achieve and how we do our work and just regular information processing and computing. All that stuff has changed so much. So when I think of the legacy, that is really what I think of. And so, it can be—uhhh, there are many careers that it, that the legacy software spans. Many different points of view. Many different missions.

While the term "patchwork" in software development discourse is often used in a more derogatory way, this engineer inverts those values and offers up a sense of patchwork code as a source of richness—a richness of history that provides diversity of work as well as a sense of continuity with the legacies left by people who worked on the Mission years before. Working with code involves working across time and building legacies and inheritances that serve as connections not only between practice and the functions of a tool but between different ways of working with systems over time.

These ways of working with code trouble the valuations that we place on current software development. As the young engineer Toby pointed out, the new software tools that are replacing the Mission's legacy systems are more "extensible" and "evolvable," but at the same time he doubts that any of these software tools will outlast the lifetimes of legacy software used at the Mission, which in some cases are over 40 years old. He points to the irony of calling a system more "extensible" than one that has already *been extended* for such long periods. These moments of software maintenance speak to the ways that the materialities of code are temporal not only in how they represent, inscribe, or mediate temporality (its rhythms or scales) in the present, but also through regimes of disciplining temporal relations to code. Methodologies for managing software change shape ideas of how software ought to evolve and in turn how subjects of software regimes should relate to time.[26]

So, while trends in software management that aim to make software more maintainable and extensible (through modularity or other strategies to make code easier to refactor) are usually considered part of best practices, they are simultaneously ways of shifting the lived temporal relations with code. As one more senior engineer described it, the new software tools make work very "cookie-cutter," making him interchangeable with any other coder, whereas the diversity of working across legacy systems at the Mission provides pleasure in retracing the work done by other engineers, brings a sense of care or duty in tending to the longevity of the mission, and is also a source of mystery and excitement, encountering older ways of working in the same way that we encounter history in a museum. This speaks to the unseen costs of newer paradigms of disciplining work with code, where code is modularized and commodified. The justifications for this are always aimed at particular anticipated futures and the need for systems to evolve.[27] But they also have impacts on the quality of work for those who live with these systems *as* they evolve.

As Bowker suggests, computing and organizations do not have separable histories: "We make an analytical error when we say that there is programming on the one side with its internal history, and organization theory on the other with its own dynamic" (Bowker 2005, 31). If this is true at the level of the field of computing, it is all the more true within an organization where the code becomes a kind of text that can be read for those entangled histories, not only by the STS analyst but by the coders whose work with old code traces back and forth across them. From this perspective, there is no such thing as code, but only "codings" and the "biograph[ies] of software . . . mingl[ing] with the lives of programmers" (Mackenzie 2006, 171).

When code is treated as a temporal object, there is a tendency to focus on issues of speed—how the infinitesimal duration of algorithmic processing or its rates of change pattern interactions in use. This seeming instantaneity is often used to define code as a medium despite all the ways that code lingers and persists. In infrastructural analyses of software, temporality is understood more broadly (Jackson et al. 2011), encompassing, for example, how software representations shape perceptions of time, how the rhythms of software upgrades shape organizational practice, or how biographical rhythms of careers present challenges to collaboration. Yet these still tend to be discussed in terms of how software's manifestations of time become problems for coordination in the present. What software maintenance brings forward is also how software is *lived with* over time. Different regimes for disciplining unruly bodies of code over time structure participation within the organization and shape "how people understand their lives and collective histories" (Bowker 2007, 22). Histories and anticipated futures of a body of code are figured together with histories and anticipated futures of work—work done together with others.

Conclusion

> *Old systems have a name. They are called "legacy systems." In the regular world, "legacy" has an aura of beneficence. . . . In computing, however, "legacy" is a curse. A legacy system is a lingering piece of junk that no one has yet figured out how to throw away. It's something to be lived with and suffered.*
>
> **—Ellen Ullman (2012, 118)**

It is a truism that someone else may have to live with the code you are writing. Legacy systems are everywhere, in long-standing institutional systems at banks, hospitals, schools, and so on but also in bits and pieces that have been cannibalized into new software. And old software, while it may become embedded as infrastructure, does not necessarily remain there. Understanding how software is lived with over time is as much about the forms of forgetting of what software an organization has, as the ways that software's histories can re-present themselves, sometimes uncannily, reminding the organization of what it has forgotten, what it once knew.

This is a mundane forgetting of code that happens in any organization over time, as the software and its performances become taken for granted, and as pieces of code are no longer actively touched by programmers.[28] Through continual updates and upgrades, new software development promises an organization eternal youth, keeping computational work current. Yet this can work only by submerging old code into irrelevance or pathologizing it as a kind of blemish on the organization, making legacy software work into "dirty work," and by doubling down on the newest tools and latest platforms and granting high status to only cutting-edge computational work. While very few software engineers would insist upon an immaterial view of software, the erasure of certain forms of computational labor is achieved in practice through the diminishment of histories and the pathologizing of the particularly messy materialities of old code.

These erasures of history are something that STS work on the digital has insufficiently addressed. While STS includes and draws upon historical perspectives on software and computing, taking a long view of software as a changing and unfolding set of relations,[29] when STS scholars examine software as a relational network of practices, they tend to leave history or the long view of software aside, despite the fact that these histories animate certain kinds of practice. When focusing on the situated enactments of software, they adopt a transhistorical view on software's materiality, neglecting the relevance of software's histories, life stories, and lived temporalities, its forms of decay, aging, and forgetting.

The tremendous challenge of keeping software present—from the work of making software organizationally accountable to the work of keeping software current—is addressed through disciplinary regimes that also impact whose work will be considered vital to the system and who can be laid off, which systems will stay and which will be replaced. If we are to understand how software's materiality shapes the politics of design of computational systems, we need to attend not only to the challenges of tracing software's material performativity in the present, but also to the challenges of keeping software present, of dealing with the continual forgetting of software's pasts and histories.[30] The question of the politics of computing may then be a question of who *can* afford to forget, or what does this forgetting cost to whom?

This chapter has shown how software and computational infrastructure are lived, enlivened, and inhabited. It has taken up the practices that make up a particular disciplinary regime that governs how software is made and what kinds of relationships to code are constituted as proper. It has dealt with the lived experience and work of translating across these different disciplinary regimes and surviving across shifting programming paradigms. The case of an aging, degrading infrastructure challenges our cultural overvaluation of software's present by looking at what happens within these entangled lifetimes.[31] The Mission presents a case where software is not considered timeless but is rather connected to particular times, events, people, and places. It is the very untimeliness of legacy software that allows it to speak back to the present, to consider the temporality of code from outside of its current forms of commodification, revealing software systems to be temporal assemblages that gather up history. Lives, careers, lifetimes, generations, paradigms, languages, lifetimes of machines are things that come and go. In their passing, a backward glance shows us the ways that these temporal relations are enacted with and through valuations of different forms of work.

Acknowledgments

I would like to acknowledge the enormous generosities of people who helped provided feedback and suggested revisions during the writing of this chapter. Janet Vertesi, Drew Paine, Laura Houston, and Elisa Arond each provided invaluable guidance. Ethnographic fieldwork was conducted in conversation with the distributed research efforts of the Spaceteams Research Group led by Janet Vertesi and including Paul Dourish, Melissa Mazmanian, Matthew Bietz, and David Reinecke, supported by NSF Socio-Computational Systems Grant 0968616 and the Consortium for the Science of Sociotechnical Systems.

Notes

1. See http://talkingpointsmemo.com/news/federal-government-ancient-technology.
2. I refer throughout this chapter to both software and code. By software I refer not only to code as a textual artifact, but also to the work that goes into making code work (following Mackenzie 2006). This includes work both to achieve operational and functioning software and also to manage software change (software change requests and review, version management, accounts and documentation of extant tools, etc.). Others have a made similar distinction; for example, Berry (2016) refers to code as a textual object and software as a broader set of tools (e.g., operating systems, programming languages, software used to write other software) that together make up software systems within the organization. As he points out, code "should be approached in its multiplicity" not only as an object but as a set of social and textual practices of writing and reading code (36). Just as a body of code can be read "as a literature, a mechanism, a spatial form (organization), and repository of

social norms, values, patterns, and processes" (65), so too can software be examined not only for its functionality, but also in terms of its social, organizational, and discursive practices, its forms of accountabilities, disciplinary regimes, and values and norms.

3. A notable exception is recent work by Drew Paine that conceptualizes "software infrastructural vision" as a specific form of "professional vision" (citing Goodwin 1994) that works by disentangling these embedded relations.
4. In her book *Beamtimes and Lifetimes* (1988), Sharon Traweek develops a notion of lifetimes to denote the time in which the lives of people and machines unfold. She notes that there is a tension between the commodified relativistic time of "beamtime"—the finite allotments of time using particle accelerators granted to individual high energy physicists—and the nonrelativistic experiential time of lifetimes. I work with a similar tension here between the performative runtime of code and the experiential time of lived relations to bodies of software. Like Traweek, I see in the jokes and stories told about code how software comes to embody this lived sense of time.
5. For unpacking software as a temporal form, I have found the work of Elisabeth Freeman useful for thinking about the way that forms of media bind people together through shared timings and shared affective histories, and how certain forms of time, particularly nonsequential time, "fold subjects into structures of belonging and duration" (Freeman 2010, 11).
6. While it is impossible to fully anonymize the organization, the name of the Mission organization and lab facility have been omitted in order to minimize the potential of implications of this chapter affecting ongoing organizational work. Likewise, while it is impossible to fully anonymize the individuals at the Mission due to the singularity of their work, I have used some elements of pastiche personas, combining interactions and statements made by different individuals at the Mission in order to make attribution of quotes ambiguous. There is therefore not an exact congruence of any one person named here with actual individuals at the organization.
7. While, at the time of writing, the mission has already entered its "grand finale" and will end in only a few weeks' time, I speak in the present tense about systems and organizational practices that are presently ongoing.
8. This is in contrast to the "on-board system," which comprises the spacecraft hardware and "flight software" and allows it to interpret incoming commands, automate safety measures, and relay data from its sensors back to earth.
9. Charlotte Linde (1993) has shown how life stories are told and retold to develop a sense of coherence about our identity or membership to a group. Life stories are used to negotiate this sense of self with others and to demonstrate one's worthiness to the social group. Life stories can change over time, organizing our understanding of the past without necessarily resolving into a single narrative. In a similar way, stories of the "life of the craft" or the "life of the mission" or the "life of the software" organize the past and orient to the future in ways that reveal what is at stake in maintaining a sense of coherence regarding the social relations of the organization.
10. In this way, the work I am speaking about could be seen as a form of "articulation work" (Star and Strauss 1999).
11. This set of concerns about the way organizational software changes over the longue durée and is remembered or forgotten resonates with concerns in infrastructure studies (Star and Ruhleder 1996), where infrastructure is defined in part by its embeddedness, allowing it to remain invisible so long as it runs smoothly, and to become visible upon break down. While I have dealt with these concerns of invisible software maintenance work elsewhere (Cohn 2016), here I focus less on the varying visibilities of different kinds of software work and more on the various temporalities of software change at play in making software organizationally accountable and tractable to change. It is not only that bodies of code or software diagrams serve as a kind of mnemonic device for how the software has evolved over the years or as archive of local organizational history (Bowker 2005), but also that the histories of both software and the organization are figured together along with speculations about their possible futures.
12. The Mission completed its first science task, capturing observations of Jupiter as it flew by on its way to Saturn. This "flyby" led to a realization of just how challenging it was to coordinate science and initiated a flurry of software development to support scientific observation design, coordination, and implementation. Sarah explained that the Jupiter flyby was the first time that the Mission fully realized the software that was needed to conduct science at Saturn. The flyby of Jupiter had given them a taste of what they needed and anxiety was high about the arrival to Saturn, prompting the creation of the diagram.

13. Level III is the highest designated engineering position at the Lab. At the time of this study, the Mission had only three level IIIs at the time of this study: Sarah in charge of software, one in charge of science, and another in charge of the spacecraft.
14. Change management review is a rigorous process through which every change to a piece of operational software is evaluated in terms of cost, benefit, and risk before it is performed.
15. Category A is mission critical in that if an error were introduced into the software tool it could lead to "mission failure," "degrade mission objects," or "result in loss of mission critical events." Category B is for software that is not mission critical but could harm the conduct of science; C is for software that might impinge on the "efficiency" of work but for which there are "operational workarounds" that still allow for timely conduct of science. While at NASA category A stands for systems on manned missions where an error could result in loss of life, at the Lab there are no manned missions.
16. That is said to mitigate risk because the individual who wrote and uses a script or hack is the person who would be impacted if it breaks, and the risk is low that it could impact others. In fact, this is not the case since these scripts, utilities, and hacks circulate throughout the organization—sometimes even crossing interinstitutional and international boundaries and becoming essential tools over time. When this happens, it is possible for software to be granted a new status. As one scientist told me about a tool he developed and shared and became commonly used, he was reprimanded for violating rules against circulating technical information across national boundaries (even within an internationally collaborative project); but by that point "the cat was out of the bag."
17. Maria Puig de la Bellacasa (2009) suggests that we might explore "meanings of touch" for the politics of knowledge production, as "to think with touch has a potential to inspire a sense of connectedness that can further problematize abstractions and disengagements of (epistemological) distances—between subjects and objects."
18. See note 5 for discussion of Beth Freeman's notion of the "bind" (Freeman 2010).
19. I further discuss a variety of these tactics of infrastructural care work in Cohn (2016).
20. This is in contrast to in situ instruments that do not need to "point" since they have no specific field of view. The orbit of the spacecraft is designed as a sequence of segments that are then divided up among the various instrument teams. For any given segment, one instrument team is designated as "primary" and will design a set of observations using SDT to model the specific geometries (shape, size, location, etc.) of the "target bodies" they wish to observe. Other instruments can also make secondary observations, "riding along" and collecting data as the spacecraft flies through the Saturn system, but these data may or may not be meaningful depending on how the spacecraft is pointed by the primary instrument.
21. Moving at tens of kilometers per second.
22. The Lab is organized in a "matrix structure" that is made up of sections and projects. Each employee belongs to a section, one for management, navigation, mission planning, etc., and to a project such as a mission.
23. Ellen Ullman writes in a memoir of her work in the 1980s in Silicon Valley of being interviewed for a job "with a man who had made peace with his own obsolescence"—a wonderful phrase for the giving into the inevitable (Ullman 2012, 118). I too met engineers who seemed to have made peace with their own obsolescence, particularly because of the gratitude that came from a long career with the Mission.
24. "An ansa is the portion of a ring that appears farthest from the disk of a planet in an image. This is the location in an image where we see the finest radial resolution on a ring. The word comes from the Latin word for 'handle,' since the earliest views of Saturn's rings suggested that the planet had two handles extending out on either side" (http://pds-rings.seti.org/glossary.html).
25. This work of configuring, understood as figuring with, is explored through other material from the same empirical case in Mazmanian et al. (2014).
26. In fact those engineers who seem to value working with legacy code too much are pathologized as backward, regressive, or nostalgic, as I explore further in a special issue on repair for *Continent* (Cohn 2017).
27. See Steinhardt and Jackson (2015) and Mackenzie (2013) for discussions of how programming and infrastructural work are animated by regimes of anticipation.
28. It is this kind of software that we might consider a form of infrastructure in the sense put forward by Star and Ruhleder as that which becomes embedded and largely invisible (Star and Ruhleder 1996). This is also a form of forgetting that Chun points to in her historical analysis of the development of

computer science as a discipline (Chun 2005) when she calls out software's alleged transparency, its ability to be treated as a neutral mediating window into other phenomena, as a "compensatory gesture," and as an ideological achievement accomplished progressively since the 1950s during which time software's layers of abstraction have been built up while simultaneously being collapsed and forgotten. Yet in the case discussed above, we can see how this form of forgetting is not performed once and for all within the discipline, but must be continuously enacted within software projects. It is interesting to note as well that the gendered nature of this forgetting within the history of the discipline (the collapsing into automation of work historically performed by women) is also repeated in these mundane forms of forgetting within a project. The work of maintaining legacy software systems is also feminized and marginalized.

29. Notable examples include work that has considered infrastructure longitudinal, for example, Williams and Pollock's consideration of the "biography" of computational artifacts (2009) or Ribes and Finholt on the "long now of infrastructure" (2009). Other examples include media archaeological modes of textual analysis of software, such as Kirschenbaum's "forensic imagination" (2008) or Terry Harpold's "ex-foliation" (2009), as ways to examine the many sedimented layers in software media. What is less well understood however is how a longitudinal perspective or forensic and historicizing view of software is taken up in organizational practice as well.
30. Rather than theorizing software through "transhistorical truths" that stand apart from its evolutions, resisting decay, we can see software as layered with histories that never fully determine its relations (Philip 2009). This has implications for what it means to work with software as a temporal medium. Instead of thinking about the technological stack, a spatial metaphor that dominates our thinking about the archaeological nature of technologies evolving over time, thinking about folds, enfoldments, and translations becomes a productive way of considering software materialities.
31. This framing of the case of legacy as one of untimely computing that can speak back to the present is inspired by Wendy Brown's discussion of Nietzsche's "untimely" in her book *Edgework* (Brown 2009).

Works Cited

Beaulieu, Anne. 2010. "Research Note: From Co-location to Co-presence: Shifts in the Use of Ethnography for the Study of Knowledge." *Social Studies of Science* 40 (3): 453–70.

Berry, David. 2016. *The Philosophy of Software: Code and Mediation in the Digital Age*. New York: Springer.

Bowker, Geoffrey C. 2005. *Memory Practices in the Sciences*. Cambridge, MA: MIT Press.

——. 2007. "The Past and the Internet." In *Structures of Participation in Digital Culture*, 20–36. New York: Social Science Research Council.

Bowker, Geoffrey C., and Susan Leigh Star. 2000. *Sorting Things Out: Classification and Its Consequences*. Cambridge, MA: MIT Press.

Brown, Wendy. 2009. *Edgework: Critical Essays on Knowledge and Politics*. Princeton University Press.

Bucher, Tania. 2012. "Programmed Sociality: A Software Studies Perspective on Social Networking Sites." Doctoral dissertation, Faculty of Humanities, UiO.

Button, Graham, and Paul Dourish. 1996. "Technomethodology: Paradoxes and Possibilities." In *Proceedings of the SIGCHI Conference on Human Factors in Computing Systems*, 19–26. New York: ACM.

Chun, Wendy Hui Kyong. 2005. "On Software, or the Persistence of Visual Knowledge." *Grey Room* 18:26–51.

Cohn, Marisa Leavitt. 2016. "Convivial Decay: Entangled Lifetimes in a Geriatric Infrastructure." In *Proceedings of the 19th ACM Conference on Computer-Supported Cooperative Work & Social Computing*, 1511–23. New York: ACM.

——. 2017. "'Lifetime Issues': Temporal Relations of Design and Maintenance." *Continent* 6 (1): 4–12.

Coleman, Gabriella. 2013. *Coding Freedom: The Ethics and Aesthetics of Hacking*. Princeton, NJ: Princeton University Press.

Crawford, Kate. 2013. "Can an Algorithm Be Agonistic? Ten Scenes about Living in Calculated Publics." *Governing Algorithms*. http://governingalgorithms.org/.

——. 2016. "Can an Algorithm Be Agonistic? Ten Scenes from Life in Calculated Publics." *Science, Technology, & Human Values* 41 (1): 77–92.

de la Bellacasa, Maria P. 2009. "Touching Technologies, Touching Visions. The Reclaiming of Sensorial Experience and the Politics of Speculative Thinking." *Subjectivity* 28 (1): 297–315.

Dourish, Paul. 2014. "NoSQL: The Shifting Materialities of Database Technology." *Computational Culture* 4. http://computationalculture.net/no-sql-the-shifting-materialities-of-database-technology/.

Ensmenger, Nathan L. 2011. "Is Chess the Drosophila of AI? A Social History of an Algorithm." *Social Studies of Science* 42:5–30.

——. 2012. *The Computer Boys Take Over: Computers, Programmers, and the Politics of Technical Expertise.* Cambridge, MA: MIT Press.

——. 2016. "When Good Software Goes Bad: The Surprising Durability of an Ephemeral Technology." *The Maintainers, Stevens Institute of Technology.* https://larlet.fr/static/david/blog/ensmenger-main tainers-v2.pdf.

Freeman, Elizabeth. 2010. *Time Binds: Queer Temporalities, Queer Histories.* Durham, NC: Duke University Press.

Galloway, Alexander R. 2004. *Protocol: How Control Exists after Decentralization.* Cambridge, MA: MIT Press.

Gillespie, Tarleton. 2014. "The Relevance of Algorithms." In *Media Technologies: Essays on Communication, Materiality, and Society,* edited by Tarleton Gillespie, Pablo Boczkowski, and Kirsten Foot, 167–95. Cambridge, MA: MIT Press.

Goodwin, C. 1994. "Professional Vision." *American Anthropologist* 96 (3): 606–33.

Graham, Stephen. 2005. "Software-Sorted Geographies." *Progress in Human Geography* 29 (5): 562–80.

Haraway, Donna. 1997. *Modest_Witness@Second Millenium. FemaleMan_Meets_OncoMouse: Feminism and Technoscience.* New York: Routledge.

Harpold, Terry. 2009. *Ex-Foliations: Reading Machines and the Upgrade Path.* Vol. 25. Minneapolis: University of Minnesota Press.

Introna, Lucas D. 2016. "Algorithms, Governance, and Governmentality on Governing Academic Writing." *Science, Technology, & Human Values* 41:17–49.

Jackson, Steven, David Ribes, Ayse Buyuktur, and Geoffrey Bowker. 2011. "Collaborative Rhythm: Temporal Dissonance and Alignment in Collaborative Scientific Work. In *Proceedings of the ACM 2011 Conference on Computer Supported Cooperative Work,* 245–54. New York: ACM.

Jasanoff, Shiela, and Sang-Hyun Kim. 2009. "Containing the Atom: Sociotechnical Imaginaries and Nuclear Power in the United States and South Korea." *Minerva* 47 (2): 119–46.

Kelty, Christopher. 2005. "Geeks, Social Imaginaries, and Recursive Publics." *Cultural Anthropology* 20 (2): 185–214.

Kirschenbaum, Matthew G. 2008. *Mechanisms: New Media and the Forensic Imagination.* Cambridge, MA: MIT Press.

Linde, Charlotte. 1993. *Life Stories: The Creation of Coherence.* New York: Oxford University Press.

Mackenzie, Adrian. 2006. *Cutting Code: Software and Sociality.* Vol. 30. New York: Peter Lang.

——. 2013. "Programming Subjects in the Regime of Anticipation: Software Studies and Subjectivity." *Subjectivity* 6 (4): 391–405.

Mazmanian, Melissa, Marisa Cohn, and Paul Dourish. 2014. "Dynamic Reconfiguration in Planetary Exploration: A Sociomaterial Ethnography." *MIS Quarterly* 38 (3): 831–48.

Neyland, Daniel. 2016. "Bearing Account-able Witness to the Ethical Algorithmic System." *Science, Technology, & Human Values* 41 (1): 50–76.

Paine, Drew, and Charlotte Lee. 2017. "'Who Has Plots?' Contextualizing Scientific Software, Practice, and Visualizations." *Proceedings of the ACM Human-Computer Interaction Conference* 1:85. https://doi .org/10.1145/3134720.

Philip, Kavita. 2009. "Postcolonial Conditions: Another Report on Knowledge." Irvine Lecture in Critical Theory, University of California, Irvine, March 18, 2009.

Ribes, David, and Thomas A. Finholt. 2009. "The Long Now of Technology Infrastructure: Articulating Tensions in Development." *Journal of the Association for Information Systems* 10 (5): 375–98.

Star, Susan Leigh. 2002. "Infrastructure and Ethnographic Practice: Working on the Fringes." *Scandinavian Journal of Information Systems* 14 (2): 6.

Star, Susan Leigh, and Karen Ruhleder. 1996. "Steps toward an Ecology of Infrastructure: Design and Access for Large Information Spaces." *Information Systems Research* 7 (1): 111–34.

Star, Susan Leigh, and Anselm Strauss. 1999. "Layers of Silence, Arenas of Voice: The Ecology of Visible and Invisible Work." *Computer Supported Cooperative Work* 8 (1–2): 9–30.

Steinhardt, Stephanie B., and Steven Jackson. 2015. *Anticipation Work: Cultivating Vision in Collective Practice.* Vancouver: ACM.

Traweek, Sharon. 1988. *Beamtimes and Lifetimes: The World of High Energy Physicists.* Cambridge, MA: Harvard University Press.

Ullman, Ellen. 2012. *Close to the Machine: Technophilia and Its Discontents.* London: Macmillan.

Vertesi, Janet. 2014. "Seamful Spaces: Heterogeneous Infrastructures in Interaction." *Science, Technology, & Human Values* 39 (2): 264–84.

Webmoor, Timothy. 2014. "Algorithmic Alchemy, or the Work of Code in the Age of Computerized Visualization." In *Visualization in the Age of Computerization*, edited by Annamaria Carusi, Aud Sissel Hoel, Timothy Webmoor, and Steve Woolgar, 19–39. New York: Taylor & Francis.

Williams, Robin, and Neil Pollock. 2009. "Beyond the ERP Implementation Study: A New Approach to the Study of Packaged Information Systems: The Biography of Artifacts Framework." In *ICIS 2009 Proceedings*, 6. New York: IEEE.

Introduction
Visualizing the Social

Yanni Loukissas

Making use of maps, timelines, trees, and networks, the authors in this section are responding to a mounting call in STS (Latour 2008a; Galison 2014) for new forms of scholarship that can visualize the social. They have found ways of creatively identifying and presenting evidence of human and nonhuman relationships imprinted in digital data: either collected through purpose-built monitoring tools or extracted from the residue of regular computer use.

Practices of visualization, which can encompass a range of representational techniques designed to "help people make sense of data" (Kennedy et al. 2017), have long been a subject of study in STS (Lynch and Woolgar 1990; Coopmans et al. 2014). Innovations in data representation play a central role in the history of science and technology. Visualization helped a Renaissance astronomer track celestial changes (Biagioli 1993), an Enlightenment economist account for the financial health of societies (Tufte 2001), and a Victorian doctor diagnose the source of an urban epidemic (Johnson 2007). The power of contemporary technoscience depends on its ability to "draw things together" through tools that allow data from across temporal and spatial divides to be jointly analyzed (Latour 1990).

In recent years, scholars have written about a new wave of digital visualization tools, which have transformed professional relationships and identities across a range of domains including engineering (Downey 1998), life science (Myers 2015), nuclear weapon design (Gusterson 1998), architecture (Loukissas 2012), and other contemporary professional cultures (Turkle 2009). But despite the long-standing interest of STS scholars in visualization, only recently has the practice itself been recognized as a potent method for the field (Vertesi et al. 2016). This portion of the book brings together a set of speculative and experimental essays that consider how visualization might not only enhance what STS scholars already do but also lend shape to the social: a subject that we know inherently resists stabilization and thus static representation (Latour 2008b).

Today, in 2019, it is a hallmark of the digital turn in STS that scholars are embracing not only new subjects—data and software, digital infrastructures, reconfigured bodies, and renewed global connections, all topics broached in this volume—but new digital methods, such as visualization. For STS, taking up digital visualization as a method means learning to identify new forms of evidence manifest in digital data. As potential traces of the social, such data offer openings for novel perspectives on both human and nonhuman action at a scale and resolution previously not possible (Rogers 2013). The abundance of digital communications, in particular, has been framed as a staggering opportunity, if also a potential ethical minefield (boyd and Crawford 2012), for social research. But digital data are not straightforward representations of social relations; they are complex "assemblages" (Kitchin and Lauriault 2014) that must be laboriously unpacked from their enveloping infrastructures (Bowker et al. 2009). Transforming digital ephemera into "alleged evidence" (Borgman 2015) for STS means acknowledging the construction of these data as well as their limits, including how they differ from traditional grounds for claims making embraced by the field, such as interviews, direct observations, or archival sources. The authors included here handily demonstrate how to do this, while also addressing a number of reflexive questions about visualization: How do different types of digital data structure social inquiry? Do visualizations, which make data legible, also transform those data? Who is empowered to make data and visualizations, and what kinds of audiences can decipher them? These questions find concrete form in the hands-on engagements with visualization in the chapters that follow.

Cardoso Llach is the most intentional in his use of visualization as a means of claims making. He has built his own custom visualization tools to illuminate the social in what he calls "digital traces" found in architectural practice: "vestiges of sociodigital transactions that are typically hidden from view and discarded from the end result." The resulting images are both aesthetically evocative and analytically precise: revealing the "sociodigital infrastructures of contemporary design practices." Meanwhile, Salamanca's use of visualization is more pragmatic. It is a means of measuring *social viscosity*, "the resistance to social action flow elicited by human and nonhuman actors acting concurrently." His visualizations are a means to an end: portraying the social as an identifiable phenomenon, which interactive artifacts can be designed to intervene into. Munk et al. delve into the context of visualization practice itself. Their chapter asks us to reflect on the ways in which visualization shapes relationships between scholars and subjects. As a counter measure, they suggest "participatory" approaches to visualization, with the promise of bringing about a kind of "common world." Finally, Venturini et al. ask us to step back even further, to consider how well digital methods—visualizations of networks in particular—actually map onto conceptions of the social developed for actor-network theory, a field-defining technique for STS that illuminates the associations between actants, human and nonhuman (Latour 2008b).

These varied approaches offer a glimpse of what the social might become, when seen through the lens of visualization. Moreover, the authors are implicitly asking what an STS researcher becomes when they adopt new forms of digital data as a way of seeing (Passi and Jackson 2017). Cardoso Llach inhabits a dual role as a designer of visualizations as well as a scholar of design: effectively demonstrating how visualization can be enrolled in participant observation. Salamanca models how data might help STS researchers intervene in social relations, by incorporating data for analysis and prediction. Munk et al. also hope to reshape social relations, but using a reflexive approach; they use visualization to break down the boundaries between scholars and the social groups they study. In contrast to the others, Venturini et al. ask how STS researchers can resist changing too much, by staying true to the deep roots of the field, even while opportunistically making use of available digital data.

These answers help us understand how a variety of ways of doing visualization might be applicable in STS. They are by no means the only answers (also see the chapters by Calvillo and Loukissas in this volume), but they deftly illustrate what it takes to successfully navigate the obstacles of methodological invention. The next generation of scholars might choose to follow one or more of these established channels or steer off on their own, with these contributions in mind as useful way-finding guides.

Works Cited

Biagioli, Mario. 1993. *Galileo, Courtier: The Practice of Science in the Culture of Absolutism*. Chicago: University of Chicago Press.

Borgman, Christine L. 2015. *Big Data, Little Data, No Data: Scholarship in the Networked World*. Cambridge, MA: MIT Press.

Bowker, Geoffrey C., Karen Baker, Florence Millerand, and David Ribes. 2009. "Toward Information Infrastructure Studies: Ways of Knowing in a Networked Environment." In *International Handbook of Internet Research*, edited by Jeremy Hunsinger, Lisbeth Klastrup, and Matthew Allen, 97–117. Dordrecht: Springer.

boyd, danah, and Kate Crawford. 2012. "Critical Questions for Big Data." *Information, Communication & Society* 15 (5): 662–79.

Coopmans, Catelijne, Janet Vertesi, Michael E. Lynch, and Steve Woolgar, eds. 2014. *Representation in Scientific Practice Revisited*. Cambridge, MA: MIT Press.

Downey, Gary Lee. 1998. *The Machine in Me: An Anthropologist Sits among Computer Engineers*. New York: Routledge.

Galison, Peter. 2014. "Visual STS." In *Visualization in the Age of Computerization*, edited by Annamaria Carusi, Aud Sissel Hoel, Timothy Webmoor, and Steve Woolgar, 197–225. New York: Routledge.

Gusterson, Hugh. 1998. *Nuclear Rites: A Weapons Laboratory at the End of the Cold War*. Berkeley: University of California Press.

Johnson, Steven. 2007. *The Ghost Map: The Story of London's Most Terrifying Epidemic—and How It Changed Science, Cities, and the Modern World*. London: Riverhead Books.

Kennedy, Helen, et al. 2017. "Seeing Data." http://seeingdata.org/.

Kitchin, Rob, and Tracey P. Lauriault. 2014. "Towards Critical Data Studies: Charting and Unpacking Data Assemblages and Their Work." Scholarly paper, Social Science Research Network.

Latour, Bruno. 1990. "Drawing Things Together." In *Representation in Scientific Practice*, edited by Michael Lynch and Stephen Woolgar, 19–68. Cambridge, MA: MIT Press.

———. 2008a. "A Cautious Prometheus? A Few Steps toward a Philosophy of Design (with Special Attention to Peter Sloterdijk)." Keynote lecture for the Networks of Design Meeting of the Design History Society Falmouth, Cornwall.

———. 2008b. *Reassembling the Social: An Introduction to Actor-Network-Theory*. Oxford: Oxford University Press.

Loukissas, Yanni Alexander. 2012. *Co-designers: Cultures of Computer Simulation in Architecture*. London: Routledge.

Lynch, Michael E., and Steve Woolgar, eds. 1990. *Representation in Scientific Practice*. Cambridge, MA: MIT Press.

Myers, Natasha. 2015. *Rendering Life Molecular*. Durham, NC: Duke University Press.

Passi, Samir, and Steven Jackson. 2017. "Data Vision: Learning to See through Algorithmic Abstraction." In *Proceedings of the 2017 ACM Conference on Computer Supported Cooperative Work and Social Computing*, 2436–47. New York: ACM.

Rogers, Richard. 2013. *Digital Methods*. Cambridge, MA: MIT Press.

Tufte, Edward R. 2001. *The Visual Display of Quantitative Information*. Cheshire: Graphics Press.

Turkle, Sherry. 2009. *Simulation and Its Discontents*. Cambridge, MA: MIT Press.

Vertesi, Janet, David Ribes, Laura Forlano, Yanni Loukissas, and Marisa Leavitt Cohn. 2016. "Engaging, Designing, and Making Digital Systems." In *The Handbook of Science and Technology Studies*, 4th ed., edited by Ulrike Felt, Rayvon Fouché, Clark A. Miller, and Laurel Smith-Doerr Felt, 169–94. Cambridge, MA: MIT Press.

Tracing Design Ecologies
Collecting and Visualizing Ephemeral Data as a Method in Design and Technology Studies

Daniel Cardoso Llach

What does the phenomenon of design look like from the perspective of data? This chapter explores ways to make visible the sociodigital infrastructures of contemporary design practices. Drawing on images of agency as relational and distributed, it proposes "design ecologies" as an analytical category to reflect on such practices, and explores it through a case study describing the collaborative coordination of a large architectural project.[1] The chapter discusses in detail my work to study this project as a participant observer defining and collecting digital traces of thousands of design conflicts reported during its coordination, and developing a series of interpretive visualizations of these data. These visualizations—a tree, a map, a field, and a network—elicit distinct images of the design process and appear as illustrations of a design-ecological condition, as elements of an inchoate visual discourse about design ecologies, and finally as constituents of this chapter's underlying argument: that the ephemeral data produced during design, and the images of practice they inscribe, configure important sites of inquiry in studies of design and technology—sites we may explore in order to trace new analytic, speculative, or critical cartographies of sociotechnical design practices.

The chapter is organized as follows. The second section describes the work of a team of computer-savvy architects as they collaborate in the process of coordinating the design of a large architectural project, chiefly by assembling a highly detailed computer simulation of the building. Different from a conventional CAD model, this simulation combines the contributions of multiple organizations. This is an increasingly dominant mode of design production—known in the building industry as building information modeling (BIM)—premised on the centrality of computer simulations to both design and construction. Tracing the figure of the design coordinator and of the digital artifacts they collaboratively assemble and manipulate, the chapter offers insight into the social, technical, and material practices of BIM.

The third section traces coordinators' negotiations leading to the definition of a common data representation for design conflicts, and shows how their circulation is essential to this type of design production. Of interest here is the contingent production of design conflict data. While dominant narratives about computerized design methods emphasize seamless collaboration, interoperability, and efficiency, the structure of design coordination data in this project resulted from multiple negotiations involving, for example, technical concerns, design sensibilities, professional hierarchies, habits of record keeping, and personal idiosyncrasies of which the data are themselves vestiges. A close examination of one such design conflict reveals it as a situated digital artifact comprising multiple media, heterogeneous and redundant fields, and visual conventions variously linked to the coordination activities taking place "outside" the simulation. I further discuss the collaborative definition for a common data structure for design conflicts and their collection as codependent processes, and reflect on my own positioning as a researcher with active roles as both an observer and active participant on this site. For example, the formalizing of a common definition of design conflict data for collection purposes is discussed as a type of research intervention that in fact challenges the very ephemerality of the data being collected.

After a brief discussion confronting the epistemic legacies of data visualization practices in surveillance and management, the fifth section presents four different data visualizations of a dataset of design conflicts. I introduce these as illustrations of a design-ecological condition and show how their different visual structures—a tree, a map, a field, and a network—elicit different types of analyses and invoke different conceptualizations of the design process they inscribe. The chapter concludes by reflecting on the notions and methods presented, proposing that tracing design ecologies might help us expand the reach of our ethnographic and historical accounts of design. It shows that by incorporating data critically as an ethnographic material susceptible of visual and computational analysis, along other types of observation and reflection, design-ecological studies can help us make visible the sociodigital infrastructures that condition the phenomenon of contemporary design, and challenge conventional narratives about individual authorial agency in design.

A brief note on method: During the period of participant observation that originated this project I had a dual role as a researcher and as a member of the design coordination team of a large architectural project. This gave me unparalleled access to the culture of design coordination I describe, and to its digital artifacts. However, the visualizations and analyses I present in this chapter were produced after this period of fieldwork for the purposes of this study, and did not support the project or any of the organizations involved. In other words, they were instruments of ethnographic inquiry in a study of design coordination and not instruments of management or coordination.[2]

Buildings and Data

Moving away from 2-D drawings, recent industry practices in architecture and engineering design overwhelmingly favor 3-D digital models as the main vehicles of design coordination. This trend, which encompasses both software systems and managerial practices, is known in the architectural and engineering worlds as building information modeling (BIM). At its core, BIM is about using software to develop a highly detailed 3-D model of the building before it is built. Assembled from the contributions of different organizations involved in the building's design (architects and engineers as well as consultants and subcontractors), these models purport to describe not only the building's shape, but also other relevant information such as the building's materials, budget, structural performance, financial viability, logistics, and more. Assembling the information from all these actors into a single, conflict-free digital model is the central challenge of design coordination. It is a reasonable proposition. In most cases, assembling a building digitally can be orders of magnitude cheaper than building it, so potential design conflicts can be addressed before they make it to the construction site. Thus, BIM is less about representation than about organization, less about the aesthetics of design than about its production and management—and its execution. Reflecting a growing consensus among architects, engineers, software companies, and governments, BIM is frequently described as a transformative technology able to collapse the boundary between design and construction and to redefine the technical, legal, and cultural frameworks of the entire building industry.[3] Accordingly—and updating a Cold War era technological imaginary of design[4]—widely circulating success stories tout BIM software and its associated process as purveyors of increased creative freedom to designers and, ambivalently, as means to maximize managerial efficiency through improved accounting and communication.

The efficiencies of BIM are often put in contrast with earlier traditions of design coordination based on 2-D drawings. In these traditions, which remain the norm in many design and building practices, teams of engineers, architects, consultants, and tradespeople gather around large tables armed with color markers, and discuss the proper course of action to solve design conflicts such as a clash between a ventilation duct and a concrete beam, a poorly calculated staircase, or a steel column disrupting a path of circulation. BIM shifts the site of coordination from paper to the screen.[5] In BIM, data produced by all trades are collected, translated, aggregated, inspected, and negotiated into a single digital model, and this endows the digital model, and those who make it, with a new kind authority (figure 1).

And yet, despite BIM's promises of seamless transits between design and construction, putting a building together remains, and will remain for the foreseeable future, a stubborn and cumbersome endeavor. It requires the alignment of numerous factors that exceed the scope of digital representations and transactions such as people—clients, engineers, designers, workers, consultants—and their professional jargons, habits of representation, and technological preferences—not to mention the proverbially obstinate stones, bricks, and glass that actually form it. Similarly, despite the familiar rhetoric of efficiency and seamlessness, building data do not come together seamlessly or by themselves. They are laboriously put together by a collective of people, organizations, and software—composing what McCann and colleagues have called, in the context of planning, "assemblage work" (2013). A protagonist of this new coordination process is the "BIM coordinator," whose role is to iteratively collect, translate, aggregate, analyze, and report on building data. Because of their important role in design coordination, these coordinators deserve further analysis.

In our site, for example, the work of BIM coordinators was organized cyclically roughly along the following sequence. First, BIM coordinators *collected* data in the form of digital models from each trade organization. Second, in order to incorporate these into the central model, they *translated* the models. As different organizations employed different software systems and modeling standards, this was often a daunting task. The diversity reflected not only different habits of representation

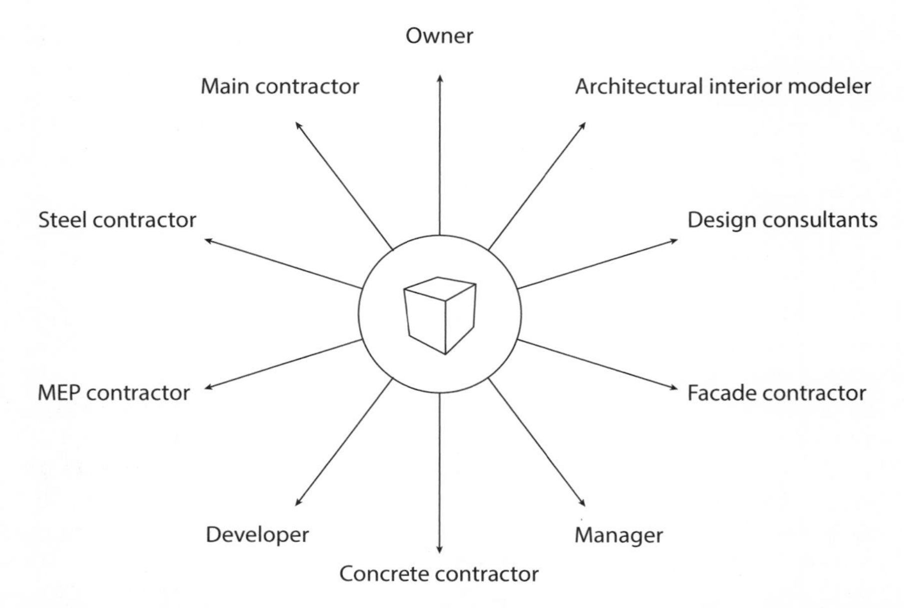

FIGURE 1: Contrasting with traditional representations emphasizing vertical contractual and legal hierarchies in design, diagrammatic representations of BIM typically show the digital model at the center of all design and building trades.

and work, but also technical literacies. For example, the concrete subcontractor and the architects used Autodesk Revit (increasingly an industry standard); the steel subcontractor used the specialized package Tekla; the mechanical systems subcontractor used AutoCAD 3-D; and so on. Grappling with this diverse landscape of software, BIM coordinators spent a significant amount of energy struggling to establish—and then enforcing—protocols of information production and exchange such as file formats, geolocation information, and modeling standards.[6]

In the project, contracts specified that all models had to be exported in IFC format, an open (nonproprietary) file format used as a BIM standard.[7] Because BIM relies on the unification of protocols for model production and exchange as a way to reduce "interoperability costs,"[8] it can be fundamentally seen as a standardization project—and a disciplining one. This rhetoric hides the often laborious and time-intensive work of translation performed by coordinators.

Third, coordinators *aggregated*. One of BIM's promises is to reduce redundancies in the design and construction process by centralizing data from different sources into a model and software platform.[9] Thus, by stitching each new update with the larger digital model, coordinators progressively composed what amounted to an aggregate model of the building from the contributions of the different organizations. Fourth, coordinators *inspected* the aggregate model for design conflicts, often with the aid of automated "clash-detection" algorithms. These algorithms, first introduced in the early CAD systems developed in the 1970s, yield exhaustive lists of clashes between sets of geometric elements. These clashes indicate discrepancies between the digital models of the different building systems, but they can be misleading. Oftentimes a modeling feature of no consequence or a slight error in the placement of the model produced hundreds of conflicts that did not in fact pose a threat to the project. Thus, when inspecting the BIM for actual conflicts, experienced coordinators relied not only on the software's clash-detecting algorithms but also on their knowledge of construction and their understanding of the modeling idiosyncrasies of the contributing organizations. Last, coordinators *reported* the conflicts resulting from these inspections to representatives of the organizations involved, who then convened in a coordination meeting to review, discuss, and determine a responsible party for each of the reported conflicts. On a typical meeting coordinators used the BIM modeling software to present each conflict to participants, who debated—sometimes heatedly—until a responsible party was identified, a plan was agreed upon, and a record was created by the coordinator, usually on an Excel spreadsheet. Actions often involved a design change—for example moving a door, changing a staircase, relocating a duct. These changes had then to be made effective by revising both the BIM and, more urgently, the 2-D drawings used to organize the construction work on site.

Throughout the process, resilient traditions of work, professional idiosyncrasies, and at times blunt skepticism toward the new process could (and often did) contest the centrality of the digital model and the authority of those commanding it. Our understanding of this context of design coordination—this "design ecology"—is what the following sections seek to enrich through the definition, collection, and visualization of data about design conflicts. What can data constructed from BIM coordinators' ephemeral, disheveled spreadsheets, screenshots, and email attachments relentlessly circulating across these sites reveal about the phenomenon of design?

On Ephemeral Data

Rather than curated to constitute a historical and navigable collection, the ephemeral data explored in this study are vestiges of sociodigital transactions that are typically hidden from view and discarded from the end result. These data are ephemeral as their chief purpose is to assist the everyday tasks of design coordination, serving as leverage to catalyze actions on site before being discarded or forgotten. Without established protocols for their archiving, they are also ephemeral in that the thing that they represent, a design conflict, is supposed to go away—to be resolved. The preservation of these data, when enforced, is a matter of contractual obligation rather than exposure or posterity. In contrast to data produced with archival and curatorial intent,[10] design conflict data are disposable: the ultimate aim of coordination is in fact to remove them from view.

And yet they are central to design. They circulated in emails and spreadsheets among the actors involved in the conflict, and their discussion paced daily coordination meetings. Their resolution punctuated the project's advance and organized the day-to-day tasks of a very large number of people across different organizations. Design conflict data were central to the sociodigital scaffolding of coordination. At the same time, design conflict data were not uniform. They were spatially and socially situated, and often reflected coordinators' professional sensibilities and personal record-keeping styles.

Design conflict data are not too distant from what recent work in STS, information science, and human-computer interaction has termed "trace data," which Geiger and Ribes usefully define as "documents and documentary traces in . . . highly technologically-mediated systems" (2011, 1).[11] However, because of the contingencies described in the above paragraph—their artisanal definition and ephemeral character—the design coordination data discussed here have a distinct quality. The relatively handcrafted, non-curated, and ephemeral data produced during messy sociodigital exchanges of design have remained relatively unexplored.[12] In what follows I return to our project to illustrate the specific quality of these data, and to show how they constitute rich sites for inquiry into situated practices of design, how we may develop instruments for their collection, analysis, and interpretation, and what their visual representation may help reveal about contemporary sociodigital design practices.

As coordination advanced at a frantic pace in the project, a small team of coordinators managed the coordination of similarly sized portions of the building, reporting dozens of new design conflicts each week. Each coordinator defined design conflicts intentionally and idiosyncratically—their record-keeping habits did not conform to a single standard and the design conflict metadata (the template establishing the categories of data to be recorded) reflected their different specialties and personal coordination styles.[13] Despite these differences, all conflicts involved certain essential details, such as the conflict's location in the building, the organizations likely to be involved in their resolution, the name of the person in charge, a description, and a date. In some cases, design conflict data incorporated multiple media, actors, and modes of representation. For example, data about a design conflict could comprise screenshots of the 3-D model, photographs of the construction site, official requests for information (RFI), as well as sketches, drawings, or excerpts of architectural plans. Often design conflicts were more sternly defined—a row in a spreadsheet with a location and a short description. In their heterogeneity design conflict data valuably trace design coordination as a sociodigital practice combining an organization's information management protocols, the affordances of software environments, and coordinators' professional inclinations and styles. As vestiges of the design coordination process, they give us access to social, technical, ideological, organizational, and material aspects of design.

This is nicely illustrated in figure 2, an image describing one of the thousands of conflicts reported during the project's design and collected for this study. The background image, an interior view of the BIM, shows two steel columns clashing with a ceiling. However, no walls are visible. The coordinator "hid" them digitally in order to create an unobstructed view of the structure. Further, the "camera" was placed and oriented to capture enough of the context to situate the conflict within the project. Annotations in red indicate the precise location of the clashes and the key measurements in the scene (the distance from floor to ceiling and the door's height). A 2-D plan of the building, itself augmented by sketches drawn by hand, provides another layer of information—a more conventional representation. Finally, a text window provides the conflict's description, index, and location expressed in grid lines. Combining different 2-D and 3-D media and multiple layers of information—including both digital and analog media—the image of the design conflict is also an illustration of the plurality of media and visual codes at play in design coordination data.

Defining a method for collecting these data for the purposes of this research posed the challenge of creating a somewhat consistent system that made conflicts comparable across coordinators, without getting rid of the data's ephemeral character, or disrupting coordinators' record-keeping habits.[14] Reaching this common definition (the conflict's "metadata") was in fact a research intervention that involved multiple conversations with the team of coordinators over a period of several weeks. Reactions varied from indifference to defensiveness and skepti-

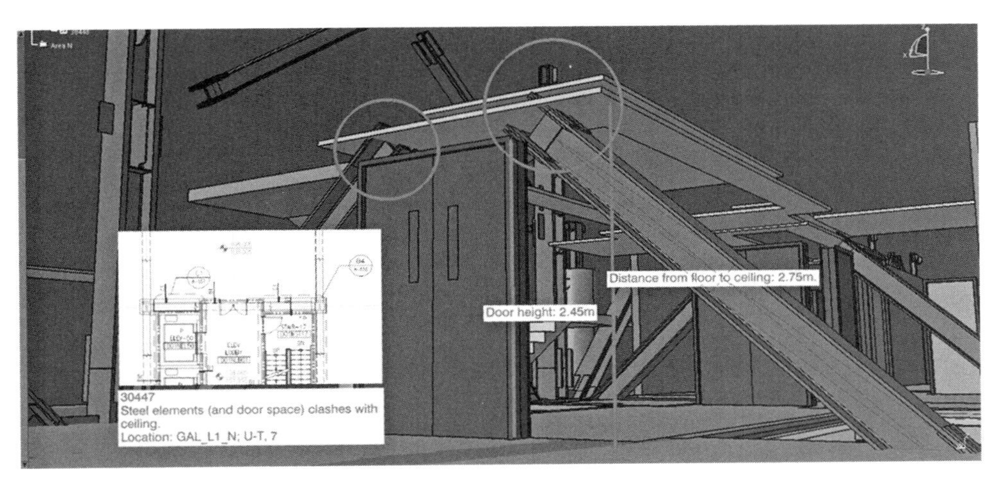

FIGURE 2: One of the thousands of design conflicts documents a clash between structural elements and an architectural ceiling.

cism. Hierarchies mattered. For example, a senior coordinator repeatedly made clear her expectation that the categories they used in their logs were preserved, and that her nomenclature was kept as the main index to the data. The final definition of conflict (C) reflected many such compromises and included the following categories:

> C = { index, building, level, trade, int code, discipline, zone, grid-lines, description, opening date, status, action, responsible person, responsible organization, RFI, next meeting date, action as agreed in meeting }

The conflict's metadata inscribe compromises and redundancies, revealing how personal, site-specific, and institutional nomenclatures converged. Redundancies and overlaps in the conflict metadata indicate that methods of spatial demarcation, professional hierarchies, and the disciplinary boundaries separating designers, builders, and other actors fluctuated during the complex process of producing a building. We might see these redundancies as an illustration of how data are always situated and contingent upon particular social, technical, and material contexts.[15] We might think of data about design conflicts as *seamful* rather than seamless.[16]

The coexistence of personal and shared nomenclatures in the definition of the design conflict reveals data as compounds of idiosyncratic knowledge, on the one hand, and larger, collective knowledge-spaces, on the other.[17] For example, the presence of two different nomenclature fields (*index* and *int-code*) to index conflicts in the database contradicts elemental good practices of information management, but was key to ease coordinators' fears that the system would force them to change their record-keeping habits. As I realized how important it was to preserve these idiosyncratic nomenclatures, a new *index* category allowed me to aggregate design conflict data from different coordinators into a larger dataset (itself an idiosyncrasy of my research design). Preserving a "key" to their prior records, the *int-code* category enabled a diversity of record-keeping styles within the new dataset. As we can see, data collection instruments shape the data they collect, and accordingly, data collection is always a form of intervention.

If redundant nomenclatures reveal the seams between personal and shared forms of record keeping in the construction of conflict data, the spatial categories locating the design conflict physically within the project reveal seams between different spatial nomenclatures and these data's specificity to the physical structure of the building itself. The categories "level" and "building," for example, were specific to the project—a building complex comprising several structures—and reflected the labeling defined by the building's designers in the construction documents. The categories "zone" and "grid-lines" reflected, by contrast, a method for spatial demarcation defined by the coordinators themselves: a rectangular grid dividing the building site into 40 alphabetically labeled square modules of about 10,000 square feet.

This grid provided an alternate abstraction to navigate the space of the project, and itself constituted a key tenet of the coordination process. While used widely by designers, builders, contractors, and subcontractors throughout the life of the project, the grid was controversial to some actors who perceived it as an unnecessary complication and as a transgression of the coordinators' professional jurisdiction. The coexistence of both spatial methods of demarcation in the design project reveals these data also as places where redundant kinds of spatial notation intersect. This is one reason spatial data are not easily abstracted or transferred—they are, by their very nature, situated.

Finally, the metadata tell us something key about BIM coordination: the collective use of a computer simulation as both an anticipatory and an accounting tool. The metadata combine fields accounting for personal and organizational responsibility with these actors' projected or requested actions. They hint at an intricate, digitally enabled bureaucracy that modulates BIM coordination while revealing design conflicts as this bureaucracy's lifeblood. Actions "as suggested by coordinators," actions as "agreed in meeting," fields for responsible person, trade, and involved organization inscribe the fundamentally contentious nature of the coordination process as well as its fragility, which—as I discuss in detail elsewhere—can engender generative forms of resistance.[18] To be sure, design conflict data were both "ephemeral" and "seamful." They reflected plurality, negotiation, site specificity, and redundancy rather than homogeneity, seamlessness, universality, and efficiency.

Once the negotiations about the conflict definition reached a solution that was acceptable for the group of coordinators, I programmed a simple data-collection widget that worked in conjunction with the Excel spreadsheets coordinators already used (figure 3). The widget offered a simple user interface for recording design conflicts as previously defined by the team and aggregated them into a common spreadsheet. Using this tool, the coordination team recorded thousands of design conflicts over a period of about four months. In what follows I describe the process I used to explore these data visually.

A Note on Data Visualization

In asking "what does design *look like* from the perspective of data?" we must consider the nature of visualizations as artifacts in their own right that, just as the data themselves, carry epistemic legacies and interpretive force. As visual theorist Johanna Drucker reminds us, by virtue of their scientific origins and their usage as instruments of management and surveillance, data and their visualizations

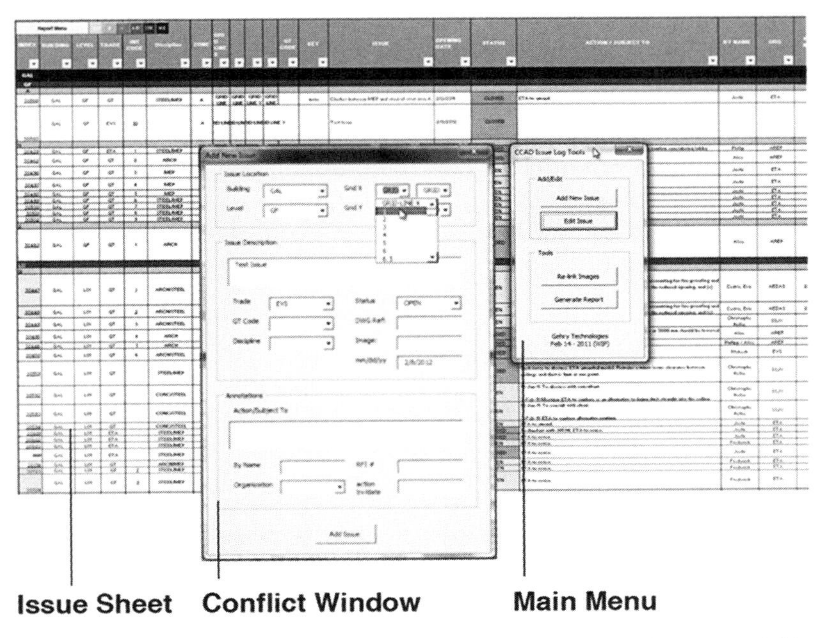

FIGURE 3: Following negotiations about the definition of a conflict, BIM coordinators used this widget to record information about design conflicts during a four-month period of data collection.

often carry with them "assumptions of knowledge as observer independent and certain" (2011). Cautioning against the uncritical use of data visualizations—and developing a critical vocabulary to probe them—Drucker usefully notes that contemporary approaches to data visualization can be traced to long-standing traditions of diagramming, charting, and mapping linked both to the appearance of the printing press and to the 18th-century demands for bureaucratic management of the emerging modern state (2014, 69).[19] Accordingly, contemporary technologies and uses of data visualization are inseparable from governmental impulses to efficiently collect, navigate, and visualize information for purposes of management, surveillance, and militaristic control.[20]

Contrasting with these approaches, recent work in STS and related fields has sought to consider visual forms of knowledge production as both subjects and vehicles of critical inquiry. They have appropriated, for example, visual and time-based media as instruments to examine sensorial and embodied aspects of technology making and use (Ratto 2011; Lehmann 2012) or film and interactive media not merely as illustrations but as constituents of scholarly arguments (Galison 2015). Scholarly efforts to use data visualizations as interpretive instruments have sought to reconcile long-standing traditions of visual communication (Tufte 2001) with an expanding landscape of computational techniques (Fry 2007; Murray 2013) to explore, for example, data visualization's capacities as instruments of artistic exploration (Viégas and Wattenberg 2007), media studies (Manovich 2012, 2016), human-machine interaction analysis (Loukissas and Mindell 2012), and humanistic inquiry (Drucker 2011). We may think of the approaches represented in this indicative sample as examples of an interpretive countertradition challenging data visualization's complicated epistemic legacies in management, surveillance, and militarism.

Extending this countertradition to sociotechnical studies of design, the following section discusses a series of four visualizations of design conflict data. Their purpose is strictly cartographical. The four visualizations—a tree, a map, a field, and a network—all invoke different understandings of the design process. They are examples of an experimental cartography of design production, illustrations of a design-ecological condition in BIM coordination, and, finally, elements of an inchoate visual discourse about design ecologies.

The Graphical Construction of Ephemeral Data

Trees: Tracing Hierarchy and Growth

Figure 4 shows a series of data visualizations where data about design conflicts are organized hierarchically as a tree.[21] The visualization maps the traces of design conflicts in time along a set of concentric rings representing the building zone and the responsible trade. To create this visualization I wrote a program that parsed all conflicts and represented them as Bezier lines inflected at points that in turn indicate the conflict's location (e.g., building 1, building 2, etc.) and trade (e.g., steel, architecture, etc.). The result is a radial tree where the leaves are conflicts and the branches provide context information about the conflict. Each leaf represents a design conflict, which in turn indexes spatial and temporal coordinates as well as graphics and annotations. However, the visualization places emphasis on two key categories: the conflict's spatial location and its trade in a hierarchical manner, while displaying the other categories in a secondary panel in the graphical user interface.

My goal with this visualization was to give a visual representation to the evolving process of design coordination. Enlivened by the data, it reveals visually the surprisingly stable growth of the tree through the four months of data collection. The visualization makes visible the explosion of conflicts related to mechanical systems (MEP) that started during the eighth week of the data collection period and that came to occupy nearly half of all the coordination efforts toward the end. This explosion reflected observations in the field to the effect that the coordination enterprise seemed at times to be solely focused on managing conflicts in the design of the mechanical systems (Cardoso Llach 2015, 121–34). The radial tree is also useful to reveal the variations in the relative importance of different trades in the coordination efforts, and on the coordination loads assigned to each of the trade organizations. As the tree gets denser, it encodes a history of coordination without losing detail—making it an interactive document indexing and providing access to a significant amount of visual and textual information.

Used for centuries to record genealogical information, trees are one of the oldest forms of information visualization. As knowledge structures, they emphasize relationships such as distance, adjacency, derivation, hierarchy, and consanguinity (Drucker 2014). Trees can grow and branch out and their elements can be rearranged, but their organization is invariably hierarchical and centralized. This is both their expressive capacity and their limitation. Visualizing design as a tree reinforces a hierarchical view of design practice placing an abstract model at the center of a ring of trades. Notably, this type of representation closely resembles the diagrammatic representation of design practice commonly used by advocates of BIM (figure 1). Placing the model at the center, the visualization makes

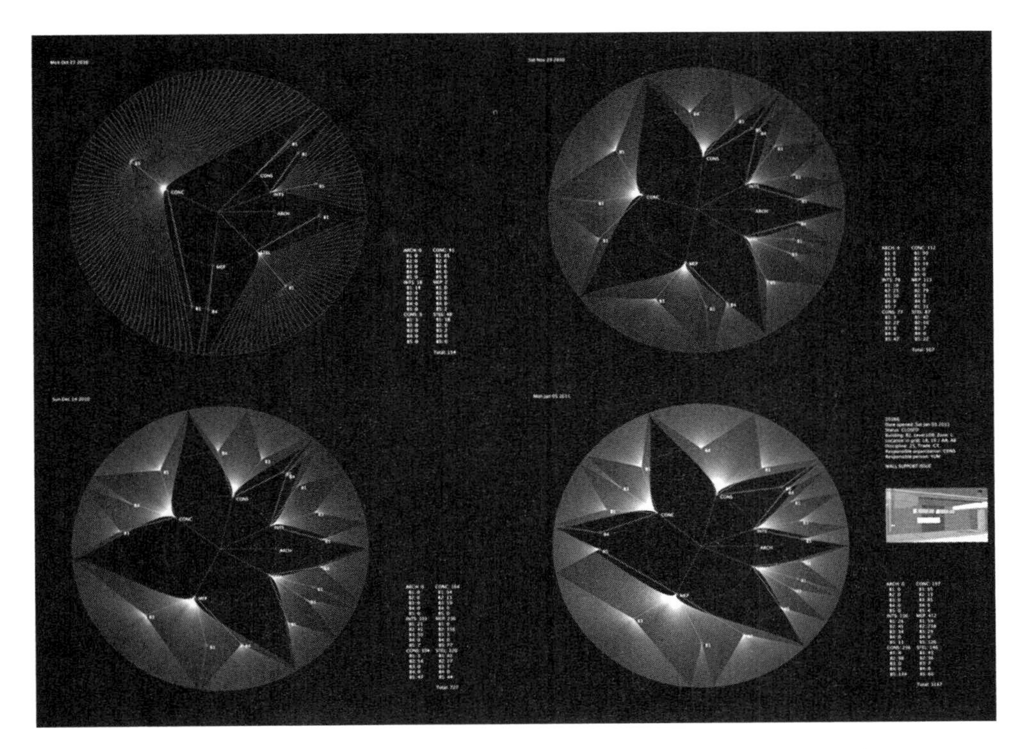

FIGURE 4: The visualization interactively represents the state of design coordination at different stages of the project by tracing the design conflicts recorded by a team of BIM coordinators.

the panoptic aspiration of BIM discourse—"enlivened" by the coordination data, it documents this aspiration in a way that reveals its often hidden and laborious production.

A design-ecological tree traces hierarchy and growth.

Fields: Tracing Clusters of Coordination Activity

In a second visualization, I sought to avoid the hierarchical organization of the data. Instead of a tree, I programmed the visualization as a rectangular grid where issues cluster by trade and building (figure 5). Design conflicts appear here not as branches or leaves but as abstract elements distributed spatially in clusters reflecting location and organization. By interactively changing the date parameter, one can see the variation of distribution of issues in time—a process that resembles (only superficially) a dynamic system where multiple elements would seem to affect each other. Without a center or root, or a reference to the physical world, the visualization evokes a view of design as a field where elements cluster around particular events (trades, indicated by colors) and subspaces (buildings, indicated by the vertical position of each conflict element), not unlike the representation of a meteorological event. The space of the representation is abstract and does not resemble the physical form of the project. It displays coordination activity as clusters of conflicts organized by building and trade, and by a series of numerical parameters in a secondary panel in the visualization's interface.

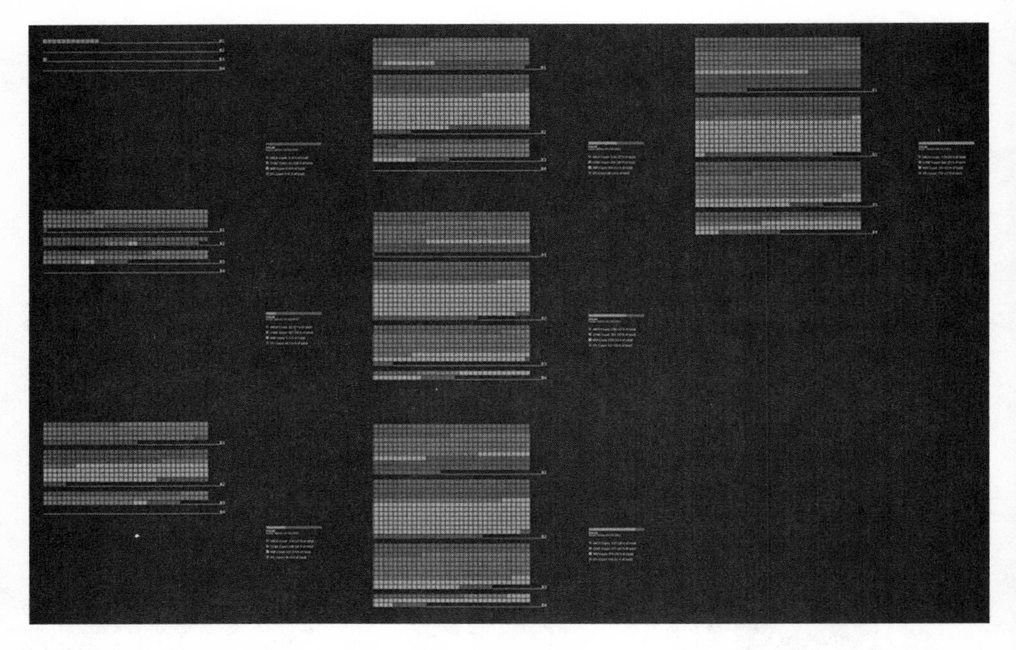

FIGURE 5: Coordination data represented as an abstract field at seven different moments during the period of observation. Different areas of the project are indicated as horizontal sections, and conflicts involving different trades are classified by color.

In order to create a view of design as a changing field, the program ignores location and reconstructs conflict data in an abstract grid. Each section of the grid dynamically displays the state of design coordination in a different area of the building, in time. This abstraction enables comparative analyses across conflicts, organizations, and trades and a bird's-eye view of the state of coordination as expressed by the distributions of conflicts and their types.

A design-ecological field traces clusters of coordination activity.

Maps: Tracing Design Spatially

A third set of visualizations explores the spatial dimension of the conflict data. Rather than a hierarchical representation of the design process as a radial tree, or an abstract one as a field, this visualization relinks the conflict data to the building's physical form and the space it occupies. The visualization takes each conflict's location data and uses them to reconstruct a plan of the building, taking the amount of conflict data as a visual unit. The resolution of the visualization is given by the project's grid of 10 feet by 10 feet, which provides enough detail to discern the building's configuration.

Tracing the spatial dimension of design conflicts in time enables new readings of the design process. The building's shape becomes apparent only to the extent that design conflicts are being reported and addressed, dematerializing the building and rematerializing it as a function of the managerial processes that modulate its production. The visualization thus defines the building literally—and visually—as a space of conflicts, relocating the managerial processes that go into design production in a virtual representation of the physical space.

What these visualizations reveal is the incidence of issues in a particular area. The intensity of conflicts is represented spatially and in time. These are explored in a number of ways. The visualizations in figures 6 and 7 (top) represent the incidence of conflicts in a particular area of the project geometrically—by scaling a shape. This produces a conflict map useful to understand the general distribution of conflicts and the critical zones in the project. The bottom set represents conflict incidence through color—as a "heatmap"—which improves the legibility of the visualization and represents conflict incidence with greater precision.

A design-ecological map rekindles the political and managerial with the spatial.

Networks: Tracing Relations

The last set of visualizations uses networks to explore and visualize different types of associations in the conflict data. These visualizations use special clustering algorithms to spatialize the concepts inscribed in the dataset and were developed using ORA-NetScene, a network analysis and visualization software developed by CASOS at Carnegie Mellon University.[22]

As Tomaso Venturini, Anders Munk, and Mathieu Jacomy usefully observe in this volume, STS scholars have engaged with network analysis methods in productive ways, at times finding a useful—if somewhat messy—convergence with actor-network theory's idioms and aims. Here, my interest is to reflect on the type of insight networks might offer a design-ecological analysis. What type of design worlds do networks describe? What types of questions might they help us address?

A design-ecological analysis seeks to make visible the sociotechnical infrastructures at play in a design process, incorporating data as ethnographic material susceptible to both visual representation and quantitative analysis alongside other forms of observation and reflection. As Drieger notes, a network analysis seeks to progressively dig into the structure of the data and the relationships they inscribe in order to offer "topological insights" (2013), which might reveal high-level relational and hierarchical patterns in the data. Thus, a network approach to a design-ecological analysis has the potential to relate social actors and collectives such as design coordinators, tradespeople, and organizations with nonhuman ones such as concepts, ducts, columns, and software in ways that reveal hidden alignments and misalignments.

Figures 8 and 9 serve as illustrations of this approach. The networks in figure 8 are directed force graphs (Eades 1984) exploring the conflicts' relationship with trades and building location. Design conflicts, represented as nodes, are connected to and clustered around their location and, depending on whether a second trade is involved in their resolution, can be linked to a second trade. A node representing a clash between a duct and a concrete beam in the "MEP" graph, for example, would be connected both to its location in the project (building 1, building 2, etc.) and to the node representing the "concrete" trade. Because conflicts involving one trade are connected only to (and clustered around) their location, these graphs show two levels of clustering.

The networks in figure 9 follow a similar principle, but explore a different type of relationship. This visualization is produced by computationally analyzing the

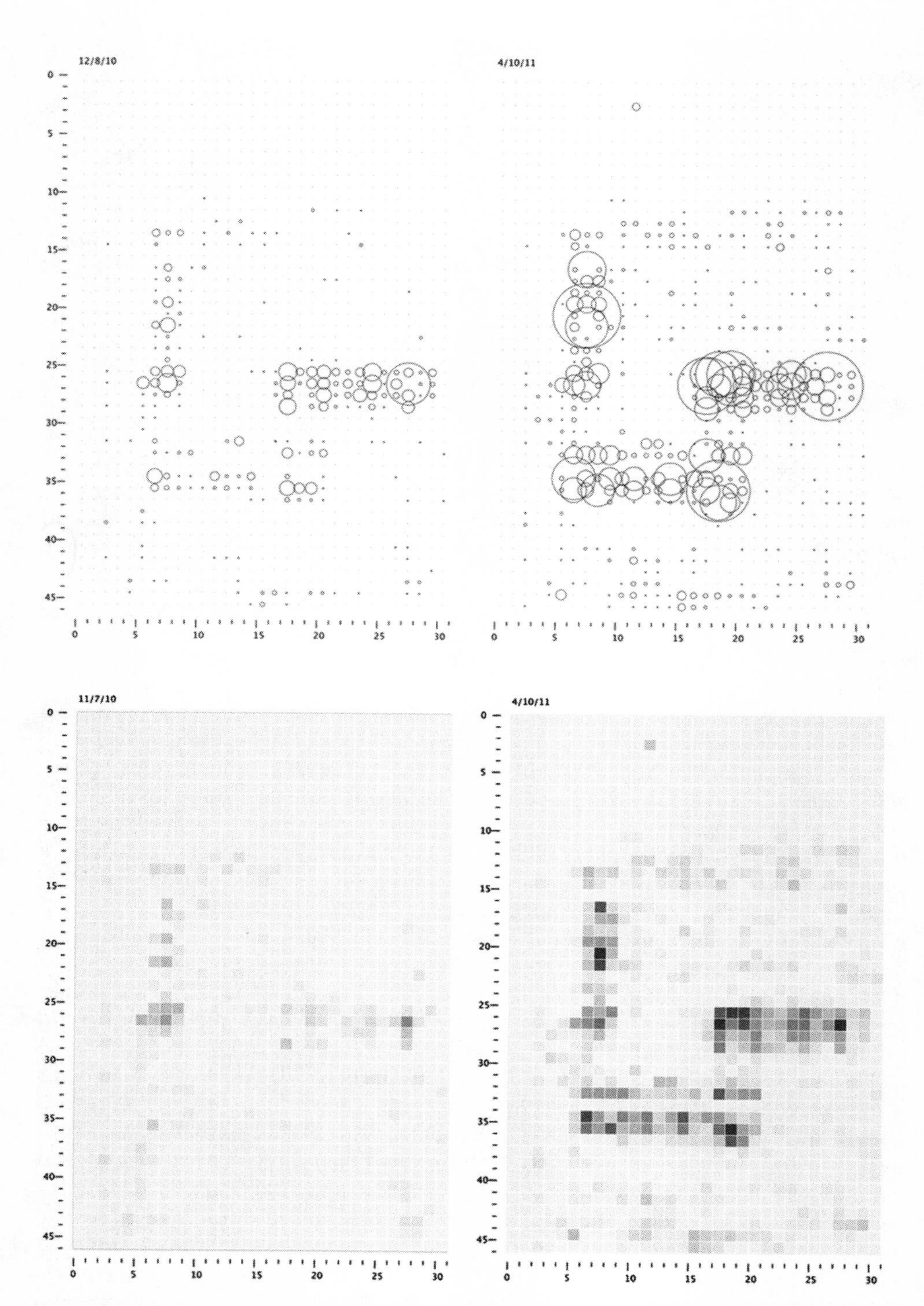

FIGURES 6 AND 7: The two maps on top represent the incidence of conflicts on a specific area geometrically by scaling a shape. The ones on the bottom do so through color. Each set comprises two different images representing different dates in the design process.

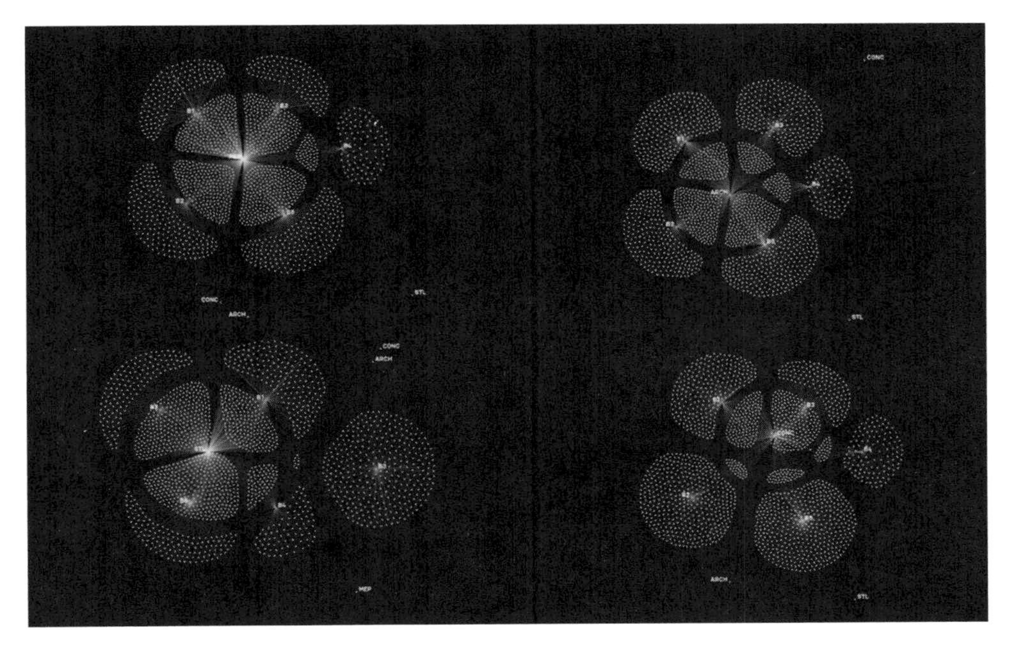

FIGURE 8: Directed force graphs representing placing design conflicts as nodes in relation to building location, main trade, and secondary trade.

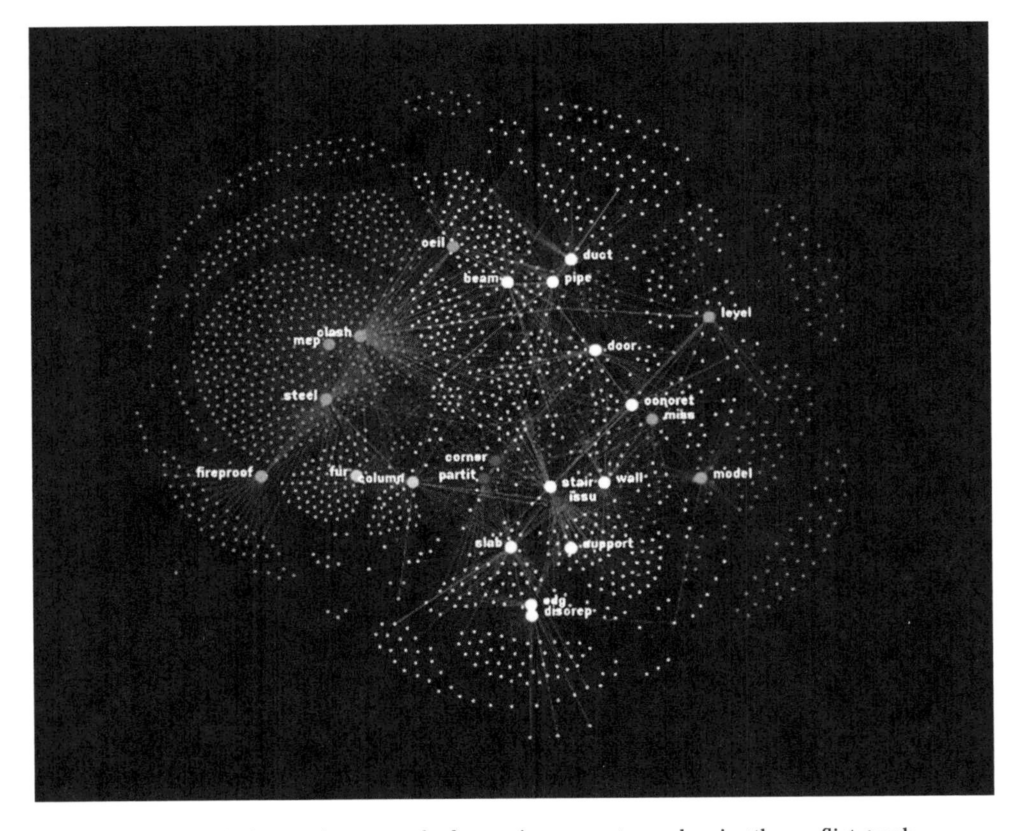

FIGURE 9: Text analysis reveals a network of recurring concepts overlapping the conflict graph.

description field of each conflict, "mining" the dataset for words that repeat over a certain threshold (50 or more appearances in the dataset) and discarding "stop-words" such as prepositions and articles. The resulting graph can be read as an elaboration of the meta-network in figure 8. Along with the trades and buildings, 24 "stem" words are represented as nodes and placed in relation to the trade and building clusters, offering insight about concepts and concerns having an impact on the process of coordination in different areas of the project and in different organizations. Relative to the examples discussed in the previous sections, network visualizations mobilize more complex methods and pack greater assumptions about the data. A detailed discussion of these experiments and their potential role in design-ecological analyses will be the subject of a forthcoming publication. Design-ecological networks explore relationality across concepts, spaces, and people.

Conclusion

A premise of this chapter has been that modes of design structured by technologies such as software, simulations, and numerically controlled machines require that we expand our descriptive and analytical capacities as scholars. The limits of design, usually circumscribing conceptual and representational work, stretch here to incorporate large collectives, sociodigital infrastructures, and relations. In this chapter, I have shown how we might go about making these infrastructures visible in ways that help challenge popular and disciplinary narratives about individual authorial agency and extend the reach of our historiography and ethnography of design. Accordingly, the chapter has outlined a method for design-ecological analysis that, as shown, incorporates the ephemeral data of design as ethnographic materials susceptible of both visual representation and computational analysis in ways that complement and rely upon other forms of observation and reflection.

My argument has hinged on a case study, the collaborative design coordination of a large architectural project, which I traced ethnographically through the observation of both social and digital transactions. In the assemblage work of creating a building's digital model, coordinators produce a design representation—a BIM—where the efforts of multiple people, systems, and organizations converge. The toil of coordination is realized in the efforts by a team of BIM coordinators to collect, translate, aggregate, inspect, report, and act upon thousands of design conflicts. Making visible a constantly changing landscape of design conflicts and indexing their coordination to various spatial, temporal, and conceptual parameters, these visualizations are illustrations of the condition I term design-ecological.

These visualizations inscribe epistemic commitments and carry interpretive force. While a visualization of design as a radial tree reinforces ideas about hierarchy and central control in design, a visualization of design as a dynamic system akin to a meteorological event suggests an open-ended, organization-agnostic reading. A spatial representation relating conflicts to locations in physical space has the capacity to situate seemingly abstract datasets within material and spatial contexts, rekindling material objects in space with the managerial processes that modulate their production. Network visualizations mobilize significant algorithmic capacities to flexibly generate high-level topological insights relating design conflicts with concepts, spaces, and organizations. We may consider these *trees*, *maps*, *fields*, and *networks* as interpretive components of a fledgling visual discourse about design ecologies—components we may tactically choose with descriptive, critical, or speculative intents.

And yet, as we have seen, their evocative power is realized in conjunction with other methods of observation and analysis, and with a critical understanding of the actions surrounding their very design and implementation. Accordingly, data collection is best understood as a type of intervention, and data visualizations as situated artifacts with their own sociomaterial histories. An effective design-ecological analysis must thus put them in conversation with different forms of observation, reflection, and self-reflection. By making the dataset open, I hope that other researchers will investigate alternative ways to visualize this particular design ecology and offer other insights and alternative readings.[23]

We are left with both questions and avenues to future work. Tracing design ecologies can help us make visible forms of technical work emerging from contemporary modes of design production. It can help us see creative practices through a lens that puts social, material, and technological infrastructures—and the labors that sustain them—in focus as integral rather than subsidiary to the phenomenon of design. For reasons including the importance of delineating a political ecology of design, this is a worthy endeavor.[24] How may we trace design ecologies in ways that reveal a broader spectrum of contentions, conflicts, and their resolutions? How may we further trace situated and performative accounts of design in ways that challenge central, panoptic views?

Acknowledgments

I am indebted to the *digitalSTS* editors, reviewers, and coauthors and to María Jose Afanador-Llach, Jedd Hakimi, Sarah Rafson, Nida Rehman, and Molly Wright Steenson for attentive reading and thoughtful comments on drafts. I am grateful to Yanni Loukissas for our many conversations, which over the years helped shape the ideas and methods I introduce here. The community of fellows at the Center for Advanced Study of Media Cultures of Computer Simulation (MECS) at the Leuphana University of Lüneburg, Germany—especially Nicole Stoecklmayr, Ricky Wichum, Dawid Kraspowicz, and Eric Winsberg—offered me and this project a supportive and collegial environment in the spring of 2016. I am grateful to Javier Argota Sánchez-Vaquerizo for his expert assistance in the network visualizations. I am most grateful to the architects, engineers, builders, and tradespeople whose practices this chapter tries to illuminate. Their amphibian work across the digital interfaces of 3D software, meeting rooms, and dusty construction sites is the spark this project tries to capture.

Notes

1. Ecology already offers what cultural geographer Matthew Gandy has termed "an ontology of interconnectedness" to a range of fields such as biology, where the term originated, to other scientific and urban fields, where it has been used, ambivalently, both because of its biophysical meaning and "as a tool for understanding the capitalist urbanisation" (2015, 150). Brought into design, ecology invites a conceptualization of its practice as a set of interdependent sociotechnical and material relations.
2. For more details and an expanded discussion of the larger study from which this project stems, see Cardoso Llach (2015).

3. For detailed introductions to BIM, see Eastman et al. (2011); for architectural perspectives highlighting its potential for architects, see Deamer (2012), Tombesi (2002), and Deutsch (2011); for critical perspectives, see Dossick and Neff (2011) and Cardoso Llach (2015).
4. See (Cardoso Llach 2015, 49–72).
5. This shift was enabled by the late 20th-century increase of computers' and software's processing, storage, and display capacities, which made interactive and highly detailed three-dimensional models possible. These advances notwithstanding, paper drawings remain the legally binding documents in the building industry and the chief means of communication between designers and builders on site.
6. File formats, software applications, and other information standards are often specified contractually. While industry and academic consortia have for years sought to establish a standard digital format, these efforts have not resulted on an industry-wide file format (Eastman et al. 2011; Deutsch 2011). For an expanded discussion of file format standardization, see Cardoso Llach (2017).
7. For a history of IFC, see Mikaeì Laakso (2012).
8. BIM is purported to greatly reduce what is often termed in this industry as "interoperability costs" (Gallaher et al. 2004). For example, a now often reviled but until recently common coordination practice was to reconstruct a given model or drawing in order to insert it into the central model.
9. Compatibility issues between different software platforms are known in the industry as "interoperability problems." According to some scholars, these problems are a major cause of inefficiency and waste in the industry. For example Gallaher and collaborators (2004) quantify the cost of this inefficiency as $15.8 billion.
10. Examples include interpretive work on data such as museum and library collections (Deschner and Dulin 2015), historical and geospatial narratives (Spatial History Project 2016; Bhawsar and Ray 2016), films (Manovich 2013), as well as, more recently, social media content (Tifentale and Manovich 2015).
11. Researchers in human-computer interaction have studied traces such as byproducts of online activity including social media exchanges (Schwanda et al. 2012; Peesapati et al. 2010), credit card transactions (Schwarz et al. 2009), and Wikipedia collaborations (Viégas et al. 2006). More generally, digital traces may include time stamps in version-controlled software projects, records of data entries in a database, (certain kinds of) sensor data, and patterns of interface usage within software environments. Geiger and Ribes propose "trace ethnography" as a methodology combining the examination of traditional ethnographic subjects with trace data (2011, 1).
12. An exception is the recent work by Albena Yaneva (2016), with which the work I present in this chapter shares some methods as well as aims.
13. Users engage software tools in different and idiosyncratic ways. Observing how scientists use PowerPoint and Photoshop to communicate visually, Janet Vertesi notes how their use of visual tools may reflect different professional sensibilities and understandings of the problem at hand (Vertesi 2009, 175).
14. In retrospect, it would have been possible to approach this problem differently, simply "cleaning the data" after collection (a process sometimes referred to as data "wrangling"). This would have likely involved a different kind of effort—entailing a significant effort to edit the data manually or semiautomatically.
15. They are an example of what digital media scholar Yanni Loukissas usefully terms, in this volume, "local data."
16. In "Seamful Spaces," Janet Vertesi usefully articulates the term "seamful" to address a concern with "multiple, coexisting, nonconforming infrastructures" in sociotechnical systems (Vertesi 2014). Human-computer interaction (HCI) researchers have used the term to acknowledge disparities and dissimilarities between the different components of a technological system (Weiser 1995; Chalmers and Maccoll 2003). I borrow the term here to refer to data's multiple, heterogeneous, and plural nature.
17. Historian and sociologist David Turnbull has used the expression "knowledge spaces" to describe situated social and material building practices. For example, communities of artisans building a Gothic cathedral configure a "knowledge space," or a "contingent assemblage of local knowledge" (Turnbull 2000, 6).
18. Many in the project confronted BIM coordination with skepticism. The worlds of building production can be ruthlessly pragmatic and little inclined to introduce technological changes into longstanding traditions of trust building, work, verbal communication, and nondigital representation

media. For an extended discussion of how engineers, architects, and tradespeople engaged in "generative forms of resistance" toward BIM in this project, see Cardoso Llach (2015, 121–48).

19. Proposing the term *capta* as a substitute for *data*, Drucker calls our attention to the codependence between observer and that which is observed, calls for site-specific interpretations, challenges direct readings of statistical generalizations, and helpfully situates data visualizations against broader cultural and historical frames (Drucker 2014).
20. In recent years work in the fields of computer science and HCI has sought to specify visual analytics as a scientific field addressing a combination of large datasets, interactive displays, and uncertain information. A premise of this approach, framing the high volume and diversity of data as a concern for cognitive task performance and human decision making, is that large volumes of data pose challenges to individuals and groups in critical scenarios such as emergency management and conflict resolution (Arias-Hernandez et al. 2011). Not surprisingly, this line of research has been largely shaped by (and funded by agencies concerned with) national security concerns such as the "threat of terror" and emergencies resulting from natural disasters, particularly by the US Department of Homeland Security (Kielman et al. 2009; Thomas and Cook 2005). Similarly, the US Office of Naval Research (ONR) has supported efforts in the related field of information visualization (Card et al. 1999, 14). As these security-minded perspectives shape the field of visual analytics, interactive data visualization tools appear chiefly as means to improve analytical efficiencies over large sets of "spatio-temporal intelligence," weather, and other applications affording the United States the capacity for *faster* and *more appropriate* action. Crucially, recent revelations that the US government routinely collects large swathes of citizen data, including phone and email metadata, raise questions about the role of these technological practices in eliciting the erosion of citizens' rights to privacy under the guise of supposedly unquestionable security concerns.
21. This visualization was first discussed in Cardoso Llach (2012b, 2012a) and subsequently elaborated in Cardoso Llach (2013).
22. I am grateful to Javier Argota Sánchez-Vaquerizo for his research assistance, crucial to using the CASOS software and to producing the network visualizations presented here.
23. An anonymized version of the dataset and the source code for the different visualizations can be downloaded at https://github.com/dcardo/Visualizing-Traces/.
24. See Latour (2004).

Works Cited

Arias-Hernandez, Richard, John Dill, Brian Fisher, and Tera Marie Green. 2011. "Visual Analytics and Human Computer Interaction." *Interactions* 18 (1): 51–55.

Bhawsar, Vaibhav, and Somnath Ray. 2016. "Timescape: Map Your World, Share Your Story." https://angel.co/timescape.

Card, Stuart K., Jock D. Mackinlay, and Ben Shneiderman. 1999. *Readings in Information Visualization*. San Francisco: Morgan Kaufmann.

Cardoso Llach, Daniel. 2012a. "Visualizing BIM Coordination." *Vimeo*. https://vimeo.com/51693897.

——. 2012b. "Builders of the Vision: Technology and the Imagination of Design." Doctoral dissertation, Massachusetts Institute of Technology.

——. 2013. "Design Conflicts: Visualizing Collective Agency in BIM." Paper presented at 4S: Society for the Social Studies of Science, San Diego.

——. 2015. *Builders of the Vision: Software and the Imagination of Design*. London: Routledge.

——. 2017. "Architecture and the Structured Image Software Simulations as Infrastructures for Building Production." In *Operative Artifacts: Imagery in the Age of Modeling*, edited by Sabine Ammon and Inge Hindterwaldner, 23–52. Berlin: Springer.

Chalmers, Matthew, and Ian Maccoll. 2003. "Seamful and Seamless Design in Ubiquitous Computing." Paper presented at the At the Crossroads: The Interaction of HCI and Systems Issues Workshop in UbiComp. http://www.techkwondo.com/external/pdf/reports/2003-chalmers.pdf.

Deamer, Peggy. 2012. "BIM and Contemporary Labor—Pidgin 15." www.peggydeamer.com/images/pid gin.pdf.

Deschner, Paul, and Kim Dulin. 2015. "Haystacks: A New Way to Look at Harvard's Library." http://hay stacks.law.harvard.edu/.

Deutsch, Randy. 2011. *BIM and Integrated Design: Strategies for Architectural Practice*. New York: John Wiley.

Dossick, Carrie S., and Gina Neff. 2011. "Messy Talk and Clean Technology: Communication, Problem-Solving and Collaboration Using Building Information Modelling." *Engineering Project Organization Journal* 1:83–93.

Drieger, Phillip. 2013. "Semantic Network Analysis as a Method for Visual Text Analytics." *Procedia—Social and Behavioral Sciences* 79:4–17. http://dx.doi.org/10.1016/j.sbspro.2013.05.053.

Drucker, Johanna. 2011. "Humanities Approaches to Graphical Display." *Digital Humanities Quarterly* 5 (1). www.digitalhumanities.org/dhq/vol/5/1/000091/000091.html.

——. 2014. *Graphesis: Visual Forms of Knowledge Production*. Cambridge, MA: Harvard University Press.

Eades, P. 1984. "A Heuristic for Graph Drawing." *Congressus Numerantium* 42:149–60.

Eastman, Chuck, Paul Teicholz, Rafael Sacks, and Kathleen Liston. 2011. *BIM Handbook: A Guide to Building Information Modeling for Owners, Managers, Designers, Engineers and Contractors*. 2nd ed. Hoboken, NJ: John Wiley.

Fry, Ben. 2007. *Visualizing Data: Exploring and Explaining Data with the Processing Environment*. Sebastopol, CA: O'Reilly Media.

Galison, Peter. 2014. "Visual STS." In *Visualization in the Age of Computerization*, edited by Annamaria Carusi, Aud Sissel Hoel, Timothy Webmoor, and Steve Woolgar, 197–225. New York: Routledge.

Gallaher, Michael, Alan O'Connor, John Dettbarn, Jr., and Linda Gilday. 2004. "Cost Analysis of Inadequate Interoperability in the U.S. Capital Facilities Industry." US Department of Commerce Technology Administration NIST GCR 04-867, Advanced Technology Program. Gaithersburg, MD: National Institute of Standards and Technology.

Gandy, Matthew. 2015. "From Urban Ecology to Ecological Urbanism: An Ambiguous Trajectory." *Area* 47 (2): 150–54. doi:10.1111/area.12162.

Geiger, R. Stuart, and David Ribes. 2011. "Trace Ethnography: Following Coordination through Documentary Practices." In *2011 44th Hawaii International Conference on System Sciences (HICSS)*, 1–10. doi:10.1109/HICSS.2011.455.

Grossman, Tovi, Justin Matejka, and George Fitzmaurice. 2010. "Chronicle: Capture, Exploration, and Playback of Document Workflow Histories." In *UIST 2010 Conference Proceedings*, 143–52. New York: ACM.

Kielman, Joe, Jim Thomas, and Richard May. 2009. "Foundations and Frontiers in Visual Analytics." *Information Visualization* 8 (4): 239–46. doi:10.1057/ivs.2009.25.

Latour, Bruno. 2004. *Politics of Nature: How to Bring the Sciences into Democracy*. Translated by Catherine Porter. Cambridge, MA: Harvard University Press.

Lehmann, Ann-Sophie. 2012. "Showing Making: On Visual Documentation and Creative Practice." *Journal of Modern Craft* 5 (1): 9–23. doi:10.2752/174967812X13287914145398.

Loukissas, Yanni A., and David Mindell. 2012. "A Visual Display of Sociotechnical Data." In *CHI '12 Extended Abstracts on Human Factors in Computing Systems*, 1103–6. New York: ACM. doi:10.1145/2212776.2212396.

Manovich, Lev. 2012. "Visualization Methods for Media Studies." http://softwarestudies.com/cultural_analytics/Manovich.Visualization_Methods_Media_Studies.pdf.

——. 2013. "Visualizing Vertov." www.softwarestudies.com.

——. 2016. "The Science of Culture? Social Computing, Digital Humanities, and Cultural Analytics." http://manovich.net/index.php/projects/cultural-analytics-social-computing.

McCann, Eugene, Ananya Roy, and Kevin Ward. 2013. "Assembling/Worlding Cities." *Urban Geography* 34 (5): 581–89. doi:10.1080/02723638.2013.793905.

Mikael Laakso, Arto Kiviniemi. 2012. "The IFC Standard: A Review of History, Development and Standardization." *ITcon* 17:134–61. www.itcon.org/cgi-bin/works/Show?2012_9.

Murray, Scott. 2013. *Interactive Data Visualization for the Web*. Sebastopol, CA: O'Reilly Media.

Peesapati, S. Tejaswi, Victoria Schwanda, Johnathon Schultz, Matt Lepage, So-yae Jeong, and Dan Cosley. 2010. "Pensieve: Supporting Everyday Reminiscence." In *Proceedings of the SIGCHI Conference on Human Factors in Computing Systems*, 2027–36. New York: ACM. doi:10.1145/1753326.1753635.

Ratto, Matt. 2011. "Critical Making: Conceptual and Material Studies in Technology and Social Life." *Information Society* 27 (4): 252–60. doi:10.1080/01972243.2011.583819.

Schwanda Sosik, Victoria, Xuan Zhao, and Dan Cosley. 2012. "See Friendship, Sort of: How Conversation and Digital Traces Might Support Reflection on Friendships." In *Proceedings of the ACM 2012 Conference on Computer Supported Cooperative Work*, 1145–54. New York: ACM. doi:10.1145/2145204.2145374.

Schwarz, Julia, Jennifer Mankoff, and H. Scott Matthews. 2009. "Reflections of Everyday Activities in Spending Data." In *Proceedings of the SIGCHI Conference on Human Factors in Computing Systems*, 1737–40. New York: ACM. doi:10.1145/1518701.1518968.

Spatial History Project. 2016. http://web.stanford.edu/group/spatialhistory/cgi-bin/site/index.php.
Thomas, James J., and Kristin A. Cook, eds. 2005. *Illuminating the Path: The Research and Development Agenda for Visual Analytics*. Los Alamitos, CA: National Visualization and Analytics Ctr.
Tifentale, Alise, and Lev Manovich. 2015. "Selfiecity: Exploring Photography and Self-Fashioning in Social Media." In *Postdigital Aesthetics: Art, Computation and Design*, 109–22. New York: Palgrave Macmillan.
Tombesi, Paolo. 2002. "Super Market: The Globalization of Architectural Production." *Arq: Architecture Research Quarterly* 6 (1): 5–11.
Tufte, Edward R. 2001. *The Visual Display of Quantitative Information*. 2nd ed. Cheshire, CT: Graphics Press.
Turnbull, David. 2000. *Masons, Tricksters and Cartographers: Comparative Studies in the Sociology of Scientific and Indigenous Knowledge*. New York: Taylor & Francis.
Vertesi, Janet. 2009. "'Seeing Like a Rover': Images in Interaction on the Mars Exploration Rover Mission." Ithaca, NY: Cornell University. http://ecommons.cornell.edu/handle/1813/13524.
——. 2014. "Seamful Spaces Heterogeneous Infrastructures in Interaction." *Science, Technology, & Human Values* 39 (2): 264–84. doi:10.1177/0162243913516012.
Viégas, Fernanda B., Scott Golder, and Judith Donath. 2006. "Visualizing Email Content: Portraying Relationships from Conversational Histories." In *Proceedings of the SIGCHI Conference on Human Factors in Computing Systems*, 979–88. New York: ACM. doi:10.1145/1124772.1124919.
Viégas, Fernanda B., and Martin Wattenberg. 2007. "Artistic Data Visualization: Beyond Visual Analytics." In *Proceedings of the 2nd International Conference on Online Communities and Social Computing*, 182–91. Berlin: Springer. http://dl.acm.org/citation.cfm?id=1784297.1784319.
Vinck, Dominique, ed. 2009. *Everyday Engineering: An Ethnography of Design and Innovation*. Cambridge, MA: MIT Press.
Weiser, Mark. 1995. "Ubiquitous Computing (Invited Talk)." Keynote at the USENIX Conference, New Orleans, January 16–20.
Yaneva, Albena. 2016. *Mapping Controversies in Architecture*. London: Routledge.

Data Sprints
A Collaborative Format in Digital Controversy Mapping

Anders Kristian Munk, Axel Meunier, and Tommaso Venturini

> *The critic is not the one who debunks, but the one who assembles. The critic is not the one who lifts the rugs from under the feet of the naive believers, but the one who offers the participants arenas in which to gather.*
>
> **—Latour (2004, 246)**

> *We don't know what a researcher who today affirms the legitimacy or even the necessity of experiments on animals is capable of becoming in an oikos that demands that he or she think "in the presence of" the victims of his or her decision. Of importance is the fact that an eventual becoming will be the researcher's own becoming; it is in that respect that it will be an event and that what I call "cosmos" can be named.*
>
> **—Stengers (2005, 997)**

The notion that researchers should think through the consequences of their knowledge claims in the presence of those affected by them, here formulated by the Belgian philosopher of science Isabelle Stengers, has long struck a chord in STS, first and foremost as the underlying credo of a genre of critical engagements with modern techno-science, but increasingly with STS researchers themselves assuming roles as caretakers and facilitators of public involvement in techno-scientific projects (Law 2009; Landström et al. 2011; Jensen 2012). At stake have been questions concerning the public trust in science and technology (Wynne 2007; Felt and Fochler 2008), the role of experts and expertise in democratic processes (Nowotny et al. 2003; Callon et al. 2009), the robustness of scientific knowledge claims (this is the position taken by Stengers in her call for a democratization of the academy; Stengers 1997, 2000), and the viability of designed and engineered solutions in everyday use practices (Woolgar 1990; Hyysalo 2006; Brandt 2006; Ehn 2008; Petersen and Munk 2013). From it has emerged a plethora of collaborative formats for involving and engaging publics (Rowe and Frewer 2005), such as participatory modeling (Yearley et al. 2003), consensus conferences (Einsiedel et al. 2001), deliberative mapping (Burgess et al. 2007), living labs (Björgvinsson et al. 2012), or competency groups (Whatmore 2009).

The EMAPS project (Electronic Maps to Assist Public Science), from which the present chapter draws its example, was in many ways conceived in this broad tradition of participatory STS. It emerged in response to a call by the European Research Council to assess "the opportunities and risks in the use of the web and social media as a meaningful information tool and for developing a participatory communication between scientists and their different publics" (ERC FP7, http://cordis.europa.eu/project/rcn/101858_en.html). But where "scientists" in such participatory experiments have usually implied someone else than the STS researcher (who has tended to assume positions somewhat on the sideline of events or as a mediator between the experts proper and their publics), the division of labor was deliberately different in EMAPS. Yes, there were experts involved, notably climate scientists and various kinds of adaptation specialists, and there was a stated ambition to render the complexities and controversies of the field of climate change adaptation navigable and interrogable by a concerned public. But the real object of the participatory experiment—the thing that was being thought through in the presence of its victims, if you will—was the mapping of these controversies itself, a practice that is, by its own accounts, distinctly STS in its origins and purposes (e.g., Latour et al. 1992; Venturini 2010, 2012; Yaneva 2012; Marres 2015).

Controversy mapping was conceived by Bruno Latour, Michel Callon, and others in the late 1980s and early 1990s as a set of techniques for charting sociotechnical debates and as a pedagogical approach to teaching STS to engineering students. Although digital methods are not an obligatory part of a controversy mapping project, the ability to harvest and analyze digital traces has become part and parcel of how the method tries to achieve its ambitions (see especially Venturini 2012 or Marres 2015). EMAPS is thus a successor project to the likewise EU-funded MACOSPOL (Mapping Controversies on Science for Politics), which posited that "citizens need to be equipped with tools to explore and visualize the complexities of scientific and technical debates" and therefore sat out to "gather and disseminate such tools through the scientific investigation and the creative use of digital technologies" (www.mappingcontroversies.net/).

As we will elaborate below, the MACOSPOL project raised a number of questions about what exactly such democratic equipment could be expected to achieve and how it could be devised in order to do so. EMAPS was founded on the realization that a careful rethink of how a controversy mapping project engages with its users in specific contexts was necessary if indeed a cosmos, as Stengers puts it, was ever to be named. What kind of common world, we asked ourselves, was a controversy mapping project trying to bring about (Venturini et al. 2015)? The core idea was that such a rethink should take place in direct collaboration with the users of the maps, which in the case of controversy mapping coincides with the objects of the cartography, namely the actors in the controversy. The data sprint format, which we present in this chapter, constitutes the eventual realization of such a participatory approach to controversy mapping.

In the Belly of the Monster

Stengers's notion that knowledge claims should be put at stake in the presence of their victims is especially pertinent in the context of a participatory approach to controversy mapping. For Stengers and others, controversies are the potent situations that make it possible to put anything authoritatively "expert" at stake in the first place. Controversies not only energize participation but also engender a world in and about which inquiries can be fruitfully undertaken. They are thus generative in the sense that it is through them that both the researcher and the researched can acquire their identities and become aware of what they have at stake in each other's practices. Importantly, it is also through controversies that these identities can be challenged and remade. "Controversies," writes Michel Callon, "establishes a brutal short circuit between specialists and laypersons" and "for a time, the relative equalization of 'rights to speak,' [affords] the opportunity for everyone to argue on his or her own account and to question the justifications of others" (Callon et al. 2009, 33). This definition of controversy as a democratic asset resonates with the pragmatist notion that "issues spark a public into being" (Marres 2007) and the proposition by Latour that it is by following controversies that we will be able to study the social in its making (Latour 2005).

For the mapmaker, acknowledging the democratic role of controversies begs the question how the cartographic instruments interfere with or contribute to these already potent becomings. When controversy mapping sees its raison d'être as that of equipping a concerned public with navigational aids, it is thus not only designing such aids but also redesigning (or attempting to) the controversy itself and thereby its emergent publics. With a term borrowed from design research, one could say that controversy mappers engage in the "infrastructuring" (Björgvinsson et al. 2012) of controversies: The controversy and its emergent publics are not (because they cannot be) staged on the cartographer's drawing board. Instead, what must be achieved is the collective mapping into knowledge of a "matter of concern" (Latour 2004), a thing that will persistently prevent any one position from reducing the others to mere fact or fiction.

We argue that this mapping into knowledge cannot be achieved unless the mapmakers acquire stakes in the controversy and, vice versa, unless the actors of the controversy acquire stakes in the mapping. We are, as Donna Haraway (1988) puts it, always and already "in the belly of the monster." We base this argument on our experiences not only with EMAPS and MACOSPOL but also with teaching and facilitating controversy mapping projects in a range of contexts. These experiences have fermented our understanding that tools and equipment do not cut the mustard alone but must be acquired—by the actors, the citizens, the users, the stakeholders, depending on the situation—for specific purposes in specific contexts.

Such an acquisition is not unproblematic. Powerful data visualization instruments put at the disposal of actors in a controversy are easily appropriated to reduce the discussion in various ways. It is therefore crucially important that the position of the controversy mapper—as one who seeks to stage the arena of debate in its complexity and heterogeneity—is also put at stake and thus hardwired into the collective. The fact that we are mapping the *controversy*—rather than trying to empower any given actor-centric position—means that the controversy mapper must find ways to deploy conflicting positions while still lending an opportunity to the individual actors to acquire their own stakes in the cartography.

On top of that comes the practical challenges of facilitating a mapping project that, in order to be agile and adaptable to user input, must bring developers, designers, and domain experts into a close and concrete dialogue. And it is of course not merely a practical challenge; more than anything STS has contributed to our understanding that robust knowledge is a distributed achievement (Latour 1993; Rheinberger 1997; Nowotny et al. 2003). Given that one of the key partners on both

MACOSPOL and EMAPS had extensive experience with facilitating such collaborative processes through their winter and summer schools in digital methods (see, for example, https://wiki.digitalmethods.net/Dmi/SummerSchool2015), we took inspiration partly from their tried-and-tested format (Berry et al. 2015), and partly from the buzzing scene of bar camps and hackathons associated with the developer and design community (Knapp et al. 2016), in appropriating the data sprint format. Before we get to that, however, we provide an overview of the experiences gained through MACOSPOL and the early stages of EMAPS that led to the idea of the sprint.

Controversy Mapping and Public Engagement

Public engagement was not always an evident course of action in controversy mapping. Cartographies of techno-scientific debates were first pursued in STS to gain analytical purchase on a set of heterogeneous objects that were difficult to represent (Callon et al. 1983; Callon et al. 1986; Latour et al. 1992), not to make them available to public scrutiny in different ways or otherwise reconstitute their implications for the democratic process. In this respect controversy mapping is aligned with a development that has taken place in the actor-network theoretical branches of STS more broadly. From its early applications as a way of crafting situated accounts of science and technology in action, and thereby as an instrument for a critique of correspondence based theories of truth (Latour and Woolgar 1979; Latour 1987, 1993; Callon 1986), to its later merits as an intervention that multiplies and interferes with singular and essentialist ontologies (Mol 2002; Moser 2008) or assembles alternative forums for public involvement (Callon 1999; Callon et al. 2009; Latour 2004; Whatmore 2009), the agnostic prescription to let actors deploy their own worlds has served a variety of interventionist strategies (Munk and Abrahamsson 2012). And just as the more general question of whether and how actor-network theory (ANT) “means business” (Woolgar et al. 2009) has been broadly debated (see also Vikkelsø 2007; Woolgar et al. 2009; Law 2009; Jensen 2012), so the specific interventions that controversy mappers engage in have developed through a series of experiments of which the EMAPS project represents one of the more recent.

The first hint of controversy mapping engaging a public can be found in the STS classroom of the 1990s, where it was often claimed that the method blossomed as a way of teaching engineering students basic insights about science and technology in society. According to legend (which is liberally recounted by those who were there, but about which little has to our knowledge yet been published), controversy mapping became popular chiefly as a didactic approach that did not require wholesale conversions of the students to social construction of technology or ANT. This is certainly still the case in the now extensive network of universities teaching controversy mapping to students in political science, geography, engineering, architecture, media studies, design, techno-anthropology, sociology, and others (see http://controverses.sciences-po.fr/archiveindex/ for an overview of student projects). By encouraging students to observe how techno-scientific controversies unfold in practice, complicated theoretical arguments can be pragmatically demonstrated rather than lectured from the blackboard. In a sense it was nothing more than a convenient remedy for a problem that almost any teacher with an STS curriculum has faced, but it did contain the nucleus of what later became controversy mapping's primary commitment to public engagement: if students could be "tooled up" with STS sensibilities, then so could issue professionals and decision makers.

At the outset of the MACOSPOL project the primary point of reference for what controversy mapping was supposed to achieve was thus firmly anchored in the classroom. Controversy mapping had been developed primarily as a pedagogical intervention, and the tools and methods associated with it had never been tested outside that context. A narrative had emerged, however, about the relevance and application of controversy mapping in democratic processes. According to this narrative, which drew on pragmatist political theory, democratic publics would need to have the right representational skills in order to assemble and engage constructively with their matters of concern. Here is how Bruno Latour formulated it at the outset of MACOSPOL:

> How you represent a river as an agent is a very interesting question. I do not know any river now that is not a contentious issue. In France, we even have a law to represent rivers politically. We actually have an institutional organ for river representation. But when you go to this parliament of rivers, which is the literal word they use, the representational tools from hydrography and geography are extremely disconnected with this question. So you have masses of maps in a traditional sense, which are critically informative but not necessarily what is needed to represent a river in this political river assembly, and that is precisely where all the questions of controversy mapping comes from. In the phrase "controversy mapping," the word "mapping" is not metaphorical but literal. We want to be able to help the citizens of this new parliament of things to have *the representational skills that are at the level of the issues* (Latour 2008, 134, emphasis added).

It is however one thing to acquire these representational skills at the level of the issues. In essence this has always been the driving ambition in controversy mapping, even in the very earliest attempts by STS scholars to develop computational ways of visualizing sociotechnical debates. But it is quite another to expect others—users, citizens, students, stakeholders—to do the same. It not only requires that a skill set (and its associated tools and methods) is available, but that those others understand what the skills are for, that is, understand what the issues are and why they "deserve more credit" (Marres 2007).

The MACOSPOL Experience and the Notion of Democratic Equipment

The notion of democratic equipment was pivotal in the MACOSPOL project. This is perhaps best illustrated by the development of the Lippmannian Device that took place in the context of MACOSPOL. Walter Lippmann, writing at the beginning of the 20th century, famously posited that the democratic ideal of a public capable of making competent decisions about the problems directly affecting it had become increasingly elusive (Lippmann 1927). Lippmann believed that, with the right tools at its disposal, the public could at best be expected to put its weight behind experts and representatives whom it considered capable of taking action on specific issues, in itself a momentous challenge considering that such tools were rarely available. Lippmann observed that the complexity of the public's problems was of such a scale and magnitude that citizens would all too easily fall prey to special interests and partisan reporting when trying to orient themselves. He thus formulated a problem that others, not least his contemporary interlocutor John Dewey (1927), have since been struggling to solve: how can a democratic public manifest itself when the issues at hand are ill defined, rife with uncertainty, and thus irreducible to questions for which expert advice can be easily and unambiguously solicited?

The Lippmannian Device (https://wiki.digitalmethods.net/Dmi/ToolGoogleScraper) scrapes Google for resonance of a set of keywords across a set of web pages. The idea is to provide the user with an indication of actors' commitment to different issues. In the fracking debate, for example, who talks about native land rights? Who talks about climate change? Who talks about natural parks? Who talks about earthquakes? And who talks about water security? The tool presupposes a user who is not an expert in, but still sufficiently implicated by, a controversy to need a temporary means of orientation, a user who has a preliminary idea of what to look for but is not a native of the terrain he or she is trying to navigate.

In doing so the Lippmannian Device offers a demonstration of how MACOSPOL imagined its commitment to public engagement. Toward the end of the project a workshop was held in Venice where potential users were invited to test tools. They were primarily journalists and decision makers—the kinds of people for whom the MACOSPOL online platform (www.mappingcontroversies.net/, now offline) had been designed and intended from the start. They were asked to map a topic of their choosing. The idea was that these types of users, as part of their profession, would routinely have stakes in a changing cast of controversies and would thus self-evidently see the point of becoming controversy mappers themselves.

The reality proved more complicated. In one of the intermediary project reports from MACOSPOL, it was noted that "what we need in the first place is to have a set of tools that targets the selected public (journalists and decision makers) as possible users of the platform." This was in many ways achieved, except that what was ignored here was the question of how exactly journalists and decision makers can be said to constitute a public. Sure enough, they are people whose job it is to act in / steer through / report on controversies, typically by trying to settle them, either by exercising editorial privileges or by acting politically. But this also implies that they are not by definition interested in opening up the controversy for exploration, especially not if they can establish criteria from which to judge up front who is right and who is wrong. They are, in other words, not necessarily inclined to become controversy mappers.

In figure 1 we see MACOSPOL researchers standing behind the users, helping them out with the tools. A frequently posed question in this situation was how the tools would compete with picking up the phone to ask a trusted expert what was important and what was not. This was how most of the users would routinely handle their encounters with controversies. The public, in this setup, was thus staged in a manner quite similar to the students learning controversy mapping—their instructors at the ready to help out where needed—but their stakes in the situation were of course completely different. In Venice we encountered users with ingrained ways of questioning the world that were not easily aligned with the tools.

It is key to understanding the MACOSPOL experience that the development and testing of tools was entirely delegated to different work packages where team members were mapping their own research topics. The experimental alpha users of the tools were thus people with a very specific, and STS informed, prior understanding

FIGURE 1: MACOSPOL final workshop in Venice. The participants are journalists and decision makers (because they are supposed to deal with controversies in their work). They sit around a table filled with computers. MACOSPOL assistants stand behind them to help the prospect mappers conduct their mappings.

of controversies and their potentials. When they conceived of the tools as "speeding up" inquiry, for instance, it was always in comparison with the extremely labor intensive work of manually and agnostically charting the arguments brought forth by actors in a controversy without a priori determining their relevance. This type of inquiry is not the standard modus operandi for journalists and decision makers. From this setup arose the fallacy that if the users only knew about the tools and how to *use* them, they would also *need* them. You could say that MACOSPOL was paradoxically and inadvertently enacting a version of the very same deficit model that it had been founded to challenge.

If MACOSPOL taught us something, it was that it does not suffice to simply make tools and maps available if they are to have public effects. In order for these things to be truly "public things" (DiSalvo et al. 2014), care and attention would have to be devoted to making them matters of concern to the users. MACOSPOL rather naïvely imagined that equipping the public would amount to offering tools that would allow concerned citizens to become cartographers of controversies themselves.

Early Experiments with Participation in EMAPS

When EMAPS succeeded MACOSPOL, it was thus decided to shift our engagements with the public in two important ways (at this point we had all become engaged

FIGURE 2: Early EMAPS workshop in London. Researchers from the project sit around the table along with other workshop participants. One assistant in each group asks questions, moderates the discussion, and takes notes. On the table are prefabricated maps that are shown and interpreted one after the other.

with the project, whereas two of us had worked on MACOSPOL as well). First, the idea of generic tools and equipment as a goal in itself was abandoned. Focus would instead be on maps tailored for specific issues. Second, the users invited to the workshops would now include people with more immediate stakes in one such issue, not just journalists and decision makers with a professional obligation toward it. We chose to split the work in two phases, first taking on a pilot case about aging in the United Kingdom, where the new approach would be tested, and then a case study on climate change adaptation, for which an online issue atlas would be published (www.climaps.eu). It was planned like this in order to have sufficient time for iterations, going back and forth between user workshops and drawing board, instead of the linear format adopted in MACOSPOL with a user workshop at the end.

MACOSPOL had no doubt contributed to the development of a common toolset for controversy mappers, but the feeling was that it was necessary to demonstrate what could be done with this toolset before one could realistically consider its affordances in actual use. The first workshop arranged in relation to the pilot study on aging thus saw EMAPS researchers bringing to the venue premade maps, which they would then explore together with care workers, advocacy groups, and other issue professionals of the aging debate. This was possible partly due to the domain expertise of one of the EMAPS partners, the Young Foundation, which had prior experience with social innovation in that sector and could thus provide necessary input on the preparation of maps prior to the workshop. The purpose was to observe how participants responded to maps that were, to the minds of the EMAPS researchers, the state of the art with the available toolset.

In figure 2 we see EMAPS researchers sitting down with users, listening in and taking stock of how they engage with their maps. The scene is notably different from the MACOSPOL workshop in Venice where they assumed the role as tool instructors helping their "students" produce maps themselves. As the pilot phase progressed, however, we gradually moved from this fairly "product-centric" mode of engagement to a more "use-centric" mode (we credit our design advisor on the project, Lucy Kimbell, for this conceptualization; see also Venturini et al. 2015). In order to make better sense of the maps, it was decided that they would have to depart from research questions derived more directly from the practices of the users.

In the first instance this was attempted by simply going back and improving the maps based on user input between the first and second workshops. No additional qualitative work was undertaken outside the workshops to get a sense of what kind of questions the users might be interested in or (perhaps especially) to situate these questions in the everyday environments of the users. The users were merely asked to respond to maps and then to respond again to a set of revised maps. During the two first workshops we were able to identify one user in particular who was engaged in advocacy work, had a sick mother herself, and seemed genuinely curious about the potentials of the mapping. She agreed to let an ethnographer from the EMAPS team work closely with her to articulate research questions that made sense in her world and in the professional lives of her colleagues. These questions were then translated into digital mapping projects, the results of which were explored with a broader selection of users at an issue safari in London. It was the first time that the EMAPS team got a sense that the cartography of controversies was being meaningfully acquired by actors in a controversy.

This taught us not only that hardwiring the mapping projects into the everyday practices of the users was essential, but also that doing so required an extremely agile and adaptable collective effort on behalf of the team. For the next phase of the project, which was supposed to build an issue atlas of climate adaptation (the eventual result is available at www.climaps.eu), we would have designers from Milan, digital methods experts from Paris and Amsterdam, domain experts from Dortmund, social innovation facilitators from London, and a range of invited issue experts involved in what was essentially a distributed exercise in collective controversy mapping. The challenge was at once to find a way of managing such a distributed work process, while efficiently translating the concerns of the actors in the controversy into feasible digital methods projects. Encouraged by our experiences at the issue safari in London, we were determined to find engaged issue experts with obvious and immediate stakes in the controversy. We had learned not to expect cartography to imbue participants with an appetite for maps. Instead we would make sure, by interviewing and spending time with potential participants in advance, that they brought such an appetite with them to the project. In a sense this reversed the roles when it came to acquiring a stake in the mapping, or at least it made us realize that it was more a matter of us learning from the issue experts how maps could be useful in their domain than a matter of them learning something about controversy mapping from us. This came with some obvious risks. There was a keen sense, for example, that we could easily end up servicing the data crunching and visualization needs of issue experts who knew quite well what kind of maps would aid their agenda, that we would, as one project partner formulated it, end up "giving too much away for the process."

The solution came from our project partners at the Digital Methods Initiative in Amsterdam who had the following program lined up for their 2013 Winter School in digital methods:

> The 2013 Digital Methods Winter School is devoted to emerging alternatives to big data. The Barcamp, Hackathon, Hack Day, Edit-a-thon, Data Sprint, Code Fest, Open Data Day, Hack the Government, and other workshop formats are sometimes thought of as "quick and dirty." The work is exploratory, only the first step, outputting indicators at most, before the serious research begins. However, these new formats also may be viewed as alternative infrastructures as well as approaches to big data in the sense of not only the equipment and logistics involved (hit and run) but also the research set-up and protocols, which may be referred to as "short-form method.
>
> The 2013 Digital Methods Winter School is dedicated to the outcomes and critiques of short-form method, and is also reflexive in that it includes a data sprint, where we focus on one aspect of the debate about short- vs. long-form method: data capture. To begin, at the Winter School the results of a data sprint from a week earlier (on counter-Jihadists, see www.hopenothate.org.uk/counter-jihad/map/) will be presented, including a specific short-form method for issue mapping (www.densitydesign.org/2012/10/visualizing-right-wing-populism-in-europe/). One outcome of the Winter School would be a comparison of short-from methods for their capacity to fit the various workshop formats (barcamp, sprint, etc.), with the question of what may be achieved in shorter (and shorter) time frames. We also will explore a variety of objects of study for sprints, including data donations, where one offers particular data sets for abbreviated analysis.

Drawing inspiration from the developer community and its tradition of hackathons (hack marathons), we decided to organize four consecutive *data sprints* (we credit Liliana Bunegru and Erik Borra at the Digital Methods Initiative for this term; see also Berry et al. 2015) where we would invite issue experts, designers, developers, and social scientists to work together for five consecutive days on a preselected theme. Over the course of a year (January–September 2014) we thus organized sprints in Paris (around the international negotiations on adaptation), Amsterdam (around the question of vulnerability assessments and indices), Oxford (around the question of financial compensation), and Milan (focusing mainly on the design of the platform for the issue atlas). Below we go through the process of organizing and executing one such sprint in order provide a feel for its gait and composition.

The Paris Sprint in Pictures

On the morning of the first day of the sprint, the invited issue experts present and discuss controversies in the international negotiations around climate change adaptation. The issue experts include (1) Farhana Yamin, associate fellow at the Royal Institute of International Affairs (London) with extensive experience as an environmental lawyer and policy expert on climate change, also a lead author for the IPCC who has served as senior adviser to the European Commission and the EU Commissioner for climate change (2) Richard Klein, senior research fellow at the Stockholm Environment Institute (SEI), lead IPCC author, and senior policy adviser to a range of governments and international organizations on the subject of climate adaptation; (3) Kees van der Geest, expert on loss and damage and researcher at the United Nations University in Bonn; (4) Alix Mazounie, policy expert on adaptation from the French branch of the NGO Climate Action Network; (5) Nicolas Bériot, secretary general of the French National Observatory for the Effects of Global Warming (ONERC) and head of climate adaptation policy at the French Ministry for the Environment; and (6) Francois Gemenne, a specialist in environmental geopolitics at the Institute for Sustainable and International Development (IDDRI) in Paris who was tasked with mediating the discussion among the experts.

While the issue experts present and discuss, the EMAPS team consisting of 25 to 30 designers, developers, and social scientists take notes in a shared Google Doc (projected on the screen on the wall; see figures 3 and 4). This is the beginning of a process to translate the concerns of the issue experts into workable digital methods projects. The team knows in advance which datasets have been collected and immediately begin discussing whether and how they might be useful to the projects taking form.

During lunch break the collective notes of the team are condensed into six project briefs that are pitched on a so-called *issue auction* (see figure 5). The team has already been divided into five working groups that can now bid on the project they want to work on. One project is deliberately left orphaned and will not be taken forward. In order to condense the collective notes as respectfully as possible, we prioritized to have meaningful and coherent project briefs even though it meant the number of briefs would be higher than the number of groups that were able to take them on.

During the afternoon of the first day of the sprint each of the working groups gather to decide how they are going to operationalize their brief. The issue experts circulate between the groups to offer their commentary and help clarify the questions they have raised in their presentations. The photo of the white board that appears as figure 6 is from a group working on the question, "who are the experts of adaptation?" It was a question raised by several issue experts who asserted that, in the international negotiations, much depends on who has privileged access to which rooms and negotiating tables. No good dataset had been prepared in advance for a question like this. Together with the issue experts, the group had to devise a mapping strategy that would scrape the participants lists year by year for all the COPs (the Conference of Parties to the United Nations Framework Convention on Climate Change, i.e., the international climate summits) and then search the meeting notes for a range of committees and subcommittees to see who had been granted the rights to speak in which forums over the course of the negotiations.

As work progresses toward the milestone presentations on the third day of the sprint, intermediary visualizations are built to explore the potentials of the mappings with the issue experts. Figure 7 displays an online interface developed for that same project on the experts of adaptation (you can explore it here: http://ladem.fr/misc/sprint2014/trajectories.php). Compared to the chaotic whiteboard of the first day, we now have a workable prototype that allows both issue experts and other sprint participants to get tactile with the datasets and explore their potentials and limitations. One of the motivations for this particular subproject was to have a hands-on tool that would allow a participant in a COP to quickly survey another participant's track record and standing in different expert groups, forums, and subsidiary boards. Another quite different motivation was to be able to gauge, on an aggregate level, how expert groups, forums, and subsidiary boards share and exchange experts. The mock-up has thus been designed in a way that enables

FIGURE 3: Issue experts pitch projects on the first day of the sprint.

FIGURE 4: Collective note taking in a shared document on the first day of the sprint.

FIGURE 5: Proposed projects are auctioned off on the afternoon of the first day of the sprint.

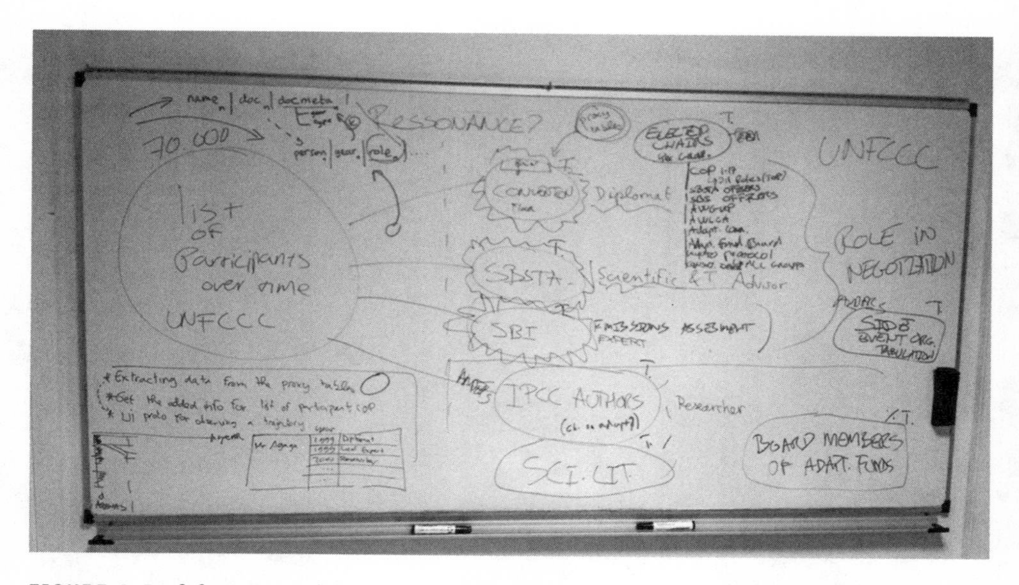

FIGURE 6: Draft for a protocol that tracks mentions of actors in meeting minutes from climate change negotiations.

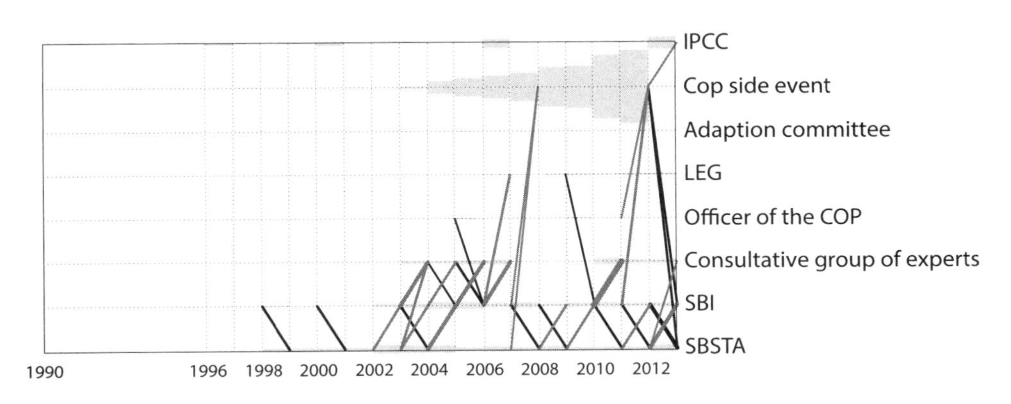

Settings

- **Normal** – All participations are displayed
- **Filtered** – Onlly with 4+ persons from one arena to another
- **Side events** – People that participated to a Side Event (once or more)
- **NO side event** – People that NEVER participated to a Side Event
- **Single person** – A person's trajectory, pick in the list, ordered by amount of participations

Carlos Fuller – 21 participations

FIGURE 7: Intermediary visualization that makes it possible for sprint participants to explore different filterings of a dataset that tracks mentions of actors in meeting minutes from climate change negotiations.

us to filter COP participants in different ways. We can for example choose to explore the trajectory of an individual participant, of all participants, of those who have been active in side events, or of those who have never been active in side events.

On the afternoon of the third day the issue experts are invited back to the conference room, but this time it is the working groups delivering the presentations of their intermediary results. It is around this point in the sprint that issue experts begin to acquire a more concrete stake in the mapping. Contrary to the presentations and discussions on the first day, when the rest of the participants attempted to get their heads around the different problems and positions of the international negotiations on climate change adaptation, the issue experts now have to get their heads around what can and cannot be done with digital methods. It has become clear that the options are not endless and that the process of crafting workable maps requires specific ways of thinking that are not always in line with what the issue experts are used to. In the project being presented in figure 8 the issue experts have been asked to help tag a dataset on adaptation funding based on the main thematic areas of intervention. According to one of them, this has been an eye-opening experience that has helped bring some of her more lofty expectations for the instant potential of data visualization healthily down to earth. It has proven much more complicated than first thought to find a logical and consistent way to tag interventions thematically. During the first three days, then, it is not only the resident controversy mappers who have become more issue-savvy; it is just as importantly the issue experts who have become more attuned to the challenges of digital controversy mapping.

During the milestone presentations on the third day the issue experts listen in and respond. They are being asked to imagine how the presented mock-ups may or

FIGURE 8: Preliminary results of data projects presented on the third day of the sprint.

may not be (made) useful for them in their own practice (see figure 9). This is also an opportunity for the working groups to deliver feedback on each other's work. The session sets the pace for the remaining days of the sprint. The job is now to finalize a set of mappings that will conceivably be of use to the issue experts and their professional communities. This involves a number of tough choices. The first three days have thrown up a range of opportunities, but the time constraints of the sprint are now acutely felt, and the need to deliver a product that will conceivably travel with the issue experts by the end of the week becomes the imperative that guides our decisions. There is excitement about the projects, but there is also a considerable amount of frustration as the EMAPS team, largely academics and accustomed to keeping questions open, has to compromise and temporarily close down avenues of exploration.

As the sprint progresses, the working groups spontaneously abandon their designated rooms and decide to gather permanently in the lunch area (see figure 10). Besides eliminating the need to break for food, this offers several other advantages. Developers have been distributed between the groups, but their specific skills are often needed in other groups; datasets that are being harvested or cleaned by one group turn out to be useful for another; results of one mapping project turn out to be interesting for the questions posed in another. This may seem like a trivial point about layout, but we found that sprints became less efficient and engaging the more subgroups were physically contained and the work organized in siloes.

On the last day of the sprint the finalized visualizations from each project are presented to the team and the issue experts (see figure 11). It provides a sense of

FIGURE 9: Issue experts provide feedback on preliminary results on the third day of the sprint.

FIGURE 10: Sprint participants work in groups to implement ideas from issue experts on the fourth day of the sprint.

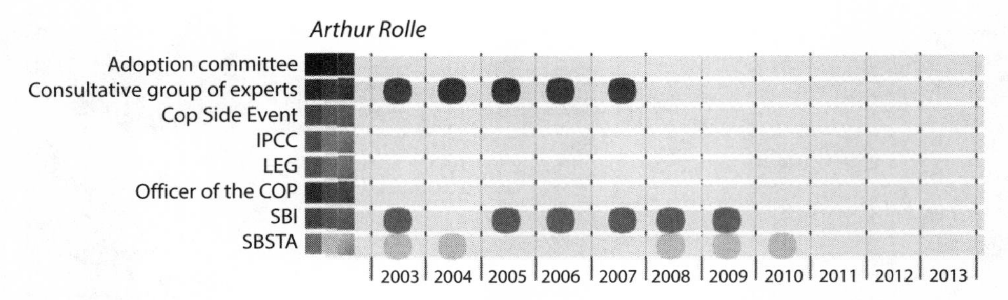

FIGURE 11: Visualization showing the career of an actor in the climate change negotiations presented on the last day of the sprint.

closure to the sprint, and a tangible target to work toward. A race toward a finish line, if you will. In figure 11 we see a further development of the interface for the subproject on "who are the experts on adaptation?" It has been designed with the scenario in mind where a negotiator on the COP floor needs to acquire a quick track record on another COP participant. It is of crucial importance that not only the finished maps and visualizations but also, to the extent possible, the datasets, the tools, and the code used are made publicly available along with a protocol that explains the projects. As Berry et al. note, "Data sprints are based on reproducibility: the work done needs to be documented and shared online in order to foster similar work and further developments" (Berry et al. 2015, 2). The results of the Paris sprint were published on the EMAPS blog (www.emapsproject.com/blog/archives/2348), and a reworked version of the results is available in the final issue atlas of the project (www.climaps.eu).

What's in a Sprint?

Data sprints in controversy mapping are extended research collectives that assemble over several days to collaboratively explore and visualize a set of pertinent questions. They comprise the necessary competencies to (1) pose these questions, (2) consider their relevance and implications for the controversy, (3) operationalize them into feasible digital methods projects, (4) procure and prepare the necessary datasets, (5) write and adapt the necessary code, (6) design and make sense of the relevant data visualizations, and (7) elicit feedback and commentary through consecutive versions of these visualizations. In practice this means that the following roles should be considered necessary to the functioning of the collective:

The Issue Experts/Alpha Users

Regardless of the subject matter of the sprint, the first order of business is always to formulate research questions. This is done with the help of people who have something at stake in the topic of the sprint (because they are affected by it, produce knowledge about it, intervene politically in it, or—highly likely—a combination of the above). They are at once the issue experts, who are able to deploy their matters of concern for the rest of the sprint participants, and the alpha users, who will be able to provide feedback and commentary on the evolving maps from the point of view of someone who might conceivably make use of them in their practice. The selection of these issue experts/alpha users does not presume to be representative (as is, for example, the case in citizen conferences) but is driven by the research collective's need to acquire stakes in the controversy. As Whatmore and Landström put it in the context of their competency groups, actors are called for who are "sufficiently affected by what is at stake to want to participate in collectively mapping it into knowledge and, thereby, in its social ordering" (2011, 583).

The Developers

Sprints are supposed to be agile. They must be able to adapt not only to the issues experts/alpha users bring to the table but also to what the research collective as a whole makes of these contributions. The one asset that more than anything ensures this agility (or hampers it if neglected) is developers. Successful sprints are fundamentally anathema to the idea that development needs can be fully anticipated much less serviced in advance. If this is possible, it almost certainly means that the labor-intensive and resource-demanding process of sprinting will have been unnecessary. The job of the developers is to adapt tools and scripts for particular analysis needs, harvest new datasets, and help the designers build applications for exploring the datasets with the issue experts when necessary.

The Project Managers

Research questions must be asked in such a way that they are amenable to the available digital methods and yet still pertinent to the issues they concern and the issue experts that asked them. This requires a translational competence. Project managers must be sufficiently knowledgeable of the controversy to understand the questions posed by the issue experts and sufficiently adept with digital methods to see the potentials and constraints flagged by the designers and the developers. This is especially critical in operationalizing the research questions where the project managers become, in a sense, the stewards of the alpha users (who often cannot be there the whole time). It is also crucial in the interpretation and exploration of the maps where the project managers can help the issue experts understand what can and cannot be claimed with digital traces.

The Designers

A sprint process relies on visualizations through several of its key stages. They are especially essential for facilitating the ongoing exploration of the datasets and the proposed analysis of them with the issue experts (which mostly happens through quick prototypes and mock-ups along the way). To the extent that the research collective is making its results public by the end of the week and is risking failure for its maps in the everyday practices of the issue experts, the design mind-set and competence are absolutely key. It is often the designers who have their eyes best trained on the final product and are able to force the pace of the sprint toward the end.

The Sprint Organizers

Besides making the necessary practical arrangements for the sprint to take place (booking rooms and accommodation, organizing food, distributing programs and practical info, etc.), the organizers play a key role in the preparatory phase leading up to the sprint. The most obvious occasion for this is the decision on the overall sprint theme. Although sprints should be agile enough to accommodate evolving research questions, thematic framing is necessary for a number of reasons. In order to invite issue experts, it is necessary not only for the organizers to know who they are looking for but also for those invited to know why they should come. Good issue experts are likely to be dedicated people with busy agendas. It falls to the sprint organizers to provide them with an incentive by giving them a sense of what their stakes in the sprint could be. Thematic framing is also necessary for preselecting datasets. Again, although sprints should be agile enough to accommodate the harvest of new datasets, if so required by the operationalized research questions, the organizers should do what they can to anticipate relevant datasets. In effect this means that important processes of scoping and foreshadowing precede the successful sprint.

In practice these roles can be filled by the same people who have crossover competences. It is our experience that the possibility to have developers with design competencies, or vice versa, can be highly beneficial to the sprint. It is in itself an important learning outcome of a sprint that participants become more attuned to each other's practices. A good example of this is provided by issue experts who in several cases have used their experiences with the sprint to formulate and launch digital methods projects of their own.

You will notice that there is no such role of "the controversy mapper." Sprints are in many ways the embodiment of what Noortje Marres calls "distributed methods." They require all the competencies in the room and draw on every available extraneous source (whether code or data) and make every attempt to contribute back to the open source/data community. To our minds, the sprint participants, if successful, are all controversy mappers in the making; they are complicit in mapping into knowledge what thus becomes a collective matter of concern.

Below we go through the different phases of the sprint in a more stylized manner.

Posing Research Questions

During the EMAPS sprints we experimented with several ways of posing research questions. Common to all of them was the fact that the invited issue experts were given time on the first day of the sprint to deploy the controversy from their respective points of view. Typically this took place as 30- to 40-minute keynotes on a preagreed theme. During the sprint in Paris these presentations were the first opportunity the other participants had to acquaint themselves with the perspectives of the issue experts. During the second sprint in Amsterdam this was changed so that issue experts were now asked to prepare written project briefs in collaboration with EMAPS researchers that could be circulated in advance. This meant that research questions were more developed at the outset of the sprint in Amsterdam, in the sense that they were more attuned to the needs and wants of a digital methods project (we elaborate in the "Operationalizing Research Questions" section). It also meant that research questions were more hardwired to the interests of specific issue experts, in the sense that initial project briefs were developed without the interventions of other issue experts (we elaborate in the "Considering the Relevance and Implications of Research Questions" section). In preparation for the sprint in Paris there had indeed been extensive consultations with issue experts about the possible forms a research question could take in a controversy mapping project, but these were mainly aimed at getting the academically trained issue experts to present their perspective on the controversy—i.e., present themselves as stakeholders—instead of what they considered to be the balanced overview of the controversy suited for a research project context.

Considering the Relevance and Implications of Research Questions

The main point of asking the issue experts to pose their questions on the first day of the sprint is to ensure that the digital methods projects, on which the sprint participants will be working for the rest of the week, are informed by, and acquire stakes in, what the actors of the controversy consider to be important matters of concern. There is an auxiliary point, which is to give the sprint participants, who cannot be expected to have prior experience with the controversy, an opportunity to get acquainted with the issues they will be working with, but both require that time is prioritized to take the problems raised by the issue experts properly into account. This can be done through conventional Q&A sessions or panel discussions following the presentations by the issue experts, but it can also, and oftentimes more fruitfully, take place as informal consultations between project groups and issue experts as part of the running feedback and commentary on data visualizations (see under "Eliciting Feedback and Commentary" below). It is a matter of sprint participants—designers and developers for instance—coming to grips with the scope of the questions. But it is also, quite importantly, a matter of issue experts questioning each other's positions. Effectively this means that no one issue expert is issued a monopoly on the mapping projects.

Operationalizing Research Questions into Feasible Digital Methods Projects

In a sense, this process begins already before the sprint where the organizer tries to gauge, in very broad terms, what type of projects the sprint might be liable to end up with. We found that an excellent way of doing this initial vetting was to ask issue experts to suggest interesting datasets in advance. This provided a chance to get back to the issue expert and explain why the dataset had the wrong structure for certain types of hypothetical projects, for example, and thus getting them attuned to what a digital methods project can and cannot achieve. The actual operationalization, however, happens in the work groups led by one or more project leaders.

Procuring and Preparing Datasets

As mentioned, it is of course desirable to have datasets available in advance, but also incompatible with the agility of the sprint to fully anticipate what kind of data will be needed. During the Paris sprint we thus invited a data provider from the NGO Climate Funds Update as an auxiliary issue expert. We also had developers putting a significant effort into procuring new datasets as a consequence of the questions raised by the issue experts. It often happened that datasets were indeed available, but had been parsed in a way that was not amenable to the projects undertaken.

Writing and Adapting Code

In stark contrast to the work done in MACOSPOL, where focus was on stand-alone tools that would be operable by a user, the sprints can potentially build customized scripts to do the kind of analysis that is required by the research questions raised by the issue experts. These projects are only as agile as the coders who are there. This does not mean, however, that old tools cannot be fruitfully adapted during the sprints, and indeed several useful developments of the controversy mapping toolset came out of the sprints.

Designing and Making Sense of Maps

The driving factor of a sprint is its tangible outcomes. In the course of five days the research collective produces a series of data visualizations that the issue experts can bring with them, that will be made openly available, and that, in the case of EMAPS, the sprint participants know they will be making public together on a website (www.climaps.eu). It is therefore obvious why data design and data narration need to be part of any data sprint in some measure. In EMAPS, however, we also took the opportunity to dedicate an entire sprint (namely the last one, in Milan) to designing the final web platform. This involved selecting and redesigning maps, publishing and organizing datasets and code, narrating stories across projects, and generally thinking through what it takes to make controversy maps public.

Eliciting Feedback and Commentary

It has almost become a cliché in controversy mapping circles that the main value is not in the map but in the mapping. Although sprints work from very concrete expectations of finished maps, it is important to remember that the main learning potential—for all parties—lies in the process of making maps together. Sprints are not reducible to five days of hard labor with a clear agenda and a fixed production deadline. They deliberately incorporate opportunities to alternate between doing things together and stopping momentarily to think through whatever is being done. This is mainly possible during the three plenary sessions on the first, third, and fifth days, but the fact that the work is colocated means that an informal conversation between, for instance, issue experts, social scientists, and developers takes place in the corridors. This should not be missed but in fact encouraged through ongoing opportunities to have casual interactions, such as eating together or being able to take breaks in nice surroundings.

Toward Controversy Mapping as Coproduction of Knowledge

Having gone through the aspects of the sprint as a participatory format in controversy mapping, we can now repose the question of what a researcher who affirms the necessity of social cartography will become in the presence of his or her victims.

Michel Callon (1999) has argued that involvement of laypeople in science and technology comes in roughly three varieties: the public education model (PEM), premised on the idea of a knowledge deficit and a fundamental opposition between lay and expert knowledge; the public debate model (PDM), which acknowledges the stakes of specific situated publics in the production of scientific knowledge, and thus accepts a need to keep knowledge claims provisional until commentary from those affected by their consequences can be obtained; and the coproduction of knowledge model (CKM), in which a public is entrusted with the competence to participate on an equal footing in all aspects of the scientific process and the lay-expert divide is thus no longer maintained.

One of the great qualities of Callon's framework is its implication that the most upstream public engagement exercise in the PDM register remains incommensurable with true coproduction of knowledge if it confines lay competence to situated and local knowledge domains. If experts remain privileged as the final arbiters of what gets to count as objective and universal, so the argument goes, the much-cited ambition to restore public trust in science and technology is unlikely to be more than an empty gesture. Here Callon resonates with a sentiment in much recent STS scholarship. The acute risk that a public involvement initiative will be perceived as a mere legitimation exercise, no matter how early or often the stakeholders are consulted, is now frequently noted and agreed upon, so much so that an author like Brian Wynne (2007) deems any attempt to engender trust in science through public engagement intrinsically futile.

For others, and arguably also for Callon, the futility is less intrinsic, or at least the means of engendering public trust are potentially available if CKM is taken seriously. The reasons given are quite pragmatic: If it is the case that scientists build trust in their own knowledge claims through a series of translations (and this is how ANT accounts for it), then something similar ought to be true for publics. Without a firsthand appreciation of the reductive choices that go into stabilizing a knowledge claim, trust cannot be expected to emerge out of the blue. One recent STS project that has taken this premise to its fullest and most radical consequence is the Pickering Flood Research Group, set up in North Yorkshire in 2008. Central to the success of this "competency group," as it dubbed itself, was the notion of apprenticeship and the maxims to be "doing things together" and "making things public" (Whatmore and Landström 2011).

To our minds, the development of the cartography of controversies through projects like MACOSPOL and EMAPS describes a movement from a quite classic PEM approach, where the public is construed as being in a navigational deficit that can be fixed by "tooling up" the citizens, to a PDM model in the early phases of EMAPS, where maps are produced in the cartographer's workshop and then solicited for commentary with the users (stakeholders in the aging debate), to the sprints as a version of CKM that allows the mapmakers to acquire stakes in the controversy and the actors of the controversy to acquire stakes in the mapping. In the end all participants became competent sprinters, sufficiently savvy about each other's domains to be mapping things together.

In this mode of engagement it becomes possible to simultaneously make controversy mapping relevant for its users and manage a necessarily distributed cartographic machinery, and also allows the research collective to render the controversy as a controversy, to deploy it in its complexity without reducing it to a single point of view.

In conclusion, however, it should be noted that *allowing* the controversy to manifest itself is not the same thing as *ensuring* that it happens. The data sprint is an occasion, characterized by specific social, temporal, and material affordances, through which it becomes possible for the controversy mapper to acquire a stake in the controversy and allow the actors of the controversy to acquire a stake in the mapmaking. But data sprinting alone will not force that to take place. It seems that one of the more serious barriers is the blackboxing of both digital methods and professional practices of issue experts that normally characterize attempts to read premade maps or use tools without a pretext. Data sprints are potential occasions precisely because they force such black boxes to be temporarily opened and scrutinized. Therein lies a *potential* for engagement, one that must be actively pursued by the organizers.

Works Cited

Berry, D. M., E. Borra, A. Helmond, J. C. Plantin, and J. Walker Rettberg. 2015. "The Data Sprint Approach: Exploring the field of Digital Humanities through Amazon's Application Programming Interface." *Digital Humanities Quarterly* 9 (3). http://digitalhumanities.org/dhq/vol/9/3/000222/000222.html.

Björgvinsson, E., P. Ehn, and P. A. Hillgren. 2012. "Design Things and Design Thinking: Contemporary Participatory Design Challenges." *Design Issues* 28 (3): 101–16.

Brandt, E. 2006. "Designing Exploratory Design Games: A Framework for Participation in Participatory Design?" In *Proceedings of the Ninth Conference on Participatory Design: Expanding Boundaries in Design*, vol. 1, 57–66. New York: ACM.

Burgess, J., A. Stirling, J. Clark, G. Davies, M. Eames, S. Staley, and S. Williamson. 2007. "Deliberative Mapping: A Novel Analytic-Deliberative Methodology to Support Contested Science-Policy Decisions." *Public Understanding of Science* 16 (3): 299–322.

Callon, M. 1986. "The Sociology of an Actor-Network: The Case of the Electric Vehicle." In *Mapping the Dynamics of Science and Technology*, edited by Michel Callon, John Law, and Arie Rip, 19–34. London: Macmillan.

——. 1998. "An Essay on Framing and Overflowing: Economic Externalities Revisited by Sociology." *Sociological Review* 46 (S1): 244–69.

——. 1999. "The Role of Lay People in the Production and Dissemination of Scientific Knowledge." *Science, Technology and Society* 4 (1): 81–94.

Callon, M., J. P. Courtial, W. A. Turner, and S. Bauin. 1983. "From Translations to Problematic Networks: An Introduction to Co-word Analysis." *Social Science Information* 22 (2): 191–235.

Callon, M., P. Lascoumes, and Y. Barthe. 2009. *Acting in an Uncertain World: An Essay on Technical Democracy.* Cambridge, MA: MIT Press.

Callon, M., J. Law, and A. Rip, eds. 1986. *Mapping the Dynamics of Science and Technology.* London: Macmillan.

Dewey, J. 1927. *The Public and Its Problems [an Essay in Political Inquiry] by John Dewey.* New York: Holt.

DiSalvo, C., J. Lukens, T. Lodato, T. Jenkins, and T. Kim. 2014. "Making Public Things: How HCI Design Can Express Matters of Concern." In *Proceedings of the 32nd Annual ACM Conference on Human Factors in Computing Systems*, 2397–2406. New York: ACM.

Ehn, P. 2008. "Participation in Design Things." In *Proceedings of the Tenth Anniversary Conference on Participatory Design 2008*, 92–101. Bloomington: Indiana University.

Einsiedel, E. F., E. Jelsøe, and T. Breck. 2001. "Publics at the Technology Table: The Consensus Conference in Denmark, Canada, and Australia." *Public Understanding of Science* 10 (1): 83–98.

Felt, U., and M. Fochler. 2008. "The Bottom-Up Meanings of the Concept of Public Participation in Science and Technology." *Science and Public Policy* 35 (7): 489–99.

Haraway, D. 1988. "Situated Knowledges: The Science Question in Feminism and the Privilege of Partial Perspective. *Feminist Studies* 14 (3): 575–99.

Horst, M., and A. Irwin. 2010. "Nations at Ease with Radical Knowledge on Consensus, Consensusing and False Consensusness." *Social Studies of Science* 40 (1): 105–26.

Hyysalo, S. 2006. "Representations of Use and Practice-Bound Imaginaries in Automating the Safety of the Elderly." *Social Studies of Science* 36 (4): 599–626.

Jensen, T. E. 2012. "Intervention by Invitation." *Science Studies* 25 (1): 13–36.

Knapp, J., J. Zeratsky, B. Kowitz. 2016. *Sprint: How to Solve Big Problems and Test New Ideas in Just Five Days.* New York: Simon & Schuster.

Landström, C., S. J. Whatmore, S. N. Lane, N. A. Odoni, N. Ward, and S. Bradley. 2011. "Coproducing Flood Risk Knowledge: Redistributing Expertise in Critical 'Participatory Modelling.'" *Environment and Planning A* 43 (7): 1617–33.

Latour, B. 1987. *Science in Action: How to Follow Scientists and Engineers through Society.* Cambridge, MA: Harvard University Press.

——. 1993. *The Pasteurization of France.* Cambridge, MA: Harvard University Press.

——. 2004. "Why Has Critique Run Out of Steam? From Matters of Fact to Matters of Concern." *Critical inquiry* 30 (2): 225–48.

——. 2005. *Reassembling the Social: An Introduction to Actor-Network-Theory.* Oxford: Oxford University Press.

——. 2008. "The Space of Controversies: Interview with Bruno Latour." *New Geographies* 0:122–35.

Latour, B., P. Mauguin, and G. Teil. 1992. "A Note on Socio-technical Graphs." *Social Studies of Science* 22 (1): 33–57.

Latour, B., and S. Woolgar. 1979. *Laboratory Life: The Social Construction of Scientific Facts.* Beverly Hills, CA: Sage.

Law, J. 2009. "The Greer-Bush Test: On Politics in STS." Draft paper, December 23.

Lippmann, W. 1927. *The Phantom Public.* New Brunswick, NJ: Transaction.

Marres, N. 2007. "The Issues Deserve More Credit Pragmatist Contributions to the Study of Public Involvement in Controversy." *Social Studies of Science* 37 (5): 759–80.

——. 2015. "Why Map Issues? On Controversy Analysis as a Digital Method." *Science, Technology, & Human Values* 40 (5): 655–86.

Mol, A. 2002. *The Body Multiple: Ontology in Medical Practice.* Durham, NC: Duke University Press.

Moser, I. 2008. "Making Alzheimer's Disease Matter: Enacting, Interfering and Doing Politics of Nature." *Geoforum* 39 (1): 98–110.

Munk, A. K., and S. Abrahamsson. 2012. "Empiricist Interventions." *Science Studies* 25 (1): 5270.

Nowotny, H., P. Scott, and M. Gibbons. 2003. "Introduction: Mode 2 Revisited: The New Production of Knowledge." *Minerva* 41 (3): 179–94.

Petersen, M. K., and A. Munk. 2013. "I Vælten: Kulturanalysens Nye Hverdag." *Kulturstudier* 4 (1): 102–17.

Rheinberger, H. J. 1997. *Toward a History of Epistemic Things: Synthesizing Proteins in the Test Tube.* Stanford, CA: Stanford University Press.

Rowe, G., and L. J. Frewer. 2005. "A Typology of Public Engagement Mechanisms." *Science, Technology, & Human Values* 30 (2): 251–90.

Stengers, I. 1997. *Power and Invention: Situating Science.* Vol. 10. Minneapolis: University of Minnesota Press.

——. 2000. *The Invention of Modern Science.* Vol. 19. Minneapolis: University of Minnesota Press.

——. 2005. "The Cosmopolitical Proposal." In *Making Things Public: Atmospheres of Democracy,* edited by Bruno Latour and Peter Weibel, 994–1003. Cambridge, MA: MIT Press.

Venturini, T. 2010. "Diving in Magma: How to Explore Controversies with Actor-Network Theory." *Public Understanding of Science* 19 (3): 258–73.

——. 2012. "Building on Faults: How to Represent Controversies with Digital Methods." *Public Understanding of Science* 21 (7): 796–812.

Venturini, T., D. Ricci, M. Mauri, L. Kimbell, and A. Meunier. 2015. "Designing Controversies and Their Publics." *Design Issues* 31 (3): 74–87.

Vikkelsø, S. 2007. "Description as Intervention: Engagement and Resistance in Actor-Network Analyses." *Science as Culture* 16 (3): 297–309.

Whatmore, S. J. 2009. “Mapping Knowledge Controversies: Science, Democracy and the Redistribution of Expertise.” *Progress in Human Geography* 33:587–98.
Whatmore, S. J., and C. Landström. 2011. “Flood Apprentices: An Exercise in Making Things Public.” *Economy and Society* 40 (4): 582–610.
Woolgar, S. 1990. “Configuring the User: The Case of Usability Trials.” *Sociological Review* 38 (S1): 58–99.
Woolgar, S., C. Coopmans, and D. Neyland. 2009. “Does STS Mean Business?” *Organization* 16 (1): 5–30.
Wynne, B. 2007. “Public Participation in Science and Technology: Performing and Obscuring a Political-Conceptual Category Mistake.” *East Asian Science, Technology and Society* 1 (1): 99–110.
Yaneva, A. 2012. *Mapping Controversies in Architecture*. Oxford: Ashgate.
Yearley, S., S. Cinderby, J. Forrester, P. Bailey, and P. Rosen. 2003. “Participatory Modelling and the Local Governance of the Politics of UK Air Pollution: A Three-City Case Study.” *Environmental Values* 12 (2): 247–62.

Smart Artifacts Mediating Social Viscosity

Juan Salamanca

Every afternoon hundreds of pedestrians scramble to cross the five intersecting streets of the Shibuya district in Tokyo each time the traffic lights grant them the right of way over motor vehicles. Astonishing collective choreographies of cooperative and collaborative interactions emerge from walkers, each with their own destination, negotiating trajectories to make it safely across the street. In doing so they encounter not only each other but also several nonhuman actors such as crosswalks, sidewalks, and signs, constituting a network of artifacts and crowds.

This kind of naturally occurring interaction in the wild exhibits recurrent forms of emergent and dissolving interactions that may inform effective methodologies for the design of "socially apt" computational artifacts. A careful analysis of how artifacts participate in *micro-social* interactions serves to elucidate a novel approach for the design of artifacts that actively mediate people's *unplanned* cooperation, collaboration, conflict, or negotiation. The method used in this research picks emerging interactions mediated by regular artifacts, then replaces the artifacts with a smart version of them programed to foster the same interactions, and finally contrasts the observations to derive conclusions about how to design such smart artifacts. To do so, this research proposes a *triadic unit of analysis* that embraces human and nonhuman actors for the examination of actors' flow of action in such mediated interactions. The triadic unit of analysis serves as a methodological instrument that unpacks action's meaning and intentionality assembled when humans and artifacts intertwine their programs-of-action (Latour 1999). Meaning and intentionality can be inferred from the expected outcomes of parties involved in arranged interactions. For example, one person collaborating with someone else to move a bulky piece of furniture can infer the meaning and intention of her partner's effort because their actions aim for the same result. But this research is interested not in such prearranged interactions but in emergent ones where parties are unaware of what others are aiming to, thus the meaning and intentionality of their actions can be inferred only from moral values while their actions unfold. In the formulation of this methodological instrument, I dislocate artifacts from being regarded as the equipment for human action and recast them as social actors with scripted programs-of-action that participate in social collectives (Callon and Law 1995; Latour [1991] 1993, 2005). This conceptual move allows the adoption of the triadic unit of analysis, in contrast to the traditional user-object dyad that figures centrally in the User-Centered Design paradigm. From early empirical studies using this analytical instrument I observed that cooperative and collaborative interactions have some sort of internal resistance to collective action flow that resembles how a fluid with certain viscosity flows. According to my analysis I define *social viscosity* as the resistance to action flow exerted by actors—whether human or nonhuman—acting concurrently. Such resistance is created by mutual disturbances that occur between humans and artifacts during the interactions that bind them together in a social practice.

The case presented herein compares the social dynamics of walkers in a regular crosswalk with those of walkers in a custom-designed smart crosswalk. It illustrates how interactive aspects of the smart crosswalk's design promote mutually beneficial interactions between pedestrians crossing two intersecting streets. The proposed smart crosswalk is not intended to be a future infrastructure available on streets. It is rather a proof of concept that helps me to explore the benefits and drawbacks of intervening unplanned collective action flows with artifacts that intend to help people do better together. From a first analysis of pedestrians in the wild, I identified the traditional crosswalk (the graphical pattern of parallel white bars painted on the ground) as a relevant actor that constitutes, together with the traffic lights and sidewalks, a stage for unplanned interactions between strangers and cliques of acquaintances. Next, I designed a smart crosswalk programmed to interpret the configurations and behaviors of cliques of walkers walking in opposite direction, and adaptively promote cooperation and collaboration. Then, I re-created observed walkers' interactions in a laboratory setting with selected participants under several smart artifact conditions and compared their action flows. From the comparison, I derive that the social viscosity of a network could be partially determined by ways in which the design of smart artifacts signifies the dominant moral agreements of the practice in which they partake. I conclude with a set of methodological considerations for the design of socially apt artifacts oriented to promote balanced social interaction by means of computational artifacts.

The Design of Smart Artifacts

A necessary starting point for my analysis is to determine to what extent an artifact can be defined as smart. I extend Maturana's (2002) notion of structured determined systems to define *smart artifact* as a tangible computational unit that reacts to external phenomena depending on the structure and organization of its constituent elements. From a posthumanistic position, artifacts can be perceived and interpreted by people as social actors that participate in—hinder or potentiate—their courses of action. In the same vein, smart artifacts incorporate technologies for "perceiving" and "interpreting" people's actions and contextual information, in order to adapt themselves to the social context (Salamanca 2012). In its most simple composition, the structure of a smart artifact has three core elements: (1) tangible components—hardware, electrical parts, sensors, and actuators; (2) software components—operational system and programs; and (3) connectivity components—ports, antennas, and communication protocols (Porter and Heppelmann 2014). I am particularly interested in smart artifacts that foster collective action in social settings such as intelligent vehicles, adaptive traffic signals, public screens, sensible floors, and crowdsourced products and services (e.g., collaborative digital cartography such as Waze traffic maps), to mention a few instances of social computing technologies.

Computational artifacts have been the subject of inquiry of the interaction design community for quite a while. In 2011, the Museum of Modern Art in New York exhibit *Talk to Me: Design and the Communication between People and Objects* highlighted nearly 200 design projects that explore communication between people and computational artifacts (Antonelli 2011). According to Antonelli, the collection portrayed how contemporary designers not only shape form, function, and meaning, but also devise algorithms that enable pancommunication between everything and everybody in all possible combinations (Rosenberg 2011). The exhibition brought forth strong evidence of how the object of inquiry in interaction design evolved from how to make computational machines actionable and understandable, to how computational artifacts embody personalities that fit unobtrusively into people's everyday practices. The *Talk to Me* exhibit urged interaction designers to look beyond the understanding of users' cognition, psychology, language, and action, central aspects of the User-Centered Design approach. This design paradigm, so popular among designers during the last three decades, proved useful to convert enigmatic computing machines into consumer products (Norman and Draper 1986; Norman 2002), but it now falls short in accounting for how pervasive computational artifacts actually mediate people's social practices. This research seeks to contribute to a better understanding of the ways that smart artifacts worsen or ameliorate social viscosity in order to create socially apt artifacts.

Hybrid Participation in Social Practices

While the User-Centered Design perspective denies the ascription of action intention to artifacts by assuming that humans—the "doers"—compose the totality of the system's agency, a practice-centered design is open to distributing the system's action—the "doing"—in collectives of actors that include artifacts. The latter not only extends the scope of the human information-processing model in interaction design, but also anchors a strand of research concerning the design of things that participate in society as assistants, agents, delegates, signifiers, mediators, or intermediaries. Forlano (2016) argues that in decentering the human and embracing the attribution of agency to nonhumans, some critical and speculative design practices (Dunne 2005; Dunne and Raby 2013; Ratto 2011) are challenging the boundaries of the corporate design brief by eliciting points of tension among design traditions. In the same vein, my proposal of recentering design practice opens into fundamental contributions that are primarily oriented toward expanding the domain of interaction design from user interface and experience to adaptive mediators in social practices. On similar grounds Kuutti and Bannon (2014) propose a turn to practice in human-computer interaction. For them, a practice perspective encompasses a broader domain beyond the human-computer dyad, and brings to center-stage bodies, environment, artifacts, human capabilities, motivations, and values. Moreover, Knorr Cetina (1997) concurs with some posthumanist thinkers (Callon and Law 1995; Latour 1988) in that the analysis of a social practice is imperfect unless it accounts for how artifacts mediate human activities. They argue that we are increasingly living in an object-centered society where artifacts take on significance through their roles not only as equipment but also as mediators, elements of cohesion, or commonplaces of identity characterization. Following this strand of thought, artifacts are not merely tools or commodities, but active partakers in human-artifact collectives.

Material Mediation of Social Practices

Material mediation in social practices occurs when artifacts participating in interactions "displace the actors' viewpoints, act as a memory support and contribute to the setting up of a work collective" (Vinck 2012, 97). In the argument presented herein, mediating smart artifacts are envisioned as "social signifiers" (Norman 2011) of the dominant form of sociality of the practice mediating the dissonances created when individuals engage in unplanned social interactions (Serrano [1977] 2008).

Even though smart artifacts might not be entitled to privileges and obligations in the same manner as humans, they can be designed to reify social patterns, motives, moral principles, and political agendas (DiSalvo 2012). Latour (1999) famously uses a speed bump as an example of how things reify such aspects of human sociality. When a speed bump located on a neighborhood street compels drivers to slow down, it reifies the desire of the neighbors to keep their children safe from cars and represents the neighbors' authority over streets, even though drivers are entitled to drive freely in the city. It is in the scope of activities interrelating several actors—parents caring for their children, children playing on the street, and drivers commuting in their cars—that the speed bump reveals its moral facet because in this moment the bump behaves as an actor mediating the actions of multiple parties. Such scope of "embodied, materially mediated arrays of human activity centrally organized around shared practical understanding" is what (Schatzki 2001, 2) defines as a social practice.

Suchman's canoeing example (Suchman 2007, 72; Hutchins 1995) helps us to illustrate the point. Situated action contends that artifacts and human actions are meaningful in relation to their moment-by-moment circumstances rather than to plans per se. A canoeist dealing with the challenges posed by rapids resorts to whatever embodied skills he or she has rather than carefully monitoring the status of a detailed plan. During the course of action the canoeist turns to the devised plan to identify the most convenient position to be able to face the challenges anticipated before boarding. In naturally occurring practices, such as walking in public spaces, people often throw themselves into acting purposefully without detailed courses of action. But, if we look at the distributed agency of the sociotechnical configuration of the street, a clique of walkers is as subscribed to the program-of-action of a sidewalk as the canoeist is subscribed to the current. Thus, both sidewalk (a designed artifact) and river (a natural object) codetermine the action of any other actor entangled in their respective programs. The action flow of each pedestrian in a clique can be estimated (by herself, an observer, or a smart artifact) in relation to the sidewalk's design. But the fluidity and consistency of their action flow depends on those of the other members of the clique; those of other humans and nonhumans acting concurrently on the sidewalk.

Unpacking Assembled Meaning and Intentionality: A Methodological Instrument

Up to this point I have argued that the analysis of artifacts participation in social practices is a source of insights for the design of socially apt smart artifacts that mediate emergent interactions. Such analysis looks at how the programs-of-action

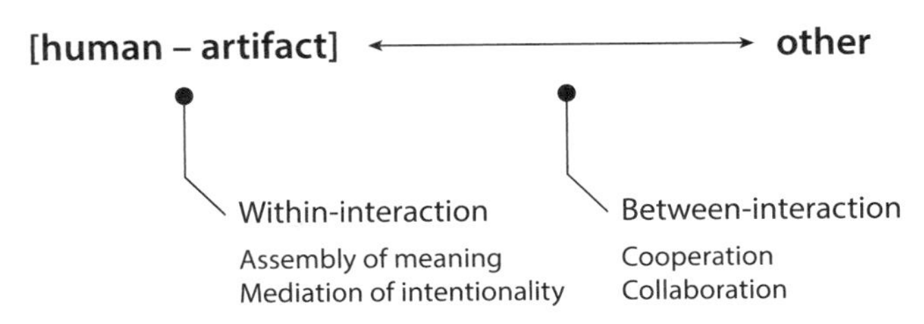

FIGURE 1: Triadic unit of analysis and domains of interaction. The within-interaction domain exhibits two facets: assembly of meaning and mediation of intentionality. The between-interaction domain comprises the forms of social interaction between collectives.

of artifacts codetermine those of humans resulting in a network of entangled actors. In this section I introduce a methodological instrument to investigate the assembling of meaning and intentionality of action in those entanglements.

As actor-network theory suggests, the methodological approach in this research is to follow the action of networked actors, whether human or nonhuman (see, e.g., (Latour 1999, 2005; Callon and Law 1995). To do so, I formulate a *triadic unit of analysis* constituted by a human actor associated with an artifact, both interacting as a whole with another human actor or collective (see figure 1).

The minimal requirement to be regarded as an actor of the triad is to get subscribed to the ongoing course of action. This means that new actors are participants in the network when they influence the course of action of current actors. While actors act on their own, the collective outcome is not necessarily a competitive game where each one maximizes one's payoffs, but a stable imbroglio of cooperation, collaboration, conflict, and negotiation in which morally laden interaction logics determine the more socially rewarding courses of action.

The triadic unit of analysis has two domains of interaction: the domain of *within-interactions* that accounts for the interactions that hold together humans and artifacts inside collectives and the domain of *between-interactions* that accounts for the interactions between collectives (e.g., cooperation, collaboration, or conflict).[1] Both domains are intrinsically related because within-interactions assemble meanings and mediate intentionalities reified by between-interactions. For simplicity, in the remaining text I use the bracketed notation [a-b] for within-interactions and a←→b for between-interactions.

A dissection of within-interactions requires a closer look at two facets of people interacting with artifacts. On facet accounts for how meaning is assembled; the other considers how artifacts mediate human intentionality. Latour (1999, 178–85) characterizes the assembly of meaning by describing how artifacts and humans translate, compose, pack, or delegate their programs-of-action.[2] His proposal has the advantage of focusing on the mechanisms of meaning articulation rather than on the meaning itself. The other facet accounts for how artifacts mediate human intentionality. It comprehends both the mental state of us and the object of such mental state. Husserl terms the former "noesis" and the latter "noema" (Dourish 2001). For Ihde (1990), people's mediated interactions with the world always occur through technologies that are present to someone in different degrees according to where it is located within our object of intentionality (Ihde 1990; Verbeek 2005, 2015). Such degrees of presence describe a continuum: On one side, the artifact's mediation of human experience of the world is unnoticeable (e.g., the car suspension operates in the background while one is driving). On the other side, the artifact's mediation is fully present in the foreground of human awareness (e.g., the speed bump hindering the car-driver flow). In the domain of within-interactions when the mediating artifact is unnoticeable, the object of intentionality targets actors outside the human-artifact collective. Conversely, when the mediating artifact is present-at-hand, the object of intentionality targets actors inside such a collective.

To illustrate how within-interactions can be dissected in relation to their two facets, picture a driver at an intersection switching on the car's left turn signal to warn a pedestrian of her intention. On the one hand, two forms of the assembled meaning are evident in that situation: *composition* because the unit `[car-driver]` turning to the left is not just the driver but a composed meaningful collective, and *delegation* at the moment the driver delegates to the car the function of warning the pedestrian of her intention. On the other hand, the mediated intentionality is explicit as the car embodies a unit with the driver, and this unit's *object of intentionality* targets the pedestrian, who is outside the domain of the collective's within-interaction.

Empirical Case Study

In order to test the proposed methodological instrument, I designed an empirical comparative study of pedestrians walking on a street crosswalk. In this section I explain the procedure and findings of an examination of raw within-interactions and between-interactions in the wild that sets the touchstone by which a custom-designed smart crosswalk was further evaluated. In the following section I describe such evaluation in terms of how the new smart crosswalk affected cooperation and collaboration in staged interactions of pedestrians. As mentioned earlier, the smart crosswalk is not intended to be a product available on streets; it is rather a research tool for the exploration of how smart artifacts affect emergent social interactions.

The case chosen for study were the naturally occurring interactions on a busy downtown crosswalk in the city of Chicago. Crosswalks are illustrative venues of nonverbal emergent cooperation, collaboration, conflict, and negotiation, where cliques and single walkers exhibit processes of participatory sense making. The sources of evidence were direct observation and two hours of video clips that captured dozens of walkers interacting on the crosswalk from a bird's-eye point of view. The video footages were analyzed with Path Analytics, a custom-made video analysis tool that generated a detailed dataset of each pedestrian path including time-stamped strides, course, displacement, and deviation from expected trajectory (see figure 2). It also visualizes forecasted positions and potential zones of conflict between pedestrians and cliques.

The analysis showed that urban pedestrians walk by themselves or in cliques of two or three. Larger groups of acquaintances regularly split in combinations of pairs or trios. The configuration of walking dynamics on sidewalks can be summarized in six combinations of walker sets (single, pair, or trio) with opposing trajectories—named "north pedestrians" and "south pedestrians." The analysis also showed that the crosswalk width and the size of the interacting cliques have a strong effect on the overall walking flow. Large cliques exhibit steadier courses, whereas small ones often divert their trajectories.

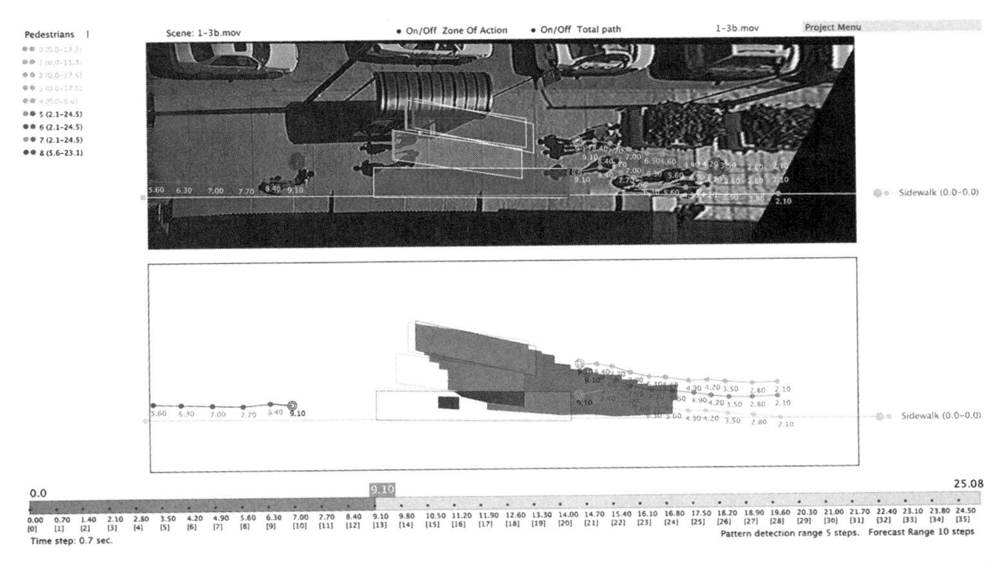

FIGURE 2: Analysis of trajectory interaction between pedestrians. The clear areas describe the zone of cooperation of a clique of three walkers that open space for another one walking in opposite direction. The hatched areas are the potential zones of conflict between them.

The moment pedestrians step on the crosswalk they subscribe to the program-of-action inscribed in its design attributes (i.e., afford walkers traveling straight across corners) constituting a collective. The representative triadic unit of analysis is defined as the collective of north pedestrians walking of the crosswalk interacting with south pedestrians or `[north pedestrians-crosswalk]` ←→`south pedestrians`. This unit is simple enough to illustrate the complexity of the triadic interaction, and rich enough to depict the human and nonhuman interactions of walkers on the street. In terms of the two facets of the within-interaction domain, the facet of assembled meaning is a composition—in Latour's terms—resulting in the articulation of pedestrians and crosswalk programs-of-action. As for its mediated intentionality, the locus of each walker's intentionality is not at the approaching pedestrian nor at the crosswalk but at the foreseen conflicting intersection of the three programs-of-action reified in a location on the pattern of white stripes. The analysis of the between-interaction clearly revealed that people walking in the same direction collaborate whereas those heading in the opposite direction cooperate. This triad of actors dissolves at the moment the pedestrians step out of the crosswalk unsubscribing from its program-of-action. Collaborative interactions differ from cooperative ones in that in the former two or more actors or collectives align their programs-of-action to achieve congruent goals, whereas in the latter they weave their discordant programs-of-action while pursuing their own best possible outcomes. Collaborative parties have selfless interest in contributing to the achievement of the goals of other actors. This was observed when cliques aligned themselves to follow the trail of unacquainted predecessors. Cooperative parties may have selfish attitudes and share a symbiotic interaction because their mutual successes benefit one another. This was frequently observed when single walkers with conflictive trajectories with large cliques gave them the right of way.

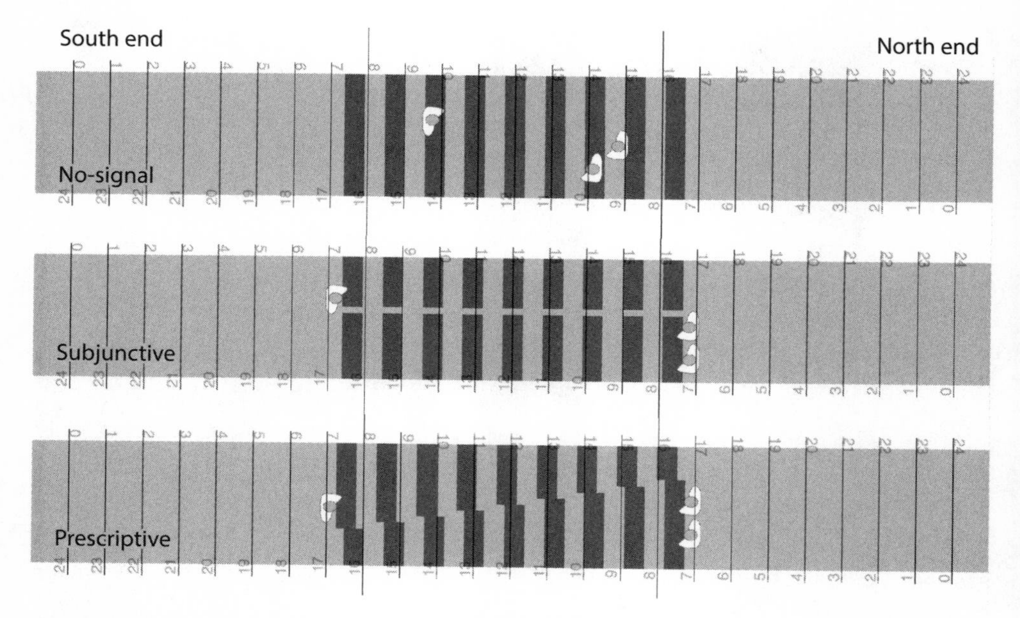

FIGURE 3: Smart crosswalk signaling modes tested in a staged experiment.

Staged Studies and Smart Crosswalk Design

Due to the fact that walkers locate their intentionality at the points of intersection foreseen by themselves in the white striped pattern, I designed an intelligent pedestrian crossing that predicts such common points of intersection and visually points out ways to avoid the potential conflict. For the prediction, a computer video processor anticipates in real time spots of conflicting interaction from a bird's-eye video source located above walkers on the smart crosswalk. Those spots became the splitting epicenters of the white striped pattern that timely prompted changes in walking courses by signaling the most balanced distribution of walking space among concurrent cliques (see figure 3). The signaling behavior had two modes: In the *subjunctive* mode the smart crosswalk signaling tone is a suggestion of the ideal distribution of its width proportionally to the size of the cliques.[3] It simply opens a gap in the striped pattern along the crosswalk longitudinal axis. In the *prescriptive* mode, the signaling tone is indicative, also displaying a balanced distribution of the walking space but emphasizing a walking direction. In this mode, the smart crosswalk splits the stripped pattern along a diagonal axis and slightly slides both halves in opposite direction indicating both parties a trajectory to get around the zone of conflict. In some of the trials the walkers' visual perception range was limited, expecting that they would heavily rely on the information signaled by the smart crosswalk's pattern.

In a staged study, 48 walkers organized in cliques of one, two, and three ran 12 trials to test a full-sized smart crosswalk deployed at the IIT Institute of Design in Chicago (see figure 4). With the help of Path Analytics, I recorded the same variables as those observed in the wild and contrasted them.

FIGURE 4: Staged experiment of smart crosswalk signaling patterns.

Results and Interpretation

For the most part, both smart crosswalk signaling modes positively impacted the consistency of pedestrians' walking course because their trajectory deviation was considerably lessened in comparison with observations in the wild. This constitutes evidence of improvement in coordination. Yet, in spite of better coordination, the pedestrians' walking speed was slower. A possible explanation is that pedestrians do not care about how much they need to deviate from their course as long as they maintain their pace. Hence, I deduce that cliques prioritize the achievement of their intended target at a consistent velocity compromising their course stability.

The subjunctive signal produced lower target deviations and clearer pedestrian walking paths than the prescriptive signal. An interpretation is that the pedestrians might improve the coordination of their actions when their attention is directed toward potential conflicts. In other words, smart crosswalk signals complement pedestrians' contextual information of the ongoing social situation. However, pedestrians' efficiency to circumvent those conflicts was diminished when prescriptive signals were suggested.

The limitation of walkers' visual perception yielded no surprising results. Pedestrians walk faster in full visual perception, and the shorter the visual perception field, the slower the walking velocity. Similarly, the larger the number of subjects participating in a trial, the slower their average speed. But data show intriguing results when we scrutinize the walkers' trajectory deviations because people with full visual perception, both in the staged experiment and on the street, tend to divert their trajectories significantly more than those with limited visual perception. A careful analysis of clique members' walking course revealed that one member usually makes a greater effort to maintain her trajectory flow. In full visual perception, the larger clique yields the right of way over the opposing party when the smart crosswalk is in subjunctive tone, whereas the opposite was observed in prescriptive tone. An interpretation is that calling the attention of pedestrians to their potential conflicts via the subjunctive signals results in improvements of the walking flow of the smaller group. Consequently, the largest party has to exert greater effort to maintain its course. The inverse situation is observed in prescriptive signal. The larger collectives benefit the most from the smart crosswalk prescriptions, shifting the burden of adapting the trajectory to the smaller collectives.

An examination of the within- and between-interactions of the triad `[pedestrian-smart crosswalk]` ⟵⟶ `other pedestrian` helps us to elucidate the creative process of augmenting a traditional artifact with autonomous and adaptive functionalities that enable it to actively intervene in social practice. Let us make first a parallel between the within-interaction domains of regular crosswalks and smart crosswalks. In either a regular crosswalk or a smart crosswalk, the assembled meaning is the result of a composition of actors, but the smart crosswalk packs more actors than the regular crosswalk because it not only includes tangible components but also entails a designer/programmer of nonhuman attitudes evident in the signaling tone of the smart crosswalk. In terms of walkers' intentionality mediation, there is a radical shift from an unnoticeable mediation of the regular crosswalk to a hermeneutic mediation of the smart crosswalk that combines visual and performing languages to overtly signify frequent courses of action for the common good. Simply put, a smart artifact is a signifier of morally accepted situated actions.

As for the between-interaction domains of regular crosswalks and smart crosswalks, cooperation and collaboration emerged indistinctively, but the coordination among walkers and the stability of their walking paths are remarkably improved while walking on smart crosswalks, especially in prescriptive mode. This means that smart crosswalks benefit the cohesion and stability of large cliques.

The Social Viscosity of Assemblages

The use of the triadic unit of analysis in the staged design study reveals that while actors collectively attempt to enact their programs-of-action, they offer resistance to one another, hindering the interaction flow of the entire assemblage. Following the naïve definition of mechanical viscosity—the resistance to flow exhibited by a viscous matter—I define *social viscosity* as the resistance to social action flow elicited by human and nonhuman actors acting concurrently.

Actor-networks exhibit social viscosity that affects the flow of independent action unevenly. While smaller collectives coordinate easily, larger collectives struggle to establish and maintain coordination by continuous acts of balancing and rebalancing. The observed accumulated disturbance reveals the internal friction of actors, as a result of their attempt to enact their programs-of-action. Such friction, which ultimately renders the actor-network viscous, seems to thicken when people act under limited access to prescient environmental information. It is under those limited conditions when smart artifacts may have a greater impact on the network's viscosity and benefit communal action flow. With the example of a skillful doorman and a mechanical door closer, Latour (1988) exemplifies how highly trained humans and better designed nonhumans could facilitate the flow of action in an actor-network. In the same vein, this study demonstrates that the design of socially apt nonhumans associated with well-behaved humans could reduce the social viscosity of the actor-network.

Conclusion

The embrace of autonomous and adaptive everyday artifacts by society highlights the social dimension of interaction design. Instead of regarding everyday artifacts as mere tools or equipment, they need to be considered as agents with capacity for action in social contexts.

This research adopted a posthumanist position to tackle such conceptual shifts. It repositions artifacts within social practices as active participants symmetrical to humans enrolled in such practices. The triadic structure of networked social interaction presented herein is a methodological instrument for the investigation of smart devices adaptively mediating people's interactions. Such a triadic unit of analysis accounts for the interactions within and between collectives in actor-networks. The within-interactions are those that hold together humans and smart artifacts inside a collective and put forward the collective's meaning and intentionality for other actors in the network. The between-interactions are those that occur among collectives and characterize the dominant model of sociality of the actor-network.

I envision that the triadic structure can be used in the interaction design process as an instrument for analysis of activities and identification of their focuses of activity. Moreover, the proposed triadic structure is an instrument for the evaluation of the social effects of artifacts in an actor-network in that the study of the between-interactions reveals the types of social interaction in the triad—cooperation and collaboration in this case—describing the directionality of actors' goals, alignment of their programs-of-action, and mutual consideration for each other's interests. The analysis of cooperation and collaboration demonstrated that coordination is a contingent condition for their emergence, regardless of the presence of coordination technologies. But our findings indicated that any mediator, at the expense of the fluidity of social action, may positively affect the degree of coordination. The challenge for smart artifact designers is to design social mediators that enhance—and do not hamper—the fluidity of social action.

This research suggests that researchers and practitioners designing smart artifacts aiming to promote cooperation and/or collaboration need to do the following:

1. Define the social practice in which the smart artifact will be involved and select the most relevant activities.
2. Understand the set of social logics and values that define the social domain of the practice and the cultural behaviors of human participants that engage and affect the practice.
3. Map the triads of networked humans and nonhumans and describe their constituent collectives. Identify which artifacts have the highest potential to become smart artifacts by determining which nonhuman actors are located at relevant focuses of activity.
4. Analyze the within-interactions of collectives to understand the assembled meanings by technological means and the mediation of intentionalities of their hybrid actors.

5. Analyze the between-interactions to understand the aspects that define the interactions between collectives: directionality of goals, compatibility of programs-of-action, and consideration for others' interests. In doing so, specify the type(s) of coordinated action as cooperative or collaborative.
6. Design smart artifacts that signify the social logics and values, and promote types of coordination that support *between-interactions.*
7. Deploy smart artifacts in the social practice and evaluate their impact in the social viscosity of the actor-network.

Notes

1. A collective is defined as an association of heterogeneous actors that forms a meaningful whole within an array of human activity (Latour 1999, 2005).
2. *Translation* is exemplified by how the moral integrity of a good citizen is degraded when she holds a gun; *composition* is exemplified by how traveling by plane is possible thanks to the association of pilots, planes, airports, launch pads, air traffic controllers, and ticket counters; *black-boxing* is exemplified by the network of repairmen, light bulbs, and lenses packed in an overhead projector (i.e., a sealed unit) that gets unpacked when the device breaks down in the middle of a lecture; and finally, *delegation* is exemplified by traffic engineers entrusting the reduction of car drivers' speed to speed bumps.
3. On subjunctive and prescriptive interaction, see Tomasello (2010) and Sennett (2012).

Works Cited

Antonelli, Paola. 2011. *Talk to Me: Design and the Communication between People and Objects.* New York: Museum of Modern Art.

Callon, Michel, and John Law. 1995. "Agency and the Hybrid Collectif." *South Atlantic Quarterly* 94 (2): 481–507.

DiSalvo, Carl. 2012. *Adversarial Design: Design Thinking, Design Theory.* Cambridge, MA: MIT Press.

Dourish, Paul. 2001. *Where the Action Is: The Foundations of Embodied Interaction.* Cambridge, MA: MIT Press.

Dunne, Anthony. 2005. *Hertzian Tales: Electronic Products, Aesthetic Experience, and Critical Design.* Cambridge, MA: MIT Press.

Dunne, Anthony, and Fiona Raby. 2013. *Speculative Everything: Design, Fiction, and Social Dreaming.* Cambridge, MA: MIT Press.

Forlano, Laura. 2016. "Decentering the Human in the Design of Collaborative Cities." *Design Issues* 32 (3): 42–54.

Hutchins, Edwin. 1995. *Cognition in the Wild.* Cambridge, MA: MIT Press.

Ihde, Don. 1990. *Technology and the Lifeworld: From Garden to Earth.* Indiana Series in the Philosophy of Technology. Bloomington: Indiana University Press.

Knorr Cetina, Karin. 1997. "Sociality with Objects." *Theory, Culture & Society* 14:1–30. doi:10.1177/026327697014004001.

Kuutti, Kari, and Liam J. Bannon. 2014. "The Turn to Practice in HCI: Towards a Research Agenda." In *Proceedings of the 32nd Annual ACM Conference on Human Factors in Computing Systems*, 3543–52. New York: ACM.

Latour, Bruno. 1988. "Mixing Humans and Nonhumans Together: The Sociology of a Door-Closer." *Social Problems* 35 (3): 298–310.

——. [1991] 1993. *We Have Never Been Modern.* Edited by Nous n'avons jamais ètè modernes. Cambridge, MA: Harvard University Press.

——. 1999. *Pandora's Hope: Essays on the Reality of Science Studies.* Cambridge, MA: Harvard University Press.

——. 2005. *Reassembling the Social: An Introduction to Actor-Network-Theory.* Oxford: Oxford University Press.

Maturana, Humberto. 2002. "Autopoiesis, Structural Coupling and Cognition." *Cybernetics & Human Knowing* 9 (3–4): 5–34.
Norman, Donald A. 2002. *The Design of Everyday Things*. New York: Doubleday.
——. 2011. *Living with Complexity*. Cambridge, MA: MIT Press.
Norman, Donald A., and Stephen W. Draper. 1986. *User Centered System Design: New Perspectives on Human-Computer Interaction*. Hillsdale, NJ: Erlbaum.
Porter, Michael E., and James E. Heppelmann. 2014. "How Smart, Connected Products Are Transforming Competition." *Harvard Business Review* 92 (11): 64–88.
Ratto, Matt. 2011. "Critical Making: Conceptual and Material Studies in Technology and Social Life." *Information Society* 27 (4): 252–60.
Rosenberg, Karen. 2011. "Art That Interacts if You Interface." *New York Times*, July 29. www.nytimes.com/2011/07/29/arts/design/momas-talk-to-me-focuses-on-interface-review.html?_r=0.
Salamanca, Juan. 2012. "Designing Smart Artifacts for Adaptive Mediation of Social Viscosity: Triadic Actor-Network Enactments as a Basis for Interaction Design." Doctoral dissertation, Illinois Institute of Technology. https://pqdtopen.proquest.com/pubnum/3570096.html.
Schatzki, Theodore. 2001. "Practice Theory." In *The Practice Turn in Contemporary Theory*, edited by Theodore Schatzki, Karin Knorr Cetina, and Eike Savigny, 1–14. Oxon: Routledge.
Sennett, Richard. 2012. *Together: The Rituals, Pleasures, and Politics of Cooperation*. New Haven, CT: Yale University Press.
Serrano, Manuel M. [1977] 2008. *La Mediación Social*. Madrid: Akal.
Suchman, Lucille Alice. 2007. *Human-Machine Reconfigurations: Plans and Situated Actions*. 2nd ed. Cambridge: Cambridge University Press.
Tomasello, M. 2010. *¿Por qué Cooperamos?* Buenos Aires: Katz Editores.
Verbeek, P. 2005. *What Things Do: Philosophical Reflections on Technology, Agency, and Design*. University Park: Pennsylvania State University Press.
——. 2015. "Cover Story: Beyond Interaction: A Short Introduction to Mediation Theory." *Interactions* 22 (3): 26–31. http://dx.doi.org/10.1145/2751314.
Vinck, Dominique. 2012. "Accessing Material Culture by Following Intermediary Objects." In *An Ethnography of Global Landscapes and Corridors*, edited by Loshini Naidoo, 89–108. London: Intech.

Actor-Network versus Network Analysis versus Digital Networks

Are We Talking about the Same Networks?

Tommaso Venturini, Anders Kristian Munk, and Mathieu Jacomy

Odi et amo. quare id faciam, fortasse requiris?
nescio, sed fieri sentio et excrucior.
Catullus 85 or Carmina LXXXV

> Professor: You should not confuse the network that is drawn by the description and the network that is used to make the description.
>
> Student: . . . ?
>
> Professor: But yes! Surely you'd agree that drawing with a pencil is not the same thing as drawing the shape of a pencil. It's the same with this ambiguous word, network. With actor-network you may describe something that doesn't at all look like a network—an individual state of mind, a piece of machinery, a fictional character; conversely, you may describe a network—subways, sewages, telephones—which is not all drawn in an "actor-networky" way. You are simply confusing the object with the method. ANT is a method, and mostly a negative one at that; it says nothing about the shape of what is being described with it.
>
> Student: This is confusing! But my company executives, are they not forming a nice, revealing, significant network?
>
> Professor: Maybe yes, I mean, surely, yes—but so what?
>
> Student: Then, I can study them with actor-network theory!
>
> Professor: Again, maybe yes, but maybe not. It depends entirely on what you yourself allow your actors, or rather your actants to do. Being connected, being interconnected, being heterogeneous, is not enough. It all depends on the sort of action that is flowing from one to the other, hence the words "net" and "work." Really, we should say "worknet" instead of "network." It's the work, and the movement, and the flow, and the changes that should be stressed. But now we are stuck with "network" and everyone thinks we mean the World Wide Web or something like that.

Student: Do you mean to say that once I have shown that my actors are related in the shape of a network, I have not yet done an ANT study?

Professor: That's exactly what I mean: ANT is more like the name of a pencil or a brush than the name of an object to be drawn or painted.

—Bruno Latour, 2005, *Reassembling the Social: An Introduction to Actor-Network Theory,* Oxford: Oxford University Press (pp. 142, 143)

From Conflation Comes Power

Say what you want, analytical dissection is not the only motive of science. Often, the desire to fit together concepts coming from different traditions and disciplines feels just as urgent. A good example is the conflation that in the last three decades has seen three different meanings of the word "network" merge in STS.

It arguably began in 1986 when Michel Callon introduced the term "actor-network" as a conceptual tool to "describe the dynamics and internal structure of actor-worlds" (Callon 1986, 28). It is worth remembering that Callon's essay appeared in the volume *Mapping the Dynamics of Science and Technology,* a book that intended to complement the traditional ethnographic techniques employed in STS with new methods derived from scientometrics and text analysis.

Three ingredients of network conflation were already there:

1. The theoretical idea that collective phenomena are best described not by the substances, but by the relations that constitute them (actor-network theory)
2. The methodological appeal for new quantitative techniques to analyze and represent the connections between social actors (network analysis)
3. The intuition that the inscriptions left by collective actions (scientific publication in the specific case) could be repurposed for social research (network data)

The ambiguity of the word "network"—which can equally refer to a conceptual topology (the space of connections as opposed to the Euclidian space of coordinates), to a set of computation techniques (the mathematics of graphs), and to the hypertextual organization of inscriptions (the relational datasets)—suggested that the conflation was possible and, indeed, desirable.

Conflating these otherwise disparate notions of "network" was more than a conceptual trick. It involved wedding the ideas of actor-network theory (ANT) to some of the methods of social network analysis (SNA). The marriage was particularly appealing because it promised a way to follow sociotechnical associations across sites (see Knorr Cetina 1995; Vinck 2012). But the wedding had appeal to social network analysts as well, who could find in it the theoretical framework that they had missed (Granovetter 1979 laments a "Theory-Gap in Social Network Analysis" and Burt 1980 argues that "the lack of network theory seems to me to be the most serious impediment to the realization of the potential value of network models in empirical research"; 134).

Yet, for quite some time, the marriage between ANT and SNA bred few progeny. ANT scholars felt the appeal of SNA techniques, but were afraid their definition of "social relations" would be too narrow. Having spent half a decade defending the role of nonhuman actors, actor-network theorists could not settle for networks restricted to human beings.

Hence the interest for scientific inscriptions and more generally for the variety of "intermediary objects" (scientific papers, technological devices, animal models, measuring instruments) producing relational data complementary to that of human relations. Many of such objects exhibited connections that could be traced and analyzed (Vinck 2012). Studying them produced the first embryo of the hybrid addressed in this article: a quali-quantitative approach to heterogeneous networks (Venturini and Guido 2012). The qualitative observations realized in science and technology studies suggested new applications for the quantitative techniques of network analysis. Callon, for example, started investigating co-occurrence in titles after observing (through ethnographic work) that the association of words was commonly used as an "interestment device."

Still, collecting traces on such hybrid networks was as demanding as traditional ethnographic work (if not more), and the shortage of relational data limited the interest of the ANT/SNA conflation. Such shortage was overcome with the advent of yet another type of networks, namely those emerging from digital mediation. Speaking at the Virtual Society? conference (Woolgar 2002), Bruno Latour (1998) suggested that social connections become more material and thereby more traceable when flowing through digital infrastructures: "Once you can get information as bores, bytes, modem, sockets, cables and so on, you have actually a more material way of looking at what happens in Society. Virtual Society thus, is not a thing of the future, it's the materialisation, the traceability of Society. It renders visible because of the obsessive necessity of materialising information into cables, into data." In the audience were two young sociologists, Richard Rogers and Noortje Marres, who, in the following years, developed a series of tools and methods to put digital traces at the service of the social sciences (see Rogers 2004, 2013; Rogers and Marres 2000, 2002; www.digitalmethods.net): "Bruno Latour (1998), argued that the Web is mainly of importance to social science insofar as it makes possible new types of descriptions of social life. According to Latour, the social integration of the Web constitutes an event for social science because the social link becomes traceable in this medium. Thus, social relations are established in a tangible form as a material network connection. We take Latour's claim of the tangibility of the social as a point of departure in our search" (Rogers and Marres 2002, 342). It is important to notice that it is not the volume of digital data that made the difference (this is *not* a "big data" argument), but its relational nature. As digital media are organized as networks at the both physical and content levels (the Internet is the interconnection of computer networks and the World Wide Web is the interconnection of online hypertexts), the inscriptions that they produce are *natively* relational. TPC/IP (Transmission Control Protocol / Internet Protocol), HTTP (Hypertext Transfer Protocol), the Relational Databases, and all major protocols and formats supporting digital communication are relations based.

By generalizing the practice of citation beyond the scientific literature (Leydesdorff 1998; Leydesdorff and Wouters 1999), digital protocols contributed to formalize collective life as a network of association, both in the sense of extending the reach of the network methods developed in scientometrics (see, for example, how Roth and Cointet 2010 employed the exact same techniques to study scientists working on the zebra fish and US political bloggers) and in the sense of encouraging collective life to organize in network-like shapes: "We took to the Web to study public debates on science and technology, but we found 'issue-networks' instead. . . . Following hyperlinks among pages dealing with a given issue, we found that these links provided a means to demarcate the network that could be said to be staging the controversy in the new medium" (Marres and Rogers 2005, 922). It would be nice here to tell the story of social sciences revealing the nature of a new medium and repurposing its formats for research. Things, however, are more complex, and while social scientists were striving to socialize web networks, computer scientists were busy engineering sociological methods—and scientometrics in particular—into digital media (Marres 2012a). The most famous example is contained in the article presenting PageRank, the algorithm that made the success of Google, where its inventors explicitly argue, "It is obvious to try to apply standard citation analysis techniques to the hypertextual citation structure of the web. One can simply think of every link as being like an academic citation" (Page et al. 1998, 2). This explains why the network conflation is so powerful: it is not just the meeting of two separate sociological schools; it is that this meeting takes place on the ground of one of the major technological (and economic) innovations of the last century. If it feels more and more natural to think of collective phenomena in relational terms, it is because digital mediation is increasingly turning them into networks. Our professional sector much more resembles a social network since our colleagues invited us on LinkedIn. Friendship has literally become a matter of connection, now that it is mediated by Facebook. And when we look at our library, we increasingly expect to see what other books "Customers Who Bought This Item Also Bought." The more it is mediated by network technologies, the more collective life can be read through the theory of networks, measured through network analysis, and captured in network data. "Sociologists of technology have long relied on methods of network and textual analysis in order to capture the unfolding of controversies. . . . Today the proliferation of digital technologies means that similar methods are deployed much more widely to analyse and visualise issues in digital networked media. . . . Indeed, network and textual analysis tools are now routinely deployed in digital culture" (Marres 2012a, 300). The (con)fusion of the four meanings of "network" (a conceptual metaphor, an analytic technique, a set of data, a sociotechnical system) is not just a product of sociology; it is a product of society. This is why the network conflation is so powerful—to the point that great is the temptation to argue not only that collective phenomena can be described and mediated through networks but also that society has in fact become a network (see Castells 2000; Van Dijk 1999) and even that everything has become a network (see Barabási 2002). And this is why the network conflation is so dangerous.

Networks Are Not Networks

As the uncle of Spiderman used to say, "With great power comes great responsibility," and the very same people who initiated the network conflation in STS, the actor-network theorists, have always been wary about its use and abuse. In particular, they were afraid that, while offering an operationalization of their relational analysis, it also risked blurring important parts of their approach. They were right.

The easiest way to answer the question asked by this chapter—"are we talking about the same networks?"—is with a simple "no, we are not." The networks captured by digital data and analyzed through the canon of graph mathematics do not resemble actor-networks in at least *four* respects.

Partiality and Bias of Digital Inscriptions

The first concerns the relational data that, as we said, catalyzed the fusion between ANT and network analysis. It is obvious but deserves to be mentioned: digital traces (like any other type of inscription) are not always representative of the phenomena that they allow to trace.

There are two main reasons for this. First, not all relevant collective actions are mediated by digital infrastructures: despite the growing extent to which digital mediation has infiltrated social life, there are still important interactions that fall beyond them. For instance, despite the advances in digitalization, the production of science and technology still relies on face-to-face interaction and direct manipulation. All the online journals and libraries will not replace the discussions in the corridors of conferences and all the computer simulations are no substitute for in vivo measure and in vitro experiments.

Second, digital technologies (as all media) do not just trace, but also translate the interactions that they support. Digital media are not the carbon paper that trace our writing, they are the paper that replace the parchment, thereby substantially affecting the nature of the books we write and read (Eisenstein 1980). This is not an abstract argument: working with digital traces entails a constant questioning of the findings obtained: What do I see when I examine the evolution of a hashtag? Public opinion, or Twitter (Marres and Gerlitz, 2015)? Digital inscriptions are not created by or for the social sciences; they are the product of vast sociotechnical systems comprising online platforms, commercial start-ups, communication protocols, fiber cables, and so forth, and bring with them the influences of such system. This is not to say, to be sure, that digital traces are more biased than other types of inscriptions, but that the conditions of their production are always to be remembered (Munk 2013; Venturini et al. 2014).

This first hitch concerns the catalyst (digital traces) that made possible the reaction between ANT and SNA, but other difficulties emerge when actor-networks and mathematical networks are compared. We describe them in the next paragraphs with reference to *conventional* graph mathematics. By *conventional*, we refer to the methods and tools that are implemented in standard network analysis. Though extensions have been proposed to overcome many of the limitations of conventional network analysis (see, for instance, Everett and Borgatti 2014 on negative connections and Chavalarias and Cointet 2013 on dynamic clustering), their experimental character has prevented them (so far at least) from entering the toolkit of social research.

Heterogeneity of Nodes and Edges

The first difference between graph mathematics and actor-networks was pointed out by Michel Callon (1986) in the very article in which he introduced the notion of the actor-network: "[An actor-network] is distinguished from a simple network because its elements are both heterogeneous and are mutually defined in the course of their association" (32). One of the ideas that aroused most interest around ANT is its extremely broad definition of social actors. According to ANT, social phenomena involve not only individuals (e.g., scientists and engineers) but also collective assemblages (e.g., laboratories and academic institutions), nonhuman actors (e.g., natural substances and technical devices), and even conceptual items (e.g., scientific theories and legal frameworks). At a first glance, this openness matches well with the agnosticism of graphs, whose elements have been used to represent almost everything (from websites to neurons, from proteins to words). Yet while actor-networks allow and even prescribe the presence of items of different nature in the same network, graph nodes tend to be of the same type.

The reason is simple: graph mathematics is hardly capable of handling qualitative differences. The items in a graph can be quantitatively different (as they may carry different "weights"), but they are all mathematically equivalent. It is possible to build networks with nodes of different type (see, for instance, Cambrosio et al. 2004), but belonging to one type or another will not affect what nodes can or cannot do.

This limitation is felt more strongly for edges than for nodes and sometimes referred to as the problem of "parallel edges." Imagine a network of Facebook accounts. As long as edges are limited to one type of connection (say friendship links), graph analysis can deliver interesting results (see Rieder 2013). But as soon as we try to project different types of relations *on the same network*, we stumble on the problem of weighting: How many "likes" should count as a comment? How much weaker does a friendship get when it is "unfollowed" (removed from the user's news feed)? Is posting a text stronger or weaker than posting an image? And, of course, combining traces coming from different platforms and media compounds the problem.

Negative relations are especially complicated. Collective life is made of opposition as much as of alliances, but conventional graph mathematics offers no convincing way to handle "negatively charged" edges. In network analysis, opposition is generally operationalized as a lack of association (see the concept of "structural hole" by Burt 1995). In citation analysis, for instance, it is commonly accepted that "there is no such thing as negative publicity." Garfield, one of the fathers of scientometrics, makes it very clear: "If scientists tend to ignore inferior work that is of little importance, then the work that they do go to the trouble of formally criticizing must be of some substance. Why, then, should negative citations be considered a sign of discredit?" (1979, 361–62).

This work-around has been successfully used to exploit network analysis for controversy mapping (Venturini 2010, 2012) and produced interesting results when applied to digital data (see, for instance, Adamic and Glance 2005). It often happens, however, that digital traces provide us information directly about opposition. For instance, studying controversies in Wikipedia, we can easily access "reverts" and other antagonistic edits, but to exploit them to detect "edit-factions" we need to turn the network around, according to the principle of "my enemy's enemy is my friend" (Borra et al. 2014).

Reversibility of Actor-Network

The second glitch in network conflation has to do with the hyphen connecting actor and network in ANT. This little typographical character is of critical importance and often misunderstood. The wrong way to read the hyphen is as a pointer to the interactions between the social actors and the system that would contain them: "the idea was never to occupy a position into the agency/structure debate, not even to overcome this contradiction. Contradictions should not be overcome, but ignored or bypassed" (Latour 1999, 15). Rather, the hyphen stands for an equal: actor=network: "To try to follow an actor-network is a bit like defining a wave-corpuscle in the 1930s: any entity can be seized either as an actor (a corpuscle) or as a network (a wave). It is in this complete reversibility—an actor is nothing but a network, except that a network is nothing but actors—that resides the main originality of this theory" (Latour 2010, 5). The hyphen is not meant to connect the two halves of the expression (actor *and* network), it is meant to deny *both* (*neither* actor *nor* network). Paradoxical as it may sound, in the world of actor-network there are no actors (entities defined by properties independent from the relations connects them) and no networks (structures defined by patterns independent from the elements that they connect).

This reversibility is absent from graph mathematics, where nodes and networks are described by different properties and measured by different metrics. It is even commonly accepted that SNA techniques can be separated in two analytic toolkits: one to study the ego networks (centered on a single node and its neighbors; see White 2000) and another to study global networks. Though such a distinction is more apparent than real (the two toolkits are based on the same graph mathematics), there is indeed a substantial difference in the way SNA conceives nodes (indivisible and impenetrable items) and networks (global and composite structures). And this difference aligns closely with the classic divides of social theory (micro/macro, interactions/structures, individuals/institutions, local/global, etc.; see Giddens 1984; Archer 1995) that ANT has always rejected (Callon and Latour 1981).

However, when looking at the actual techniques of network analysis, the separation between nodes and networks appears less significant. All the key properties of nodes (authority, centrality, betweenness, etc.) depend on the overall topology of the network in which they are located and, conversely, all the key properties of networks (diameter, modularity, clustering, etc.) depend on the local arrangements between nodes. In graph mathematics, nothing can be calculated about networks without considering each node and little can be calculated about nodes without considering the network it its entirety.

This is particularly more visible in the digital implementations of social networks (Latour et al. 2012). Consider, for instance, how Facebook breached earlier WWW conventions by developing a website without a homepage and without individual pages. And Facebook is no exception. All the homepages of the main Web2.0 platforms (Twitter, Flickr, Tumblr, Pinterest, etc.) are remarkably empty and systematically deserted by their users (how many times have you visited the homepage of Wikipedia?). But what is most striking about Facebook is that even the individual pages are of little importance. Yes, users can choose their name, edit their description, and upload a cover photo, but what core of their account is the "wall" in which the users' posts are mixed with (often drown in) the contents published by their "friends." The largest online social network is not a global structure lodging an ensemble of individuals (actors *and* network). It is a constant flux of recombinable contents relentlessly clotting and dissolving (actor=network) (see a similar analysis of Flickr by Boullier and Crépel 2012).

Dynamics of Relational Change

The last and possibly the most serious divergence between ANT and network analysis concerns time. ANT is essentially a theory of change. Its focus is not on the structure of associations, but on their dynamics. "Reality," writes Michel Callon, "is a process. Like a chemical body, it passes through successive states" (Callon 1984, 207). The difficulty in accounting for time as networks is not only a problem for ANT. According to Mustafa Emirbayer (1997), time remains one of the main obstacles in the operationalizing all relational sociologies: "Paradoxically (for a mode of study so intently focused upon processuality), relational sociology has the greatest difficulty in analyzing, not the structural features of static networks, whether these be cultural, social structural, or social psychological, but rather, the dynamic processes that transform those matrices of transactions in some fashion. Even studies of 'processes-in-relations,' in other words, too often privilege spatiality (or topological location) over temporality and narrative unfolding" (305). The difficulty graphs have in rendering dynamics is probably the reason why none of the diagrams appearing in the foundational texts of ANT are networks (see, for instance, Callon 1986; Latour et al. 1992; Law and Callon 1992). To be sure, it is not that graph mathematics is not interested in dynamics. On the contrary, movement has always been one of the major preoccupations of network analysts. After all, Euler (1736) invented graph mathematics precisely to solve the problem of moving through the city of Konigsberg and the core application of network theory is the management of flows (the routing of trains first and of communication soon after). Yet, movement in graph theory is usually movement *through* networks and not movement *of* networks. Rooted deep in graph mathematics is the separation between what flows (ideas, goods, signals, etc.) and what stays (the structure of connections) (Madsen 2015).

This separation is highly problematical for ANT, which has always denied the existence of a "context" in which action will take place. In ANT (which, it is worth to remember, is a sociology of translation, not of transport), networks are *not* conceptualized as systems of routes through which actors drive their way. Quite the opposite: they are the maze of trails left by children running through the uncut grass. It is the runner who makes the trail, not the other way around. This is yet another reason why actor-network theorists have been uncomfortable with the graph topography and why, for instance, John Law and Annemarie Mol (Mol and Law 1994; Law and Mol 2001) propose to replace networks with "fluid spaces" and "fire spaces," respectively characterized by the constant transformation and the constant overflowing of boundaries.

Being Sensitive to the Difference in the Density of Association

So is this it? Should we declare the case closed, divorce network analysis from ANT, and renounce exploiting the traceability of digital networks? We think not. We believe that there is a more positive (though admittedly riskier) answer to the question posed in the title of this chapter. To formulate it, one must gauge the potential equivalence among the three notions of "network" in a less literal way. No, graphs do not resemble actor-networks. As the pipe painted by Magritte does not resemble its referent (Foucault 1983), so the relations between the Bush and the bin

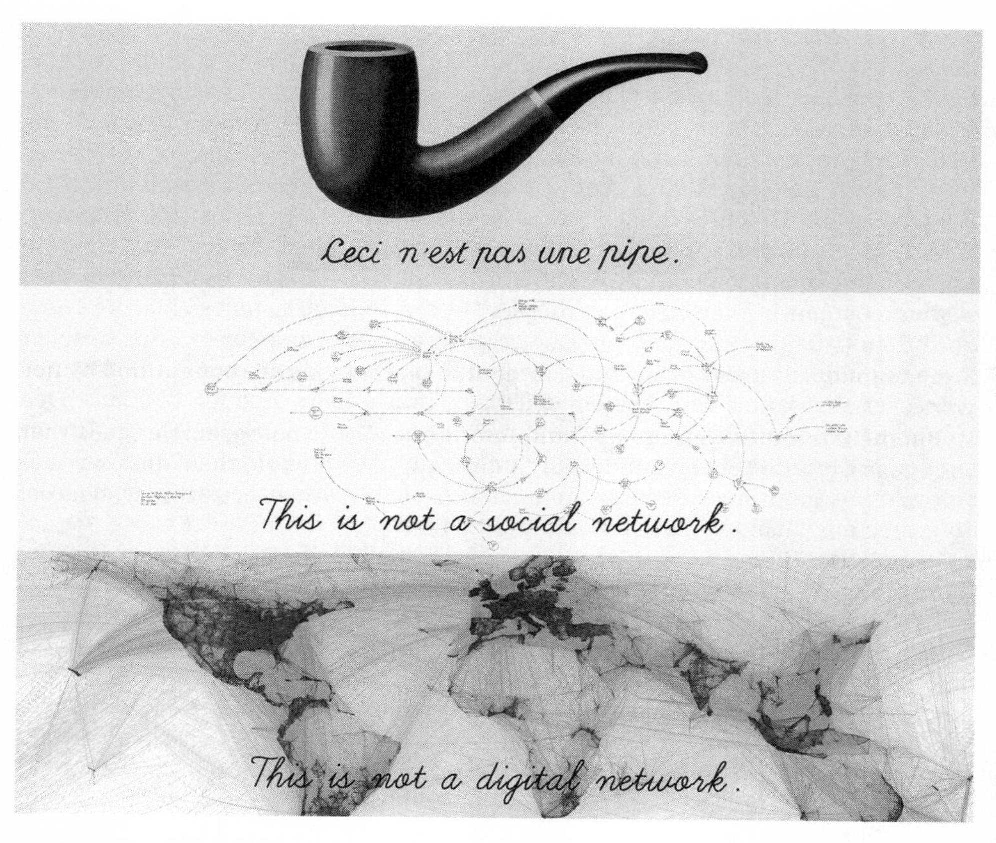

FIGURE 1: a. René Magritte, *La Trahison des images* (1928). b. Marc Lombardi, "George W. Bush, Harken Energy, and Jackson Stephens, ca. 1979–90" (1999). c. Paul Butler, "Visualizing Friendship" (2010). Captions for b and c added by the authors.

Laden families designed by Mark Lombardi or the Facebook connections designed by Paul Butler do not resemble the phenomena that they portray (see figure 1).

Social networks are not made of lines on canvas, digital networks are not made of pixels, and neither is made of data. Actor-networks are made of flesh and fabric, of words and memories, of contracts and laws, of money and transactions, and, increasingly, of cables and protocols. It is not surprising that graphs do not resemble them. And yet, this does not mean that graphs cannot help us understand collective topologies. If there is something that STS observed over and over, it is that scientific representations do not have to resemble to their referent to be useful.

Abandoning the benchmark of resemblance is important because it allows us to put aside the differences between graphs and actor-networks and consider their similarities. A first reason for SNA and ANT to be good friends is that they both fight the same assumptions of classic sociology. Their "networks" may not be synonyms, but their antonym is the same.

Both of these approaches reject a priori reifications such as "the social" or "society"; instead, these notions are constructions out of social enmeshing and become observable only ex post. Both resist reference to the representational or the symbolic; instead, they focus their empirical analyses on material reality and the meanings actors themselves ascribe to it in struggles and controversies. Both of these approaches consider the production of meaning as an activity of connecting/ disconnecting and analyze how actors come to be created through collaborations of other actors in different contexts. The stories actors tell make the links between them explicit. For both approaches, the ties precede the nodes (Mutzel 2009, 878).

ANT and network analysis are both inspired by the same relational thinking (Emirbayer 1997), whose first tenet is the refusal of any form of substantialism (Robinson 2014). For both ANT and SNA, associations (and dissociations) are the only things that matter. John Law (1999; but see also Blok 2010) described this opposition by contrasting "topographical" and "topological" approaches and suggested to "imagine actor-network theory as a machine for waging war on Euclideanism: as a way of showing, inter alia, that regions are constituted by networks" (7).

But there is more. While graphs and actor-networks do not resemble each other, they bear a distinct correspondence: "A diagram of a network, then, does not look like a network but maintain the same qualities of relations—proximities, degrees of separation, and so forth—that a network also requires in order to form. Resemblance should here be considered a resonating rather than a hierarchy (a form) that arranges signifiers and signified within a sign" (Munster 2013, 24). The easiest way to understand the way in which networks resonate with collective phenomena is to consider the drawing of social networks. Of all the techniques associated with graph analysis, the ones developed to *visualize* networks are those that most closely resonate with ANT. It is not accidental that while graphs had been around for more than two centuries (Euler 1736), it was only when sociologists seized upon them that visualization joined computation as an analytical tool.

Jacob Moreno, the founder of SNA (see Moreno 1934), is formal about the importance of visualization:

> If we ever get to the point of *charting* a whole city or a whole nation, we would have an intricate maze of psychological reactions which would *present* a *picture* of a vast solar system of intangible structures powerfully influencing conduct, as gravitation does bodies in space. Such an *invisible* structure underlies society and has in influence in determining the conduct of society as a whole. . . . Until we have at least determined the nature of these fundamental structures which form the networks, we are working *blindly* in a hit-or-miss effort to solve problems which are caused by group attraction, repulsion and indifference. (*New York Times* 1933, emphasis added)

The interest for network visualization has recently surfaced again in both academic and popular culture. Images of networks are sprouting everywhere. They decorate buildings and objects; they are printed on T-shirts and posters; they colonize the desktop of our computers and the walls of our airports. Networks have become the emblem of modernity, the very form of its imagination. In part, of course, this is linked to the success of digital networks, but there is something else. Something connected to the *figurative power* of network visualization.

This *something*, we believe, is directly connected to the way networks are designed. Although several techniques for "network spatialisation" exist, a family of algorithms has progressively emerged as a standard for graph visualization: the so-called "force-directed layouts" (or "force-vectors"). A force-vector layout works following a physical analogy: nodes are given a repulsive force that drives them apart, while edges work as springs binding the nodes that they connect. Once the algorithm is launched, it changes the disposition of nodes until reaching the equilibrium that guarantees the balance of forces.

At equilibrium, force-vectors not only minimize line crossings, but also give sense to the disposition of nodes in space. In a force-spatialized network, spatial distribution becomes meaningful: groups of nodes are closer the more they are directly or indirectly connected (Jacomy et al. 2014). As proved in Noack (2009), visual clustering in force-spatialized networks is directly equivalent to clustering by modularity. In force-directed layout, "centrality," "betweenness," "diameter," "density," "structural separation," all these concepts (and many others) found a graphical equivalent (Venturini et al. 2014). They can be not only calculated, but also *seen*. This is where the figurative power of networks, their *un-resembling resonance*, comes from. This is also where the deepest bond between SNA and ANT is to be found.

Looking at a force-spatialized network provides a visual experience of *both* the metrics of network analysis *and* the notions of ANT—thus revealing their *elective affinity* (Jensen et al. 2014). Consider, for example, the notion of "boundary," which has long been a puzzle for SNA (Laumann et al. 1989). "Networks are interesting but difficult to study because since real-world network lack convenient natural boundaries. When a network as a whole is impracticably large, the usual procedure is to arbitrarily delimit a subgraph and treat it as a representative sample of the whole network. Unfortunately, this procedure is hazardous not only qualitatively . . . but quantitatively as well" (Barnes 1979, 416). On the other hand, ANT has been often accused of dissolving all the classic distinctions of social theory (micro/macro, science/politics, science/technology, nature/culture, etc.), without replacing them with any clear analytic framework. Yet ANT is not a night where all cows are black. If it is true that, following the actors and their relations, we rarely encounter clear-cut boundaries, it also true that we do experience *variations in the density of association*. Our collective existence is a "small world" (Milgram 1967; Watts and Strogatz 1998) where everything is connected. And yet, the density of associations is not homogeneous. This *in*homogeneity is manifest in spatialized networks: nodes and edges do not dispose regularly—some of them flock together, while others repulse each other. The visual space of graphs as the conceptual space of actor-network *is continuous but not uniform*. It is because of this similarity that networks can be used to represent actor-networks, despite the differences that separate them.

After all, we might be talking about the same networks.

Works Cited

Adamic, L. A., and Natalie Glance. 2005. "The Political Blogosphere and the 2004 US Election: Divided They Blog." In *Proceedings of the 3rd International Workshop on Link Discovery*, 36–43. New York: ACM. doi:10.1145/1134271.1134277.

Archer, Margaret S. 1995. *Realist Social Theory: The Morphogenetic Approach*. Cambridge: Cambridge University Press.

Barabási, Albert-László. 2002. *Linked: The New Science of Networks. How Everything Is Connected to Everything Else and What It Means for Business, Science, and Everyday Life*. Cambridge, MA: Perseus Books.

Barnes, John A. 1979. "Network Analysis: Orienting Notion, Rigorous Technique, or Substantive Field of Study." In *Perspectives on Social Network Research*, edited by W. Paul Holland and Samuel Leinhardt, 403–23. New York: Academic Press.

Barney, Darin. 2004. *The Network Society*. Cambridge: Polity.
Bateson, Gregory. 1972. "Form, Substance, and Difference." In *Steps to an Ecology of Mind*, 454–71. Chicago: University of Chicago Press.
Blok, Anders. 2010. "Topologies of Climate Change: Actor-Network Theory, Relational-Scalar Analytics, and Carbon-Market Overflows." *Environment and Planning D: Society and Space* 28 (1): 896–912. doi:10.1068/d0309.
Borra, Erik, Esther Weltevrede, Paolo Ciuccarelli, Andreas Kaltenbrunner, David Laniado, Giovanni Magni, Michele Mauri, Richard Rogers, and Tommaso Venturini. 2014. "Contropedia—The Analysis and Visualization of Controversies in Wikipedia Articles." In *OpenSym 2014 Proceedings*, 34. New York: ACM.
Boullier, Dominique, and Maxime Crépel. 2012. "Biographie D'une Photo Numérique et Pouvoir Des Tags: Classer/Circuler." *Anthropologie Des Connaissances* 7 (4): 785–813.
Burt, Ronald S. 1980. "Models of Network Structure." *Annual Review of Sociology* 6:79–141. doi:10.1146/annurev.so.06.080180.000455.
——. 1995. *Structural Holes: The Social Structure of Competition*. Cambridge, MA: Harvard University Press.
Callon, Michel. 1984. "Some Elements of a Sociology of Translation: Domestication of the Scallops and the Fishermen of St Brieuc Bay." *Sociological Review* 32 (S1): 196–233.
——. 1986. "The Sociology of an Actor-Network: The Case of the Electric Vehicle." In *Mapping the Dynamics of Science and Technology*, edited by Michel Callon, John Law, and Arie Rip, 19–34. London: Macmillan.
Callon, Michel, and Bruno Latour. 1981. "Unscrewing the Big Leviathans: How Do Actors Macrostructure Reality?" In *Advances in Social Theory and Methodology. Toward an Integration of Micro and Macro Sociologies*, edited by Karin Knorr Cetina and Aron Cicourel, 277–303. London: Routledge.
Cambrosio, Alberto, Peter Keating, and Andrei Mogoutov. 2004. "Mapping Collaborative Work and Innovation in Biomedicine: A Computer Assisted Analysis of Antibody Reagent Workshops." *Social Studies of Science* 34:325–64.
Castells, Manuel. 2000. *The Rise of the Network Society*. Oxford: Blackwell.
Chavalarias, David, and Jean Philippe Cointet. 2013. "Phylomemetic Patterns in Science Evolution—The Rise and Fall of Scientific Fields." *PLOS ONE* 8 (2). doi:10.1371/journal.pone.0054847.
Eisenstein, Elizabeth L. 1980. *The Printing Press as an Agent of Change*. Cambridge: Cambridge University Press.
Emirbayer, Mustafa. 1997. "Manifesto for a Relational Sociology." *American Journal of Sociology* 103 (2): 281–317. doi:10.1086/231209.
Euler, Leonhard. 1736. "Solutio Problematis Ad Geometriam Situs Pertinentis." *Commentarii Academiae Scientiarum Petropolitanae*, no. 8: 128–40.
Everett, Martin, and Stephen Borgatti. 2014. "Networks Containing Negative Ties." *Social Networks* 38:111–20. doi:10.1016/j.socnet.2014.03.005.
Foucault, Michel. 1983. *This Is Not a Pipe*. Berkeley: University of California Press.
Garfield, Eugene. 1979. "Citation Analysis." *Scientometrics* 1 (4): 359–75.
Giddens, Anthony. 1984. *Constitution of Society: Outline of the Theory of Structuration*. Berkeley: University of California Press.
Granovetter, Mark. 1979. "The Theory-Gap in Social Network Analysis." In *Perspectives on Social Research*, edited by P. Holland and S. Leinhardt, 501–18. New York: Academic Press.
Jacomy, Mathieu, Tommaso Venturini, Sebastien Heymann, and Mathieu Bastian. 2014. "ForceAtlas2, a Continuous Graph Layout Algorithm for Handy Network Visualization Designed for the Gephi Software." *PLOS ONE* 9 (6): e98679. doi:10.1371/journal.pone.0098679.
Jensen, T. E., A. K. Munk, A. K. Madsen, and A. Birkbak. 2014. *Expanding the Visual Registers of STS. Visualization in the Age of Computerization*. New York: Routledge.
Knorr Cetina, Karin. 1995. "Laboratory Studies, the Cultural Approach to the Study of Science." In *Handbook of Science and Technology Studies*, edited by Sheila Jasanoff, Gerald E. Markle, James C. Peterson, and Trevor J. Pinch, 140–66. London: Sage.
Latour, Bruno. 1998. "Thought Experiments in Social Science: From the Social Contract to Virtual Society." Virtual Society? Annual Public Lecture, Brunel University, London, April 1.
——. 1999. "On Recalling ANT." In *Actor Network and After*, edited by John Law and John Hassard, 15–25. Oxford: Blackwell.
——. 2010. "Networks, Societies, Spheres: Reflections of an Actor-Network Theorist." Keynote address to the International Seminar on Network Theory: Network Multidimensionality in the Digital Age,

Annenberg School for Communication and Journalism, Los Angeles. https://hal-sciencespo.archives-ouvertes.fr/hal-00972865.

Latour, Bruno, Pablo Jensen, Tommaso Venturini, Sébastian Grauwin, and Dominique Boullier. 2012. "'The Whole Is Always Smaller Than Its Parts': A Digital Test of Gabriel Tardes' Monads." *British Journal of Sociology* 63 (4): 590–615. doi:10.1111/j.1468-4446.2012.01428.x.

Latour, Bruno, Philippe Mauguin, and Geneviève Teil. 1992. "A Note on Socio-technical Graphs." *Social Studies of Science* 22 (1): 33–57.

Laumann, Edward, Peter Marsden, and David Prensky. 1989. "The Boundary Specification Problem in Network Analysis." In *Research Methods in Social Network Analysis*, edited by Linton C. Freeman, Douglas R. White, and Antone Kimball Romney, 61–79. Fairfax, VA: George Mason University Press.

Law, John. 1999. "After ANT: Complexity, Naming and Topology." In *Actor Network and After*, edited by John Law and John Hassard, 1–14. Oxford: Blackwell.

Law, John, and Michel Callon. 1992. "The Life and Death of an Aircraft: A Network Analysis of Technical Change." In *Shaping Technology/Building Society: Studies in Sociotechnical Change*, edited by Wiebe Bijker and John Law, 21–52. Cambridge, MA: MIT Press.

Law, John, and Annemarie Mol. 2001. "Situating Technoscience: An Inquiry into Spatialities." *Environment and Planning D: Society and Space* 19:609–21. doi:10.1068/d243t.

Leydesdorff, Loet. 1998. "Theories of Citation?" *Scientometrics* 43 (1): 5–25. doi:10.1007/BF02458394.

Leydesdorff, L., and P. Wouters. 1999. "Between Texts and Contexts: Advances in Theories of Citation? (A Rejoinder)." *Scientometrics* 44:169–82. doi:10.1007/BF02457378.

Madsen, A. K. 2015. "Tracing Data—Paying Attention." In *Making Things Valuable*, edited by M. Kornberger, L. Justesen, A. K. Madsen, and J. Mouritsen, 257–78. Oxford: Oxford University Press.

Marres, Noortje. 2012a. "On Some Uses and Abuses of Topology in the Social Analysis of Technology." *Theory, Culture & Society* 29 (4–5): 288–310.

——. 2012b. "The Redistribution of Methods: On Intervention in Digital Social Research, Broadly Conceived." *Sociological Review* 60:139–65. doi:10.1111/j.1467-954X.2012.02121.x.

Marres, Noortje, and Caroline Gerlitz. 2015. "Interface Methods: On Some Confluence between Sociology, STS and Digital Research." *Information, Communication & Society* 64:21–46.

Marres, Noortje, and Richard Rogers. 2005. "Recipe for Tracing the Fate of Issues and Their Publics on the Web." In *Making Things Public*, edited by Latour Bruno and Peter Weibel, 922–35. Karlsruhe: ZKM.

Milgram, Stanley. 1967. "The Small World Problem." *Psychology Today* 2 (1): 60–67.

Mol, A., and J. Law. 1994. "Regions, Networks and Fluids: Anaemia and Social Topology." *Social Studies of Science* 24 (4): 641–71. doi:10.1177/030631279402400402.

Moreno, Jacob. 1934. *Who Shall Survive?* Washington, DC: Nervous and Mental Disease Publishing.

——. 1953. *Who Shall Survive?* 2nd ed. New York: Beacon House.

Munk, Anders Kristian. 2013. "Techno-Anthropology and the Digital Natives." In *What Is Techno-Anthropology?*, edited by L. Botin and T. Borsen. Aalborg: Aalborg University Press. http://forskningsbasen.deff.dk/Share.external?sp=S2eb954fb-fb8c-46df-b850-fc1ffa3bb4e4&sp=Saau.

Munster, Anna. 2013. *An Aesthesia of Networks: Conjunctive Experience in Art and Technology*. Cambridge, MA: MIT Press.

Mutzel, S. 2009. "Networks as Culturally Constituted Processes: A Comparison of Relational Sociology and Actor-Network Theory." *Current Sociology* 57:871–87. doi:10.1177/0011392109342223.

New York Times. 1933. "Emotions Mapped by New Geography." April 3.

Noack, Andreas. 2009. "Modularity Clustering Is Force-Directed Layout." *Physical Review E* 79 (2). doi:10.1103/PhysRevE.79.026102.

Page, Lawrence, Sergey Brin, Motwani Rajeev, and Winograd Terry. 1998. "The PageRank Citation Ranking: Bringing Order to the Web." Stanford, CA: Stanford University Press.

Rieder, Bernhard. 2013. "Studying Facebook via Data Extraction: The Netvizz Application." In *Proceedings of WebSci '13: The 5th Annual ACM Web Science Conference*, 346–55. New York: ACM. doi:10.1145/2464464.2464475.

Robinson, Howard. 2014. "Substance." In *The Stanford Encyclopedia of Philosophy*, edited by Edward N. Zalta. https://plato.stanford.edu/entries/substance/.

Rogers, Richard. 2000. *Preferred Placement: Knowledge Politics on the Web*. Maastricht: Jan Van Eyck Edition.

——. 2004. *Information Politics on the Web*. Cambridge, MA: MIT Press.

——. 2013. *Digital Methods*. Cambridge, MA: MIT Press.

Rogers, Richard, and Noortje Marres. 2000. "Landscaping Climate Change: A Mapping Technique for Understanding Science and Technology Debates on the World Wide Web." *Public Understanding of Science* 9:141–63. doi:10.1088/0963-6625/9/2/304.

——. 2002. "French Scandals on the Web, and on the Streets: A Small Experiment in Stretching the Limits of Reported Reality." *Asian Journal of Social Science* 66:339–53.

Roth, Camille, and Jean-Philippe Cointet. 2010. "Social and Semantic Coevolution in Knowledge Networks Epistemic Networks." *Social Networks* 32 (1): 16–29.

Van Dijk, Jan. 1999. *The Network Society*. London: Sage.

Venturini, Tommaso. 2010. "Diving in Magma: How to Explore Controversies with Actor-Network Theory." *Public Understanding of Science* 19 (3): 258–73. doi:10.1177/0963662509102694.

——. 2012. "Building on Faults: How to Represent Controversies with Digital Methods." *Public Understanding of Science* 21 (7): 796–812. doi:10.1177/0963662510387558.

Venturini, Tommaso, and Daniele Guido. 2012. "Once Upon a Text: An ANT Tale in Text Analysis." *Sociologica* 3. doi:10.2383/72700.

Venturini, Tommaso, Mathieu Jacomy, and Débora Carvalho Pereira. 2015. "Visual Network Analysis." www.tommasoventurini.it/wp/wp-content/uploads/2014/08/Venturini-Jacomy_Visual-Network-Analysis_WorkingPaper.pdf.

Venturini, Tommaso, Nicolas Baya Laffite, Jean-Philippe Cointet, Ian Gray, Vinciane Zabban, and Kari De Pryck. 2014. "Three Maps and Three Misunderstandings: A Digital Mapping of Climate Diplomacy." *Big Data & Society* 1 (2). doi:10.1177/2053951714543804.

Vinck, Dominique. 2012. "Accessing Material Culture by Following Intermediary Objects." In *An Ethnography of Global Landscapes and Corridors*, edited by Loshini Naidoo, 89–108. London: Intech.

Watts, Duncan J., and Steven H. Strogatz. 1998. "Collective Dynamics of 'Small-World' Networks." *Nature* 393 (6684): 440–42.

White, Harrison. 2000. "Toward Ego-Centered Citation Analysis." In *The Web of Knowledge*, edited by B. Cronin and H. B. Atkins, 475–96. Medford, NJ: Information Today.

Woolgar, Steve. 2002. *Virtual Society. Technology, Cyberbole, Reality*. Oxford: Oxford University Press.

Acknowledgments

We are grateful to all members of the digitalSTS community, past and present, who shaped the contents of this volume and provided through many conversations an infusion of fresh intellectual ideas into our field. These individuals attended events and workshops, participated in online open peer review, came to conferences and meetings, and in some cases organized or assisted in organizing panels, workshops, social events, and more. A list of participants follows. We are thankful for their time, energy, and intellectual generosity.

In addition, we thank the Society for Social Studies of Science, especially past president Trevor Pinch and secretary Wes Shrum; the NSF Sociotechnical Studies Research Coordination Network led by Steve Sawyer, Brian Butler, and colleagues; the Sloan Foundation and Josh Greenberg; Microsoft Research; the IT University of Copenhagen; and the Harvard University Arboretum. All believed in our vision and offered generous support of our workshops, our community, and our activities throughout the course of the project. Many thanks also to Paul Dourish, Christina Dunbar-Hester, Camilla Hawthorne, Stephanie Steinhardt, and our co-editors for helpful commentary on prior versions of the introduction, to Leah Reisman for assisting in assembling the final collection, to Al Bertrand and colleagues at Princeton University Press for their unwavering support of this project, and to our dedicated and indefatigable co-editorial team for shaping this project and its community every step of the way.

Contributors

Carlo Allende, Arizona State University
Doris Allhutter, Austrian Academy of Sciences
Morgan Ames, University of California, Berkeley
Mike Annany, University of Southern California
Smiljana Antonijevic, University of Copenhagen
Margy Avery, MIT Press/University of Amherst Press
David Banks, Rensselaer Polytechnic Institute
Matthew Battles, Harvard Unviersity
Erin Whitney Boesel, University of California at Santa Cruz
Heidi Boisvert, Rensselaer Polytechnic Institute
Geof Bowker, Unviersity of California, Irvine
Angie Boyce, Harvard University
Andrea Brennen, Massachusetts Institute of Technology
Hronn Brynjarsdottir, Cornell Unviersity
Tania Bucher, University of Copenhagen
Matt Burton, University of Michigan

Nerea Calvillo, University of Warwick
Alexandre Camus, University of Lausanne
Anita Say Chan, University of Illinois Urbana-Champaign
Padma Chirumamilla, University of Michigan
Jean Ho Chu, Georgia Institute of Technology
Marisa Leavitt Cohn, IT University of Copenhagen
Peter Sachs Collopy, University of Southern California
Stéphane Couture, York University
Dharma Dailey, University of Washington
Ben Dalton, Royal College of Art, London
Jenny Davis, Texas A&M
Deanna Day, Chemical Heritage Foundation
Carl DiSalvo, Georgia Institute of Technology (Editor, Mentor)
Kelly Dobson, Rhode Island School of Design (Mentor)
Vincent Duclos, McGill University
Christina Dunbar-Hester, University of Southern California
Ingrid Erickson, Rutgers University
Anne Catherine Feldman, Hubspot
Megan Finn, University of Washington
Laura Forlano, Illinois Institute of Technology (Editor, Workshop Co-organizer)
Stuart Geiger, University of California, Berkeley
Amanda Geppart, Illinois Institute of Technology
Shad Gross, Indiana University
Camilla A. Hawthorne, University of California, Berkeley
Chris Hesselbein, Cornell University
Lara Houston, Lancaster University
Carla Ilten, University of Illinois at Chicago
Lilly Irani, University of California, San Diego
Steven J. Jackson, Cornell University (Editor)
Mathieu Jacomy, Sciences Po Médialab
Mohammad Hossein Jarrahi, University of North Carolina Chapel Hill
Steve Jones, University of Illinois at Chicago
Nadja Kanellopoulou, The University of Edinburgh
Jelena Karanovic, New York University
Christian Katzenbach, Freie Universität Berlin
Shreeharsh Kelkar, MIT
Xaroula (Charalampia) Kerasidou
Duong Tam Kien, Institute for Research and Innovation in Society
Joseph Klett, Yale University
Cory Knobel, University of California at Irvine
N. A. Knouf, Wellesley College
Guillaume Latzko-Toth, Université Laval
Lucian Leahu, Royal Technical Institute Stockholm, Sweden
Max Liboiron, Memorial University
Silvia Lindtner, University of Michigan
Daniel Cardoso Llach, Carnegie Mellon University
Yanni Loukissas, Georgia Institute of Technology (Editor, Workshop Co-organizer)
Jonathan Lukens, Georgia Institute of Technology
Michael Lynch, Cornell University

James Maguire, IT University of Copenhagen
Melissa Mazmanian, University of California, Irvine
Paul-Brian McInerney, University of Illinois at Chicago
Axel Meunier, Sciences Po Médialab
Florence Millerand, Université du Québec à Montréal
Eric Monteiro, Norwegian University of Science and Technology
Anders Kristian Munk, TANTLab, Aalborg University Copenhagen
Sumitra Nair, Virginia Institute of Technology
David Nemer, University of Kentucky
Kasper Ostrowski, University of Aarhus
Elena Parmiggiani, Norwegian University of Science and Technology
Kyle Parry, Harvard University
Winifred R. Poster, Washington University
Jessica Price, Cornell University
Bryce Renninger, Rutgers University
David Ribes, University of Washington (Editor, Workshop Co-organizer)
Daniela K. Rosner, University of Washington (Editor, Mentor)
Juan Salamanca, Universidad Icesi
Stephanie Santoso, Cornell University
Steve Sawyer, Syracuse University
Caterina Scaramelli, Massachusetts Institute of Technology
Nick Seaver, Tufts University
Karen Schrier Shaenfield, Marist College
Kalpana Shankar, University College Dublin
Laura Sheble, University of North Carolina, Chapel Hill
Hanna Rose Shell, Massachusetts Institute of Technology (Editor, Mentor)
Katie Shilton, University of Maryland
Wes Shrum, Louisiana University
Ranjit Singh, Cornell University
Johan Söderberg, University of Gothenberg
Luke Stark, New York University
Molly Wright Steenson, Carnegie Mellon University (Mentor)
Maurizio Teli, Madeira Interactive Technologies Institute
Mitali Thakor, Massachusetts Institute of Technology
Teresa Velden, University of Michigan
Tommaso Venturini, Institut national de recherche en informatique et en automatique
Janet Vertesi, Princeton University (Editor, Workshop Co-organizer)
Dominique Vinck, University of Lausanne
Laura Watts, University of Edinburgh
Brooke Foucault Welles, Northeastern University
Kaiton Williams, Cornell University
Robin Williams, University of Edinburgh
Heather Wiltse, Umeå Institute of Design
Brit Ross Winthereik, IT University Copenhagen
Sara Wylie, Northeastern University (Mentor)
Malte Ziewitz, Cornell University

Contributors

Doris Allhutter is Elise Richter Senior Researcher at the Austrian Academy of Sciences and lecturer in STS at the University of Vienna. Currently she is working on an onto-epistemological conceptualization of "infrastructural power" based on ethnographic research in various computer science fields. This research focuses on the co-emergence of digital infrastructures with ideologies and hegemonies in technoscientific practices and society by tracing their affective and material-discursive performativities. She was a member of the COST Action (European Co-operation in Science and Technology) on "New Materialism. How Matter comes to Matter" and is co-editing a forthcoming special issue on "Materiality-Critique-Transformation" in Feminist Theory. She holds a PhD in political science and an MA in business, economics, and social sciences.

Nerea Calvillo is Assistant Professor at the Centre for Interdisciplinary Methodologies (University of Warwick), and founder of the architecture practice C+ arquitectos and the visualization project In the Air. She is an architect, researcher, and curator, working on the urbanisms of the air, environmental sensing, and collaborative practices at the intersection between architecture and STS.

Alexandre Camus is a PhD candidate in the STS Lab of the Institute for Social Sciences of the University of Lausanne and in the Centre de Sociologie de l'Innovation of Mines ParisTech. He studies the digitization of culture and humanities through ethnography in engineering projects investing in the future of cultural heritage. His work intersects with innovation and engineering studies, digital materiality, sound studies, and material practices.

Daniel Cardoso Llach is Assistant Professor of Architecture at Carnegie Mellon University. His first book, *Builders of the Vision: Software and the Imagination of Design* (2015), is an intellectual history of computer-aided design that documents the emergence of technological imaginaries of design in postwar research projects in the US, and traces their socio-technical repercussions in architecture. An interdisciplinary scholar and researcher, his work contributes to the fields of architecture, design, and science and technology studies. He holds a PhD and an MS (with honors) in design and computation from MIT and has been a visiting scholar at Leuphana, Germany, and the University of Cambridge, UK.

Anita Say Chan is Associate Professor in the Department of Media and Cinema Studies and Faculty Affiliate at the National Center for Supercomputing Applications and the School of Information Sciences at the University of Illinois, Urbana-Champaign. Her research and teaching interests include globalization and digital cultures, innovation networks and the "periphery", science and technology studies in Latin America, and data ethics and cultures. She is the author of *Networking Peripheries: Technological Futures and the Myth of Digital Universalism* (2014), which explores the competing imaginaries of global connection and information technologies in network-age Peru.

Padma Chirumamilla is a PhD candidate at the University of Michigan School of Information. Her dissertation examines how the work of television repairmen was crucial to transformation of the television set into an everyday media object in south India. Her broader research interests are in histories of media infrastructures in South Asia, theories of everyday life and temporality, material culture, ordinary ethics, and science and technology studies.

Marisa Leavitt Cohn is Associate Professor at the IT University of Copenhagen, where she is a member of the Technologies in Practice and Interaction Design research groups and co-founder of the ETHOS Lab. She is also a Research Fellow in Critical Design and Engineering at the Madeira Interactive Technologies Institute. With a PhD in information and computer science from University of California, Irvine, she combines approaches from HCI, anthropology, and STS to examine the politics of computational work in the design, maintenance, and repair of sociotechnical systems.

Stéphane Couture is Assistant Professor in Communications at Glendon Campus, York University in Toronto, Canada. He is also affiliated with the STS graduate program in this same university. Initially trained in computer science, he holds a joint PhD in communication and sociology. His research is concerned with the values and practices of free and open source software developers, digital and media activism, and the articulation of political dimensions in the design of digital technologies and Internet infrastructures. His PhD and postdoctoral work, on which his chapter in this collection is based, addressed the collective making of computer source code and its articulation with cultural and political dimensions.

Carl DiSalvo is Associate Professor in the Digital Media Program in the School of Literature, Media, and Communication at the Georgia Institute of Technology. At Georgia Tech he directs the Public Design Workshop: a design research studio that explores socially engaged design and civic media. He publishes regularly in design, science and technology studies, and human-computer interaction journals and conference proceedings. His first book, *Adversarial Design*, is part of the Design Thinking, Design Theory series. He is also a co-editor of *Design Issues*. He holds a PhD in design from Carnegie Mellon University (2006).

Christina Dunbar-Hester is an ethnographer who studies the politics of technology. She is the author of *Low Power to the People: Pirates, Protest, and Politics in FM Radio Activism* (MIT Press, 2014) and the forthcoming *Hacking Diversity: The Politics of Inclusion in Open Technology Cultures* (Princeton University Press). She is a faculty member in communication at the University of Southern California's Annenberg School for Communication and Journalism, and she holds a PhD in science and technology studies from Cornell University.

Ingrid Erickson is Assistant Professor at the School of Information Studies at Syracuse University in Syracuse, New York, where she teaches in the Information

Management program. She received her PhD from the Center for Work, Technology, and Organization in the Department of Management Science and Engineering at Stanford University. Her research centers on the way that mobile devices and ubiquitous digital infrastructures are influencing how we work and communicate with one another, navigate and inhabit spaces, and engage in new types of sociotechnical practices. Her most recent work explores the infrastructuring skills and practices that individual workers need in order to be both highly mobile and appropriately professional—a concept that she and her collaborators call "infrastructural competence."

Laura Forlano, a Fulbright Award–winning and National Science Foundation–funded scholar, is a writer, social scientist, and design researcher. She is Associate Professor of Design at the Institute of Design and Affiliated Faculty in the College of Architecture at Illinois Institute of Technology, where she is Director of the Critical Futures Lab. Her research is focused on the aesthetics and politics at the intersection between design and emerging technologies. Over the past ten years, she has studied the materialities and futures of sociotechnical systems such as autonomous vehicles and smart cities; 3D printing, local manufacturing, and innovation ecosystems; automation, distributed labor practices, and the future of work; and computational fashion, smart textiles, and wearable medical technologies. She is co-editor with Marcus Foth, Christine Satchell, and Martin Gibbs of *From Social Butterfly to Engaged Citizen* (2011). She received her PhD in communications from Columbia University.

Camilla A. Hawthorne is Assistant Professor of Sociology at UC Santa Cruz and also coordinates the Black Europe Summer School in Amsterdam, the Netherlands. She received her PhD in geography and science and technology studies from UC Berkeley. Her research addresses the cultural politics of Blackness among youth of African descent in Italy. Broadly, her work intersects with postcolonial STS, new media studies, diaspora and race-critical theory, and Black European studies.

Chris Hesselbein is a PhD candidate in science and technology studies at Cornell University whose research aligns with feminist studies of technoscience, anthropology of the senses, and studies of everyday life. Their current work focuses on the relationship between mundane technologies, embodied skills, and gendered comportment at the intersection of fashion and technology. Ongoing projects focus on the co-construction of (digital) technology and sensory experience, the materiality of authenticity and realness, and the performance of gender in/through electronic music.

Carla Ilten is a sociology PhD candidate at the University of Illinois Chicago. She studies how people use organizations and technologies to structure social and economic relations and engages with organization theory, economic sociology, STS, and media studies. She holds a diplom (MA) in sociology and technology studies from Technische Universität Berlin. Her thesis on sociotechnical innovation by civil society actors was published in German.

Steven J. Jackson is Associate Professor of Information Science and Science and Technology Studies and Chair of Information Science at Cornell University.

His work centers on questions of value and power in contemporary technoscience, with special emphasis on problems of time, infrastructure, maintenance, repair, and hope in complex sociotechnical systems. Current empirical projects include work on computational development and infrastructural change in the sciences; ethics, law, and policy in emerging media environments; collaboration, creativity, and improvisation in science, music, and new media arts; and problems of computation and social change in postcolonial environments. His work has been supported by the Ford Foundation, the World Bank, Intel Research, and the U.S. National Science Foundation. For more information, see https://sjackson.infosci.cornell.edu.

Mathieu Jacomy has been a research engineer at the Sciences Po Paris Médialab since 2010. As part of the Dime Web Team he develops digital tools for the social sciences and provides support and advice in digital methods to scholars. His current research focuses on visual network analysis, digital methods, and issue mapping. He initiated the Gephi network analysis software and is still developing it, notably through algorithms like ForceAtlas2, and he is currently developing the web crawler Hyphe. He was also strongly involved in the e-Diasporas Atlas program with Dana Diminescu from 2006 to 2010. He tweets @jacomyma.

Mohammad Hossein Jarrahi is Assistant Professor in the School of Information and Library Science at the University of North Carolina at Chapel Hill. He received his PhD from the School of Information Studies at Syracuse University. His research focuses on the use and consequences of information and communication technologies (ICTs) and the accompanying social and organizational changes that these bring to knowledge-intensive organizational contexts.

Steve Jones is UIC Distinguished Professor of Communication and Adjunct Professor of Computer Science at the University of Illinois at Chicago, where he is also a research associate in the Electronic Visualization Laboratory (EVL). Co-founder of the Association of Internet Researchers and editor of the international journal *New Media & Society,* he has studied the social and behavioral aspects of new media, the Internet, and technology since the late 1980s. He earned his PhD in communications at the University of Illinois at Urbana-Champaign, where he began using and authoring on the PLATO system in the late 1970s.

Xaroula (Charalampia) Kerasidou is an independent researcher in the fields of feminist science and technology studies and media and cultural studies with a special interest in exploring how new technologies challenge us to reconceptualize the ontologies and power relations between the human and the machine, and understanding what this means for ethics, politics, and policy. Currently, she has joined the Centre for Mobilities Research at Lancaster University as a Wellcome Research Fellow investigating ethical AI configurations in healthcare. She holds a PhD in sociology from Lancaster University (UK), an MA in digital media studies from Goldsmiths College (UK), and a BSc in informatics from Aristotle University (GR).

Guillaume Latzko-Toth is Associate Professor in the Department of Information and Communication at Laval University (Quebec City, Canada) and the Codirector of the Laboratory for Communication and the Digital (LabCMO). Rooted in a science and technology studies perspective, his research and publications address users' contribution to the development of digital media; the role of artifacts in digitally supported communities; transformations in publics and publicness; and methodological and ethical issues related to research in digital contexts. He is a member of the Interuniversity Research Center on Science and Technology (CIRST) and co-founded the Technology and Emerging Media interest group within the Canadian Communication Association.

Yanni Loukissas is Assistant Professor of Digital Media in the School of Literature, Media, and Communication at Georgia Tech. He holds a PhD and an SM in design and computation from MIT, where he also completed postdoctoral work in the Program in Science, Technology and Society. His research is focused on helping creative people think critically about the social implications of emerging technologies. He is the author of two books, *All Data Are Local: Thinking Critically in a Data-Driven Society* (2019) and *Co-Designers: Cultures of Computer Simulation in Architecture* (2012).

Michael Lynch is Emeritus Professor of Science and Technology Studies at Cornell University. He takes an ethnomethodological approach toward discourse and practice in scientific and legal settings, as well as in the intersection between the two. His current preoccupation is with "radical ethnomethodology": a critical intervention into recent developments in ethnomethodology and conversation analysis.

James Maguire is an anthropologist and Assistant Professor at the IT University of Copenhagen. His research focuses on the complex interrelationship between the digital and the environmental and pays particular attention to the production of digital infrastructures through the appropriation of environmental forms. He has published on questions of energy, infrastructures, environmental resources, digitalization, as well as on the relationship between the state and Big-Tech.

Paul-Brian McInerney is Associate Professor of Sociology at the University of Illinois at Chicago. He is the author of *From Social Movement to Moral Market: How the Circuit Riders Sparked an IT Revolution and Created a Technology Market* (2015). His research interests include social studies of technology, economic and organizational sociology, and social movements. He holds a PhD and MPhil in sociology from Columbia University and an MA and BA in sociology from St. John's University.

Axel Meunier is an independent researcher, cartographer, and collaborative workshop designer. He has degrees in science and technology studies from EHESS, and in experimentation in arts and politics from Sciences Po. At Sciences Po's Médialab his research concerns the integration of digital and design methods in social sciences. He also does social practice in the art field, through experimentations with mixed human/nonhuman communities. He is a contributor to the webradio *DUUU and a member of the art collective Arts Visuels Debout.

Florence Millerand is Full Professor in the Department of Public and Social Communication at the University of Quebec at Montreal (UQAM). She is a member of the Interuniversity Research Center on Science and Technology (CIRST), is chairholder of the Research Chair on Digital Technology Uses and Changes in Communication, and coleads the Laboratory for Communication and the Digital (LabCMO).

Her research interests include research infrastructures, their design and use in relation to changes in scientific work and data practices, citizen and participatory science, social media, and digital culture. She is a regular contributor to the fields of science and technology studies and communication studies. For more information, please visit http://florencemillerand.uqam.ca/index.php/en/.

Eric Monteiro is Professor of Information Systems at the Norwegian University of Science and Technology (NTNU) and Adjunct Professor at the University of Oslo. He is interested in the interplay between technology and the social, especially in health care and the energy/oil sector. His work draws on insights from STS, CSCW, and information systems. His work has been published in *STHV*, *CSCW*, *MISQ*, and *EJIS*.

Anders Kristian Munk is Associate Professor in Techno-Anthropology and Director of the Techno-Anthropology Lab at the University of Aalborg in Copenhagen. His research interests include controversy mapping, science and technology studies (STS), public engagement with science (PES), pragmatism, actor-network theory (ANT), and new digital methods for the social sciences and humanities. He holds a PhD in human geography from the University of Oxford and an MA in European ethnology from the University of Copenhagen. He has worked as a senior visiting researcher at the Sciences Po Médialab and received research funding from the ESRC and the Carlsberg Foundation.

David Nemer is Assistant Professor in the School of Information Science at the University of Kentucky. His research and teaching interests cover the intersection of ICT for development (ICT4D), science and technology studies (STS), postcolonial STS, and human-computer interaction (HCI). He is an ethnographer who is specifically interested in studying ICTs in less industrialized parts of the world to understand the effects of ICTs on the development and empowerment of marginalized communities. He is the author of *Favela Digital: The Other Side of Technology* (2013). He holds a PhD in informatics from Indiana University and an MSc in computer science from Saarland University.

Elena Parmiggiani is Postdoctoral Researcher in Information Systems at the Norwegian University of Science and Technology (NTNU). She contributes to information systems, STS, and CSCW and is interested in studying the sociotechnical challenges of implementing, integrating, and maintaining information infrastructures and in the methodological stakes of studying such distributed and long-term arrangements. Empirically, she has focused primarily on environmental monitoring and the energy/oil industry. She holds a PhD in information technology from NTNU (2015) and an MSc in computer engineering from the University of Modena and Reggio Emilia.

Winifred R. Poster teaches in international affairs at Washington University, St. Louis, and has held visiting positions in India, Sweden, Germany, and Canada. Her research interests are in digital globalization, feminist labor theory, and Indian outsourcing. Under several grants from the National Science Foundation, she has been following ICT industries from the United States to India, in both earlier waves of computer manufacturing and software and later waves of back-office data processing and call centers. Recent projects explore the labors of surveillance, crowdsourcing, cybersecurity, and artificial intelligence. Her latest books are *Invisible Labor* (2016) and *Borders in Service* (2016).

Jessica Price is a PhD candidate in science and technology studies at Cornell University. She researches European knowledge practices in the early modern period. Her dissertation examines the intersections of demonology and English colonialism in South Asia. She holds a BA in history from the University of York and an MSc in the history of science, technology, and medicine jointly from University College London and Imperial College London.

David Ribes is Associate Professor in the Department of Human Centered Design and Engineering (HCDE) and Director of the Data Ecologies Lab (deLAB) at the University of Washington. He is a sociologist of science and technology who focuses on the development and sustainability of research infrastructures (i.e., networked information technologies for the support of interdisciplinary science), their relation to long-term changes in the conduct of science, and transformations in objects of research. His current research investigates the emerging institutions of data science at multiple scales, such as changing scientific practices, budding regional or national organizations, and novel public-private partnerships. He is a regular contributor to the fields of science and technology studies (STS) and information studies. His methods are ethnographic, archival-historical, and comparative. See davidribes.com or dataecologi.es for more.

Daniela K. Rosner is Assistant Professor in Human Centered Design and Engineering (HCDE) at the University of Washington and Codirector of the Tactile and Tactical Design (TAT) Lab. Her research critically investigates ways people work with materials to imagine more socially responsible technological futures, focusing on practices historically marginalized within engineering cultures such as electronics maintenance and needlecraft. Her work has generated several best paper nominations and awards and appeared in *Public Culture*, *New Media & Society*, and other journals, conference proceedings, and edited volumes. She is the author of *Critical Fabulations: Reworking the Methods and Margins of Design* (2018). Her work has been supported by multiple awards from the US National Science Foundation, including an NSF CAREER award. She earned her PhD from the University of California, Berkeley. She also holds a BFA in graphic design from the Rhode Island School of Design and an MS in computer science from the University of Chicago. She serves on the Editorial Board of *Artifact: Journal of Design Practice* and as the editor of the Design as Inquiry forum for *Interactions* magazine, a bimonthly publication of ACM SIGCHI.

Juan Salamanca is Assistant Professor in the Department of Graphic Design at University of Illinois at Urbana-Champaign. His research in social computing explores the design of digitally enabled nonhuman agents and their sociality. He has worked as data visualization leader for the Big Data Analytics Center of Excellence (CAOBA) in Colombia and actively contributes to the Design and Emotion Society and CSCW community as conference organizer and reader. He holds a PhD in design from Illinois Institute of Technology and an MA in design direction from Domus Academy, Italy. Visit smartartifact.com to learn more.

Steven Sawyer is Professor in the School of Information Studies at Syracuse University. He received his doctorate from Boston University's School of Management. He conducts research in the social informatics tradition with particular attention to the ways in which people organize to work together and use information and communication technologies. He leads courses that focus students' attention to the design, development, and implementation of information systems, to managing projects and systems, and to the roles of information and communication technologies relative to organizational and social change.

Nick Seaver is Assistant Professor of Anthropology at Tufts University, where he also teaches in the Science, Technology, and Society Program. His research examines how culturally significant concepts are operationalized in technical systems and understood by engineers and scientists. He has conducted ethnographic research with the developers of algorithmic recommender systems for music and is currently working on a project investigating the valences of "attention" in machine learning practice. He holds a PhD in anthropology from the University of California, Irvine and an SM in comparative media studies from the Massachusetts Institute of Technology.

Hanna Rose Shell is Associate Professor of Cinema Studies and Moving Image Arts, and of Art and Art History, at the University of Colorado Boulder. She holds an MA from Yale in American Studies and a PhD in the History of Science from Harvard, where she also completed her postdoctoral work as Junior Fellow at the Society of Fellows. Previously Associate Professor of STS at MIT, she is the author of *Hide and Seek: Camouflage, Photography, and the Media of Reconnaissance* (2012), audio-visual editor at *Technology & Culture*, and creator of multiple films, public installations and multimodal scholarly projects. Her research focuses on the interwoven histories of science, technology, aesthetics, and material culture in relation to practices of self-fashioning, surveillance and practices of recycling.

Ranjit Singh is a PhD candidate in the Department of Science and Technology Studies at Cornell University. His dissertation project examines the legal, administrative, and technological challenges in the implementation of India's biometrics-based national identification project, Aadhaar. He is also involved in a research project that traces the conceptualization, design, and implementation of updating the National Register of Citizens (NRC) in Assam, India. Both projects are geared toward elucidating the rapidly changing understandings, practices, and evaluations of the state-citizen relationship mediated by information infrastructures.

Johan Söderberg is Associate Professor in Theory of Science at the Department of Philosophy, Linguistics and Theory of Science at Göteborg University. His two most recent research projects were on hobbyists building self-replicating 3D printers and DIY drug chemistry. In his current work he is exploring the implications of the "moment of post-truth" for the study of science and scientific expertise.

Luke Stark is Researcher at Microsoft Research Montréal and an Affiliate of the Berkman Klein Center for Internet & Society at Harvard University. His work explores the histories, cultures, and ethics of computing, digital privacy, and artificial intelligence. His work has appeared in venues including *Social Studies of Science*, *History of the Human Sciences*, and *International Journal of Communication*, and he is a frequent contributor to the fields of media studies, science and technology studies, and critical HCI. He holds a PhD in Media, Culture, and Communication from New York University and an MA in history from the University of Toronto. He tweets @luke_stark; learn more at https://starkcontrast.co.

Tommaso Venturini (www.tommasoventurini.it) is Researcher at the Institut des Systèmes Complexes Rhône-Alpes and the École Normale Supérieure of Lyon and cofounder of the Public Data Lab. He is recipient of the Advanced Research fellowship of the French Institute for research in computer science and automation. He has been Lecturer at the Department of Digital Humanities of King's College London and Researcher at the Médialab of Sciences Po Paris, with which he is still associated.

Janet Vertesi is Assistant Professor of Sociology at Princeton University. She is author of *Seeing Like a Rover* (2015) and coeditor of *Representation in Scientific Practice Revisited* (2014), and her primary project is a long-term ethnography of NASA's robotic spacecraft mission teams. A frequent contributor to CHI and CSCW and past chair of alt.CHI, she is also passionate about critical design. She holds a PhD in science and technology studies from Cornell University and an MPhil in history and philosophy of science from the University of Cambridge.

Dominique Vinck is Full Professor at the University of Lausanne (UNIL) and also teaches at the Collège des Humanités of the Ecole Polytechnique Fédérale of Lausanne. He is a member of the UNIL Institute for Social Sciences and directs the STS Lab. His investigations focus on science and innovation studies. He is currently working on the engineering of digital cultures and humanities. He has notably published *Everyday Engineering: An Ethnography of Design and Innovation* (2003), *Comment les acteurs s'arrangent avec l'incertitude* (2009), *The Sociology of Scientific Work: The Fundamental Relationship between Science and Society* (2010), *Les Masques de la convergence* (2012), *Sciences et technologies émergentes. Pourquoi tant de promesses* (2015), *Humanités numériques. La culture face aux nouvelles technologies* (2016), and *Critical Studies of Innovation Bias: Alternative Approaches to the Pro-Innovation Bias* (2017). He directs the *Revue d'Anthropologie des Connaissances.*

Laura Watts is an ethnographer of futures, poet, writer, and Senior Lecturer in Energy and Society at Institute of Geography, Geosciences, University of Edinburgh. As an STS scholar, her research is focused on how futures are imagined and made, as well as methods for writing futures otherwise. Her recent book, *Energy at the End of the World: An Orkney Islands Saga,* is published by MIT Press, and she is co-author of *Ebban an' Flowan,* a poetic guide to marine renewable energy. Her work is published at www.sand14.com.

Brit Ross Winthereik is Professor at the IT University of Copenhagen in the Technologies in Practice research group, and PI of the collective research project "Data as Relation: Governance in the Age of Big Data." She has published widely in STS and anthropology journals on public sector digitalization, information infrastructures, and inventive methods.

Index

Unpack
Sag
Stabilized